D1034315

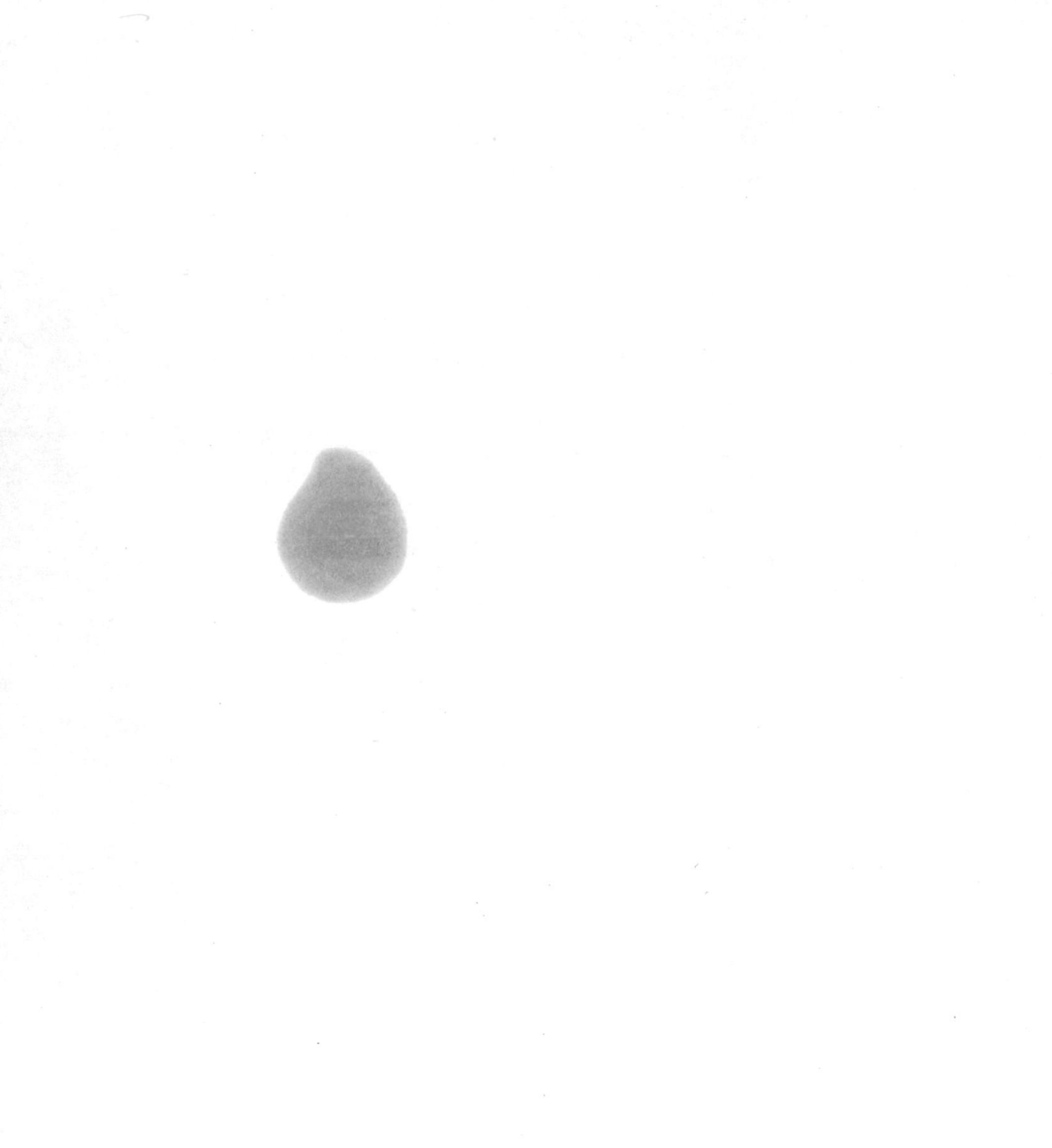

HANDBOOK OF CONVEX GEOMETRY

Volume B

HANDBOOK OF CONVEX GEOMETRY

Volume B

edited by

P.M. GRUBER,
Technische Universität Wien, Austria

J.M. WILLS,
Universität Siegen, Germany

1993

NORTH-HOLLAND
AMSTERDAM · LONDON · NEW YORK · TOKYO

ELSEVIER SCIENCE PUBLISHERS B.V.
Sara Burgerhartstraat 25
P.O. Box 211, 1000 AE Amsterdam, The Netherlands

Library of Congress Cataloging-in-Publication Data

Handbook of convex geometry / edited by P.M. Gruber, J.M. Wills.
p. cm.
Includes bibliographical references and index.
ISBN 0-444-89598-1 (set of vols A and B). – ISBN 0-444-89596-5
(vol A: alk. paper). – ISBN 0-444-89597-3 (vol B: alk. paper)
1. Convex geometry. I. Gruber, Peter M., 1941– II. Wills,
Jörg M., 1937–
QA639.5.H36 1993
516'.08–dc20 93–6696
CIP

ISBN: 0 444 89596 5 (Vol A)
ISBN: 0 444 89597 3 (Vol B)
ISBN: 0 444 89598 1 (Set of Vols A and B)

This book is printed on acid-free paper.

Printed in The Netherlands

Preface

One aim of this Handbook is to survey convex geometry, its many ramifications and its relations with other areas of mathematics. We believe that it will be a useful tool for the expert. A second aim is to give a high level introduction to most branches of convexity and its applications, showing the major ideas, methods and results. We hope that because of this feature the Handbook will act as an appetizer for future researchers in convex geometry. For them the many explicitly or implicitly stated problems should turn out to be a valuable source of inspiration. Third, the Handbook should be useful for mathematicians working in other areas as well as for econometrists, computer scientists, crystallographers, phycisists and engineers who are looking for geometric tools for their own work. In particular, mathematicians specializing in optimization, functional analysis, number theory, probability theory, the calculus of variations and all branches of geometry should profit from the Handbook.

The famous treatise "Theorie der konvexen Körper" by Bonnesen and Fenchel presented in 164 pages an almost complete picture of convexity as it appeared around 1930. While a similarly comprehensive report today seems to be out of reach, the Handbook deals with most of the more important topics of convexity and its applications. By comparing the Handbook with the survey of Bonnesen and Fenchel and with more recent collections of surveys of particular aspects of geometric convexity such as the AMS volume edited by Klee (1963), the Copenhagen Colloquium volume edited by Fenchel (1967), the two green Birkhäuser volumes edited by Tölke and Wills (1979) and Gruber and Wills (1983), respectively, and the New York Academy volume edited by Goodman, Lutwak, Malkevitch and Pollack (1985), the reader may see where progress was made in recent years.

During the planning stage of the Handbook, which started in 1986, we got generous help from many prominent convex geometers, in particular Peter McMullen, Rolf Schneider and Geoffrey Shephard. The discussion of the list of contents and of prospective authors turned out to be difficult. Both of us are obliged to the authors who agreed to contribute to the Handbook. In the cooperation with them we got much encouragement and the professional contacts furthered our good personal relations with many of them. The manuscripts which we finally received turned out to be much more diverse than we had anticipated. They clearly exhibit the most different characters and scientific styles of the authors and this should make the volume even more attractive.

There are several researchers in geometric convexity whom we invited to contribute to the Handbook but who for personal, professional or other reasons – regretfully – were not able to participate. The reader will also note that one area or another is missing in the list of contents. Examples are elementary geometry of normed planes, axiomatic convexity, and Choquet theory, but this should not diminish the usefulness of the Handbook.

Some fields such as computational and algorithmic aspects or lattice point results are dealt with in several chapters. The subjects covered are organized in five parts which in some sense reflect the fact that convexity is situated between analysis, geometry and discrete mathematics. This organization clearly has some disadvantages but we think that it will be helpful for the reader who wants to orient himself.

In the editing of the Handbook we received much help, in particular from Dr. M. Henk, Dr. J. Müller, Professor S. Hildebrandt, Professor L. Payne and Professor F. Schnitzer. We are most grateful to Ms. S. Clarius and Ms. E. Rosta who typed about 1000 letters to the authors, to colleagues whom we asked for advice and to the Publishers. We most gratefully acknowledge the friendly cooperations and expert support of Dr. A. Sevenster and Mr. W. Maas from Elsevier Science Publishers.

Finally we should like to express our sincere hope that the readers will appreciate the great effort of so many prominent authors and will find the Handbook useful for their scientific work.

Peter M. Gruber
Jörg M. Wills

Contents

Volume A

Volume B

List of Contributors

Bayer, M.M., *University of Kansas, Lawrence, KS* (Ch. 2.3).
Bokowski, J., *TH Darmstadt* (Ch. 2.5).
Brechtken-Manderscheid, U., *Universität Würzburg* (Ch. 4.3).
Brehm, U., *Technische Universität Dresden* (Ch. 2.4).
Burkard, R.E., *Technische Universität Graz* (Ch. 2.8).
Connelly, R., *Cornell University, Ithaca, NY* (Ch. 1.7).
Eckhoff, J., *Universität Dortmund* (Ch. 2.1).
Edelsbrunner, H., *University of Illinois, Urbana, IL* (Ch. 2.9).
Engel, P., *Universität Bern* (Ch. 3.7).
Ewald, G., *Universität Bochum* (Ch. 2.6).
Fejes Tóth, G., *Hungarian Academy of Sciences, Budapest* (Ch. 3.3).
Florian, A., *Universität Salzburg* (Ch. 1.6).
Goodey, P., *University of Oklahoma, Norman, OK* (Ch. 4.9).
Gritzmann, P., *Universität Trier* (Chs. 2.7, 3.2 and 3.4).
Groemer, H., *University of Arizona, Tucson, AZ* (Chs. 1.4 and 4.8).
Gruber, P.M., *Technische Universität Wien* (Chs. 0, 1.9, 1.10, 3.1 and 4.10).
Heil, E., *TH Darmstadt* (Chs. 1.11 and 4.3).
Klee, V., *University of Washington, Seattle, WA* (Ch. 2.7).
Kuperberg, W., *Auburn University, Auburn, AL* (Ch. 3.3).
Lee, C.W., *University of Kentucky, Lexington, KY* (Ch. 2.3).
Leichtweiß, K., *Universität Stuttgart* (Ch. 4.1).
Lindenstrauss, J., *The Hebrew University, Jerusalem* (Ch. 4.5).
Lutwak, E., *Polytechnic University, Brooklyn, NY* (Ch. 1.5).
Mani-Levitska, P., *Universität Bern* (Ch. 1.1).
Martini, H., *Technische Universität Chemnitz* (Ch. 1.11).
McMullen, P., *University College, London* (Ch. 3.6).
Milman, V.D., *Tel-Aviv University* (Ch. 4.5).
Papini, P.L., *Università degli Studi, Bologna* (Ch. 4.6).
Pták, V., *Czechoslovak Academy of Sciences, Prague* (Ch. 4.7).
Roberts, A.W., *Macalester College, St. Paul, MN* (Ch. 4.2).
Sangwine-Yager, J.R., *Saint Mary's College of California, Moraga, CA* (Ch. 1.2).
Schmitt, P., *Universität Wien* (Ch. 2.2).
Schneider, R., *Universität Freiburg* (Chs. 1.8 and 5.1).
Schulte, E., *Northeastern University, Boston, MA* (Ch. 3.5).
Talenti, G., *Università di Firenze* (Chs. 1.3 and 4.4).
Weil, W., *TH Karlsruhe* (Chs. 4.9 and 5.2).
Wieacker, J.A., *Universität Freiburg* (Chs. 5.1 and 5.2).
Wills, J.M., *Universität Siegen* (Chs. 2.4, 3.2 and 3.4).

Part 3
Discrete Aspects of Convexity

CHAPTER 3.1

Geometry of Numbers

Peter M. GRUBER

Abteilung für Analysis, Technische Universität Wien, Wiedner Hauptstraße 8–10, A-1040 Wien, Austria

Contents

HANDBOOK OF CONVEX GEOMETRY
Edited by P.M. Gruber and J.M. Wills

1. Introduction

Cum grano salis one may say that classical geometry of numbers forms a bridge between convexity, Diophantine approximation and the theory of quadratic forms. Today it is an independent problem-oriented field of mathematics having relations with such diverse areas as coding theory, modular functions, numerical integration and complexity of algorithms. During the last two decades it has again obtained its original geometric flavour.

First traces of the geometry of numbers can be found in the work of Kepler, Lagrange, Gauss, Hermite, Korkin and Zolotarev (lattice packing of balls), of Dirichlet and Fedorov (lattice tiling) and of Klein (geometric theory of continued fractions). The systematic general study was initiated by Minkowski (who also coined the name of this field) in the last years of the nineteenth century. His achievements were complemented in the first decade of the twentieth century by the work of Voronoi on the geometric theory of positive quadratic forms. Important contributors in this century were, amongst others, Blichfeldt, Siegel, Furtwängler, Delone, Venkov, Davenport, Mahler, Mordell, Hajós, their students and the schools in Vienna, Manchester–London–Cambridge, Moscow–Leningrad–Samarkand, Chandigarh, Columbus. The last two decades are characterized by the emergence of new fertile areas and the introduction of powerful tools from other parts of mathematics. Examples of such newer developments are lattice polytopes, the zeta-function, effective dense packings of balls, algorithmic problems.

In the following we omit almost completely results dealing with lattice packing and covering, particularly with Euclidean balls and lattice polytopes. Very little is said about star bodies and indefinite quadratic forms. Lattice tilings are rarely mentioned, nor applications to Diophantine approximation.

For more detailed information and references the reader is referred to the books and surveys of Minkowski (1896, 1907), Keller (1954), Rogers (1964), Cassels (1972), Fejes Tóth (1972), Gruber (1979), Lovász (1986), Kannan (1987a), Conway and Sloane (1987), Gruber and Lekkerkerker (1987), Lovász (1988), Grötschel, Lovász and Schrijver (1988), Erdős, Gruber and Hammer (1989), Pohst and Zassenhaus (1989), and to the collected or selected works of Minkowski (1911), Voronoi (1952), Davenport (1977), and Hlawka (1990). See also the sections on packing, covering, tiling, lattice points and crystallography in this volume.

2. Lattices and the space of lattices

A *lattice* $\mathbb{L}$ (of full rank d) in d-dimensional Euclidean space $\mathbb{E}^d$ is the set of all integer linear combinations of d linearly independent vectors $b_1, \dots, b_d \in \mathbb{E}^d$. (We do not distinguish between points and vectors.) $\{b_1, \dots, b_d\}$ is often identified with the $d \times d$-matrix B having columns $b_1, \dots, b_d$. $\{b_1, \dots, b_d\}$ or B is called a *basis* of $\mathbb{L}$. Clearly $\mathbb{L} = BZ^d$ where Z^d is the lattice of all points with integer coordinates. Given a basis B of $\mathbb{L}$ all others bases of $\mathbb{L}$ are of the form BU where

U is an arbitrary integral unimodular $d \times d$ matrix, i.e., it has integer entries and determinant ± 1. The *determinant* $d(\mathbb{L})$ of $\mathbb{L}$ is the absolute value of the determinant of a basis $B = \{b_1, \ldots, b_d\}$ of $\mathbb{L}$ or, geometrically expressed, the volume of the *fundamental parallelepiped* $\{\alpha_1 b_1 + \cdots + \alpha_d b_d : 0 \leqslant \alpha_i < 1\}$ of $\mathbb{L}$.

An inequality of Hadamard says that $d(\mathbb{L}) \leqslant |b_1| \cdots |b_d|$ where equality holds precisely if the basis vectors $b_1, \ldots, b_d$ of $\mathbb{L}$ are pairwise orthogonal. ($|\cdot|$ stands for the ordinary Euclidean norm.) Schnorr (1985) has called the quotient $|b_1| \cdots |b_d|/d(\mathbb{L})$ ($\geqslant 1$) the *orthogonality defect* of the basis $\{b_1, \ldots, b_d\}$. Bases with small orthogonality defect are of importance for the calculation of "short" basis vectors. The search for bases with small orthogonality defect has recently become of importance for reduction theory, see section 6.

A subset of $\mathbb{E}^d$ is *discrete* if it has only finitely many points in any bounded set. The following simple result is well known.

Theorem 1. *A subset $\mathbb{L}$ of $\mathbb{E}^d$ is a lattice precisely when it is a discrete subgroup of $\mathbb{E}^d$ containing d linearly independent vectors.*

If a subset $\mathbb{M}$ of a lattice $\mathbb{L}$ is also a lattice it is called a *sublattice* of $\mathbb{L}$. In this case one may choose bases of $\mathbb{L}$ and $\mathbb{M}$ which are related to each other in particularly simple ways. In the following result (i) and (ii) are due to Hermite and Minkowski, (iii) follows from the theory of elementary divisors.

Theorem 2. *Let $\mathbb{M}$ be a sublattice of a lattice $\mathbb{L}$. Then*

(i) *For any basis $\{b_1, \ldots, b_d\}$ of $\mathbb{L}$ there is a basis $\{c_1, \ldots, c_d\}$ of $\mathbb{M}$ such that*

$$\begin{aligned} c_1 &= u_{11} b_1 \\ c_2 &= u_{21} b_1 + u_{22} b_2 \\ &\vdots \\ c_d &= u_{d1} b_1 + \cdots + u_{dd} b_d \end{aligned} \tag{1}$$

where $u_{11}, \ldots, u_{dd} \in Z$.

(ii) *For any basis $\{c_1, \ldots, c_d\}$ of $\mathbb{M}$ there is a basis $\{b_1, \ldots, b_d\}$ of $\mathbb{L}$ such that* (1) *holds.*

(iii) *There are bases $\{b_1, \ldots, b_d\}$ of $\mathbb{L}$ and $\{c_1, \ldots c_d\}$ of $\mathbb{M}$ such that* (1) *holds and $u_{ij} = 0$ for $i \neq j$.*

Given a basis of $\mathbb{M}$, the problem arises to actually determine a basis $\{c_1, \ldots, c_d\}$ of $\mathbb{M}$ as in (i). Frumkin (1976), and Kannan and Bachem (1979) first described *polynomial time algorithms* for this purpose. More precisely, there is a polynomial $p(\cdot, \cdot)$ such that given $\mathbb{M}$ one can determine $\{c_1, \ldots, c_d\}$ in at most $p(d, s)$ arithmetic operations where d is the dimension and s the largest binary size of the coordinates of the vectors of the given basis of $\mathbb{M}$ with respect to $\{b_1, \ldots, b_d\}$. (The *binary size* of a rational number is the sum of the lengths of the binary representations of its numerator and its denominator.)

Let $\mathbb{L}$ be a lattice and let $a_1, \ldots, a_{d-1} \in \mathbb{E}^d$ be linearly independent. Then a result of Davenport (1955) says that for each sufficiently large $\lambda > 0$ there is a basis $\{b_1, \ldots, b_d\}$ of $\mathbb{L}$ such that $|b_i - \lambda a_i|$, $i = 1, \ldots, d-1$, is "small" relative to λ. For refinements see Lekkerkerker (1961) and Wáng (1975).

An important concept is that of the *polar* or *dual lattice* $\mathbb{L}^*$ of a lattice $\mathbb{L} = BZ^d$:

$$\mathbb{L}^* = \{m\colon l \cdot m \in Z \text{ for all } l \in \mathbb{L}\} = (B^{-1})^{\mathrm{tr}} Z^d$$

where the dot $\cdot$ indicates the ordinary inner product in $\mathbb{E}^d$. Clearly $d(\mathbb{L}^*)d(\mathbb{L}) = 1$.

The *natural topology* on the space $\mathscr{L}$ of all lattices in $\mathbb{E}^d$ can be introduced in different ways, for example by the following definition: a sequence $\mathbb{L}_i \in \mathscr{L}, i = 1, 2, \ldots$, is *convergent* if there are a lattice $\mathbb{L} \in \mathscr{L}$ and bases $\{b_{i1}, \ldots, b_{id}\}$ and $\{b_1, \ldots, b_d\}$ of $\mathbb{L}_i$ and $\mathbb{L}$, respectively, such that $b_{i1} \to b_1, \ldots, b_{id} \to b_d$ as $i \to \infty$. Let $\mathscr{L}$ (1) denote the subspace $\{\mathbb{L} \in \mathscr{L}\colon d(\mathbb{L}) = 1\}$ of $\mathscr{L}$.

The following *selection* or *compactness theorem of Mahler* (1946) guarantees the existence of lattices having certain extremal properties. For example, given a (proper, compact) convex body K it yields the existence of a lattice $\mathbb{L} \in \mathscr{L}$ such that the translates $K + l\colon l \in \mathbb{L}$ form a *packing* of K (i.e., any two distinct translates have disjoint interiors) with *packing lattice* $\mathbb{L}$ for which the *density* $V(K)/d(\mathbb{L})$ is maximal. (V stands for volume.) One may think of the density as the proportion of space covered by the bodies $K + l\colon l \in \mathbb{L}$.

Theorem 3. *Let $\mathbb{L}_i \in \mathscr{L}, i = 1, 2, \ldots$, be a sequence of lattices such that the determinants $d(\mathbb{L}_i)$ are bounded above and which are all admissible for a fixed neighborhood of the origin o. Then $\mathbb{L}_1, \mathbb{L}_2, \ldots$ contains a convergent subsequence.*

(A lattice is *admissible* for a set M or *M-admissible* if it contains no interior point of it, except, possibly, o.)

Outline of Proof (*Following* Groemer 1971). For each $\mathbb{L}_i$ consider its *Dirichlet–Voronoi cell*

$$D(\mathbb{L}_i, o) = \{x\colon |x| \leqslant |x - l| \text{ for all } l \in \mathbb{L}\}.$$

These cells are o-symmetric convex polytopes and the assumptions on the $\mathbb{L}_i$'s imply that they all contain a fixed ball with center o and are contained in another fixed ball with center o. Thus the Blaschke selection theorem (see chapter 1.9) assures the existence of a subsequence $D(\mathbb{L}_{i_k}, o)$ which converges to a convex body D, say. It turns out that D is the Dirichlet–Voronoi cell of a lattice $\mathbb{L}$ and $\mathbb{L}_{i_k} \to \mathbb{L}$. □

Since $\mathscr{L}$ and its closed subspace $\mathscr{L}(1)$ are both locally compact there are many *Borel measures* on these spaces. It is of interest to specify such measures which are of geometric significance. Following Siegel (1945) we describe a "natural" construction of such a measure on $\mathscr{L}(1)$. Let $\mathcal{M}(1)$ be the (locally compact) multiplicative group of all real $d \times d$-matrices of determinant ± 1 where each matrix

$M \in \mathcal{M}(1)$ is considered as a point of $\mathbb{E}^{d^2}$ (take lexicographic ordering). There is a Borel subset $\mathcal{R}$ of $\mathcal{M}(1)$ which contains precisely one element of each left coset of the subgroup $\mathcal{U}(1)$ of all integral unimodular matrices. It can be shown that $0 < v_d = V(\bigcup\{\lambda\mathcal{R} : 0 \leqslant \lambda \leqslant 1\}) < +\infty$ where V denotes ordinary Lebesgue measure in $\mathbb{E}^{d^2}$. Now, for any Borel set $\mathcal{B} \subset \mathcal{R}$ let $\mu(\mathcal{B}) = V(\bigcup\{\lambda\mathcal{B} : 0 \leqslant \lambda \leqslant 1\})/v_d$. This defines a Borel measure on $\mathcal{R}$; actually μ is the restriction to $\mathcal{R}$ of the suitably normalized Haar measure on $\mathcal{M}(1)$. By letting correspond to each lattice $\mathbb{L} \in \mathcal{L}(1)$ its system of bases we obtain a one-to-one correspondence between the lattices $\mathbb{L} \in \mathcal{L}(1)$ and the left cosets $B\mathcal{U}$ of $\mathcal{U}$ and thus between $\mathcal{L}(1)$ and $\mathcal{R}$. Thus we may consider μ a Borel measure on $\mathcal{L}(1)$.

There are several *proposals for classifications of the lattices* in $\mathcal{L}$. Some of these are connected with reduction or with particular tilings related to a lattice $\mathbb{L}$, for example the tiling $D(\mathbb{L}, o) + l$: $l \in \mathbb{L}$; see section 7. Another classification (into so-called Bravais types) stems from crystallography and is connected with the groups of rigid motions mapping lattices onto themselves; see, e.g., Erdős, Gruber and Hammer (1989).

3. The fundamental theorems

A basic problem in geometry of numbers is to investigate whether a given set in $\mathbb{E}^d$ contains a point (possibly $\neq o$) of a fixed lattice. In arithmetic terms this amounts to the problem whether an inequality or a system of inequalities in d variables has a (non-trivial) integer solution.

In the following we will present several simple but far reaching results, each having a wide range of mainly number-theoretic applications. We begin by stating *Minkowski's first* or *(first) fundamental theorem* of 1891.

Theorem 4. *Let K be an o-symmetric convex body and $\mathbb{L}$ a lattice in $\mathbb{E}^d$. If $V(K) \geqslant 2^d d(\mathbb{L})$ then K contains at least one pair of points $\pm l \neq o$ of $\mathbb{L}$.*

Outline of Proof (*Via a density argument for lattice packings*)**.** It is sufficient to show the following:

$$\text{Let } K \text{ contain no point } l \in \mathbb{L}\backslash\{o\}. \text{ Then } V(K) < 2^d d(\mathbb{L}). \tag{1}$$

The convexity and the symmetry of K yield that any two distinct bodies of the form $\frac{1}{2}K + l$: $l \in \mathbb{L}$ are disjoint. Hence they form a packing with packing lattice $\mathbb{L}$ and the property that any two translates of $\frac{1}{2}K$ are disjoint. Since the density of such a packing is < 1, i.e., $V(\frac{1}{2}K)/d(\mathbb{L}) < 1$, we obtain $V(K) < 2^d d(\mathbb{L})$. This proves (1), concluding the proof of Theorem 4. □

For any K, except for so-called *parallelohedra* (see section 7), one may replace the constant 2^d by a smaller number (depending on K) with the conclusion still holding. The smallest such constant is $V(K)/\Delta(K)$ where $\Delta(K)$ is the *lattice constant* of K, which is defined by

$$\Delta(K) = \inf\{d(\mathbb{L}) \colon \mathbb{L} \in \mathscr{L} \text{ admissible for } K\}.$$

That the density of the densest lattice packing of K equals $V(K)/2^d\Delta(K)$ is easy to see. The fundamental theorem is equivalent to the inequality $2^d\Delta(K) \geqslant V(K)$.

Of particular interest is $\Delta(\mathbb{B}^d)$ where $\mathbb{B}^d$ is the solid Euclidean unit ball. Sometimes (for example in the theory of positive quadratic forms) one uses *Hermite's constant* $\gamma_d = \Delta(\mathbb{B}^d)^{-2/d}$ instead of $\Delta(\mathbb{B}^d)$. A result of Kabatjanskii and Levenštein (1978) together with the Minkowski–Hlawka theorem (see section 4) shows that

$$2^{-0.401d+o(d)}V(\mathbb{B}^d) \leqslant \Delta(\mathbb{B}^d) \leqslant V(\mathbb{B}^d) \quad (d \to \infty).$$

The *first successive* or *homogeneous minimum* $\lambda_1 = \lambda_1(K, \mathbb{L})$ of K with respect to $\mathbb{L}$ is defined by

$$\lambda_1 = \inf\{\lambda > 0 \colon \lambda K \text{ contains a point of } \mathbb{L}\backslash\{o\}\}.$$

The fundamental theorem is equivalent to the inequality

$$\lambda_1(K, \mathbb{L})^d \leqslant 2^d d(\mathbb{L})/V(K). \tag{2}$$

Applying (2) to the polar body $K^* = \{y \colon x \cdot y \leqslant 1 \text{ for all } x \in K\}$ and to $\mathbb{L}^*$ we obtain

$$\lambda_1(K, \mathbb{L})\lambda_1(K^*, \mathbb{L}^*) \leqslant 4/(V(K)V(K^*))^{1/d}. \tag{3}$$

A conjecture of Mahler (1939) and the inequality of Blaschke–Santaló (see Santaló 1949) say that

$$\frac{4^d}{d!} \leqslant V(K)V(K^*) \leqslant V(\mathbb{B}^d)^2.$$

Bourgain and Milman (1985) proved the existence of a constant $\alpha > 0$ such that $\alpha^d V(\mathbb{B}^d)^2 \leqslant V(K)V(K^*)$, thus improving an earlier result of Bambah (1954). Hence $4/(V(K)V(K^*))^{1/d} \leqslant \beta d$ for some constant $\beta > 0$. On the other hand a result of Conway and Thompson (see Milnor and Husemoller 1973) shows that the inequality $\lambda_1(\mathbb{B}^d, \mathbb{L})\lambda_1(\mathbb{B}^d, \mathbb{L}^*) \leqslant \beta d$ thus obtained is best possible – up to the constant. See also Banaszczyk (1989).

Results such as (3) relating successive minima and the inhomogeneous minimum of a lattice and its polar lattice now all go under the name of *transference theorems*. For more examples and definitions see section 5 below.

We next consider computational problems. First, let d be fixed. Let $\mathbb{L}$ be a lattice with a basis consisting of rational vectors. The problem is to determine a vector in $\mathbb{L}\backslash\{o\}$ of minimal length (i.e., of length $\lambda_1(\mathbb{B}^d, \mathbb{L})$). Kannan (1987) described a polynomial time algorithm for obtaining such a vector. That is, there is a fixed polynomial p such that for any $\mathbb{L}$ the algorithm computes the desired vector in $p(s)$ arithmetic operations where s is the largest binary size of the coordinates of the vectors of the given basis of $\mathbb{L}$. If instead of the Euclidean norm the maximum norm is used, i.e., $\mathbb{B}^d$ is replaced by the unit cube, the analog of Kannan's result is due to Lenstra (1983).

For variable d the situation is worse: improving upon an earlier result of Lenstra, Lenstra and Lovász (1982), Schnorr (1987) showed the following: for any $\alpha > 1$ there is a polynomial time algorithm which computes a vector of $\mathbb{L}\backslash\{o\}$ of length $\leqslant \alpha^d \lambda_1(\mathbb{B}^d, \mathbb{L})$. (Replace now $p(s)$ by $p(d,s)$.) See also the discussion and additional results in Lagarias, Lenstra and Schnorr (1990). More generally, as a consequence of a result of Lenstra, Lenstra and Lovász (1982) on properties of LLL-reduced bases (see section 6 below) and of John's theorem on ellipsoids circumscribed to an o-symmetric convex body, Grötschel, Lovász and Schrijver (1988) obtain an algorithmic version of the fundamental theorem. (A *weak membership oracle* for a convex body K is defined by the following property: Given any rational vector y and any rational number $\varepsilon > 0$ then it answers that the ε-neighborhood of y intersects K or is not contained in K. If y is near the boundary of K, then either answer is legal.)

Theorem 5. *There are a polynomial $p(\cdot,\cdot)$ and an algorithm (which can ask an oracle and each such question and answer requires* 1 *unit of time) with the following property: let an o-symmetric convex body K be given by means of a weak membership oracle such that for suitable rationals $r, R > 0$ we have $r\mathbb{B}^d \subset K \subset R\mathbb{B}^d$ and let a lattice $\mathbb{L}$ be given by means of a rational basis $\{b_1, \ldots, b_d\}$. If*

$$V(K) \geqslant 2^{(d^2/4)+f(d)} d(\mathbb{L}),$$

the algorithm finds a point $l \in \mathbb{L}\backslash\{o\}$ in K in time at most $p(d,s)$ where s is the maximum of the binary sizes of r, R and the components of $b_1, \ldots, b_d$. $f(d) = \mathrm{O}(d \log d)$ can be given explicitly.

In spite of its simplicity the fundamental theorem as well as some of its refinements have important applications in algebraic number theory and Diophantine approximations. We cite the *linear form theorem of Minkowski* (1896).

Theorem 6. *Let $l_1, \ldots, l_d$ be d real linear forms in d variables of determinant* 1. *If $\tau_1, \ldots, \tau_d > 0$ are chosen such that $\tau_1 \cdots \tau_d = 1$ then the system of inequalities*

$$|l_1(u)| \leqslant \tau_1, \ldots, |l_d(u)| \leqslant \tau_d \tag{4}$$

has an integer solution $u \in Z^d\backslash\{o\}$.

Proof. The o-symmetric parallelepiped defined by (4) has volume $2^d \tau_1 \cdots \tau_d = 2^d$. Thus the fundamental theorem implies that it contains a point $u \in Z^d\backslash\{o\}$. □

There are many *refinements and extensions* of the first fundamental theorem. Below some classical results are quoted which, in part, admit more sophisticated number-theoretic applications. For another series of modern results on lattice points in convex bodies we refer to chapter 3.2.

There are numerous criteria that a convex body contains points of a given lattice. Many of these results suffer from the defect that they are proved only in the case $d = 2$ or the conditions given are not invariant with respect to (volume preserving) linear transformations. Notable exceptions are a result of Chalk (1967) and Yates (1970), and a recent remark of Wills (1991).

Verifying a statement of Blichfeldt (1921), van der Corput (1935/6) derived the following extension of Minkowski's theorem (and a corresponding refinement of Blichfeldt's Theorem 9 below). His proof is based on Dirichlet's "pigeon hole principle" which was used in this context already by Scherrer (1922) and by Mordell.

Theorem 7. *Let K be an o-symmetric convex body, $\mathbb{L}$ a lattice in $\mathbb{E}^d$ and $k \geqslant 1$ an integer. If $V(K) \geqslant 2^d k d(\mathbb{L})$ then K contains k distinct pairs of points $\pm l_1, \ldots, \pm l_k \neq o$ of $\mathbb{L}$.*

Lattices are periodic sets. Hence a natural idea is to apply (multidimensional) Fourier series. The following *formula of Siegel* (1935) is of importance for the investigation of discriminants in algebraic number fields. It is based on Parseval's theorem.

Theorem 8. *Let K be an o-symmetric convex body and $\mathbb{L}$ a lattice in $\mathbb{E}^d$ such that $\mathbb{L}$ is admissible for K. Then*

$$V(K) + 4^d V(K)^{-1} \sum_{m \in \mathbb{L}^* \setminus \{o\}} \left| \int_{(1/2)K} \mathrm{e}^{-2\pi i m \cdot x} \, \mathrm{d}x \right|^2 = 2^d d(\mathbb{L}). \tag{5}$$

The integrals in (5) are generalizations of Fraunhofer's integrals; they are difficult to calculate, see, e.g., Hlawka (1950). In a similar way as Minkowski's theorem has been refined by Siegel's formula, the theorem of Blichfeldt (see Theorem 9) has been refined by several authors, for example by Uhrin (1981).

The following *theorem of Blichfeldt* (1914) has been proved independently by Scherrer (1922). An important application is Blichfeldt's (1929) upper estimate $2^{-d/2}(1 + d/2)^{-1}$ for the density of the densest lattice packing of balls. (The smallest upper estimate known is $2^{-0.599d+o(d)}$ as $d \to \infty$; it is due to Kabatjanskii and Levenštein (1978).)

Theorem 9. *Let M be a measurable set and $\mathbb{L}$ a lattice in $\mathbb{E}^d$. If $V(M) > d(\mathbb{L})$ or if $V(M) = d(\mathbb{L})$ and M is compact, then M contains two distinct points p, q with $p - q \in \mathbb{L}$.*

By applying Blichfeldt's theorem to $\frac{1}{2}K$ the first fundamental theorem follows. A proof of Blichfeldt's theorem can be modelled along the outline of the proof of Minkowski's theorem given above by replacing $\frac{1}{2}K$ by M.

4. The Minkowski–Hlawka theorem

Minkowski's theorem shows that a lattice $\mathbb{L}$ which is admissible for an o-symmetric convex body K satisfies the inequality $2^{-d}V(K) \leqslant d(\mathbb{L})$. On the other hand the following result of Hlawka (1944) which verifies a conjecture of Minkowski implies that we may always choose a K-admissible lattice $\mathbb{L}$ for which $d(\mathbb{L}) \leqslant V(K)$.

Theorem 10. *Let M be a bounded Jordan measurable set in $\mathbb{E}^d$ with $V(M) < 1$. Then there is a lattice $\mathbb{L}$ with $d(\mathbb{L}) = 1$ such that M contains no point of $\mathbb{L}\backslash\{o\}$.*

Proof (*Following* Rogers 1947). Let $p = 2,3,5,\ldots$ be a prime. For $u = (1, u_2, \ldots, u_d)^{\mathrm{tr}} \in Z^d$ with $0 \leqslant u_2, \ldots, u_d < p$ let $\mathbb{L}(p,u)$ be the lattice with basis $u, (0,p,0, \ldots, 0)^{\mathrm{tr}}, (0,0,p,0,\ldots,0)^{\mathrm{tr}}, \ldots, (0,0,\ldots,0,p)^{\mathrm{tr}}$. Then

for any $v \in Z^d$ with $v_1 \neq 0, \pm p, \pm 2p, \ldots$ there is a unique lattice $\mathbb{L}(p,u)$ containing v. (1)

For the proof of (1) it is sufficient to show the following: there are unique integers $u_2, \ldots, u_d$ with $0 \leqslant u_2, \ldots, u_d < p$ such that for suitable integers $k, k_2, \ldots, k_d$ the following conditions hold:

$$\begin{array}{lcl} v_1 = k & & v_1 = k \\ v_2 = ku_2 + k_2 p & & v_2 \equiv ku_2 \bmod p \\ \vdots & \text{or} & \vdots \\ v_d = ku_d + k_d p & & v_d \equiv ku_d \bmod p. \end{array}$$

Since $v_1 = k$ is not an integral multiple of p, these conditions uniquely determine integers $u_2, \ldots, u_d$ with $0 \leqslant u_2, \ldots, u_d < p$, concluding the proof of (1). Hence

$Y = \{v \in Z^d \colon v_1 \neq 0, \pm p, \ldots\}$ is the disjoint union of the p^{d-1} sets $\mathbb{L}(p,u) \cap Y \colon 0 \leqslant u_2, \ldots, u_d < p$. (2)

Since M is bounded and has Jordan measure < 1 one may choose p so large that

$$M \subset \{x \colon |x_i| < p^{1/d}\}, \tag{3}$$

and

$$\frac{1}{p^{d-1}} \cdot \#\{M \cap \frac{1}{p^{(d-1)/d}} Y\} \leqslant \frac{1}{p^{d-1}} \cdot \#\{M \cap \frac{1}{p^{(d-1)/d}} Z^d\} < 1. \tag{4}$$

(# stands for cardinal number and $(1/p^{(d-1)/d})Z^d$ is a lattice of determinant $1/p^{d-1}$.) (2) and (4) imply the existence of a lattice $\mathbb{L}(p,u)$ such that

$$\#\{M \cap \frac{1}{p^{(d-1)/d}} (\mathbb{L}(p,u) \cap Y)\} < 1,$$

or

$$(p^{(d-1)/d} M) \cap (\mathbb{L}(p,u) \cap Y) = \emptyset. \tag{5}$$

Now noting that $\mathbb{L}(p,u) \cap \{x : x_1 = 0, |x_2|, \ldots, |x_d| < p\} = \{o\}$, (3) and (5) imply

$$(p^{(d-1)/d} M) \cap \mathbb{L}(p,u) \subseteq \{o\} \quad \text{or} \quad M \cap \frac{1}{p^{(d-1)/d}} \mathbb{L}(p,u) \subseteq \{o\}.$$

Since $(1/p^{(d-1)/d})\mathbb{L}(p,u)$ is a lattice of determinant 1, the proof of Theorem 10 is complete. □

The above proof is based on an averaging argument. This is also true for other proofs and is clearly recognizable in the derivation of the Minkowski–Hlawka theorem from the following *mean value theorem of Siegel* (1945) (take for f the characteristic function of M).

Theorem 11. *Let $f : \mathbb{E}^d \to \mathbb{R}$ be Riemann integrable with bounded support. Then*

$$\int_{\mathcal{L}(1)} \sum \{f(l) : l \in \mathbb{L} \backslash \{o\}\} \, \mathrm{d}\mu(\mathbb{L}) = \int_{\mathbb{E}^d} f(x) \, \mathrm{d}x.$$

(The measure μ was defined in section 2.)

There are many *refinements* of the Minkowski–Hlawka theorem, the sharpest one is due to Schmidt (1963): there are $\alpha_d > 0, \alpha_d \to \log \sqrt{2}$ as $d \to \infty$ ($\alpha_2 = \frac{8}{15}$) such that for any M with $V(M) \leqslant \alpha_d d$ there is a lattice $\mathbb{L}$ with $d(\mathbb{L}) = 1$ which is admissible for M.

In contrast to earlier beliefs it now seems plausible that one might not be able to essentially improve the Minkowski–Hlawka theorem, not even in the case of Euclidean balls. This is due to the belief that it might be possible to lower the upper bound $2^{-0.599d+o(d)}$ for the maximum density of lattice packings of balls by Kabatjanskii and Levenštein (1978) to $2^{-d+o(d)}$. (The Minkowski–Hlawka theorem implies that for any o-symmetric convex body there is a lattice packing having density at least 2^{-d}.)

In principle the original proof of Hlawka as well as other proofs and, in particular, the proof presented above can be used to determine in finitely many steps admissible lattices of small determinant for M. Unfortunately the large amount of calculations makes this procedure useless in practice. *Explicit constructions* of dense lattice and non-lattice packings of balls based on binary error-correcting codes were first given by Leech and Sloane (1971). So far the best result in a long series of more and more sophisticated constructions is due to Rush (1989). It gives a rather explicit lattice packing of density $2^{-d+o(d)}$ for any convex body K which is symmetric through each coordinate hyperplane. See also Quebbemann (1991). For earlier references see Conway and Sloane (1987).

5. Successive minima

Let K be a convex body in $\mathbb{E}^d$ with o in its interior and let L be a lattice. For $i = 1, \ldots, d$, the ith *successive minimum* of K with respect to L is defined by

$$\lambda_i = \lambda_i(K, \mathbb{L}) = \inf\{\lambda > 0\colon \lambda K \text{ contains } i \text{ linearly independent vectors of } \mathbb{L}\}.$$

Clearly $0 < \lambda_1 \leqslant \lambda_2 \leqslant \cdots \leqslant \lambda_d < +\infty$. There are linearly independent vectors $l_1, \ldots, l_d \in \mathbb{L}$ (not necessarily a basis) such that l_i is a boundary point of $\lambda_i K$. (According to Mahler (1938) there is a basis $\{b_1, \ldots, b_d\}$ of $\mathbb{L}$ such that $b_i \in i\lambda_i K$.)

Minkowski's (1896) *second (fundamental) theorem* is the following result.

Theorem 12. *Let K be an o-symmetric convex body and $\mathbb{L}$ a lattice in $\mathbb{E}^d$. Then*

$$(2^d/d!)d(\mathbb{L}) \leqslant \lambda_1(K, \mathbb{L}) \cdots \lambda_d(K, \mathbb{L}) V(K) \leqslant 2^d d(\mathbb{L}). \tag{1}$$

The equality case on the right-hand side was discussed by Jarnik (1949). There is no particularly simple proof available. In essence Minkowski's original proof is surprisingly simple, see Henk (1992). Unfortunately it is hidden behind a bulk of details. More easily to grasp are the proofs of Bambah, Woods and Zassenhaus (1964). Danicic (1969) gave a proof in the spirit of Scherrer's (1922) proof of Blichfeldt's theorem.

If K is a convex body with o in its interior and $\mathbb{L}$ any lattice then a *conjecture* cited by Davenport (1946) states that

$$\lambda_1(K, \mathbb{L}) \cdots \lambda_d(K, \mathbb{L}) \Delta(K) \leqslant d(\mathbb{L}). \tag{2}$$

If K is o-symmetric, then $V(K) \leqslant 2^d \Delta(K)$ by the first fundamental theorem; see section 3. Hence (2) generalizes (1). For $d = 2$, (2) is true as shown by Woods (1958) and in (1956) he showed that it is true for $d = 3$ and K o-symmetric. Minkowski (1896) proved (2) for general d if K is an o-symmetric ellipsoid.

Mahler (1939) applied (1) to K^*, L^* to obtain the inequalities

$$1 \leqslant \lambda_i(K, \mathbb{L}) \lambda_{d+1-i}(K^*, \mathbb{L}^*) \leqslant 4^d / V(K) V(K^*) \quad \text{for } i = 1, \ldots, d. \tag{3}$$

By a result of Bourgain and Milman (1985) the right-hand side is $\leqslant \beta^d d^d$ for a constant $\beta > 0$; compare the discussion in section 3 above. Using reduction in the sense of Korkin and Zolotarev it was proved by Lagarias, Lenstra and Schnorr (1990) that there is a constant $\gamma > 0$ such that the right-hand side of (3) can be replaced by $\gamma d^{5/2}$ and that in case $K = \mathbb{B}^d$ it may be replaced by δd^2 where $\delta > 0$ is a suitable constant. In a recent letter Banaszczyk pointed out that for $K = \mathbb{B}^d$ the exponent 2 may be replaced even by 1.

Besides the transference result (3) and its refinements there are other transference theorems connecting the successive minima of an o-symmetric convex body K with respect to a lattice $\mathbb{L}$ and the *inhomogeneous minimum* or *covering radius* $\mu(K, \mathbb{L})$ of K with respect to $\mathbb{L}$. It is defined by

$$\mu(K, \mathbb{L}) = \inf\{\mu > 0 \colon \bigcup\{\mu K + l \colon l \in \mathbb{L}\} = \mathbb{E}^d\}.$$

A simple example of this type of transference theorem is due to Jarnik (1941), namely

$$\lambda_d(K, \mathbb{L}) \leqslant 2\mu(K, \mathbb{L}) \leqslant \lambda_1(K, \mathbb{L}) + \cdots + \lambda_d(K, \mathbb{L}).$$

Further classical results of this type are due to Mahler, Hlawka, Kneser, Scherk, Birch and others, see Gruber and Lekkerkerker (1987).

By means of Korkin–Zolotarev reduction Lagarias, Lenstra and Schnorr (1990) proved that there is a constant $\delta > 0$ such that for any o-symmetric convex body K and any lattice L

$$\mu(K, \mathbb{L})\lambda_1(K^*, \mathbb{L}^*) \leqslant \delta d^2$$

and

$$\mu(\mathbb{B}^d, \mathbb{L})\lambda_1(\mathbb{B}^d, \mathbb{L}^*) \leqslant \delta d^{3/2}.$$

Due to a written communication of Banaszczyk in 1991 one may replace $\frac{3}{2}$ by 1 in the last inequality. For not necessarily o-symmetric convex bodies and for so-called *covering minima* $[\mu(K, \mathbb{L})$ then is $\mu_d(K, \mathbb{L})]$ covering results of this type have been found by Kannan and Lovász (1986). For algorithmic problems related to $\mu(\mathbb{B}^d, L)$ see Lagarias, Lenstra and Schnorr (1990) and Babai (1986).

Relations to algebraic number theory are discussed by Chalk (1983).

6. Reduction theory

The *basic problem* of reduction theory can be stated in two ways. First, given a lattice $\mathbb{L} = BZ^d$, determine (by means of a suitable algorithm) a basis (the "reduced" basis) having "nice" geometric or arithmetic properties. Second, given a positive quadratic form $q(x) = |Bx|^2 = x^{\mathrm{tr}}B^{\mathrm{tr}}Bx = x^{\mathrm{tr}}Ax$, find an equivalent form $q(Ux)$ having "nice" arithmetic properties.

Aside from its impact on (locally or globally) densest packings of balls, the elaborated and elegant results of classical reduction theory were mainly of intrinsic interest. Surprisingly enough the picture has changed completely in the last decade as the following exposition will show. The first definitions of reductions are due to Lagrange (1773), Seeber (1831), Gauss (1831), and Dirichlet (1850) for $d = 2, 3$. For general d definitions were proposed by Hermite (1850); Korkin and Zolotarev (1873); Selling (1874) ($d = 2, 3$), Charve (1882), Hofreiter (1933) ($d = 4$), Ryškov (1974), Baranovskii (1980) (general d); Minkowski (1911, Abh. II, pp. 53–100), Voronoi (1908a), see also Ryškov (1974), and Ryškov and Baranovskii (1979); Venkov (1940) specified a reduction depending on a given quadratic form; finally we mention the LLL-basis or Lovász reduction, see Lenstra, Lenstra and Lovász (1982) and the total reduction of Pohst and Zassenhaus (1989) which refines Minkowski's definition.

In the following we will consider Korkin–Zolotarev and LLL-reduction. General references for the other types have already been mentioned in the Introduction to which we add Delone (1937/8), van der Waerden (1956), van der Waerden and Gross (1968), Delone and Ryškov (1971), Ryškov and Baranovskii (1979), Ryškov (1980), Lovász (1986), Pohst and Zassenhaus (1989), and Siegel (1989).

6.1. *Korkin–Zolotarev reduction*

Let $B = \{b_1, \ldots, b_d\}$ be a basis of the lattice $\mathbb{L} = BZ^d$. By the Gram–Schmidt orthogonalization we have the following representation of d orthogonal vectors $\hat{b}_1, \ldots, \hat{b}_d$:

$$\begin{aligned} \hat{b}_1 &= b_1 \\ \hat{b}_2 &= b_2 - \mu_{21}\hat{b}_1 \\ &\vdots \\ \hat{b}_d &= b_d - \mu_{d1}\hat{b}_1 - \cdots - \mu_{d\ d-1}\hat{b}_{d-1} \end{aligned} \tag{1}$$

where

$$\mu_{ij} = \frac{b_i \cdot \hat{b}_j}{|\hat{b}_j|^2}.$$

For $0 \leqslant i < d$ consider the orthogonal projection (i.e., the projection parallel to the subspace generated by $b_1, \ldots, b_i$ or, equivalently, by $\hat{b}_1, \ldots, \hat{b}_i$) of $\mathbb{L}$ onto the subspace generated by $\hat{b}_{i+1}, \ldots, \hat{b}_d$. The image of $\mathbb{L}$ is a lattice $\mathbb{L}^{d-i}$ in this subspace with basis $\hat{b}_{i+1}, \hat{b}_{i+2} + \mu_{i+2\ i+1}\hat{b}_{i+1}, \ldots, \hat{b}_d + \mu_{d\ i+1}\hat{b}_{i+1} + \cdots + \mu_{d\ d-1}\hat{b}_{d-1}$. The basis $\{b_1, \ldots, b_d\}$ is said to be *reduced in the sense of Korkin–Zolotarev* if the following conditions hold:

(i) $|\mu_{ij}| \leqslant \frac{1}{2}$ for $1 \leqslant j < i \leqslant d$,

(ii) $\hat{b}_{i+1}$ is a shortest vector of $\mathbb{L}^{d-i}$ for $0 \leqslant i < d$.

There exists always a Korkin–Zolotarev reduced basis of $\mathbb{L}$. Schnorr (1987) showed that the orthogonality defect (see section 2) of a Korkin–Zolotarev reduced basis is at most d^d. This seems to be the smallest upper bound for the orthogonality defect known for any definition of reduction. Lagarias, Lenstra and Schnorr (1990) proved several structure theorems for Korkin–Zolotarev reduced basis. Kannan (1987) specified for fixed d a polynomial time algorithm to find Korkin–Zolotarev reduced bases for lattices with rational basis vectors. For other properties of Korkin–Zolotarev reduction see Ryškov and Baranovskii (1979).

The first application of Korkin–Zolotarev reduction was the derivation of the densest lattice packings of $\mathbb{B}^d$ for $d \leqslant 4$; see Korkin and Zolotarev (1872, 1877), and Ryškov and Baranovskii (1979). Lagarias, Lenstra and Schnorr (1990) used it for transference theorems; see the previous subsection. The problem of finding in polynomial time (in d and the maximum of the binary sizes of the coordinates of the given basis) a short vector of a lattice and the problem of finding in polynomial time a basis with small orthogonal defect are equivalent. This was proved by Lenstra and Schnorr (1984) and Schnorr (1985), compare the discussion in Lovász (1986, pp. 24–25). The problem of finding a shortest vector in a lattice $\mathbb{L}$ with rational basis vectors and that of finding a Korkin–Zolotarev reduced basis of it are "polynomial time equivalent"; see Lagarias, Lenstra and Schnorr (1990), but no polynomial time algorithm for these problems is known. The best known algorithms require time which depends on d in the form $d^{O(d)}$; see Kannan (1987b) and Schnorr (1987). For the importance of Korkin–Zolotarev reduction for the problem of finding for a (rational) $x \in \mathbb{E}^d$ a closest point of a lattice (with rational basis) and the relation of this problem to the shortest vector problem we refer to Lagarias, Lenstra and Schnorr (1990) and Kannan (1987a).

6.2. LLL *or Lovász reduction*

Let $B = \{b_1, \ldots, b_d\}$ be a basis of a lattice $\mathbb{L} = BZ^d$ and consider the representation (1). Then $\{b_1, \ldots, b_d\}$ is LLL or *Lovász reduced* if

(i) $|\mu_{ij}| \leqslant \frac{1}{2}$ for $1 \leqslant j < i \leqslant d$,

(ii') $(2/\sqrt{3})|\hat{b}_{i+2} + \mu_{i+2\ i+1}\hat{b}_{i+1}| \geqslant |\hat{b}_{i+1}|$ for $0 \leqslant i < d-1$;

see Lenstra, Lenstra and Lovász (1982).

This type of reduction can be considered as a sort of weakened version of the Korkin–Zolotarev reduction. (i) means that the lattice vectors are pairwise "close to orthogonal". (ii') means that the lengths of the basis vectors $\hat{b}_{i+1}, \ldots,$ of the projected lattice $\mathbb{L}^{d-1}$ do not decrease too fast.

Results of Lenstra, Lenstra and Lovász (1982) show that every lattice has an LLL-reduced basis. If $\{b_1, \ldots, b_d\}$ is such a basis then its orthogonality defect is at most $2^{d(d-1)/4}$ (which is much larger than that for Korkin–Zolotarev reduced bases). Furthermore, $|b_1| \leqslant 2^{(d-1)/2}\lambda_1(\mathbb{B}^d, \mathbb{L})$. The following result is due to Lenstra, Lenstra and Lovász (1982), for refinements see Schönhage (1984) and Schnorr (1988).

Theorem 13. *There is an algorithm having the following property: Let $\mathbb{L}$ be a lattice in $\mathbb{E}^d$ with a basis consisting of rational vectors. Then the algorithm produces a* LLL-*reduced basis in* $O(d^4 s)$ *arithmetic operations using only integers of binary length* $O(ds)$ *where s is the largest binary length of the coordinates of the vectors of the given basis.*

This algorithm has many applications. Examples are polynomial time algorithms for factorization of polynomials with integer coefficients (Lenstra, Lenstra and Lovász 1982), simultaneous Diophantine approximation (Lenstra, Lenstra and Lovász 1982) and integer programming for fixed d (Lenstra 1983). Further applications appear in the disproof of the Mertens conjecture (Odlyzko and te Riele 1985) and cryptography. For references see, e.g., Lovász (1986, 1988), Kannan (1987a) and, in particular, Grötschel, Lovász and Schrijver (1988).

7. Selected special problems and results

General references are Gruber and Lekkerkerker (1987) (and Erdős, Gruber and Hammer 1989).

7.1. Lattice constants

A *star body* is a closed subset of $\mathbb{E}^d$ with o in its interior such that any ray starting at o meets the boundary of S in at most one point. The *lattice constant* $\Delta(S)$ of S is defined by

$$\Delta(S) = \inf\{d(\mathbb{L}) : \mathbb{L} \in \mathcal{L} \text{ admissible for } S\}.$$

(If S has no admissible lattices then $\Delta(S) = \infty$.) If $\Delta(S) < \infty$ then Mahler's compactness theorem yields the existence of *critical lattices* of S; these are admissible lattices with determinant $\Delta(S)$. A major problem of the geometry of numbers is the determination of the lattice constants and the critical lattices of star bodies and convex bodies.

If K is a convex body then $V(K)/\Delta(K-K)$ is the maximum density of a lattice packing of K and the critical lattices of $K - K = \{x - y : x, y \in K\}$ are the corresponding packing lattices.

For the (small number of) known lattice constants of o-symmetric convex bodies consult the general references mentioned before.

For $1 \leqslant r < d$ let $S_{r,d-r}$ be the star body

$$S_{r,d-r} = \{x : |(\xi_1)^2 + \cdots + (\xi_r)^2 - (\xi_{r+1})^2 - \cdots - (\xi_d)^2| \leqslant 1\}.$$

The determination of the lattice determinants $\Delta(S_{r,d-r})$ or, in arithmetic terms, the homogeneous minima of indefinite quadratic forms has attracted some interest. Markov (1878, 1879, 1903) determined $\Delta(S_{11}), \Delta(S_{12}) = \Delta(S_{21})$ and Oppenheim (1929) found the values for $\Delta(S_{13}) = \Delta(S_{31})$ and $\Delta(S_{22})$. A *conjecture of Oppenheim*

(1929) which implies that $\Delta(S_{r,d-r}) = \infty$ for $1 \leqslant r < d$, $d \geqslant 5$, was proved by Margulis (1987) (in an even stronger form). Related inhomogeneous problems were investigated in detail by Dumir, Grover, Hans-Gill, Raka.

7.2. Minima of the Epstein zeta-function

Given a lattice $\mathbb{L} \in \mathscr{L}$ the *zeta-function* $\zeta(\mathbb{L}, \cdot)$ (introduced by Epstein and rediscovered by Sobolev) is defined in the following way:

$$\zeta(\mathbb{L}, s) = \sum\{\frac{1}{|l|^{2s}} : l \in \mathbb{L}\backslash\{o\}\} \quad \text{for } s > \tfrac{1}{2}d.$$

The main problem is to determine those lattices $\mathbb{L} \in \mathscr{L}(1)$, say, at which for a given s or for all sufficiently large s the function $\zeta(\cdot, s)$ has a local or global minimum. For $d = 2, 3$ substantial steps towards a solution have been made; for general d some extremal criteria are known. Delone and Ryškov (1967) pointed out that any lattice in $\mathscr{L}(1)$ which provides a local minimum of $\zeta(\cdot, s)$ for all sufficiently large s provides a lattice packing of Euclidean balls for which the density has a local maximum among all lattice packings of balls. The converse of this result holds for $d < 6$ but not for $d \geqslant 6$; see Ryškov (1973).

Sobolev (1974) considered the space of all real functions on a bounded open subset G of $\mathbb{E}^d$ with continuous partial derivatives up to order s and posed the following problem: in order to approximate

$$\int_G f \, \mathrm{d}x$$

by

$$\frac{V(G)}{n} \sum_{i=1}^{n} f(x_i),$$

how should one choose "knots" $x_1, \ldots, x_n \in G$ such as to minimize the error. If the knots are chosen from a lattice of given determinant δ, say, the best choice is a lattice $\mathbb{L}$ such that $\zeta(\cdot, s)$ attains its minimum on $\mathscr{L}(1)$ at $\delta^{1/d}\mathbb{L}^*$.

7.3. Moment problem

Motivated by several applications we propose the following *problem*: let $\omega : [0, +\infty[\to [0, +\infty[$ be a continuous non-decreasing function with $\omega(0) = 0$. Determine those lattices $\mathbb{L} \in \mathscr{L}(1)$ for which

$$\int_{D(\mathbb{L}, o)} \omega(|x|) \, \mathrm{d}x$$

is minimal where $D(\mathbb{L}, o)$ is the Dirichlet–Voronoi cell of $\mathbb{L}$; see section 2.

This problem was solved by Fejes Tóth (1972, p. 81), for $d = 2$ and Barnes and Sloane (1983) found the solution (conjectured by Bambah) in case $d = 3$ and $\omega(\tau) = \tau^2$. For $d > 3$ and $\omega(\tau) = \tau^2$ they conjecture that the minimizing lattices are the polar lattices of the lattices which provide densest packings of Euclidean balls.

The solution of the moment problem has important implications for numerical integration (see Babenko 1977) and the quantization of data (see Conway and Sloane 1987).

7.4. *Parallelohedra*

A system of compact subsets of $\mathbb{E}^d$ with non-empty interior is a *tiling* of $\mathbb{E}^d$ if their union equals $\mathbb{E}^d$ and any two distinct sets have disjoint interiors. The sets are called *tiles*. If the sets are all translates of a fixed set (called *prototile*) where the translation vectors are precisely the vectors of a lattice $\mathbb{L}$, we speak of a *lattice tiling* and $\mathbb{L}$ is called its *tiling lattice*. A *parallelohedron* P is a convex polytope which is the prototile of a lattice tiling. Given a lattice, the cells

$$D(\mathbb{L}, m) = \{x: |x - m| \leqslant |x - l| \text{ for all } l \in \mathbb{L}\}, \quad m \in \mathbb{L}$$

form the *Dirichlet–Voronoi tiling* corresponding to $\mathbb{L}$. It is a lattice tiling with prototile $D(\mathbb{L}, o)$ and tiling lattice $\mathbb{L}$. Call a parallelohedron P *primitive* if there is a lattice tiling of it such that each vertex of any tile belongs to and is a vertex of precisely d other tiles.

A *problem of Voronoi* asks to show that any parallelohedron is an affine image of a suitable Dirichlet–Voronoi cell. Voronoi (1908b, 1909) gave an affirmative answer for primitive parallelohedra and Delone (1929) did so for $d \leqslant 4$. A substantial improvement of Voronoi's result is due to Žitomirskii (1929).

A tiling with convex polytopes is *facet-to-facet* if each facet (i.e., a $(d - 1)$-dimensional face) of a tile is a facet of precisely one other tile.

It is *face-to-face* if the intersection of any two distinct tiles is a common face. By a result of Gruber and Ryškov (1989) a facet-to-facet tiling with parallelohedra is face-to-face and thus gives rise to an (infinite) polyhedral complex. We say that two parallelohedra are of the same *(combinatorial) type* if their (up to rigid motions) unique facet-to-facet tilings determine combinatorially isomorphic polytopal complexes. For $d = 2, 3$ the two, respectively five types were determined by Fedorov (1885). For $d = 4$ there are 52 types according to Delone (1929) and Štogrin (1973). For $d \geqslant 5$ there are contributions of Ryškov and Baranovskii (1976), but the general problem remains open.

7.5. *Conjecture on the product of non-homogeneous linear forms*

Let $l_1, \ldots, l_d$ be d real linear forms in d variables of determinant 1 and let $\alpha_1, \ldots, \alpha_d \in \mathbb{R}$ be arbitrary. Then there are integral values of the variables, say $u \in Z^d$, such that

$$|(l_1(u) - \alpha_1) \cdots (l_d(u) - \alpha_d)| \leqslant 1/2^d. \tag{1}$$

Here equality occurs if and only if after a suitable integral unimodular transformation of the variables the linear forms are of the form $\beta_1 u_1, \ldots, \beta_d u_d$ while $\alpha_1 \equiv \beta_1/2 \bmod \beta_1, \ldots, \alpha_d \equiv \beta_d/2 \bmod \beta_d$. For equivalent (more geometric) formulations and a comprehensive survey see Gruber and Lekkerkerker (1987).

There are several lines of attack: so far the conjecture has been proved for $d \leqslant 5$; see Skubenko (1976). For general d there are upper estimates for the product in (1), all of the form $2^{-d/2+o(d)}$. More recent ones are due to Andriyasyan, Ilin and Malyshev (1986) and Malyshev (1987). For "many" (coefficient) matrices (of the linear forms) the conjecture is true, where "many" can be interpreted as "for any DOTU-matrix"; see Macbeath (1961) or in the measure sense; see Gruber (1967).

7.6. Mordell's converse problem for the linear form theorem

Find the supremum κ_d of the $\kappa > 0$ having the following property: Let $l_1, \ldots, l_d$ be d real linear forms of determinant 1. Then there are reals $\tau_1, \ldots, \tau_d > 0, \tau_1 \cdots \tau_d = \kappa$ such that the inequalities

$$|l_1(u)| \leqslant \tau_1, \ldots, |l_d(u)| \leqslant \tau_d$$

have no solution $u \in Z^d$ except for $u = o$.

κ_2 is known and there is a conjecture for κ_3. Estimates for κ_d for large values of d are due to Hlawka (1950, Selecta 118–124), and to Gruber and Ramharter (1982). So-called isolation phenomena were considered by Ramharter (1981). For a corresponding inhomogeneous problem see Bambah, Dumir and Hans-Gill (1986).

Acknowledgements

For many discussions and suggestions I am obliged to H. Groemer, M. Grötschel, E. Hlawka, F.J. Schnitzer and J.M. Wills.

References

Andriyasyan, A.K., I.V. Ilin and A.V. Malyshev
[1986] Application of computers to estimates of Chebotarev type in the non-homogeneous Minkowski conjecture, *Zap. Naučn. Sem. Leningrad. Otdel. Mat. Inst. Steklov. (LOMI)* **51**, 9–25, 195.

Babai, L.
[1986] On Lovász's lattice reduction and the nearest lattice point problem, *Combinatorica* **6**, 1–14.

Babenko, V.F.
[1977] On the optimal error bound for cubature formulae on certain classes of continuous functions, *Anal. Math.* **3**, 3–9.

Bambah, R.P.
[1954] Polar reciprocal convex bodies, *Proc. Cambridge Philos. Soc.* **51**, 377–378.

Bambah, R.P., A.C. Woods and H.J. Zassenhaus
[1964] Three proofs of Minkowski's second inequality in the geometry of numbers, *J. Austral. Math. Soc.* **5**, 453–462.

Bambah, R.P., V.C. Dumir and R.J. Hans-Gill
[1986] An analogue of a problem of Mordell, *Studia Sci. Math. Hungar.* **21**, 135–142.

Banaszczyk, W.
[1989] Polar lattices from the point of view of nuclear spaces, *Rev. Mat. Univ. Compl. Madrid* **2**(suppl.), 35–46.
[1990] On the lattice packing–covering ratio of finite dimensional normed spaces, *Colloq. Math.* **59**, 31–33.
[1991] Additive subgroups of topological vector spaces, in: *Lecture Notes in Math.*, Vol. 1466 (Springer, Berlin).

Baranovskiĭ, E.P.
[1980] The reduction domain in the sense of Selling of positive quadratic forms of five variables, *Trudy Mat. Inst. Steklov* **152**, 5–33 [*Proc. Steklov Math. Inst.* **152**, 5–35].

Barnes, E.S., and N.J.A. Sloane
[1983] The optimal lattice quantizer in three dimensions, *SIAM J. Algebraic Discrete Methods* **4**, 30–41.

Blichfeldt, H.F.
[1914] A new principle in the geometry of numbers with some applications, *Trans. Amer. Math. Soc.* **15**, 227–235.
[1921] Note on geometry of numbers, *Bull. Amer. Math. Soc.* **27**, 152–153.
[1929] The minimum values of positive quadratic forms and closest packing of spheres, *Math. Ann.* **101**, 605–608.

Bourgain, J., and V.D. Milman
[1985] Sections euclidiennes et volume des corps symétriques convexes dans R^n, *C.R. Acad. Sci. Paris Sér. I Math.* **300**, 435–438.

Cassels, J.W.S.
[1972] *An Introduction to the Geometry of Numbers* (Springer, Berlin, 2nd ed.).

Chalk, J.H.H.
[1967] A note on coverings of E^d by convex sets, *Canad. Math. Bull.*, 668–673.
[1983] Algebraic lattices, in: *Convexity and its Applications* (Birkhäuser, Basel) pp. 97–110.

Charve, H.F.
[1882] De la réduction des formes quadratiques quaternaires positives, *Ann. Sci. École Norm. Sup. (2)* **11**, 119–134.

Conway, J.H., and N.J.A. Sloane
[1987] *Sphere Packings, Lattices and Groups* (Springer, Berlin).

Dani, S.G., and G.A. Margulis
[1989] Values of quadratic forms at primitive integral points, *C.R. Acad. Sci. Paris Sér. I Math.* **308**, 199–203.

Danicic, V.I.
[1969] An elementary proof of Minkowski's second inequality, *J. Austral. Math. Soc.* **10**, 177–181.

Davenport, H.
[1946] The product of n homogeneous linear forms, *Indag. Math.* **8**, 525–531.
[1955] On a theorem of Furtwangler, *J. London Math. Soc.* **30**, 185–195.
[1977] *The Collected Works of Harold Davenport,* Vols. 1, 2 (Academic Press, New York).

Delone, B.N.
[1929] Sur la partition régulière de l'espace à 4 dimensions, *Izv. Akad. Nauk SSSR Otdel. Fiz.-Mat. Nauk* **7**, 79–110; 147–164.
[1937] Geometry of positive quadratic forms, *Uspekhi Mat. Nauk* **3**, 16–62.
[1938] Geometry of positive quadratic forms, *Uspekhi Mat. Nauk* **4**, 102–164.

Delone, B.N., and S.S. Ryškov
[1967] A contribution to the theory of the extrema of a multidimensional S-function, *Dokl. Akad. Nauk SSSR* **8**, 499–503.
[1971] Extremal problems in the theory of positive quadratic forms, *Trudy Mat. Inst. Steklov* **112**, 203–223 [*Proc. Steklov Inst. Math.* **112**, 211–231].

Dirichlet, P.G.L.
[1850] Über die Reduktion der positiven quadratischen Formen mit drei unbestimmten ganzen Zahlen, *J. Reine Angew. Math.* **40**, 209–227 [*Werke*, Vol. 2 (Reimer, Berlin, 1897) pp. 27–48].

Engel, P.
[1988] Mathematical problems in modern crystallography. Crystal symmetries, *Comput. Math. Appl.* **16**, 425–436.

Erdős, P., P.M. Gruber and J. Hammer
[1989] *Lattice Points* (Longman Scientific, Harlow, Essex/Wiley, New York).

Fedorov, E.S.
[1885] Elements of the study of figures, *Zap. Mineralog. Obsc. (2)* **21**, 1–279. Reprinted: Izdat. Akad. Nauk SSSR, Moscow, 1953.

Fejes Tóth, L.
[1972] *Lagerungen in der Ebene, auf der Kugel und im Raum* (Springer, Berlin, 2nd ed.).

Frumkin, M.A.
[1976] An algorithm for the reduction of a matrix of integers to triangular form with complexity of the computations, *Èkonom. i Mat. Metody* **12**, 173–178.

Gauss, C.F.
[1831] Untersuchungen über die eigenschaften der positiven ternären quadratischen Formen von Ludwig August Seeber, Göttingische gelehrte Anzeigen. Reprinted: *Werke*, Vol. 2 (Kgl. Ges. Wiss., Göttingen, 1836) pp. 188–196; *J. Reine Angew. Math.* **20** (1840) 312–320.

Groemer, H.
[1971] Continuity properties of Voronoi domains, *Monatsh. Math.* **74**, 423–431.

Grötschel, M., L. Lovász and A. Schrijver
[1988] *Geometric Algorithms and Combinatorial Optimization* (Springer, Berlin).

Gruber, P.M.
[1967] Über das Produkt inhomogener Linearformen, *Acta Arith.* **13**, 9–27.
[1979] Geometry of numbers, in: *Contributions to Geometry, Proc. Symp. Siegen, 1978* (Birkhäuser, Basel) pp. 186–225.
[1986] Typical convex bodies have surprisingly few neighbours in densest lattice packings, *Studia Sci. Math. Hungar.* **21**, 163–173.

Gruber, P.M., and C.G. Lekkerkerker
[1987] *Geometry of Numbers* (North-Holland, Amsterdam, 2nd ed.).

Gruber, P.M., and G. Ramharter
[1982] Beiträge zum Umkehrproblem für den Minkowskischen Linearformensatz, *Acta Math. Acad. Sci. Hungar.* **39**, 135–141.

Gruber, P.M., and S.S. Ryškov
[1989] Facet-to-facet implies face-to-face, *European J. Combin.* **10**, 83–84.

Håstad, J.
[1988] Dual vectors and lower bounds for the nearest lattice point problem, *Combinatorica* **8**, 75–81.

Henk, H.
[1992] Minkowski's second theorem, Manuscript.

Hermite, Ch.
[1850] Extraits de lettre de M. ch. Hermite à M. Jacobi súr differents objets de la théorie des nombres. Première lettre, *J. Reine Angew. Math.* **40**, 261–278 [*Oeuvres*, Vol. 1 (Gauthier-Villars, Paris, 1905) pp. 100–121].

Hlawka, E.
[1944] Zur Geometrie der Zahlen, *Math. Z.* **49**, 285–312 [*Selecta* (Springer, Heidelberg) pp. 38–65].
[1950] Über Integrale auf konvexen Körpern I, *Monatsh. Math.* **54**, 1–36 [*Selecta* (Springer, Heidelberg) pp. 125–160].
[1980] 90 Jahre Geometrie der Zahlen, in: *Jahrb. Überbl. Math.* (Bibliographisches Institut, Mannheim) pp. 9–41 [*Selecta* (Springer, Heidelberg) pp. 398–430].
[1990] *Selecta* (Springer, Heidelberg).

Hofreiter, N.
[1933] Verallgemeinerung der Sellingschen Reduktionstheorie, *Monatsh. Math. Phys.* **40**, 393–406.

Jarnik, V.
[1941] Zwei Bemerkungen zur Geometrie der Zahlen, *Věstnik Kråloveské České Společn. Nauk*, 12pp.
[1949] On Estermann's proof of a theorem of Minkowski, *Časopis Pěst Mat. Fyz.* **73**, 131–140.

Kabatjanskii, G.A., and V.I. Levenštein
[1978] Bounds for packings on a sphere and in space, *Problemy Peredachi Informatsii* **14**, 3–25 [*Problems Inform. Transmission* **14**, 1–17].

Kannan, R.
[1987a] Algorithmic geometry of numbers, *Ann. Rev. Comput. Sci.* **2**, 231–267.
[1987b] Minkowski's convex body theorem and integer programming, *Math. Oper. Res.* **12**, 415–440.

Kannan, R., and A. Bachem
[1979] Polynomial time algorithms for computing the Smith and Hermite normal forms of an integer matrix, *SIAM J. Comput.* **8**, 499–507.

Kannan, R., and L. Lovász
[1986] Covering minima and lattice point free convex bodies, in: *Lecture Notes Comput. Sci.*, Vol. 241 (Springer, Berlin) pp. 193–213.

Keller, D.-H.
[1954] Geometrie der Zahlen, in: *Enzyklopedie Math. Wiss.*, Vol. 1(2) (Teubner, Leipzig) p. 27.

Korkin, A., and G. Zolotarev
[1872] Sur les formes quadratiques positives quaternaires, *Math. Ann.* **5**, 581–583.
[1873] Sur les formes quadratiques, *Math. Ann.* **5**, 366–389.
[1877] Sur les formes quadratiques positives, *Math. Ann.* **11**, 242–292.

Lagarias, J.C., H.W. Lenstra Jr and C.P. Schnorr
[1990] Korkine–Zolotarev bases and successive minima of a lattice and its reciprocal lattice, *Combinatorica* **10**, 333–348.

Lagrange, J.L.
[1773] Recherches d'arithmétique, *Nouv. Mém. Acad. Roy. Sc. Belles Lettres Berlin*, 265–312 [*Oeuvres*, Vol. 3 (Gauthier–Villars, Paris, 1869) pp. 693–758].

Leech, J., and N.J.A. Sloane
[1971] Sphere packings and error-correcting codes, *Canad. J. Math.* **23**, 718–745.

Lekkerkerker, C.G.
[1961] A theorem on the distribution of lattices, *Nederl. Akad. Wetensch. Indag. Math.* **23**, 197–210.

Lenstra, A.K., H.W. Lenstra Jr and L. Lovász
[1982] Factoring polynomials with rational coefficients, *Math. Ann.* **261**, 515–534.

Lenstra Jr, H.W.
[1983] Integer programming with a fixed number of variables, *Math. Oper. Res.* **8**, 538–548.

Lovász, L.
[1986] *An Algorithmic Theory of Numbers, Graphs and Convexity* (SIAM, Philadelphia).
[1988] Geometry of numbers: an algorithmic view, in: *ICIAM'87: Proc. 1st Int. Conf. Industr. Appl. Math., Paris, 1987* (SIAM, Philadelphia, PA) pp. 144–152.

Macbeath, A.M.
[1961] Factorization of matrices and Minkowski's conjecture, *Proc. Glasgow Math. Assoc.* **5**, 86–89.

Mahler, K.
[1938] A theorem on inhomogeneous diophantine inequalities, *Proc. Kon. Ned. Akad. Wet.* **41**, 634–637.
[1939a] Ein Übertragungsprinzip für konvexe Körper, *Časopis Pěst. Mat. Fyz.* **68**, 85–92.
[1939b] Ein Übertragungsprinzip für konvexe Körper, *Časopis Pěst. Mat. Fyz.* **68**, 93–102.
[1946] On lattice points in n-dimensional star bodies I. Existence theorems, *Proc. Roy. Soc. London A* **187**, 151–187.

Malyshev, A.V.
[1987] Estimates for the inhomogeneous arithmetic minimum of a product of linear forms, *Zap. Naučn. Sem. Leningrad. Otdel. Mat. Inst. Steklov. (LOMI)* **160**, 138–150, 299.

Margulis, G.A.
[1987] Formes quadratiques indéfinies et flots unipotents sur les espaces homogènes, *C.R. Acad. Sci. Paris Sér. I Math.* **304**, 249–253.

Markov, A.
[1878] Sur les formes quadratiques binaires indéfinies, *Math. Ann.* **15**, 381–406.
[1879] Sur les formes quadratiques binaires indéfinies, *Math. Ann.* **17**, 379–399.
[1903] Sur les formes quadratiques ternaires indéfinies, *Math. Ann.* **56**, 233–251.

Milnor, J., and D. Husemoller
[1973] *Symmetric Bilinear Forms* (Springer, Berlin).

Minkowski, H.
[1896] *Geometrie der Zahlen* (Teubner, Leipzig). Reprinted: Johnson, New York, 1968.
[1907] *Diophantische Approximationen* (Teubner, Leipzig). Reprinted: Physika, Würzburg, 1961.
[1911] *Gesammelte Abhandlungen*, Vols. 1, 2 (Teubner, Leipzig).

Odlyzko, A.M., and H. te Riele
[1985] Disproof of the Mertens conjecture, *J. Reine Angew. Math.* **357**, 138–160.

Oppenheim, A.
[1929] The minima of indefinite quaternary quadratic forms, *Proc. Nat. Acad. Sci. U.S.A.* **15**, 724–727.

Pohst, M., and H. Zassenhaus
[1989] *Algorithmic Algebraic Number Theory* (Cambridge Univ. Press, Cambridge).

Quebbemann, H.-G.
[1991] A note on lattices of rational functions, *Mathematika* **38**, 10–13.

Ramharter, G.
[1981] Über ein Problem von Mordell in der Geometrie der Zahlen, *Monatsh. Math.* **92**, 143–160.

Rogers, C.A.
[1947] Existence theorems in the geometry of numbers, *Ann. of Math. (2)* **48**, 994–1002.
[1964] *Packing and Covering* (Cambridge Univ. Press, Cambridge).

Rush, J.A.
[1989] A lower bound on packing density, *Invent. Math.* **98**, 499–509.

Ryškov, S.S.
[1973] On the question of the final ζ-optimality of lattices providing the closest lattice packing of n-dimensional spheres, *Sibirsk. Mat. Zh.* **14**, 1065–1075, 1158 [*Siberian Math. J.* **14** (1973/74) 743–750].
[1974] The Venkov reduction of positive quadratic forms, *Ivanov. Gos. Univ. Ucen. Zap.* **89**, 5–36.
[1977] The geometry of positive quadratic forms, in: *Proc. Int. Congr. Math., Vancouver, 1974*, Vol. 1 (Canad. Math. Congress, Montreal, 1975) pp. 501–506 [*Amer. Math. Soc. Transl. (2)* **109**, 27–32].

Ryškov, S.S., ed.
[1980] *The Geometry of Quadratic Forms,* Trudy Mat. Inst. Steklov., Vol. 152 [*Proc. Steklov. Inst. Math.* 1982].

Ryškov, S.S., and E.P. Baranovskii
[1976] *C-Types of* n*-dimensional Lattices and 5-dimensional Primitive Parallelohedra (with Applications to the Theory of Coverings),* Akad. Nauk SSSR Trudy Otdel. Leningrad. Mat. Inst. Steklov., Vol. 137 [*Proc. Steklov. Math. Inst.*, 1978].
[1979] Classical methods in the theory of lattice packings, *Uspekhi Mat. Nauk* **34**, 3–63, 356 [*Russian Math. Surveys* **34**, 1–68].

Santaló, L.A.
[1949] Un invariante afin para las cuerpos convexos del espacio de n dimensiones, *Portugal. Math.* **8**, 155–161.

Scherrer, W.
[1922] Ein Satz über Gitter und Volumen, *Math. Ann.* **86**, 99–107.

Schmidt, W.M.
[1963] On the Minkowski–Hlawka theorem, *Illinois J. Math.* **7**, 18–23.

Schnorr, C.P.
[1985] A hierarchy of polynomial time basis reduction algorithms, in: *Colloq. Math. Soc. János Bolyai* **44**, 375–386.
[1987] A hierarchy of polynomial time lattice basis reduction algorithms, *Theoret. Comput. Sci.* **53**, 201–224.
[1988] A more efficient algorithm for lattice basis reduction, *J. Algorithms* **9**, 47–62.

Schönhage, A.
[1984] Factorization of univariate integer polynomials by Diophantine approximation and an improved basis reduction algorithm, in: *Lecture Notes Comput. Sci.*, Vol. 172 (Springer, Berlin) pp. 436–447.

Seeber, L.A.
[1831] *Untersuchungen über die Eigenschaften der positiven ternären quadratischen Formen* (Freiburg).

Selling, E.
[1874] Über die binären und ternären quadratischen Formen, *J. Reine Angew. Math.* **77**, 143–229.

Siegel, C.L.
[1935] Über Gitterpunkte in konvexen Körpern und ein damit zusammenhängendes Extremalproblem, *Acta Math.* **65**, 307–323 [*Abhandlungen*, Vol. 1 (Springer, Berlin) pp. 311–325].
[1945] A mean value theorem in geometry of numbers, *Ann. of Math. (2)* **46**, 340–347 [*Abhandlungen*, Vol. 3 (Springer, Berlin) pp. 39–46].
[1989] *Lectures on the Geometry of Numbers* (Springer, Berlin).

Skubenko, B.F.
[1976] A proof of Minkowski's conjecture on the product of n linear inhomogeneous linear forms in n variables for $n \leqslant 5$, *J. Soviet Math.* **6**, 627–650.

Sobolev, S.L.
[1974] *Introduction to the Theory of Cubature Formulae* (Izdat Nauka, Moscow).

Štogrin, M.I.
[1973] Regular Dirichlet–Voronoi partitions for the second triclinic group, *Akad. Nauk SSSR Trudy Mat. Inst. Steklov.* **123** [*Proc. Steklov. Inst. Math.* **123** (1975)].

Uhrin, B.
[1981] On a generalization of Minkowski's convex body theorem, *J. Number Theory* **13**, 192–209.

Van der Corput, J.G.
[1935] Verallgemeinerung einer Mordellschen Beweismethode in der Geometrie der Zahlen, *Acta Arith.* **1**, 62–66.
[1936] Verallgemeinerung einer Mordellschen Beweismethode in der Geometrie der Zahlen, *Acta Arith.* **2**, 145–146.

Van der Waerden, B.L.
[1956] Die Reduktionstheorie der positiven quadratischen Formen, *Acta Math.* **96**, 265–309. *Berichtigung* **98** (1957) 3–4.

Van der Waerden, B.L., and H. Gross, eds
[1968] *Studien zur Theorie der Quadratischen Formen* (Birkhäuser, Basel).

Venkov, B.A.
[1940] On the reduction of positive quadratic forms, *Izv. Akad. Nauk SSSR Ser. Mat.* **4**, 37–52.

Voronoi, G.
[1908a] Nouvelles applications des paramètres continus à la théorie des formes quadratiques I, *J. Reine Angew. Math.* **134**, 198–287.
[1908b] Propriétés des formes quadratiques positives parfaites, *J. Reine Angew. Math.* **133**, 97–178.
[1909] Nouvelles applications des paramètres continus à la théorie des formes quadratiques II, *J. Reine Angew. Math.* **136**, 97–178.
[1952] *Collected Works*, Vols. 1–3 (in Russian) (Izdat. Akad. Nauk Ukrain SSSR, Kiev).

Wáng, Y.
[1975] A note on a theorem of Davenport, *Acta Math. Sinica* **18**, 286–289.

Wills, J.M.
[1991] Affine surface area and a lattice point inequality, Manuscript.

Woods, A.C.
[1956] The anomaly of convex bodies, *Proc. Cambridge Philos. Soc.* **52**, 406–423.
[1958] On two-dimensional convex bodies, *Pacific J. Math.* **8**, 635–640.

Yates, D.L.
[1970] Suprema for a class of finite sums with applications to lattice coverings, *Proc. London Math. Soc. (3)* **21**, 731–756.

Žitomirskii, O.K.
[1929] Verschärfung eines Satzes von Woronoi, *Z. Leningrad. Fiz.-Mat. Obsc.* **2**, 131–151.

CHAPTER 3.2

Lattice Points

Peter GRITZMANN*

Universität Trier, Fb IV, Mathematik, Postfach 3825, D-54228 Trier, Germany

Jörg M. WILLS

Fachbereich Mathematik, Universität Siegen, Hölderlinstrasse 3, D-57068 Siegen, Germany

Contents

* Research of the first author supported in part by the Deutsche Forschungsgemeinschaft.

HANDBOOK OF CONVEX GEOMETRY
Edited by P.M. Gruber and J.M. Wills

1. Preliminaries

1.1. Introduction

This survey deals with various lattice point problems, both, from a theoretical and from an algorithmic point of view. Lattice point problems trace back to Lagrange, Gauß, Dirichlet and Hermite. Most important contributions came from Minkowski, when he created in 1891 the field "Geometry of Numbers" by applying geometric tools to number-theoretic problems. The basic idea was to interpret integer solutions of equations or inequalities as points of the integer lattice. This approach was very fruitful and many eminent mathematicians – Weyl, Siegel, Mordell, Blichfeldt, van der Corput, Mahler, Davenport, Hlawka, W.M. Schmidt among them – contributed to this field; for historic outlines see Hlawka (1980) and Schwermer (1991).

Lattice points are also used in other areas including numerical analysis, computer science and, in particular, integer programming.

The central notions in this paper are, first, the *lattice point enumerator* $G_{\mathbb{L}}$, a functional that is defined for a given point lattice $\mathbb{L}$ in $\mathbb{R}^d$ and that associates with a given subset S of $\mathbb{R}^d$ the number $\mathrm{card}(S \cap \mathbb{L})$ of $\mathbb{L}$-points in S and, second, *convex lattice polytopes*, polytopes that are the convex hull of their $\mathbb{L}$-points. In particular, we are interested in properties of $\mathbb{L}$-polyhedra $I_{\mathbb{L}}(K)$ that are associated with given convex bodies K via $I_{\mathbb{L}}(K) = \mathrm{conv}\,(\mathbb{L} \cap K)$.

The integer programming problem is the task to maximize a linear functional over $I_{\mathbb{Z}^d}(F)$ where F is the region of points that satisfy a given system of rational linear equalities and linear inequalities. A basic question of interest is to give criteria for $I_{\mathbb{Z}^d}(F)$ to be nonempty. A related but more general problem is to give upper and lower bounds for $G_{\mathbb{L}}(I_{\mathbb{L}}(F))$, bounds on the number of such solutions for a given lattice $\mathbb{L}$.

In the geometry of numbers there has been some interest in bounds of the discrete functional $G_{\mathbb{L}}$ in terms of continuous functionals like the volume, the surface area and other quermassintegrals. For algorithmic purposes other functionals are of some relevance.

All these connections with other areas have been fruitful and some of the more recent contacts will be addressed in this article, in particular some of the computational aspects of the field. We will, however, not focus on aspects that are being covered by various books and survey articles like Cassels (1971), Gruber and Lekkerkerker (1987), Erdős, Gruber and Hammer (1989), Schrijver (1986, 1993a,b), Gruber (1979), Hlawka (1980), Keller (1954), Lang (1990) and Lagarias (1993) and, particularly, limit the overlap with other chapters of this Handbook as much as possible. These chapters are: 3.1 *Geometry of Numbers* by Gruber, 3.3 *Packing and Covering with Convex Sets* by G. Fejes Tóth and Kuperberg, 3.4 *Finite Packing and Covering* by Gritzmann and Wills, 2.8 *Convexity and Discrete Optimization* by Burkard and 2.7 *Mathematical Programming and Convex Geometry* by Gritzmann and Klee. In particular, we will not cover the following topics:

weighted lattice point enumerator;
lattice points on the boundary of convex bodies;
lattice points in "large" convex bodies; asymptotic results.

For surveys on these subjects see Betke and Wills (1979), Fricker (1982) and Walfisz (1957). Further there are far too many ramifications of 2-dimensional results to be mentioned here in detail; we restrict our attention to those which we deemed most closely related to the purpose of this article.

1.2. General definitions and notation

Let $\mathbb{R}^d$ denote the d-dimensional vector space over the reals; $\mathbb{E}^d$ is the d-dimensional Euclidean space. The Euclidean scalar product and unit ball are denoted by $\langle\cdot,\cdot\rangle$ and $\mathbb{B}^d$, respectively.

A *convex body* in $\mathbb{R}^d$ is a nonempty compact convex subset of $\mathbb{R}^d$. Let $\mathcal{K}^d$ denote the set of convex bodies of $\mathbb{R}^d$, let $\mathcal{K}_0^d$ be the set of convex bodies with nonempty interior and let $\hat{\mathcal{K}}_0^d$ denote the subset of convex bodies in $\mathcal{K}_0^d$ that are symmetric about the origin.

A *convex polytope* in $\mathbb{R}^d$ is a convex body that can be presented as the convex hull of finitely many points (or equivalently, as the intersection of finitely many closed halfspaces). The set of all convex polytopes in $\mathbb{R}^d$ will be denoted by $\mathcal{P}^d$; $\mathcal{P}_0^d$ and $\hat{\mathcal{P}}_0^d$ are abbreviations for $\mathcal{P}^d \cap \mathcal{K}_0^d$ and $\mathcal{P}^d \cap \hat{\mathcal{K}}_0^d$, respectively. Given $P \in \mathcal{P}^d$, let $\mathcal{F}_i(P)$ denote the set of all i-dimensional faces of P. For comprehensive treatments of combinatorial properties of convex polytopes see Grünbaum (1967) and chapter 2.3 by Bayer and Lee in this Handbook.

For computational purposes it is sometimes relevant to distinguish between the two different ways how polytopes may be given.

A $\mathcal{V}$*-presentation* of a polytope $P \subset \mathbb{R}^d$ consists of positive integers d and m, and m points $v_1, \ldots, v_m$ in $\mathbb{Q}^d$ such that $P = \operatorname{conv}\{v_1, \ldots, v_m\}$. An $\mathcal{H}$*-presentation* of P consists of positive integers d and m, a rational $m \times d$ matrix A, and a vector $b \in \mathbb{Q}^m$ such that $P = \{x \in \mathbb{R}^d\colon Ax \leqslant b\}$ Sometimes we use the shorter form $\mathcal{V}$- or $\mathcal{H}$-polytope when dealing with a polytope that is $\mathcal{V}$- or $\mathcal{H}$-presented, respectively, and we will also speak of an $\mathcal{H}$-polyhedron F to indicate that $F = \{x \in \mathbb{R}^d\colon Ax \leqslant b\}$ with rational A and b.

For $K \in \mathcal{K}^d$ let $V(K)$ denote K's volume; in particular, we set $\kappa_d = V(\mathbb{B}^d)$, the volume of the Euclidean unit ball. An important set of functionals on $\mathcal{K}^d$ are Minkowski's quermassintegrals $W_0, \ldots, W_d$ or their renormalization, the intrinsic volumes $V_0, \ldots, V_d$, given for $j = 0, 1, \ldots, d$ by $V_j = \kappa_{d-j}^{-1} \binom{d}{j} W_{d-j}$; (see Hadwiger 1957, McMullen 1975, 1977, McMullen and Schneider 1983, and chapter 1.8 by Schneider in this Handbook). The intrinsic volume V_i is continuous, additive, monotone and positive, invariant under rigid motions, homogeneous of degree i and independent of the dimension of the space in which K is embedded. In particular, $V_d = V$ is the volume, $V_{d-1} = \frac{1}{2}F$ is half the surface area and $V_0 = 1$.

A lattice $\mathbb{L}$ in $\mathbb{R}^d$ is a discrete subset $\mathbb{L}$ of $\mathbb{R}^d$ of the following form: there is a basis $\{v_1,\ldots,v_d\}$ such that

$$\mathbb{L}=\left\{\sum_{j=1}^{d}\nu_i v_i:\ \nu_1,\ldots,\nu_d\in\mathbb{Z}\right\}.$$

The set of all lattices in $\mathbb{R}^d$ will be denoted by $\mathscr{L}^d$. For a lattice $\mathbb{L}$ with basis $\{v_1,\ldots,v_d\}$, $\det(\mathbb{L})$ denotes the determinant of $\mathbb{L}$, the volume of the parallelotope $\sum_{i=1}^{d}[0,1]v_i$, and $C(\mathbb{L})$ denotes the *Dirichlet*-cell of 0, the set of all points of $\mathbb{R}^d$ which are not farther away from 0 than from any other lattice point. Observe that $C(\mathbb{L})$ is the closure of a fundamental region of $\mathbb{L}$; see chapters 3.1 by Gruber and 3.3 by G. Fejes Tóth and Kuperberg.

For a given lattice $\mathbb{L}$ let $\mathscr{P}^d(\mathbb{L})$ and $\mathscr{P}_0^d(\mathbb{L})$ denote the subsets of $\mathscr{P}^d$ and $\mathscr{P}_0^d$, respectively, of convex polytopes P with $\mathscr{F}_0(P)\subset\mathbb{L}$. The polytopes $P\in\mathscr{P}^d(\mathbb{L})$ are called convex $\mathbb{L}$-polytopes (or $\mathbb{L}$-integer polytopes). A convex $\mathbb{L}$-polyhedron P is the intersection of finitely many closed halfspaces such that $P=\operatorname{conv}(P\cap\mathbb{L})$. $\mathbb{L}$-rational polytopes and polyhedra are defined analogously. For $\mathbb{L}\in\mathscr{L}^d$ and an arbitrary convex set C the $\mathbb{L}$-hull (or the $\mathbb{L}$-integer hull) of C is given by $I_{\mathbb{L}}(C)=\operatorname{conv}(\mathbb{L}\cap C)$.

For a subset $S\subset\mathbb{R}^d$ the lattice point enumerator $G_{\mathbb{L}}(S)$ is defined by $G_{\mathbb{L}}(S)=\operatorname{card}(S\cap\mathbb{L})$. Typically, S will be a convex body but we will also consider cases where S is the boundary or the interior of a convex body. For $\mathbb{L}=\mathbb{Z}^d$ we write G instead of $G_{\mathbb{Z}^d}$.

As usual, the polar lattice $\mathbb{L}^*$ of a lattice $\mathbb{L}$ is defined by

$$\mathbb{L}^*=\{x\in(\mathbb{E}^d)^*\colon y\in\mathbb{L}\Longrightarrow\langle x,y\rangle\in\mathbb{Z}\}.$$

$\mathbb{L}^*$ is a lattice in the conjugate space $(\mathbb{E}^d)^*$. Since $\mathbb{E}^d$ is self-conjugate we will regard $\mathbb{L}^*$ as a lattice in $\mathbb{E}^d$. Then, clearly, $(\mathbb{L}^*)^*=\mathbb{L}$ and, in particular, $(\mathbb{Z}^d)^*=\mathbb{Z}^d$. For various interactions between $\mathbb{L}$ and $\mathbb{L}^*$ see McMullen (1984), Kannan and Lovász (1988) and Schnell (1992).

In sections 5 and 6 we will use the standard notation of complexity theory. We use the binary Turing machine model, hence, the size of the input of a problem – usually denoted by L – is the number of binary digits needed to encode the input data. The complexity classes that are relevant here are the classes P, NP-hard, NP-complete, #P-hard and #P-complete. For underlying concepts of theoretical computer science, definitions and for numerous results see Aho, Hopcroft and Ullman (1974) and Garey and Johnson (1979).

2. Centrally symmetric convex bodies

2.1. *Minkowski's fundamental theorem*

The starting point of the geometry of numbers and the predominant result of this section is Minkowski's fundamental theorem, Minkowski (1896).

Let $\mathbb{L} \in \mathscr{L}^d$, $K \in \hat{\mathscr{K}}_0^d$ *and suppose* $G_{\mathbb{L}}(\operatorname{int} K) = 1$. *Then*

$$V(K) \leqslant 2^d \det \mathbb{L}. \tag{2.1}$$

This estimate is tight; in particular, it holds with equality for lattice parallelotopes. In addition, the slack in (2.1) can be expressed explicitly:

$$\frac{1}{V(K)} \sum_{u \in \mathbb{L}^* \setminus \{0\}} \left| \int_{(1/2)K} \mathrm{e}^{-\pi i u x}\, \mathrm{d}x \right|^2 = 2^d \det \mathbb{L} - V(K).$$

This identity holds under the same assumptions as (2.1); it was derived by Siegel (1935) as a special multiple Fourier series instance of Parseval's identity from functional analysis.

Minkowski's theorem has numerous applications in the geometry of numbers (see chapter 3.1). Its impact shows in the number of ramifications, refinements and generalizations that it led to. In particular, Blichfeldt (1914) and van der Corput (1935, 1936) proved that for $\mathbb{L} \in \mathscr{L}^d$ and $K \in \hat{\mathscr{K}}^d$:

$$2 \left\lfloor \frac{V(K)}{2^d (\det \mathbb{L})} \right\rfloor + 1 \leqslant G_{\mathbb{L}}(K). \tag{2.2}$$

Equality occurs for suitable "quasi-1-dimensional prisms" of the form $\alpha C(\mathbb{L}) + \beta v_1$ with $0 < \alpha, \beta$ and $\alpha < 2$.

The following inequalities for the lattice point enumerator are closely related to (2.1); they are also due to Minkowski (1896).

Let $\mathbb{L} \in \mathscr{L}^d$, $K \in \hat{\mathscr{K}}_0^d$ *and suppose* $G_{\mathbb{L}}(\operatorname{int} K) = 1$. *Then*

$$G_{\mathbb{L}}(K) \leqslant 3^d. \tag{2.3}$$

If, in addition, K is strictly convex then $G_{\mathbb{L}}(K) \leqslant 2^{d+1} - 1$.

Further analogues of (2.1) can be found in Cassels (1971), Gruber and Lekkerkerker (1987), and Erdős, Gruber and Hammer (1989).

2.2. *Successive minima*

The most important improvement of (2.1) is Minkowski's theorem on successive minima. Given a lattice $\mathbb{L}$ and a convex body $K \in \hat{\mathscr{K}}_0^d$, then the successive minima $\lambda_1, \ldots, \lambda_d$ are defined by

$$\lambda_i(K, \mathbb{L}) = \inf\{\lambda > 0\colon \dim \operatorname{aff}(\lambda K \cap \mathbb{L}) = i\}, \quad i = 1, \ldots, d.$$

For brevity we write $\lambda_i(\mathbb{L})$ and λ_i for $\lambda_i(K, \mathbb{L})$, $\lambda_i(K, \mathbb{Z}^d)$, respectively. Then, (2.1) reads

$$\lambda_1(\mathbb{L})^d V(K) \leqslant 2^d \det \mathbb{L},$$

while Minkowski's (1896) generalization is the following result:

Given $\mathbb{L} \in \mathcal{L}^d$ *and* $K \in \mathcal{K}_0^d$, *then*

$$\frac{2^d}{d!} \det \mathbb{L} \leqslant \lambda_1(\mathbb{L}) \times \cdots \times \lambda_d(\mathbb{L}) V(K) \leqslant 2^d \det \mathbb{L}. \tag{2.4}$$

The left inequality is simple, but the right one is a deep improvement of (2.1). For a generalization of (2.4) see Woods (1966) and for various proofs see Bambah, Woods and Zassenhaus (1965), Cassels (1971, p. 208) and Gruber and Lekkerkerker (1987, p. 59).

There are some recent analogues of (2.1), (2.2), (2.3) and (2.4), Henk (1990), Betke, Henk and Wills (1993):

$$\begin{aligned}
&\lambda_{i+1} \times \cdots \times \lambda_d V(K) \leqslant 2^{d-i} V_i(K), \quad i = 1, \ldots, d, \\
&\frac{2^i}{i!} \leqslant \lambda_1 \times \cdots \times \lambda_i V_i(K), \quad i = 1, \ldots, d, \\
&G_{\mathbb{L}}(K) \leqslant \left\lfloor \frac{2}{\lambda_1(\mathbb{L})} + 1 \right\rfloor^d, \\
&\frac{1}{d!} \prod_{i=1}^{d} \left(\frac{2}{\lambda_i(\mathbb{L})} - 1 \right) \leqslant G_{\mathbb{L}}(K).
\end{aligned} \tag{2.5}$$

The estimate (2.5) extends Minkowski's inequality (2.3). Betke, Henk and Wills (1993) conjecture that the stronger bound

$$G_{\mathbb{L}}(K) \leqslant \prod_{i=1}^{d} \left\lfloor \frac{2}{\lambda_i(\mathbb{L})} + 1 \right\rfloor$$

holds and prove it for $d = 2$; along with analogous results for strictly convex bodies.

For details on reduction theory of quadratic forms and other classical results we refer to Cassels (1971), Gruber and Lekkerkerker (1987), Erdős, Gruber and Hammer (1989) and chapter 3.1 of this Handbook.

3. General convex bodies

3.1. Lattice points and intrinsic volumes

The classical theory of geometry of numbers deals mainly with centrally symmetric convex bodies; hence there are only a few classical results on general convex bodies. One such result is the simple mean value theorem:

$$V(K) = \int_{C(\mathbb{L})} G_{\mathbb{L}}(x + K) \, \mathrm{d}x,$$

which holds for arbitrary $\mathbb{L} \in \mathcal{L}^d$ and $K \in \mathcal{K}^d$.

Mahler (1939) generalized Minkowski's fundamental theorem by relaxing the symmetry assumption on K. As a measure of asymmetry Mahler defined

$$\sigma(K) = \min\{\sigma \geqslant 1\colon -K \subset \sigma K\},$$

for all $K \in \mathcal{K}^d$ with $0 \in \operatorname{int} K$ and derived an inequality that was later improved by Sawyer (1954) to the following result:

Let $K \in \mathcal{K}^d$ *and suppose* $(\operatorname{int} K) \cap \mathbb{L} = \{0\}$. *Then*

$$V(K) \leqslant (1+\sigma(K))^d \left(1 - \left(1 - \frac{1}{\sigma(K)}\right)^d\right) \det \mathbb{L}. \tag{3.1}$$

Observe that when $K = -K$, (3.1) is just Minkowski's theorem (2.1). If $z(K)$ denotes the centroid of K then $z(K) = 0$ implies $\sigma(K) \leqslant d$ (cf. Bonnesen and Fenchel 1934, p. 53). Together with (3.1) this gives

$$V(K) \leqslant (d+1)^d \left(1 - \left(1 - \frac{1}{d}\right)^d\right) \det \mathbb{L}$$

for all $K \in \mathcal{K}^d$ with $z(K) = 0$ and $(\operatorname{int} K) \cap \mathbb{L} = \{0\}$. Ehrhart (1955a) conjectured, however, that the much stronger inequality

$$V(K) \leqslant \frac{(d+1)^d}{d!} \det \mathbb{L}$$

holds for all $K \in \mathcal{K}^d$ with $z(K) = 0$. This estimate would be tight but has only been verified for $d = 2$, Ehrhart (1955a).

In the remainder of this section we will focus on bounds for the lattice-point enumerator $G_{\mathbb{L}}$ in terms of quermassintegrals and closely related functionals. Observe that for finding upper bounds for $G_{\mathbb{L}}(K)$ in terms of monotone functionals it is enough to just consider convex lattice polytopes since $I_{\mathbb{L}}(K)$ has the same number of $\mathbb{L}$-lattice points as K.

The oldest general upper bound for the lattice point enumerator is due to Blichfeldt (1921).

Let $\mathbb{L} \in \mathcal{L}^d$, *let* $K \in \mathcal{K}^d$ *and suppose* $I_{\mathbb{L}}(K) \in \mathcal{P}_0^d$. *Then*

$$G_{\mathbb{L}}(K) \leqslant \frac{d!}{\det \mathbb{L}} V(K) + d. \tag{3.2}$$

Equality holds in (3.2) for instance for lattice simplices of volume $(1/d!) \det \mathbb{L}$. If $R_{\mathbb{L}}$ denotes the circumradius of $C(\mathbb{L})$ then

$$V(K) - R_{\mathbb{L}} F(K) \leqslant G_{\mathbb{L}}(K) \cdot \det \mathbb{L} \leqslant V\big(K + C(\mathbb{L})\big); \tag{3.3}$$

Gritzmann (1984, pp. 52, 88), Wills (1990a, p. 37).

Schnell (1992) and Wills (1991) gave the bounds

$$\frac{V_d(K)}{D_d(\mathbb{L})} - \frac{1}{2}d^{3/2}\frac{V_{d-1}(K)}{D_{d-1}(\mathbb{L})} \leqslant G_{\mathbb{L}}(K) \leqslant \sum_{i=0}^{d} i!\frac{V_i(K)}{D_i(\mathbb{L})}, \tag{3.4}$$

which involve the lattice functionals $D_i(\mathbb{L})$, $i = 0, \ldots, d$ defined by

$$D_0(\mathbb{L}) = 1$$

and for $i = 1, \ldots, d$

$$D_i(\mathbb{L}) = \min\{|\det(\mathbb{L}_i)| : \mathbb{L}_i \text{ is an } i\text{-dimensional sublattice of } \mathbb{L}\}.$$

Note that, in particular, $D_d(\mathbb{L}) = \det(\mathbb{L})$ and that $D_1(\mathbb{L})$ is the length of a shortest nonzero lattice vector. Neither of the two inequalities of (3.4) is tight. For $d = 2$, the factors $\frac{1}{2}d^{3/2}$ and $i!$ can be replaced by 1, and this is best possible; Oler (1961) [upper bound; see (3.9)] and Schnell and Wills (1991) (lower bound). Further, for $\mathbb{L} = \mathbb{Z}^d$ there is the asymptotically tight inequality

$$V(K) - \frac{1}{2}F(K) \leqslant G(K) \tag{3.5}$$

of Bokowski, Hadwiger and Wills (1972) for arbitrary $K \in \mathcal{K}^d$. The proof of (3.5) is based on previous work of Hadwiger (1972). Partial results can also be found in Nosarzewska (1948), Bender (1962), Wills (1968, 1970), Hadwiger (1970), Schmidt (1972) and Bokowski and Wills (1974). Some consequences of (3.5) were given by Hammer (1971) and Bokowski and Odlyzko (1973).

The right-hand inequality in (3.3) can be developed into mixed volumes (or via $C(\mathbb{L}) \subset R_{\mathbb{L}}\mathbb{B}^d$ and Steiner's formula into quermassintegrals). For $\mathbb{L} = \mathbb{Z}^d$, we obtain Davenport's (1951) inequality.

Let for $K \in \mathcal{K}^d$, K_j^i be the orthogonal projection of K into the i-dimensional coordinate subspace $\mathbb{E}_j^i \subset \mathbb{E}^d$, $i = 1, \ldots, d$; $j = 1, \ldots, \binom{d}{j}$. Then

$$G(K) \leqslant \sum_{i=0}^{d} \sum_{j=1}^{\binom{d}{i}} V_i(K_j^i). \tag{3.6}$$

(See Betke 1979.) Equality holds for lattice boxes. A simple consequence of (3.6) is the estimate

$$G(K) \leqslant V(K) + \sum_{i=0}^{d-1} V_i(Q), \quad K \in \mathcal{K}^d,$$

where Q denotes the smallest lattice box containing K. Ehrhart (1977) conjectured that the bound can be improved to

$$G(K) \leqslant V(K) + \tfrac{1}{2}F(K) + \sum_{i=0}^{d-2} V_i(Q), \quad K \in \mathcal{K}^d;$$

which holds for $d \leqslant 3$ (Ehrhart 1977), but is open for $d \geqslant 4$.

Another simple consequence of (3.6) is

$$G(K) \leqslant \sum_{i=0}^{d} \binom{d}{i} V_i(K), \quad K \in \mathcal{K}^d, \tag{3.7}$$

which holds with equality if and only if K is a lattice point. Obviously, the coefficients of V_0 and V_d are best possible. There were, however, many attempts to improve the other coefficients. Wills (1973) conjectured that

$$G(K) \leqslant \sum_{i=0}^{d} V_i(K). \tag{3.8}$$

This estimate holds with equality for lattice boxes. Further, it has been verified for dimensions $d = 2, 3$ (Nosarzewska 1948, Overhagen 1975), for rotation bodies when $d \leqslant 20$ (Hadwiger and Wills 1974) and for arbitrary lattice zonotopes (Betke and Gritzmann 1986). Moreover, Hadwiger (1975) gave integral representations of the functional $W = \sum_{i=0}^{d} V_i$ which inspired some work from the viewpoint of valuations (McMullen 1975; see also chapter 3.6 by McMullen in this Handbook). However, the conjecture turned out to be false. Hadwiger (1979) showed that for $d \geqslant 441$ there are lattice simplices for which (3.8) does not hold. Later Betke and Henk (1993) showed that it is false for suitable lattice cross polytopes already when $d \geqslant 207$. In fact, even the much weaker estimate

$$G(K) \leqslant V(K + \kappa_d^{-1/d}\mathbb{B}^d),$$

conjectured and proved for $d \leqslant 5$ by Bokowski (1975), is false for $d \geqslant 3.7 \cdot 10^{159}$, Höhne (1980). Another attempt to fill the gap between (3.7) and (3.8) is the following conjecture of Wills (1990b) which is closely related to a covering theorem by Santaló (1976, p. 274):

$$G(K) \leqslant \sum_{i=0}^{d} \frac{\kappa_i \kappa_{d-i}}{\kappa_d} V_i(K), \quad K \in \mathcal{K}^d.$$

3.2. *Classes of lattices*

Lattice point problems are closely related to lattice packing and lattice covering. In order to facilitate the transition between these areas the following classes of lattices are introduced for $C \in \mathcal{K}_0^d$.

$\mathcal{L}_p(C)$ is the subset of $\mathcal{L}^d$ of all lattices $\mathbb{L}$ which pack C, i.e., the translates of C in $\{g + C\colon g \in \mathbb{L}\}$ do not have interior points in common;

$\mathcal{L}_c(C)$ is the subset of $\mathcal{L}^d$ of all lattices $\mathbb{L}$ for which $\mathbb{L} + C$ covers $\mathbb{R}^d$, i.e., $\bigcup_{g\in\mathbb{L}}(g + C) = \mathbb{R}^d$.

Since the main interest focuses on bounds for the lattice point enumerator in terms of quermassintegrals only the case $C = \mathbb{B}^d$ has been studied thoroughly.

For $\mathbb{L} \in \mathcal{L}_p(\mathbb{B}^d)$ and $K \in \mathcal{K}^d$ there is the simple but useful upper bound

$$G_{\mathbb{L}}(K) \leqslant \kappa_d^{-1} V(K + \mathbb{B}^d) = \sum_{i=0}^{d} \frac{\kappa_{d-i}}{\kappa_d} V_i(K).$$

With $\delta_{\mathcal{L}}(\mathbb{B}^d)$ and $\vartheta_{\mathcal{L}}(\mathbb{B}^d)$ denoting the lattice packing and lattice covering density of $\mathbb{B}^d$, respectively, it is easy to see that the inequality

$$G_{\mathbb{L}}(K) \leqslant \sum_{i=0}^{d} \frac{\delta_{\mathcal{L}}(\mathbb{B}^d)}{\kappa_i} V_i(K)$$

would be the best possible bound of this kind. It has been shown by Gritzmann and Wills (1986) that this inequality holds for $\mathbb{L}$-zonotopes but is false in general for suitably high dimension. However, in dimension 2 it holds for general $K \in \mathcal{K}^2$ and $\mathbb{L} \in \mathcal{L}_p(\mathbb{B}^d)$ even in the Minkowski plane (Oler 1961). In the Euclidean case Oler's result reads as follows:

$$G_{\mathbb{L}}(K) \leqslant \frac{1}{2\sqrt{3}} V(K) + \frac{1}{2} V_1(K) + 1. \tag{3.9}$$

Observe that (3.9) follows from Pick's (1899) identity (4.1). In fact, (4.1) implies that for $K \in \mathcal{K}^2$ and $\mathbb{L} \in \mathcal{L}^2$

$$G_{\mathbb{L}}(K) \leqslant \frac{V(K)}{\det \mathbb{L}} + \frac{V_1(K)}{D_1(\mathbb{L})} + 1.$$

Now, (3.9) follows since for each $\mathbb{L} \in \mathcal{L}_p(\mathbb{B}^2)$ we have $\det \mathbb{L} \geqslant \kappa_2/\delta_{\mathcal{L}}(\mathbb{B}^2) = 2\sqrt{3}$ and $D_1(\mathbb{L}) \geqslant 2$. The inequality (3.9) is equivalent to the "lattice version" of a packing theorem of Groemer (1960); another proof is due to Folkman and Graham (1969); cf. also Zassenhaus (1961). Oler's result might possibly be the 2-dimensional case of the general inequality for $\mathbb{L} \in \mathcal{L}_p(\mathbb{B}^d)$ and $K \in \mathcal{K}^d$:

$$G_{\mathbb{L}}(P) \leqslant \sum_{i=0}^{d} \frac{\sigma_i}{\kappa_i} V_i(P). \tag{3.10}$$

Here σ_i denotes Rogers's (1964) packing constants, i.e. the ratio of the sum of the volumes of the intersection of $d+1$ unit balls centered at the vertices of a regular simplex of side 2 to the volume of the simplex. Gritzmann and Wills (1986) showed that (3.10) is valid for $\mathbb{L}$-zonotopes and conjecture that it holds in general.

Gritzmann (1984, 1986) showed that

$$G_{\mathbb{L}}(K) \leqslant \left((2+\sqrt{2})\sqrt{\frac{\pi}{2d}} + \frac{2}{\sqrt{d(d-1)}} \right) \sum_{i=0}^{d} \frac{\kappa_{d-i}}{\kappa_d} V_i(K);$$

again, for arbitrary $\mathbb{L} \in \mathcal{L}_p(\mathbb{B}^d)$ and $K \in \mathcal{K}^d$. (For a "finite packing interpretation" of this result see chapter 3.4.)

As in section 3.1, the lower bound problem is simpler and essentially solved by Wills (1989):

Given $\mathbb{L} \in \mathcal{L}_c(\mathbb{B}^d)$ *and* $K \in \mathcal{K}^d$, *then* $\vartheta^d_{\mathcal{L}} \kappa_d^{-1}\{V(K) - F(K)\} \leqslant G_{\mathbb{L}}(K)$.

3.3. Nonlinear inequalities

In the previous two sections the bounds for G were linear combinations of the V_i or related functionals. We now consider nonlinear inequalities for G in terms of functionals which are homogeneous of degree 1.

Let $D(K)$, $R(K)$, $r(K)$ and $w(K)$ denote the diameter, circumradius, inradius and width of $K \in \mathcal{K}^d$, respectively. We begin with the following trivial inequalities.

$$G(K) \leqslant (D(K)+1)^d,$$
$$G(K) < \kappa_d(R(K) + \tfrac{1}{2}\sqrt{d})^d,$$
$$\kappa_d(r(K) - \tfrac{1}{2}\sqrt{d})^d < G(K), \quad \text{if } r(K) \geqslant \tfrac{1}{2}\sqrt{d}.$$

There are various inequalities for lattice-point-free convex bodies which involve the width. Results of this kind are of some relevance for reducing a d-dimensional integer programming problem to lower-dimensional ones; see Lenstra (1983). Let

$$\omega_d = \max\{w(K): K \in \mathcal{K}^d \wedge G(K) = 0\}.$$

McMullen and Wills (1981) showed

$$(\sqrt{2}+1)(\sqrt{d-1} - \alpha) < \omega_d < d+1,$$

where $\alpha \approx 1.018$ is a constant. For $K \in \mathcal{K}^d$ let $d_i(K)$ denote the length of the projection of K onto the x_i-axis ("outer quermass") and let $s_i(K)$ denote the length of a maximal segment of K parallel to the x_i-axis ("inner quermass"). Then Scott (1979) (cf. also Wills 1990a) showed that for $K \in \mathcal{K}^d$ with $G(K) = 0$

$$\frac{1}{d_i(K)} + \sum_{j \neq i} \frac{1}{s_j(K)} \geqslant 1, \quad i = 1, \ldots, d,$$

with equality for suitable cross polytopes. McMullen and Wills (1981) proved under the same assumptions that

$$\frac{\sqrt{2}\omega_{d-1}}{w(K)} + \frac{1}{d_i(K)} \geqslant 1, \quad i = 1, \ldots, d,$$

$$\frac{\sqrt{2}\omega_{d-1}}{w(K)} + \frac{\sqrt{d}}{D(K)} \geqslant 1.$$

Possibly the last inequality can be improved to

$$\frac{\omega_{d-1}}{w(K)} + \frac{1}{D(K)} \geqslant 1,$$

a result confirmed by Scott (1973) for $d = 2$. Some sharp nonlinear inequalities for special convex bodies can be found in Erdős, Gruber and Hammer (1989) and Xu and Yau (1992).

A simple but useful inequality is

$$G(K) - V(K) \leqslant \prod_{i=1}^{d} \bigl(d_i(K) + 1\bigr) - \prod_{i=1}^{d} d_i(K); \tag{3.11}$$

it holds with equality for lattice boxes. For applications of (3.11) and some closely related inequalities see Chalk (1980) and Niederreiter and Wills (1975).

The following results use the covering minimum μ_i, introduced by Kannan and Lovász (1988), which is defined for $\mathbb{L} \in \mathcal{L}^d$, $K \in \mathcal{K}^d$ and $i = 1, \ldots, d$ by

$$\mu_i(K, \mathbb{L}) = \inf\{t\colon tK + \mathbb{L} \text{ meets every } (d-i)\text{-flat of } \mathbb{R}^d\}.$$

The μ_i correspond to Minkowski's successive minima λ_i and, in fact, μ_d is the classical inhomogeneous minimum μ in geometry of numbers. $\mu_1(K, \mathbb{L})^{-1}$ is called the $\mathbb{L}$-width of K – or, when the lattice is specified by the context, simply *lattice width* of K. Kannan and Lovász (1988) prove that there is a positive constant γ such that for $\mathbb{L} \in \mathcal{L}^d$ and $K \in \mathcal{K}_0^d$

$$\left\lfloor \frac{1}{\mu_1(K, \mathbb{L})\gamma d^2} \right\rfloor^d - 1 \leqslant G_{\mathbb{L}}(K), \tag{3.12}$$

and if $K \in \hat{\mathcal{K}}_0^d$

$$\left\lfloor \left(\frac{1}{\mu_1(K, \mathbb{L})\gamma d} - d \right)^d \right\rfloor \leqslant G_{\mathbb{L}}(K). \tag{3.13}$$

The constant γ comes from the constant in the following theorem.

There is a positive constant β such that for $K \in \mathcal{K}_0^d$

$$\beta^d \kappa_d^2 \leqslant V(K)V(K^*) \leqslant \kappa_d^2. \tag{3.14}$$

The upper bound in (3.14) is due to Santaló (1949), the lower bound was given by Bourgain and Milman (1985). Using the known formula for κ_d one sees that $V(K)V(K^*)$ is bounded below by $(\alpha d)^{-d}$, where α is a positive absolute constant. As a consequence of (3.14) and Minkowski's theorem (2.4) on successive minima Bourgain and Milman (1987) obtain

$$\prod_{i=1}^{d} \lambda_i(K, \mathbb{L})\lambda_i(K^*, \mathbb{L}^*) \leqslant \gamma^d d^d.$$

Their paper contains also various other results on the λ_i and μ_i.

Betke, Henk and Wills (1993) prove some inequalities for the μ_i and the V_i; an example is the result that there is a constant β with $0 < \beta \leqslant 1/d!$ such that for every $K \in \mathcal{K}_0^d$

$$\mu_1(K, \mathbb{Z}^d) \cdots \mu_d(K, \mathbb{Z}^d)V(K) \geqslant \beta.$$

4. Lattice polytopes

4.1. General lattice polytopes

The methods in the theory of lattice polytopes are mainly from combinatorics and linear algebra and several results do not require convexity. For the purpose of this subsection only, a polytope in $\mathbb{R}^d$ (or a polygon in $\mathbb{R}^2$) is the underlying point set of a simplicial cell complex (in the sense of Grünbaum 1967). Equivalently, a polytope can be defined as a finite union of convex polytopes of dimension at most d. A polytope is called proper, if it is the closure of its interior or, equivalently, if it can be represented as the finite union of convex d-polytopes. Given a lattice $\mathbb{L} \in \mathcal{L}^d$, a polytope P is called $\mathbb{L}$-polytope if there is a simplicial cell complex with underlying point set P whose vertices are all in $\mathbb{L}$. As usual the Euler-characteristic $\chi(P)$ of a polytope is the sum $\sum_{i=0}^{d}(-1)^i|\mathcal{F}_i(\mathcal{C})|$ where $\mathcal{C}$ is any simplicial cell complex with underlying point set P. The first result on lattice polygons is Pick's identity, Pick (1899).

If $P \subset \mathbb{E}^2$ is a lattice polygon whose boundary is a closed Jordan curve, then

$$G_{\mathbb{L}}(P) = \frac{V(P)}{\det \mathbb{L}} + \tfrac{1}{2}G_{\mathbb{L}}(\operatorname{bd} P) + 1. \tag{4.1}$$

(Compare also Varberg 1985.) Observe that Nosarzewska's inequality (3.8, for $d = 2$) is a simple consequence of (4.1). There are various proofs of (4.1) and many generalizations. For instance, Reeve (1957) showed

$$G_{\mathbb{L}}(P) = \frac{V(P)}{\det \mathbb{L}} + \tfrac{1}{2} G_{\mathbb{L}}(\operatorname{bd} P) + \chi(P) - \tfrac{1}{2}\chi(\operatorname{bd} P), \tag{4.2}$$

and Hadwiger and Wills (1975) proved

$$G_{\mathbb{L}}(P) = \frac{V(P)}{\det \mathbb{L}} + \tfrac{1}{2} E(P) + \chi(P),$$

where $E(P)$ denotes the number of segments between two consecutive lattice points on $\operatorname{bd} P$, and 1-dimensional parts of P are counted twice.

Ding and Reay (1987a) give a generalization of (4.1) to hexagonal tilings and indicate possible applications to computer graphics; for further results along these lines see Ding and Reay (1987b) and Ding, Kolodziejczyk and Reay (1988). Grünbaum and Shephard (1992) extend (4.1) to more general oriented polygons, a result that implies many of the previously known (plane) variants of (4.1).

Hadwiger and Wills (1975) showed the following.

Let $\mathbb{L} \in \mathcal{L}^2$, let P be an $\mathbb{L}$-polygon and $x \in \mathbb{R}^2 \setminus \mathbb{L}$. Then

$$G_{\mathbb{L}}(P) - G_{\mathbb{L}}(x + P) \geqslant \chi(P); \tag{4.3}$$

and for each $\mathbb{L} \in \mathcal{L}^2$ and $\chi \in \mathbb{Z}$ there are an $\mathbb{L}$-polygon P with $\chi(P) = \chi$ and an $x \in \mathbb{R}^2 \setminus \mathbb{L}$ such that (4.3) *holds with equality.*

No analogue of (4.3) is known for $d \geqslant 3$; but there are remarkable generalizations of (4.1) and (4.2) for arbitrary $\mathbb{L} \in \mathcal{L}^d$ and proper $\mathbb{L}$-polytopes P:

$$\begin{aligned}\frac{(d-1)d!}{2\det \mathbb{L}} V(P) &= (-1)^{d-1}\left(\chi(P) - \tfrac{1}{2}\chi(\operatorname{bd} P)\right) \\ &\quad + \sum_{j=0}^{d-2} \binom{d-1}{j} (-1)^j \left(G_{\mathbb{L}}((d-1-j)P) - \tfrac{1}{2} G_{\mathbb{L}}(\operatorname{bd}((d-1-j)P))\right),\end{aligned}$$

$$\frac{d!}{\det \mathbb{L}} V(P) = (-1)^d \chi(P) + \sum_{j=0}^{d-1} \binom{d}{j} (-1)^j G_{\mathbb{L}}((d-1-j)P).$$

These identities are due to Macdonald (1963); the case $d = 3$ was first proved by Reeve (1957, 1959).

4.2. Convex lattice polytopes: Equalities

Ehrhart (1967, 1968) discovered the "polynomiality" of the lattice point enumerator $G_{\mathbb{L}} : \mathcal{P}^d(\mathbb{L}) \to \mathbb{N}_0$:

For each $\mathbb{L} \in \mathcal{L}$ *there are functionals* $G_{\mathbb{L},i} : \mathcal{P}^d(\mathbb{L}) \to \mathbb{N}_0$ *such that for every* $P \in \mathcal{P}^d(\mathbb{L})$ *and* $n \in \mathbb{N}$

$$G_{\mathbb{L}}(nP) = \sum_{i=0}^{d} n^i G_{\mathbb{L},i}(P) \tag{4.4}$$

and

$$G_{\mathbb{L}}\big(\mathrm{relint}(nP)\big) = (-1)^{\dim P} \sum_{i=0}^{d} (-n)^i G_{\mathbb{L},i}(P). \tag{4.5}$$

Equation (4.5) is called the *reciprocity law*, and the polynomial in (4.4) is often referred to as *Ehrhart-polynomial*. It is quite important for various questions. In particular, the question when the dual of an integer polytope P (with $0 \in \mathrm{int}(P)$) is again an integer polytope can be answered in terms of conditions on the Ehrhart-polynomial, Hibi (1992). See the book by Stanley (1986) for more facts on the Ehrhart polynomial. For $\mathbb{L} = \mathbb{Z}^d$ we will use the abbreviation G_i rather than $G_{\mathbb{L},i}$.

In case of lattice zonotopes one can give an explicit formula for the Ehrhart polynomial; Stanley (1980, 1986). Stanley (1991a) used this fact to find a generating function for the number of degree sequences of simple n-vertex graphs; in fact, there is a one-to-one correspondence between these degree sequences and the integer points of a suitable zonotope.

The $G_{\mathbb{L},i}$ are valuations (cf. McMullen 1977 and McMullen's chapter 3.6 in this Handbook). Further they are invariant under unimodular transformations, i.e., transformations U of $\mathbb{R}^d$ with $U(x) = Ax + b$, where A is an integer $d \times d$-matrix, $\det A = \pm 1$ and $b \in \mathbb{L}$. These properties provide the framework of an analogy between the $G_{\mathbb{L},i}$ and the intrinsic volumes V_i which shows in the following two important theorems.

(Hadwiger 1951) *Every continuous and additive functional on* $\mathcal{K}^d$ *which is invariant under rigid motions is a linear combination of the* $d+1$ *intrinsic volumes* $V_0, \ldots, V_d$.

(Betke and Kneser 1985) *Every additive and unimodular invariant functional on* $\mathcal{P}^d(\mathbb{L})$ *is a linear combination of the* $d+1$ *functionals* $G_{\mathbb{L},0}, \ldots, G_{\mathbb{L},d}$.

In addition, for $i = 0, d-1, d$, V_i and G_i are very similar

$$G_d = V_d = V, \qquad G_0 = V_0 = 1$$

and

$$G_{d-1}(P) = \tfrac{1}{2} \sum_{F \in \mathcal{F}_{d-1}} \mu_F V_{d-1}(F) \leqslant \tfrac{1}{2} F(P),$$

where $1/\mu_F = \det(\mathbb{Z}^d \cap \mathrm{aff}\, F)$ is the determinant of the sublattice of $\mathbb{Z}^d$ induced by $\mathrm{aff}\, F$. As opposed to the corresponding intrinsic volumes, $G_1, \ldots, G_{d-2}$, however, do not admit a simple geometric interpretation. In particular they are neither monotone nor nonnegative. Let us point out that Stanley (1991b) introduced a different basis $h_0^*, \ldots, h_d^*$ via

$$(1-x)^{d+1} \sum_{n\geqslant 0} G_{\mathbb{L}}(nP)x^n = h_0^* + h_1^* x + \cdots + h_d^* x^d$$

for the same space whose elements are monotone and, hence, nonnegative; see also Hibi (1991).

As Wills (1982) pointed out there is a simple analogue of Minkowski's lattice point theorem (2.1) for G_{d-1}. In fact,

$$G_{d-1}(P) \leqslant d2^{d-1},$$

for all $P \in \hat{\mathcal{P}}_0^d$ with $G(\operatorname{int} P) = 1$.

It is open whether

$$G_i(P) \leqslant \binom{d}{i} 2^i$$

holds for $i = 1, \ldots, d-2$ under the same assumptions. Betke and McMullen (1985) showed that for each $i = 1, \ldots, d$ there are constants α_i, β_i such that for all $P \in \mathcal{P}^d$

$$G_i(P) \leqslant \alpha_i G_d(P) + \beta_i.$$

The following identity resembles the polynomial expansion of the Minkowski sums of convex bodies into mixed volumes and generalizes Ehrhart's polynomial expansion (4.4).

(Bernstein 1976, McMullen 1977) *Let* $\mathbb{L} \in \mathcal{L}^d$, $k \in \mathbb{N}$, $P_1, \ldots, P_k \in \mathcal{P}^d(\mathbb{L})$ *and* $n_1, \ldots, n_k \in \mathbb{N}$. *Then there are coefficient functionals* $G_{\mathbb{L}}(P_1, n_1, \ldots, P_k, n_k)$ *such that*

$$G_{\mathbb{L}}\left(\sum_{i=1}^{k} n_i P_i\right) = \sum_{\substack{j_1,\ldots,j_k=0 \\ j_1+\cdots+j_k\leqslant d}}^{d} n_1^{j_1} \cdots n_k^{j_k} G_{\mathbb{L}}(P_1, j_1, \ldots, P_k, j_k).$$

The number (of equivalence classes under the group of unimodular transformations) of convex lattice polygons and polytopes has been studied by Arnold (1980), Bárány and Pach (1992), and Bárány and Vershik (1992).

4.3. *Convex lattice polytopes: Inequalities*

Unlike the bounds in section 3 which hold for general $K \in \mathcal{K}^d$, the following inequalities are tailored to the case of convex lattice polytopes.

Using methods of Ehrhart (1955a) and Stanley (1976) on formal power series and polyhedral cell complexes Betke and McMullen (1985) extended Blichfeldt's inequality (3.2) as follows:

$$G_{\mathbb{L}}(nP) \leqslant \left(\frac{V(nP)}{\det \mathbb{L}} + \frac{n^{d-1}}{(d-1)!}\right) \prod_{i=1}^{d-1} \left(1 + \frac{i}{n}\right),$$

$$\left(\frac{V(nP)}{\det \mathbb{L}} - \frac{n^d}{d!}\right) \prod_{|i|<d/2} \left(1 + \frac{i}{n}\right) + \binom{n+d}{d} \leqslant G_{\mathbb{L}}(nP).$$

These bounds hold for all $\mathbb{L} \in \mathscr{L}^d$ and $P \in \mathscr{P}_0^d(\mathbb{L})$.

Ehrhart (1955b) and Scott (1976) proved that for $\mathbb{L}$-polygons P with $G_{\mathbb{L}}(\operatorname{int} P) = k \geqslant 1$,

$$V(P) \leqslant \begin{cases} 4.5 \cdot \det \mathbb{L}, & \text{if } k = 1, \\ 2(k+1) \cdot \det \mathbb{L}, & \text{if } k \geqslant 2. \end{cases}$$

The bounds are sharp. There is no direct analogue for $k = 0$. Perles, Wills and Zaks (1982) showed that there is a constant $\alpha \approx 0,5856$ such that for each $d \geqslant 3$ and $k \geqslant 1$ there is a $P \in \mathscr{P}^d(\mathbb{L})$ with $G_{\mathbb{L}}(\operatorname{int} P) = k$ and

$$V(P) \geqslant \frac{k+1}{d!} 2^{2^{d-\alpha}} \det \mathbb{L}. \tag{4.6}$$

The much harder problem of the existence of an upper bound for V was solved by Hensley (1983). Subsequently, Lagarias and Ziegler (1991) improved his bound and showed that

$$V(P) \leqslant k(7(k+1))^{2^{d+1}} \det \mathbb{L}, \tag{4.7}$$

whenever $P \in \mathscr{P}^d(\mathbb{L})$ and $G_{\mathbb{L}}(\operatorname{int} P) = k \geqslant 1$. Further, Lagarias and Ziegler (1991) conjecture that the examples for (4.6) are optimal and show that (4.6) and (4.7) can be generalized to rational convex polytopes. Via Blichfeldt's inequality (3.2) one obtains similar results for $G_{\mathbb{L}}(P)$.

Rabinowitz (1989) determined all convex lattice polygons (up to unimodular transformations) with at most one interior lattice point. Lattice simplices in $\mathbb{E}^3$ containing no lattice points except their vertices were studied among others by Reeve (1957), White (1964), Scarf (1985) and Reznick (1986). A relation of the Frobenius problem to "maximal lattice free bodies" was given by Scarf and Shallcross (1990) (cf. also Kannan 1989).

5. Lattice polyhedra in combinatorial optimization

As we will see in the last two sections combinatorial optimization problems are naturally related to some special lattice point problems which are usually formulated for $\mathbb{Z}^d$. Of course, most problems can also be phrased in terms of other lattices. However, from a theoretical point of view most problems studied in sections 5 and 6 are, indeed, affinely invariant. The same is true from a computational point of view if we assume that the occurring lattices $\mathbb{L}_d$ are related to $\mathbb{Z}^d$ by affine transformations A_d of size that is bounded by a polynomial in d. Hence it seems unnecessarily clumsy to formulate the following results for lattices other than the integer lattice. Therefore, with the exception of section 6.1 the results in sections 5 and 6 will be phrased in terms of $\mathbb{Z}^d$ only.

5.1. The combinatorics of associated lattice polyhedra

In this section we deal with the integer hull $I_{\mathbb{Z}^d}(F)$ of polyhedra F. It is easy to see that $I_{\mathbb{Z}^d}(F)$ is not a polyhedron in general. However, Meyer (1974) showed that for a rational polyhedron F, $I_{\mathbb{Z}^d}(F)$ is again a polyhedron. Hence, we will in the following only deal with rational polyhedra. In fact, we have the following stronger statement (cf. Schrijver 1986, p. 237).

Let A be an integer $m \times d$ matrix, $b \in \mathbb{Z}^m$ and let $F = \{x \in \mathbb{R}^d\colon Ax \leqslant b\}$. Further, let Δ be the maximum of the absolute values of the subdeterminants of the matrix $[A\ b]$. Then there are integer vectors $x_1, \ldots, x_n, y_1, \ldots, y_s$ with all components at most $(d+1)\Delta$ such that $I_{\mathbb{Z}^d}(F) = \operatorname{conv}\{x_1, \ldots, x_n\} + \operatorname{pos}\{y_1, \ldots, y_s\}$.

This result implies that if a rational system $Ax \leqslant b$ has an integer solution then it has one of size that is bounded by a polynomial in the size of the input A and b. Further, if c is rational and $\max\{\langle c, x\rangle\colon x \in I_{\mathbb{Z}^d}(F)\}$ is finite then the maximum is attained by a vector of polynomial size. This means that one can restrict all considerations in integer programming to polytopes P and associated lattice polytopes $I_{\mathbb{Z}^d}(P)$ and that the integer programming problem is in the class NP. For various other results along this line see Schrijver (1986).

Doignon (1973) and, independently, Bell (1977) and Scarf (1977) showed the Helly-type theorem that if each set of at most 2^d of the constraints in $Ax \leqslant b$ has an integer solution then there is an integer solution for the complete set $Ax \leqslant b$ of constraints.

The importance of integer hulls of $\mathcal{H}$-polytopes in combinatorial optimization stems from the fact that linear functionals can be maximized (or minimized) over rational polyhedra $F = \{x\colon Ax \leqslant b\}$ in polynomial time; indeed, this is the linear programming problem, Khachiyan (1979), Karmarkar (1984). Hence, if we could find in polynomial time a presentation of $I_{\mathbb{Z}^d}(F)$ in terms of linear inequalities we could solve the integer programming problem in polynomial time. This is particularly easy if b is integer and if A is totally unimodular, i.e., if each subdeterminant of A is in $\{-1, 0, 1\}$ hence, then, $F = \{x\colon Ax \leqslant b\}$ is already a lattice polyhedron, all vertices are integer. Therefore, in this case the integer programming problem is solved by any linear programming algorithm. This result is essentially a characterization of total unimodularity: Hoffman and Kruskal (1956) show the following theorem, for which a short proof was later provided by Veinott and Dantzig (1968).

An integer $m \times d$ matrix A is totally unimodular if and only if for each vector $b \in \mathbb{Z}^m$ the polyhedron $\{x\colon Ax \leqslant b \wedge x \geqslant 0\}$ is integer.

Note that this property holds, in particular, for network matrices; see Schrijver (1986, p. 272). As a consequence of Seymour's (1980) decomposition theorem total unimodularity can be tested in polynomial time; for more details see Schrijver (1986, p. 290).

Unfortunately, the situation is much worse, in general. Edmonds (1965) showed that there is no polynomial p such that for each $\mathcal{H}$-polyhedron $F = \{x \in \mathbb{R}^d\colon Ax \leqslant b\}$, the associated polyhedron $I_{\mathbb{Z}^d}(F)$ has at most $p(\text{size}(A, b))$ facets.

Let for $j \in \mathbb{N}$, T_j denote the triangle in $\mathbb{R}^2$ given by

$$T_j = \{x = (\xi_1, \xi_2) \in \mathbb{R}^2\colon \phi_{2j}\xi_1 + \phi_{2j+1}\xi_2 \leqslant \phi_{2j+1}^2 - 1 \wedge \xi_1, \xi_2 \geqslant 0\},$$

where ϕ_k denotes the kth Fibonacci-number. Rubin (1970) showed that $I_{\mathbb{Z}^d}(T_j)$ has $j+3$ vertices and therefore also $j+3$ facets. Hence, there is no bound on the number of facets of the polytope $I_{\mathbb{Z}^d}(P)$ in terms of the number of facets of a polytope P.

Hayes and Larman (1983) showed that the number of vertices of the knapsack polytope $P = \{x \in \mathbb{R}^d\colon \langle a, x\rangle \leqslant \beta \wedge x \geqslant 0\}$, with $a \in \mathbb{N}^d$, $\beta \in \mathbb{N}$, is at most $(\log_2(2 + 2\beta/\alpha))^d$, where α is the smallest component of a. Extension of their arguments yields an $\mathrm{O}(m^d L^d)$ upper bound for arbitrary rational polyhedra of size L with m facets in fixed $\mathbb{R}^d$. (For some related results see Shevchenko 1981.) Strengthening this result Cook et al. (1992) showed that a rational polyhedron F in $\mathbb{R}^d$ presented as the set of solutions to a system of m linear inequalities of total size L can have at most $2m^d(6d^2L)^{d-1}$ vertices.

Rubin's (1970) result shows that the order in L of the above result is best possible for polygons. Recently, Bárány, Howe and Lovász (1992) proved that for any fixed $d \geqslant 2$ and for any $L \in \mathbb{N}$ there exists a rational polyhedron $F \in \mathbb{R}^d$ of size at most L and with at most $2d^2$ facets such that the number of vertices of $I_{\mathbb{Z}^d}(F)$ is at least γL^{d-1}, where γ is a constant depending only on n.

Similarly sharp results for the number of facets of associated lattice polyhedra are not known.

5.2. Polyhedral combinatorics

Associated lattice polyhedra and lattice polytopes play an important role for the algorithmic solution of combinatorial optimization problems. Here we will only give some paradigms to outline the concept. For more details we refer to the relevant literature on combinatorial optimization, particularly to Lawler et al. (1985) for results on the traveling salesman polytope, to Schrijver (1993a) and to Schrijver's (1993b) forthcoming book on polyhedral combinatorics. The general approach of polyhedral combinatorics is to apply linear programming techniques to combinatorial optimization problems by studying the structure of corresponding lattice polytopes. These polytopes are usually given as the convex hull of a finite set of points. Hence, in order to apply linear programming techniques we have to find a (possibly small) system of linear inequalities that presents the polytope P. This might seem a rather strange detour since, obviously, a linear functional can be maximized over a finite point set by evaluating it for each such point and taking the maximum. However, typically the number of points is exponential. This is also true for the number of facets but on the other hand an optimal vertex v of P is already characterized by $\dim P$ of its facets that are incident with v. Hence the underlying philosophy is that it might be possible to find an optimum solution without having to consider too many facets. Focusing on its theoretical aspects, polyhedral combinatorics is mainly concerned with the combinatorial and geometric study of special lattice polytopes. In many cases these polytopes evolve as follows.

Let $E = \{e_1, \dots, e_n\}$, and let $\mathcal{S}$ be a subset of 2^E, the set of all subsets of E. With every $S \in 2^E$ we associate the incidence vector $x^S = (\xi_1^S, \dots, \xi_n^S)$, where

$$\xi_i^S = \begin{cases} 0 & \text{if } e_i \notin S, \\ 1 & \text{if } e_i \in S, \end{cases} \qquad i = 1, \dots, n.$$

Then we set

$$P_{\mathcal{S}} = \operatorname{conv}\{x^S \in \mathbb{R}^n \colon S \in \mathcal{S}\}.$$

To give a concrete example suppose $V = \{1, \dots, d\}$ and let E denote the set of all edges of the complete graph $K_{|V|}$ on the vertex set V. Hence $n = \binom{d}{2}$. Further, let $\mathcal{S}^d$ denote the subset of 2^E of all Hamiltonian cycles (tours) in $K_{|V|}$. Then the polytope $P_{\mathcal{S}^d}$ is called the symmetric traveling salesman polytope. If we do the same in the complete directed graph on V and consider the set $\mathcal{D}^d$ of all directed Hamiltonian cycles then $P_{\mathcal{D}^d}$ is the asymmetric traveling salesman polytope. In principle, solving the symmetric or asymmetric traveling salesman problem is just a linear programming problem over $P_{\mathcal{S}^d} \subset \mathbb{R}^{\binom{d}{2}}$ or $P_{\mathcal{D}^d} \subset \mathbb{R}^{d(d-1)}$. Clearly, these polytopes are the convex hull of a suitable subset of the vertices of the cube $[0,1]^{\binom{d}{2}}$, $[0,1]^{d(d-1)}$, respectively.

The dimensions of the traveling salesman polytopes are for $d \geqslant 3$:

$$\dim P_{\mathcal{S}^d} = \tfrac{1}{2} d(d-3), \qquad \dim P_{\mathcal{D}^d} = (d-1)^2 - d,$$

cf. Grötschel and Padberg (1985).

Padberg and Rao (1974) further showed that for $d \geqslant 6$, the (graph-theoretic) diameter of the 1-skeleton of $P_{\mathcal{D}^d}$ is 2; for $3 \leqslant d \leqslant 5$ it is 1. (This does not imply that two pivot operations of the simplex algorithm would suffice; however, $2d - 1$ pivot steps do suffice, Padberg and Rao 1974.)

There are many facets known for the traveling salesman polytopes – too many to be described here, see Grötschel and Padberg (1985). Some classes can be used in cutting plane approaches. However, there are also facets which are defined by properties that – unless P = NP – cannot be checked in polynomial-time. For a survey on algorithmic implications of polyhedral theory see Grötschel and Padberg (1985).

6. Computational complexity of lattice point problems

6.1. Algorithmic problems in geometry of numbers

Essentially any result in the geometry of numbers can be studied from an algorithmic point of view. Here, we only give some examples and refer to Schrijver (1986), Kannan (1987b), Grötschel, Lovász and Schrijver (1988) and chapter 3.1 by Gruber for further studies.

Recall, first, the definition of a reduced basis of a given lattice $\mathbb{L}$. Let $(v_1,\dots,v_d)$ be an ordered basis of $\mathbb{L}$, let $(v_1^*,\dots,v_d^*)$ be its Gram–Schmidt orthogonalization and let

$$v_i = \sum_{j=1}^{i} \mu_{ij} v_j^*, \quad i = 1,\dots,d.$$

$(v_1,\dots,v_d)$ is called *reduced* if the following two conditions hold:

$$|\mu_{ij}| \leqslant \tfrac{1}{2} \quad \text{for } 1 \leqslant j < i \leqslant d,$$

$$\|v_{i+1}^* + \mu_{i+1,i} v_i^*\|^2 \geqslant \tfrac{3}{4}\|v_i^*\|^2 \quad \text{for } i = 1,\dots,d-1.$$

One of the fundamental results in the algorithmic geometry of numbers is Lovász' basis reduction algorithm which first appeared in Lenstra, Lenstra and Lovász (1982), (cf. also Grötschel, Lovász and Schrijver 1988, Gruber and Lekkerkerker 1987, and Kannan 1987b).

There is a polynomial-time algorithm that, for any given linearly independent vectors $v_1,\dots,v_d \in \mathbb{Q}^d$, finds a reduced basis of the lattice $\mathbb{L}$ spanned by $v_1,\dots,v_d$.

Needless to say how important basis reduction is in geometry of numbers, hence this result has numerous implications. For example, it leads to a polynomial-time algorithm for factorization of polynomials and to an approximate polynomial-time algorithm for simultaneous Diophantine approximation, Lenstra, Lenstra and Lovász (1982), but it can also be used in cryptography, Shamir (1984). A generalization of the basis reduction algorithm to Minkowski geometry was given by Lovász and Scarf (1990).

The basis reduction algorithm was, in particular, applied by Grötschel, Lovász and Schrijver (1988, p. 149) to give the following algorithmic version of Minkowski's fundamental theorem (2.1).

Let $v_1,\dots,v_d \in \mathbb{Q}^d$ *and let* $\mathbb{L} \in \mathscr{L}^d$ *be the lattice spanned by* $v_1,\dots,v_d$. *Let* $K \in \hat{\mathscr{K}}_0^d$, *let* $r, R \in \mathbb{Q}$ *such that* $r\mathbb{B}^d \subset K \subset R\mathbb{B}^d$ *and suppose*

$$V(K) \geqslant \frac{2^{d(d-1)/4}\pi^{d/2}}{\Gamma(d/2+1)}(d+1)^d \det \mathbb{L}.$$

Further, let $\varepsilon \in \mathbb{Q}, \varepsilon > 0$. *Then there is a rational arithmetic algorithm for finding a nonzero lattice point in* K *that uses as a subroutine a procedure that, given a point* $x \in \mathbb{Q}^d$ *either asserts that* $x \in K + \varepsilon\mathbb{B}^d$ *or asserts that*

$$x \notin \operatorname{cl}\Big(\mathbb{R}^d \setminus \big((\mathbb{R}^d \setminus K) + \varepsilon\mathbb{B}^d\big)\Big).$$

Assuming that a call of the subroutine has unit complexity, the algorithm runs in time that is bounded by a polynomial in $d, \log r, \log R$ *and* $\log \varepsilon$.

Observe that the requirement for the volume of K is much stronger than in (2.1), and it is not known whether there is a similar algorithmic version of Minkowski's theorem with its original bound 2^d.

For more results on the algorithmic theory of convex bodies see Grötschel, Lovász and Schrijver (1988) and chapter 2.7 by Gritzmann and Klee. For some interesting recent algorithmic results concerning successive minima see Kannan, Lovász and Scarf (1990).

6.2. NP-hard problems

Karp (1972) showed that the feasibility problem of integer programming is NP-complete:

> Given a rational $m \times d$ matrix A and a vector $b \in \mathbb{Q}^m$, is there a point $x \in \mathbb{Z}^d$ such that $Ax \leqslant b$?

The problem remains NP-hard if all entries of A and b are in $\{0,1\}$ and x is required to be a 0–1-vector. Hence, integer programming is NP-hard in the strong sense, even over rational polytopes. Even the following variant of the knapsack problem is NP-hard.

> Given $a \in \mathbb{Q}^d$ with $a \geqslant 0$ and $\beta \in \mathbb{Q}$; does there exist a vector $x \in \{0,1\}$ such that $\langle a, x\rangle = \beta$?

A transformation from this problem can be used to show that the problem of deciding whether a given $\mathscr{V}$-polytope contains an integer point is also NP-complete (Freund and Orlin 1985)

However, the situation for $\mathscr{H}$-polytopes is even worse: Papadimitriou and Yannakakis (1982) (cf. also Schrijver 1986, p. 253) showed that the following problem is NP-complete.

Given an $\mathcal{H}$-polytope P and a vector $y \in \mathbb{Q}^d$; is $y \in I_{\mathbb{Z}^d}(P)$?

The following problems are also NP-complete; Schrijver (1986, p. 254), Papadimitriou (1978):

Let P be an $\mathcal{H}$-polytope. Given $y \in P \cap \mathbb{Z}^d$, is y not a vertex of $I_{\mathbb{Z}^d}(P)$? Given $x_1, x_2 \in P \cap \mathbb{Z}^d$, are x_1 and x_2 non-adjacent vertices of $I_{\mathbb{Z}^d}(P)$?

The last problem remains NP-complete even when $P \subset (0,1)^d$. In this case x_1, x_2 are vertices of P. Similar difficulties arise when a hyperplane is to be tested for being a facet of $I_{\mathbb{Z}^d}(F)$. Even the problem of deciding whether an $\mathcal{H}$-polytope has only integer vertices is NP-hard: Papadimitriou and Yannakakis (1990) show that the following problem is NP-complete.

Given an $\mathcal{H}$-polytope P, is $P \neq I_{\mathbb{Z}^d}(P)$?

Clearly, computing $G_{\mathbb{Z}^d}(P)$ for an $\mathcal{H}$-polytope P is NP-hard, even #P-complete (see Valiant 1979). This does, however, not directly imply that the problem of counting the number of lattice points of a lattice polytope is also #P-hard since the restriction to polytopes P with integer vertices changes the problem. However, as it is known (Valiant 1979) the problem of determining the number of perfect matchings in a bipartite graph is #P-complete and on the other hand the node–edge incidence matrix of a bipartite graph is totally unimodular. This implies that the corresponding polytope has 0–1-vertices. Hence the problem of counting the number of lattice points of a lattice $\mathcal{H}$-polytope is, indeed, #P-complete. For integer $\mathcal{V}$-polytopes the #P-hardness can be inferred from Ehrhart's result (4.4) and the fact that, by Dyer and Frieze (1988), computing the volume of a rational polytope is #P-hard, Dyer, Gritzmann and Hufnagel (1993). The same is true for the problem of counting the number of lattice points of a lattice zonotope, Dyer, Gritzmann and Hufnagel (1993).

The problem of counting lattice points of polytopes in fixed dimensions was studied by various authors; it can be solved in polynomial time for $d \leqslant 4$ (Zamanskii and Cherkasskii 1983, 1985, Dyer 1991).

For various additional results on the computational complexity of integer programming etc. see Schrijver (1986) and chapter 2.8 by Burkard.

6.3. *Polynomial time solvability*

As we have seen, counting the number of lattice points in an $\mathcal{H}$-polytope is a hard problem. Cook et al. (1992), however, describe an algorithm that determines the number of integer points in a polyhedron $\{x\colon Ax \leqslant b\}$ to within a multiplicative factor of $1+\varepsilon$ in time polynomial in m, L and $1/\varepsilon$ when the dimension d is fixed. This result has to be seen in connection with results of Zamanskii and Cherkasskii (1985) but also in conjunction with Lenstra's (1983) polynomial-time algorithm for integer programming in fixed dimension. Lenstra (1983) proved the following remarkable theorem.

For fixed $d \in \mathbb{N}$ there is a polynomial-time algorithm for the following problem: given $m \in \mathbb{N}$, a rational $m \times d$ matrix A and $b \in \mathbb{Q}^m$, find an integer solution of the system $Ax \leqslant b$ or decide that there is no such solution.

This result implies, in particular, that the integer programming problem can be solved in polynomial time when the dimension is fixed. Based on work of Lenstra, Lenstra and Lovász (1982) and making crucial use of Minkowski's theorem (2.1) Kannan (1987a) improved Lenstra's complexity bound by giving an integer programming algorithm whose complexity depends on the dimension d as $d^{O(d)}$. The problem of counting the number of lattice points of lattice polytopes in fixed $\mathbb{R}^d$ has some relevance for problems in computer algebra, cf. Gritzmann and Sturmfels (1993).

6.4. The complexity of computing upper and lower bound functionals

In sections 2, 3 and 4.3 we have stated various inequalities involving $G_{\mathbb{L}}$ and some other functionals. In view of the hardness of counting lattice points, it is natural to ask whether these functionals can be computed easily. It turns out that functionals like the volume, the surface area, the diameter and the width can be computed for rational ($\mathscr{V}$- or $\mathscr{H}$-presented) polytopes in polynomial-time, if the dimension is fixed. For d being part of the input, computing the volume or surface area is #P-hard, Dyer and Frieze (1988) (see also Khachiyan 1992). The complexity of inner and outer radii like diameter, width, inradius and circumradius in finite-dimensional normed spaces has been studied by Gritzmann and Klee (1993) – and we refer to their paper for the precise statement (and some applications) of the following results. In Euclidean space the situation is roughly as follows: the diameter (and hence the circumradius) problem is NP-hard even for $\mathscr{H}$-presented parallelotopes (centered at the origin); see also Bodlaender et al. (1990). The width problem is NP-hard already for ($\mathscr{H}$- or $\mathscr{V}$-presented) simplices, the inradius problem is NP-hard already for $\mathscr{V}$-presented cross-polytopes. The following radii can be computed or at least approximated in polynomial time: the inradius for $\mathscr{H}$-polytopes, the width for symmetric $\mathscr{H}$-polytopes, the diameter and the circumradius for $\mathscr{V}$-polytopes. In ℓ_1 and in ℓ_∞ spaces some of the radius computations become easier.

The complexity of computing the lattice width of a polytope as used in (3.12) and (3.13) has not been determined, yet.

Let us close with the discouraging result of Cook et al. (1992) that for any polynomial $p : \mathbb{Z} \to \mathbb{Z}$ the following problem is NP-hard:

> Given an $\mathscr{H}$-polytope P in $\mathbb{R}^d$; find positive integers α, β such that $\alpha \leqslant G_{\mathbb{Z}^d}(P) + 1 \leqslant \beta$ and $\beta \leqslant 2^{p(d)}\alpha$.

This result shows a dilemma for the problem of computing lower and upper bound functionals for the lattice point enumerator in variable dimensions: either the gap between the lower and the upper bound grows super-exponentially in the dimension or at least one of the two functionals is itself hard to compute.

Acknowledgment

We would like to express our gratitude to the following friends and colleagues for their valuable comments on a previous version of this article: Martin Grötschel, Peter M. Gruber, Victor Klee, Vitali Milman, Herbert E. Scarf, Alexander Schrijver, Richard P. Stanley and Günter M. Ziegler.

References

Aho, A.V., J.E. Hopcroft and J.D. Ullman
[1974] *The Design and Analysis of Computer Algorithms* (Addison-Wesley, Reading, MA).

Arnold, V.I.
[1980] Statistics of integral lattice polygons (in Russian), *Funkcional. Anal. i Priložen.* **14**, 1–3.

Bambah, R.P., A.C. Woods and H. Zassenhaus
[1965] Three proofs of Minkowski's second inequality in the geometry of numbers, *J. Austral. Math. Soc.* **5**, 453–462.

Bárány, I., and J. Pach
[1992] On the number of convex lattice polygons, Manuscript.

Bárány, I., and A.M. Vershik
[1992] On the number of convex lattice polytopes, Manuscript.

Bárány, I., R. Howe and L. Lovász
[1992] On integral points in polyhedra: A lower bound, *Combinatorica*, **12**, 135–141.

Bell, D.E.
[1977] A theorem concerning the integer lattice, *Stud. Appl. Math.* **56**, 187–188.

Bender, E.A.
[1962] Area–perimeter relations for two-dimensional lattices, *Amer. Math. Monthly* **69**, 742–744.

Bernstein, D.N.
[1976] The number of integral points in integral polyhedra, *Functional Anal. Appl.* **10**(3), 223–224.

Betke, U.
[1979] *Gitterpunkte und Gitterpunktfunktionale*, Habilitationsschrift (Universität Siegen, Siegen).

Betke, U., and P. Gritzmann
[1986] An application of valuation theory to two problems of discrete geometry, *Discrete Math.* **58**, 81–85.

Betke, U., and M. Henk
[1993] Intrinsic volumes and lattice points of crosspolytopes, *Monatsh. Math.*, in press.

Betke, U., and M. Kneser
[1985] Zerlegung und Bewertungen von Gitterpolytopen, *J. Reine Angew. Math.* **358**, 202–208.

Betke, U., and P. McMullen
[1985] Lattice points in lattice polytopes, *Monatsh. Math.* **99**, 253–265.

Betke, U., and J.M. Wills
[1979] Stetige und diskrete Funktionale konvexer Körper, in: *Contributions to Geometry*, Proc. Geom. Sympos., Siegen, eds J. Tölke and J.M. Wills (Birkhäuser, Basel) pp. 226–237.

Betke, U., M. Henk and J.M. Wills
[1993] Successive minima type inequalities, *Discrete Comput. Geom.* **9**, in press.

Blichfeldt, H.F.
[1914] A new principle in the geometry of numbers with some applications, *Trans. Amer. Math. Soc.* **15**, 227–235.
[1921] Note on geometry of numbers, *Bull. Amer. Math. Soc.* **27**, 150–153.

Bodlaender, H.L., P. Gritzmann, V. Klee and J. van Leeuwen
[1990] Computational complexity of norm-maximization, *Combinatorica* **10**, 203–225.

Bokowski, J.
[1975] Gitterpunktanzahl und Parallelkörpervolumen von Eikörpern, *Monatsh. Math.* **79**, 93–101.

Bokowski, J., and A.M. Odlyzko
[1973] Lattice points and the volume area ratio of convex bodies, *Geom. Dedicata* **2**, 249–254.

Bokowski, J., and J.M. Wills
[1974] Eine Ungleichung zwischen Volumen, Oberfläche und Gitterpunktanzahl konvexer Mengen im R^3, *Acta Math. Acad. Sci. Hungar.* **25**, 7–13.

Bokowski, J., H. Hadwiger and J.M. Wills
[1972] Eine Ungleichung zwischen Volumen, Oberfläche und Gitterpunktanzahl konvexer Körper im n-dimensionalen euklidischen Raum, *Math. Z.* **127**, 363–364.

Bonnesen, T., and W. Fenchel
[1934] *Theorie der konvexen Körper* (Springer, Berlin).

Bourgain, J., and V.D. Milman
[1985] Sectiones euclidiennes et volume des corps symétriques convexes dans R^n, *C.R. Acad. Sci. Paris 300, Ser. 1* **13**, 435–438.
[1987] New volume ratio properties for convex symmetric bodies in R^n, *Invent. Math.* **88**, 319–340.

Cassels, J.W.S.
[1971] *An Introduction to Geometry of Numbers* (Springer, New York).

Chalk, J.H.
[1980] The Vinogradoff–Mordell–Titäväinen inequalities, *Indag. Math.* **42**, 367–374.

Cook, W.J., M. Hartmann, R. Kannan and C. McDiarmid
[1992] On integer points in polyhedra, *Combinatorica* **12**, 27–37.

Davenport, H.
[1951] On a principle of Lipschitz, *J. London Math. Soc.* **26**, 179–183.

Ding, R., and J.R. Reay
[1987a] Areas of lattice polygons, applied to computer graphics, *Appl. Math.* **19**, 547–556.
[1987b] The boundary characteristic and Pick's theorem in the Archimedean planar tilings, *J. Combin. Theory Ser. A* **44**, 110–119.

Ding, R., K. Kolodziejczyk and J.R. Reay
[1988] A new Pick-type theorem on the hexagonal lattice, *Discrete Math.* **68**, 171–177.

Doignon, J.P.
[1973] Convexity in crystallographical lattices, *J. Geom.* **3**, 71–85.

Dyer, M.
[1991] On counting lattice points in polyhedra, *SIAM J. Comput.* **20**, 695–707.

Dyer, M., P. Gritzmann and A. Hufnagel
[1993] On the complexity of computing mixed volumes, Manuscript.

Dyer, M.E., and A.M. Frieze
[1988] The complexity of computing the volume of a polyhedron, *SIAM J. Comput.* **17**, 967–974.

Edmonds, J.
[1965] Maximum matching and a polyhedron with 0, 1-vertices, *J. Res. Nat. Bur. Standards (B)* **69**, 125–130.

Ehrhart, E.
[1955a] Sur les ovales et les ovoides, *C.R. Acad. Sci. Paris* **258**, 573–575.
[1955b] Une généralisation du théorème de Minkowski, *C.R. Acad. Sci. Paris* **258**, 483–485.
[1967] Sur un problème de géometrie diophantienne linéaire, *J. Reine Angew. Math.* **226**, 1–29; **227**, 25–49.
[1968] Démonstration de la loi de réciprocité, *C.R. Acad. Sci. Paris* **265**, 5–9, 91–94; **266** (1969) 696–697.
[1977] *Polynômes Arithmétiques et Méthode des Polyèdres en Combinatoire* (Birkhäuser, Basel).

Erdős, P., P.M. Gruber and J. Hammer
[1989] *Lattice Points* (Longman, Essex).

Folkman, J.H., and R.L. Graham
[1969] A packing inequality for compact convex subsets of the plane, *Canad. Math. Bull.* **12**, 745–752.

Freund, R.M., and J.B. Orlin
[1985] On the complexity of four polyhedral set containment problems, *Math. Programming* **33**, 139–145.

Fricker, F.
[1982] *Einführung in die Gitterpunktlehre* (Birkhäuser, Basel).

Garey, M.R., and D.S. Johnson
[1979] *Computers and Intractability: A Guide to the Theory of NP-completeness* (Freeman, San Francisco).

Gritzmann, P.
[1984] *Finite Packungen und Überdeckungen,* Habilitationsschrift (Universität Siegen, Siegen).
[1986] Finite packing of equal balls, *J. London Math. Soc.* **33**, 543–553.

Gritzmann, P., and V. Klee
[1993] Computational complexity of inner and outer *j*-radii of polytopes in finite dimensional normed spaces, *Math. Programming,* to appear.

Gritzmann, P., and B. Sturmfels
[1993] Minkowski addition of polytopes: Computational complexity and applications to Gröbner bases, *SIAM J. Discrete Math.*, to appear.

Gritzmann, P., and J.M. Wills
[1986] An upper estimate for the lattice point enumerator, *Mathematika* **33**, 197–203.

Groemer, H.
[1960] Über die Einlagerung von Kreisen in einen konvexen Bereich, *Math. Z.* **73**, 285–294.

Grötschel, M., and M. Padberg
[1985] in: *The Traveling Salesman Problem,* eds E.L. Lawler, J.K. Lenstra, A.H.G. Rinnooy Kan and D.B. Shmoys (Wiley, New York) Polyhedral theory, pp. 251–360; Polyhedral computations, pp. 307--360.

Grötschel, M., L. Lovász and A. Schrijver
[1988] *Geometric Algorithms and Combinatorial Optimization* (Springer, Berlin).

Gruber, P.M.
[1979] Geometry of numbers, in: *Contributions to Geometry,* eds J. Tölke and J.M. Wills (Birkhäuser, Basel).

Gruber, P.M., and C.G. Lekkerkerker
[1987] *Geometry of Numbers* (North-Holland, Amsterdam).

Grünbaum, B.
[1967] *Convex Polytopes* (Wiley, London). New edition with an additional chapter by V. Klee and P. Kleinschmidt: 1993, Springer, Berlin.

Grünbaum, B., and G.C. Shephard
[1992] Pick's theorem revisited, preprint.
Hadwiger, H.
[1951] Beweis eines Funktionalsatzes für konvexe Körper, *Abh. Math. Sem. Univ. Hamburg* **17**, 69–76.
[1957] *Vorlesungen über Inhalt, Oberfläche und Isoperimetrie* (Springer, Berlin).
[1970] Volumen und Oberfläche eines Eikörpers, der keine Gitterpunkte überdeckt, *Math. Z.* **116**, 191–196.
[1972] Gitterperiodische Punktmengen und Isoperimetrie, *Monatsh. Math.* **76**, 410–418.
[1975] Das Wills'sche Funktional, *Monatsh. Math.* **79**, 213–221.
[1979] Gitterpunktanzahl im Simplex und Wills'sche Vermutung, *Math. Ann.* **239**, 271–288.
Hadwiger, H., and J.M. Wills
[1974] Gitterpunktanzahl konvexer Rotationskörper, *Math. Ann.* **208**, 221–232.
[1975] Neuere Studien über Gitterpolygone, *J. Reine Angew. Math.* **280**, 61–69.
Hammer, J.
[1971] Volume–surface area relations for n-dimensional lattices, *Math. Z.* **123**, 219–222.
Hayes, A.C., and D.G. Larman
[1983] The vertices of the knapsack polytope, *Discrete Appl. Math.* **6**, 135–138.
Henk, M.
[1990] Inequalities between successive minima and intrinsic volumes of a convex body, *Monatsh. Math.* **110**, 279–282.
Hensley, D.
[1983] Lattice vertex polytopes with few interior lattice points, *Pacific J. Math.* **105**, 183–191.
Hibi, T.
[1991] Ehrhart polynomials of convex polytopes, h-vectors of simplicial complexes and non-singular projective toric varieties, preprint.
[1992] Dual polytopes of rational convex polytopes, *Combinatorica* **12**, 237–240.
Hlawka, E.
[1980] 90 Jahre Geometrie der Zahlen, in: *Jahrbuch Überblicke Mathematik* (Bibliografisches Institut, Mannheim) pp. 9–41.
Hoffman, A.J., and J.B. Kruskal
[1956] Integral boundary points of convex polyhedra, in: *Linear Inequalities and Related Systems*, eds H.W. Kuhn and A.W. Tucker (Princeton University Press, Princeton, NJ) pp. 223–246.
Höhne, R.
[1980] Gitterpunktanzahl und Parallelkörpervolumen von Eikörpern, *Math. Ann.* **251**, 269–276.
Kannan, R.
[1987a] Minkowski's convex body theorem and integer programming, *Math. Oper. Res.* **12**, 415–440.
[1987b] Algorithmic geometry of numbers, *Ann. Rev. Comput. Sci.* **2**, 231–267.
[1989] The Frobenius problem, *Lecture Notes in Comput Sci.* **405** (summary) 242–251.
Kannan, R., and L. Lovász
[1988] Covering minima and lattice point free convex bodies, *Ann. Math.* **128**, 577–602.
Kannan, R., L. Lovász and H.E. Scarf
[1990] The shapes of polyhedra, *Math. Oper. Res.* **15**, 364–380.
Karmarkar, N.
[1984] A new polynomial-time algorithm for linear programming, *Combinatorica* **4**, 373–395.
Karp, R.M.
[1972] Reducibility among combinatorial problems, in: *Complexity of Computer Computations*, eds R.E. Miller and J.W. Thatcher (Plenum Press, New York) pp. 85–103.
Keller, O.H.
[1954] *Geometrie der Zahlen* (Enzykl. Wiss. Bd. I 2, 27) (Teubner, Leipzig).

Khachiyan, L.G.
[1979] A polynomial algorithm in linear programming (in Russian), *Dokl. Akad. Nauk. SSSR* **244**, 1093–1096 [Engl. translation: *Soviet Math. Dokl.* **20**, 191–194].
[1992] Complexity of polytope volume computation, in: *New Trends in Discrete and Computational Geometry*, ed. J. Pach (Springer, Berlin).

Lagarias, J.C.
[1993] Point lattices, in: *Handbook of Combinatorics*, eds R.L. Graham, M. Grötschel and L. Lovász (North-Holland, Amsterdam), in press.

Lagarias, J.C., and G.M. Ziegler
[1991] Bounds for lattice polytopes containing a fixed number of interior points in a sublattice, *Canad. J. Math.* **43**, 1022–1035.

Lang, S.
[1990] Old and new conjectured diophantine inequalities, *Bull. Amer. Math. Soc.* **23**, 37–75.

Lawler, E.L., J.K. Lenstra, A.H.G. Rinnooy Kan and D.B. Shmoys, eds
[1985] *The Traveling Salesman Problem. A Guided Tour of Combinatorial Optimization* (Wiley, New York).

Lenstra, H.W.
[1983] Integer programming with a fixed number of variables, *Math. Oper. Res.* **8**, 538–548.

Lenstra, J.K., H.W. Lenstra and L. Lovász
[1982] Factoring polynomials with rational coefficients, *Math. Ann.* **261**, 513–534.

Lovász, L., and H.E. Scarf
[1990] The generalized basis reduction algorithm, preprint.

Macdonald, I.G.
[1963] The volume of a lattice polyhedron, *Proc. Cambridge Philos. Soc.* **59**, 719–726.

Mahler, K.
[1939] Ein Übertragungsprinzip für konvexe Körper, *Časopis Pěst. Mat. Fyz.* **68**, 93–102.

McMullen, P.
[1975] Non-linear angle-sum relations for polyhedral cones and polytopes, *Math. Proc. Cambridge Philos. Soc.* **78**, 247–261.
[1977] Valuations and Euler-type relations on certain classes of convex polytopes, *Proc. London Math. Soc. (3)* **35**, 113–135.
[1984] Determinants and lattices induced by rational subspaces, *Bull. London Math. Soc.* **16**, 275–277.

McMullen, P., and R. Schneider
[1983] Valuations on convex bodies, in: *Convexity and its Applications*, eds P. Gruber and J.M. Wills (Birkhäuser, Basel) pp. 170–247.

McMullen, P., and J.M. Wills
[1981] Minimal width and diameter of lattice-point-free convex bodies, *Mathematika* **28**, 255–264.

Meyer, R.R.
[1974] On the existence of optimal solutions to integer and mixed integer programming problems, *Math. Programming* **7**, 223–235.

Minkowski, H.
[1896] *Geometrie der Zahlen* (Teubner, Leipzig).

Niederreiter, H., and J.M. Wills
[1975] Diskrepanz und Distanz von Maßen bezüglich konvexer und Jordanscher Mengen, *Math. Z.* **144**, 125–134.

Nosarzewska, M.
[1948] Evaluation de la difference entre l'aire d'une région plane convexe et le nombre des points aux coordinières couverts par elle, *Coll. Math.* **1**, 305–311.

Oler, N.
[1961] An inequality in the geometry of numbers, *Acta Math.* **105**, 19–48.

Overhagen, T.
[1975] Zur Gitterpunktanzahl konvexer Körper im 3-dimensionalen euklidischen Raum, *Math. Ann.* **216**, 217–224.

Padberg, M., and M.R. Rao
[1974] The travelling salesman problem and a class of polyhedra of diameter two, *Math. Programming* **7**, 32–45.

Papadimitriou, C.H.
[1978] The adjacency relation on the traveling salesman polytope is NP-complete, *Math. Programming* **14**, 312–324.

Papadimitriou, C.H., and M. Yannakakis
[1982] The complexity of facets (and some facets of complexity), in: *Proc. 14th Ann. ACM Symp. Theory of Comput.*, San Francisco (Ass. Comput. Mach., New York) pp. 255–260; also in: *J. Computer Syst. Sci.* **28** (1984) 244–259.
[1990] On recognizing integer polyhedra, *Combinatorica* **10**, 107–109.

Perles, M., J.M. Wills and J. Zaks
[1982] On lattice polytopes having interior lattice points, *Elem. Math.* **37**, 44–46.

Pick, G.
[1899] Geometrisches zur Zahlenlehre, *Naturwiss. Z. Lotos, Prag.*, 311–319.

Rabinowitz, S.
[1989] A census of convex lattice polygons with at most one interior lattice point, *Ars Combin.* **28**, 83–96.

Reeve, J.E.
[1957] On the volume of lattice polyhedra, *Proc. London Math. Soc. (3)* **7**, 378–395.
[1959] A further note on the volume of lattice polyhedra, *J. London Math. Soc.* **34**, 57–62.

Reznick, B.
[1986] Lattice point simplices, *Discrete Math.* **60**, 219–242.

Rogers, C.A.
[1964] *Packing and Covering* (Cambridge University Press, Cambridge).

Rubin, D.S.
[1970] On the unlimited number of faces in integer hulls of linear programs with a single constraint, *Oper. Res.* **18**, 940–946.

Santaló, L.A.
[1949] Una invariante afin pasa os cuerpas convexos del espacio du n-dimensiones, *Portugal. Math.* **8**, 155–161.
[1976] *Integral Geometry and Geometric Probability* (Addison-Wesley, London).

Sawyer, D.B.
[1954] The lattice determinants of asymmetric convex regions, *J. London Math. Soc.* **29**, 251–254.

Scarf, H.E.
[1977] An observation on the structure of production sets with invisibilities, *Proc. Nat. Acad. Sci. U.S.A.* **74**, 3637–3641.
[1985] Integral polyhedra in three-space, *Math. Oper. Res.* **10**, 403–438.

Scarf, H.E., and D.F. Shallcross
[1990] The Frobenius problem and maximal lattice free bodies, preprint.

Schmidt, W.M.
[1972] Volume, surface area and the number of integer points covered by a convex set, *Arch. Math.* **33**, 537–543.

Schnell, U.
[1992] Minimal determinants and lattice inequalities, *Bull. London Math. Soc.* **24**, 606–612.

Schnell, U., and J.M. Wills
[1991] Two isoperimetric inequalities with lattice constraints, *Monatsh. Math.* **112**, 227–233.

Schrijver, A.
[1986] *Theory of Linear and Integer Programming* (Wiley, Chichester).
[1993a] Polyhedral combinatorics, in: *Handbook of Combinatorics*, eds R. Graham, M. Grötschel and L. Lovász (North-Holland, Amsterdam), in press.
[1993b] *Polyhedral Combinatorics*, forthcoming.

Schwermer, J.
[1991] Räumliche Anschauung und Minima positiver quadratischer Formen, *Jber. Deutsch. Math.-Vereinig.* **93**, 49–105.

Scott, P.R.
[1973] A lattice problem in the plane, *Mathematika* **20**, 247–252.
[1976] On convex lattice polygons, *Bull. Austral. Math. Soc.* **15**, 395–399.
[1979] Two inequalities for convex sets with lattice point constraints in the plane, *Bull. London Math. Soc.* **11**, 273–278.

Seymour, P.D.
[1980] Decomposition of regular matroids, *J. Combin. Theory Ser. B* **28**, 305–359.

Shamir, A.
[1984] A polynomial time algorithm for breaking the Merkle–Hellman crypto system, *IEEE Trans. Inform. Theory* **30**, 699–704.

Shevchenko, V.N.
[1981] On the number of extreme points in integer programming, *Kibernetika* **2**, 133–134.

Siegel, C.L.
[1935] Über Gitterpunkte in konvexen Körpern und ein damit zusammenhängendes Extremalproblem, *Acta Math.* **65**, 307–323.

Stanley, R.P.
[1976] Magic labelings of graphs, symmetric magic squares, systems of parameters and Cohen–Macauley rings, *Duke Math. J.* **43**, 511–531.
[1980] Decompositions of rational convex polytopes, *Ann. Discrete Math.* **6**, 333–342.
[1986] *Enumerative Combinatorics*, Vol. 1 (Wadsworth & Brooks/Cole, Pacific Grove).
[1991a] A zonotope associated with graphical degree sequences, in: *Applied Geometry and Discrete Mathematics: "The Victor Klee Festschrift"*, eds P. Gritzmann and B. Sturmfels, DIMACS Series on Discrete Mathematics and Computer Science, Vol. 4 (Amer. Math. Soc & Ass. Comput. Mach., Providence, RI) pp. 555–570.
[1991b] A monotonicity property of h-vectors and h^*-vectors, preprint.

Valiant, L.G.
[1979] The complexity of enumeration and reliability problems, *SIAM J. Comput.* **8**, 410–421.

Van der Corput, J.G.
[1935] Verallgemeinerung einer Mordellschen Beweismethode in der Geometrie der Zahlen I, *Acta Arith.* **1**, 62–66.
[1936] Verallgemeinerung einer Mordellschen Beweismethode in der Geometrie der Zahlen II, *Acta Arith.* **2**, 145–146.

Varberg, D.E.
[1985] Pick's theorem revisited, *Amer. Math. Monthly* **92**, 584–587.

Veinott, A.F., and G.B. Dantzig
[1968] Integer extreme points, *SIAM Rev.* **10**, 371–372.

Walfisz, A.
[1957] *Gitterpunkte in mehrdimensionalen Kugeln* (Warschau).

White, G.K.
[1964] Lattice tetrahedra, *Canad. J. Math.* **16**, 389–396.

Wills, J.M.
[1968] Ein Satz über konvexe Mengen und Gitterpunkte, *Monatsh. Math.* **72**, 451–463.
[1970] Ein Satz über konvexe Körper und Gitterpunkte, *Abh. Math. Sem. Univ. Hamburg* **35**, 8–13.

[1973] Zur Gitterpunktanzahl konvexer Mengen, *Elem. Math.* **28**, 57–63.
[1982] On an analog to Minkowski's lattice point theorem, in: *The Geometric Vein: The Coxeter Festschrift,* eds C. Davies, B. Grünbaum and F.A. Sherk (Springer, New York) pp. 285–288.
[1989] A counterpart to Oler's lattice point theorem, *Mathematika* **36**, 216–220.
[1990a] Kugellagerungen und Konvexgeometrie, *Jber. Deutsch. Math.-Vereinig.* **92**, 21–46.
[1990b] Minkowski's successive minima and the zeros of a convexity-function, *Monatsh. Math.* **109**, 157–164.
[1991] Bounds for the lattice point enumerator, *Geom. Dedicata* **40**, 237–244.

Woods, A.C.
[1966] A generalization of Minkowski's second inequality in the geometry of numbers, *J. Austral. Math. Soc.* **6**, 148–152.

Xu, Y.J., and S.S.T. Yau
[1992] A sharp estimate of the number of integral points in a tetrahedron, *J. Reine Angew. Math.* **423**, 199–219.

Zamanskii, L.Y., and V.L. Cherkasskii
[1983] Determination of the number of integer points in polyhedra in R^3: Polynomial algorithms, *Dokl. Akad. Nauk Ukrain. USSR. Ser. A* **4**, 13–15.
[1985] Generalization of the Jacobi–Perron algorithm for determining the number of integer points in polyhedra, *Dokl. Akad. Nauk. USSR A* **10**, 10–13.

Zassenhaus, H.
[1961] Modern developments in the geometry of numbers, *Bull. Amer. Math. Soc.* **67**, 427–439.

CHAPTER 3.3

Packing and Covering with Convex Sets

Gábor FEJES TÓTH

Mathematical Institute, Hungarian Academy of Sciences, Reáltanoda u. 13–15, H-1053 Budapest, Hungary

Wlodzimierz KUPERBERG

Auburn University, Division of Mathematics F.A.T., Parker Hall, Auburn AL 36849, USA

Contents

HANDBOOK OF CONVEX GEOMETRY
Edited by P.M. Gruber and J.M. Wills

Overview

In this chapter we will discuss arrangements (families) of sets in a space E which should have a structure admitting the notions of congruence, measure (volume) and convexity. We will concentrate mostly on the Euclidean d-dimensional space E^d, but spherical and hyperbolic spaces will also be considered. Members of the arrangements will be *convex bodies* (compact convex sets with non-void interior) unless otherwise specifically assumed. Given a domain in E, a *packing in the domain* is an arrangement whose members are all contained in the domain and have mutually disjoint interiors, and a *covering of the domain* is an arrangement whose union contains the domain. A packing in or a covering of the whole space E is simply called a *packing* or a *covering*, respectively. An arrangement which is a packing and a covering at the same time is called a *tiling* (see chapter 3.5).

It is possible to consider these concepts in a more general setting. There are in the literature problems and results concerning packing and covering in discrete spaces, for example in coding theory and combinatorics, but those will be mentioned here only as far as they have connections with the continuous type of geometry discussed in this book.

1. Introduction

The most important notion associated with packings and coverings is density. For a packing of E, density represents, intuitively speaking, the ratio between the sum of the measures (areas in E^2, volumes in E^3, etc.) of the bodies being packed and that of the space in which they are packed. Analogously, for a covering, density is the ratio of the sum of measures of the bodies and that of the space being covered. Precise definitions require use of limits and will be given in the next section, where the natural questions of existence and uniqueness also will be addressed.

The problems that play a central role in the theory of packings and coverings are of the following types: among a given family of packings select a *densest* one (i.e., one whose density is maximum) and among a given family of coverings select a *thinnest* one (i.e., one whose density is minimum). For example, for a given convex body K, consider the family of all packings of E consisting of congruent copies of K. Which of the packings is of maximum density? (It has been shown that this maximum density is attained in the family.) This maximum density is denoted by $\delta(K)$ and is called the *packing density of* K. The *covering density of* K, $\vartheta(K)$, is defined analogously.

If the family of packings is restricted to translates of K, then we obtain $\delta_T(K)$, the *translation packing density of* K or, analogously for coverings, the *translation covering density of* K, $\vartheta_T(K)$. While $\delta(K)$ and $\vartheta(K)$ can be defined in more general spaces E, the translation densities $\delta_T(K)$ and $\vartheta_T(K)$ are restricted to the Euclidean spaces E^d. In these spaces, even further restrictions are possible. One considers so-called *lattice arrangements* (packings, coverings, etc.) whose members are translates of each other in such a way that the corresponding translation vectors form a lattice (see chapters 3.1 and 3.2 of this Handbook). In this way, we obtain the *lattice packing density* and *lattice covering density* of K denoted by $\delta_L(K)$ and $\vartheta_L(K)$, respectively.

Determining the values of $\delta(K)$, $\delta_T(K)$, $\delta_L(K)$, $\vartheta(K)$, $\vartheta_T(K)$ and $\vartheta_L(K)$ is usually a different problem for a different body K, and usually a very difficult one. Only a few solutions in some very special cases exist, mostly in dimension 2. In dimension 3, the determination of these quantities for a given body K remains, in Roger's words (1964*, p. 7), "a formidable task". In particular, the problem of maximum density sphere packings in E^3 still remains open, despite the well-known, long-standing conjecture and the common belief in it, and despite the fact that the lattice version of the problem has been solved a long time ago (Gauss 1831). The conjecture is that the maximum density in general is the same as that in the tightest lattice packing, which is $\pi/\sqrt{18}$. In view of these difficulties, the main problem of the theory of packings and coverings shifts to obtaining good estimates of the densities in terms of upper and lower bounds for fairly general classes of convex bodies.

Directly from definitions, one obtains the following inequalities:

$$\delta_L(K) \leqslant \delta_T(K) \leqslant \delta(K) \leqslant 1 \leqslant \vartheta(K) \leqslant \vartheta_T(K) \leqslant \vartheta_L(K) \tag{1.1}$$

for every K.

Obviously, there exist constants $p > 0$ and c (depending on d only) such that $p \leqslant \delta_L(K)$ and $\vartheta_L(K) \leqslant c$ for all K in E^d. Some of the leading problems concern the best possible such constants for each of the corresponding densities δ, δ_T, δ_L, ϑ, ϑ_T, ϑ_L and for some special classes of convex bodies, e.g., centrally symmetric bodies. Also, one can ask which of the inequalities in (1.1) can be replaced by equalities under certain restrictions on K, or, more specifically, to describe classes of convex bodies K for which a certain inequality in (1.1) can be replaced by equality.

Early results related to packing and covering can be found in the works of Lagrange (1773) who was interested in the theory of quadratic forms and who implicitly found the lattice packing density of the circular disk in E^2, and in the discoveries of Gauss (1831) who explicitly considered lattice packings of spheres in E^3. It was Minkowski who systematized these topics into a separate theory which he named *geometry of numbers*. The early development of the theory of packings and coverings was stimulated by its connection to number theory and crystallography, but more recently it developed into a broad theory of its own which includes a multitude of problems besides those mentioned above. Some of them are concerning the number of neighbors, some deal with various concepts of efficiency of arrangements other than density, and some are related to special types of convex bodies.

In the following sections of this chapter we give an outline of the classical part of the theory as well as of its more recent development, including main results and often sketching various techniques used to obtain them.

The limitations of this chapter do not allow us to discuss all of the topics related to the theory of packings and coverings. Some of the topics, such as tilings and finite arrangements are discussed in detail in other chapters of this Handbook (see chapters 3.4 and 3.5, respectively). Some other have been treated in recent books and surveys. For problems concerning packing and covering with sequences of (not necessarily congruent) convex bodies we refer the reader to Groemer (1985*). Covering a convex body with smaller, homothetic copies of itself is described in K. Bezdek (1993*). Arrangements of circular disks in the plane are extensively treated in G. Fejes Tóth (1983*). The literature given at the end of the chapter lists virtually all relevant books and expository articles separately from research papers. Each reference to an item in the "books and survey papers" part of the literature is marked with an asterisk, as above.

2. Density bounds in d-dimensional Euclidean space

2.1. Densities

As mentioned in the introduction, E will denote one of the following d-dimensional spaces: the Euclidean space E^d, the spherical space S^d or the hyperbolic space H^d, although several of the concepts and theorems presented here can be stated in a more general setting.

The measure of a (measurable) set A in E will be denoted by $V(A)$. By the measure we will usually understand the Lebesgue or Jordan measure, but since we will be dealing with sets of a simple geometric structure, such as convex sets, polytopes, or, most generally, open or closed domains, the question of measurability or of the type of measure used will seldom concern us. By a *region* we shall mean a Jordan-measurable set in E^d homeomorphic to a closed ball.

Given an *arrangement*, i.e., a family of sets, $\mathcal{A}$, and a bounded domain G in E, we define the *inner density of $\mathcal{A}$ relative to G*, $d_{\mathrm{inn}}(\mathcal{A}|G)$, the *outer density of $\mathcal{A}$ relative to G*, $d_{\mathrm{out}}(\mathcal{A}|G)$, and the *density of $\mathcal{A}$ relative to G*, $d(\mathcal{A}|G)$, as follows:

$$d_{\mathrm{inn}}(\mathcal{A}|G) = \frac{1}{V(G)} \sum_{A\in\mathcal{A},\, A\subset G} V(A),$$

$$d_{\mathrm{out}}(\mathcal{A}|G) = \frac{1}{V(G)} \sum_{A\in\mathcal{A},\, A\cap G\neq\emptyset} V(A),$$

and

$$d(\mathcal{A}|G) = \frac{1}{V(G)} \sum_{A\in\mathcal{A}} V(A\cap G).$$

Concerning density of an arrangement $\mathcal{A}$ relative to the whole space E, observe that it is already defined for $E = S^d$, since S^d is bounded. For $E = H^d$ attempts to define density in a natural way lead to substantial difficulties. We shall discuss these difficulties in section 3.2. For $E = E^d$, the concept is defined as follows:

Start with a region G and its interior point o; we will call the pair (G, o) a *gauge for density*. Then let

$$d_-(\mathcal{A}, G, o) = \liminf_{\lambda \to \infty} d_{\text{inn}}(\mathcal{A}|\lambda G),$$

and

$$d_+(\mathcal{A}, G, o) = \limsup_{\lambda \to \infty} d_{\text{out}}(\mathcal{A}|\lambda G),$$

where λG means the homothetic image of G with coefficient λ and with o as the center of homothety.

The number $d_-(\mathcal{A}, G, o)$ [$d_+(\mathcal{A}, G, o)$], possibly infinite, is called the *lower* [resp. the *upper*] *density of* $\mathcal{A}$ *with gauge* (G, o). If these two densities are equal, we call their common value the *density of* $\mathcal{A}$ *with gauge* (G, o) and denote it by $d(\mathcal{A}, G, o)$. In case the gauge is the unit ball centered at o, in the notation for the densities of an arrangement $\mathcal{A}$ we omit the gauge and simply write $d_+(\mathcal{A})$, $d_-(\mathcal{A})$ and $d(\mathcal{A})$. One can prove that these densities remain the same if the gauge (G, o) is replaced by its translate $(G + v, o + v)$ by any vector v, but otherwise they depend on the choice of gauge.

However, for certain crucial types of arrangements density does not depend on gauge. One of such types consists of lattice arrangements. It is easy to prove that if $\mathcal{A}$ is a lattice arrangement of a body K and if P is a basic parallelepiped of the lattice, then density of $\mathcal{A}$ exists, is independent from gauge, and is equal to the ratio $V(K)/V(P)$. More generally, the independence from gauge occurs for every periodic arrangement (see remark below).

It turns out that every convex body K admits a packing and a covering whose density is extreme and does not depend on gauge. Define the *packing density* $\delta(K)$ for any convex body K in E^d as the supremum of the densities $d_+(\mathcal{P}, G, o)$ over all packings $\mathcal{P}$ with congruent copies of K and with all possible gauges (G, o) used to measure density. Similarly, the *covering density* $\vartheta(K)$ of K is defined as the infimum of the densities $d_-(\mathcal{C}, G, o)$ among all coverings $\mathcal{C}$ with congruent copies of K and with all gauges (G, o). Although in the above definitions the supremum and the infimum is taken over all gauges, it turns out that the choice of gauge is irrelevant. Specifically, the following holds true: For every convex body K in E^d there exists a packing $\mathcal{P}$ and a covering $\mathcal{C}$ with congruent copies of K such that

$$d_+(\mathcal{P}, G, o) = d_-(\mathcal{P}, G, o) = \delta(K)$$

and

$$d_-(\mathcal{C}, G, o) = d_+(\mathcal{C}, G, o) = \vartheta(K)$$

for every gauge (G, o).

For a proof of this theorem in a considerably more general setting see Groemer (1963, 1968, 1986b). In view of the above theorem, for the purpose of finding or estimating the size of $\delta(K)$ or $\vartheta(K)$ a suitable gauge may be chosen, such as the unit ball or the unit cube with its center at the origin. Moreover, since the extreme arrangements possess density, an upper bound for $d_{-}(\mathcal{P}, G, o)$ for all packings $\mathcal{P}$ with congruent copies of K and a specific gauge (G, o) will provide an upper bound for $\delta(K)$. Similarly, any lower bound for the upper density of all coverings with congruent copies of K yields a lower bound for $\vartheta(K)$.

It should be noted that the above theorem and its proof do not require the assumption of convexity of K. All that is needed is that K is a compact region. Furthermore, the theorem and the definitions of $\delta(K)$ and $\vartheta(K)$ can be easily adapted to accommodate arrangements consisting of translates of K. This modification produces the *translation packing density of* K, $\delta_T(K)$, and the *translation covering density of* K, $\vartheta_T(K)$. The definitions of the corresponding *lattice packing density*, $\delta_L(K)$ and *lattice covering density* $\vartheta_L(K)$ do not require this elaboration.

In the definition of density of an arrangement other measures can be considered instead of the volume of the members of the arrangement, or a different measure can be assigned to the gauge (or both). For example, the so called *number density* (or *point density*) can be defined by assigning to each member of the arrangement the constant 1, and retaining the usual volume for the gauge. Existence and uniqueness theorems in such general settings can be found in Groemer (1963, 1968), including packings and coverings with not necessarily congruent bodies.

If K is a region in E^d whose congruent copies can be arranged so as to form a tiling, then K is called a *tile*. Obviously, the packing density as well as the covering density of a tile is equal to 1. Moreover, the converse, included in the following statement, is also true: Let K be a convex body in E^d. Then the following conditions are equivalent:

(i) $\delta(K) = 1$,
(ii) $\vartheta(K) = 1$,
(iii) K is a tile.

This theorem was proved by Schmidt (1961); we include its easy proof below for completeness' sake. We prove here the implication (i) $\Rightarrow$ (iii); the implication (ii) $\Rightarrow$ (iii) can be seen quite analogously, and, as mentioned above, the converse implications are obvious. The assumption $\delta(K) = 1$ implies that for every n and every $\varepsilon > 0$ there exists a packing $\mathcal{P}_n^\varepsilon$ with congruent copies of K such that

$$d(\mathcal{P}_n^\varepsilon | Q_n) \geqslant 1 - \varepsilon,$$

where Q_n is a cube of edge length n, centered at the origin. Set

$$\varepsilon_n = \frac{V(K)}{n^{d+1}}.$$

From the sequence of packings $\mathcal{P}_n = \mathcal{P}_n^{\varepsilon_n}$ a convergent subsequence $\mathcal{P}_{n_k}$ can be selected by a diagonal procedure: first select a subsequence convergent on Q_1, then from it select a subsequence convergent on Q_2, etc., and then take a subsequence which is co-terminal with all of these subsequences by taking the ith term from the ith subsequence. Let $\mathcal{P}$ be the limit packing. Since $n_k \geqslant k$, the kth term of the subsequence is a packing whose members occupy so much volume of Q_k, that the total volume not covered by $\mathcal{P}_k$ is at most $V(K)/k$. Therefore $\mathcal{P}$ covers the entire cube Q_m, for every $m = 1, 2, \ldots$, thus $\mathcal{P}$ is a tiling.

An arrangement $\mathcal{A}$ of sets in E^d is *periodic* if it is obtained by translates of a finite arrangement by the vectors of a lattice. It is easy to prove that for every region K and for every number $\varepsilon > 0$ there exists a periodic packing $\mathcal{P}$ and a periodic covering $\mathcal{C}$ with copies of K such that $d(\mathcal{P}) > \delta(K) - \varepsilon$ and $d(\mathcal{C}) < \vartheta(K) + \varepsilon$ (compare Rogers 1964*, pp. 29, 32). Whether for each convex body K, $\delta(K)$ and $\vartheta(K)$ can be attained in periodic arrangements remains an open problem. However, already in E^2 there exists a non-convex disk which tiles the plane, but not periodically (see chapter 3.5). Also, there exists a non-tiling star-shaped disk K for which $\delta(K)$ is not attained in any periodic packing (see Schmitt 1991).

A packing $\mathcal{P}$ of E with congruent copies of a region K is *saturated* if it cannot be augmented with any additional copy of K without overlapping with a member of $\mathcal{P}$, and it is *completely saturated* if no finite set of its members can be replaced with a more numerous set of copies of K. Similarly, a covering $\mathcal{C}$ of E with congruent copies of K is *reduced* if none of its members can be removed from it without uncovering a portion of E, and it is called *completely reduced* if no finite set of its members can be replaced with a less numerous set of copies of K. Obviously, every packing can be augmented to become saturated and every covering can be cut back to a reduced one. Surprisingly, the existence of completely saturated packings and completely reduced coverings with copies of a region K is, in general, unknown. It is easily seen that complete saturation implies maximum density and complete reduction implies minimum density. With the additional assumption of periodicity, the converse is also true. Specifically, if a periodic packing [covering] with copies of K is of density $\delta(K)$ [respectively $\vartheta(K)$], then the packing [covering] is completely saturated [reduced].

2.2. *Existence of efficient arrangements*

Recall inequalities (1.1) from the introduction:

$$\delta_L(K) \leqslant \delta_T(K) \leqslant \delta(K) \leqslant 1 \leqslant \vartheta(K) \leqslant \vartheta_T(K) \leqslant \vartheta_L(K).$$

Since a parallelepiped P containing [contained in] K generates a lattice packing [covering] with copies of K whose density is $V(K)/V(P)$, an approximation result of Hadwiger (1955) yields immediately additional inequalities:

$$1/d! \leqslant \delta_L(K); \tag{2.2.1}$$

$$d^d \geqslant \vartheta_L(K). \tag{2.2.2}$$

Significant improvements of these inequalities and inequalities of this type for the other densities are discussed in this section.

First, observe that a mere saturation of any packing of E^d with translates of a centrally symmetric convex body K produces a lower bound for $\delta_T(K)$ and an upper bound for $\vartheta_T(K)$. If every member of such a saturated packing is homothetically enlarged about its center by a factor of 2, then the resulting arrangement is a covering. Since the expansion of the members of the packing causes an increase of density by a factor of 2^d, we get

$$2^d \delta_T(K) \geqslant \vartheta_T(K). \tag{2.2.3}$$

In particular,

$$\delta_T(K) \geqslant 2^{-d} \tag{2.2.4}$$

and

$$\vartheta_T(K) \leqslant 2^d. \tag{2.2.5}$$

It is remarkable that, for large values of d, inequality (2.2.3), obtained in such a simple way, yields the best known lower bound for the packing density for spheres in E^d. Even inequality (2.2.4) has been only slightly improved by considerably more elaborate methods. On the other hand, those more elaborate methods produce lattice packings (not just packings with translates) of reasonably high density. One of the first such results is the theorem of Minkowski–Hlawka which states that

$$\delta_L(K) \geqslant \zeta(d)/2^{d-1} \tag{2.2.6}$$

for every centrally symmetric convex body K in E^d, where $\zeta(d) = \sum_{k=1}^{\infty} k^{-d}$.

Inequality (2.2.6) was announced by Minkowski in (1893) who gave a proof in (1905) in the special case when K is a ball. The first complete proof was published by Hlawka (1944). Since then many alternative proofs and refinements were given [see, e.g., Rogers (1964*, p. 8), and Gruber and Lekkerkerker (1987*, p. 202)]. The best refinement is due to Schmidt (1963) who established the inequality

$$\delta_L(K) \geqslant \frac{cd}{2^d} \tag{2.2.7}$$

for every convex centrally symmetric K and sufficiently large d, provided $c < \log 2$.

All the known proofs of the theorem of Minkowski–Hlawka and its refinements are non-constructive. The main idea is to introduce a measure on the set of lattices, and estimate the average density of lattice packings of a given body according to this measure. For details we refer the reader to chapter 3.1. Constructive methods for obtaining dense packings will be discussed in section 2.4.

Concerning packings of translates of an arbitrary (not necessarily centro-symmetric) convex body K, Minkowski (1904) observed a connection between such packings and corresponding packings with the difference body of K. More precisely, for any arrangement $\{K + v_i\}$ of translates of K by vectors v_i, Minkowski considered the corresponding arrangement $\{\frac{1}{2}(K-K)+v_i\}$, and he noticed the simple fact that if one of these arrangements is a packing, then so is the other one. This observation leads immediately to the identity

$$\begin{aligned}\delta_T(K) &= \delta_T\left(\tfrac{1}{2}(K-K)\right)V(K)/V\left(\tfrac{1}{2}(K-K)\right)\\ &= \delta_T(K-K)2^d V(K)/V(K-K).\end{aligned}$$

Rogers and Shephard (1957) proved the inequality

$$V(K-K)/V(K) \leqslant \binom{2d}{d}$$

for an arbitrary convex body K in E^d, which, combined with (2.2.7) yields the lower bound

$$\delta_T(K) \geqslant c\frac{d^{3/2}}{4^d} \tag{2.2.8}$$

for each K, where c is a suitable constant, independent from K and d.

Each of the results mentioned here yields the existence of a relatively dense lattice packing of E^d. It seems that by considering arbitrary packings, especially those in which arbitrary congruent copies of a convex body are allowed without restriction to translates only, one should be able to obtain better estimates for the corresponding packing densities. Unfortunately, this apparent advantage has not been exploited yet in dimensions above 2, perhaps due to the increase in complexity of the problem.

Additional information on the topic of dense packings can be found in section 2.4, devoted to packings produced by constructive methods.

Turning to coverings, we would like first to point out some essential contrasts with the corresponding properties of packings. In many cases, despite apparent analogy in definitions, there is a surprising lack of duality.

To begin with, recall the inequality

$$2^d\delta_T(K) \geqslant \vartheta_T(K)$$

for every centro-symmetric convex body K, which was obtained by a saturation of an arbitrary packing with translates of K. It turns out that a reduction of a covering with translates of K produces no bound for densities, as the following example shows: The lattice covering of E^2 with unit circles, with base vectors

$$\boldsymbol{u} = \left(\frac{1}{n}, 0\right) \quad \text{and} \quad \boldsymbol{v} = \left(1+\sqrt{1-\frac{1}{4n^2}}, \frac{1}{2n}\right)$$

is reduced, but its density tends to ∞ as $n \to \infty$.

Secondly, in contrast to packings, the assumption of central symmetry seems to give no advantage for coverings (Minkowski's idea involving the difference body has no analogue here). On the other hand, the advantage of allowing translation coverings other than just lattice ones, has been successfully exploited. Rogers (1957) proved that

$$\vartheta_T(K) \leqslant d \ln d + d \ln \ln d + 4d \tag{2.2.9}$$

for an arbitrary convex body K in E^d, while the best known upper bound for $\vartheta_L(K)$, also given by Rogers (1959), is:

$$\vartheta_L(K) \leqslant d^{\log_2 \log_2 d + c}. \tag{2.2.10}$$

Neither of inequalities (2.2.9) and (2.2.10) has a constructive proof. In both cases the proofs consist of two major steps. First a mean value argument is used to establish the existence of arrangements with low density yet covering a great portion of space. Based on the arrangement so obtained, an economical covering is constructed in the second step. Of the two, the proof of (2.2.9) is the easier one. Here the mean value argument establishes for any positive number ϱ the existence of arrangements of translates of K with density ϱ and leaving uncovered a portion of space not greater than $\mathrm{e}^{-\varrho}$. By augmenting this arrangement with additional translates of K and subsequently replacing each member of the arrangement with a slightly enlarged homothetic copy of K we obtain a covering whose density does not exceed the quantity on the right side of (2.2.9).

The proof of inequality (2.2.10) is more involved. A mean value argument similar to that which is used in the proof of the Minkowski–Hlawka theorem (see chapter 3.1) yields the existence of a lattice arrangement with translates of a given body H which has density 1 and covers nearly one-half of space. We note that the density of the arrangement and the part of space covered are independent of the dimension of the underlying space. A further mean value argument is used to show that by doubling the density one can obtain from any lattice arrangement of a body H in E^k a lattice arrangement of a cylinder with base H in E^{k+1} leaving a much smaller portion of space uncovered. By repeated applications of this result, one obtains the existence of lattice arrangements of generalized cylinders with relatively small density covering most of space. In order to obtain a similar result for a given convex body K in E^d, Rogers approximates K by generalized cylinders contained in K. Generalizing a result of Macbeath (1951), he shows that there is an affine image of K with the same volume $V(K)$ containing the Cartesian product $H \times C$ of a k-dimensional convex body H and a $(d-k)$-dimensional cube C such that

$$V(H \times C) \geqslant \frac{k^k}{d^d} V(K).$$

The last step of the proof of inequality (2.2.10) is to show that from a lattice arrangement of a given convex body covering most of space one obtains a covering of a slightly higher density by enlarging the bodies homothetically.

The result mentioned above concerning approximation by generalized cylinders can be improved for special classes of convex bodies yielding a considerable improvement of inequality (2.2.10). Rogers already pointed out that there is a constant c such that the inequality

$$\vartheta_L(B^d) \leqslant cd(\log_e d)^{(1/2)\log_2 2\pi e} \tag{2.2.11}$$

holds. Gritzmann (1985) obtained a similar bound for a larger class of convex bodies:

$$\vartheta_L(K) \leqslant cd(\log_e d)^{1+\log_2 e} \tag{2.2.12}$$

for every convex body K in E^d which has an affine image symmetric about at least $\log_2 \log_e d + 4$ coordinate hyperplanes.

2.3. *Upper bounds for $\delta(B^d)$ and lower bounds for $\vartheta(B^d)$*

There are only a few special classes of convex bodies K in E^d for which a non-trivial upper bound for $\delta(K)$ and non-trivial lower bound for $\vartheta(K)$ have been found. Finding such bounds just for spheres is a central problem in the theory of packings and coverings. The complexity of the problem makes it difficult to expect that similar results be obtained for an arbitrary convex body K. We begin this section with a review of the bounds for sphere packings and coverings.

Chronologically, the first upper bound for $\delta(B^d)$ was given by Blichfeldt in (1929):

$$\delta(B^d) \leqslant (d+2)2^{-(d+2)/2}. \tag{2.3.1}$$

Rogers (1958) gave a better bound:

$$\delta(B^d) \leqslant \sigma_d, \tag{2.3.2}$$

where σ_d is the ratio between the total volume of the sectors of $d+1$ unit balls centered at the vertices of a regular simplex of edge 2 and the volume of the simplex. It should be noted that asymptotically, the bound of Rogers is better than the bound of Blichfeldt by a factor of $2/\mathrm{e}$. In addition, Roger's bound is sharp for $d=2$, as it implies Thue's result $\delta(B^2) = \pi/\sqrt{12}$ (see section 4.2 below). Rogers' simplex bound (2.3.2) was rediscovered by Baranovskiĭ (1964).

Many mathematicians believed that the (asymptotic) order of magnitude of Blichfeldt's bound (2.3.1) was the best possible, until the break-through result of Sidelnikov (1973, 1974):

$$\delta(B^d) \leqslant 2^{-(0.509+o(1))d} \quad (\text{as } d \to \infty). \tag{2.3.3}$$

Subsequently, this bound was improved; first by Levenštein (1975) to

$$\delta(B^d) \leqslant 2^{-(0.523+o(1))d} \quad (\text{as } d \to \infty) \tag{2.3.4}$$

and then by Kabatjanskiĭ and Levenštein (1978) to

$$\delta(B^d) \leqslant 2^{-(0.599+o(1))d} \quad (\text{as } d \to \infty). \tag{2.3.5}$$

Among upper bounds of an asymptotic nature, this last one is currently the best known. Inequality (2.2.7), giving the best known asymptotic bound from the other side, leaves a sizable gap between the two. For low dimensions Rogers' bound remains the best known, except for dimension 3, where it has been improved recently by Lindsey (1986) to $\delta(B^3) \leqslant 0.77844\ldots$ and, even more recently, by Muder (1988) to

$$\delta(B^3) \leqslant \left(6\arccos\left(\frac{\sqrt{3}}{2}\sin\frac{\pi}{5}\right) - \frac{9\pi}{5}\right)\cot\frac{\pi}{5} = 0.77836\ldots. \tag{2.3.6}$$

For the analogous problem on coverings a more satisfactory stage has been reached. Coxeter, Few and Rogers (1959) proved that

$$\vartheta(B^d) \geqslant \tau_d, \tag{2.3.7}$$

where τ_d is defined analogously to σ_d in the bound (2.3.2) of Rogers: it is the ratio between the total volume of the intersections of $d+1$ unit balls with the regular simplex of side $\frac{1}{2}\sqrt{2}(n+1)/n$ whose centers lie on the vertices of the simplex, and the volume of the simplex. Asymptotically, $\tau_d \sim d/e^{3/2}$. This bound appears to be quite strong in view of the upper bound (2.2.9) for $\vartheta(K)$. Moreover, for $d=2$ the bound is sharp as it implies Kershner's result (1939) $\vartheta(B^2) = 2\pi/\sqrt{27}$ (see section 4.2 below).

Certain ideas used in the proofs of these results constitute useful tools in estimating the densities in several more specific problems. For this reason we shall describe them in some detail here.

Blichfeldt's enlargement method. Given a packing $\{B_i\}$ of E^d with unit balls. Replace each ball with a concentric one of radius $\varrho > 1$, and furnish each enlarged ball with a mass distribution, variable from point to point, but translation invariant from ball to ball. If the total density at each point of E^d contributed by the enlarged balls containing the point is bounded above by a constant C, then the density of the original packing is at most $C\omega_d/M(d)$, where $M(d)$ is the mass of an enlarged ball, and ω_d is the volume of a unit ball. A suitable choice of ϱ and the mass distribution can produce a meaningful bound for $\delta(B^d)$. In his original work, Blichfeldt took $\varrho = \sqrt{2}$ and he defined a mass distribution m by

$$m(p) = 2 - \operatorname{dist}^2(p, c_i),$$

where c_i is the center of B_i. He then estimated that the mass density at each point is bounded above by 2, which resulted in (2.3.1). The upper bound of 2 for the mass density is easily obtained from the inequality

$$k \sum_{i=1}^{k} x_i^2 \geqslant \sum_{1 \leqslant i \leqslant j \leqslant k} (x_i - x_j)^2 \tag{2.3.8}$$

for any set $\{x_1, x_2, \ldots, x_k\}$ of vectors in E^d, called Blichfeldt's inequality.

The Dirichlet cell (Voronoi region) method. Again, let $\{B_i\}$ be a packing of E^d with unit balls. Each ball B_i is assigned the set D_i of points of E^d whose distance from c_i, the center of B_i, is less than or equal to the distance from the centers of each of the other balls. The set D_i is called the *Dirichlet cell of* B_i. Observe that $B_i \subset D_i$ and that the family $\{D_i\}$ forms a tiling of E^d with convex cells. If a constant $C < 1$ can be found such that the density of B_i relative to its Dirichlet cell D_i is smaller than or equal to C for each i, then the inequality $\delta(B^d) \leqslant C$ follows. It should be pointed out that this implication is less obvious than it seems, for it depends substantially on special properties of the Euclidean space E^d, and it does not generalize to the hyperbolic space H^d (see section 3.2). Rogers (1958) finds a value for C based on the following:

Lemma. *Suppose that a d-dimensional ball B centered at the origin* 0 *is contained in a convex polytope* $P \subset E^d$ *and let* $a_0 > a_1 > \cdots > a_{d-1}$ *be numbers such that the distance from any i-dimensional face of P to* 0 *is at least* a_i. *Let* Δ *be the simplex with vertices* $0, v_1, v_2, \ldots, v_d$ *such that* $\operatorname{dist}(0, v_i) = a_{d-i}$ *and the subspaces* $A(0, v_1, v_2, \ldots, v_i)$ *and* $A(v_i, v_{i+1}, v_{i+2}, \ldots, v_d)$ *are orthogonal for each* $i = 1, 2, \ldots, d-1$. *Then the density of B relative to P is greater than or equal to the density of B relative to* Δ.

As a consequence of Blichfeldt's inequality (2.3.8) it easily follows that in a packing with unit balls each ball and its Dirichlet cell satisfy all conditions of Rogers' Lemma with

$$a_i \geqslant \sqrt{\frac{2(d-i)}{d-i+1}},$$

and the density bound (2.3.2) follows. Moreover, consider the intersection of a d-dimensional (affine) hyperplane E^d with a ball and its Dirichlet cell from a packing with congruent balls in E^{d+k}. We obtain a ball of radius r in E^d and a convex polyhedral cell containing it. It can be noticed that this ball and the cell satisfy the assumptions of Rogers' Lemma with

$$a_i \geqslant r\sqrt{\frac{2(d-i)}{d-i+1}}.$$

By this observation, G. Fejes Tóth (1980) extended the simplex bound of Rogers over any packing with (not necessarily congruent) balls which arises as the section with a d-dimensional hyperplane of a higher-dimensional packing with congruent balls. This bound, as well as Rogers' bound, is sharp for $d = 2$, and it yields that the density of a packing of the plane with circular disks which is the plane section of a congruent ball packing in any higher dimension, is at most $\delta(B^2)$. Whether the analogous statement is true for any $d \geqslant 3$ remains an open problem. The problem concerning relations between the density of an arrangement and the density of its linear sections was investigated by Groemer (1966b, 1986a) and A. Bezdek (1984).

The spherical geometry method. The proofs of the more recent bounds (2.3.3)–(2.3.5) make use of a connection between the problem of the densest ball packing in E^d and the corresponding problem in the spherical space S^d. Let $\mathcal{P}$ be a packing of unit balls in E^d with upper density $d_+(\mathcal{P}, B^d, o) = \delta(B^d)$. A simple mean value argument implies that every region G in E^d has a translate containing at least $V(G)\delta(B^d)/V(B^d)$ centers of balls from $\mathcal{P}$. In particular, for every positive number λ there is a ball of radius λ containing at least $\lambda^d\delta(B^d)$ centers. Without loss of generality we assume that the ball λB^d contains a set C of centers of balls from $\mathcal{P}$ with cardinality at least $\lambda^d\delta(B^d)$.

Consider the mapping which sends the point $c = (c_1, \ldots, c_d) \in C$ into the point $c^* = (c_1^*, \ldots, c_{d+1}^*)$, where $c_i^* = \lambda^{-1}c_i$ for $i = 1, \ldots, d$ and

$$c_{d+1}^* = \left(1 - \sum_{i=1}^{d} \lambda^{-2}c_i^2\right)^{1/2}.$$

The image points lie in the spherical space S^d, and it is easy to check that the angular separation between any two of them is at least $\varphi = 2\arcsin(1/\lambda)$. Denoting by $M(d, \varphi)$ the maximum cardinality of a set of points on S^d with mutual angular distances greater than or equal to φ, we obtain

$$\delta(B^d) \leqslant \lim_{\varphi \to 0} \left(\tfrac{1}{2}\right)^d (1 - \cos\varphi)^{d/2} M(d, \varphi). \tag{2.3.9}$$

Thus, upper bounds for $\delta(B^d)$ can be found by estimating $M(d, \varphi)$. This latter problem will be addressed in section 3.1.

The Delone triangulation method. Let S be a discrete set of points spanning E^d, i.e., not contained in one $(d-1)$-hyperplane. We have described above the collection of Dirichlet cells associated with S, which form a tiling of E^d. Another tiling, dual to the Dirichlet cell tiling, can be obtained by constructing the convex hull of each maximal subset of S whose corresponding closed Dirichlet regions intersect. The resulting convex cells are called the *Delone cells associated with S*. These cells can be constructed in another, equivalent way: consider a ball whose interior does not contain any point of S and whose boundary intersects S in a set spanning E^d. The convex hull of the intersection of the ball with S is a Delone cell, and every Delone cell can be obtained in this way. It is easily verified that if

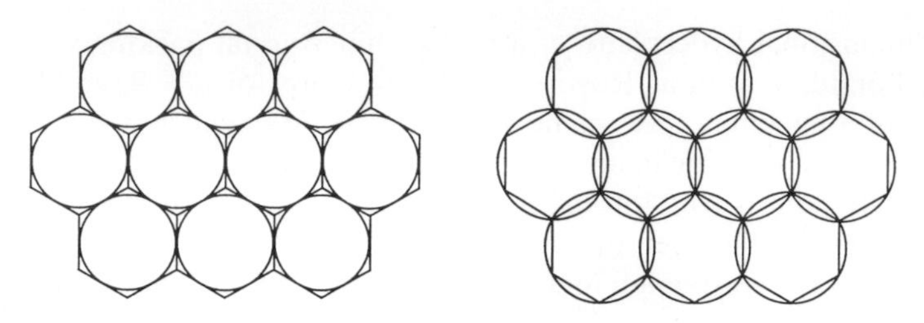

Figure 1.

S is the set of centers of congruent balls in a reduced covering, then the Delone cells form a tiling. If each Delone cell which is not a simplex is subdivided into simplices without introducing new vertices, then we obtain a partition of E^d into simplices, called a *Delone triangulation*, associated with the covering. Observe that if the members of the covering are unit balls, then each simplex of the Delone triangulation is contained in a unit ball.

In the proof of inequality (2.3.7), Coxeter, Few and Rogers consider a single simplex of the Delone triangulation associated with a covering with unit balls and they assign to it the *local density* of the covering as the ratio between the sum of measures of the solid angles of the simplex and the volume of the simplex. The crucial part of the proof is a lemma which states that the minimum of the local density occurs when the simplex is regular. Density bound (2.3.7) follows directly.

2.4. Constructive methods for dense packings

The densest lattice packings with balls in E^d are known for $d \leqslant 8$ only. Case $d = 2$ was settled by Lagrange (1773); $d = 3$ by Gauss (1831); $d = 4$ and $d = 5$ by Korkin and Zolotarev (1872, 1877); and the remaining cases by Blichfeldt (1925, 1926, 1934). Even less is known about lattice coverings. The thinnest lattice covering of E^d is known for $d \leqslant 5$. Case $d = 2$ was implicitly solved by Kershner (1939); $d = 3$ by Bambah (1954); $d = 4$ by Delone and Ryškov (1963) and $d = 5$ by Ryškov and Baranovskiĭ (1975, 1976).

Among the lattice arrangements related to the above results, only the two-dimensional ones are known to be of extreme density among all, not just lattice, arrangements [circle packings: Thue (1910); circle coverings: Kershner (1939)]. In the remaining dimensions, the arrangements are conjectured to be of such extreme density, and the conjecture remains open even in dimension 3. In the plane, both extreme lattice arrangements of circular disks form the familiar hexagonal pattern. In the densest lattice packing, the circles are inscribed in regular hexagons tiling the plane, and in the thinnest lattice covering, they are circumscribed about them (see fig. 1).

In E^3, the sphere centers in the densest lattice packing are the vertices and the face-centers of a cubic lattice. This arrangement of balls resembles the usual pattern in which congruent spherical objects such as cannonballs, melons or oranges are stacked for display, and it occurs quite often in nature (see fig. 2).

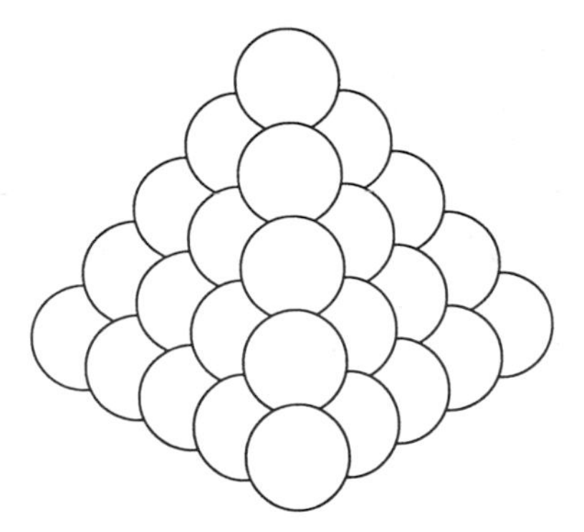

Figure 2.

In contrast with the plane, the arrangement of sphere centers in the thinnest lattice covering of E^3 is quite different: they are the vertices and the cube-centers of a cubic lattice. Some dimensions above 3 provide other, startling surprises.

Section 2.2 of this chapter quotes some theorems about existence of dense sphere packings and thin sphere coverings, results of an asymptotic nature, giving rather poor estimates in low dimensions. Also, it should be noted that these theorems establish the existence of such arrangements without actually constructing them. The need for constructive methods could be sufficiently motivated by purely mathematical curiosity alone, but the greatest contributing factor was the realization of their applicability in information processing and transmission. An extremely fruitful period of research followed the discovery by Leech (1964) of a connection between efficient error-correcting codes and dense sphere packings. The topic of sphere packings quickly gained a great number of significant results and became a vast, separate area of research, with its own combinatorial and algebraic techniques. Here we will mention only some of the basic ideas of this theory, referring the interested reader to the nearly encyclopedic account of the subject in Conway and Sloane (1988*).

A *binary code* is a set of d-dimensional binary vectors (or $\{0,1\}$-*words* of length d). The *weight* of a word is the number of its coordinates that differ from zero. The *Hamming distance* between two such vectors (or *codewords*) is the number of coordinates at which they differ. The greater the distance between two codewords, the easier it is to distinguish one from the other. Thus, reliable and efficient storage and transmission of information requires an explicitly constructed code with a large number of codewords as far from one another as possible. This can be interpreted as the sphere packing problem in the Hamming space.

A code consisting of M words of length d and with minimum Hamming distance k between any two of its words is called a (d,M,k)-code. We point out that in the literature on coding theory, the length of codewords is denoted by n and the minimum distance is denoted by d. For consistency of notation throughout this Handbook, we reserve "d" for dimension, and we hope that this will cause no confusion. The set of all binary words of length d is a linear space over the binary field GF(2). A code which is a subspace of that linear space is called a *linear code*, which codes are of special importance in coding theory and turn out to be interesting from a geometric point of view.

Geometrically, a (d,M,k)-code is a subset of the set of vertices of the d-dimensional unit cube. The balls of radius $\frac{1}{2}\sqrt{k}$ centered at the codewords form a packing in E^d. In a suitably chosen manner, this local sphere packing can be used to generate a periodic packing in the whole space E^d. Three basic constructions were devised by Leech and Sloane (1971) to generate such packings. We briefly describe them as follows.

Construction A. The point $x=(x_1,x_2,\dots,x_d)$ is a center of a sphere if x is congruent (modulo 2) to a codeword. Geometrically, this means that the unit cube, with the spheres attached to it, is translated to new positions, filling a "checkerboard" pattern in E^d. In this construction, the radii of the spheres are $\min\{1,\frac{1}{2}\sqrt{k}\}$.

Construction B. This construction applies to codes in which every codeword is of even weight. The point $x=(x_1,x_2,\dots,x_d)$ is a center of a sphere if x is congruent (modulo 2) to a codeword and $\sum_{i=1}^{d} x_i$ is divisible by 4. The radius of the spheres is $\min\{\sqrt{2},\frac{1}{2}\sqrt{k}\}$.

Construction C. To describe this construction, begin with assigning to each integer i an (infinite) sequence of 0's and 1's called its *binary representation*. For a nonnegative integer, the assigned sequence consists of the integer's expansion in the binary system in the increasing order of powers of 2, followed by a string of 0's. For a negative integer, the sequence ends with a string of 1's and is assigned to the integer in such a manner that the formal "carry-over" addition algorithm is preserved. For instance, $4 \leftrightarrow \{0,0,1,0,0,\dots\}$ and $-4 \leftrightarrow \{0,0,1,1,1,\dots\}$, so that $\{0,0,1,0,0,\dots\}+\{0,0,1,1,1,\dots\}=\{0,0,0,\dots\} \leftrightarrow 0$. For a point $x=(x_1,x_2,\dots,x_d)$ with integer coordinates, the *coordinate array* of x is defined as the matrix whose d columns are the binary representations of the integers $x_1,x_2,\dots,x_d$.

Now, Construction C produces a sphere packing generated by a collection of codes C_i, where C_i is a (d,M_i,k_i)-code, $i=1,2,\dots,a$, with $k_i=\gamma\cdot 4^{a-i}$ and $\gamma=1$ or 2. A point $x=(x_1,x_2,\dots,x_d)$ is a center of a sphere if the ith row (corresponding to the 2^i power in the binary representation) of the coordinate array of x is a codeword in C_i for $i=1,2,\dots,a$. The radius of the spheres is $\sqrt{\gamma}\cdot 2^{a-1}$.

Each of the Constructions A and B yields a lattice packing if and only if the code on which it is based is linear. Construction C does not possess this property. Most of the examples of sphere packings of greatest known density have been obtained by one of these constructions or their modifications.

The densest lattice packing of balls in E^d for $d=3,4$ and 5 is obtained by applying Construction A to the $(d,2^{d-1},2)$-code consisting of all words of even weight. The densest lattice packing of balls in E^7 and E^8 can be obtained by Construction A using codes obtained from an Hadamard matrix. An *Hadamard matrix* H is an $n\times n$ matrix with entries ± 1 such that $HH^{\mathrm{T}}=nI$. It is known that an Hadamard matrix can exist only for $n = 1$, 2 and multiples of 4. In the case when n is a power of 2, an $n \times n$ Hadamard matrix H_n can be easily obtained by induction, setting $H_1 = (1)$ and

$$H_{2k} = \begin{pmatrix} H_k & H_k \\ H_k & -H_k \end{pmatrix}$$

for $k = 1, 2, 4, \ldots$. We consider the matrix H_8 so obtained, and observe that its first row and column contain only 1's. Next we consider the matrix $\widetilde{H}_8$ obtained from H_8 upon replacing the +1's by 0's and the −1's by 1's. The rows of $\widetilde{H}_8$ with the first column deleted form a $(7, 8, 4)$-code. The codewords of this code are the vertices of a 7-dimensional regular simplex. We obtain the densest lattice packing of balls in E^7 by applying Construction A to this code. The densest lattice packing of balls in E^8 is obtained through Construction A used on the $(8, 16, 4)$-code consisting of the rows of H_8 together with the complements of the rows. The same lattice of balls can be obtained also by Construction B applied to the trivial $(8, 2, 8)$-code consisting of the all 0's and all 1's words. For future reference, we note that in this packing each ball is tangent to 240 other balls. We also mention that the densest lattice packing of balls in E^6 cannot be obtained directly by Construction A or B, but is obtained as an appropriate section of the densest lattice ball packing in E^7.

A remarkably dense lattice packing of balls in E^{24}, playing a fundamental role not only in the theory of ball packings but in group theory as well, was constructed by Leech (1964). The construction of the Leech lattice is based on the *Golay code* G_{24}, a $(24, 2^{12}, 8)$-code, which can be described as follows: let $C_{12} = (c_{ij})$, $i, j = 1, \ldots, 12$, be defined by

$$c_{ij} = \begin{cases} 0, & \text{if } i = j = 1 \text{ or } i, j \geqslant 2 \text{ and } i + j - 4 \text{ is a quadratic residue mod } 11, \\ 1, & \text{otherwise,} \end{cases}$$

and let I_{12} be the unit matrix of order 12. Then G_{24} is the 12-dimensional linear space [over GF(2)] spanned by the rows of the matrix (I_{12}, C_{12}). Each codeword in the Golay code is of even weight; in fact the weight of each codeword is divisible by 4.

An application of Construction B to G_{24} results in a lattice packing of balls of radius $\sqrt{2}$, which makes possible to join this packing, without overlap, with a translate of itself through the vector $(-\frac{5}{2}, \frac{1}{2}, \ldots, \frac{1}{2})$ thereby to form a new lattice packing, called, after its inventor, the *Leech lattice*. The density of this packing is $0.001930\ldots$, which compares quite favorably to the Rogers' bound $\delta(B^{24}) \leqslant \sigma_{24} = 0.002455\ldots$. Each ball in the Leech lattice is touched by 196560 other balls, a number which happens to be the maximum possible contact number in any packing of congruent balls in E^{24} (see section 5.3). Each of the densest known lattice packings in dimensions below 24 can be obtained as an appropriate section of the Leech lattice.

It is conjectured (Rogers 1964*, p. 15) that for sufficiently high dimensions, $\delta(B^d) > \delta_L(B^d)$. Thus far, this inequality has not been confirmed in any dimension at all, but in some dimensions, for instance $d = 10$, non-lattice ball packings were found denser than any *known* lattice packing (Best 1980, Conway and Sloane 1988*, pp. 140–141). These examples were produced by applying Construction A to certain non-linear codes.

There were several attempts to give effective constructions of sphere packings in E^d for large values of d. Early attempts, for which we refer to Conway and Sloane (1988*), produced lattice sphere packings in E^d of order of magnitude c^d with $c < \frac{1}{2}$, and thus they fell short the (non-constructive) asymptotic bound of the Minkowski–Hlawka theorem (see section 2.2). Quite recently, Rush (1989) succeeded in constructing lattice packings with B^d of density $(1/2)^{d+o(d)}$ through an application of Construction A to codes over larger alphabets. Moreover, in his construction, Rush assumed a more general setting, considering an arbitrary convex body which is symmetric about each of the coordinate hyperplanes in place of B^d, and he still obtained equally high density. In this setting, the Hamming metric was replaced by a metric generated by the convex body. The idea of furnishing the space of words with a metric suited to the convex body was introduced earlier by Rush and Sloane (1987). There the bound of Minkowski–Hlawka was exceeded in lattice packings of "superballs" $\{(x_1, x_2, \ldots, x_d) \in E^d\colon \sum_{i=1}^{d} x_i^\sigma \leqslant 1\}$ for the sequence $d_j = p_j^\sigma$ of dimensions, where p_j is an odd prime. Further generalizations in the direction of constructive dense lattice packings with special types of convex bodies were obtained by Elkies, Odlyzko and Rush (1991).

2.5. *Lattice arrangements*

The concepts of lattices and lattice arrangements are intimately related to the geometry of numbers, therefore the basic topics of that area are treated extensively in chapter 3.1 of this Handbook. Here we discuss some special problems concerning arrangements and their densities, for which the restriction to lattice arrangements is self-imposed by the nature of the problem.

An arrangement $\mathscr{A}$ of sets is *point trapping* if every component of the complement of the union of members of $\mathscr{A}$ is bounded. It is natural to ask: what is the minimum density of a point trapping lattice arrangement of any convex body in E^d? Confirming the "checkerboard conjecture" of L. Fejes Tóth (1975a), Böröczky et al. (1986) proved that the minimum is $\frac{1}{2}$, attained in the "checkerboard" lattice arrangement of cubes.

We sketch here the proof of the "checkerboard theorem" of Böröczky et al. which uses a result on polytopes of interest on its own.

Let P be an arbitrary polytope in E^d with facets $F_1, \ldots, F_m$. Let v_i denote the outward normal vector of F_i (symbolically, $v_i = v\bot_{\text{out}} F_i$) whose length is equal to the surface area of F_i. Then we have $\sum_{i=1}^{m} v_i = 0$, and a well-known theorem of Minkowski implies that there exists a unique convex polytope P^* with m facets $F_1^*, \ldots, F_m^*$ such that v_i is an outward normal vector of F_i^* and $\mathrm{Vol}_{d-1}(F_i) = \mathrm{Vol}_{d-1}(F_i^*)$. The result to which we referred above is that

$$V(P) \leqslant V(P^*)$$

for any polytope in E^d with equality if and only if P is convex. For the proof of the checkerboard theorem its authors use the following, seemingly stronger, but in fact equivalent, statement. Let $P = \bigcup_{i=1}^{r} P_i$ be a finite union of some d-dimensional polytopes P_i ($d \geqslant 2$) such that the intersection of any two of them is at most $(d-2)$-dimensional. Let F_j, $j = 1, \dots, m$, be the facets of P, and let P' be a convex polytope with facets F'_k, $k = 1, \dots, m'$ such that

$$\sum_{v = v \perp_{\text{out}} F_j} \mathrm{Vol}_{d-1}(F_j) \leqslant \sum_{v = v \perp_{\text{out}} F'_k} \mathrm{Vol}_{d-1}(F'_k)$$

for every vector $v \in E^d$. Then

$$V(P) \leqslant V(P'),$$

with equality only if $r = 1$ and $P = P_1$ is convex.

Using this result, the proof of the checkerboard theorem is quite easy. Obviously, it suffices to prove the theorem for arrangements of polytopes. Consider the connected components of a point trapping lattice arrangement $\mathscr{A}$ of a convex polytope K. Two components are called equivalent if there is a lattice translation carrying one onto the other. It is easy to see that there are only finitely many, say s, different equivalence classes. Choose a representative P_i, $i = 1, \dots, s$ from each equivalence class, and consider the set $P = \bigcup_{i=1}^{s} P_i$. Each facet of P can be subdivided into finitely many openly disjoint $(d-1)$-dimensional simplices contained in the boundary of some member of $\mathscr{A}$. That is, we can choose lattice vectors v_j and $(d-1)$-dimensional simplices S_j, $j = 1, \dots, m$, such that

$$\mathrm{bd}\, P = \bigcup_{j=1}^{m} S_j \quad \text{and} \quad S_j \subset \mathrm{bd}(K + v_j).$$

One can check that the simplices $S'_j = S_j - v_j$ are openly disjoint pieces of $\mathrm{bd}\, K$. Thus, an application of the theorem mentioned above yields

$$V(P) \leqslant V(-K) = V(K).$$

Now we consider an elementary cell E of the lattice and denote by E_i the set of those points of E which are covered by exactly i members of $\mathscr{A}$. We observe that

$$V(E_0) = V(P) \quad \text{and} \quad \sum_{i=1}^{\infty} V(E_i) = V(K).$$

Hence we obtain for the density $d(\mathscr{A})$ of $\mathscr{A}$ the inequality

$$d(\mathscr{A}) = \frac{VK)}{V(E)} = \frac{V(K)}{V(P) + V(K)} \geqslant \frac{V(K)}{2V(K)} = \frac{1}{2}.$$

A similar problem can be stated for any specific convex body K. Bleicher (1975) proved that the minimum density of a point trapping lattice arrangement of balls in E^3 equals to

$$32\sqrt{(7142+1802\sqrt{17})^{-1}} = 0.265\ldots;$$

moreover, the extreme lattice is generated by three vectors of length $\frac{1}{2}\sqrt{7+\sqrt{17}}$ any two of which make an angle of $\arccos\frac{1}{8}(\sqrt{17}-1) = 67.021\ldots^\circ$.

An arrangement of sets is said to be *connected* if the union of the sets is connected. The problem of the thinnest connected lattice arrangement with a convex body in E^d has been solved by Groemer (1966a). For a convex body K, let $c(K)$ denote the minimum density of a connected lattice arrangement of copies of K. Groemer has proved the sharp inequalities

$$\frac{1}{d!} \leqslant c(K) \leqslant \frac{\pi^{d/2}}{2^d\Gamma((d/2)+1)}.$$

The value $c(K) = 1/d!$ is attained when K is a simplex or a cross-polytope, and the other extreme of $c(K)$ is attained when K is a ball. Extending Groemer's investigation of such extreme bodies restricted to the centro-symmetric case, L. Fejes Tóth (1973b) characterizes all convex K in E^d for which the first one of the above two inequalities turns into an equality: they are the limiting figures of topological isomorphs of the regular cross-polytope.

An arrangement $\mathscr{A}$ of sets in E^d is said to be *non-separable* if every $(d-1)$-dimensional flat intersects the interior of a member of $\mathscr{A}$. For a given convex body K, let $\varrho(K)$ denote the infimum of the density of a non-separable lattice arrangement of copies of K, and let $\widehat{K}$ denote the polar body of the symmetrization of K, $\widehat{K} = \frac{1}{2}(K-K)^*$. A surprising connection between $\varrho(K)$ and $\delta_L(\widehat{K})$ was found by Makai (1978), expressed in the formula

$$\varrho(K)\delta_L(\widehat{K}) = V(K)V(\widehat{K}).$$

The result of Makai includes a connection between the lattices which produce the extreme densities $\varrho(K)$ and $\delta_L(\widehat{K})$: the two lattices are polar to each other. Makai's formula allows to determine the values of $\varrho(B^d)$ for $d \leqslant 8$ since the corresponding values of $\delta_L(B^d)$ are known (see section 2.4). The problems of the minimum value of $\varrho(K)$ over all K and all centrally symmetric K remain open. Makai conjectures that the minimum over all K is $(d+1)/(2^d d!)$ attained by the d-simplex, and that the minimum over all centrally symmetric K is $1/d!$ attained by the d-cross polytope.

The notion of non-separability can be generalized as follows. Let $k \leqslant d-1$ be a non-negative integer and let $\mathscr{A}$ be an arrangement of sets in E^d. Say that $\mathscr{A}$ is *k-impassable* if every k-flat intersects the interior of a member of $\mathscr{A}$. Observe that $\mathscr{A}$ is 0-impassable if and only if $\mathscr{A}$ is a covering, and that $(d-1)$-impassibility means non-separability. The infimum density problem discussed above can be generalized to the case of k-impassable lattice arrangements, but little is known in that regard. One conjectures that the thinnest 1-impassable lattice arrangement of balls in E^3 is obtained by suitably enlarging the balls of the maximum density lattice packing. Heppes (1960) proved that a lattice packing of balls in E^3 cannot be 1-impassable. This holds true for all dimensions $d \geqslant 3$. To see this, consider a 1-impassable lattice packing $\mathcal{P}$ of unit balls in E^d and a lattice vector v of minimum length. Without loss of generality we assume that $|v| = 2$. Projecting $\mathcal{P}$ onto a hyperplane orthogonal to v yields a lattice covering of E^{d-1}. Comparing the densities we get $\delta(B^d) \geqslant d(\mathcal{P}) \geqslant \vartheta(B^{d-1})V(B^d)/2V(B^{d-1})$ which is impossible for $d > 3$. The fact that each lattice packing of balls in E^d $(d \geqslant 3)$ leaves room for lines disjoint with the balls has been exploited by A. Bezdek and W. Kuperberg (1991a) in a construction of dense non-lattice packings with certain congruent ellipsoids in E^d, producing densities higher than in any lattice packing.

2.6. *Multiple arrangements*

The concepts of packing and covering have the following natural generalizations. Let k be a positive integer. A family $\mathcal{F}$ of sets is said to be a *k-fold packing* if each point belongs to the interior of at most k members of $\mathcal{F}$, and $\mathcal{F}$ is called a *k-fold covering* if each point belongs to at least k members of $\mathcal{F}$. The 1-fold packings and coverings coincide with the usual (simple) packings and coverings. Many of the questions asked about simple packings and coverings can be as well asked about multiple packings and coverings. The classical problems concerning extreme densities and the arrangements that produce them play as important roles here as in the context of simple packings and coverings.

We define densities of multiple packings and coverings in the same way as those of the (simple) packings and coverings. Thus, to every convex body K we assign the corresponding k-fold packing and k-fold covering densities: general, restricted to translations, and restricted to lattice-like arrangements, denoted by $\delta^k(K)$, $\delta^k_T(K)$, $\delta^k_L(K)$ and $\vartheta^k(K)$, $\vartheta^k_T(K)$, $\vartheta^k_L(K)$, respectively. The natural problems concerning lower bounds for multiple packing densities and upper bounds for multiple covering densities for arbitrary convex bodies form a vast, almost untouched research area. Among the few obtained results, we mention here the following. Erdős and Rogers (1962) proved that E^d can be covered by translates of any convex body K with density at most $d \log d + d \log(\log d) + 4d$ so that no point is covered more than $\mathrm{e}(d \log d + d \log(\log d) + 4d)$ times. This implies that for any dimension d there exists a constant c_d such that

$$\delta^k_T(K) \geqslant c_d k$$

for every convex body K in E^d.

Cohn (1976) proved that

$$\vartheta^k_L(K) \leqslant \left((k+1)^{1/d} + 8d\right)^d$$

for each d-dimensional convex body K. This inequality should be regarded as an asymptotic estimate as k or d tends to infinity. For instance, for $d = 2$, Cohn's bound is greater than the trivial bound of $\frac{3}{2}k$ for all $k \leqslant 1914$. The trivial bound of $\frac{3}{2}k$ is derived from the obvious inequality $\vartheta_L^k(K) \leqslant k\vartheta_L(K)$ and Fáry's (1950) inequality $\vartheta_L(K) \leqslant \frac{3}{2}$ for convex disks K.

Asymptotic bounds for $\delta_L^k(K)$ and $\vartheta_L^k(K)$ as $k \to \infty$ and K remains a fixed convex disk are given by Bolle (1989): For every convex disk K there exist positive constants c_1 and c_2 depending on K only, such that

$$\delta_L^k(K) \geqslant k - c_1 k^{2/5}$$

and

$$\vartheta_L^k(K) \leqslant k + c_2 k^{2/5}.$$

For convex disks K of a special kind which Bolle calls "acute" disks, and which include all convex polygons, the above inequalities are strengthened by replacing the exponent $\frac{2}{5}$ with $\frac{1}{3}$.

Concerning k-fold packings with congruent balls in E^d, Few (1964) observed that if each ball in a (simple) packing with unit balls is replaced with a concentric ball of radius $2k/(k+1)$, then the resulting arrangement of balls is a k-fold packing. This observation yields

$$\delta^k(B^d) \geqslant [2k/(k+1)]^{d/2}\delta(B^d)$$

and

$$\delta_L^k(B^d) \geqslant [2k/(k+1)]^{d/2}\delta_L(B^d).$$

Few's observation, by its very nature, cannot yield a good lower bound for $\delta^k(B^d)$ when k is large, but, at least for balls, it supplements the bound of Erdős and Rogers which requires k to be large.

We now turn to problems on density bounds from the opposite side: for a given convex body K or a given class of convex bodies K, find a reasonable upper bound for the k-fold packing density and a reasonable lower bound for the k-fold covering density. Florian (1978), generalizing a result of Schmidt (1961), proved that if the boundary of K is smooth, then $\delta^k(K) < k$ and $\vartheta^k(K) > k$. Obviously, for each specific smooth body K, one expects specific bounds $\alpha < k$ and $\beta > k$ such that $\delta^k(K) < \alpha$ and $\vartheta^k(K) > \beta$, but Florian's method does not yield such specific bounds, even when K is a ball. Such specific bounds for $\delta^k(B^d)$, at least for certain pairs of k and d, have been obtained first by Few (1964). Adapting Blichfeldt's idea (see section 2.3 above) to multiple packings, he proved that

$$\delta^k(B^d) \leqslant (1 + d^{-1})[(d+1)^k - 1][k/(k+1)]^{d/2}.$$

Observe that this inequality is significant (i.e., it gives an upper bound better than the trivial bound of k) only if d is much greater than k. By a further elaboration of this idea for $k = 2$, in (1968), Few obtained a stronger inequality

$$\delta^2(B^d) \leqslant \tfrac{4}{3}(d+2)\left(\tfrac{2}{3}\right)^{d/2},$$

which is better than the trivial one for all $d \geqslant 11$. By means of a suitable generalization and modification of the Dirichlet cell method (see section 2.3 above) some upper bounds for $\delta^k(B^d)$ and for $\vartheta^k(B^d)$ have been obtained, non-trivial for all k and d (see G. Fejes Tóth 1976 and 1979).

There is extensive literature devoted to lattice-like multiple arrangements of balls in E^d. In particular, the values of $\delta^k(B^d)$ have been determined for the following pairs (k, d): $d = 2$, $2 \leqslant k \leqslant 8$, and $d = 3$, $k = 2$ [see the survey of G. Fejes Tóth (1983*) and the more recent publications of Temesvári, Horváth and Yakovlev (1987) and Temesvári (1988)]. The latter two papers give algorithms determining the densest k-fold lattice packing and the thinnest k-fold lattice covering with circles. Earlier, Linhart (1983) constructed and implemented algorithms which determine the values of $\delta^k_L(B^2)$ and $\vartheta^k_L(B^2)$, but not the corresponding lattices.

In some cases the extreme density multiple arrangements are not better than repeated simple arrangements. For instance, a theorem of Dumir and Hans-Gill (1972a) states that $\vartheta^2_L(K) = 2\vartheta_L(K)$ for every centrally symmetric convex disk K. Analogously, for packings, the equality $\delta^k_L(K) = k\delta_L(K)$ holds true for $k = 2$ (Dumir and Hans-Gill 1972b) and for $k = 3$ and 4 (G. Fejes Tóth 1983) for every centrally symmetric disk K. In fact, the equalities $\delta^k_L(K) = k\delta_L(K)$ for $k = 3$ and 4 were proved by observing that each 3-fold lattice-like packing of the plane with a centrally symmetric K is the union of 3 simple lattice packings, and each such 4-fold packing is the union of two 2-fold ones.

This last observation belongs to the topic concerning decompositions of multiple arrangements into simple ones, results and problems which focus on the combinatorial structure of such arrangements. Following is a basic problem of this type: decompose a k-fold packing [covering] into as few [as many] as possible simple packings [coverings]. Typically, this leads to finding common bounds for the number of simple arrangements into which every k-fold arrangement from a given class can be decomposed. Research in this direction was initiated by Pach (1980). He proved that any 2-fold packing with positively-homothetic copies of a convex disk (in E^2) can be decomposed into 4 (simple) packings. For a convex disk K, let $r(K)$ denote the circumradius of K and let $l(K) = \pi r^2(K)/A(K)$. In the same paper, Pach (1980) proved that if $\mathcal{F}$ is a k-fold packing with (not necessarily congruent) convex disks such that $l(K) \leqslant L$ for each $K \in \mathcal{F}$, then $\mathcal{F}$ can be decomposed into $n \leqslant 9Lk$ packings.

Concerning coverings, Mani-Levitska and Pach (1990) showed that every 33-fold covering of the plane with congruent circles can be decomposed into 2 coverings. For any polygonal centrally symmetric convex disk P, Pach (1986) proved that if r is a positive integer, then there exists an integer $k = k(P, r)$ such that every k-fold covering with translates of P can be decomposed into r coverings. Unfortunately, the number $k(P,r)$ in Pach's theorem increases with the number of sides of P, thus the "natural" attempt to extend this result to all centrally symmetric convex disks through polygonal approximation fails. Also, as examples of Mani-Levitska and Pach (1990) show, in 3 dimensions statements analogous to the above two theorems are not true.

3. Packing in non-Euclidean spaces

The definitions of packing and covering are not restricted to Euclidean spaces, and in spherical and hyperbolic spaces they seem just as natural. However, problems involving density create here certain difficulties. Lack of similarities in these spaces changes the nature of such problems considerably. In addition, in the spherical geometry most of the problems assume a finite, combinatorial form, and in the hyperbolic space, as we mentioned in section 2.1, there is no natural way to define density. Moreover, the curvature of the non-Euclidean spaces amplifies the contrast between packings and coverings. These may be the reasons why in the non-Euclidean case relatively few results have been obtained and most of them are about sphere packings. We discuss these results here for the interest of their own, but we emphasize the packing problem on the sphere because of its role in the ball packing problem in E^d (see section 2.3).

3.1. Spherical space

The spherical space S^d can be viewed as the unit sphere in E^{d+1}. For a pair of points $x, y \in S^d$, we define the *angular* (geodesic) *distance* $\varphi(x,y)$ between x and y on S^d as the angle between the radii ox and oy. The *angular diameter* of a set $A \subset S^d$ is defined as $\sup\{\varphi(x,y)\colon x,y \in A\}$. We will be concerned with packings on S^d with congruent spherical caps of angular diameter $\varphi < \pi$. Instead of density, we will consider the number of caps in a packing, representing the concept of density in a more natural way. Let $M(d,\varphi)$ denote the maximum number of caps of diameter φ in a packing on S^d. Observe that this number is, at the same time, the maximum number of points on S^d all of whose mutual angular distances are greater than or equal to φ.

Davenport and Hajós (1951) proved that

$$M(d,\varphi) = d+2 \quad \text{for } \tfrac{1}{2}\pi < \varphi < \tfrac{1}{2}\pi + \arcsin\frac{1}{d+1}$$

and

$$M(d, \tfrac{1}{2}\pi) = 2(d+1).$$

This implies that if $d+3$ congruent caps can be packed on S^d, then there is enough room for as many as $2(d+1)$ of caps of the same size. Rankin (1955) determined the exact values of $M(d,\varphi)$ for all $\varphi > \frac{1}{2}\pi$ and gave an upper bound for $M(d,\varphi)$

for all $\varphi < \frac{1}{2}\pi$ (see also Bloh 1956). Lower bounds for $M(d,\varphi)$ were given by Shannon (1959) and Wyner (1965). The order of magnitude of these bounds is

$$\left(\frac{1}{\sin\varphi} + \mathrm{o}(1)\right)^d \leqslant M(d,\varphi) \leqslant (1-\cos\varphi + \mathrm{o}(1))^{-d/2} \quad (\text{as } d\to\infty). \tag{3.1.1}$$

While the lower bound remained the best known up to now, significant improvements of the upper bound were obtained by Sidelnikov (1973, 1974), Levenštein (1975) and Kabatjanskiĭ and Levenštein (1978). Currently, the (asymptotically) best known upper bound for $M(d,\varphi)$ is given by Kabatjanskiĭ and Levenštein (1978):

$$M(d,\varphi) \leqslant (1-\cos\varphi)^{-d/2}2^{-d(0.099+\mathrm{o}(1))} \quad (\text{as} \quad d\to\infty) \tag{3.1.2}$$

for all $\varphi \leqslant \varphi^* = 62.9974\ldots$. This bound was obtained by a method exploiting a surprising connection between $M(d,\varphi)$ and the coefficients of the expansion of a real-variable polynomial in terms of certain Jacobi polynomials. The Jacobi polynomials, $P_i^{(\alpha,\beta)}(x)$, $i = 0, 1, 2, \ldots$, $\alpha > -1$, $\beta > 1$, form a complete system of orthogonal polynomials on $[-1,1]$ with respect to the weight function $(1-x)^\alpha(1+x)^\beta$. For the application, set $\alpha = \beta = \frac{1}{2}(d-2)$. The connection between $M(d,\varphi)$ and the Jacobian polynomials is expressed as follows: Let

$$f(t) = \sum_{i=0}^{k} f_i P_i^{(\alpha,\alpha)}(t)$$

be a real polynomial such that $f_0 > 0$, $f_i \geqslant 0$ $(i = 1, 2, \ldots, k)$ and $f(t) \leqslant 0$ for $-1 \leqslant t \leqslant \cos\varphi$. Then

$$M(d,\varphi) \leqslant f(1)/f_0. \tag{3.1.3}$$

This theorem, commonly referred to as the *linear programming bound* actually goes back to Delsarte (1972, 1973). Alternative proofs were given by Delsarte, Goethals and Seidel (1977), Kabatjanskiĭ and Levenštein (1978), Odlyzko and Sloane (1979), Lloyd (1980) and Seidel (1984).

Kabatjanskiĭ and Levenštein show that the polynomials of degree $2k+1$

$$f(t) = \frac{1}{t-s}\{P_{k+1}^{(\alpha,\alpha)}(t)P_k^{(\alpha,\alpha)}(s) - P_k^{(\alpha,\alpha)}(t)P_{k+1}^{(\alpha,\alpha)}(s)\}^2,$$

where $s = \cos\varphi$, satisfy the conditions of the linear programming bound. These polynomials are analogous to the ones used previously by McEliece et al. (1977) for the study of codes (or, equivalently, sphere packings, see section 2.4) in Hamming spaces. Their use in the linear programming bound yields that

$$M(d,\varphi) \leqslant 4\binom{d+k-2}{k}(1-t_{1,k+1}^{\alpha})$$

for $\cos\varphi \leqslant t^{\alpha}_{1,k}$, where $1 > t^{\alpha}_{1,k}$ is the greatest root of the Jacobi polynomial $P^{(\alpha,\alpha)}_k$. ($P^{(\alpha,\alpha)}_k$ has k simple real roots.) By a detailed investigation of the values of $t^{\alpha}_{1,k}$ Kabatjanskiĭ and Levenštein show that

$$\frac{1}{d}\log_2 M(d,\varphi) \leqslant \frac{1+\sin\varphi}{2\sin\varphi}\log_2\frac{1+\sin\varphi}{2\sin\varphi} - \frac{1-\sin\varphi}{2\sin\varphi}\log_2\frac{1-\sin\varphi}{2\sin\varphi} + \mathrm{o}(d).$$

For $\varphi \leqslant \varphi^* = 62.9974\ldots$ this implies the simpler bound (3.1.2), which, combined with (2.3.9) yields the inequality (2.3.5) for the packing density of the d-dimensional ball.

The proof of the linear programming bound outlined below follows the presentation given by Seidel (1984). We consider the unit sphere S^d in E^{d+1}, and the linear space L^2 of square integrable real functions on S^d. Throughout the proof d will be a fixed positive integer, and often we will omit it in our notation. For two functions f_1 and f_2 in L^2 we define the inner product $\langle f_1, f_2\rangle$ by

$$\langle f_1, f_2\rangle = \frac{1}{\sigma_d}\int_{S^d} f_1(x)f_2(x)\,\mathrm{d}\omega(x).$$

Here σ_d denotes the surface area of S^d. For distinction, we denote the inner product of the vectors x and y in E^{d+1} by $x \cdot y$. Let $\mathrm{Pol}(k)$ denote the subspace of L^2 consisting of functions which are restrictions of a polynomial of degree at most k in $d{+}1$ variables to S^d. Of course, different polynomials, when restricted to S^d, may yield the same element of $\mathrm{Pol}(k)$. A polynomial p in $d{+}1$ variables of degree at most k is a linear combination of monomials of the form $m(x_1,\ldots,x_{d+1}) = x^{k_1}\cdots x^{k_{d+1}}$ with $k_1 + \cdots + k_{d+1} = l \leqslant k$. We observe that for $x = (x_1,\ldots,x_{d+1}) \in S^d$ we have $m(x) = (x\cdot x)^i m(x)$. For $i = \left[\frac{1}{2}(k-l)\right]$, the polynomial $(x{\cdot}x)^i m(x)$ is homogeneous, of degree k if l is even and of degree $k-1$ if l is odd. Denote by $\mathrm{Hom}(k)$ the space of homogeneous polynomials of degree k in $d+1$ variables restricted to S^d, and observe that $\mathrm{Pol}(k)$ is the direct sum of the spaces $\mathrm{Hom}(k)$ and $\mathrm{Hom}(k-1)$. Moreover, the spaces $\mathrm{Hom}(k)$ and $\mathrm{Hom}(k-1)$ are orthogonal, since the integral of an odd function over S^d vanishes. Thus

$$\mathrm{Pol}(k) = \mathrm{Hom}(k) \perp \mathrm{Hom}(k-1). \tag{3.1.4}$$

In order to obtain a decomposition of $\mathrm{Hom}(k)$ we consider the space $\mathrm{Harm}(k)$ consisting of homogeneous harmonic polynomials of degree k in $d+1$ variables, restricted to S^d. A polynomial p is harmonic if it satisfies Laplace's equation

$$\frac{\partial^2 p}{\partial x_1^2} + \cdots + \frac{\partial^2 p}{\partial x_{d+1}^2} = 0.$$

The elements of $\mathrm{Harm}(k)$ are called *spherical harmonics*. The reader can learn more about these functions and their use in convexity from chapter 4.8. The Laplace operator acts linearly on $\mathrm{Hom}(k)$, its kernel is, by definition, $\mathrm{Harm}(k)$, and its image is Hom$(k-2)$. The spaces Harm(k) and Hom$(k-2)$ are orthogonal as well, which is a direct consequence of Green's theorem. For details we refer to Stein and Weiss (1971*, p. 141). Thus

$$\mathrm{Hom}(k) = \mathrm{Harm}(k) \perp \mathrm{Hom}(k-2). \tag{3.1.5}$$

From (3.1.4) and (3.1.5) we immediately obtain the direct sum decomposition

$$\mathrm{Pol}(k) = \mathrm{Harm}(k) \perp \cdots \perp \mathrm{Harm}(0). \tag{3.1.6}$$

Next we introduce a special class of spherical harmonics, which play a particularly important role. We observe that since Harm(k) is a finite dimensional linear space equipped with a non-trivial inner product, therefore to any real valued linear function $L = L(h)$ defined on the space Harm(k) there is a unique element $\widetilde{L}$ of Harm(k) such that $L(h) = \langle h, \widetilde{L}\rangle$. In particular, for each element x of S^d the function $L(h) = h(x)$ is linear on Harm(k), and thus we can associate with x a unique function $\widetilde{x} = Q_k(x,y) \in \mathrm{Harm}(k)$ such that

$$h(x) = \langle h, \widetilde{x}\rangle = \frac{1}{\sigma_d}\int_{S^d} h(y)Q_k(x,y)\,\mathrm{d}\omega(y).$$

$Q_k(x,y)$ is called the *kth zonal spherical harmonic with pole x*. One of the important features of $Q_k(x,y)$ is that it is invariant under the orthogonal group $O(d+1)$. That is, if σ is an isometry of E^{d+1} leaving the origin fixed, then

$$Q_k(\sigma x, \sigma y) = Q_k(x,y). \tag{3.1.7}$$

To see this we first observe that it is easily seen that $h(\sigma x) \in \mathrm{Harm}(k)$ if $h(x) \in \mathrm{Harm}(k)$. In particular, $Q_k(\sigma x, \sigma y)$, $y \in S^d$, is a spherical harmonic. For an arbitrary spherical harmonic h consider the integral

$$\frac{1}{\sigma_d}\int_{S^d} h(y)Q_k(\sigma x,\sigma y)\,\mathrm{d}\omega(y).$$

Substituting $z = \sigma y$, we get

$$\frac{1}{\sigma_d}\int_{S^d} h(y)Q_k(\sigma x,\sigma y)\,\mathrm{d}\omega(y) = \frac{1}{\sigma_d}\int_{S^d} h(\sigma^{-1}z)Q_k(\sigma x,z)\,\mathrm{d}\omega(z).$$

The integral on the right side is, by the definition of $Q_k(\sigma x, z)$, the value of the spherical harmonic $h(\sigma^{-1}y)$ at $y = \sigma x$. Thus we have

$$\frac{1}{\sigma_d}\int_{S^d} h(y)Q_k(\sigma x,\sigma y)\,\mathrm{d}\omega(y) = h(\sigma^{-1}\sigma x) = h(x),$$

from which (3.1.7) follows by the uniqueness of $Q_k(x,y)$.

It immediately follows from (3.1.7) that $Q_k(x,y)$ is constant on the "zone" consisting of points $y \in S^d$ for which $x \cdot y$ is constant, which explains the name of $Q_k(x,y)$. Another consequence of (3,1.7) is that $Q_k(x,y)$ depends only on $x \cdot y$. Thus, writing $t = x \cdot y$, we can define the function

$$G_k(t) = Q_k(x,y),$$

which is a polynomial (in the variable t) of degree k.

We consider the integral

$$\int_{S^d} Q_k(x,y)Q_l(x,y)\,\mathrm{d}\omega(y),$$

and observe that it vanishes for $k \neq l$ because of the orthogonality of Harm(k) and Harm(l). On the other hand, substituting $t = x \cdot y$ and writing the area element of the zone $x \cdot y = t$ as $\sigma_{d-1}(1-t^2)^{(d-2)/2}\,\mathrm{d}t$, we get

$$\int_{S^d} Q_k(x,y)Q_l(x,y)\,\mathrm{d}\omega(y) = \sigma_{d-1}\int_{-1}^{1} G_k(t)G_l(t)(1-t^2)^{(d-2)/2}\,\mathrm{d}t.$$

Hence

$$\int_{-1}^{1} G_k(t)G_l(t)(1-t^2)^{(d-2)/2}\,\mathrm{d}t = 0 \quad \text{for } k \neq l.$$

This means that the polynomials $G_k(t)$, $k = 0,1,\ldots$, form a complete system of orthogonal polynomials on the interval $-1 \leqslant t \leqslant 1$ with the weight function given by $(1-t^2)^{(d-2)/2}$. A system of polynomials is determined by this property up to constant factors. Thus there are constants c_k, $k = 0,1,\ldots$, such that

$$G_k(t) = c_k P_k^{(\alpha,\alpha)}(t), \quad \alpha = \frac{d-2}{2}, \quad k = 0,1,\ldots.$$

Let $f_{k,i}$, $i = 1,\ldots,\mu_k$, be an orthonormal basis of Harm(k). Any element $h \in$ Harm(k) can be expanded as

$$h = \sum_{i=1}^{\mu_k} \langle f_{k,i}, h\rangle f_{k,i}.$$

In particular, we have

$$Q_k(x,y) = \sum_{i=1}^{\mu_k} \langle f_{k,i}(y), Q_k(x,y)\rangle f_{k,i}(y).$$

We observe that $\langle f_{k,i}(y), Q_k(x,y)\rangle = f_{k,i}(x)$, by the definition of $Q_k(x,y)$. Hence we obtain the following, so-called "addition formula", which is of fundamental importance.

If $f_{k,1},\dots,f_{k,\mu_k}$ is an orthonormal basis of Harm(k), then

$$Q_k(x,y)=\sum_{i=1}^{\mu_k} f_{k,i}(x)f_{k,i}(y).$$

The addition formula enables us to determine the value of $G_k(1)$. For we have $G_k(1)=Q_k(x,x)=\sum_{i=1}^{\mu_k} f_{i,k}^2(x)$. Taking for both sides the mean value over S^d we get

$$\begin{aligned}G_k(1)&=\frac{1}{\sigma_d}\int_{S^d} G_k(1)\,\mathrm{d}\omega(x)=\frac{1}{\sigma_d}\int_{S^d}\sum_{i=2}^{\mu_k} f_{k,i}^2(x)\,\mathrm{d}\omega(x)\\&=\frac{1}{\sigma_d}\sum_{i=2}^{\mu_k}\int_{S^d} f_{k,i}^2(x)\,\mathrm{d}\omega(x)=\sum_{i=2}^{\mu_k}1=\mu_k=\dim\mathrm{Harm}(k).\end{aligned}$$

For further reference, we note that, obviously, $\dim\mathrm{Harm}(0)=1$, so we have

$$G_0(1)=1.$$

The Jacobi polynomials $P_k^{(\alpha,\alpha)}$, $\alpha=\frac{1}{2}(d-2)$, $k=0,1,\dots$ are normalized by the equation $P_k^{(\alpha,\alpha)}(1)=\binom{k+\alpha}{k}$. In the proof of the linear programming bound the normalization plays role only inasmuch as

$$G_0(t)=P_0^{(\alpha,\alpha)}(t)\quad\text{and}\quad\frac{G_k(t)}{P_k^{(\alpha,\alpha)}(t)}=c_k>0,\ k=0,1,\dots.\tag{3.1.8}$$

We also observe the following simple consequence of the addition formula. For an arbitrary finite subset S of S^d we have

$$\sum_{x,y\in S} Q_k(x,y)=\sum_{i=1}^{\mu_k}\left(\sum_{x\in S} f_{k,i}(x)\right)^2\geqslant 0.\tag{3.1.9}$$

After all these preparations the proof of the linear programming bound is quite easy. Let S be a finite subset of S^d such that the caps of angular diameter φ centered at the points of S form a packing. This is equivalent to saying that

$$x\cdot y\leqslant\cos\varphi\quad\text{for } x,y\in S,\ x\neq y.$$

Let

$$f(t)=\sum_{k=1}^{n} f_k P_k^{(\alpha,\alpha)}(t)$$

be a real polynomial such that $f_0 > 0$, $f_k \geqslant 0$ $(k = 1,\dots,n)$ and $f(t) \leqslant 0$ for $-1 \leqslant t \leqslant \cos\varphi$. We will use the expansion

$$f(t) = \sum_{k=1}^{n} g_k G_k(t),$$

rather than the expansion into the Jacobi polynomials. We can do this, since, in view of (3.1.8), we have $g_0 = f_0 > 0$ and $g_k = c_k f_k \geqslant 0$ $(k = 1,\dots,n)$.

We consider the sum $\sum_{x,y\in S} f(x\cdot y)$ and evaluate it in two different ways. Let m denote the cardinality of S. Let A be the set of numbers a for which $a = x\cdot y$, $x,y \in S$, $x \neq y$. For $a \in A$ let $n(a)$ denote the number of ordered pairs (x,y), $x,y \in S$ for which $x\cdot y = a$. Then we have

$$\sum_{x,y\in S} f(x\cdot y) = mf(1) + \sum_{a\in A} n(a)f(a).$$

On the other hand, in view of the definition of G_k and the relation $G_0 = 1$,

$$\sum_{x,y\in S} f(x\cdot y) = \sum_{k=0}^{n} g_k \sum_{x,y\in S} G_k(x\cdot y) = \sum_{k=0}^{n} g_k \sum_{x,y\in S} Q_k(x,y)$$
$$= g_0 m^2 + \sum_{k=1}^{n} g_k \sum_{x,y\in S} Q_k(x,y).$$

Observing that $f(a) \leqslant 0$ for $a \in A$ and using (3.1.9), the comparison of the two expressions for $\sum_{x,y\in S} f(x\cdot y)$ yields $mf(1) \geqslant m^2 g_0 = m^2 f_0$.

The significance of the linear programming bound is additionally underscored by the fact that for certain special values of d and φ the estimates obtained by this method turn out to be exact. Following is a list of such cases given by Levenštein (1979):

$M(2, \arccos 1/\sqrt{5}) = 12$, $\quad M(21, \arccos\frac{1}{11}) = 100$,
$M(4, \arccos\frac{1}{5}) = 16$, $\quad M(21, \arccos\frac{1}{6}) = 275$,
$M(5, \arccos\frac{1}{4}) = 27$, $\quad M(21, \arccos\frac{1}{4}) = 891$,
$M(6, \arccos\frac{1}{3}) = 56$, $\quad M(22, \arccos\frac{1}{5}) = 552$,
$M(7, \frac{1}{3}\pi) = 240$, $\quad M(22, \arccos\frac{1}{3}) = 4600$,
$M(20, \arccos\frac{1}{9}) = 112$, $\quad M(23, \frac{1}{3}\pi) = 196560$.
$M(20, \arccos\frac{1}{7}) = 162$,

The values of $M(7, \frac{1}{3}\pi)$ and $M(23, \frac{1}{3}\pi)$ prove to be of particular importance for ball packings in E^8 and E^{24}, respectively (see section 5.3).

The bound of (3.1.3) cannot be considered an explicit bound for $M(d,\varphi)$ because of the compound difficulties related to the choice of polynomials $P_i^{(\alpha,\alpha)}$ and coefficients f_i. Moreover, for some small values of d and specific φ, the results of this method are superseded by the "simplex" bound of Böröczky (1978), analogous to the bound (2.3.2) of Rogers for sphere packings in E^d.

The problem of determining the values of $M(2, \varphi)$ and the corresponding extremal arrangements was first raised by the Dutch biologist Tammes (1930) who found a connection between this problem and the highly symmetric distributions of orifices on spherical pollen grains. Since the work of Tammes, the problem has attracted interest of many scientists. For example, connections to information theory (van der Waerden 1961) to virulogy (Goldberg 1967, Molnár 1975), chemistry (Melnyk, Knop and Smith 1977) and architecture (Tarnai 1984) have been established. There is a comprehensive discussion of this subject in the book of L. Fejes Tóth (1972*). For some of the recent developments we refer to Böröczky (1983), Danzer (1986), Hárs (1986) and Tarnai and Gáspár (1983). The latter pair of authors utilized techniques of rigidity theory to establish lower bounds for $M(2, \varphi)$.

3.2. Hyperbolic space

The concepts of packing and covering, as well as the concept of density relative to a bounded domain are well-defined in hyperbolic geometry. However, when attempting to define the notion of density of an arrangement relative to the whole hyperbolic space H^d, one encounters a still unresolved challenge. If we try to adopt here the definition used for E^d (see section 2.1), we notice the first difficulty, which is the absence of similarity transformations in hyperbolic geometry. Therefore defining density with various gauges is not possible.

Even if we restrict the gauges to spheres with a fixed center, a further obstacle remains: In the hyperbolic space, the volume and surface area of a sphere of radius r are of the same order of magnitude for large r. Therefore, for a fixed positive number a, the quotient of the volume of a ball of radius r to the volume of the spherical shell between concentric spheres of radii r and $r + a$ approaches a positive number as $r \to \infty$. It follows that if $\mathscr{A}$ is an arrangement of bounded domains evenly distributed in H^d, the outer and inner density of $\mathscr{A}$ relative to balls differ in a constant factor. Thus, while it is possible to define lower and upper density of an arrangement by taking the lower limit of the inner density and the upper limit of the outer density of the arrangement relative to balls of radius r centered at a fixed point when $r \to \infty$, the lower and upper densities will not be equal for any reasonable arrangement. Moreover, in contrast with the case of Euclidean geometry, densities defined in this way would not be independent from the choice of origin taken as the balls' center.

We could hope that density can be defined at least for "densest packings" and "thinnest coverings". But Groemer's theorem mentioned in section 2.1 does not generalize to hyperbolic geometry. In view of the above, it seems that in the hyperbolic space, any attempt to introduce density of an arbitrary arrangement in analogy with the process in the Euclidean space, by an appropriate limit, is bound to fail.

Recall that in the Euclidean space, for lattice arrangements, or, more generally, for periodic arrangements, density relative to the whole space could be equivalently defined as density relative to a fundamental domain. If any hope remained that for fairly regular arrangements in the hyperbolic space density can be similarly defined, it was effectively dashed by a series of ingenious constructions by Böröczky (1974). These constructions are published by Böröczky in Hungarian and in a rather inaccessible periodical, therefore we describe here one of his main ideas in detail.

In the hyperbolic plane of curvature -1 let L be an oriented line (see fig. 3). Let m be an integer greater than 1, and place points o_i on L, $i = 0, \pm 1, \pm 2, \ldots$, so that the length of the oriented segment $o_{i+1}o_i$ is $\ln m$. Let $\{H_i\}$, $i = 0, \pm 1, \pm 2, \ldots$, be concentric horocycles with the common axis L such that $o_i \in H_i$. Orient H_i so that at o_i the direction of H_i is obtained from the direction of L by a rotation through the angle $-\frac{1}{2}\pi$. Now choose points o_i^j, $i, j = 0, \pm 1, \pm 2, \ldots$, on H_i so that $o_i^0 = o_i$ and the length of the oriented arc of H_i from o_i^j to o_i^{j+1} is a given positive number l. It follows from the construction that the axis of H_i through o_i^j intersects H_{i+1} at o_{i+1}^{mj}.

Next, assign to each point o_i^j two cells, $C_i^j(a)$ and $C_i^j(b)$ as follows: choose a number x, $1 < x < m$ and place points a_i and b_i on the segment o_io_{i+1} so that $o_ia_i = b_io_{i+1}$ and $a_ib_i = \ln x$. Here we use the convention that ab denotes the segment with endpoints a and b, as well as the length of this segment. Let A_i and B_i be the horocycles with axis L, through a_i and b_i, respectively. Denote by $R_i(a)$ the ring between the horocycles A_{i-1} and A_i and by $R_i(b)$ the ring between B_{i-1} and B_i. Let S_i^j be the parallel strip bounded by the perpendicular bisectors of the segments $o_i^{j-1}o_i^j$ and $o_i^jo_i^{j+1}$. Then we can define the cells $C_i^j(a)$ and $C_i^j(b)$ as $C_i^j(a) = S_i^j \cap R_i^j(a)$ and $C_i^j(b) = S_i^j \cap R_i^j(b)$.

It follows from the construction that the cells $C_i^j(a)$, $i, j = 0, \pm 1, +2, \ldots$, as well as the cells $C_i^j(b)$, $i, j = 0, \pm 1, \pm 2, \ldots$, are congruent among each other, and each of the two sets of cells forms a tiling. However, the different types of cells have different areas, and the ratio of the area of $C_i^j(b)$ to the area of $C_i^j(a)$ is $\ln x$. Thus, considering the "density" of the family of congruent circles around the points o_i^j, the "densities" defined by the two types of cells can differ by an arbitrary factor.

In this construction the cells are not convex, but by a suitable modification, Böröczky succeeded in constructing similar examples with convex cells. He also generalized the construction to higher dimensions. Further examples of Böröczky show that the Dirichlet cell subdivision and Delone triangulation can yield different "densities".

Only a few arrangements in hyperbolic space admit a reasonable definition of density. Obviously, tilings are special in this respect: if $\mathcal{A}$ is a tiling, then $d(\mathcal{A}) = \lim_{r\to\infty} d(\mathcal{A}|B(r)) = 1$ as expected, where $B(r)$ is the ball of radius r centered at a fixed point O and the relative density $d(\mathcal{A}|G)$ is defined as in section 2.1. Arrangements $\mathcal{A}$ with a very large symmetry group constitute less obvious examples for which density is well-defined in the "natural" way as $\lim_{r\to\infty} d(\mathcal{A}|B(r))$.

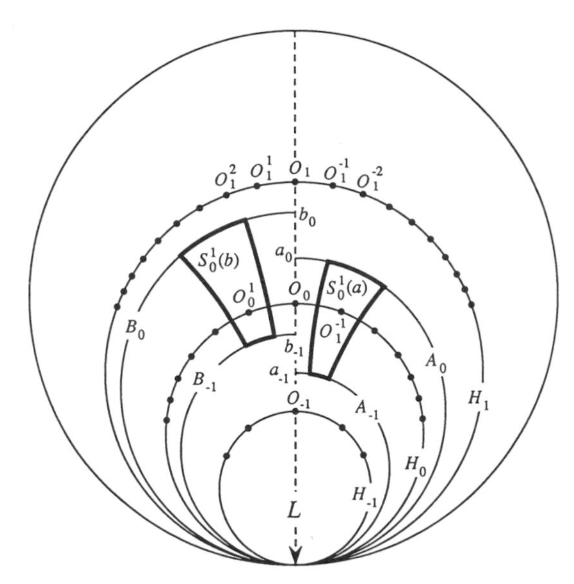

Figure 3.

Specifically, as G. Kuperberg and C. McMullen (1991) point out, if the symmetry group of $\mathscr{A}$ has a fundamental domain of a finite volume, then the above limit exists, is independent of the choice of O, and is equal to $d(\mathscr{A}|F)$ for any fundamental domain F. This fact is a consequence of the so-called "mixing theorem" for geodesic flows on manifolds of negative curvature (see Eberlein 1973).

In absence of a general definition of density, two major problems arise: (1) find bounds for the density of packings and coverings relative to bounded domains, and (2) give substitutes for the concept of densest packing and thinnest covering. In the remaining part of this section we review results of this type.

K. Bezdek (1982) proved that if at least two congruent circular disks are packed in a circular domain in H^2, then the density of the packing relative to the domain is smaller than $\pi/\sqrt{12}$. He generalized this result in (K. Bezdek 1984): the same bound holds for the density of packings of at least two, but finitely many circular disks of radius r relative to the parallel domain of radius r of the convex hull of the set of the circles' centers. One should note here that the density of such a finite packing in its convex hull can be arbitrarily close to 1 as $r \to \infty$. Also, by inscribing a circular disk in each tile of a 3-valent regular tiling of H^2 one can obtain a packing whose density relative to each of its Dirichlet cells is greater than $\pi/\sqrt{12}$.

Now let us examine these regular arrangements of congruent disks in H^2 more closely. Suppose that we started with the tiling $\{p, 3\}$ of H^2. The incircles of the tiles form a very regular packing of H^2 with circular disks of radius r. It turns out that in any packing with circular disks of radius r, the density relative to each Dirichlet cell is smaller than or equal to that in the original, regular packing. A general result on packings with congruent balls has been obtained by Böröczky

(1978) who proved that the density of any such packing in H^d relative to each of its Dirichlet cells has an upper bound analogous to the Rogers' "simplex" bound. Of course, this local bound should not be interpreted as a global density bound. For this reason, the regular packing described above could be regarded as if it were the densest packing with circular disks of radius r, even though density is not defined. This brings us to a substitute notion for the maximum density packing.

A packing $\mathcal{P}$ is *solid* (see L. Fejes Tóth 1968) if no finite number of its members can be rearranged so as to form together with the rest of the members a packing not congruent to $\mathcal{P}$. This notion applies to any space of constant curvature, and in the Euclidean case, solidity implies maximum density. Specifically, if $\mathcal{P}$ is a solid packing of E^d with congruent copies of a body K, then $d(\mathcal{P}) = \delta(K)$. One can notice some connection between solidity and complete saturation (see section 2.1). With few exceptions, a solid packing is completely saturated. L. Fejes Tóth (1968) proved that the packing with the incircles of the tiling $\{p,3\}$ is solid. He also proved the analogous theorem for coverings, with an analogous definition of solid coverings.

Some results have been obtained on solid packings and coverings with incongruent circular disks. A general conjecture is that the incircles [circumcircles] of any trihedral Archimedean tiling form a solid packing [covering]. The conjecture has been confirmed for many special cases (see L. Fejes Tóth 1968, G. Fejes Tóth 1974, and Heppes 1992). In particular, Heppes settled it for packings of the Euclidean plane.

Returning to the packings with congruent circular disks, associated with the regular tiling $\{p,3\}$, we mention a conjecture stating that, for $p \geqslant 5$, the packing remains solid even after the removal of one of its members. The conjecture has been confirmed for $p = 5$ by Danzer (1986) and Böröczky (1983) and for all $p \geqslant 8$ by A. Bczdck (1979). For $p - 6$ and $p = 7$ the conjecture remains open.

4. Problems in the Euclidean plane

Guided by the nature of the Handbook, in previous sections we emphasized the general results (in all dimensions). In this part, motivated by the abundance of sharp inequalities and complete solutions of problems in the plane, we present an array of results concerning two dimensions. We consider here problems on existence and non-existence of economical packings and coverings (in the spirit of sections 2.2 and 2.3) and a variety of special problems on arrangements in the plane. A region in E^2 will be called a *disk* and a convex body in E^2 will be consistently referred to as a *convex disk*. The measure (area) of a convex disk K will be denoted by $A(K)$.

4.1. Existence of efficient arrangements

Observe that the general bounds from section 2.2, mostly asymptotic in nature, are rather poor for low values of d, especially for $d = 2$. Moreover, those general

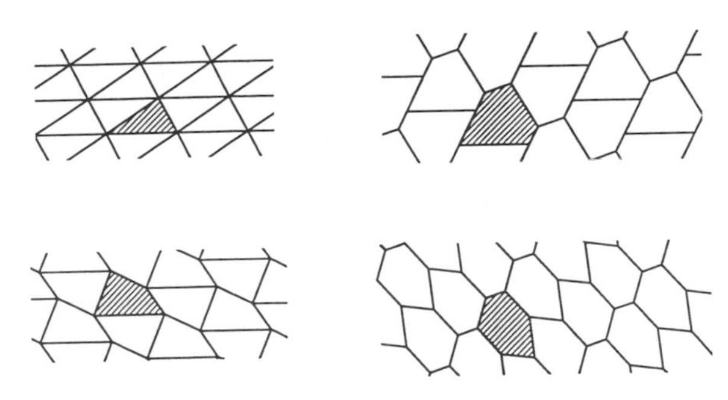

Figure 4.

existence theorems have been obtained through non-constructive methods. In two dimensions, a number of good estimates, in some cases sharp, have been obtained through explicit constructions of economical arrangements.

A very natural idea for constructing a dense packing [a thin covering] with congruent copies of a convex disk K is to enclose K in [enclose in K] a reasonably small [large] (by area) polygon T which tiles the plane. The density of an arrangement of copies of K obtained by tiling the plane with copies of T equals to $A(K)/A(T)$. This method resulted in the inequalities

$$\delta(K) \geqslant \tfrac{25}{32} \tag{4.1.1}$$

and

$$\vartheta(K) \leqslant \tfrac{8}{3}(2\sqrt{3} - 3) \tag{4.1.2}$$

by W. Kuperberg (1987a, 1989). Also, a sharp inequality linking $\delta(K)$ with $\vartheta(K)$, namely

$$\frac{\delta(K)}{\vartheta(K)} \geqslant \frac{3}{4} \tag{4.1.3}$$

was obtained by that method by W. Kuperberg (1987b). We outline this technique by showing the proof of inequality (4.1.3).

By a *p-hexagon* we understand a convex hexagon with a pair of parallel opposite sides of equal length (degenerate p-hexagons are allowed). It is easily noticed that each p-hexagon tiles the plane (see fig. 4). For an arbitrary direction (unit vector) v, let l_1 and l_2 be the support lines of K, parallel to v, and let A_1 and A_2 be the points at which l_1 and l_2 touch K, respectively. Another pair of lines, m_1 and m_2, can be found, parallel to v, such that the distance between them is one-half of the distance between l_1 and l_2 and such that the chords $m_1 \cap K$ and $m_2 \cap K$ are of equal length. The four end points of the chords together with A_1 and A_2 are vertices of a p-hexagon h inscribed in K. The support lines of K at the end points of the chords $m_1 \cap K$ and $m_2 \cap K$ together with the lines l_1 and l_2 define a hexagon H circumscribed about K. The sides of H lying on l_1 and l_2 are opposite and parallel

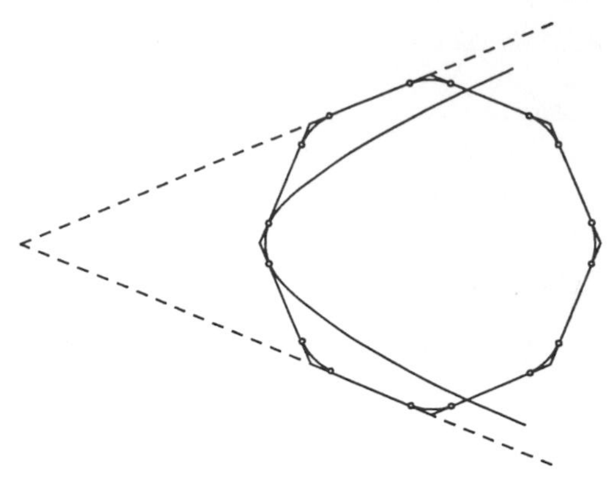

Figure 5.

(but not necessarily of equal length), and it is easy to verify that $A(H)/A(h) \leqslant \frac{3}{4}$. If the direction v is chosen suitably, then H is a p-hexagon, since a continuous change of direction from v to $-v$ results in the interchange of lines l_1 and l_2. The resulting p-hexagons h and H with $h \subset K \subset H$ produce corresponding arrangements with copies of K, whose density ratio yields inequality (4.1.3).

The inequality (4.1.2) shows the presently best known upper bound for $\vartheta(K)$ (for all convex disks K). It is conjectured that the lowest bound is $2\pi/\sqrt{27}$, i.e., that each convex disk admits a covering of the plane with density at most $2\pi/\sqrt{27}$. The bound of (4.1.1) has been improved to

$$\delta(K) \geqslant \tfrac{1}{2}\sqrt{3} \tag{4.1.4}$$

by G. Kuperberg and W. Kuperberg (1990) by an application of a different technique. It seems that inequality (4.1.4) can still be improved. Under the restriction to centrally symmetric convex disks K, some better bounds are known. Tammela (1970) proved that

$$\delta(K) \geqslant 0.892656\ldots$$

for every centrally symmetric convex disk K. Tammela obtained this inequality by an elaboration of a method of Ennola (1961) who suggested this bound as an improvement of his own result. It is conjectured (see Reinhardt 1934 and Mahler 1947) that the greatest lower bound for $\delta(K)$ over all centrally symmetric disks K is

$$\frac{8-\sqrt{32}-\ln 2}{\sqrt{8}-1} = 0.902414\ldots.$$

This is the packing density of the "smoothed octagon" obtained by rounding off each corner of a regular octagon with an arc of a hyperbola tangent to each of the adjacent sides and whose asymptotes contain the second-adjacent sides (see fig. 5).

Concerning coverings of the plane with congruent centrally symmetric convex disks K, the least upper bound for $\vartheta(K)$ is known to be $2\pi/\sqrt{27}$ (the same as the conjectured bound for all convex disks). This result follows directly from the way in which Sas (1939) proved a theorem on inscribed polygons. The method of Sas implies that every centrally symmetric convex disk K contains a centrally symmetric polygon P with at most six sides such that

$$A(K)/A(P) \leqslant 2\pi/\sqrt{27}.$$

4.2. Upper bounds for packing densities and lower bounds for covering densities

We begin with a couple of theorems of Dowker (1944) concerning polygons inscribed in and circumscribed about a convex disk. These theorems are useful tools in obtaining several of the density bounds discussed in this section and proving sharpness of some of the bounds.

For a given convex disk K and integer $n \geqslant 3$, let $p_n(K)$ denote the maximum area of an n-gon contained in K; analogously, let $P_n(K)$ denote the minimum area of an n-gon containing K. Dowker's theorems are the following:

$$2p_n(K) \geqslant p_{n-1}(K) + p_{n+1}(K) \quad \text{for } n \geqslant 4 \tag{4.2.1}$$

and

$$2P_n(K) \leqslant P_{n-1}(K) + P_{n+1}(K) \quad \text{for } n \geqslant 4. \tag{4.2.2}$$

Moreover, if K is symmetric about a point o and n is even, then each of $p_n(K)$ and $P_n(K)$ is attained by an n-gon symmetric about o.

The inscribed and circumscribed hexagons play a particularly important role in density bounds for packings and coverings, thus the maximum area of an inscribed hexagon ($p_6(K)$) is denoted by $h(K)$ and the minimum area of a circumscribed hexagon ($P_6(K)$) is denoted by $H(K)$. A theorem of L. Fejes Tóth (1950) states that

$$\delta(K) \leqslant A(K)/H(K) \tag{4.2.3}$$

for all convex disks K. By the above theorem of Dowker, if K is centrally symmetric, then there exists a centrally symmetric hexagon containing K with area $H(K)$. Since every centrally symmetric hexagon tiles the plane in a lattice manner, inequality (4.2.3) yields

$$\delta(K) = \delta_T(K) = \delta_L(K) = A(K)/H(K) \tag{4.2.4}$$

for each centrally symmetric convex disk K.

The above statement, combined with the "difference-body trick" of Minkowski (see section 2.2 above), implies that

$$\delta_T(K) = \delta_L(K) = A(K)/H\left(\tfrac{1}{2}(K-K)\right) \tag{4.2.5}$$

for every convex disk K. The identity $\delta_T(K) = \delta_L(K)$ (for every convex disk K) was first discovered by Rogers (1951). These results allow an explicit computation of $\delta_T(K)$ and $\delta(K)$ in case when $H(\frac{1}{2}(K-K))$ and $H(K)$ can be determined. In particular, for the unit circular disk B^2, $H(B^2) = \sqrt{12}$ is the area of the circumscribed regular hexagon. Thus

$$\delta(B^2) = \pi/\sqrt{12}, \tag{4.2.6}$$

as discovered by Thue (1910) in a different way. The method of circumscribed hexagons also produced the packing density for the parallel domain of a rectangle (see L. Fejes Tóth 1967b) and for the intersection of two congruent circular disks (L. Fejes Tóth 1971).

For coverings, the inequality $\vartheta(K) \geqslant A(K)/h(K)$, analogous to (4.2.3), has been conjectured, but only partial results support the conjecture. We say that two convex disks, K and L, *cross* each other, if each of the sets $K \setminus L$ and $L \setminus K$ is disconnected. A theorem of L. Fejes Tóth (1950) states that if $\mathscr{C}$ is a covering with congruent copies of K which do not cross, then

$$d(\mathscr{C}) \geqslant A(K)/h(K). \tag{4.2.7}$$

In particular, since translates of K do not cross, it follows that

$$\vartheta_T(K) \geqslant A(K)/h(K) \tag{4.2.8}$$

for every convex disk K. Again, as it was the case for packings, for centrally symmetric K, the above inequality turns into the equality

$$\vartheta_T(K) = \vartheta_L(K) = A(K)/h(K) \tag{4.2.9}$$

by an application of the corresponding theorem of Dowker on inscribed polygons. Kershner's theorem (1939) stating that

$$\vartheta(B^2) = 2\pi/\sqrt{27} \tag{4.2.10}$$

follows directly from (4.2.9).

Mount and Silverman (1990) constructed algorithms based on relations (4.2.5) and (4.2.9) to produce a densest packing and a thinnest covering with translates of a convex polygon. In each case, the time complexity of the algorithm is bounded above by a linear function of the number of sides of the polygon.

The technique which L. Fejes Tóth used in proving (4.2.3) and (4.2.7) has been utilized in attacking other, related problems, therefore we include an outline of this technique below.

In order to obtain a general density bound, it is sufficient to consider the density of an arrangement of a finite number of convex disks $\{K_1, K_2, \ldots, K_N\}$ relative to an arbitrary convex hexagon D. Suppose that the arrangement $\mathscr{A} =$

$\{K_1, K_2, \ldots, K_N\}$ is a packing in D or a crossing-free covering of D. To each K_i a convex polygon D_i with n_i sides is assigned so that

(i) $K_i \subset D_i$ if $\mathscr{A}$ is a packing and $D_i \subset K_i$ if $\mathscr{A}$ is a covering of D;

(ii) $\{D_1, D_2, \ldots, D_N\}$ is a packing in D if $\mathscr{A}$ is a packing and it is a tiling of D if $\mathscr{A}$ is a covering of D; and

(iii) $\sum n_{i-1}^N \leqslant 6N$.

In case $\mathscr{A}$ is a packing in D, the polygons D_i are obtained by simultaneously "inflating" maximally each K_i, and in case $\mathscr{A}$ is a covering of D, they are obtained by "trimming" the K_i's in intersecting pairs. The trimming procedure is made possible by the absence of crossings. Properties (i) and (ii) are direct consequences of the construction, and inequality (iii) follows by an application of Euler's formula for planar cell complexes.

As before, let $P_n(K)$ denote the minimum area of an n-gon containing K. Then, assuming that $\mathscr{A}$ is a packing with congruent copies of a convex disk K, the following inequalities hold:

$$A(D) \geqslant \sum_{i-1}^{N} A(D_i) \geqslant \sum_{i-1}^{N} P_{n_i}(K) \geqslant NP_6(K).$$

The last inequality is a consequence of (iii) and Theorem (4.2.2) of Dowker. This implies

$$d(\mathscr{A}|D) = NA(K)/A(D) \leqslant A(K)/P_6(K) = A(K)/H(K).$$

In case $\mathscr{A}$ is a covering of D with congruent copies of K, the inequality $d(\mathscr{A}|D) \geqslant A(K)/h(K)$ is derived in much the same way.

Observe that the above argument holds even if $\mathscr{A}$ is an arrangement of equiareal affine images of K. Moreover, as Böröczky noticed, the assumption of equal areas can be relaxed by requiring only that the areas of the disks K_i satisfy the inequalities

$$\frac{A(K_i)}{A(K_j)} \geqslant \frac{P_6(K) - P_7(K)}{P_5(K) - P_6(K)} \qquad (i, j = 1, \ldots, N)$$

for packings, and

$$\frac{A(K_i)}{A(K_j)} \geqslant \frac{p_7(K) - p_6(K)}{p_6(K) - p_5(K)} \qquad (i, j = 1, \ldots, N)$$

for coverings.

Bambah, Rogers and Zassenhaus (1964) proved that for every convex disk K the inequality

$$\vartheta_T(K) \geqslant A(K)/2t(K) \tag{4.2.11}$$

holds, where $t(K)$ denotes the maximum area of a triangle inscribed in K. This inequality and (4.2.9) coincide for the centrally symmetric disks K, but otherwise are quite independent.

To derive inequality (4.2.11), its authors construct a special triangulation associated with a covering with translates of a convex disk K. The vertices of the triangulation are the translates of an arbitrarily selected point in K and the triangles are constructed in a way somewhat similar to that in which Delone triangulation is obtained. The crucial property of the triangulation is that the vertices of each triangle correspond to three translates of K which intersect, therefore each of the triangles can be covered by a translate of $-K$. In a large circle, the number of triangles is asymptotically twice the number of vertices (by Euler's formula), and since each vertex corresponds to a translate of K, the density bound (4.2.11) follows. G. Fejes Tóth (1988) observed that the proof of Bambah, Rogers and Zassenhaus actually allows a stronger conclusion:

$$\vartheta_T(K) \geqslant A(K)/h_p(K), \tag{4.2.12}$$

where $h_p(K)$ is the maximum area of a p-hexagon contained in K (p-hexagons are defined in section 4.1). It is conjectured that $\vartheta_T(K) = A(K)/h_c(K)$, where $h_c(K)$ is the maximum area of a centrally symmetric hexagon contained in K. This conjecture, if true, would imply that $\vartheta_T(K) = \vartheta_L(K)$ for every convex disk K.

Several of the density relations mentioned above have corresponding versions for finite arrangements. Results of this type have been obtained by Rogers (1951), Bambah and Rogers (1952), Oler (1961), Bambah, Rogers and Zassenhaus (1964), Folkman and Graham (1969), Graham, Witsenhausen and Zassenhaus (1972), and G. Fejes Tóth (1987a) (see chapter 3.4 for a review of these results).

4.3. Additional results on density bounds

The problems of densest packing and thinnest covering have a common generalization (see L. Fejes Tóth 1972*, p. 80) obtained in the following way: for an arrangement $\mathscr{A}$ with congruent convex disks, let $s(\mathscr{A})$ denote the number expressing the ratio between the part of the plane covered by the members of $\mathscr{A}$ and the whole plane. The number $s(\mathscr{A})$ can be defined rigorously in a similar way as density (see section 1.1 above). The general problem is: Given a convex disk K and a number $\alpha > 0$, among all arrangements $\mathscr{A}$ of congruent copies of K with $d(\mathscr{A}) = \alpha$, find one for which $s(\mathscr{A})$ is maximum. This maximum value of $s(\mathscr{A})$ is denoted by $\sigma(K, \alpha)$. Since $s(\mathscr{A}) = d(\mathscr{A})$ if $\mathscr{A}$ is a packing, the inequality $\delta(K) \leqslant \sup\{\alpha: \sigma(K, \alpha) = \alpha\}$ is immediate. The opposite inequality can be proved by a compactness argument, very much in the same way as certain theorems stated in section 2.1 are proved. Thus

$$\delta(K) = \sup\{\alpha: \sigma(K, \alpha) = \alpha\}.$$

Analogously,

$$\vartheta(K) = \inf\{\alpha: \sigma(K, \alpha) = 1\}.$$

These two relations justify the problem of finding $\sigma(K,\alpha)$ as a common generalization for the problems of finding $\delta(K)$ and $\vartheta(K)$. Assume $A(K)=1$ and define $f(K,x)$ as the maximum area of the intersection between K and a hexagon of area x. Observe that $f(K,x)=x$ for $x \leqslant h(K)$, $f(K,x)=1$ for $x \geqslant H(K)$ and $f(K,x)$ is increasing (but not necessarily concave down) with x. Let $\bar{f}(K,x)$ be the least among all concave down functions $g(x) \geqslant f(K,x)$. G. Fejes Tóth (1973) proved that if $\mathcal{A}$ is an arrangement of congruent copies of K with $d(\mathcal{A})=\alpha$ and with no crossings, then

$$s(\mathcal{A}) \leqslant \alpha \bar{f}(K,1/\alpha). \tag{4.3.1}$$

Since crossings cannot occur in packings, this generalizes each of the inequalities (4.2.3) and (4.2.7).

For centrally symmetric convex disks K, inequality (4.3.1) is sharp. In the case when $\bar{f}(K,1/\alpha)=f(K,1/\alpha)$, there exists a lattice arrangement $\mathcal{A}$ with density α for which $s(\mathcal{A})$ attains the upper bound of (4.3.1). However, as an example constructed by G. Fejes Tóth (1977) shows, there exist centrally symmetric convex disks K for which $f(K,x)$ is not concave down. When $\bar{f}(K,1/\alpha) \neq f(K,1/\alpha)$, the upper bound in (4.3.1) is attained by a suitably chosen combination of two different lattice arrangements: in a certain part of the plane the disks are chosen from one lattice arrangement, and in the rest of the plane they are chosen from another lattice arrangement. G. Fejes Tóth and L. Fejes Tóth (1989) observe that this pattern of the optimal arrangement shows a remarkable analogy with the presence of two different forms of crystals during phase transition of certain materials from one crystalline form to another.

Let us return to the theorem of Rogers (1951) stating that $\delta_T(K)=\delta_L(K)$ for every convex disk K (see (4.2.5) above). A disk D (not necessarily convex) for which $\delta_T(D)=\delta_L(D)$ is called a *Rogerian disk* or an *R-disk*, for short. In (1983a) L. Fejes Tóth gave an alternative proof of the fact that every convex disk is Rogerian. In a subsequent paper (1985) he generalized the method introduced in (1983a) and exhibited a class of Rogerian disks, called limited semiconvex disks. A disk D is said to be *semiconvex* if there exists a pair of parallel lines of support of D at points a and b such that one of the arcs into which a and b partition the boundary of D is convex and has D on its convex side. If in addition there exists a point c on the convex arc ab such that the translates of the arc ab by the vector $\overrightarrow{ca}$ and $\overrightarrow{cb}$, together with the arc ab bound a region containing D, then D is *limited semiconvex*. The problem of a complete characterization of R-disks, posed by L. Fejes Tóth in the same paper, remains open. An example found by A. Bezdek and Kertész (1987) shows that not every semiconvex disk is Rogerian. Heppes (1990) simplified the example of A. Bezdek and Kertész to obtain a non-Rogerian semiconvex disk which is the union of 3 convex disks. On the other hand, L. Fejes Tóth conjectures that disks that are the union of two translates of a convex disk having a point in common are Rogerian. In (1986) he proved this conjecture in the special case when the two disks are congruent circles. For each of the above examples of non-Rogerian disks D, $\delta_T(D) \neq 1$ (i.e., D does not tile

E^2 by translations). Grünbaum conjectures that if D tiles E^2 by translates, then it also tiles in a lattice manner.

Some bounds for $\delta(K)$ and $\vartheta(K)$ are given by Groemer (1961a) for convex disks K of a special kind. Groemer proves that if the radius of curvature of (the boundary of) K is bounded below by $r > 0$, then

$$\delta(K) \leqslant [1 + (2\sqrt{3} - \pi)r^2/A(K)]^{-1}$$

and

$$\vartheta(K) \geqslant [1 - (\pi - 3\sqrt{3}/2)r^2/A(K)]^{-1}$$

and if the radius of curvature of K is bounded above by $R > 0$, then

$$\delta(K) \leqslant A(K)/6R^2 \tan(A(K)/6R^2)$$

and

$$\vartheta(K) \geqslant A(K)/3R^2 \sin(A(K)/3R^2).$$

Each of the above pairs of inequalities generalizes the results of Thue (4.2.6) and Kershner (4.2.10) on $\delta(B^2)$ and $\vartheta(B^2)$, respectively. The first of the four inequalities above follows from a much more general result of G. Fejes Tóth and L. Fejes Tóth (1983).

L. Fejes Tóth (1982, 1983b) considers packings and crossing-free coverings with *r-convex disks*, which are disks that can be presented as intersections of circular disks of radius r. He proves that

$$d_+(\mathcal{P}) \leqslant \left(1 + \frac{6r^2}{a}\left(\tan\frac{p}{12r} - \frac{p}{12r}\right)\right)^{-1}$$

and

$$d_-(\mathcal{C}) \geqslant \left(1 - \frac{3r^2}{a}\left(\frac{p}{6r} - \sin\frac{p}{6r}\right)\right)^{-1}$$

for every packing $\mathcal{P}$ and every crossing-free covering $\mathcal{C}$ with r-convex disks K of area a and perimeter p. Again, these inequalities generalize the results of Thue and Kershner, respectively.

Other generalizations of the theorems of Thue and Kershner are obtained by considering problems on packings and coverings consisting of convex disks which are not necessarily congruent, but selected from a special class. For a pair (a, p) of positive numbers satisfying the isoperimetric inequality $p^2/a \geqslant 4\pi$, let $\mathcal{K}(a, p)$ be the class of convex disks of area at least a and perimeter at most p. Observe that if $p^2/a \geqslant 8\sqrt{3}$, then a tiling of E^2 with members of $\mathcal{K}(a, p)$ is possible, since then

$\mathcal{K}(a,p)$ contains a regular hexagon. Otherwise, the ratio of p^2/a can be attained by a "*vertex-rounded*" regular hexagon $H_v(a,p)$, i.e., a convex disk obtained by rounding off the corners of a regular hexagon with circular arcs of a suitably chosen radius. L. Fejes Tóth (1957) proved that the density of any packing of the plane with members of $\mathcal{K}(a,p)$ cannot exceed $\delta(H_v(a,p))$. The packing density $\delta(H_v(a,p))$ can be computed, based on (4.2.4), as the ratio between the area of $H_v(a,p)$ and the area of the regular hexagon before it was vertex-rounded. Heppes (1963) observed that this density bound remains valid even if the class $\mathcal{K}(a,p)$ is extended to all (non-convex as well as convex) disks satisfying the same restrictions on area and perimeter.

The corresponding result for coverings is not completely analogous, for it requires the absence of crossings (compare (4.2.7) as well as other results discussed above). Again, we can assume $p^2/a \geqslant 8\sqrt{3}$. This time, to attain the ratio of p^2/a, an "*edge-rounded*" regular hexagon $H_e(a,p)$ is constructed by replacing each edge of a regular hexagon with a circular arc of a suitably chosen radius (the vertices of the hexagon remain unchanged). G. Fejes Tóth and Florian (1984) proved that a crossing-free covering of the plane with members of $\mathcal{K}(a,p)$ is of density greater than or equal to $\vartheta_T(H_e(a,p))$, which can be computed, based on (4.2.9) as the ratio between the area of $H_e(a,p)$ and the area of the regular hexagon before it was edge-rounded. Similar problems, in which the perimeter p was replaced by affine perimeter, were considered by L. Fejes Tóth (1980), L. Fejes Tóth and Florian (1982) and Florian (1986).

There is an abundance of problems and results concerning circle packings and circle coverings. These results constitute an essential part of the theory of packing and covering, but their role in convexity is not central. For lack of space we omit a detailed discussion of this subject and refer the interested reader to the relevant sections of the book of L. Fejes Tóth (1972*) and the survey paper of G. Fejes Tóth (1983*).

5. Additional topics

5.1. *Other concepts of efficiency of arrangements*

While high density packings and low density coverings can be considered efficient, the definition of density allows some undesired local deviations to occur which go contrary to the intuitive concept of efficiency. For instance, a packing can be of maximum density and still leave arbitrary large empty spots, and globally thin coverings can be locally very thick. Notions other than density have been introduced to express local efficiency, often at the expense of global efficiency.

We define the *gap size* of an arrangement $\mathcal{A}$ of convex sets as the supremum of the radius of a ball disjoint with all members of $\mathcal{A}$. Similarly, the *overlap size* of $\mathcal{A}$ is the supremum of the radius of a ball contained in the intersection of two members of $\mathcal{A}$. These concepts are closely related to the notions of closeness of a packing and looseness of a covering (compare L. Fejes Tóth 1976, 1978, Linhart

1978, and Böröczky 1986) and to the notion of an (r, R)-system of points (see Ryškov 1974). Observe that if the gap size of an arrangement $\mathcal{A}$ of convex bodies is finite, then it is equal to the minimum number r such that the (outer) parallel domains of radius r of members of $\mathcal{A}$ form a covering. Similarly, if the overlap size of $\mathcal{A}$ is smaller than the inradius of all members of $\mathcal{A}$, then it is equal to the minimum number r such that the inner parallel domains of radius r of members of $\mathcal{A}$ form a packing.

In a modified version of the (still unsolved) ball packing problem in E^3, minimum gap size replaces maximum density. Confirming a conjecture of L. Fejes Tóth (1976), Böröczky (1986) proved that the packing with unit balls in E^3 of minimum gap size is unique and is obtained by placing the centers of the balls at the vertices and at the centers of all cubes of a cubic lattice of edge-length $4/\sqrt{3}$. The gap size so obtained equals to $\sqrt{5/3}-1 = 0.29099444\ldots$. Since for unit balls the problems of minimum gap size of a packing and minimum overlap size of a covering are equivalent, Böröczky's solution settles both of them. The result provides some insight in the problems of densest packing and thinnest covering with congruent balls in E^3. It implies that the configurations of centers of balls in such two extreme density arrangements cannot coincide.

Utilizing a technique of Delone and Ryškov (1963) and Ryškov and Baranovskiĭ (1975, 1976), Horváth (1980, 1986a) solved the problem of minimum gap size of a lattice packing with unit balls in E^4 and E^5. In (1986b), Horváth considers cylinders of various kinds (i.e., parallel domains of k-flats in E^d, $k = 0, 1, \ldots, d-1$) and gives bounds for the gap size of a packing and for the overlap size of a covering with congruent cylinders. In particular, if $\mathcal{P}$ is a packing of E^3 with unit circular cylinders (infinite in both directions), then the gap size of $\mathcal{P}$ is greater than or equal to $2/\sqrt{3} - 1$, and equality occurs only if all cylinders of $\mathcal{P}$ are parallel and each of them touches six others. Furthermore, if no two cylinders of $\mathcal{P}$ are parallel, then, according to A. Bezdek and W. Kuperberg (1991b), for any pair of balls of radius $2/\sqrt{3} - 1$, not overlapping with any of the cylinders, each of the balls can move between the cylinders to assume the other ball's place, without overlapping with any of the cylinders during the motion.

Turning to arrangements in the plane, we recall the theorem of Rogers included in (4.2.5) above, which states that the maximum density of a packing in E^2 with translates of a convex disk is attained in a lattice packing. L. Fejes Tóth (1978) obtained a theorem analogous to Rogers' theorem, in which maximum density was replaced by minimum gap size. However, this resemblance between maximum density and minimum gap size does not extend over other theorems of a similar nature. According to a theorem of L. Fejes Tóth given in (4.2.4), for any centrally symmetric convex disk K, the maximum density of a packing of E^2 with congruent copies of K is attained in a lattice packing. On the other hand, centrally symmetric disks were found by L. Fejes Tóth (1978) and A. Bezdek (1980) with the property that the minimum gap size of a packing with congruent copies of the disk require non-lattice arrangements.

For arrangements $\mathcal{A} = \{a_i + K\}$ of translates of a convex body K, it is natural to modify the concepts of gap and overlap size by using homothetic images of K

and $-K$ as gauges rather than balls. Linhart (1978) considers the numbers

$$\lambda(\mathscr{A}) = \inf\{\lambda > 0: \{a_i + (1+\lambda)K\} \text{ is a covering}\}$$

and

$$\mu(\mathscr{A}) = \inf\{\mu > 0: \{a_i + (1-\mu)K\} \text{ is a packing}\},$$

and he shows that if $\mathscr{P}$ is a packing of translates of K, then

$$\lambda(\mathscr{P}) = \sup\{\lambda > 0: \text{there is a translate of } -\lambda K \text{ disjoint to all disks of } \mathscr{P}\}$$

and if $\mathscr{C}$ is a covering with translates of K, then

$$\mu(\mathscr{C}) = \sup\{\mu > 0: \text{there is a translate of } \mu K \text{ contained in the intersection of two disks of } \mathscr{C}\}.$$

With a convex disk K Linhart associates the quantities

$$C(K) = \sup\{\frac{1}{\lambda(\mathscr{P})} > 0: \mathscr{P} \text{ is a packing of translates of } K\}$$

and

$$c(K) = \sup\{\frac{1}{\mu(\mathscr{C})} > 0: \mathscr{C} \text{ is a covering with translates of } K\}.$$

It turns out that these quantities satisfy the relation

$$C(K) - c(K) = 1.$$

Linhart also shows that

$$c(K) \geqslant 2 \quad \text{and} \quad C(K) \geqslant 3$$

for any convex disk K and in both inequalities equality characterizes triangles.

5.2. Density bounds for arrangements with restricting conditions

A packing $\mathscr{P}$ is said to be *totally separable* if each pair of its members can be separated by a hyperplane not intersecting the interior of any member of $\mathscr{P}$. Given a convex body K in E^d, what is the maximum density of a totally separable packing with congruent copies (translates) of K? G. Fejes Tóth and L. Fejes Tóth (1973) consider this question for $d = 2$ and they prove that if $\mathscr{P}$ is a totally separable packing with congruent copies of a convex disk K, then

$$d_+(\mathscr{P}) \leqslant A(K)/P_4(K), \tag{5.2.1}$$

where $P_4(K)$ denotes, as before, the minimum area of a quadrilateral containing K (see section 4.2). If K is centrally symmetric, the number $A(K)/P_4(K)$ is actually the maximum density of such a packing since, by Dowker's theorem (see section 4.2), $P_4(K)$ can be realized by a parallelogram, and congruent parallelograms admit a totally separable tiling. If K is a circular disk, then (5.2.1) states that $d_+(\mathcal{P}) \leqslant \frac{1}{4}\pi$. A. Bezdek (1983) proves this last inequality under an assumption weaker than totally separability, requiring only that the packing $\mathcal{P}$ with congruent circles satisfies the following *local separability* condition: for every triple of members of $\mathcal{P}$ there is a line separating one from the other two. He also shows that, in general, for non-circular K local separability of $\mathcal{P}$ does not imply (5.2.1).

The interesting problem of maximum density of a totally separable packing of E^2 with (not necessarily congruent) circular disks remains open. One can rephrase this problem in the following way: start with a configuration of lines in the plane which partition the plane into bounded regions and place a circle in each of the regions. What is the maximum density of a circle packing so obtained? In this phrasing, the analogous question about the minimum density of a covering can be asked. It is conjectured that the configurations of lines providing each these extreme densities is that which partitions the plane into the Archimedean tiling $(3,6,3,6)$. G. Fejes Tóth (1987b) proves that neither of the extreme densities in question equals 1.

The problem of the densest totally separable packing of E^3 with congruent balls was solved by Kertész (1988) who proved that if a cube of volume V contains N unit balls forming a totally separable packing, then $V \geqslant 8N$. Consequently, the cube-lattice packing is the densest one among all totally separable packings of E^3 with congruent balls, and the maximum density is $\frac{1}{6}\pi$.

The *homogeneity* of an arrangement of bodies $\{K_i\}$ is defined as $\inf V(K_i)/\sup V(K_i)$. Two intersecting members of an arrangement are each other's *neighbors*. A packing of E^2 is said to be *compact* if each member K of the packing satisfies the following three conditions:

(i) K has a finite number of neighbors,

(ii) all neighbors of K can be ordered cyclically so that each of them touches its successor,

(iii) the union of the neighbors of K contains a polygon enclosing K.

L. Fejes Tóth (1984) originated the study of compact packings by proving that the lower density of any positive-homogeneity compact packing of the plane with circular disks is at least $\pi/\sqrt{12}$, and the lower density of any positive-homogeneity compact packing of the plane with homothetic centro-symmetric convex disks is at least $\frac{3}{4}$, where equality occurs only for packings with affinely regular hexagons.

Florian (1985) considers compact packings with circular disks on the sphere S^2 and in the hyperbolic plane H^2, defined in the same manner as for E^2. In each case, let $d(r)$ denote the density of three mutually tangent circular disks of radius r each, with respect to the triangle spanned by their centers (it is assumed here that $r > 0$, and in the spherical case, additionally $r \leqslant \frac{1}{3}\pi$). Florian proves that the density of any finite compact packing of the sphere with circular disks with radii at most r is greater than or equal to $d(r)$. For compact circle packings of H^2 with radii $r_i \leqslant r$ the same lower bound established by Florian is for density of a "local" character, namely with respect to triangles of a triangulation whose vertices are the centers of the circles.

Bezdek, Bezdek and Böröczky (1986) prove that the density of a positive-homogeneity compact packing of E^2 with homothetic convex disks is at least $\frac{1}{2}$, and that equality occurs for various packings with homothetic triangles.

The concept of compact packing has been generalized to higher dimensions by K. Bezdek (1987) and K. Bezdek and Connelly (1991). They say that a member of a packing is *enclosed* by its neighbors if any curve connecting a point of the member with a point sufficiently far from it must intersect one of the neighbors. A packing is compact if each of its members is enclosed by its neighbors. K. Bezdek proved that the density of any compact lattice packing of E^d $(d \geqslant 3)$ with translates of a centrally symmetric convex body is at least $2^{1/d-1}/\left(2^{1/d-1}+1\right)$. K. Bezdek and Connelly consider positive-homogeneity compact packings of E^d $(d \geqslant 2)$ consisting of homothetic centrally symmetric convex bodies and they establish a lower density bound of $(d+1)/2d$ for such packings, sharp for $d \leqslant 3$. Also, they observe that compactness of packings in E^d, defined as above, is a very strong condition. For instance, in dimensions $d \geqslant 3$ strict convexity of members of a packing precludes compactness. Therefore they introduce an alternative concept, at least for packings of E^d with centrally symmetric convex bodies. Such a packing is said to be *triangulated* if there exists a triangulation of E^d whose vertices are the centers of the bodies and whose edges are determined by pairs of touching bodies. In dimension two, triangulated packings and compact packings are the same. Of course, in dimensions above two, triangulated packings can be far less dense than compact ones have to be. The sharp lower density bound for positive-homogeneity triangulated packings of E^d with homothetic centrally symmetric convex bodies, established by K. Bezdek and Connelly, is $(d+1)/2^d$.

The natural restrictions on the number of neighbors in a packing lead to several important problems, especially concerning ball packings. Most of these problems are of a rather combinatorial or structural nature and we address them in section 5.3 below; here we discuss only those which involve density bounds.

If every member of a packing has exactly [at least] n neighbors, we call it an *n-neighbor* [n^+*-neighbor*] packing. An easy construction shows that there exists a zero density 5-neighbor packing of the plane with translates of a parallelogram. It turns out that this property characterizes parallelograms (see L. Fejes Tóth 1973a). Let K be a convex disk other than a parallelogram. Makai (1987) proved that every 5^+-neighbor packing with translates of K is of density greater than or equal to $\frac{3}{7}$ and equality can occur only if K is a triangle. For centrally symmetric disks K, Makai proved that the corresponding lower bound is $\frac{9}{14}$, attained only if K is an affine regular hexagon. Concerning 6^+-neighbor packings, the following is known: L. Fejes Tóth (1973a) showed that the density of a 6^+-neighbor packing with translates of a convex disk is at least $\frac{1}{2}$, and Makai (1987) obtained the corresponding lower bound of $\frac{3}{4}$ for centrally symmetric disks. Each of these bounds is sharp, and, again, the extreme values are produced only by triangles and affine regular hexagons, respectively. Chvátal (1975) considered 6^+-neighbor packings with translated parallelograms and proved that the density of such a packing is at least $\frac{11}{15}$.

Next, we consider n-neighbor packings of E^3 with congruent balls. For $n = 9$, density zero can be attained by arranging the balls in two adjacent hexagonal layers. On the other hand, any 10^+-neighbor packing with congruent balls must have positive density, as G. Fejes Tóth (1981), Sachs (1986) and A. Bezdek and K. Bezdek (1988) proved. Each of them used an idea of L. Fejes Tóth who reduced the problem to proving that the maximum number of points on a closed unit hemisphere no two of which are closer than 1 to each other is 9, and each of them proved just that. The problem of the minimum densities for 10^+- and 11^+-neighbor packings of E^3 with congruent balls is open, as well as a multitude of similar problems in higher dimensions and for other convex bodies are.

In the densest lattice packing with balls in E^3 (see section 2.4), each ball has exactly 12 neighbors, which is the maximum number of neighbors of a single ball in any packing of E^3 with congruent balls (see section 5.3). L. Fejes Tóth (1969a, 1989) conjectures that any 12-neighbor packing of E^3 with congruent balls is *laminated*, i.e., it is composed of parallel layers of the same hexagonal structure as the layers in the densest lattice packing, and thus has density $\pi/\sqrt{18}$, conjectured to be equal to $\delta(B^3)$.

In relation to the sphere packing conjecture in E^3, Böröczky (1975) suggested to prove a special case of the conjecture by restricting it to packings which consist of translates of a string of spheres (a *string of spheres* is a collection of congruent balls whose centers are collinear and such that each of them touches two others). Böröczky's conjecture was confirmed by Bezdek, Kuperberg and Makai (1991). The analogous conjectures for non-parallel strings and for finite strings remain open.

An arrangement $\mathcal{A}$ of translates of homothetic copies of a convex body K is said to be *saturated* if there exists a number $r > 0$ such that each coefficient of homothety is at least r and every translate of rK overlaps with a member of $\mathcal{A}$ (compare section 2.1). Suppose $\mathcal{C}$ is a covering with translates of a centrally symmetric convex body K and $t > \frac{1}{2}$. Then replacing each member of $\mathcal{C}$ by its concentric homothetic copy with coefficient t produces a saturated arrangement. As a consequence, every centrally symmetric convex body K admits a saturated arrangement with congruent copies of K and of density arbitrarily close to $\vartheta_T(K)/2^d$. Dumir and Khassa (1973a, b) address the problem of the infimum of densities of saturated arrangements in E^2 of translates of homothetic copies of a given centrally symmetric convex disk K and they prove that the infimum is equal to $\vartheta_T(K)/4$. Their result generalizes earlier results by Eggleston (1965) and Bambah and Woods (1968a, b) inspired by a conjecture of L. Fejes Tóth (1967*) in which the density bound was proposed for saturated arrangements of circular disks.

Certain stronger forms of saturation have been considered by L. Fejes Tóth and Heppes (1980) and A. Bezdek (1990) who found some density bounds for corresponding circle packings in the plane.

5.3. Newton number and related problems

For every convex body K, let $N(K)$ denote the maximum number of neighbors of a member of any packing with congruent copies of K. Obvously, $N(K)$ is always a finite number. This number is called the *Newton number of* K because of the dispute between Sir Isaac Newton and David Gregory concerning $N(B^3)$, which Newton conjectured to be 12 and Gregory thought to be 13 [see Coxeter (1963) for an interesting account of this dispute]. One hundred and eighty years elapsed before the problem was settled by Hoppe (1874) in favor of Newton's conjecture.

Observe that $N(B^d) = M(d-1, \frac{1}{3}\pi)$ where $M(d, \varphi)$ denotes the maximum number of spherical caps of angular diameter φ in a packing of S^d, as defined in section 3.1. Thus, the results quoted in that section yield upper bounds for the numbers $N(B^d)$ which, in cases $d = 8$ and $d = 24$ are sharp: $N(B^8) = 240$ and $N(B^{24}) = 196560$.

In dimension 3, the arrangement of 12 unit balls touching one is not unique, which may explain why the problem of determining $N(B^3)$ was so hard to solve. The simplest known solution belongs to Leech (1956) and is not very simple. On the other hand, the arrangements providing $N(B^8)$ and $N(B^{24})$ were found to be unique (Bannai and Sloane 1981). It is not completely understood what distinguishes dimensions $d = 8$ and $d = 24$ and it is not known in which other dimensions $d > 3$ this uniqueness phenomenon occurs. For a detailed account of this topic see Conway and Sloane (1988*).

For convex bodies other than balls, the problem of finding or estimating $N(K)$ was considered only in dimension 2. Böröczky (1971) determined the Newton number of the regular n-gon for all $n \neq 5$ and Linhart (1973) found the number for the regular pentagon. Several upper bounds for $N(K)$ were found in terms of certain parameters associated with a convex disk K. L. Fejes Tóth (1967a) established that if K is a convex disk of diameter d and width w, then

$$N(K) \leqslant (4+2\pi)\frac{d}{w} + 2 + \frac{w}{d}.$$

This inequality is sharp in many cases. For instance, it implies that the Newton number of the isosceles triangle with $w/d = \sin\frac{1}{9}\pi$ is 64. Other general bounds for the Newton number of a convex disk involving different parameters were given by Hortobágyi (1972, 1975) and Wegner (1992). Each of them considered a generalization of the Newton number, produced by counting the number of congruent copies of a convex disk K that can touch another convex disk C. In particular, Wegner's result includes the computation of the Newton number of the 30°–30°–120° triangle, which, as conjectured, turns out to be 21. Together with a previous result of Linhart (1977) this implies that in a tiling with congruent convex disks in which each tile's number of neighbors is its Newton number, the maximum number of neighbors is 21.

A modification of the Newton number by considering translates of a convex body K results in the number $H(K)$, called the *Hadwiger number of* K, defined as the maximum number of non-overlapping translates of K that can touch K.

The term was introduced by L. Fejes Tóth (1970) in appreciation of the elegant theorem of Hadwiger (1957) stating that

$$H(K) \leqslant 3^d - 1$$

for all convex bodies K in E^d. The equality occurs only if K is a parallelotope (see Groemer 1961b and Grünbaum 1961). On the other hand, Swinnerton-Dyer (1953) showed that $H(K) \geqslant d^2 + d$ for all convex bodies K in E^d. This inequality was generalized by Smith (1975) to all compact sets K for which $K - K$ is a neighborhood of the origin. K. Kuperberg and W. Kuperberg (1993) consider a related problem. They prove that if n non-overlapping translates of a star-shaped disk have a point in common, then $n \leqslant 4$. They conjecture that $H(K) \leqslant 8$ for every star-shaped disk.

The concepts of the Newton number and the Hadwiger number can be generalized by considering the kth order neighbors in a packing. We say that the members A and B of a packing are *kth order neighbors* if there exist $k - 1$ members A_1, A_2,...,A_{k-1} of the packing but not fewer than $k - 1$, such that in the sequence $A = A_0$, A_1, A_2,...,A_{k-1}, $A_k = B$, A_{k-1} is a neighbor of A_i for $i = 1, \dots, k$. The kth *Newton number* $N_k(K)$ and the kth *Hadwiger number* $H_k(K)$ of K, are defined as the maximum number of neighbors of order at most k of a single member of any packing with congruent copies of K or with translates of K, respectively. As was shown by L. Fejes Tóth and Heppes (1967),

$$N_2(B^2) = H_2(B^2) = 18.$$

For $k > 2$ the exact value of $N_k(B^2) = H_k(B^2)$ is not known. However, asymptotically, the magnitude of $H_k(K)$ is of order k^2 for every convex disk K (L. Fejes Tóth 1975b), and, in particular,

$$\lim_{k\to\infty} H_k(B^2)/k^2 = 2\pi/\sqrt{3},$$

as proved by L. Fejes Tóth (1969b).

In addition, let us mention in this section a result concerning plane packings with non-congruent circular disks in which the number of neighbors plays a crucial role. Bárány, Füredi and Pach (1984), confirming a conjecture of L. Fejes Tóth (1977), proved that if in a packing of the plane with circular disks every disk has at least 6 neighbors, then either all disks are congruent or arbitrarily small disks occur. Their proof combines a geometric idea with a combinatorial one, each of interest on its own.

References

Books and survey papers

Bambah, R.P.
[1960] Some problems in the geometry of numbers, *J. Indian Math. Soc. (N.S.)* **24**, 157–172.

[1970] Packing and covering, *Math. Student* **38**, 133–138.
[1971] Geometry of numbers, packing and covering and discrete geometry, *Math. Student* **39**, 117–129.

Baranovskiĭ, E.P.
[1969] Packings, coverings, partitionings and certain other distributions in spaces of constant curvature (in Russian), *Itogi Nauki – Ser. Mat., Vol. 14 (Algebra, Topologiya, Geometrya)* pp. 189–225 [*Progr. Math.* **9** (1971) 209–253].

Bezdek, K.
[1993] Hadwiger–Levi's covering problem revisited, in: *New Trends in Discrete and Computational Geometry*, ed. J. Pach (Springer, Berlin) to appear.

Bos, A.
[1979] Sphere-packings in euclidean space, in: *Packing and Covering in Combinatorics, Study week, 1978,* Math. Centre Tracts, Vol. 106, pp. 161–177.

Cassels, J.W.S.
[1971] *An Introduction to the Geometry of Numbers* (Springer, Berlin, 2nd edition).

Conway, J.H., and N.J.A. Sloane
[1988] *Sphere Packings, Lattices and Groups* (Springer, Berlin).

Erdős, P., P.M. Gruber and J. Hammer
[1989] *Lattice Points* (Longman, Harlow and John Wiley, New York).

Fejes Tóth, G.
[1983] New results in the theory of packing and covering, in: *Convexity and its Applications*, eds P.M. Gruber and J.M. Wills (Birkhäuser, Basel) pp. 318–359.

Fejes Tóth, G., and W. Kuperberg
[1993] A survey of recent results in the theory of packing and covering, in: *New Trends in Discrete and Computational Geometry*, ed. J. Pach (Springer, Berlin) to appear.

Fejes Tóth, L.
[1960] Neuere Ergebnisse in der diskreten Geometrie, *Elem. Math.* **17**, 25–36.
[1964] *Regular Figures* (Pergamon Press, Oxford).
[1967] Packings and coverings in the plane, in: *Proc. Coll. Convexity, Copenhagen, 1965* (Københavns Univ. Math. Inst., Copenhagen) pp. 78–87.
[1972] *Lagerungen in der Ebene, auf der Kugel und im Raum*, 2. Aufl. (Springer, Berlin).
[1977] Dichteste Kugelpackung. Eine Idee von Gauss, *Festschrift zur 200. Wiederkehr des Geburtstages von Gauss, Abh. Braunschweig. Wiss. Ges.* **27**, 311–321.
[1984] Density bounds for packing and covering with convex discs, *Exposition. Math.* **2**, 131–153.
[1986] Symmetry induced by economy, *Comput. Math. Appl. Part B* **12**, 83–91.

Few, L.
[1967] Multiple packing of spheres: A survey, in: *Proc. Coll. Convexity, Copenhagen, 1965* (Københavns Univ. Mat. Inst., Copenhagen) pp. 88–93.

Florian, A.
[1987] Packing and covering with convex discs, in: *Intuitive Geometry, Siófok, 1985,* Colloq. Math. Soc. János Bolyai, Vol. 48 (North-Holland, Amsterdam) pp. 191–207.

Gritzmann, P., and J.M. Wills
[1986] Finite packing and covering, *Studia Sci. Math. Hungar.* **21**, 149–162.

Groemer, H.
[1985] Coverings and packings by sequences of convex sets, in: *Discrete Geometry and Convexity*, eds J.E. Goodman, E. Lutwak, J. Malkevitch and R. Pollack, Ann. New York Acad. Sci., Vol. 440 (New York Acad. Sci, New York) pp. 262–278.

Gruber, P.M.
[1979] Geometry of numbers, in: *Contributions to Geometry, Proc. Geom. Symp., Siegen, 1978* (Birkhäuser, Basel) pp. 186–225.

Gruber, P.M., and C.G. Lekkerkerker
[1987] *Geometry of Numbers* (North-Holland, Amsterdam).

Grünbaum, B., and G.C. Shephard
[1986] *Tilings and Patterns* (Freeman, New York).

Levenštein, V.I.
[1983] Bounds for packings of metric spaces and some of their applications (in Russian), *Problemy Kibernet.* **40**, 43–110.

Moser, W.O.J., and J. Pach
[1986] 100 Research problems in discrete geometry, Mimeograph.
[1993] Recent developments in combinatorial geometry, in: *New Trends in Discrete and Computational Geometry,* ed. J. Pach (Springer, Berlin) to appear.

Pach, J., and P.K. Agarwal
[1991] *Combinatorial Geometry,* Courant Lecture Notes Series (New York University, New York).

Rogers, C.A.
[1964] *Packing and Covering* (Cambridge Univ. Press, Cambridge).

Saaty, T.L., and J.M. Alexander
[1975] Optimization and the geometry of numbers: packing and covering, *SIAM Rev.* **17**, 475–519.

Stein, E.M., and G. Weiss
[1971] *Introduction to Fourier Analysis on Euclidean Spaces* (Princeton Univ. Press, Princeton).

Wills, J.M.
[1990] Kugellagerungen und Konvexgeometrie, *Jber. Deutsch. Math.-Vereinig.* **92**, 21–46.

Research papers

Bambah, R.P.
[1954] On lattice coverings by spheres, *Proc. Nat. Inst. Sci. India* **20**, 25–52.

Bambah, R.P., and C.A. Rogers
[1952] Covering the plane with convex sets, *J. London Math. Soc.* **27**, 304–314.

Bambah, R.P., and A.C. Woods
[1968a] On the minimal density of maximal packings of the plane by convex bodies, *Acta Math. Acad. Sci. Hungar.* **19**, 103–116.
[1968b] On the minimal density of plane coverings by circles, *Acta Math. Acad. Sci. Hungar.* **19**, 337–343.

Bambah, R.P., C.A. Rogers and H. Zassenhaus
[1964] On coverings with convex domains, *Acta Arith.* **9**, 191–207.

Bannai, E., and N.J.A. Sloane
[1981] Uniqueness of certain spherical codes, *Canad. J. Math.* **33**, 437–449.

Baranovskiĭ, E.P.
[1964] On packing of n-dimensional Euclidean space by equal spheres (in Russian), *Izv. Vyssh. Uchebn. Zved. Mat.* **39**, 14–24.

Bárány, I., Z. Füredi and J. Pach
[1984] Discrete convex functions and proof of the six circle conjecture of Fejes Tóth, *Canad. J. Math.* **36**, 569–576.

Best, M.R.
[1980] Binary codes with minimum distance of four, *IEEE Trans. Inform. Theory* **26**, 738–742.

Bezdek, A.
[1979] Solid packing of circles in the hyperbolic plane, *Studia Sci. Math. Hungar.* **14**, 203–207.
[1980] Remark on the closest packing of convex discs, *Studia Sci. Math. Hungar.* **15**, 283–285.
[1983] Locally separable circle packings, *Studia Sci. Math. Hungar.* **18**, 371–375.
[1984] On the section of lattice-coverings of balls, *Rend. Circ. Math. Palermo Ser. II, Suppl. 3*, 23–45.
[1990] Double-saturated packing of unit disks, *Period. Math. Hungar.* **21**, 189–203.

Bezdek, A., and K. Bezdek
[1988] A note on the ten-neighbour packings of equal balls, *Beiträge Algebra Geom.* **27**, 49–53.

Bezdek, A., and G. Kertész
[1987] Counter-examples to a packing problem of L. Fejes Tóth, in: *Intuitive Geometry, Siófok, 1985,* Colloq. Math. Soc. János Bolyai, Vol. 48 (North-Holland, Amsterdam) pp. 29–36.

Bezdek, A., and W. Kuperberg
[1991a] Packing Euclidean space with congruent cylinders and with congruent ellipsoids, in: *Applied Geometry and Discrete Mathematics: The Victor Klee Festschrift*, eds P. Gritzmann and B. Sturmfels, DIMACS Series on Discrete Mathematics and Computer Science (Amer. Math. Soc. & ACM, New York) pp. 71–80.
[1991b] Placing and moving spheres in the gaps of a cylinder packing, *Elem. Math.* **46**, 47–51.

Bezdek, A., K. Bezdek and K. Böröczky
[1986] On compact packings, *Studia Sci. Math. Hungar.* **21**, 343–346.

Bezdek, A., W. Kuperberg and E. Makai Jr
[1991] Maximum density space packing with parallel strings of spheres, *Discrete Comput. Geom.* **6**, 277–283.

Bezdek, K.
[1982] Ausfüllungen eines Kreises durch kongruente Kreise in der hyperbolischen Ebene, *Studia Sci. Math. Hungar.* **17**, 353–366.
[1984] Ausfüllungen in der hyperbolischen Ebene durch endliche Anzahl kongruenter Kreise, *Ann. Univ. Sci. Budapest. Eötvös Sect. Math.* **27**, 113–124.
[1987] Compact packing in Euclidean space, *Beiträge Algebra Geom.* **25**, 79–84.

Bezdek, K., and R. Connelly
[1991] Lower bound for packing densities, *Acta Math. Hungar.* **57**, 291–311.

Bleicher, M.N.
[1975] The thinnest three dimensional point lattice trapping a sphere, *Studia Sci Math. Hungar.* **10**, 157–170.

Blichfeldt, H.F.
[1925] On the minimun value of positive real quadratic forms in 6 variables, *Bull. Amer. Math. Soc.* **31**, 386.
[1926] The minimum value of positive quadratic forms in seven variables, *Bull. Amer. Math. Soc.* **32**, 99.
[1929] The minimum value of quadratic forms, and the closest packings of spheres, *Math. Ann.* **101**, 605–608.
[1934] The minimum value of positive quatratic forms in six, seven and eight variables, *Math. Z.* **39**, 1–15.

Bloh, E.L.
[1956] On the most dense arrangement of spherical segments on a hypersphere (in Russian), *Izv. Akad. Nauk SSSR Ser. Mat.* **20**, 707–712.

Bolle, U.
[1989] On the density of multiple packings and coverings of convex discs, *Studia Sci. Math. Hungar.* **24**, 119–126.

Böröczky, K.
[1971] Über die Newtonsche Zahl regulärer Vielecke, *Period. Math. Hungar.* **1**, 113–119.
[1974] Sphere packing in spaces of constant curvature I (in Hungarian), *Mat. Lapok* **25**, 265–306.
[1975] Research problem no. 12, *Period. Math. Hungar.* **6**, 109.
[1978] Packing of spheres in spaces of constant curvature, *Acta Math. Acad. Sci. Hungar.* **32**, 243–261.
[1983] The problem of Tammes for $n = 11$, *Studia Sci. Math. Hungar.* **18**, 165–171.
[1986] Closest packing and loosest covering of the space with balls, *Studia Sci. Math. Hungar.* **21**, 79–89.

Böröczky, K., I. Bárány, E. Makai Jr and J. Pach
[1986] Maximal volume enclosed by plates and proof of the chessboard conjecture, *Discrete Math.* **60**, 101–120.

Chvátal, V.
[1975] On a conjecture of Fejes Tóth, *Period. Math. Hungar.* **6**, 357–362.

Cohn, M.J.
[1976] Multiple lattice covering of space, *Proc. London Math. Soc. (3)* **32**, 117–132.

Coxeter, H.S.M.
[1963] An upper bound for the number of equal nonoverlapping spheres that can touch another of the same size, in: *Convexity*, Proc. Sympos. Pure Math., Vol. 7 (Amer. Math. Soc., Providence, RI) pp. 53–71.

Coxeter, H.S.M., L. Few and C.A. Rogers
[1959] Covering space with equal spheres, *Mathematika* **6**, 147–157.

Danzer, L.
[1986] Finite point-sets on S^3 with minimum distance as large as possible, *Discrete Math.* **60**, 3–66.

Davenport, H., and G. Hajós
[1951] Problem 35 (in Hungarian), *Mat. Lapok* **2**, 68.

Delone, B.N., and S.S. Ryškov
[1963] Solution of the problem of the least dense lattice covering of a 4-dimensional space by equal spheres (in Russian), *Dokl. Akad. Nauk SSSR* **152**, 523–524 [*Soviet Math. Dokl.* **4**, 1333–1334].

Delsarte, P.
[1972] Bounds for unrestricted codes by linear programming, *Philips Res. Rep.* **27**, 272–289.
[1973] An algebraic approach to the association schemes of coding theory, *Philips Res. Rep. Suppl.* **27**, 1–97.

Delsarte, P., J.M. Goethals and J.J. Seidel
[1977] Spherical codes and designs, *Geom. Dedicata* **6**, 363–388.

Dowker, C.H.
[1944] On minimum circumscribed polygons, *Bull. Amer. Math. Soc.* **50**, 120–122.

Dumir, V.C., and R.J. Hans-Gill
[1972a] Lattice double coverings in the plane, *Indian J. Pure Appl. Math.* **3**, 466–480.
[1972b] Lattice double packings in the plane, *Indian J. Pure Appl. Math.* **3**, 481–487.

Dumir, V.C., and D.S. Khassa
[1973a] Saturated systems of symmetric convex domains; results of Eggleston, Bambah and Woods, *Proc. Cambridge Philos. Soc.* **74**, 107–116.
[1973b] A conjecture of Fejes Tóth on saturated systems of circles, *Proc. Cambridge Philos. Soc.* **74**, 453–460.

Eberlein, P.
[1973] Geodesic flows on negatively curved manifolds II, *Trans. Amer. Math. Soc.* **178**, 57–82.

Eggleston, H.G.
[1965] A minimal density plane covering problem, *Mathematika* **12**, 226–234.

Elkies, N.D., A.M. Odlyzko and J.A. Rush
[1991] On the packing densities of superballs and other bodies, preprint.

Ennola, V.
[1961] On the lattice constant of a symmetric convex domain, *J. London Math. Soc.* **36**, 135–138.

Erdős, P., and C.A. Rogers
[1962] Covering space with convex bodies, *Acta Arith.* **7**, 281–285.

Fáry, I.
[1950] Sur la densité des reseaux de domaines convexes, *Bull. Soc. Math. France* **78**, 152–161.

Fejes Tóth, G.
[1973] Covering the plane by convex discs, *Acta Math. Acad. Sci. Hungar.* **23**, 263–270.
[1974] Solid sets of circles, *Studia Sci. Math. Hungar.* **9**, 101–109.
[1976] Multiple packing and covering of the plane with circles, *Acta Math. Acad. Sci. Hungar.* **27**, 135–140.
[1977] On the intersection of a convex disc and a polygon, *Acta Math. Acad. Sci. Hungar.* **29**, 149–153.
[1979] Multiple packing and covering of spheres, *Acta Math. Acad. Sci. Hungar.* **34**, 165–176.
[1980] On the section of a packing of equal balls, *Studia Sci. Math. Hungar.* **15**, 487–489.
[1981] Ten-neighbour packing of equal balls, *Period. Math. Hungar.* **12**, 125–127.
[1983] Multiple lattice packings of symmetric convex domains in the plane, *J. London Math. Soc.(2)* **29**, 556–561.
[1987a] Finite coverings by translates of centrally symmetric convex domains, *Discrete Comput. Geom.* **2**, 353–363.
[1987b] Totally separable packing and covering with circles, *Studia Sci. Math. Hungar.* **22**, 65–73.
[1988] Note to a paper of Bambah, Rogers and Zassenhaus, *Acta Arith.* **50**, 119–122.

Fejes Tóth, G., and L. Fejes Tóth
[1973] On totally separable domains, *Acta Math. Acad. Sci. Hungar.* **24**, 229–232.
[1983] Packing the plane with Minkowskian sums of convex sets, *Studia Sci. Math. Hungar.* **18**, 461–464.
[1989] A geometrical analogue of the phase transformation of crystals, *Comput. Math. Appl.* **17** 251–254.

Fejes Tóth, G., and A. Florian
[1984] Covering of the plane by discs, *Geom. Dedicata* **16**, 315–333.

Fejes Tóth, L.
[1950] Some packing and covering theorems, *Acta Sci. Math. Szeged* **12/A**, 62–67.
[1957] Filling of a domain by isoperimetric discs, *Publ. Math. Debrecen* **5**, 119–127.
[1967a] On the number of equal discs that can touch another of the same kind, *Studi Sci. Math. Hungar.* **2**, 363–367.
[1967b] On the arrangement of houses in a housing estate, *Studia Sci. Math. Hungar.* **2**, 37–42.
[1968] Solid circle-packings and circle-coverings, *Studia Sci. Math. Hungar.* **3**, 401–409.
[1969a] Remarks on a theorem of R.M. Robinson, *Studia Sci. Math. Hungar.* **4**, 441–445.
[1969b] Über die Nachbarschaft eines Kreises in einer Kreispackung, *Studia Sci. Math. Hungar.* **4**, 93–97.
[1970] Über eine affininvariante Masszahl bei Eipolyedern, *Studia Sci. Math. Hungar.* **5**, 173–180.
[1971] Densest packing of lenses on the plane (in Hungarian), *Mat. Lapok* **22**, 209–213.
[1973a] Five-neighbour packing of convex discs, *Period. Math. Hungar.* **4**, 221–229.
[1973b] On the density of a connected lattice of convex bodies, *Acta Math. Acad. Sci. Hungar.* **24**, 373–376.
[1975a] Research Problem no. 14, *Period. Math. Hungar.* **6**, 277–278.
[1975b] On the Hadwiger numbers and Newton numbers of a convex body, *Studia Sci. Math. Hungar.* **10**, 111–115.
[1976] Close packing and loose covering with balls, *Publ. Math. Debrecen* **23**, 323–326.
[1977] Research problem no. 21, *Period. Math. Hungar.* **8**, 103–104.
[1978] Remarks on the closest packing of convex discs, *Comment. Math. Helv.* **53**, 536–541.
[1980] Approximation of convex domains by polygons, *Studia Sci. Math. Hungar.* **15**, 133–138.
[1982] Packing of r-convex discs, *Studia Sci. Math. Hungar.* **17**, 449–452.
[1983a] On the densest packing of convex discs, *Mathematika* **30**, 1–3.
[1983b] Packing and covering with r-convex discs, *Studia Sci. Math. Hungar.* **18**, 69–75.
[1984] Compact packing of circles, *Studia Sci. Math. Hungar.* **19**, 103–107.
[1985] Densest packing of translates of a domain, *Acta Math. Hungar.* **45**, 437–440.

[1986] Densest packing of translates of the union of two circles, *Discrete Comput. Geom.* **1**, 307–314.
[1989] Research problem no. 44, *Period. Math. Hungar.* **20**, 89–91.

Fejes Tóth, L., and A. Florian
[1982] Packing and covering with convex discs, *Mathematika* **29**, 181–193.

Fejes Tóth, L., and A. Heppes
[1967] A variant of the problem of the thirteen spheres, *Canad. J. Math.* **19**, 1092–1100.
[1980] Multi-saturated packings of circles, *Studia Sci. Math. Hungar.* **15**, 303–307.

Few, L.
[1964] Multiple packing of spheres, *J. London Math. Soc.* **39**, 51–54.
[1968] Double packing of spheres: a new upper bound, *Mathematika* **15**, 88–92.

Florian, A.
[1978] Mehrfache Packung konvexer Körper, *Österreich. Akad. Wiss. Math.- Natur. Kl. Sitzungsber. II* **186**, 238–247.
[1985] On compact packing of circles, *Studia Sci. Math. Hungar.* **20**, 473–480.
[1986] Packing and covering with convex discs, *Studia Sci. Math. Hungar.* **21**, 251–265.

Folkman, J.H., and R.L. Graham
[1969] A packing inequality for compact convex subsets of the plane, *Canad. Math. Bull.* **12**, 745–752.

Gauss, C.F.
[1831] Untersuchungen über die Eigenschaften der positiven ternären quadratischen Formen von Ludwig August Seber, *Gőttingische gelehrte Anzeigen, Juli 9* [*J. Reine Angew. Math.* **20** (1840) 312–320; *Werke,* Vol. 2 (Königliche Gesellschaft der Wissenschaften, Göttingen, 1876) pp. 188–196].

Goldberg, M.
[1967] Viruses and a mathematical problem, *J. Mol. Biol.* **24**, 337–338.

Graham, R.L., H.S. Witsenhausen and H.J. Zassenhaus
[1972] On tightest packings in the Minkowski plane, *Pacific J. Math.* **41**, 699–715.

Gritzmann, P.
[1985] Lattice covering of space with symmetric convex bodies, *Mathematika* **32**, 311–315.

Groemer, H.
[1961a] Lagerungs- und Überdeckungseigenschaften konvexer Bereiche mit gegebener Krümmung, *Math. Z.* **76**, 217–225.
[1961b] Abschätzungen für die Anzahl der konvexen Körper, die einen konvexen Körper berühren, *Monatsh. Math.* **65**, 74–81.
[1963] Existenzsätze für Lagerungen im Euklidischen Raum, *Math. Z.* **81**, 260–278.
[1966a] Zusammenhängende Lagerungen konvexer Körper, *Math. Z.* **94**, 66–78.
[1966b] Über ebene Schnitte von Lagerungen, *Monatsh. Math.* **70**, 213–222.
[1968] Existenzsätze für Lagerungen in metrischen Räumen, *Monatsh. Math.* **72**, 325–334.
[1986a] On linear sections of lattice packings, *Monatsh. Math.* **102**, 199–216.
[1986b] Some basic properties of packing and covering constants, *Discrete Comput. Geom.* **1**, 183–193.

Grünbaum, B.
[1961] On a conjecture of H. Hadwiger, *Pacific J. Math.* **11**, 215–219.

Hadwiger, H.
[1955] Volumenabschätzungen für die einen Eikörper überdeckenden und unterdeckenden Parallelotope, *Elem. Math.* **10**, 122–124.
[1957] Über Treffenzahlen bei translationsgleichen Eikörpern, *Arch. Math.* **8**, 212–213.

Hárs, L.
[1986] The Tammes problem for $n = 10$, *Studia Sci. Math. Hungar.* **21**, 439–451.

Heppes, A.
[1960] Ein Satz über gitterförmige Kugelpackungen, *Ann. Univ. Sci. Budapest Eötvös, Sect. Math.* **3–4**, 89–90.
[1963] Filling of a domain by discs, *Magyar Tud. Akad. Mat. Kutató Int. Közl.* **8**, 363–371.
[1990] On the packing density of translates of a domain. *Studia Sci. Math. Hungar.* **25**, 117–120.
[1992] Solid circle-packings in the Euclidean plane, *Discrete Comput. Geom.* **1**, 29–43.

Hlawka, E.
[1944] Zur Geometrie der Zahlen, *Math. Z.* **49**, 285–312.

Hoppe, R.
[1874] Bemerkung der Redaktion, *Arch. Math. Phys. (Grunert)* **56**, 307–312.

Hortobágyi, I.
[1972] The Newton number of convex plane regions (in Hungarian), *Mat. Lapok* **23**, 313–317.
[1975] Über die auf Scheibenklassen bezügliche Newtonsche Zahl der konvexen Scheiben, *Ann. Univ. Sci. Budapest Eötvös Sect. Math.* **18**, 123–127.

Horváth, J.
[1980] Close lattice packing of unit balls in the space E^n (in Russian), in: *Geometry of Positive Quadratic Forms,* Trudy Mat. Inst. Steklov., Vol. 152, pp. 216–231 [*Proc. Steklov. Math. Inst.* **152** (1982) 237–254].
[1986a] Some problems in discrete geometry in higher-dimensional spaces (in Russian), Doctoral dissertion, Steklov Institute, Moscow.
[1986b] Über die Enge von Zylinderpackungen und die Lockerheit von Zylinderüberdeckungen im n-dimensionalen Euklidischen Raum, *Studia Sci. Math. Hungar.* **21**, 219–225.

Kabatjanskiĭ, G.A., and V.I. Levenštein
[1978] Bounds for packings on a sphere and in space (in Russian), *Problemy Peredachi Informatsii* **14**, 3–25 [*Problems Inform. Transmission* **14**, 1–17].

Kershner, R.
[1939] The number of circles covering a set, *Amer. J. Math.* **61**, 665–671.

Kertész, G.
[1988] On totally separable packings of equal balls, *Acta Math. Hungar.* **51**, 363–364.

Korkin, A., and G. Zolotarev
[1872] Sur les formes quadratiques positives quaternaires, *Math. Ann.* **5**, 581–583.
[1877] Sur les formes quadratiques positives, *Math. Ann.* **11**, 242–292.

Kuperberg, G., and W. Kuperberg
[1990] Double-lattice packings of convex bodies in the plane, *Discrete Comput. Geom.* **5**, 389–397.

Kuperberg, G., and C. McMullen
[1991] Private communication.

Kuperberg, K., and W. Kuperberg
[1993] Translates of a starlike plane region with a common point, in: *Intuitive Geometry, Szeged 1991,* Colloq. Math. Soc. János Bolyai, Vol. 63 (North-Holland, Amsterdam) to appear.

Kuperberg, W.
[1987a] On packing the plane with congruent copies of a convex body, in: *Intuitive Geometry, Siófok, 1985,* Colloq. Math. Soc. János Bolyai, Vol. 48 (North-Holland, Amsterdam) pp. 317–329.
[1987b] An inequality linking packing and covering densities of plane convex bodies, *Geom. Dedicata* **23**, 59–66.
[1989] Covering the plane with congruent copies of a convex body, *Bull. London Math. Soc.* **21**, 82–86.

Lagrange, J.L.
[1773] Recherches d'arithmetique, *Nouv. Mem. Acad. Roy. Sci. Belles Lettres Berlin*, 265–312 [*Oeuvres,* Vol. III (Gauthier–Villars, Paris, 1869) pp. 693–758].

Leech, J.
[1956] The problem of the thirteen spheres, *Math. Gazette* **40**, 22–23.
[1964] Some sphere packings in higher space, *Canad. J. Math.* **16**, 657–682.

Leech, J., and N.J.A. Sloane
[1971] Sphere packings and error-correcting codes, *Canad. J. Math.* **23**, 718–745.

Levenštein, V.I.
[1975] Maximal packing density of equal spheres in n-dimensional Euclidean space (in Russian), *Mat. Zametki* **18**, 301–311 [*Math. Notes Acad. Sci. USSR* **18**, 765–771].
[1979] On bounds for packings in n-dimensional Euclidean space (in Russian), *Dokl. Akad. Nauk. SSSR* **245**, 1299–1303 [*Soviet Math. Dokl.* **20**, 417–421].

Lindsey, J.H.
[1986] Sphere packing in R^3, *Mathematika* **33**, 137–147.

Linhart, J.
[1973] Die Newtonsche Zahl von regelmässigen Fünfecken, *Period. Math. Hungar.* **4**, 315–328.
[1977] Scheibenpackungen mit nach unten beschränkter Nachbarnzahlen, *Studia Sci. Math. Hungar.* **12**, 281–293.
[1978] Closest packings and closest coverings by translates of a convex disc, *Studia Sci. Math. Hungar.* **13**, 157–162.
[1983] Eine Methode zur Berechnung der Dichte einer dichtesten gitterförmigen k-fachen Kreispackung, Arbeitsber. Math. Inst. Univ. Salzburg.

Lloyd, S.P.
[1980] Hamming association schemes and codes on spheres, *SIAM J. Math. Anal.* **11**, 488–505.

Macbeath, A.M.
[1951] A compactness theorem for affine equivalence-classes of convex regions, *Canad. J. Math.* **3**, 54–61.

Mahler, K.
[1947] On the minimum determinant and the circumscribed hexagons of a convex domain, *Nederl. Akad. Wetensch. Proc. Ser. A* **50**, 692–703.

Makai Jr, E.
[1978] On the thinnest non-separable lattice of convex bodies, *Studia Sci. Math. Hungar.* **13**, 19–27.
[1987] Five-neighbour packing of convex plates, in: *Intuitive Geometry, Siófok, 1985,* Colloq. Math. Soc. János Bolyai, Vol. 48 (North-Holland, Amsterdam) pp. 373–381.

Mani-Levitska, P., and J. Pach
[1990] Decomposition problems for multiple coverings with unit balls, preprint.

McEliece, R.J., E.R. Rodemich, H. Rumsey Jr and L.R. Welch
[1977] New upper bounds on the rate of a code via the Delsarte–MacWilliams inequalities, *IEEE Trans. Inform. Theory* **23**, 157–166.

Melnyk, T.W., O. Knop and W.R. Smith
[1977] Extremal arrangements of points and unit charges on a sphere: equilibrium configurations revisited, *Canad. J. Chem.* **55**, 1745–1761.

Minkowski, H.
[1893] Extrait d'une lettre adressée à M. Hermite, *Bull. Sci. Math.* **17**(2), 24–29 [*Ges. Abh.*, Vol. 1 (Teubner, Leipzig, 1911) pp. 266–270].
[1904] Dichteste gitterförmige Lagerung kongruenter Körper, *Nachr. Ges. Wiss. Göttingen*, 311–355 [*Ges. Abh.*, Vol. 2 (Teubner, Leipzig, 1911) pp. 3–42].
[1905] Diskontinuitätsbereich für arithmetische Äquivalenz, *J. Reine Angew. Math.* **129**, 220–274 [*Ges. Abh.*, Vol. 2 (Teubner, Leipzig, 1911) pp. 53–100].

Molnár, J.
[1975] On a generalisation of the Tammes problem, *Publ. Math. Debrecen* **22**, 109–114.

Mount, D.M., and R. Silverman
[1990] Packing and covering the plane with translates of a convex polygon, *J. Algorithms* **11**, 564–580.

Muder, D.J.
[1988] Putting the best face on a Voronoi polyhedron, *Proc. London Math. Soc. (3)* **56**, 329–348.

Odlyzko, A.M., and N.J.A. Sloane
[1979] New bounds on the number of unit spheres that can touch a unit sphere in n dimensions, *J. Combin. Theory Ser. A* **26**, 210–214.

Oler, N.
[1961] An inequality in the geometry of numbLrs, *Acta Math.* **105**, 19–48.

Pach, J.
[1980] Decomposition of multiple packing and covering, in: *Diskrete Geometrie,* Vol. 2., Kolloq. Inst. Math. Univ. Salzburg, pp. 169–178.
[1986] Covering the plane with convex polygons, *Discrete Comput. Geom.* **1**, 73–81.

Rankin, R.A.
[1955] The closest packing of spherical caps in n dimensions, *Proc. Glasgow Math. Assoc.* **2**, 139–144.

Reinhardt, K.
[1934] Über die dichteste gitterförmige Lagerung kongruenter Bereiche in der Ebene und eine besondere Art konvexer Kurven, *Abh. Math. Sem. Hansischer Univ.* **10**, 216–230.

Rogers, C.A.
[1951] The closest packing of convex two-dimensional domains, *Acta Math.* **86**, 309–321.
[1957] A note on coverings, *Mathematika* **4**, 1–6.
[1958] The packing of equal spheres, *Proc. London Math. Soc. (3)* **8**, 609–620.
[1959] Lattice coverings of space, *Mathematika* **6**, 33–39.

Rogers, C.A., and G.C. Shephard
[1957] The difference body of a convex body, *Arch. Math.* **8**, 220–233.

Rush, J.A.
[1989] A lower bound on packing density, *Invent. Math.* **98**, 499–509.

Rush, J.A., and N.J.A. Sloane
[1987] An improvement to the Minkowski–Hlawka bound for packing superballs, *Mathematika* **34**, 8–18.

Ryškov, S.S.
[1974] Density of an (r, R)-system (in Russian), *Mat. Zametki* **16**, 447–454 [*Math. Notes* **16**, 855–858].

Ryškov, S.S., and E.P. Baranovskiĭ
[1975] Solution of the problem of the least dense lattice covering of five-dimensional space by equal spheres (in Russian), *Dokl. Akad. Nauk. SSSR* **222**, 39–42 [*Soviet Math. Dokl.* **16**, 586–590].
[1976] S-types of n-dimensional lattices and five-dimensional primitive parallelohedra (with application to covering theory) (in Russian), *Trudy Mat. Inst. Steklov.* **137** [*Proc. Steklov. Inst. Math.* **127**].

Sachs, H.
[1986] No more than nine unit balls can touch a closed unit hemisphere, *Studia Sci. Math. Hungar.* **21**, 203–206.

Sas, E.
[1939] Über eine Extremumeigenschaft der Ellipsen, *Compositio Math.* **6**, 468–470.

Schmidt, W.M.
[1961] Zur Lagerung kongruenter Körper im Raum, *Monatsh. Math.* **65**, 154–158.
[1963] On the Minkowski–Hlawka theorem, *Illinois J. Math.* **7**, 18–23.

Schmitt, P.
[1991] Discs with special properties of densest packings, *Discrete Comput. Geom.* **6**, 181–190.

Seidel, J.J.
[1984] Harmonics and combinatorics, in: *Special functions: Group Theoretical Aspects and Applications* (Reidel, Dordrecht) pp. 287–303.

Shannon, C.E.
[1959] Probability of error for optimal codes in a Gaussian channel, *Bell System Tech. J.* **38**, 611–656.

Sidelnikov, V.M.
[1973] On the densest packing of balls on the surface of the n-dimensional Euclidean sphere, and the number of vectors of a binary code with prescribed code distances (in Russian), *Akad. Nauk SSSR* **231**, 1029–1032 [*Soviet Math. Dokl.* **14**, 1851–1855].
[1974] New estimates for the closest packing of spheres in n-dimensional Euclidean space (in Russian), *Mat. Sb.* **95**, 148–158 [*USSR Sb.* **24**, 147–157].

Smith, M.J.
[1975] Packing translates of a compact set in Euclidean space, *Bull. London Math. Soc.* **7**, 129–131.

Swinnerton-Dyer, H.P.F.
[1953] Extremal lattices of convex bodies, *Proc. Cambridge Philos. Soc.* **49**, 161–162.

Tammela, P.
[1970] An estimate of the critical determinant of a two-dimensional convex symmetric domain (in Russian), *Izv. Vyš. Učebn. Zaved. Mat.* **12**(103), 103–107.

Tammes, P.M.L.
[1930] On the origin of number and arrangement of the places of exit on the surface of pollen grains, *Rec. Trav. Bot. Neerl.* **27**, 1–84.

Tarnai, T.
[1984] Spherical circle-packing in nature, practice and theory, *Structural Topology* **9**, 39–58.

Tarnai, T., and Zs. Gáspár
[1983] Improved packing of equal circles on a sphere and rigidity of its graph, *Math. Proc. Cambridge Philos. Soc.* **93**, 191–218.

Temesvári, A.H.
[1988] Eine Methode zur Bestimmung der dünnsten gitterförmigen k-fachen Kreisüberdeckungen, *Studia Sci. Math. Hungar.* **23**, 23–35.

Temesvári, A.H., J. Horváth and N.N. Yakovlev
[1987] A method for finding the densest lattice k-fold packing of circles (in Russian), *Mat. Zametki* **41**, 625–636, 764 [*Math. Notes* **41**, 349–355].

Thue, A.
[1910] Über die dichteste Zusammenstellung von kongruenter Kreisen in der Ebene, *Norske Vid. Selsk. Skr.* **1**, 1–9.

Van der Waerden, B.L.
[1961] Pollenkörner, Punktverteilungen auf der Kugel und Informationstheorie, *Naturwiss.* **48**, 189–192.

Wegner, G.
[1992] Relative Newton numbers, *Monatsh. Math.*, to appear.

Wyner, J.M.
[1965] Capabilities of bounded discrepancy decoding, *Bell System Tech. J.* **44**, 1061–1122.

CHAPTER 3.4

Finite Packing and Covering

Peter GRITZMANN*

Universität Trier, Fb IV, Mathematik, Postfach 3825, D-54228, Trier, Germany

Jörg M. WILLS

Fachbereich Mathematik, Universität Siegen, Hölderlinstrasse 3, D-57068 Siegen, Germany

Contents

* Research of the first author supported in part by the Deutsche Forschungsgemeinschaft.

HANDBOOK OF CONVEX GEOMETRY
Edited by P.M. Gruber and J.M. Wills

1. Preliminaries

1.1. Introduction

The problem of packing finitely many objects into some container is an every day task in real world. Usually there is an underlying concept of efficiency of a packing and hence one is in effect confronted with an optimization problem. The "dual" problem of covering a given set with finitely many objects is also a basic task in real world applications. In many cases this is less apparent since the objects involved are not "physical". Examples are a covering of jobs by workers or a covering of streets by mailmen.

In mathematics, (infinite) packing and covering problems have a long history, particularly since lattice packings and lattice coverings are closely related to problems in number theory and crystallography. More recently, there has been growing interest in the theory of *finite* packing and covering.

The present paper gives a survey of these more recent developments concerning finite packing and covering problems with special emphasis on problems involving convex bodies. Two basic questions considered here are *sausage-problems* which came up in the context of discrete geometry and *bin packing* problems which have been studied extensively from the point of view of mathematical programming. Sausage packing and bin packing problems have in common that convex bodies are packed into a convex body of smallest possible volume among all bodies of a given family. Sausage covering problems fall into a corresponding category. We will deal with these problems from a theoretical but also from an algorithmic point of view and we will also mention some applications.

1.2. General packing and covering problems

When speaking of finite packing and covering we are in effect dealing with an instance of the following problem. Given a ground set X, let $\mathcal{M}$ be a subset of 2^X and let Ξ denote a family of finite subsets of 2^X. For $M \in \mathcal{M}$ and $\mathcal{F} \in \Xi$, $\mathcal{F}$ is called a *finite packing* into M if

$$F \in \mathcal{F} \implies F \subset M,$$
$$F_1, F_2 \in \mathcal{F} \implies F_1 \cap F_2 = \emptyset.$$

$\mathcal{F}$ is called a *finite covering* of M if

$$M \subset \bigcup_{F \in \mathcal{F}} F.$$

Now, a *finite packing problem* consists of a set $\mathcal{M} \subset 2^X$, a family Ξ of finite subsets of 2^X and a *density function*

$$\pi : \Xi \times \mathcal{M} \to [0,1].$$

The task is to solve the maximization problem

$$\begin{aligned}&\max \pi(\mathcal{F}, M)\\&\mathcal{F} \in \Xi, \quad M \in \mathcal{M},\\&\mathcal{F} \text{ is a finite packing into } M.\end{aligned} \tag{1.1}$$

A *finite covering problem* consists again of a set $\mathcal{M} \subset 2^X$, a family Ξ of finite subsets of 2^X and a *density function*

$$\gamma : \Xi \times \mathcal{M} \to [1, \infty[,$$

and the task is to solve the minimization problem

$$\begin{aligned}&\min \gamma(\mathcal{F}, M)\\&\mathcal{F} \in \Xi, \quad M \in \mathcal{M},\\&\mathcal{F} \text{ is a finite covering of } M.\end{aligned} \tag{1.2}$$

For each pair $(\mathcal{F}, M)$ the values $\pi(\mathcal{F}, M)$ and $\gamma(\mathcal{F}, M)$ are called (packing or covering) *density* of $(\mathcal{F}, M)$, while the solutions of (1.1) and (1.2) are referred to as (packing or covering) *density* of $(\Xi, \mathcal{M})$.

Of course, this is a very general theme, far too general for our purposes. In particular, subjects like Tarski's plank problem, or Borsuk's and some Helly-type problems would fit into this framework, as well as the problem of packing flexible objects with joints, a special case of tasks that come up in *robotics*.

Hence, we will fill the definitions with life by restricting $\mathcal{M}$, Ξ and the density functions to instances that are relevant in our context. Here we are particularly interested in $\mathcal{M}$ being a family of convex bodies (compact convex subset of $\mathbb{E}^d$ with nonempty interior) in some Euclidean space $\mathbb{E}^d$, Ξ being a family of finite subsets of the set $\mathcal{K}^d$ of convex bodies and π, γ measuring a volume relation. In a slight (but reasonable) abuse of notation we will speak of packings of convex bodies even if we are really dealing with packings of their interiors.

1.3. The scope of this survey

As pointed out before we will restrict our considerations mainly to geometric finite packing and covering problems that satisfy the following informal criteria:

(i) their "natural formulation" involves only packings of (coverings with) translates of (at least two) convex bodies K_i into (of) convex bodies M of a given family $\mathcal{M} \subset \mathcal{K}^d$, and, if a density function is involved, in addition:

(ii) the density function is the volume ratio $\sum_i V(K_i)/V(M)$,

where, for a Lebesgue measurable set S, $V(S)$ denotes the *Lebesgue measure* of S. Most of the problems that belong to this category fall into the following two classes:

Let $k \in \mathbb{N}$, $K_1, \ldots, K_k \in \mathcal{K}^d$, $\mathcal{M} \subset \mathcal{K}^d$ and let Ξ be the family of subsets $\{b_1 + K_1, \ldots, b_k + K_k\}$ of $2^{\mathbb{R}^d}$; (1.3)

Let $\mathcal{K}$ be a finite subfamily of $\mathcal{K}^d$, $m \in \mathbb{N}$, $M_1, \ldots, M_m \in \mathcal{K}^d$ pairwise disjoint, $M = M_1 \cup \cdots \cup M_m$ and let Ξ denote the family of all subsets of $2^{\mathbb{R}^d}$ of the form $\{b + K\colon b \in \mathbb{R}^d \wedge K \in \mathscr{C}\}$, where $\mathscr{C} \subset \mathcal{K}$. (1.4)

In the first case, for packings, we want to minimize the volume of the sets $M \in \mathcal{M}$ into which translates of $K_1, \ldots, K_k$ can be packed. For $\mathcal{M} = \mathcal{K}^d$ we get the *sausage packing problem* from discrete geometry, for $\mathcal{M}$ being the set of all boxes $[0, \alpha_1] \times \cdots \times [0, \alpha_d]$ we obtain the general *bin packing* problem from combinatorial optimization. We will deal with the former topic in section 2 and with the latter problem in section 3.

Observe that problems like the classical "dictator problem" of determining the smallest radius of a sphere that contains a packing of k spherical caps of given radius can be formulated in this framework by allowing the sets M to be (the closure of) set-theoretic differences of convex bodies. In fact, with $\mathbb{B}^d$ denoting the Euclidean unit ball, the dictator problem is modeled by $K_1 = \cdots = K_k = \mathbb{B}^d$ and $\mathcal{M} = \{(\lambda + 2)\mathbb{B}^d \setminus \mathrm{int}(\lambda\mathbb{B}^d)\colon \lambda \in]0, \infty[\}$.

Similar kinds of examples can be stated for coverings.

In the second class (1.4) of problems containers $M_1, \ldots, M_m$ are given and, for packings, the task is to pack items of a given list into the containers in such a way that the "wasted space" is as small as possible. Examples are the classical problem of determining the maximal number of unit disks that can be packed into a square but also some variants of bin packing problems.

The present survey focuses mainly on sausage and bin packing problems; see, however, section 4. For surveys on other kinds of packing and covering problems we refer to the books of L. Fejes Tóth (1965, 1972b) and to the survey articles by Coxeter (1954), Saaty and Alexander (1975) and G. Fejes Tóth (1983).

Naturally, we try to limit the overlap with other chapters of this handbook as much as possible. However, there are aspects of finite packing and covering problems treated in other chapters in this Handbook. These chapters are: chapter 3.3. Packing and Covering with Convex Sets, by G. Fejes Tóth and Kuperberg, chapter 3.7. Geometric Crystallography, by Engel, chapter 3.1. Geometry of Numbers, by Gruber, and chapter 3.2. Lattice Points, by Gritzmann and Wills. Further there are far too many ramifications of 2-dimensional results to be mentioned here in detail; we restrict our attention to those which we deemed most closely related to the purpose of this article. Moreover, we do not include *m-fold packings and coverings*. For some results in this area in the context of sausage problems see G. Fejes Tóth, Gritzmann and Wills (1990); a survey on other results involving m-fold packings and coverings is contained in chapter 3.3.

2. Sausage problems

Let $K \in \mathcal{K}^d$ and $k \in \mathbb{N}$. Suppose we want to pack k translates of the convex body K into a container $M \in \mathcal{K}^d$. The density of such a packing is defined as the ratio of the total volume of all translates of K and the volume of M. Now suppose we regard this ratio as a function of M and set $\nu(M) = kV(K)/V(M)$. Then the supremum of $\nu(M)$ taken over all convex bodies M that admit a packing of k translates of K is an upper bound for the densities of any such packing into any container $M \in \mathcal{K}^d$. Hence, determining $\sup \nu(M)$ is a fundamental task in the theory of finite packing of convex bodies. In asking for this supremum, we are in effect dealing with the following question:

> Given $k \in \mathbb{N}$ and $K \in \mathcal{K}^d$; how can we arrange k nonoverlapping translates of K so as to minimize the volume of their convex hull?

The corresponding question for coverings which, in turn, yields lower bounds for finite covering densities is as follows:

> Given $k \in \mathbb{N}$ and $K \in \mathcal{K}^d$; how can we arrange k translates of K so as to maximize the volume of a largest convex body covered by their union?

Clearly, here Ξ consists of all sets $\{b_1 + K, \ldots, b_k + K\}$, $\mathcal{M} = \mathcal{K}^d$ and the value of the density function for a packing (covering) $\{b_1 + K, \ldots, b_k + K\}$ into (of) M is given by the ratio $kV(K)/V(M)$.

These problems will be studied in the present section; the heading "sausage problems" was coined after the conjectured extremal configurations (2.3) and (2.4).

2.1. Densities

Formally, we are dealing with the following problems: Given $d, k \in \mathbb{N}$ and $K \in \mathcal{K}^d$; find vectors $b_1^*, \ldots, b_k^* \in \mathbb{R}^d$ and a convex body C^* with the following properties:

(i) Sausage packing problem:

$$\operatorname{int}(b_i^* + K) \cap \operatorname{int}(b_j^* + K) = \emptyset \quad \text{for all } i, j = 1, \ldots, n,\ \ i \neq j,$$

$$\bigcup_{i=1}^{n} (b_i + K) \subset C^*$$

and

$$\delta_k(K) = \frac{kV(K)}{V(C^*)}$$

$$= \max\left\{ \frac{kV(K)}{V(C)} : \exists b_1, \ldots, b_k \in \mathbb{R}^d, C \in \mathcal{K}^d : \bigcup_{i=1}^{k} (b_i + K) \subset C \right.$$

$$\left. \wedge\, (i, j = 1, \ldots, n \wedge i \neq j \Rightarrow \operatorname{int}(b_i + K) \cap \operatorname{int}(b_j + K) = \emptyset) \right\}.$$

(ii) Sausage covering problem:

$$C^* \subset \bigcup_{i=1}^{n} (b_i + K)$$

and

$$\vartheta_k(K) = \frac{kV(K)}{V(C^*)}$$
$$= \min \left\{ \frac{kV(K)}{V(C)} \colon \exists b_1, \ldots, b_k \in \mathbb{R}^d, C \in \mathcal{K}^d \colon C \subset \bigcup_{i=1}^{k} (b_i + K) \right\}.$$

Let us first remark that δ_k, ϑ_k are well defined since (as a consequence of Blaschke's selection theorem) maximal or minimal elements of the corresponding families do exist.

Let us further point out that for any k translates $K_1, \ldots, K_k$ of K, $\operatorname{conv}(K_1 \cup \cdots \cup K_k)$ is the unique minimal convex body that contains $K_1 \cup \cdots \cup K_k$. Maximal convex bodies which are contained in the union of k translates of K, however, have not yet been characterized. This "asymmetry" leads, in particular, to different algorithmic behavior (see section 2.4).

Obviously, both problems are trivial for $d = 1$ or $k = 1$. Hence we suppose in the following that $d, k \geqslant 2$.

As functionals of K the numbers $\delta_k(K), \vartheta_k(K)$ are nonnegative, invariant under translations, motions and dilations, continuous, and, of course, we have

$$\delta_k(K) \leqslant 1 \leqslant \vartheta_k(K).$$

Therefore $\delta_k(K)$ and $\vartheta_k(K)$ are called (*sausage*) *packing density* and (*sausage*) *covering density*, respectively.

2.2. Sphere packings and sphere coverings

Sausage packing and covering problems have first been introduced as an aid to deal with general infinite packing and covering problems. We will use the notation $\delta(K)$, $\vartheta(K)$ and $\delta_{\mathcal{L}}(K)$, $\vartheta_{\mathcal{L}}(K)$ for the "classical" (infinite) *packing* and *covering* and *lattice packing* and *lattice covering densities* for translates of a given convex body K (with respect to the whole space); for a detailed account of this theory see Conway and Sloane (1988), Gruber and Lekkerkerker (1987) and chapters 3.1 and 3.3 in this Handbook.

It is, of course, clear that for all $d \in \mathbb{N}$ and $K \in \mathcal{K}^d$

$$\delta(K) \leqslant \limsup_{k \to \infty} \delta_k(K) \quad \text{and} \quad \liminf_{k \to \infty} \vartheta_k(K) \leqslant \vartheta(K).$$

L. Fejes Tóth (1949) showed that for the unit ball in the Euclidean plane infinite packings and coverings really behave like the limit of finite packings and coverings: given $k \in \mathbb{N}$ with $k \geqslant 2$, then

$$\delta_k(\mathbb{B}^2) < \delta(\mathbb{B}^2) = \frac{\pi}{2\sqrt{3}} \quad \text{and} \quad \frac{2\pi}{3\sqrt{3}} = \vartheta(\mathbb{B}^2) < \vartheta_k(\mathbb{B}^2). \tag{2.1}$$

For some explicit bounds for the slack between $\delta(\mathbb{B}^2)$ and $\delta_k(\mathbb{B}^2)$ and between $\vartheta(\mathbb{B}^2)$ and $\vartheta_k(\mathbb{B}^2)$ see Groemer (1960) (cf. also Wegner 1984, 1986), Gritzmann and Wills (1985, 1986) and G. Fejes Tóth (1987).

Wills (1987) conjectured that a result analogous to (2.1) holds for $d = 3, 4$ when $k \geqslant 4$. Anyway, it turns out that L. Fejes Tóth's result (2.1) does not extend to any dimension $d \geqslant 5$; in fact, linear arrangements seem to be best possible.

Before stating the corresponding conjectures and known results let us introduce some notation and give some computations to motivate these statements. Let $K \in \mathcal{K}^d$ and $k \in \mathbb{N}$. The set $\{b_1 + K, \ldots, b_k + K\}$ of translates of K is called a *sausage arrangement* (or short a *sausage*) if $\dim(\operatorname{conv}\{b_1, \ldots, b_k\}) = 1$, i.e., if all b_i are collinear. Further, let

$$\Delta_k(K) = \max\left\{\frac{kV(K)}{V(C)} : \exists b_1, \ldots, b_k \in \mathbb{R}^d, C \in \mathcal{K}^d \colon \{b_1 + K, \ldots, b_k + K\} \text{ is a sausage and a packing into } C\right\};$$

$$\Theta_k(K) = \min\left\{\frac{kV(K)}{V(C)} : \exists b_1, \ldots, b_k \in \mathbb{R}^d, C \in \mathcal{K}^d \colon \{b_1 + K, \ldots, b_k + K\} \text{ is a sausage that covers } C\right\}.$$

The numbers $\Delta_k(K)$, $\Theta_k(K)$ are called *sausage arrangement densities*.

The name "sausage" comes from the fact that the optimal bodies C^* for sausage arrangements resemble d-dimensional sausages.

It is not trivial to compute the sausage densities for an arbitrary convex body K (see section 2.4). However, for sausage packings of $\mathbb{B}^d$ this is easy. In fact, using the abbreviation $\kappa_d = V(\mathbb{B}^d)$ for the volume of the Euclidean unit sphere $\mathbb{B}^d$ we have

$$\sqrt{\frac{\pi}{2}}\sqrt{\frac{1}{d+1}} < \Delta_k(\mathbb{B}^d) = \frac{k\kappa_d}{\kappa_d + 2(k-1)\kappa_{d-1}} < \frac{k}{k-1}\sqrt{\frac{\pi}{2}}\sqrt{\frac{1}{d+1}}.$$

In case of sausage coverings we readily obtain

$$1 < \Theta_k(\mathbb{B}^d) < \sqrt{\frac{\pi \mathrm{e}}{2}} \approx 2.066.$$

These bounds are (asymptotically) much better than the (asymptotic) estimates for $\delta(\mathbb{B}^d)$ and $\vartheta(\mathbb{B}^d)$:

$$\begin{aligned} &\frac{d\zeta(d)}{\mathrm{e}(1-\mathrm{e}^{-d})2^{d-1}} \leqslant \delta(\mathbb{B}^d) \leqslant \frac{1}{2^{0.5990d+o(d)}}, \\ &\mathrm{e}^{-3/2}\, d \sim \tau_d \leqslant \vartheta(\mathbb{B}^d) \leqslant d\log_{\mathrm{e}} d + d\log_{\mathrm{e}}\log_{\mathrm{e}} d + 5d. \end{aligned} \tag{2.2}$$

Here τ_d denotes the ratio of the sum of the volumes of the intersection of $d+1$ unit balls centered at the vertices of a regular simplex of side $\sqrt{2(d+1)d^{-1}}$ to the volume of the simplex. [The lower bound for $\delta(\mathbb{B}^d)$ is an improvement obtained by Rogers (1947) of the Minkowski–Hlawka theorem, while the upper bound is due to Kabatjanskii and Levenštein (1978). The lower bound for $\vartheta(\mathbb{B}^d)$ was proved by Coxeter, Few and Rogers (1959) while the upper bound (which holds for $d \geqslant 3$ and arbitrary convex bodies) is due to Rogers (1957).]

In fact, in comparing the sausage densities with the known bounds for the densities $\delta(\mathbb{B}^d)$, $\vartheta(\mathbb{B}^d)$ of (infinite) sphere packings and coverings in $\mathbb{E}^d$ (as listed in Conway and Sloane 1988) we see that we have for all $k \in \mathbb{N}$ and for $d \geqslant 5$

$$\delta(\mathbb{B}^d) < \Delta_k(\mathbb{B}^d) \leqslant \delta_k(\mathbb{B}^d)$$

and

$$\vartheta_k(\mathbb{B}^d) \leqslant \Theta_k(\mathbb{B}^d) < \vartheta(\mathbb{B}^d).$$

This observation leads to the *sausage conjectures* for finite sphere packing and covering: for all $k \in \mathbb{N}$ and for $d \geqslant 5$,

$$\delta_k(\mathbb{B}^d) = \Delta_k(\mathbb{B}^d); \quad \text{(L. Fejes Tóth 1975)} \tag{2.3}$$

$$\vartheta_k(\mathbb{B}^d) = \Theta_k(\mathbb{B}^d). \quad \text{(Wills 1983)} \tag{2.4}$$

We will now give results showing the status of these conjectures; the intermediate dimensions 3 and 4 will be dealt with at the end of this section.

We will say for a subclass $\mathcal{B}_k$ of all families $\{b_1+\mathbb{B}^d, \ldots, b_k+\mathbb{B}^d\}$ of k translates of $\mathbb{B}^d$ that the sausage conjecture for finite sphere packings (finite sphere coverings) holds for $\mathcal{B}_k$ if for all $\{b_1+\mathbb{B}^d, \ldots, b_k+\mathbb{B}^d\} \in \mathcal{B}_k$ and for all $C \in \mathcal{K}^d$ such that $\{b_1+\mathbb{B}^d, \ldots, b_k+\mathbb{B}^d\}$ is a finite packing into C (a finite covering of C) we have

$$\frac{k\kappa_d}{V(C)} \leqslant \Delta_k(\mathbb{B}^d) \quad \left(\Theta_k(\mathbb{B}^d) \leqslant \frac{k\kappa_d}{V(C)}\right).$$

Further, for a given $\{b_1+\mathbb{B}^d, \ldots, b_k+\mathbb{B}^d\} \in \mathcal{B}_k$ we set $C_k = \operatorname{conv}\{b_1, \ldots, b_k\}$, the *center polytope* of the packing.

The sausage conjecture for finite sphere packings of the unit ball holds in the following cases:

$$\dim C_k \leqslant \tfrac{7}{12}(d-1) \tag{2.5}$$

(Betke, Gritzmann and Wills 1982)

$$\dim C_k \leqslant 9 \quad \text{and} \quad d \geqslant \dim C_k + 1 \tag{2.6}$$

(Betke and Gritzmann 1984)

C_k is a lattice zonotope

(Betke and Gritzmann 1986)

C_k is a regular simplex

(G. Fejes Tóth, Gritzmann and Wills 1989)

$$1 + r(C_k) > \sqrt{2}\left(1 + \frac{R(C_k)}{2}\right)\left(\frac{(d+2)\sqrt{d+1}}{\sqrt{2\pi}}\right)^{1/d} \tag{2.7}$$

(G. Fejes Tóth, Gritzmann and Wills 1989)

In (2.7), r, R denote the inradius and the circumradius, respectively. Observe that none of the above results places any restriction on the dimension d of the space we are working in. This is also true for the following result of Gritzmann (1986):

$$\sqrt{1 - \frac{1}{d+1}} < \sqrt{\frac{2}{\pi}}\sqrt{d}\,\delta_k(\mathbb{B}^d) < 2 + \sqrt{2} + \frac{2}{\sqrt{d-1}},$$

which in particular shows that the sausage-conjecture is correct up to a constant factor, essentially $2+\sqrt{2}$. Finally we mention two very recent results on the sausage conjecture for sphere packings: Böröczky and Henk (1992) show that (2.3) holds whenever C_k is not "too flat" and not "too fat", hence they partially fill the gap between (2.5), (2.6) and (2.7).

Betke, Henk and Wills (1993) develop a new approach of "variable radii" which leads to a proof of (2.3) for all sufficiently large dimensions.

For finite sphere coverings, Gritzmann and Wills (1985) show that

$$\frac{V(C)}{F(C)} \leqslant \left(1 - \frac{\tau}{\tau_d}\right)^{-1},$$

whenever $0 < \tau < \tau_d$, $C \in \mathcal{K}^d$ and $\{b_1 + \mathbb{B}^d, \ldots, b_k + \mathbb{B}^d\}$ is a covering of C with $k\kappa_d \leqslant \tau V(C)$. τ_d is again the constant that occurs in (2.2) and $F(C)$ denotes the surface area of C. Thus, for $\tau = \Theta_k(\mathbb{B}^d)$, by Daniels' asymptotic formula for τ_d (see Rogers' 1964), the densities of coverings of a convex body C which (essentially) satisfies the inequality

$$V(C) \geqslant \left(1 + \frac{1}{d}\sqrt{\frac{\pi}{2}}\,\mathrm{e}^2\right)F(C)$$

are greater than the density of the optimal sausage arrangement, i.e., such coverings cannot provide counterexamples to the sausage conjecture for finite sphere coverings.

For the densities G. Fejes Tóth, Gritzmann and Wills (1984) gave the following bounds:

$$1 + \frac{1}{2^{1.21d+6.04}} < \vartheta_k(\mathbb{B}^d) < \sqrt{\frac{\pi \mathrm{e}}{2}}.$$

Let us now deal with sphere packings in dimensions $d = 3$ and $d = 4$. It is these "intermediate" dimensions where the transition occurs between L. Fejes Tóth's (1949) result (2.1) which says that for growing k extremal packings of $\mathbb{B}^2$ spread out in all directions and the sausage conjecture supported by results that indicate that for $d \geqslant 5$ 1-dimensional arrangements are best possible.

The farthest reaching partial result for these intermediate dimensions follows from (2.5) and (2.6): for $d = 3, 4$ the density of packings of $\mathbb{B}^d$ with the convex hull of the centers of the balls being at most $(d-1)$-dimensional cannot exceed $\Delta_k(\mathbb{B}^d)$.

This result seems to indicate the following phenomenon in dimensions 3 and 4: for small k sausage arrangements are best possible, then, for a certain number k_d^p of balls the extremal configurations become full-dimensional without going through intermediate dimensional arrangements first. This phenomenon was first mentioned in Wills (1985) and referred to as *sausage catastrophe*. Computer-aided computations seem to indicate that $k_3^p \sim 56$ and $k_4^p \sim 75000$. For a detailed investigation of k_3^p see Gandini and Wills (1992) and Böröczky (1992).

For coverings with $\mathbb{B}^d$ the same phenomenon occurs in dimensions 2, 3, 4. Here, computer-aided experiments indicate $k_2^c \sim 10, k_3^c \sim 490$ and $k_4^c \sim 5 \cdot 10^5$.

2.3. *General convex bodies*

We begin again with results for problems in the plane.

Given $K \in \mathcal{K}^2$, $K = -K$, and $k \geqslant 2$, then

$$\begin{aligned}(\delta(K))^{-1} &\leqslant \left(\left(1 - \frac{1}{k}\right)\delta_k(K)\right)^{-1} - \frac{1}{k-1},\\ (\vartheta(K))^{-1} &\geqslant \left(\left(1 - \frac{1}{k}\right)\vartheta_k(K)\right)^{-1} - \frac{1}{k-1}.\end{aligned} \tag{2.8}$$

The first inequality is due to Rogers (1951) while the second is contained in Bambah and Rogers (1952). Both results are asymptotically tight. There are variants of (2.8) by Bambah, Rogers and Zassenhaus (1964) and Bambah and Woods (1971). Further, there is an improved version of (2.8) due to G. Fejes Tóth (1987) which implies that for a centrally symmetric $K \in \mathcal{K}^2$ and all $k \geqslant 26$

$$\vartheta(K) \leqslant \vartheta_k(K). \tag{2.9}$$

If K is not a parallelogram this inequality is strict. G. Fejes Tóth (1987) conjectured that (2.9) holds already for $k \geqslant 2$.

Let us now deal with arbitrary dimensions.

It is not too surprising that even in high dimensions for arbitrary convex bodies sausage arrangements are not always optimal. Take for example the Cartesian product of the regular hexagon and the ball $\mathbb{B}^{d-2}$. For sufficiently many translates a 2-dimensional arrangement is, then, obviously better than any 1-dimensional arrangement. Similar examples can be constructed with any convex body which tiles $\mathbb{R}^d$ by translation and whose finite packing and covering densities differ from 1. (For a survey on tilings see chapter 3.5 in this Handbook.) But in any case sausage arrangements yield rather surprising bounds. In fact, Gritzmann (1985a) showed

$$\begin{aligned} \frac{1}{d} < \frac{k}{d(k-1)+1} &\leqslant \delta_k(K), \\ \vartheta_k(K) \leqslant \left(1+\frac{1}{d-1}\right)^{d-1} &< \mathrm{e}, \end{aligned} \tag{2.10}$$

for arbitrary $K \in \mathcal{K}^d$.

2.4. Algorithmic aspects

There are numerous interesting algorithmic problems related to finite packing and covering. For problems in combinatorial optimization it is of course standard fare to deal with their computational complexity. The algorithmic theory of geometric packing and covering on the other hand is (to say the least) not highly developed yet. This is partly due to the fact that the occurring mathematical theory is not developed far enough. There are, however, many algorithmic problems which are relevant to geometric packing and covering whose computational complexity seems worthwhile studying in some detail. The purpose of this section is to give some first results and state some open problems. For the most part we follow the more extensive study Gritzmann (1993).

In the following we will make use of the standard notation of complexity theory. We use the binary Turing machine model, hence, the size of the input of a problem – usually denoted by L – is the number of binary digits needed to encode the input data. The complexity classes that are relevant here are the classes P, NP-hard, NP-complete, #P-hard and #P-complete. For underlying concepts of theoretical computer science, definitions and for numerous results see Aho, Hopcroft and Ullman (1974) and Garey and Johnson (1979).

To keep the exposition simple we restrict our considerations here mainly to polytopes. We end this subsection, however, with some remarks concerning an algorithmic theory of convex bodies.

Recall that from a theoretical point of view a convex polytope in $\mathbb{R}^d$ is equivalently a convex body that can be presented as the convex hull of finitely many points or a convex body that is given as the intersection of finitely many closed halfspaces. (For comprehensive treatments of combinatorial properties of convex polytopes see Grünbaum (1967) and chapter 2.3 in this Handbook.) For computational purposes it is, however, often relevant to distinguish between the two different ways polytopes may be given.

A $\mathscr{V}$*-presentation* of a polytope $P \subset \mathbb{R}^d$ consists of positive integers d and m and points $v_1, \ldots, v_m$ in $\mathbb{Q}^d$ such that $P = \operatorname{conv}\{v_1, \ldots, v_m\}$. An $\mathscr{H}$*-presentation* of P consists of positive integers d and m, a rational $m \times d$ matrix A, and a vector $b \in \mathbb{Q}^m$ such that $P = \{x \in \mathbb{R}^d \colon Ax \leqslant b\}$. Sometimes we use the shorter form $\mathscr{V}$- or $\mathscr{H}$-polytope when dealing with a polytope that is $\mathscr{V}$- or $\mathscr{H}$-presented, respectively.

Let us begin with the following fundamental task for finite packings of polytopes: the problem

> Given $d \in \mathbb{N}$ and ($\mathscr{V}$- or $\mathscr{H}$-) polytopes $P_1, \ldots, P_k$ in $\mathbb{R}^d$; determine whether $\{P_1, \ldots, P_k\}$ is a finite packing.

can be reduced to linear programming and can hence (Khachiyan 1979) be solved in polynomial time. Things become already more subtle when we deal with packings into a polytope:

> Given $d \in \mathbb{N}$ and ($\mathscr{V}$- or $\mathscr{H}$-) polytopes $P_1, \ldots, P_k, Q$ in $\mathbb{R}^d$; determine whether $\{P_1, \ldots, P_k\}$ is a finite packing into Q.

This problem can be solved in polynomial time in the cases that $P_1, \ldots, P_k, Q$ are all $\mathscr{V}$-presented or all $\mathscr{H}$-presented and when $P_1, \ldots, P_k$ are $\mathscr{V}$-polytopes while Q is an $\mathscr{H}$-polytope. The problem,

> Given $d \in \mathbb{N}$, an $\mathscr{H}$-polytope P and a $\mathscr{V}$-polytope Q in $\mathbb{R}^d$; determine whether $P \subset Q$,

however, is NP-complete. This result is due to Freund and Orlin (1985). It follows from Bodlaender et al. (1990) and Gritzmann and Klee (1993) that the NP-hardness persists even if P is a (regular) cube and Q is (the affine image of) a cross-polytope. The same kind of difficulties occur for finite coverings.

The following problem is closely related to the problem of determining the complexity of Minkowski-addition of polytopes studied by Gritzmann and Sturmfels (1993) since the convex hull of k translates $b_1 + P, \ldots, b_k + P$ of a polytope P is the Minkowski sum of P and the center polytope $\operatorname{conv}\{b_1, \ldots, b_k\}$.

> Given $d \in \mathbb{N}$ and ($\mathscr{V}$- or $\mathscr{H}$-) polytopes $P_1, \ldots, P_k$; find the (unique) irredundant ($\mathscr{V}$- or $\mathscr{H}$-) presentation of the smallest polytope P that contains $P_1, \ldots, P_k$.

This problem can be solved in polynomial time when all polytopes are $\mathscr{V}$-presented. In the other cases there is, however, no algorithm that runs in time which is bounded by a polynomial in the size of the input since the size of the output can already be exponential.

The corresponding problem for coverings, finding a "largest" polytope in $P_1 \cup \cdots \cup P_k$ seems even harder since it is not clear that a largest convex body that is contained in $P_1 \cup \cdots \cup P_k$ is, indeed, a polytope with sufficiently few vertices or facets.

For computing the density of a packing it would be useful to be able to compute the volume of a polytope. Dyer and Frieze (1988) (see also Khachiyan 1992) showed, however, that computing the volume of a ($\mathcal{V}$- or $\mathcal{H}$-) polytope is #P-hard. Hence the following problems are #P-hard:

> Given $d \in \mathbb{N}$ and ($\mathcal{V}$- or $\mathcal{H}$-) polytopes $P_1, \ldots, P_k$; compute the volume of $\operatorname{conv}\{P_1, \ldots, P_k\}$.

Lawrence (1991) even gave an example that showed that the volume of an $\mathcal{H}$-polytope may be a number of size that is exponential in the size of the presentation. For a recent survey on the problem of volume computation with special emphasis on the use of randomized algorithms see Dyer and Frieze (1991).

Even though the sausage packing density is a ratio of volumes Dyer and Frieze's (1988) #P-hardness result for volume computation does not directly imply that the problem of computing the densities of finite P-packings is hard. It is not even clear how hard it is to compute the sausage arrangement densities or even just to find a "best sausage direction":

> Given $d \in \mathbb{N}$ and a ($\mathcal{V}$- or $\mathcal{H}$-) polytope P; find collinear vectors $b_1, \ldots, b_k$ such that $\{b_1 + P, \ldots, b_k + P\}$ is a sausage packing arrangement of density $\Delta_k(P)$.

First, it is easy to see that the volume of the convex hull of a minimal sausage packing arrangement of k translates of a polytope P in direction $z \in \mathbb{S}^{d-1}$ is

$$V(P) + (k-1) \cdot \overline{V}_{d-1}(P, z) \cdot \underline{V}_1(P, z),$$

where $\overline{V}_{d-1}(P, z)$ is the *outer* $(d-1)$*-quermass* of P in direction z, i.e., the $(d-1)$-volume of the orthogonal projection of P on a hyperplane perpendicular to z, while $\underline{V}_1(P, z)$ is the *inner* 1*-quermass* of P in direction z, i.e., the length of a maximal chord of P in direction z. Hence, for computing $\Delta_k(P)$ we have to minimize $\overline{V}_{d-1}(P, z) \cdot \underline{V}_1(P, z)$ over $\mathbb{S}^{d-1}$. Let us point out that there are some lurking NP-hard problems that seem to be closely related. Clearly, $\min_{z \in \mathbb{S}^{d-1}} \underline{V}_1(P, z)$ is just the width of P. By Gritzmann and Klee (1993) even the problem of upper-bounding the width of ($\mathcal{V}$- or $\mathcal{H}$-presented) simplices is NP-complete. It has also been shown that the problem of maximizing the volume of orthogonal projections of a simplex on hyperplanes is NP-hard, Gritzmann and Klee (1992); the complexity of minimizing the volume of projections of arbitrary ($\mathcal{V}$- or $\mathcal{H}$-presented) polytopes has not been determined yet.

On the other hand, for simplices S

$$\overline{V}_{d-1}(S, z) \cdot \underline{V}_1(S, z) = dV(S);$$

this is actually a characterization of simplices, see Martini (1990). Thus, the sausage packing density for a simplex is

$$\frac{kV(S)}{V(S)+(k-1)dV(S)} = \frac{k}{d(k-1)+1},$$

and, hence, easy to compute (this is actually the lower bound given in (2.10)). The complexity of computing or approximating an optimal sausage direction for a given arbitrary polytope has not been determined, yet. Again, for coverings the problem seems even harder.

All the above problems can be posed for translates of a given polytope, for special polytopes, for simplices, for zonotopes, etc. Further, it is interesting to investigate also the case of fixed dimension. For fixed d many of the problems mentioned above become easier. In particular, from the point of view of polynomial time solvability versus NP-hardness the distinction between $\mathcal{V}$- and $\mathcal{H}$-polytopes becomes irrelevant since one presentation can be converted into the other in polynomial time. In addition, the volume of a polytope can be computed in polynomial time. We mention here only one problem in $\mathbb{R}^3$ that is particularly relevant for being able to produce further numerical evidence for a relation between packings of unit balls and the phenomenon of chemical whiskers (see 2.5.4).

> Given $k \in \mathbb{N}$ and a lattice $\mathbb{L}$ that packs $\mathbb{B}^3$ find $b_1, \ldots, b_k \in \mathbb{L}$ such that $\{b_1 + \mathbb{B}^3, \ldots, b_k + \mathbb{B}^3\}$ is a finite sphere packing of largest density among all such packings with translation vectors restricted to $\mathbb{L}$.

Even if we assume that we could evaluate the density of a given arrangement at unit cost, a simple enumeration procedure would require the checking of $k^{O(k)}$ arrangements and is hence not feasible for the relevant order of magnitude of k. Is there a substantially better algorithm?

Clearly, for a general algorithmic theory of packing and covering it is desirable to be able to deal with arbitrary convex bodies (see, e.g., the paragraph on chemical whiskers in 2.5.4.). For this reason we end this subsection with some short remarks concerning an algorithmic theory of convex bodies. For details see Lovász (1986) and Grötschel, Lovász and Schrijver (1988).

The underlying idea how to deal with convex bodies from an algorithmic point of view is to augment the Turing machine with a device, called *oracle* that answers certain questions about the bodies under consideration. A well-known example is the *weak membership oracle*, an algorithm which solves the following problem for $K \in \mathcal{K}^d$:

> Given $y \in \mathbb{Q}^d$ and a rational number $\varepsilon > 0$; assert that $y \in K + \varepsilon\mathbb{B}^d$ or that $y \notin K \ominus \varepsilon\mathbb{B}^d$.

Here, for $K, C \in \mathcal{K}^d$, the notation $K \ominus C$ is used to denote the set $\mathbb{R}^d \setminus ((\mathbb{R}^d \setminus K) + C)$.

Usually, it is important to have some a priori information, for instance a lower bound on the inradius of K, an upper bound on K's circumradius, etc. This data enters the size of the input of K and the question is whether a given problem can be solved in oracle-polynomial time, in time that is polynomial in the size of the input and in the number of calls to the oracle. Grötschel, Lovász and Schrijver (1988) study different oracles and their relationship and derive oracle-polynomial time algorithms for numerous problems, many of which rely on the *ellipsoid-algorithm*.

Some algorithmic results for (infinite) lattice packings and coverings can be found in Conway and Sloane (1988) and Mount and Silverman (1987).

2.5. Some applications

In the following we are going to outline some relations of the theory surveyed in sections 2.1–2.3 to other fields of mathematics; and we will also mention some possible connections with phenomena in chemistry, physics and engineering.

2.5.1. Lattice covering

Compared to the upper estimate $\vartheta_k(\mathbb{B}^d) < \sqrt{(\pi\,\mathrm{e})/2} \approx 2.066$ the general bound $\vartheta_k(K) < \mathrm{e} \approx 2.718$ has only increased by about 0.65. This shows that every convex body provides a rather efficient finite covering. The proof of this result is based on an improvement given in Gritzmann (1985a) of a theorem of Macbeath (1951) on the approximation of convex bodies by cylinders. Macbeath's (1951) result has been used earlier by Rogers (1959, 1964) to give the upper bound

$$\vartheta_{\mathcal{L}}(K) \leqslant d^{\log_2 \log_{\mathrm{e}} d + c}$$

for the density $\vartheta_{\mathcal{L}}(K)$ of lattice coverings of space with convex bodies K. Here, c is a suitable constant which does not depend on d and K. Using Gritzmann's (1985a) approximation theorem rather than Macbeath's (1951) result Rogers' bound can be sharpened for a certain class of convex bodies. In fact, Gritzmann (1985b) proved the following theorem.

Let $K \in \mathcal{K}^d$, $d > 1$, and let k be an integer satisfying $k > \log_2 \log_e d + 4$. Furthermore, let $H_1, \ldots, H_k$ denote k hyperplanes with normals being mutually perpendicular and let Y be an affine transformation.

(i) *If $Y(K)$ is symmetrical with respect to $H_1, \ldots, H_k$, respectively, then there is a constant c such that*

$$\vartheta_{\mathcal{L}}(K) < cd(\log_{\mathrm{e}} d)^{1+\log_2 \mathrm{e}}.$$

(ii) *If, for $i = 1, \ldots, k$, $Y(K) \cap H_i$ is equal to the orthogonal projection of $Y(K)$ onto H_i then there is a constant c such that*

$$\vartheta_{\mathcal{L}}(K) \leqslant cd(\log_{\mathrm{e}} d)^{2+\log_2 \mathrm{e}}.$$

In both cases, the constant c does not depend on d and K.

2.5.2. The lattice point enumerator

There is a close relation between lattice point problems and lattice packing and covering. After all, this is the heart of geometry of numbers; see Gruber's survey (chapter 3.1). However, since lattice point problems are dealt with in chapter 3.2 in this handbook, we will only give one simple but important transition lemma.

Let for a given lattice $\mathbb{L}$

$$G_{\mathbb{L}}: \mathcal{K}^d \to \mathbb{N} \cup \{0\}, \quad G_{\mathbb{L}}(K) = \operatorname{card}(K \cap \mathbb{L}),$$

be the *lattice point enumerator*. Then there is the following relation for all $K \in \mathcal{K}^d$.

If $\{K + l\colon\ l \in \mathbb{L}\}$ is a packing then we have for each $C \in \mathcal{K}^d$

$$G_{\mathbb{L}}(C)V(K) \leqslant \delta_{G_{\mathbb{L}}(C)}(K)V(C+K).$$

If $\{K + l\colon\ l \in \mathbb{L}\}$ covers $\mathbb{R}^d$ then we have for each $C \in \mathcal{K}^d$

$$\vartheta_{G_{\mathbb{L}}(C)}V(C \ominus K) \leqslant G_{\mathbb{L}}(C)V(K).$$

Here, again, $C \ominus K$ is an abbreviation for $\mathbb{R}^d \setminus ((\mathbb{R}^d \setminus C) + K)$.

Although both results seem very similar they are, in fact, quite different with respect to the problems stated at the beginning of this paragraph. The first bound can be expanded into *mixed volumes* (see chapter 1.2 by Sangwine-Yager). Further, proofs can be restricted to lattice polytopes. Each of these two statements is different for lower bounds. The main problem is that there is no analogon of the mixed volume expansion for $V(C \ominus K)$.

2.5.3. Dispersion

For a bounded real-valued function f on a compact subset A of $\mathbb{E}^d$, one wants to give an approximate algorithm to calculate the (say) supremum of f on A. Particularly, for nondifferentiable functions one might want to use Monte Carlo methods, i.e., search algorithms based on a stochastical choice of points. Choosing such points deterministically, one obtains *quasi-Monte Carlo methods*. A measure of the uniformity of the distribution of the points $x_1, \ldots, x_k$ in A is the so-called *dispersion*

$$d_k(A) = \max_{x \in A} \min_{1 \leqslant i \leqslant k} |x - x_k|.$$

If ω_A denotes the *modulus of continuity*, we simply have

$$y_k \leqslant \sup_{x \in A} f(x) \leqslant y_k + \omega_A\bigl(d_k(A)\bigr),$$

where $y_k := \max\{f(x_1), \ldots, f(x_k)\}$ (compare Niederreiter 1983 and Niederreiter and Wills 1975; see also Niederreiter 1992).

For efficient algorithms, one has to work with points of small dispersion.

As Niederreiter remarks the dispersion of $x_1, \ldots, x_k$ is the minimal radius ρ, such that $\sum_{i=1}^{k}(x_i + \rho\mathbb{B}^d)$ covers A. A general inequality applied in numerical analysis is then obtained from the trivial remark that the k balls must at least have total volume $V(A)$.

Since for $A \in \mathcal{K}^d$ we obviously have

$$d_k(A) \geqslant \left(\frac{1}{k}\frac{V(A)}{\omega_d}\vartheta_k(\mathbb{B}^d)\right)^{1/d},$$

the results on finite sphere coverings yield improvements, cf. Gritzmann (1984).

Let us finally remark that the special role of $\mathbb{B}^d$ arises from using the Euclidean distance to define the dispersion. For other norms, i.e., in general Minkowski spaces, this problem is related to finite covering problems with other convex bodies.

2.5.4. Crystallography (see also chapter 3.7 by Engel)
As is well known, the study of densely arranged configurations of 3-dimensional balls gives some insight into the behavior of solids and liquids. For example, molecular properties of many crystals which bear, of course, a lattice structure can be described, at least approximately, as the effect of short range and large range forces on a huge number of closely packed balls.

Such a model fails, however, to explain the phenomenon of "degenerate" molecular arrangements (as they occur in chemical whiskers, iceflowers or dendrites), of quasicrystals or of molecular clusters in metallic glasses or Buckminster-Fullerites.

In the following we deal briefly with some of these phenomena that seem to be related to sausage packings. Some relations of other finite sphere packing problems to quasicrystals, metallic glasses or Buckminster-Fullerites will be discussed in sections 4.1 and 4.2.

It turns out that dense finite sphere packings "mimic" some of the molecular phenomena occurring in whiskers, iceflowers and dendrites – even though (or maybe, because) the density function is defined with respect to the volume, rather than (as usual) in terms of energies. It is not clear at this point if the striking resemblance between the occurring phenomena is more than "phenomenological" and, hence, if finite sphere packings provide a reasonable, though limited model.

Chemical whiskers. Unlike the usual lattice-like 3-dimensional atomic growth of metals, under certain conditions some metals develop whiskers, microscopic filaments of a certain maximum length that is dependent on the metals that are involved; see Britton (1974). This phenomenon is similar to the sausage catastrophe in $\mathbb{E}^3$ as described in section 2.2. There is even a certain quantitative correlation if one takes into account that – for chemical reasons – molecular growth is constrained to subsets of specific lattice packings. For instance iron atoms are usually arranged in the *space centered cubical lattice*. Hence, a possible quantitative relation between these whiskers and the sausage catastrophe will only show if we restrict our packings to this lattice.

So, let for an arbitrary lattice $\mathbb{L}$ that packs $\mathbb{B}^3$, $k(\mathbb{L})$ denote the largest number k such that sausage arrangements are best possible for all finite packings of less than k translates whose centers are restricted to $\mathbb{L}$. As pointed out in G. Fejes Tóth, Gritzmann and Wills (1989) and Gritzmann and Wills (1993) these critical numbers $k(\mathbb{L})$ which depend on the molecular lattice of the specific metal under consideration seem very closely related to the experimentally measured ratio of length and width of its maximal whiskers.

Iceflowers and dendrites. Iceflowers and dendrites are forced to grow in a plane. However, they do not expand uniformly into all planar directions but grow (first) essentially in one dimension. This corresponds to the "iceflower theorem" (2.5) of Betke, Gritzmann and Wills (1982) which says that among all finite packings of $\mathbb{B}^3$ with centers restricted to a plane sausage arrangements are best-possible.

3. Bin packing

In the following we are dealing with a general bin packing problem, a generalization of the classical problem of packing given intervals into as few as possible unit intervals. This problem is highly practical and has hence been studied extensively from an algorithmic point of view. The main purpose of this section is not to give a comprehensive survey of algorithms for the bin packing problem and its ramifications but to link this problem that has been the subject of research mainly by those working in combinatorial optimization and computer science to similar problems worked on by research groups in geometry, number theory or coding theory.

3.1. Bin packing problems and their relatives

Even though bin packing has been studied mostly in dimensions $d \leqslant 3$ we will introduce a general bin packing problem in arbitrary dimension.

Let $d \in \mathbb{N}$, let $\mathcal{I}$ denote the set of all intervals $[0, \alpha]$ with $\alpha \in \mathbb{N}$, and set $\mathcal{I}^d = \mathcal{I} \times \cdots \times \mathcal{I}$. Further, let $\mathcal{M}_1, \ldots, \mathcal{M}_d$ denote subsets of $\mathcal{I}$, and set $\mathcal{M}^d = \mathcal{M}_1 \times \cdots \times \mathcal{M}_d$. $\mathcal{I}^d$ and $\mathcal{M}^d$ are the sets of all boxes in the positive orthand of $\mathbb{R}^d$ with edges parallel to the coordinate axes and vertex 0 and, in case of $\mathcal{M}^d$ with ith edge length belonging to $\mathcal{M}_i$, respectively. Then our basic *general bin packing problem* is the following task.

> Given $k \in \mathbb{N}$, boxes $R_1, \ldots, R_k \in \mathcal{I}^d$; find $M^* \in \mathcal{M}^d$ and vectors $b_1^*, \ldots, b_k^* \in \mathbb{Z}^d$ such that $\{b_1^* + R_1, \ldots, b_k^* + R_k\}$ is a packing into M^* and such that the density $\sum_{i=1}^{k} V(R_i)/V(M^*)$ is minimal among all such packings into all such subsets M; or determine that such a packing does not exist. (3.1)

The boxes $R_1, \ldots, R_k$ are called *items*. Let us point out that the requirement that all intervals considered here have integer vertices is just a tribute to the binary

Turing machine model that underlies our statements dealing with the algorithmic side of the problem.

Observe, that as compared to the sausage problems bin packing is more restrictive since the items are restricted to boxes, but on the other hand is more general since the items are allowed to vary in size and shape. Furthermore, the density functions are different. The restriction of items to boxes is motivated by the main applications of bin packing. Let us point out, however, that it is also of practical interest to study bin packings of more general convex or even nonconvex bodies. Examples contain cloth cutting problems in fashion industry. We will, however, not address this problem here.

Let us now show how two problems that have been considered before fit into this framework. The first deals with the task of *packing boxes into strips.*

> Let $r \in \mathbb{N}$, $r \leqslant d-1$, $\alpha_1, \dots, \alpha_r \in \mathbb{N}$.
> Given a positive integer k, boxes $R_1, \dots, R_k \in \mathcal{I}^d$; find positive integers $\beta^*_{r+1}, \dots, \beta^*_d$ and vectors $b^*_1, \dots, b^*_k \in \mathbb{Z}^d$ such that $\{b^*_1 + R_1, \dots, b^*_k + R_k\}$ is a packing into $[0, \alpha_1] \times \dots \times [0, \alpha_r] \times [0, \beta^*_{r+1}] \times \dots \times [0, \beta^*_d]$ and $\prod_{i=r+1}^{d} \beta^*_i$ is minimal for all such packings; or determine that such a packing does not exist. (3.2)

Clearly, this is just the special case of our general problem where $\mathcal{M}_1 = \{[0, \alpha_1]\}, \dots, \mathcal{M}_r = \{[0, \alpha_r]\}$ and $\mathcal{M}_{r+1} = \dots = \mathcal{M}_d = \mathcal{I}$, and the density function is again just a ratio of volumes.

The second task is that of *packing items into a minimum number of bins*:

> Given $k \in \mathbb{N}$, boxes $R_1, \dots, R_k \in \mathcal{I}^d$ and $M \in \mathcal{M}^d$, find the smallest $m \in \mathbb{N}$ such that there is a partition $J_1, \dots, J_m$ of $\{1, \dots, k\}$ and vectors $b^*_1, \dots, b^*_k \in \mathbb{R}^d$ such that for each $i = 1, \dots, m$, $\{b^*_j + R_j\colon\ j \in J_i\}$ is a packing into M; or determine that such a partition does not exist. (3.3)

Clearly, the classical 1-dimensional bin packing problem is just the 1-dimensional case of (3.3). To see that the problem of packing items into a minimum number of bins fits into the framework of the general bin packing problem (3.1) (with the dimension raised by one) just replace each box R_i by $R'_i = R_i \times [0, 1]$ and set $\mathcal{M}_1 \times \dots \times \mathcal{M}_d = \{M\}$, $\mathcal{M}_{d+1} = \mathcal{I}$. In particular, this means in view of (3.2) that for the classical bin packing problem we can replace the 1-dimensional items by a unit prism over themselves and the bins by the strip $[0, \alpha] \times [0, \infty[$ where α is the old bin size.

Let us mention some general results dealing with the density of bin packing.

Clearly, for $\mathcal{M}^d = \mathcal{I}^d$ the maximum of the densities of bin packings taken over all possible choices of k items is 1. However, unlike (2.10) for sausage problems, there is not even a nontrivial lower bound when $d, k \geqslant 2$. In fact, for $\varepsilon \in]0, \infty[$, the density of any packing of the two items $R_1 = [0, 1] \times \dots \times [0, 1] \times [0, \varepsilon]$ and $R_2 = [0, \varepsilon] \times \dots \times [0, \varepsilon] \times [0, 1]$ is less than 2ε.

However, the problem of finding nontrivial lower bounds becomes interesting for restricted classes of items. Consider, for example the following cube packing problem, a special case of (3.2):

> Given $k \in \mathbb{N}$, and cubes $C_1, \ldots, C_k \in \mathcal{J}^d$ (of possibly different sizes), find positive integers $\beta_1^*, \ldots, \beta_d^*$ and vectors $b_1^*, \ldots, b_k^* \in \mathbb{Z}^d$ such that $\{b_1^* + C_1, \ldots, b_k^* + C_k\}$ is a packing into $[0, \beta_1^*] \times \cdots \times [0, \beta_d^*]$ and $\prod_{i=1}^d \beta_i^*$ is minimal for all such packings.

It is not hard to see that there is a nontrivial lower bound for the corresponding density which holds for an arbitrary choice of the sizes of the cubes. A tight lower bound is, however, only known for $d = 2$. In fact, Kleitman and Krieger (1975) show that for a family of squares there is always a rectangle into which all the squares can be packed such that the density of the packing is at least $\sqrt{3}/(2\sqrt{2})$, and this bound is best possible. Furthermore, if the total area of the squares is normalized to 1, there is an (essentially unique) universal bin that always suffices: $[0, 2/\sqrt{3}] \times [0, \sqrt{2}]$.

Erdős and Graham (1975) studied packings of isometric copies of the unit square into a square of side α. They obtained the rather suprising upper bound $O(\alpha^{7/11})$ for the minimum possible uncovered area. This shows that, when α is bounded away from an integer, the possibility of using isometric copies rather than translates of the squares results in a considerable decrease in uncovered area. Lower bounds for this problem were given by Roth and Vaughan (1978).

It seems to be quite interesting to study the problem of finding good or even optimal lower bounds for the bin packing density for various classes of items or bins.

Groemer (1982) and Makai and Pach (1983) study conditions under which a (finite or infinite) sequence of convex bodies permits a packing into or a covering of a given convex body; see Groemer (1985) for a corresponding survey. In particular, Groemer (1982) gives the following theorem (which we state only for a finite set of bodies).

Let $K_1, \ldots, K_k \in \mathcal{K}^2$ with diameters uniformly bounded by 1*. Further, let $\gamma \in]3, \infty[$ and $\sum_i V(K_i) \leqslant \frac{1}{4}(\gamma - 1)(\gamma - 3)$. Then there are vectors $b_1, \ldots, b_k$ such that $\{b_1 + K_1, \ldots, b_k + K_k\}$ is a packing into $[0, 2\gamma] \times [0, \gamma]$.*

Furthermore, a similar result is given for coverings, and – for arbitrary dimension – there are much further reaching results when the group of isometries is allowed to act on the bodies. See Groemer (1982, 1985), Makai and Pach (1983), Lassak and Zhang (1991), and the survey by Göbel (1982) for more details and further results.

Let us close this section with mentioning some more variants of the bin packing problem which have been studied from an algorithmic point of view.

A first ramification is to enlarge the group of motions that act on the items. For example, depending on the kind of applications one has in mind it might make sense to allow not only translations but also arbitrary rotations of the items.

Furthermore, an extremely practical and algorithmically very important variant is to study on-line packings where one has to pack the items in a given order and the packing of the ith item R_i is independent of the items R_j with $j > i$ but only dependent on the packing of the items $R_1, \dots, R_{i-1}$ produced so far. Once R_i is packed it cannot be removed or reordered again. Of course, there is an abundance of ramifications where partial information on the sequence of items is available. Examples that have been studied include packings of 2-dimensional rectangles which are preordered in increasing or decreasing width or height.

While the general bin packing problem and the above mentioned variants fall into the general category (1.3) there are other ramifications which are of the form (1.4). An example is the container problem to maximize the number of items that can be packed into a given number of bins. For a survey of algorithmic results in this context see Coffman, Garey and Johnson (1984). For some theoretical results on some container problems see section 4.1.

3.2. *Approximation algorithms*

There is a vast literature on approximation algorithms for bin packing problems and its relatives and we refer to Garey and Johnson (1981) and Coffman, Garey and Johnson (1984) for a detailed survey. Here we concentrate on some basic results that – as we think – serve best the purpose of this handbook article.

The classical bin packing problem, (3.3) with $d = 1$, of packing 1-dimensional items $R_1, \dots, R_k \in \mathcal{I}$ into as few as possible bins $[0, \alpha]$ is already NP-hard in the strong sense (in fact, it contains as a special case the problem 3-PARTITION). The corresponding decision version persists to be NP-complete when the number m of bins is fixed and at least 2; it becomes then, however, solvable in pseudopolynomial time.

Hence, many authors focused on developing efficient approximate algorithms in hopes of guaranteeing near-optimal results.

For the classical 1-dimensional bin packing problem the currently best asymptotic performance ratio for a "single heuristic", is due to Garey and Johnson (1985). They show that an $O(k \log k)$ algorithm which they call MODIFIED FIRST FIT DECREASING uses at most

$$\tfrac{71}{60} \cdot \mathrm{Opt} + O(1)$$

bins where Opt denotes the optimum number of bins for the given instance.

Fernandez de la Vega and Lueker (1981) showed that for every fixed $\varepsilon \in]0, \infty[$ there is a linear time algorithm that uses at most

$$(1 + \varepsilon) \cdot \mathrm{Opt} + \frac{1}{\varepsilon^2}$$

bins. Subsequently, Karmarkar and Karp (1982) devised an approximation scheme for which the running time is polynomial both in L and $1/\varepsilon$ and the additive constant is also a polynomial in $1/\varepsilon$. These approximation results are, unfortunately, not practical to implement; they do, however, imply the theoretically important corollary that the asymptotic worst-case performance ratio is 1. This result should be contrasted with the fact that unless P = NP there is no fully polynomial approximation scheme that uses at most

$$(1+\varepsilon)\cdot \mathrm{Opt}$$

bins, cf. Garey and Johnson (1979). See Johnson (1982b) and Coffman, Garey and Johnson (1984) for a brief history of 1-dimensional bin packing.

Algorithms for the problem of packing 1-dimensional items into bins of different sizes have been given by Friesen and Langston (1986) (off-line) and Kinnersley and Langston (1989) (on-line) and an asymptotic analysis is due to Murgolo (1987).

For results on 1-dimensional bin packing with items of random sizes see Coffman, Garey and Johnson (1984) and for some more recent results see, e.g., Rhee (1990) and the papers quoted there. A lot of work has been done on the analysis of approximative algorithms when further restrictions are imposed on the item sizes, see Coffman, Garey and Johnson (1984).

There is also an extensive literature on the problem (3.2) for $d=2$ of packing rectangles into a strip $[0,\alpha]\times[0,\infty[$ since this problem has a variety of applications (see section 3.3). A first algorithmic result which gives performance guarantees is due to Baker, Coffman and Rivest (1980). Coffman et al. (1980) develop a "first-fit-decreasing-height" level algorithm that focuses around 1-dimensional techniques. In fact, as we have seen the 1-dimensional problem may be regarded as a special case of the 2-dimensional (rectangle into strip packing) problem with all heights of the items being equal. The basic idea of their algorithm is to presort the items according to nonincreasing height and to suitably apply horizontal "cuts" to the strip; hereby obtaining a set of rectangles into which the items are packed. The cuts are made in such a way that each item that is packed into a rectangle has an edge in the bottom segment and one of the items has also an edge lying in the top segment of the rectangle.

It turns out that the above-mentioned algorithm (Coffman et al. 1980) has an asymptotic worst-case error ratio of 1.7.

Baker, Brown and Katseff (1981) give an $\mathrm{O}(k\log k)$ algorithm with asymptotic worst-case error ratio 1.25.

If a preordering of the items is impossible since the applications require a genuine on-line algorithm one can still apply successfully the basic idea of the level algorithm and devise so-called *shelf algorithms*, introduced by Baker and Schwartz (1983). Here the heights of the rectangles that can be created by the cuts is restricted (and largely independent of the items). A popular shelf algorithm determines (under the assumption that an upper bound κ on the heights of the items is known) the positive integer s such that the height $h(R)$ of the next item R that has to be packed lies in the interval $\kappa]\rho^{s+1},\rho^{s}]$ where ρ is a suitable constant chosen beforehand. If there is already a rectangle of height $\kappa\rho^{s}$ where R still fits, R is placed there, otherwise a cut is performed to create a rectangle of this height. A variant of this algorithm obtains an asymptotic worst-case error ratio of $1.7/\rho$; the best absolute worst-case error ratio is obtained for $\rho\sim 0.62$ and is around 6.98.

Bartholdi, Van de Vate and Zhang (1989), analyse the expected performance of the shelf heuristic for packing a finite collection of rectangles into a strip $[0,\alpha] \times [0,\infty[$. The heights and widths are assumed to be independently and identically distributed. For a survey on the probabilistic analysis of the 2-dimensional bin packing problem see Coffman, Lueker and Rinnooy Kan (1988); see Coffman and Lagarias (1989) for a probabilistic analysis of algorithms for packing squares into a strip or a set of bins.

Clearly, for the off-line version of the bin packing problem there is a trade-off between the worst-case error performance and the worst-case time complexity of the applied algorithms. In particular there are exact algorithms if we do not insist that they run in polynomial time. For the on-line version we have an additional problem. In fact, Brown, Baker and Katseff (1982) show that any on-line algorithm has absolute worst-case error ratio at least 2.

There are not many results on d-dimensional bin packing problems. However, the problem of packing boxes into identical cubes has been considered by some authors. Chung, Garey and Johnson (1982) consider this problem in the plane while Karp, Luby and Marchetti-Spaccamela (1984) study some relevant algorithms for the d-dimensional problem from a probabilistic point of view, and Coppersmith and Raghavan (1989) deal with an on-line version of the problem in the plane and discuss extensions to higher dimensions. In particular, they show that in the plane their algorithm uses at most 3.25Opt+8 bins. This result has to be contrasted with work of Liang (1980) who showed that no on-line algorithm has a worst-case error ratio better than ~ 1.536.

3.3. *Some applications*

The bin packing problem is highly practical and has numerous applications. These include the obvious examples of packing items into trucks with a given weight limit, cutting-stock problems where demands for pieces have to be satisfied from a minimum number of standard lengths and various problems from VLSI technology.

In the following we give three further examples (for which it might be less obvious how to formulate them as bin packing problems). For additional examples see, e.g., Coffman, Garey and Johnson (1984).

There is a close connection between bin packing and the multiprocessor scheduling problem. The multiprocessor scheduling problem involves scheduling jobs with given execution time on a number m of identical processors. The goal is to minimize the overall execution time. Usually the problem involves an additional structure, precedence constraints, stating that certain jobs can be executed only after some other jobs have already been processed. A "bin packing formulation" of the multiprocessor scheduling problem without precedence constraints asks, for a fixed given number m of bins, for the smallest bin size α such that the given set of items can be packed into m bins of size α. A higher dimensional variant of this problem corresponds to schedulings of jobs which use several different resources. Here each item R_i is identified with a vector a_i and the task is to find the smallest number m of bins – identified with a vector $e = \alpha(1,\ldots,1)$ – such that $\sum_{j\in J_i} a_i \leqslant e$, for $i = 1, \ldots, m$, where $J_1, \ldots, J_m$ gives the partition of items corresponding to the bins. For a discussion of approximate algorithms for this problem see Coffman, Garey and Johnson (1984).

A motivation for the rectangle packing problem comes from computer job scheduling. Suppose we are given k jobs that have to be executed on a computer. Each job requires a certain portion of the computer's RAM, which is of course limited, and runs for a certain time. The task is to schedule these jobs in such a way that the total time needed to execute all of them is minimum.

If the required capacities are regarded as intervals $[0, \rho]$ on the x-axis, and the required time is depicted as an interval $[0, \tau]$ on the y-axis, this task is just the problem of packing rectangles into a strip $[0, \alpha] \times [0, \infty[$ where $[0, \alpha]$ corresponds to the available RAM capacity.

In some practical applications packings may be required to satisfy additional constraints. For instance, in aircraft cargo loading the weight distribution is relevant. In view of this application, Amiouny et al. (1992) study the problem of constructing *balanced* packings: the items have specified weights and the center of gravity of the packing is "close" to the center of gravity of the bin.

4. Miscellaneous packing and covering problems

This final section contains a collection of various packing and covering problems which belong to the general theme of our survey but are not central for our purposes. Since some of these aspects and results are very nice (and would in fact justify a much broader treatment in a less restricted framework) we decided to deal with them at least briefly. Unlike the other sections the following subsections make little contact with each other.

4.1. Some container problems

In the following we mention some classical container problems, give references and state some results which are paradigmatic for this area. For surveys covering this subject see L. Fejes Tóth (1965, 1972b), Saaty and Alexander (1975) and G. Fejes Tóth (1983). Saaty and Alexander (1975) give also a variety of applications. Let us begin with the following problems for convex bodies.

> Given $K, M \in \mathbb{R}^d$, how many translates of K can be packed into M; how few translates of K are needed to cover M?

Groemer's (1960) inequality for the slack between $\delta(\mathbb{B}^2)$ and $\delta_k(\mathbb{B}^2)$ was already cited in section 2. It is equivalent to the following result of Oler (1961a): a convex body M of $\mathbb{E}^2$ contains at most

$$\left\lfloor \frac{2}{\sqrt{3}} V(M) + \tfrac{1}{2} F(M) + 1 \right\rfloor$$

points of mutual distances at least 1. This result can be generalized to the Minkowski plane, Oler (1961b, 1962); see also Folkman and Graham (1969) and Graham, Witsenhausen and Zassenhaus (1972).

Here is a corresponding result for coverings with unit circles: a convex body M of $\mathbb{E}^2$ can be covered by

$$\left\lfloor \frac{2}{3\sqrt{3}} V(M) + \frac{2}{\pi\sqrt{3}} F(M) + 1 \right\rfloor$$

translates of $\mathbb{B}^2$, see Saaty and Alexander (1975, p. 488).

Another result of this kind is due to Santaló (1976), see also Santaló (1949): a convex body M of $\mathbb{E}^d$ can be covered by

$$\left\lfloor \sum_{i=0}^{d} \frac{\kappa_i \kappa_{d-i}}{\kappa_d} V_i(M) \right\rfloor$$

translates of $[0,1]^d$. Here $V_0, \ldots, V_d$ denote the intrinsic volumes, a renormalization of Minkowski's quermassintegrals; see Hadwiger (1957), McMullen (1975, 1977), McMullen and Schneider (1983) and chapter 1.8.

Let us now turn to a famous container problem of 1694: suppose $M = 2\mathbb{B}^d \setminus \operatorname{int}(\mathbb{B}^d)$ and consider the container problem of packing as many translates of $\mathbb{B}^d$ into M as possible. Obviously, this container packing problem is equivalent to the *kissing number problem*, the question of how many nonoverlapping unit balls can touch a given one. Clearly the answer is six in the plane and (as Newton argued in his famous quarrel with Gregory in 1694) twelve for $d = 3$. The only other known values for the kissing number are 240 for $d = 8$ and 196560 attained in the famous Leech-lattice for $d = 24$. For references and further results on this subject see Conway and Sloane (1988).

As pointed out in section 1.3, a closely related kind of problem is the question of the maximum radius of a fixed number of congruent spherical caps that can be packed on the unit sphere. Because of an obvious interpretation this problem is often called the "dictator problem". There is an abundant literature on this problem and we refer to L. Fejes Tóth (1965, 1972b), Danzer (1963), G. Fejes Tóth (1983) and chapter 3.3 for details.

Let us point out, however, that these kinds of problems are related to some interesting observations in nature. For instance, there is a relation to the distribution of holes on the surface of pollen grains, see Tammes (1930) and Tarnai (1984). The recent striking discovery of the C_{60}-molecules, the so-called *Buckminster-Fullerites*, gives another example of such configurations in the physical world; see Krätschmer et al. (1990). In the C_{60}-molecule the 60 atoms are arranged as the vertices of an Archimedian solid (the "soccerball") which has icosahedral rotational symmetry. (There are other arrangements of 60 balls with the same kind of symmetry even denser and each ball touching even five (rather than just three) neighboring balls. They seem, at present however, not to be of any chemical significance.)

Let us close this subsection by mentioning the remarkable discovery of L. Fejes

Tóth (1953) and Coxeter (1954) of a packing of 120 3-balls on a 3-sphere with density 0.77412... as opposed to the density 0.74048... of the densest sphere packing of $\mathbb{E}^3$.

4.2. *Other density functions*

We have dealt with density functions that are defined in terms of volume ratios that measure the relative slack of a packing or the relative surplus of a covering. There are, of course, other measures that may be relevant. In the following we give some results that involve some other density functions that come up naturally in their context.

4.2.1. *The volume of the center polytope*

Let us consider a finite sphere packing $\{b_1+\mathbb{B}^d, \dots, b_k+\mathbb{B}^d\}$. In general, it does not make sense to regard the volume of the center polytope $C_k = \operatorname{conv}\{b_1, \dots, b_k\}$ as a measure of the density of the packing. However, under some additional assumptions this measure might be reasonable. Problems of this kind have been considered in $\mathbb{E}^3$ from the point of view of applications to crystallography. Motivation stems from the study of *metallic glasses* (cf., e.g., Frank and Kasper 1958, Chaudhary, Giessen and Turnbull 1980) which contain small densely packed clusters of atoms whose centers do not belong to the face-centered cubical lattice, the lattice that provides the densest lattice packing of $\mathbb{B}^3$. An interesting open problem is to find a (in this sense) densest packing of thirteen unit balls in $\mathbb{E}^3$, one touching all twelve others. It is still a widespread belief that an icosahedral arrangement is optimal even though an arrangement with centers at the vertices and at the center of the cuboctahedron is known to be better, see Mackay (1962). Bagley (1970) found even a series of such "Newton–Gregory" packings with a five-fold symmetry axis, and still denser than the icosahedral packing. For some additional results in this vein and applications to quasicrystals which contain (local) arrangements of molecules with "forbidden" five-fold or icosahedral symmetry see Boerdijk (1952), Nelson (1986), Olamy and Alexander (1988), Olamy and Kléman (1989), Wills (1990a,b).

A notion of a covering density based on the central polytope has been introduced and studied by Gritzmann and Wills (1985); see also Bambah and Woods (1971).

4.2.2. *Minimum energy packing*

An interesting variant of our finite sphere packing problem (as studied in section 2.2) occurs if we replace the objective function by a "minimal energy criterion". More precisely, the task is the following nonlinear optimization problem:

$$U(k) = \min \sum_{i=1}^{k} \|b_i - c\|_{(2)}^2$$

$$\|b_i - b_j\|_{(2)} \geqslant 2 \qquad \text{for } i, j = 1, \dots, k,\ i \neq j,$$

$$kc = \sum_{i=1}^{k} b_i.$$

Observe that the constraints just say that $\{b_1 + \mathbb{B}^d, \ldots, b_k + \mathbb{B}^d\}$ is a packing and that c is the centroid of $\{b_1, \ldots, b_k\}$ (or, equivalently, of the center polytope $C_k = \operatorname{conv}\{b_1, \ldots, b_k\}$). The objective function is just the second moment of $\{b_1, \ldots, b_k\}$. This problem is studied in Graham and Sloane (1990) for $d = 2$. The use of the second moment is well suited for applications in coding theory (see, e.g., Conway and Sloane 1983, 1988); in fact the objective function measures the energy in the code. Furthermore, Graham and Sloane (1990) point out that this packing criterion avoids the sausage phenomenon that was observed in section 2.2 when minimizing the convex hull of the balls. Graham and Sloane (1990) construct such "penny packings" in $\mathbb{E}^2$ for all $k \leqslant 500$, study the performance of the greedy algorithm which adds one penny at a time and show that asymptotically

$$U(k) \sim \frac{\sqrt{3}k^2}{4\pi} \quad \text{as } k \to \infty.$$

The lower bound is based on Oler's (1961a) result stated in section 4.1.

4.2.3. Sausage-skin problems

Density functions that are defined in terms of quermassintegrals or intrinsic volumes other then the ordinary volume come up naturally in the following generalization of the sausage covering problem:

> Given $j, k \in \mathbb{N}$, $2 \leqslant j \leqslant d$, $K \in \mathcal{K}^d$, maximize $V_j(C)$ for all convex bodies C whose j-skeleton can be covered by k translates of K.

Note that the case $j = d$ is just the sausage covering problem of section 2. The general problem has been studied in G. Fejes Tóth, Gritzmann and Wills (1984) and Gritzmann (1987). We just mention one "sausage-skin" result obtained in the former paper: Let U_k denote the maximum of the perimeters of all plane convex bodies whose boundary can be covered by k translates of $\mathbb{B}^2$. Then

$$U_k = 4\left(\sqrt{k^2 - 1} + \arcsin\frac{1}{k}\right).$$

The optimal arrangement is unique (up to motions), the centers of the balls are equally spaced at distance $2\sqrt{1 - 1/k^2}$.

4.2.4. Insphere, circumsphere and related problems

An insphere (a circumsphere) of a convex body K in some Minkowski-space $(\mathbb{R}^d, \|\cdot\|)$ with unit ball B is characterized as a solution to the finite packing (finite covering) problem with $\Xi = \{\{b + \lambda \mathrm{B}\}\colon b \in \mathbb{R}^d \wedge \lambda \in]0, \infty[\}$, $\mathcal{M} = \{K\}$ and where the density is measured as the ratio of the volumes $V(b + \lambda\mathrm{B})/V(K)$. Other radii of convex bodies fit also into this context.

Problems of this kind have many applications in mathematical programming and computer science. For a comprehensive study of the computational complexity of such problems for $\mathcal{V}$- and $\mathcal{H}$-presented polytopes we refer to Gritzmann and Klee (1993).

The following covering problems of points by k unit cubes or Euclidean unit balls are ramifications of the circumradius problem in ℓ_2 and ℓ_∞ space:

> Given $d, m \in \mathbb{N}$, a set $V = \{v_1, \ldots, v_m\} \subset \mathbb{Q}^d$ and $K \in \{[0,1]^d, \mathbb{B}^d\}$; find vectors $b_1, \ldots, b_k \in \mathbb{Q}^d$ such that $V \subset \bigcup_{i=1}^k (b_i + K)$ or recognize that such vectors do not exist.

Megiddo (1990) showed that the cube covering problem can be solved in polynomial time for $k = 2$ but is already NP-complete for $k = 3$. The latter is shown via a transformation of 3-COLORABILITY of graphs. It follows also that for the problem of finding the minimum number of translates of the unit cube that cover a given point set there is no polynomial time approximation algorithm with an error ratio less than 1 unless P=NP.

The covering problem by k translates of $\mathbb{B}^d$ is NP-complete already for $k = 2$; Megiddo (1990) gave a reduction from 3-SAT. Again, he also showed that for finding the minimum number of translates of $\mathbb{B}^d$ that cover a given point set there is no polynomial time approximation algorithm with an error ratio less than $\frac{1}{2}$ unless P=NP.

While the minimum covering problem both, for $[0,1]^d$ and $\mathbb{B}^d$, persists to be NP-hard even in the plane (Fowler, Paterson and Tanimoto 1981, Megiddo and Supowit 1984), there exists a polynomial time approximation scheme for this problem in any fixed dimension, Hochbaum and Maass (1985).

The complexity of the problem of covering m points in $\mathbb{R}^d$ with a single cube whose edges are not necessarily parallel to the axes is conjecture by Megiddo (1990) to be NP-complete in unbounded dimension.

See Johnson (1982a) for some further results along these lines and various applications.

4.3. The Koebe–Andreev–Thurston theorem and its relatives

Let K be an arbitrary convex body in $\mathbb{R}^d$, $b_1, \ldots, b_k \in \mathbb{R}^d$, $\rho_1, \ldots, \rho_k \in]0, \infty[$ and suppose that $K_1 = b_1 + \rho_1 K, \ldots, K_k = b_k + \rho_k K$ form a packing. We will speak of $\mathcal{K} = \{K_1, \ldots, K_k\}$ as a homothetic K-packing. The *nerve* of $\mathcal{K}$ is the *intersection graph* $G(\mathcal{K}) = (V, E)$, where the vertices of $G(\mathcal{K})$ correspond to the bodies K_i and two vertices of $G(\mathcal{K})$ are joined by an edge if and only if the corresponding bodies have a point in common. Given a convex body in $\mathbb{R}^d$, a graph G is called a homothetic K-*nerve* if there exists a homothetic K-packing $\mathcal{K}$ whose nerve is isomorphic to G. A centrally symmetric convex body K gives rise to a Minkowski space with unit ball K. Hence, in this situation a K-packing is, in fact, a packing of unit balls. There is a *natural embedding* of the nerve of such packings: V is the set of centers of the balls and two vertices v_1, v_2 are joined by the edge $\operatorname{conv}\{v_1, v_2\}$ if and only if the corresponding balls have a point in common.

A famous result of Koebe (1936), Andreev (1970a,b) and Thurston (1979) says that every planar graph is a homothetic $\mathbb{B}^2$-nerve and vice versa. Schramm (1989) extended this result and showed that, given a smooth planar convex body K in $\mathbb{R}^2$, then every planar graph is a homothetic K-nerve and vice versa. These results have various applications in complex analysis.

There are many interesting questions related to nerves of packings, problems from a combinatoric, geometric, algebraic or algorithmic point of view. There are even some aspects that relate to dual interpretations of some algorithms in combinatorial optimization. For various related results see Thurston (1979), Schramm (1989, 1990) and Gritzmann, Odor and Schulte (1993).

Rigidity questions of finite packings have been studied by Connelly (1988), Tarnai (1984), Schramm (1991); see chapter 1.7.

4.4. *Packing and covering in combinatorial optimization*

There are many problems in combinatorics and combinatorial optimization which can be formulated as finite packing or covering problems. Usually the density functions facilitate counting procedures. For a survey of such problems and their applications we refer to Schrijver (1982), Balas and Padberg (1975) and Frank (1990). Here we give only a few examples.

The *set-packing* and *set-covering* problem are basic problems in combinatorial optimization:

> Given a finite set E and subsets $S_1, \dots, S_k$ of E and suppose there is a cost c_i associated with each S_i; find an index set $I \subset \{1, \dots, k\}$ such that $\sum_{i \in I} c_i$ is maximum and $S_i \cap S_j = \emptyset$ for all $i \in I$ and $j \in I \setminus \{i\}$; $\sum_{i \in I} c_i$ is minimum and $E = \bigcup_{i \in I} S_i$.

These problems are extremely practical and have many applications to problems like airline crew scheduling, facility location problems etc; for a bibliography of such applications see Balas and Padberg (1975).

Special cases of these problems include node-packing and node-covering problems in finite graphs or the *weighted matching problem*:

> Given a finite graph G with vertex set V and edge set E and a weight function $\omega : E \to \mathbb{N}$; find a subset $\hat{E}$ of E such that no two edges of $\hat{E}$ have a vertex in common and such that $\sum_{e \in \hat{E}} \omega(e)$ is maximal for all such subsets.

Let us point out that this problem can easily be transformed into a "geometric" packing problem involving convex bodies by "fattening" the edges of G. However, methods of solving the weighted matching problem do not seem to benefit from this "geometrization" of the problem and hence we do not regard this formulation as "natural".

While the general set-packing and set-covering problem is NP-complete the weighted matching problem can be solved in polynomial time, Edmonds (1965).

Acknowledgment

We would like to express our gratitude to the following friends and colleagues for their valuable comments on a previous version of this article: J.J. Bartholdi, J. Eckhoff, G. Fejes Tóth, R. Graham, H. Niederreiter.

References

Aho, A.V., J.E. Hopcroft and J.D. Ullman
[1974] *The Design and Analysis of Computer Algorithms* (Addison-Wesley, Reading, MA).

Amiouny, S.V., J.J. Bartholdi, J.H. Van de Vate and J. Zhang
[1992] Balanced loading, *Oper. Res.*.

Andreev, E.M.
[1970a] On convex polyhedra in Lobacevskii spaces, *Math. USSR-Sb.* **10**, 413–440.
[1970b] On convex polyhedra of finite volume in Lobacevskii space, *Math. USSR-Sb.* **12**, 255–259.

Bagley, B.G.
[1970] Fivefold pseudosymmetry, *Nature* **225**, 1040–1041.

Baker, B.S., and J.S. Schwartz
[1983] Shelf algorithms for two-dimensional bin packing problems, *SIAM J. Comput.* **12**, 508–525.

Baker, B.S., E.G. Coffman and R.L. Rivest
[1980] Orthogonal packings in two dimensions, *SIAM J. Comput.* **9**, 846–855.

Baker, B.S., D.J. Brown and H.P. Katseff
[1981] A 5/4 algorithm for two-dimensional packing, *J. Algorithms* **2**, 348–368.

Balas, E., and M.W. Padberg
[1975] Set partitioning, in: *Combinatorial Programming: Methods and Applications*, ed. B. Roy (Reidel, Dordrecht) pp. 205–258.

Bambah, R.P.
[1971] Geometry of numbers, packing and covering and discrete geometry, *Math. Student* **39**, 117–129.

Bambah, R.P., and C.A. Rogers
[1952] Covering the plane with convex sets, *J. London Math. Soc.* **27**, 304–314.

Bambah, R.P., and A.C. Woods
[1971] On plane coverings with convex domains, *Mathematika* **18**, 91–97.

Bambah, R.P., C.A. Rogers and H. Zassenhaus
[1964] On coverings with convex domains, *Acta Arith.* **9**, 191–207.

Bartholdi, J.J., J.H. Van de Vate and J. Zhang
[1989] Expected performance of the shelf heuristic for 2-dimensional packing, *Oper. Res. Lett.* **8**, 11–16.

Betke, U., and P. Gritzmann
[1984] Über L. Fejes Tóths Wurstvermutung in kleinen Dimensionen, *Acta Math. Hungar.* **43**, 299–307.
[1986] An application of valuation theory to two problems of discrete geometry, *Discrete Math.* **58**, 81–85.

Betke, U., P. Gritzmann and J.M. Wills
[1982] Slices of L. Fejes Tóth's sausage conjecture, *Mathematika* **29**, 194–201.

Betke, U., M. Henk and J.M. Wills
[1993] Finite and infinite packings, Manuscript.

Bodlaender, H.L., P. Gritzmann, V. Klee and J. van Leeuwen
[1990] Computational complexity of norm-maximization, *Combinatorica* **10**, 203–225.

Boerdijk, A.H.
[1952] Some remarks concerning close packing of equal spheres, *Philips Res. Rep.* **7**, 303–313.

Böröczky Jr, K.
[1992] Four-ball packings in the three-space, Manuscript.

Böröczky Jr, K., and M. Henk
[1992] Radii and the sausage conjecture, Manuscript.

Britton, S.C.
[1974] Spontaneous growth of whiskers on tin coating; 20 years of observation, *Trans. Inst. Met. Finish.* **52**, 95–102.

Brown, D.J., B.S. Baker and H.P. Katseff
[1982] Lower bounds for the on-line two-dimensional packing algorithm, *Acta Inform.* **18**, 207–225.

Chaudhary, P., B.C. Giessen and B. Turnbull
[1980] Metallic glasses, *Sci. Amer.*, April.

Chung, F.R.K., M.R. Garey and D.S. Johnson
[1982] On packing two-dimensional bins, *SIAM J. Algebraic Discrete Methods* **3**, 66–76.

Coffman, E.G., and J.C. Lagarias
[1989] Algorithms for packing squares: A probabilistic analysis, *SIAM J. Comput.* **18**, 166–185.

Coffman, E.G., J.Y. Leung and D.W. Ting
[1978] Bin packing: Maximizing the number of pieces packed, *Acta Inform.* **9**, 263–271.

Coffman, E.G., M.R. Garey, D.S. Johnson and R.E. Tarjan
[1980] Performance bounds for level-oriented two-dimensional packing algorithms, *SIAM J. Comput.* **9**, 808–826.

Coffman, E.G., M.R. Garey and D.S. Johnson
[1984] Approximation algorithms for bin-packing – An updated survey, in: *Algorithm Design for Computer Systems Design*, eds G. Ausiello, M. Lucertini and P. Serafini (Springer, New York) pp. 49–106.

Coffman, E.G., G.S. Lueker and A.H.G. Rinnooy Kan
[1988] Asymptotic methods in the probabalistic analysis of sequencing and packing heuristics, *Management Sci.* **34**, 266–290.

Connelly, R.
[1988] Rigid circle and sphere packings I, finite packings, *Structural Topology* **14**, 43–60.

Conway, J.H., and N.J.A. Sloane
[1983] A fast encoding method for lattice codes and quantizers, *IEEE Trans. Inform. Theory* **29**, 820–824.
[1988] *Sphere Packings, Lattices and Groups* (Springer, New York).

Coppersmith, D., and P. Raghavan
[1989] Multidimensional on-line bin packing: Algorithms and worst case analysis, *Oper. Res. Lett.* **8**, 17–20.

Coxeter, H.S.M.
[1954] Arrangements of equal spheres in non-euclidean spaces, *Acta Math. Hungar.* **5**, 263–274.

Coxeter, H.S.M., L. Few and C.A. Rogers
[1959] Covering space with equal spheres, *Mathematika* **6**, 147–157.

Danzer, L.
[1963] Endliche Punktmengen auf der 2-Sphäre mit möglichst großem Minimalabstand, Habilitationsschrift (Göttingen) [English translation: Finite point-sets on S^2 with minimum distance as large as possible, *Discrete Math.* **60** (1986) 3–66].

Dauenhauer, M.H., and H. Zassenhaus
[1987] Local optimality of the critical lattice sphere-packing of regular tetrahedra, *Discrete Math.* **64**, 129–146.

Dyer, M.E., and A.M. Frieze
[1988] The complexity of computing the volume of a polyhedron, *SIAM J. Comput.* **17**, 967–974.
[1991] Computing the volume of convex bodies: A case where randomness provably helps, in: *Probabilistic Combinatorics and its Applications*, ed. B. Bollobás, Proc. Symposia in Pure Math., Vol. 44 (Amer. Math. Soc., Providence, RI) pp. 123–169.

Edmonds, J.
[1965] Maximum matching and a polyhedron with 0, 1-vertices, *J. Res. Nat. Bur. Standards (B)* **69**, 125–130.

Erdős, P., and R.L. Graham
[1975] On packing squares with equal squares, *J. Combin. Theory Ser. A* **19**, 119–123.

Erdős, P., P.M. Gruber and J. Hammer
[1989] *Lattice Points* (Longman, Essex).

Fejes Tóth, G.
[1983] New results in the theory of packing and covering, in: *Convexity and its Applications*, eds P. M. Gruber and J. M. Wills (Birkhäuser, Basel) pp. 318–359.
[1987] Finite coverings by translates of centrally symmetric convex domains, *Discrete Comput. Geom.* **2**, 353–364.

Fejes Tóth, G., P. Gritzmann and J.M. Wills
[1984] Sausage-skin problems for finite coverings, *Mathematika* **31**, 118–137.
[1989] Finite sphere packing and sphere covering, *Discrete Comput. Geom.* **4**, 19–40.
[1990] On finite multiple packings, *Arch. Math.* **54**, 407–411.

Fejes Tóth, L.
[1949] Über die dichteste Kreislagerung und dünnste Kreisüberdeckung, *Comment Math. Helv.* **23**, 342–349.
[1953] On close packings of spheres in spaces of constant curvature, *Publ. Math. Debrecen* **3**, 158–167.
[1965] *Reguläre Figuren* (Ungar. Acad. Wiss., Budapest).
[1972a] Some packing and covering theorems, *Acta Sci. Math. (Szeged.)* **12** (A), 62–67.
[1972b] *Lagerungen in der Ebene, auf der Kugel und im Raum* (Springer, Berlin, 2nd ed.).
[1975] Research problem 13, *Period. Math. Hungar.* **6**, 197–199.

Fejes Tóth, L., and J.M. Wills
[1984] Enclosing a convex body by homothetic copies, *Geom. Dedicata* **15**, 279–384.

Fernandez de la Vega, W., and G.S. Lueker
[1981] Bin packing can be solved within $1+\varepsilon$ in linear time, *Combinatorica* **1**, 349–355.

Florian, A.
[1985] On compact packing of circles, *Studia Sci. Math. Hungar.* **20**, 473–480.

Folkman, J.H., and R.L. Graham
[1969] A packing inequality for compact convex subsets of the plane, *Canad. Math. Bull.* **12**, 745–752.

Fowler, R.J., M.S. Paterson and S.L. Tanimoto
[1981] Optimal packing and covering in the plane are NP-complete, *Inform. Process. Lett.* **12**, 133–137.

Frank, A.
[1990] Packing paths, circuits and cuts, in: *Paths, Flows, and VLSI Layout*, eds B. Korte, L. Lovász, H.J. Prömel and A. Schrijver (Springer, Berlin) pp. 47–100.

Frank, F.C., and J.S. Kasper
[1958] Complex alloy structures regarded as sphere packings I, *Acta Cryst.* **11**, 184–190.

Freund, R.M., and J.B. Orlin
[1985] On the complexity of four polyhedral set containment problems, *Math. Programming* **33**, 139–145.

Friesen, D.K., and M.A. Langston
[1986] Variable sized bin packing, *SIAM J. Comput.* **15**, 222–230.

Füredi, Z.
[1991] The densest packing of equal circles into a parallel strip, *Discrete Comput. Geom.* **6**, 95–106.

Gandini, P.M., and J.M. Wills
[1992] On finite sphere packings, *Math. Pannonica,* **3**, 19–29.

Garey, M.R., and D.S. Johnson
[1979] *Computers and Intractability: A Guide to the Theory of NP-completeness* (Freeman, San Francisco).
[1981] Approximation algorithms for bin packing problems: A survey, in: *Analysis and Design of Algorithms in Combinatorial Optimization,* eds G. Ausiello and M. Lucertini, CISM Courses and Lectures, Vol. 266 (Springer, Berlin) pp. 147–172.
[1985] A 71/60 theorem for bin packing, *J. Complexity* **1**, 65–106.

Göbel, F.
[1982] Geometrical packing and covering problems, in: *Packing and Covering in Combinatorics,* ed. A. Schrijver, Math. Centre Tracts, Vol. 106 (Math. Centre, Amsterdam) pp. 179–199.

Graham, R.L., and N.J.A. Sloane
[1990] Penny-packing and two-dimensional codes, *Discrete Comput. Geom.* **5**, 1–11.

Graham, R.L., H.S. Witsenhausen and H.J. Zassenhaus
[1972] On tightest packings in the Minkowski plane, *Pacific J. Math.* **41**, 699–715.

Gritzmann, P.
[1984] Finite Packungen und Überdeckungen, Habilitationsschrift, Siegen.
[1985a] Ein Approximationssatz für konvexe Körper, *Geom. Dedicata* **19**, 277–286.
[1985b] Lattice covering of space with symmetric convex bodies, *Mathematika* **32**, 311–315.
[1986] Finite packing of equal balls, *J. London Math. Soc.* **33**, 543–553.
[1987] Über die *j*-ten Überdeckungsdichten konvexer Körper, *Monatsh. Math.* **103**, 207–220.
[1993] On the computational complexity of some packing and covering problems, in preparation.

Gritzmann, P., and V. Klee
[1992] Finding optimal shadows of polytopes, in preparation.
[1993] Computational complexity of inner and outer *j*-radii of polytopes in finite dimensional normed spaces, to appear in *Math Prog.*

Gritzmann, P., and B. Sturmfels
[1993] Minkowski-addition of polytopes: Computational complexity and applications to Gröbner bases, to appear in *SIAM J. Discrete Math.*

Gritzmann, P., and J.M. Wills
[1985] On two finite covering problems of Bambah, Rogers, Woods and Zassenhaus, *Monatsh. Math.* **99**, 279–296.
[1986] Finite packing and covering, *Studia Sci. Math. Hungar.* **21**, 149–162.
[1993] On chemical whiskers, in preparation.

Gritzmann, P., T. Odor and E. Schulte
[1993] On nerve graphs of packings: Some relatives of the Koebe–Andreev–Thurston theorem, in preparation.

Groemer, H.
[1960] Über die Einlagerung von Kreisen in einen konvexen Bereich, *Math. Z.* **73**, 285–294.
[1982] Covering and packing properties of bounded sequences of convex sets, *Mathematika* **29**, 18–31.

[1985] Coverings and packings by sequences of convex sets, in: *Discrete Geometry and Convexity*, eds J.E. Goodman, E. Lutwak, J. Malkevitch and R. Pollack, Ann. New York Acad. Sci., Vol. 440 (New York Acad. Sci., New York) pp. 262–278.

Grötschel, M., L. Lovász and A. Schrijver
[1988] *Geometric Algorithms and Combinatorial Optimization* (Springer, Berlin).

Gruber, P.M.
[1979] Geometry of numbers, in: *Contributions to Geometry*, eds J. Tölke and J.M. Wills (Birkhäuser, Basel).

Gruber, P.M., and C.G. Lekkerkerker
[1987] *Geometry of Numbers* (North-Holland, Amsterdam).

Grünbaum, B.
[1967] *Convex Polytopes* (Wiley, London). New edition with an additional chapter by V. Klee and P. Kleinschmidt (Springer, New York, 1993).

Hadwiger, H.
[1957] *Vorlesungen über Inhalt, Oberfläche und Isoperimetrie* (Springer, Berlin).

Hoare, M.R., and J.A. McInnes
[1983] Morphology and statistical statics of simple microclusters, *Adv. in Phys.* **32**, 791–821.

Hochbaum, D.S., and W. Maass
[1985] Approximation schemes for covering and packing problems in image processing and VLSI, *J. Assoc. Comput. Mach.* **32**, 130–136.

Johnson, D.S.
[1982a] The NP-completeness column: An ongoing guide, *J. Algorithms* **3**, 182–195.
[1982b] The NP-completeness column: An ongoing guide, *J. Algorithms* **3**, 288–300.

Kabatjanskii, G.A., and V.I. Levenštein
[1978] Bounds for packings on the sphere and in space, *Problems Inform. Transmission* **14**, 1–17.

Karmarkar, N., and R.M. Karp
[1982] An efficient approximation scheme for the one-dimensional bin packing problem, in: *Proc. 23rd Ann. Symp. Found. of Comput. Sci.*, pp. 312–320.

Karp, R.M., M. Luby and A. Marchetti-Spaccamela
[1984] A probabilistic analysis of multidimensional bin-packing problems, in: *16th ACM Symp. on Theory of Comput.*, pp. 289–298.

Khachiyan, L.G.
[1979] A polynomial algorithm in linear programming (in Russian), *Dokl. Akad. Nauk. SSSR* **244**, 1093–1096 [English translation: *Soviet Math. Dokl.* **20**, 191–194].
[1992] Complexity of polytope volume computation, in: *New Trends in Discrete and Computational Geometry*, ed. J. Pach (Springer, Berlin).

Kinnersley, N.G., and M.A. Langston
[1989] On-line variable sized bin packing, *Discrete Appl. Math.* **22**, 143–146.

Kleinschmidt, P., U. Pachner and J.M. Wills
[1984] On L. Fejes Tóth's sausage conjecture, *Israel J. Math.* **47**, 216–226.

Kleitman, D.J., and M.K. Krieger
[1975] An optimal bound for two dimensional bin packing, in: *Proc. 16th Ann. Symp. Found. of Comput. Sci.*, pp. 163–168.

Koebe, P.
[1936] Kontaktprobleme der konformen Abbildung, *Ber. Verh. Sächs. Akad. Wiss. Leipzig, Math.-Phys. Kl.* **88**, 141–164.

Krätschmer, W., L.D. Lamb, K. Festiopoulos and D.R. Haffman
[1990] Solid C_{60}: A new form of carbon, *Nature* **347**, 354–358.

Lassak, M., and J. Zhang
[1991] An on-line potato sack theorem, *Discrete Comput. Geom.* **6**, 1–7.

Lawrence, J.
[1991] Polytope volume computation, *Math. Comput.* **57**, 259–271.

Liang, F.M.
[1980] A lower bound for on-line bin-packing, *Inform. Process. Lett.* **10**, 76–79.

Lovász, L.
[1986] *An Algorithmic Theory of Numbers, Graphs and Convexity*, AMS–SIAM Reg. Conf. Ser., Vol. 50 (SIAM, Philadelphia).

Macbeath, A.M.
[1951] A compactness theorem for affine equivalence classes of convex regions, *Canad. J. Math.* **3**, 54–61.

Mackay, A.L.
[1962] A dense non-crystallographic packing of equal spheres, *Acta Cryst.* **15**, 916–918.

Makai Jr, E., and J. Pach
[1983] Controlling function classes and covering euclidean space, *Studia Sci. Math. Hungar* **18**, 435–459.

Martini, H.
[1990] A new view on some characterizations of simplices, *Arch. Math.* **55**, 389–393.

McMullen, P.
[1975] Non-linear angle-sum relations for polyhedral cones and polytopes, *Math. Proc. Cambridge Philos. Soc.* **78**, 247–261.
[1977] Valuations and Euler-type relations on certain classes of convex polytopes, *Proc. London Math. Soc. (3)* **35**, 113–135.

McMullen, P., and R. Schneider
[1983] Valuations on convex bodies, in: *Convexity and its Applications*, eds P. Gruber and J.M. Wills (Birkhäuser, Basel) pp. 170–247.

Megiddo, N.
[1990] On the complexity of some geometric problems in unbounded dimension, *J. Symbolic Comput.* **10**, 327–334.

Megiddo, N., and K.J. Supowit
[1984] On the complexity of some geometric location problems, *SIAM J. Comput.* **13**, 182–196.

Mount, D.M., and R. Silverman
[1987] Packing and covering the plane with translates of a convex polygon, Manuscript.

Murgolo, F.D.
[1987] An efficient approximation scheme for variable-sized bin packing, *SIAM J. Comput.* **16**, 149–161.

Nelson, D.R.
[1986] Quasicrystals, *Sci. Amer.* **255**, 32–41.

Niederreiter, H.
[1983] *A Quasi-Monte-Carlo Method for Approximate Computation of the Extreme Values of a Function* (Akad. Kiadó, Budapest) pp. 523–529.
[1992] *Random Number Generation and Quasi-Monte Carlo Methods* (SIAM, Philadelphia, PA).

Niederreiter, H., and J.M. Wills
[1975] Diskrepanz und Distanz von Maßen bezüglich konvexer und Jordanscher Mengen, *Math. Z.* **144**, 125–134.

Olamy, Z., and S. Alexander
[1988] Quasiperiodic packing densities, *Phys. Rev.* **37**, 3973–3978.

Olamy, Z., and M. Kléman
[1989] A 2-dimensional aperiodic dense tiling, *J. Phys. (Paris)* **50**, 19–33.

Oler, N.
[1961a] An inequality in the geometry of numbers, *Acta Math.* **105**, 19–48.

[1961b] A finite packing problem, *Canad. Math. Bull.* **4**, 153–155.
[1962] The slackness of finite packings in E^2, *Amer. Math. Monthly* **69**, 511–514.

Rhee, W.T.
[1990] A note on optimal bin packing and optimal bin covering with items of random size, *SIAM J. Comput.* **19**, 705–710.

Rogers, C.A.
[1947] Existence theorems in the geometry of numbers, *Ann. of Math.* **48**, 994–1002.
[1951] The closest packing of convex two-dimensional domains, *Acta Math.* **86**, 309–321.
[1957] A note on coverings, *Mathematika* **4**, 1–6.
[1959] Lattice coverings of space, *Mathematika* **6**, 33–39.
[1964] *Packing and Covering* (Cambridge Univ. Press, Cambridge).

Roth, K.F., and R.C. Vaughan
[1978] Inefficiency in packing squares with unit squares, *J. Combin. Theory Ser. A* **24**, 170–186.

Saaty, T.L., and J.M. Alexander
[1975] Optimization and the geometry of numbers: packing and covering, *SIAM Rev.* **17**, 475–519.

Santaló, L.A.
[1949] Una invariante afin pasa os cuerpas convexos del espacio du n-dimensiones, *Portugal. Math.* **8**, 155–161.
[1976] *Integral Geometry and Geometric Probability* (Addison-Wesley, London).

Schramm, O.
[1989] Packing 2-dimensional bodies with prescibed combinatorics and applications to the construction of conformal and quasiconformal mappings, Ph.D. Thesis, Princeton University.
[1990] How to cage an egg, Manuscript.
[1991] Rigidity of infinite (circle) packings, *J. Amer. Math. Soc.* **4**, 127–149.

Schrijver, A., ed.
[1982] *Packing and Covering in Combinatorics,* Math. Centre Tracts, Vol. 106 (Math. Centre, Amsterdam).

Sloane, N.J.A., and B.K. Teo
[1985] Theta series and magic numbers for close-packed spherical clusters, *J. Chem. Phys.* **83**, 6520–6534.

Tammes, R.M.L.
[1930] On the number and arrangement of the places of exit on the surface of pollen grains, *Rec. Trav. Bot. Neerl.* **27**, 1–84.

Tarnai, T.
[1984] Spherical circle packing in nature, practice and theory, *Structural Topology* **9**, 39–58.

Thurston, W.
[1979] The geometry and topology of 3-manifolds, Princeton University Notes.

Wegner, G.
[1984] Extremale Groemerpackungen, *Studia Sci. Math. Hungar.* **19**, 299–302.
[1986] Über endliche Kreispackungen in der Ebene, *Studia Sci. Math. Hungar.* **21**, 1–28.

Wills, J.M.
[1983] Research problem 33, *Period. Math. Hungar.* **14**, 189–191.
[1985] On the density of finite packings, *Acta Math. Hungar.* **46**, 205–210.
[1987] Research problem 41, *Period. Math. Hungar.* **18**, 251–252.
[1990a] Kugellagerungen und Konvexgeometrie, *Jber. Deutsch. Math.-Vereinig.* **92**, 21–46.
[1990b] A quasi-crystalline sphere-packing with unexpected high density, *J. Phys. (Paris)* **5**, 1061–1064.

CHAPTER 3.5

Tilings

Egon SCHULTE

Mathematics Department, Northeastern University, Boston, MA 02115, USA

Contents

HANDBOOK OF CONVEX GEOMETRY
Edited by P.M. Gruber and J.M. Wills

1. Introduction

Tiling problems for Euclidean or other spaces have been investigated throughout the history of mathematics, leading to a vast literature on the subject. Almost all variants of the question "How can a space be tiled by copies of one (or more) sets?" has been studied in some form or another.

Though many tiling problems have a strong intuitive geometric appeal, it is surprising that historically a systematic study of the theory of tilings started only very late. In this and the last century many achievements result from the work of crystallographers connecting tilings with the theory of the crystallographic groups (Fedorov 1885, 1899, Voronoi 1908, Schoenflies 1891, Delone 1929); for a more detailed account see Engel (1986) and chapter 3.7 in this Handbook.

At the beginning of this century the theory of tilings was stimulated by Hilbert's (1900) 18th problem which relates to discrete groups and their fundamental regions. Especially noteworthy is the pioneering work on plane tilings by Reinhardt (1918, 1928) and Heesch (1935, 1968) who both provided solutions to the 2nd part of Hilbert's problem. Another important impetus on the subject of tiling, especially in higher dimensions, came from the geometry of numbers and started with Minkowski's (1897) work. See also chapter 3.1 in this Handbook.

In the 1970s Grünbaum and Shephard started their comprehensive work on tilings that resulted in a beautiful book on plane tilings; see Grünbaum and Shephard (1986). This provided a systematic approach to the theory of plane tilings and highly stimulated this area of mathematics. More recently Dress (1987) and co-workers developed a powerful technique to deal with certain classification problems for plane and higher-dimensional tilings.

One of the most exciting developments in the theory of tilings is the relatively recent discovery of the phenomenon of aperiodicity for tilings. Here, fantastic achievements have been obtained over the last 20 years, and it is likely that this challenging area of research will remain active for a longer time. Part of the motivation for studying aperiodic tilings stems from the very recent discovery of their connection with quasi-crystals.

The purpose of this work is to give a short survey on some topics from the theory of tilings, emphasizing aspects which are closely related to convex geometry in Euclidean spaces of any dimension d. This limits the discussion essentially to relatively well behaved tilings and tiles, and necessarily must leave aside the discussion of tilings with topologically strange properties.

Our discussion of plane tilings will largely follow the exposition in Grünbaum and Shephard (1986); the reader is often referred to this book for further details.

For tilings in higher dimensions no complete account on the present knowledge is available in the literature. We will attempt to survey the main results as far as possible, but clearly this cannot come anywhere near a satisfactory comprehensive exposition of the subject.

2. Basic notions

Following the terminology of Grünbaum and Shephard (1986) a *tiling* $\mathcal{T}$ of Euclidean d-space E^d is a countable family of closed subsets T of E^d, the *tiles* of $\mathcal{T}$, which cover E^d without gaps and overlaps; that is to say, the union of all tiles of $\mathcal{T}$ is E^d, and any two distinct tiles do not have interior points in common. Throughout this paper we are always assuming that the tiles of $\mathcal{T}$ are *closed topological d-cells*. In fact, in many cases they will be convex d-polytopes. For properties of convex polyhedra and polytopes we refer to Grünbaum (1967) and chapter 2.3 in this Handbook.

Theorem 1. *If $\mathcal{T}$ is a tiling of E^d with (compact) convex tiles, then each tile in $\mathcal{T}$ is a convex d-polyhedron (d-polytope, respectively).*

Trivially, any tiling $\mathcal{T}$ is a packing and a covering of E^d, both with density 1; see chapter 3.3 in this Handbook.

Two tilings $\mathcal{T}_1$ and $\mathcal{T}_2$ of E^d are called *congruent* if there is a Euclidean motion of E^d which maps (the tiles of) $\mathcal{T}_1$ onto (the tiles of) $\mathcal{T}_2$. In dealing with classification problems for certain kinds of tilings we usually do not distinguish two tilings of E^d which are obtained from each other by a similarity transformation (a Euclidean motion followed by a homothety); such tilings are said to be *equal*, or *the same*.

A tiling $\mathcal{T}$ is called *locally finite* if each point of E^d has a neighbourhood meeting only finitely many tiles. To avoid pathological situations, from now on *all tilings are taken to be locally finite*. Note for this general assumption that in this article the underlying space of a tiling will always be a finite-dimensional Euclidean space. For results on tilings of topological vector spaces see, e.g., Klee (1986).

A tiling $\mathcal{T}$ of E^d by convex polytopes is *normal* if its tiles are uniformly bounded; that is, there exist positive real numbers r and R such that each tile contains a Euclidean ball of radius r and is contained in a Euclidean ball of radius R. We remark that the definition of normality is more complicated in case the tiles of $\mathcal{T}$ are allowed to be non-convex; see section 3.2.

It follows from the definition of a tiling that the intersection of any finite set of (at least two) tiles necessarily has measure zero. For an arbitrary (locally finite) tiling of E^d (by topological d-cells) the intersection pattern of tiles can be very complicated. It seems that except for plane tilings and special kinds of tilings in higher dimensions there is no generally accepted terminology in the literature which captures the various possibilities.

The best behaved tilings of E^d are the tilings by convex polytopes which respect the facial structure of the tiles; these are the face-to-face tilings. More precisely, a tiling $\mathcal{T}$ by convex d-polytopes is called *face-to-face* if the intersection of any two tiles is a face of each tile, possibly the (improper) empty face. For a face-to-face tiling the intersection of any number of tiles is a face of each of the tiles.

Let $\mathcal{T}$ be a face-to-face tiling of E^d by convex d-polytopes. For $i = 0, \ldots, d$, the *i-faces* (faces of dimension i) of the tiles of $\mathcal{T}$ are called the *i-faces* of $\mathcal{T}$. Then the d-faces of $\mathcal{T}$ are precisely the tiles of $\mathcal{T}$. The faces of dimension 0 and 1 are also called the *vertices* and *edges* of $\mathcal{T}$, respectively. Also, the empty set and E^d itself are considered as (improper) faces of $\mathcal{T}$ of dimension -1 and $d+1$, respectively. The face-to-face property implies that the set of all faces of $\mathcal{T}$, ordered by set-theoretic inclusion, is a lattice called the *face-lattice* of $\mathcal{T}$. By an abuse of notation, if $\mathcal{T}$ is a face-to-face tiling of E^d by convex polytopes, then we often do not distinguish between $\mathcal{T}$ and its face-lattice.

The notion of "face-to-face" carries over to more general tilings in which the tiles are topological d-polytopes, that is, homeomorphic images of convex d-polytopes (equipped with the corresponding boundary complex).

Let $\mathcal{T}$ and $\tilde{\mathcal{T}}$ be face-to-face tilings by convex (or more generally, topological) polytopes in E^d. A mapping $\Phi:\mathcal{T} \to \tilde{\mathcal{T}}$ of the face-lattice of $\mathcal{T}$ onto the face-lattice of $\tilde{\mathcal{T}}$ is called a *(combinatorial) isomorphism* if it is one-to-one and inclusion preserving. If such a mapping Φ exists, then $\mathcal{T}$ and $\tilde{\mathcal{T}}$ are said to be *isomorphic*, or *combinatorially equivalent*. An isomorphism of $\mathcal{T}$ onto itself is called an *automorphism*. The set of all automorphisms of $\mathcal{T}$ forms a group, the *automorphism group* $A(\mathcal{T})$ of $\mathcal{T}$.

A mapping $\Phi:\mathcal{T} \to \tilde{\mathcal{T}}$ between the face-lattices of $\mathcal{T}$ and $\tilde{\mathcal{T}}$ is called a *duality* if it is one-to-one and inclusion reversing. If such a mapping exists, then $\tilde{\mathcal{T}}$ is said to be a *dual* of $\mathcal{T}$. A tiling can have many (metrically distinct) duals, but any two of these are isomorphic.

Contrary to a general misbelief there seems to be no satisfactory theory of metrical duality which extends the well-known concept of duality for highly symmetrical polytopes such as the regular polytopes (cf. Coxeter 1973). It is not even known if any face-to-face tiling $\mathcal{T}$ by convex d-polytopes has a dual whose tiles are also convex d-polytopes.

Let $\mathcal{T}$ be a (locally finite) plane tiling (by topological discs), where a priori there need not be any facial structure on the tiles. We will impose a facial structure on $\mathcal{T}$ and its tiles as follows. A point in the plane is called a *vertex* of $\mathcal{T}$ if it is contained in at least 3 tiles. Since the tiles are topological discs, each simple closed curve which forms the boundary of a tile T is divided into a finite number of closed arcs by the vertices of $\mathcal{T}$, where any two arcs are disjoint except for possibly vertices of $\mathcal{T}$. These arcs are called the *edges of the tile* T and are also referred to as *edges of* $\mathcal{T}$. Each vertex x of $\mathcal{T}$ is contained in finitely many edges of $\mathcal{T}$, the number being referred to as the *valence* $v(x)$ of x; clearly, $v(x) \geqslant 3$ for all vertices x. Note that two tiles can intersect in several edges, where pairs of edges may or may not have vertices in common. Two tiles with a common edge are called *adjacents* of each other.

If the tiles of $\mathcal{T}$ are planar convex or non-convex polygons, then generally the above $\mathcal{T}$-induced notion of vertices and edges of the tiles will not coincide with the standard notion of vertices and edges of polygons. To avoid confusion we shall refer to the latter as *corners* and *sides* of the polygons. In a tiling $\mathcal{T}$ by polygons, the vertices and edges of the tiles may or may not coincide with the corners and sides of the polygons, respectively; if they do coincide, then we call $\mathcal{T}$ an *edge-to-edge* tiling.

The notions of isomorphism, automorphism and duality carry over to arbitrary plane tilings $\mathcal{T}$. The corresponding mapping Φ is now an inclusion preserving or inclusion reversing one-to-one correspondence between the sets of all vertices, edges and tiles of the tilings.

There is a close relationship between combinatorial and topological equivalence of tilings. Two tilings of E^d are said to be of the *same topological type*, or to be *topologically equivalent*, if there is a homeomorphism of E^d which maps one onto the other. For normal plane tilings the two concepts of topological equivalence and combinatorial equivalence turn out to be the same, but an analogous result for higher dimensions seems to be unknown. This result can be used to prove that for every normal plane tiling $\mathcal{T}$ there exists a normal tiling which is dual to $\mathcal{T}$ (cf. Grünbaum and Shephard 1986, pp. 169, 174).

We remark that a more general notion of (plane) tiling deals with so-called *marked tilings* (cf. Grünbaum and Shephard 1986, p. 28). These are tilings in our sense, in which there is a *marking*, or a *motif*, on each tile. By thinking of each tile of a tiling as a motif on itself we can regard each tiling trivially as a marked tiling. Unless specified differently, all tilings are supposed to be unmarked, or more precisely, marked trivially. We will discuss marked tilings in connection with non-periodic tilings.

It seems that in higher dimensions no general terminology has been introduced which deals with the distinction of an a priori-facial-structure and the $\mathcal{T}$-induced facial structure on the tiles of $\mathcal{T}$.

There are other (equivalent) approaches to the notion of "tiling". For example, in Dress and Huson (1987) a plane tiling is defined in terms of its 1-skeleton (that is, its set of vertices and edges) rather than in terms of its set ot tiles.

A *monohedral* tiling $\mathcal{T}$ of E^d is a tiling in which all tiles are congruent to one fixed set T, the (*metrical*) *prototile* of $\mathcal{T}$. We say that T *admits* the tiling $\mathcal{T}$. More generally, a tiling $\mathcal{T}$ is *n-hedral* ($n \geqslant 1$) if each tile in $\mathcal{T}$ is congruent to one *prototile* from an n-element set $\mathcal{S}$, with each prototile actually representing one congruence class of tiles of $\mathcal{T}$; as above, we say that $\mathcal{S}$ *admits* the tiling $\mathcal{T}$.

The simplest examples of monohedral tilings $\mathcal{T}$ are those in which the tiles are translates of a fixed set T. If T admits such a tiling, then we say that T *tiles by translation*. In a tiling $\mathcal{T}$ by translates of T, if the corresponding translation vectors form a d-dimensional lattice L in E^d, then $\mathcal{T}$ is called a *lattice tiling with lattice* L. If $\mathcal{T}$ is a lattice tiling of E^d with convex d-polytopes as tiles, then the prototile T is called a *parallelohedron*.

A central notion is that of a symmetry of a tiling. A Euclidean motion σ of E^d is a *symmetry* of a tiling $\mathcal{T}$ if σ maps (each tile of) $\mathcal{T}$ onto (a tile of) $\mathcal{T}$. The set of all symmetries of $\mathcal{T}$ forms a group, the *symmetry group* $S(\mathcal{T})$ of $\mathcal{T}$.

The subgroup of all translations in $S(\mathcal{T})$, the *translation subgroup* of $S(\mathcal{T})$, is in one-to-one correspondence with the k-dimensional lattice in E^d consisting of all corresponding translation vectors; here, $k = 0, 1, \ldots$ or d. A tiling $\mathcal{T}$ in E^d is called *periodic* if $k = d$, that is, if $S(\mathcal{T})$ contains translations in d linearly independent directions. Trivially, each lattice tiling of E^d is periodic. A tiling $\mathcal{T}$ is called *non-periodic* if $S(\mathcal{T})$ contains no other translation than the identity; here $k = 0$. Note that a tiling which is not periodic need not be non-periodic.

An important problem in the theory of tilings is the classification of tilings with respect to certain transitivity properties of the symmetry group.

A tiling $\mathcal{T}$ of E^d is called *isohedral* if $S(\mathcal{T})$ acts transitively on the tiles of $\mathcal{T}$. Clearly, any isohedral tiling is monohedral, but the converse is not true. More generally, $\mathcal{T}$ is said to be *k-isohedral* $(k \geqslant 1)$ if there are precisely k transitivity classes for the action of $S(\mathcal{T})$ on the tiles of $\mathcal{T}$.

Analogously one can consider transitivity properties for other elements of the tiling. Let $\mathcal{T}$ be a plane tiling, so that there are well-defined notions of vertices and edges of $\mathcal{T}$. Then $\mathcal{T}$ is called *isogonal* (*isotoxal*) if $S(\mathcal{T})$ acts transitively on the vertices (edges, respectively) of $\mathcal{T}$. In analogy to the notion of "k-isohedral" we can define "*k-isogonal*" and "*k-isotoxal*" by replacing tiles by vertices and edges of $\mathcal{T}$, respectively. The notions of isogonality and isotoxality carry over to face-to-face tilings $\mathcal{T}$ in E^d.

There are various other ways to study tilings with respect to transitivity properties of their symmetry group. A face-to-face tiling $\mathcal{T}$ of E^d is called *regular* if $S(\mathcal{T})$ acts transitively on the flags (maximal chains of mutually incident faces). In the plane there are only three regular tilings, namely those by regular triangles, squares and hexagons. In E^4 there are two exceptional regular tilings, with Schläfli-symbols $\{3,3,4,3\}$ and $\{3,4,3,3\}$ (cf. Coxeter 1973); their tiles are 4-crosspolytopes and 24-cells, respectively. The only other regular tilings are the well-known cubical tessellations in E^d, $d \geqslant 1$.

The structure of the symmetry group $S(\mathcal{T})$ of a tiling $\mathcal{T}$ in E^d depends essentially on the number of independent translations in $S(\mathcal{T})$. First note that $S(\mathcal{T})$ cannot contain arbitrarily small translations, since by our general assumptions the tiles in $\mathcal{T}$ are topological d-cells. This shows that $S(\mathcal{T})$ is a discrete group of Euclidean motions in E^d. Recall that a group G of Euclidean motions is *discrete* (or, *acts discretely*) if for each point x in E^d the set of its transforms by G is a discrete set in E^d.

Given two tilings $\mathcal{T}_1$ and $\mathcal{T}_2$ of E^d, we call $\mathcal{T}_1$ and $\mathcal{T}_2$ of the *same symmetry type* if $S(\mathcal{T}_1)$ and $S(\mathcal{T}_2)$ are *isomorphic* in the following geometric sense: there exists an affinity (non-singular affine transformation) α of E^d such that $\sigma \mapsto \alpha\sigma\alpha^{-1}$ defines an (abstract) isomorphism of $S(\mathcal{T}_1)$ onto $S(\mathcal{T}_2)$. In other words, $S(\mathcal{T}_1)$ and $S(\mathcal{T}_2)$ are *isomorphic* if and only they are conjugate subgroups of the group of all affinities of E^d. Note that this definition of isomorphism deviates from the standard definition of abstract isomorphism, but is more useful for geometric purposes.

In classifying (up to isomorphism) symmetry groups $S(\mathcal{T})$ of plane tilings $\mathcal{T}$, three cases have to be distinguished (cf. Grünbaum and Shephard 1986, p. 43, Burckhardt 1966, L. Fejes Tóth 1964, Coxeter and Moser 1980, Engel 1986).

(1) If $S(\mathcal{T})$ contains no translations, then $S(\mathcal{T})$ is (geometrically) isomorphic to the cyclic group C_n of order n or the dihedral group D_n of order $2n, n \geqslant 1$.

(2) If $S(\mathcal{T})$ contains only translations in one independent direction, then $S(\mathcal{T})$ is isomorphic to one of 7 groups called *strip groups*, or *frieze groups*.

(3) If $S(\mathcal{T})$ contains translations in two independent directions, then $\mathcal{T}$ is periodic and $S(\mathcal{T})$ is isomorphic to one of 17 groups known as (*plane*) *crystallographic groups*, *periodic groups* or *wallpaper groups*.

A group G of isometries in E^d is called a *crystallographic group*, or *space group*, if it is discrete and contains translations in d independent directions; for a detailed discussion of these groups see Engel (1986) and chapter 3.7 in this Handbook. See also Milnor (1976). By a result of Bieberbach (1910) the crystallographic groups are precisely the discrete groups of isometries in E^d whose fundamental region is compact; see also Buser (1985) for a new proof of this result. Similar as above, two space groups are said to be (*geometrically*) *isomorphic*, or of the *same space group type*, if they are conjugate subgroups in the group of all affinities of E^d. A theorem of Bieberbach (1912) shows that two space groups are geometrically isomorphic if and only if they are isomorphic in the usual abstract sense. In the plane there are 17 types of crystallographic groups, first enumerated by Fedorov. In 3-space there are 219 types of crystallographic groups, first enumerated by Fedorov (1892) and Schoenflies (1891). Of these types, 11 split further into enantiomorphic pairs. The number of types in 4-space is precisely 4783 (of which 112 split); these were classified in Brown et al. (1978). For $d \geqslant 5$ the number of types is not known but is finite, by theorems of Bieberbach (1910) and Frobenius (1911); see also Buser (1985).

For a discussion of several interdisciplinary aspects of the subject of tiling the reader is referred to Senechal and Fleck (1988).

3. Plane tilings

Much of the attraction of tilings comes from their appearance in nature and art. Thus it is only natural that in any detailed exposition of plane tilings figures should play an important role. However, in this paper lack of space prevents us from illustrating most results by appropriate figures, so that we must refer the reader to other references, usually to Grünbaum and Shephard (1986).

3.1. Tilings by regular polygons

Historically, tilings by regular polygons were the first kinds of tilings to be the subject of mathematical investigations. For a detailed discussion see Grünbaum and Shephard (1986, chapter 2). We begin by observing the obvious fact that the only monohedral edge-to-edge tilings by regular polygons are the three regular tilings by triangles, squares and hexagons.

The situation becomes more interesting if we allow regular polygons of various kinds. An edge-to-edge tiling $\mathcal{T}$ by regular polygons is said to be of *type* $(n_1.n_2.\ \cdots\ .n_r)$ if each vertex x of $\mathcal{T}$ is of *type* $n_1.n_2.\ \cdots\ .n_r$; that is to say, in a cyclic order, x is surrounded by an n_1-gon, an n_2-gon, and so on. Here, and in similar situations below, the mere reversal of cyclic order is not counted as distinct. To obtain a unique type symbol for $\mathcal{T}$ and its vertices we will always choose the one which comes lexicographically first, and we do so in similar situations below. For convenience we will also denote $\mathcal{T}$ itself by $(n_1.n_2.\ \cdots\ .n_r)$. For the next theorem recall our definition of equality of tilings.

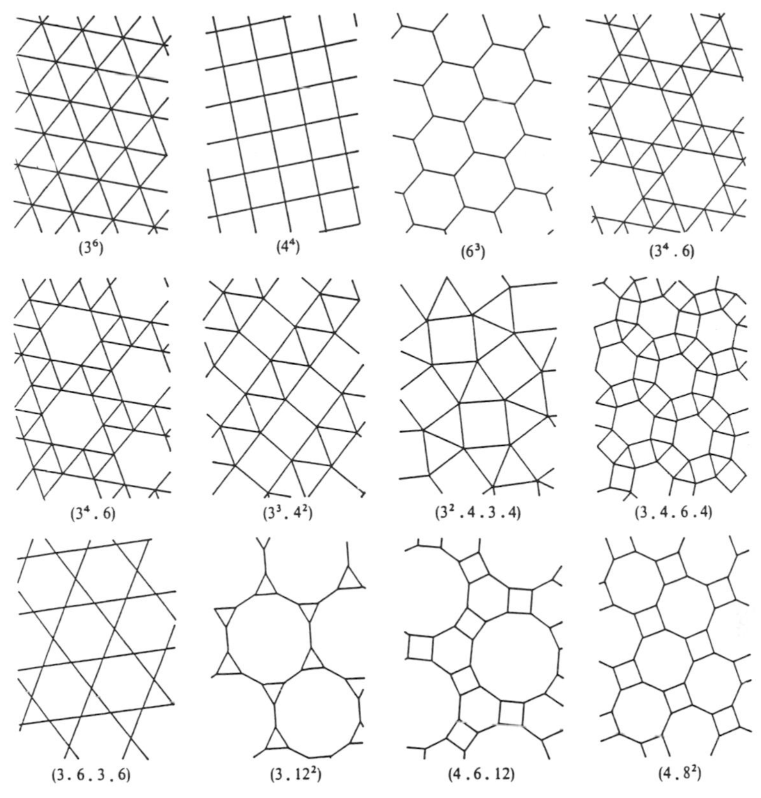

Figure 1. Archimedean plane tilings.

Theorem 2. *There exist precisely* 11 *distinct edge-to-edge tilings by regular polygons such that all vertices are of the same type. These are* (3^6), $(3^4.6)$, $(3^3.4^2)$, $(3^2.4.3.4)$, $(3.4.6.4)$, $(3.6.3.6)$, (3.12^2), (4^4), $(4.6.12)$, (4.8^2) *and* (6^3).

The 11 tilings of Theorem 2 are known as the *Archimedean tilings* and were already enumerated by Kepler (1619); see Grünbaum and Shephard (1986, p. 59). The tilings are shown in fig. 1. The tilings (3^6), (4^4) and (6^3) are the regular tilings; the tiling $(3^4.6)$ occurs in two enantiomorphic (mirror image) forms and is the only Archimedean tiling with this property.

An edge-to-edge tiling $\mathcal{T}$ is called *uniform* if it is isogonal and its tiles are convex regular polygons.

Theorem 3. *The uniform plane tilings are precisely the* 11 *Archimedean tilings.*

The uniform tilings of the plane have as their dual counterpart the so-called *Laves tilings*. A vertex x of an edge-to-edge plane tiling $\mathcal{T}$ is called *regular* if the edges meeting at x dissect a small neighbourhood of x into equiangular parts, the angles being $2\pi/v$ if v is the valence of x. Then, in some sense (which can be made precise) the classification of monohedral edge-to-edge tilings $\mathcal{T}$ with regular vertices is dual to the classification of edge-to-edge tilings by regular polygons with all vertices of the same type.

Let $\mathcal{T}$ be edge-to-edge and monohedral, with prototile an r-gon whose vertices have valences $v_1, \ldots, v_r$ in $\mathcal{T}$. If all vertices of $\mathcal{T}$ are regular, we denote $\mathcal{T}$ by the symbol $[v_1.v_2. \cdots .v_r]$, with appropriate conventions to assure uniqueness of the symbol for $\mathcal{T}$ (cf. Grünbaum and Shephard 1986, p. 97).

Theorem 4. (a) *If $\mathcal{T}$ is a monohedral edge-to-edge plane tiling with only regular vertices, then its symbol is one of the* 11 *symbols mentioned in* (b).

(b) *To each of the symbols* $[3^4.6]$, $[3^2.4.3.4]$, $[3.6.3.6]$, $[3.4.6.4]$, $[3.12^2]$, $[4.6.12]$, $[4.8^2]$ *and* $[6^3]$ *corresponds a unique such tiling $\mathcal{T}$; to* $[3^3.4^2]$ *and* $[4^4]$ *correspond families of such tilings depending on one real-valued parameter; and to* $[3^6]$ *corresponds a family of tilings with two such parameters.*

The tilings of Theorem 4 (or more precisely, representatives for each symbol) are known as the 11 *Laves tilings*, after the crystallographer Laves; see fig. 2. Again, the tiling $[3^4.6]$ occurs in two enantiomorphic forms. All Laves tilings are isohedral; this is analogous to the fact that the Archimedean tilings are uniform.

The Laves tilings and uniform tilings are related by duality, with the tilings $[v_1.v_2. \cdots .v_r]$ and $(v_1.v_2. \cdots .v_r)$ corresponding to each other.

If we relax the condition that an arbitrary (not necessarily transitive) plane tiling $\mathcal{T}$ by regular convex polygons be edge-to-edge, then the number of further possibilities is enormous. For more details see Grünbaum and Shephard (1986, section 2.4).

A plane tiling $\mathcal{T}$ by regular polygons is called *equitransitive* if each set of mutually congruent tiles forms one transitivity class with respect to $S(\mathcal{T})$. All uniform edge-to-edge tilings are equitransitive except for $(3^4.6)$ which is dihedral (2-hedral) but 3-isohedral. However there are many further equitransitive tilings of both kinds, edge-to-edge or not edge-to-edge; see Grünbaum and Shephard (1986, pp. 70, 73), and Danzer, Grünbaum and Shephard (1987).

3.2. Some general results on well-behaved plane tilings

For most results in this section the reader is referred to Grünbaum and Shephard (1986, chapter 3). Throughout, $\mathcal{T}$ will be a (locally finite) plane tiling (by topological discs). By a *patch* of $\mathcal{T}$ we mean a finite collection of tiles of $\mathcal{T}$ whose union is a topological disc. For a connected plane set D, we denote by $A(D,\mathcal{T})$ the *patch generated by D*; this patch is derived from the union of all tiles of $\mathcal{T}$ which meet D by adjoining just enough tiles to fill up the "holes" of this union and turn it into a topological disc.

Extending the definition of normality given in section 2, $\mathcal{T}$ is called *normal* if both the intersection of every two tiles in $\mathcal{T}$ is connected and the tiles of $\mathcal{T}$ are uniformly bounded in the sense described in section 2.

For $r > 0$ and a point P in E^2 we write $D(r,P)$ for the closed circular disc of radius r centered at P. By $t(r,P)$, $e(r,P)$ and $v(r,P)$ we denote the numbers

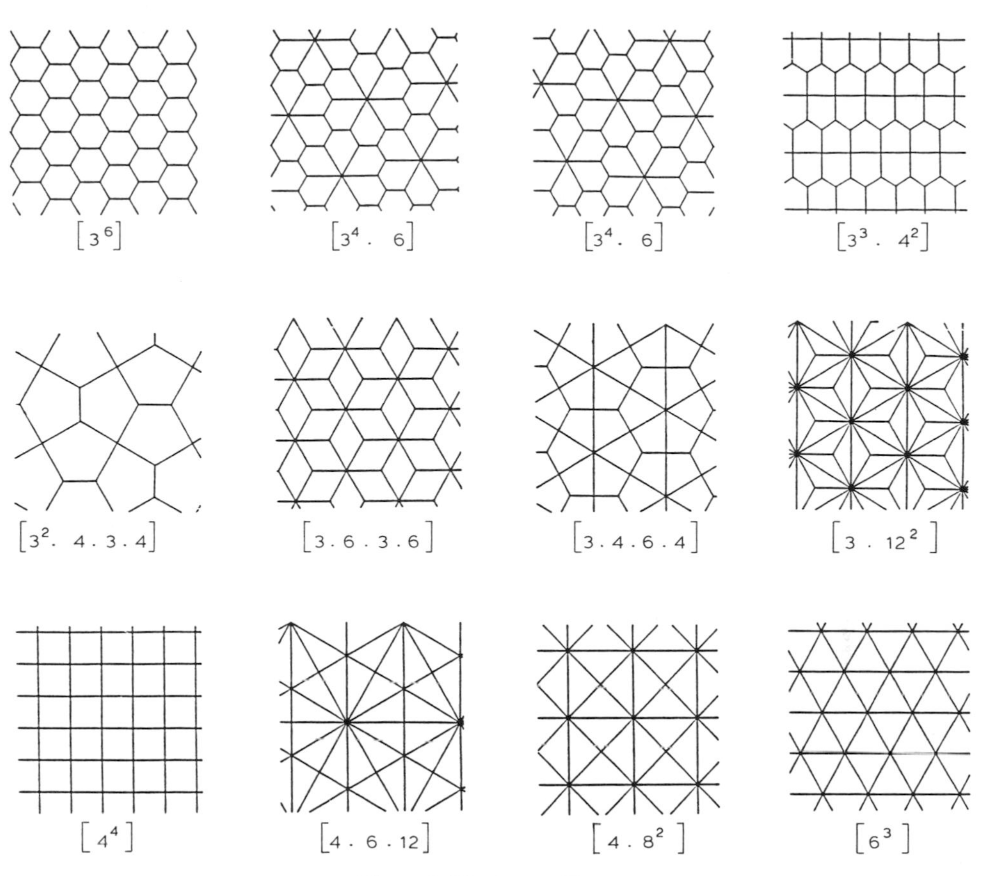

Figure 2. The Laves tilings.

of tiles, edges and vertices in the patch $A(r,P) := A(D(r,P),\mathcal{T})$ generated by $D(r,P)$, respectively. The following basic result implies that normality prevents the existence of "singularities" in $\mathcal{T}$ both at finite points and at infinity. Historically, Reinhardt's (1918) work was the first to include a statement resembling that of Theorem 5.

Theorem 5 (Normality Lemma). *If $\mathcal{T}$ is a normal plane tiling, then for every $s > 0$*

$$\lim_{r\to\infty} \frac{t(r+s,P) - t(r,P)}{t(r,P)} = 0.$$

We remark that there is also a d-dimensional analogue of Theorem 5 for normal tilings of E^d by convex polytopes. The following theorem is a consequence of Theorem 5.

Theorem 6. *If $\mathcal{T}$ is a normal plane tiling in which each tile has the same number k of edges, then $k = 3,4,5$ or 6.*

In the literature Euler's Theorem on Tilings is often illegally used in that no restrictions are imposed on the tilings to which it is applied. To formulate the correct statement we introduce the following notation.

A normal plane tiling $\mathcal{T}$ is said to be *balanced* (*with respect to a point* P) if both limits

$$\lim_{r\to\infty} \frac{v(r,P)}{t(r,P)}$$

and

$$\lim_{r\to\infty} \frac{e(r,P)}{t(r,P)}$$

exist and are finite. If for a tiling $\mathcal{T}$ and a point P the two limits exist and are finite, then these limits exist and have the same value $v(\mathcal{T})$ and $e(\mathcal{T})$, respectively, for any other choice of P in the plane.

Theorem 7 (Euler's Theorem for Tilings). *For any normal plane tiling $\mathcal{T}$, if one of the limits $v(\mathcal{T})$ or $e(\mathcal{T})$ exists and is finite, then so does the other. Thus $\mathcal{T}$ is balanced and $v(\mathcal{T}) = e(\mathcal{T}) - 1$.*

The following kinds of normal plane tilings can be shown to be balanced: tilings in which each tile has the same number of adjacents; tilings in which each vertex has the same valence; each periodic tiling (cf. Grünbaum and Shephard 1986, section 3.3).

A more refined notion is that of a strongly balanced tiling. For a plane tiling $\mathcal{T}$ we write $v_j(r,P)$ for the number of vertices of valence j in the patch $A(r,P)$ and $t_k(r,P)$ for the number of tiles with k adjacents in $A(r,P)$. Then a normal tiling $\mathcal{T}$ is called *strongly balanced* if for one point P all the limits

$$v_j(\mathcal{T}) := \lim_{r\to\infty} \frac{v_j(r,P)}{t(r,P)}$$

and

$$t_k(\mathcal{T}) := \lim_{r\to\infty} \frac{t_k(r,P)}{t(r,P)}$$

exist ($j,k \geqslant 3$); by normality they must be finite. Again, this definition is independent of P. In any strongly balanced tiling $\mathcal{T}$, $v(\mathcal{T}) = \sum_{j\geqslant 3} v_j(\mathcal{T}) < \infty$ and $\sum_{k\geqslant 3} t_k(\mathcal{T}) = 1$; in particular, $\mathcal{T}$ is necessarily balanced. The following kind of normal tilings are strongly balanced: tilings in which each tile has precisely k vertices and these vertices have valences $j_1,\ldots,j_k$ (in some order); tilings in which each vertex has valence j and is incident with tiles that have $k_1,\ldots,k_j$ adjacents (in some order); each periodic tiling.

The following theorems can be derived from Euler's Theorem for Tilings (cf. Grünbaum and Shephard 1986, sections 3.4 and 3.5). They have counterparts in the theory of convex polyhedra (cf. Grünbaum 1967).

Theorem 8. *In every strongly balanced plane tiling $\mathcal{T}$, we have*

(a) $$2\sum_{j\geqslant 3}(j-3)v_j(\mathcal{T})+\sum_{k\geqslant 3}(k-6)t_k(\mathcal{T})=0,$$

(b) $$\sum_{j\geqslant 3}(j-4)v_j(\mathcal{T})+\sum_{k\geqslant 3}(k-4)t_k(\mathcal{T})=0,$$

(c) $$\sum_{j\geqslant 3}(j-6)v_j(\mathcal{T})+2\sum_{k\geqslant 3}(k-3)t_k(\mathcal{T})=0.$$

Theorem 9. *For each strongly balanced plane tiling $\mathcal{T}$ we have*

$$\frac{1}{\sum_{j\geqslant 3} jw_j(\mathcal{T})}+\frac{1}{\sum_{k\geqslant 3} kt_k(\mathcal{T})}=\frac{1}{2},$$

with

$$w_j(\mathcal{T}) = \frac{v_j(\mathcal{T})}{v(\mathcal{T})}.$$

Note that the sum $\sum_{j\geqslant 3} jw_j(\mathcal{T})$ can be interpreted as the average valence taken over all vertices of $\mathcal{T}$; similarly, $\sum_{k\geqslant 3} kt_k(\mathcal{T})$ is the average number of edges of the tiles, taken over all tiles in $\mathcal{T}$.

Theorems 8 and 9 are concerned with equations involving the relative frequences of vertices of various valences and tiles with various number of adjacents. There exist results which deal with the problem of preassigning such frequences for periodic tilings; by analogy with results of a similar nature for convex 3-polytopes (cf. Grünbaum 1967), they are called Eberhard-type-theorems. Further refinements of the notion of balanced tilings deal with so-called *prototile balanced* tilings and *metrically balanced* tilings. For more details see Grünbaum and Shephard (1986, sections 3.4 and 3.9).

There are also stronger versions of Euler's Theorem for tilings on which for specific values of k certain restrictions on the distribution of the k-gonal tiles are imposed (Stehling 1988). To give an example, if a tiling has only finitely many triangles, quadrangles and pentagons, but infinitely many heptagons as tiles, then it cannot be normal no matter how the heptagons are distributed.

An important class of plane tilings is given by the periodic tilings.

Theorem 10. *Let $\mathcal{S}$ be a finite set of polygonal prototiles. If $\mathcal{S}$ admits an edge-to-edge plane tiling $\mathcal{T}_1$ which possesses a (non-trivial) translational symmetry, then $\mathcal{S}$ also admits a periodic tiling.*

Let $\mathcal{T}$ be a periodic tiling and L the 2-dimensional lattice generated by the translation vectors corresponding to translations in $S(\mathcal{T})$. A *fundamental parallelogram* $\mathcal{P}$ for L is a parallelogram whose translates by L cover the plane without gaps and overlaps. We will always assume that $\mathcal{P}$ is properly chosen such that no vertex of $\mathcal{T}$ lies on a side of $\mathcal{P}$, and no corner of $\mathcal{P}$ lies on an edge of $\mathcal{T}$. By V, E and T we denote the number of vertices, edges and tiles in $\mathcal{P}$, respectively, where fractions of edges and tiles are counted appropriately. The following statement is known as *Euler's Theorem for Periodic Plane Tilings* and corresponds to Euler's Theorem for the torus.

Theorem 11. *For every normal periodic plane tiling $\mathcal{T}$ the numbers V, E and T do not depend on the (proper) choice of the fundamental parallelogram $\mathcal{P}$, and moreover,*

$$V - E + T = 0.$$

In constructing plane tilings one often comes along the problem of extending patches of tiles to larger patches or to a tiling of the whole plane. A set $\mathcal{S}$ of prototiles is said to *tile over a subset* D of E^2 if $\mathcal{S}$ admits a patch A such that the union of the tiles in A contains D.

Theorem 12 (Extension Theorem)**.** *Let $\mathcal{S}$ be any finite set of planar prototiles, each of which is a topological disc. If $\mathcal{S}$ tiles over arbitrary large circular discs D, then $\mathcal{S}$ admits a tiling $\mathcal{T}$ of the whole plane.*

Note that there are patches of tiles covering a circular disc, which cannot be extended so as to cover larger circular discs. That is, in constructing a global tiling it may be necessary to "rearrange" the tiles after each step. See Grünbaum and Shephard (1986, section 3.8).

3.3. Classification with respect to topological transitivity properties

The results discussed in this section can be found in Grünbaum and Shephard (1986, chapter 4). As remarked in section 2, for normal plane tilings $\mathcal{T}$ the concepts of topological equivalence and combinatorial equivalence are the same. Any automorphism in $A(\mathcal{T})$ can be realized by a homeomorphism of the plane that preserves $\mathcal{T}$, and vice versa.

Let $\mathcal{T}$ be a plane tiling and T a tile of $\mathcal{T}$. Similar as in section 3.1 we associate with T the *valence-type* $j_1.j_2.\cdots.j_k$ provided T has k vertices which, in cyclic order, have valences $j_1, j_2, \ldots, j_k$. We call $\mathcal{T}$ *homogeneous of type* $[j_1.j_2.\cdots.j_k]$ if $\mathcal{T}$ is normal and each tile of $\mathcal{T}$ has valence-type $j_1.j_2.\cdots.j_k$. To obtain a unique symbol for $\mathcal{T}$ we choose among the various symbols again the one which comes lexicographically first, and do so in similar situations below.

A plane tiling $\mathcal{T}$ is called *homeohedral*, or *topologically tile-transitive*, or *combinatorially tile-transitive*, if it is a normal tiling and $A(\mathcal{T})$ acts transitively on the tiles of $\mathcal{T}$. Clearly, each homeohedral tiling is homogeneous. The classification of homeohedral tilings has been carried out by many authors; see Laves (1931) or Bilinski (1949).

Theorem 13. (a) *If $\mathcal{T}$ is a homogeneous plane tiling, then it is of one of the eleven types* $[3^6], [3^4.6], [3^3.4^2], [3^2.4.3.4], [3.4.6.4], [3.6.3.6], [3.12^2], [4^4], [4.6.12], [4.8^2], [6^3]$.

(b) *Each homogeneous plane tiling is homeohedral. All homogeneous plane tilings of the same type are topologically equivalent to each other, and each type is represented by one of the Laves tilings in Theorem* 4 *and fig.* 2.

A tiling $\mathcal{T}$ is called *homeogonal*, or *topologically vertex-transitive*, or *combinatorially vertex-transitive*, if $\mathcal{T}$ is normal and $A(\mathcal{T})$ acts transitively on the vertices of $\mathcal{T}$. Similar as for uniform tilings we associate with a homeogonal $\mathcal{T}$ its *type* $(k_1.k_2.\cdots.k_j)$; here, j is the valence of any vertex x, and the tiles that contain x have, in a suitable cyclic order, k_1 vertices, k_2 vertices, and so on. If $\mathcal{T}$ and $\tilde{\mathcal{T}}$ are normal dual tilings, then $\mathcal{T}$ is homeogonal if and only if $\tilde{\mathcal{T}}$ is homeohedral. See Šubnikov (1916) for the historically first account on homeogonal tilings.

Theorem 14. (a) *If $\mathcal{T}$ is a homeogonal plane tiling, then it is of one of the eleven types* (3^6), $(3^4.6)$, $(3^3.4^2)$, $(3^2.4.3.4)$, $(3.4.6.4)$, $(3.6.3.6)$, (3.12^2), (4^4), $(4.6.12)$, (4.8^2), (6^3).

(b) *All homeogonal tilings of the same type are mutually topologically equivalent, and each type is represented by one of the uniform tilings of Theorems* 2, 3 *and fig.* 1.

We remark that there is an analogous result for *homeotoxal* tilings, that is, normal tilings $\mathcal{T}$ for which $A(\mathcal{T})$ is edge-transitive. There are five "types" of homeotoxal tilings; see Grünbaum and Shephard (1986, p. 180).

If the requirement of normality is dropped, then many further possibilities arise. For example, each of the regular tessellations $\{p,q\}$ of the hyperbolic plane can be realized in the Euclidean plane by an edge-to-edge tiling with convex p-gonal tiles; this tiling is necessarily non-normal.

3.4. Classification with respect to symmetries

The classification of plane tilings with respect to symmetries is a special case of the more general concept of classifying geometric objects with respect to symmetries, or, as we will say, *by homeomerism*. This general approach fits into Klein's Erlangen Program and has further applications in the classification of other geometric objects such as coloured tilings, various kinds of patterns and line arrangements (cf. Grünbaum and Shephard 1986, chapter 7).

Let $\mathcal{R}$ and $\mathcal{R}'$ be geometric objects in Euclidean space E^d, with symmetry groups $S(\mathcal{R})$ and $S(\mathcal{R}')$, respectively. Let $\Phi : E^d \to E^d$ be a homeomorphism which maps $\mathcal{R}$ onto $\mathcal{R}'$. Then Φ is said to be *compatible with a symmetry* σ of $\mathcal{R}$ if there exists a symmetry σ' of $\mathcal{R}'$ such that $\sigma'\Phi = \Phi\sigma$. We call Φ *compatible with* $S(\mathcal{R})$ if Φ is compatible with each σ in $S(\mathcal{R})$.

Now, two geometric objects $\mathcal{R}$ and $\mathcal{R}'$ in E^d are said to be *homeomeric*, or *of the same homeomeric type*, if there exists a homeomorphism $\Phi : E^d \to E^d$ which maps $\mathcal{R}$ onto $\mathcal{R}'$ such that Φ is compatible with $S(\mathcal{R})$ and Φ^{-1} is compatible with $S(\mathcal{R}')$. It follows that $S(\mathcal{R})$ and $S(\mathcal{R}')$ are (geometrically) isomorphic in the sense of section 2.

This general concept applies to the classification of normal plane tilings with certain transitivity properties of their symmetry group (cf. Grünbaum and Shephard 1986, sections 6.2 and 7.2, Dress 1987).

Theorem 15. *There exist precisely* 81 *homeomeric types of normal isohedral plane tilings. Precisely* 47 *of these can be realized by a normal isohedral edge-to-edge tiling with convex polygonal tiles.*

In proving Theorem 15 one comes along the problem of actually enumerating the different types of tilings. One way to solve this problem is to introduce so-called *incidence symbols* which encode data about the local structure of an isohedral plane tiling. Then two isohedral plane tilings are said to be *of the same isohedral type* if they are of the same topological type (in the sense of section 2, or Theorem 13) and their incidence symbols are (essentially) the same. Then it can be shown that for an isohedral plane tiling the classification according to homeomerism coincides with the classification according to isohedral type. It turns out that incidence symbols provide an algorithmic approach to the problem by which it can finally be solved. It is worth mentioning that the actual classification is carried out for marked tilings; in fact, there are precisely 12 isohedral types of marked tilings which are not realizable by (unmarked) tilings (cf. Grünbaum and Shephard 1986, pp. 283, 344). Also it is worth remarking that isohedral tilings by polygons admit a finer classification according to their *polygonal isohedral type* (Grünbaum and Shephard 1986, pp. 474, 489).

The situation is similar for the classification of normal isogonal and normal isotoxal plane tilings. Again, suitable incidence symbols can be used for the enumeration of the types (cf. Grünbaum and Shephard 1986, sections 6.3 and 6.4).

Theorem 16. *There exist precisely* 91 *homeomeric types of normal isogonal plane tilings. Precisely* 63 *types can be realized by normal isogonal edge-to-edge tilings with convex polygonal tiles.*

Theorem 17. *There exist precisely* 26 *homeomeric types of normal isotoxal plane tilings. Precisely* 6 *types can be realized by a normal isotoxal edge-to-edge tiling with convex polygonal tiles.*

The classification results of Theorems 15, 16 and 17 have a long history; see Grünbaum and Shephard (1986, sections 6.6 and 7.8) for a detailed discussion. In full generality and rigor the concept of classification by homeomerism was developed by Grünbaum and Shephard. Various notions of "types" of isohedral or isogonal tilings have been used prior to Grünbaum and Shephard (1977, 1978,

1986); see, e.g., Delone (1959), and Delone, Galiulin and Štogrin (1979). The classification presented in Delone, Dolbilin and Štogrin (1978) seems to be equivalent to that covered by Theorem 15.

Besides these solutions to the classification problems there is yet another approach which was recently made by Dress and co-workers. In its general form this approach gives a device to deal with classification problems for equivariant chamber systems on (simply connected) manifolds of any dimension; see Dress (1987). For an exposition of this method for plane tilings see,e.g., Dress and Huson (1987).

To explain the planar case let $\Sigma := \langle \sigma_0, \sigma_1, \sigma_2 \mid \sigma_0{}^2 = \sigma_1{}^2 = \sigma_2{}^2 = 1 \rangle$ denote the free Coxeter group with the three generators $\sigma_0, \sigma_1, \sigma_2$. For a plane tiling $\mathcal{T}$ its barycentric subdivision $\mathcal{C}$ (thin chamber system of rank 2) has the natural structure of a *Σ-set*; that is, Σ acts on $\mathcal{C}$ (on the right) in such a way that for each chamber $C \in \mathcal{C}$ the chamber $C\sigma_i$ is i-adjacent to C (that is, differs from C in its vertex given by an i-dimensional element of $\mathcal{T}$). An *equivariant plane tiling* $(\mathcal{T}, \Gamma)$ is a pair consisting of a plane tiling $\mathcal{T}$ and a distinguished discrete group Γ of automorphisms of $\mathcal{T}$. With each equivariant tiling $(\mathcal{T}, \Gamma)$ is associated its *Delaney symbol* (generalized *Schläfli symbol*) $(\mathcal{D}; m_{01}, m_{12})$ consisting of the set $\mathcal{D}$ of orbits of Γ on $\mathcal{C}$ and two integer-valued functions m_{01}, m_{12} on $\mathcal{D}$ encoding local data of $\mathcal{T}$. The crucial observation is that up to isomorphism each equivariant tiling is completely determined by its Delaney symbol. This one-to-one correspondence of equivariant tilings and Delaney symbols reduces the classification problem for a specific kind of tilings to the purely combinatorial problem of classifying all corresponding Delaney symbols. The latter can be done either by hand or by appropriate computer programs.

For applications of this method see Dress and Scharlau (1984, 1986), Dress and Huson (1992), Dress and Franz (1992), Franz (1988) and Huson (1992), and other references quoted there. See also Dress, Huson and Molnar (1991) for an application of the general technique to the classification of all marked tilings $\mathcal{T}$ in E^3 (by topological polytopes) whose symmetry group is transitive on the 2-faces of $\mathcal{T}$. It seems that at present the classification of all isohedral (tile transitive) tilings in E^3 is hopeless; see Delone and Sandakova (1961) and Molnar (1983).

4. Monohedral tilings

This section deals with tilings of Euclidean spaces of any dimension $d \geqslant 2$. One of the main problems in tiling theory is the classification of all convex d-polytopes T which are prototiles of monohedral tilings of E^d. This problem is not even solved for the plane, where necessarily T must be an n-gon with $n = 3, 4, 5$ or 6; see Theorem 6. In its full generality the classification problem in higher dimensions seems to be intractable, so that suitable restrictions must be imposed on the tiles or on the kind of tilings. In this section we discuss various ramifications of the classification problem, ranging from an account on lattice tilings and space-fillers to combinatorial analogues of the classification problem.

4.1. Lattice tilings

One of the best studied class of tilings are the lattice tilings in E^d by convex d-polytopes, $d \geqslant 2$. The interest in such tilings stems from the interaction with crystallography and the geometry of numbers. Early contributions on the subject can be found in the work of Dirichlet, Fedorov, Minkowski, Voronoi and Delone.

An important concept is that of a *Voronoi region*, or *Dirichlet region*, or *Wirkungsbereich*. Let L be a discrete set in E^d, not necessarily a lattice, and let φ be a positive definite quadratic form on E^d. For $x \in L$ the Voronoi region $V(\varphi, L, x)$ is defined by

$$V(\varphi, L, x) = \{y \in E^d \mid \varphi(y - x) \leqslant \varphi(y - z) \text{ for all } z \in L - \{x\}\}.$$

Then the family of all Voronoi regions $V(\varphi, L, x)$ with $x \in L$ forms a face-to-face tiling $\mathcal{T}$ of E^d, with further properties depending on the choice of L. This tiling is a suitable affine image of another tiling whose tiles are the Voronoi-regions defined with respect to the standard quadratic form and another suitable discrete set. If L is a lattice in E^d, then the Voronoi regions are translates of $V(\varphi, L) := V(\varphi, L, 0)$ and $\mathcal{T}$ is a lattice tiling with $V(\varphi, L)$ as prototile; in particular, $V(\varphi, L)$ is a parallelohedron in the sense of section 2. Note that in applications the term "Voronoi region" is often used only for Voronoi regions which are defined with respect to the standard quadratic form.

The structure of the Voronoi regions of a lattice is intimately related to the structure of the lattice itself. For a description of the Voronoi regions of the root lattices see Conway and Sloane (1988, chapter 21).

The convex d-polytopes T which tile E^d by translation can be characterized by properties involving among other things central symmetry for T and its facets; see Venkov (1954), Aleksandrov (1954), McMullen (1980). This characterization can be used to prove the following theorem, which impinges on Hilbert's 18th problem; see section 4.2 below.

Theorem 18. *Let T be a convex d-polytope. If T tiles E^d by translation, then T admits (uniquely) a face-to-face lattice tiling of E^d and thus is a parallelohedron.*

Call a parallelohedron T *primitive* if in its face-to-face tiling $\mathcal{T}$ each vertex of $\mathcal{T}$ is contained in exactly $d+1$ tiles. The problem of classifying all parallelohedra is completely solved only for $d \leqslant 4$. Table 1 lists the total number of combinatorial types of (primitive) parallelohedra together with corresponding references. The types of parallelohedra in E^2 and E^3 are shown in fig. 3. Minkowski (1897) proved that parallelohedra in E^d have at most $2(2^d - 1)$ facets. As a consequence, for each d there are only finitely many combinatorial types of parallelohedra in E^d. For $d = 3$ the maximum number of facets is 14, which is attained for the last parallelohedron of fig. 3. For a more detailed exposition on parallelohedra see chapter 3.7 in this Handbook. For a short survey see also Gruber (1979), Gruber and Lekkerkerker (1987), and Erdős, Gruber and Hammer (1989).

Table 1

d	Total number of parallelohedra	Number of primitive parallelohedra	References
2	2	1	Fedorov (1885)
3	5	1	Voronoi (1908)
4	52	3	Delone (1929), Štogrin (1973)
5	?	$\geqslant 223$	Baranovskiĭ and Ryškov (1973), Ryškov and Baranovskiĭ (1976), Engel (1989)

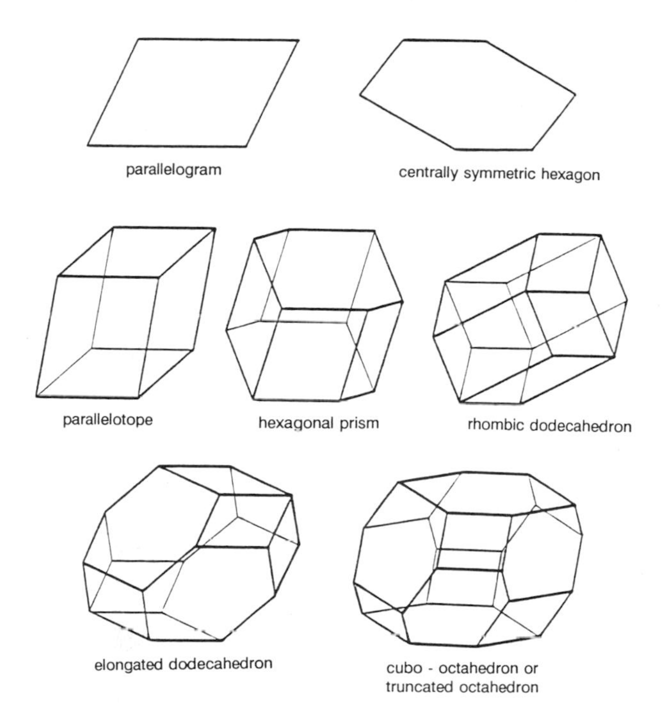

Figure 3. Parallelohedra in the plane and 3-space.

It is an unsettled question of Voronoi (1908) whether each parallelohedron is in fact (a translate of) a Voronoi region for some lattice. This is true for primitive parallelohedra and in general for $d \leqslant 4$; see Voronoi (1908, 1909), Delone (1929) and Žitomirskiĭ (1929). It also holds for zonotopes (Minkowski-sums of line segments); see Shephard (1974), Coxeter (1962), McMullen (1975) and Schneider and Weil (1983). All five parallelohedra in E^3 are zonotopes, but the analogous fact is not true for parallelohedra in E^d with $d \geqslant 4$. See also Groemer (1978/79) for interesting results about multiple tilings by zonotopes.

It is interesting to remark that the result of Theorem 18 is not true for star-shaped polytopes. Stein (1972) has shown that there are non-convex star-shaped polyhedral sets in E^d with $d \geqslant 5$, even centrally symmetric ones if $d \geqslant 10$, which tile space by translation, but do not admit any lattice tiling; see also Stein (1986). Recently Szabó (1985) has constructed centrally symmetric examples in spaces of dimension $d = 2^m - 1$, $m \geqslant 2$, thereby settling the question for ordinary 3-space. See also Bezdek and Kuperberg (1990) for other examples in E^3.

The theory of lattice tilings is, of course, closely related to the theory of lattice packings and lattice coverings. The parallelohedra are precisely those centrally symmetric convex bodies for which both the lattice packing density and the lattice covering density equal 1; see Gruber and Lekkerkerker (1987), G. Fejes Tóth (1983) and chapter 3.3 in this Handbook.

A central problem in the geometry of numbers is that of finding the critical lattices of a convex body; see chapter 3.1 in this Handbook. For the d-cube this amounts essentially to the following problem posed by Minkowski (1907, p. 74).

Given a lattice tiling of E^d by congruent d-cubes, does there always exist a "stack" of cubes in which each two adjacent cubes meet in a whole facet? This question was answered in the affirmative by Hajós (1942) who determined all critical lattices of the cube; for partial results see also Perron (1940b). Generalizing Minkowski's problem Keller (1930) asked if in every monohedral tiling of E^d by congruent d-cubes there exists a pair of tiles which meet in a common facet. Perron (1940a) established that the answer is affirmative for $d \leqslant 6$, but only recently Lagarias and Shor (1992) proved that the answer is negative for large d. For a discussion of these and related problems see also Stein (1974) and Grünbaum and Shephard (1980).

4.2. Prototiles of monohedral tilings

As mentioned in the introduction to section 4, the classification of prototiles of monohedral tilings of the plane or higher-dimensional Euclidean spaces is far from being complete. This classification problem got an important impetus from the 18th of Hilbert's famous problems which concerns crystallographic groups and their fundamental regions (cf. Hilbert 1900, Milnor 1976). In part it asks whether there exists a 3-dimensional polyhedral tile that admits a monohedral, but no isohedral, tiling of E^3. Tiles with this property (for E^d) are called *anisohedral*. Another way to phrase the problem is whether there are polyhedral prototiles of monohedral tilings of E^3 which are not fundamental regions for a discrete group of Euclidean motions. From the context it appears that Hilbert believed that the corresponding planar problem had a negative solution. Note that by taking prisms one can easily construct anisohedral prototiles in dimensions $d \geqslant 3$ from anisohedral planar prototiles.

Hilbert's opinion about the planar case was shared by Reinhardt (1918) who made the first attempt to classify all polygonal prototiles of monohedral plane tilings, but later his enumeration turned out to be incomplete. Hilbert's question was settled by Reinhardt (1928) who discovered an anisohedral polyhedron in E^3.

The planar case was finally solved by Heesch (1935) who found a non-convex anisohedral polygon which admits a periodic monohedral tiling; see also Heesch (1968). Kershner (1968) improved on Heesch's example by producing anisohedral convex pentagons which admit periodic monohedral tilings.

Theorem 19. *For each dimension $d \geqslant 2$ there exist anisohedral convex d-polytopes.*

Many authors who attempted the classification for the planar case believed, or even stated explicitly, that their enumeration is complete. Often this was generally accepted until finally a new prototile was discovered by someone else. For a more detailed account see Grünbaum and Shephard (1986, sections 9.3 and 9.6) and Schattschneider (1978). At present the list of convex prototiles comprises all triangles and quadrangles, 14 "types" of pentagons, and 3 "types" of hexagons. Among these prototiles, the triangles, quadrangles and hexagons and exactly 5 "types" of the pentagons admit isohedral tilings; there are no further convex polygons which admit isohedral plane tilings. The various prototiles were discovered by Reinhardt (1918), Kershner (1968), M. Rice, R. James, and Stein (1985).

See also Grünbaum and Shephard (1986, section 9.4), and the references mentioned there, for monohedral tilings by polyiamonds, polyominos and polyhexes. Noteworthy is also the existence of monohedral *spiral* tilings whose prototile has the remarkable property that two copies of it can completely surround a third (cf. Voderberg 1936, and Grünbaum and Shephard 1986, p. 512).

As soon as we raise the dimension of the space, the classification problem becomes increasingly difficult. Not even the easiest case of tilings by translates (parallelohedra) is solved for $d \geqslant 5$; see section 4.1. Many investigations on monohedral tilings concentrate on special kinds of prototiles to obtain partial classification results. Examples are the monohedral tilings of higher-dimensional spaces by simplices discussed in Baumgartner (1971), Danzer (1968) and Debrunner (1985). In a series of papers Goldberg obtained 3-dimensional prototiles with few facets, such as pentahedra, hexahedra or heptahedra; see, e.g., Goldberg (1978). In higher dimensions there are various isolated examples of prototiles, often with many symmetries.

The problem has especially attracted attention in three dimensions. Here, the prototiles for monohedral tilings are called *space-fillers*. Many examples of space-fillers were discovered by crystallographers and were constructed by taking Voronoi regions for suitable discrete point sets (dot patterns) in E^3. There have been several contradictory claims as to how many facets a space-filler can have; for a detailed discussion on this see Grünbaum and Shephard (1980). Early examples have 16, 17 or 18 facets and were discovered by Föppl (1914), Nowacki (1935), and Löckenhoff and Hellner (1971). More recent examples, with the number of facets ranging from 20 to 26, were found by Smith (1965), Štogrin (1973), Koch and Fischer (1972), and Fischer (1979). All conjectures on the true size of the bound were completely upset by the spectacular discovery in Engel (1981) of space-fillers with up to 38 facets. These and the above mentioned examples of space-fillers all admit isohedral tilings, with the tiles Voronoi regions defined by a suitable dot pattern.

In the investigations of Engel (and in other publications) these Voronoi regions are defined from the corresponding dot pattern by using a computer.

A more modest problem is the classification of all combinatorial types of convex polytopes which are prototiles of monohedral tilings of E^d. This problem is trivial for the plane, but open for dimensions $d \geqslant 3$. For $d \geqslant 3$ it is not even known if there are only finitely many such combinatorial types. Equivalently, it is not known whether there is an upper bound on the number of facets of convex prototiles of monohedral tilings in E^d. The only general result available in the literature is the following theorem of Delone (1961), which provides an upper bound for stereohedra; see also Štogrin (1973). *Stereohedra* are convex polytopes which are prototiles of isohedral tilings. By an *aspect* of an isohedral tiling $\mathcal{T}$ of E^d we mean a translation class of the tiles, that is, a transitivity class of tiles with respect to the translation subgroup of the symmetry group of $\mathcal{T}$.

Theorem 20. *The number of facets of a stereohedron T in E^d is at most $2^d(1+a)-2$, where a is the number of aspects of T in an isohedral tiling of E^d.*

For general d the number a of aspects is bounded by the maximum order of the point groups (stabilizer of the origin) of crystallographic groups; by Theorem 20, this gives an upper bound for the number of facets of a stereohedron. If $d = 3$, this maximum is 48, so that the bound takes the value 390; it is likely that the true bound is considerably lower, possibly as low as 38.

4.3. Monotypic tilings

The strong requirement of congruence of the tiles considerably restricts the various possibilities for constructing monohedral tilings. If this requirement is relaxed to combinatorial equivalence, there is generally much freedom for choosing the metrical shape of the tiles and arrive at a tiling of the whole space.

Call a (locally finite) tiling $\mathcal{T}$ of E^d by convex polytopes *monotypic* if all the tiles are combinatorially isomorphic to a convex polytope T; then T is said to be the *combinatorial prototile* of $\mathcal{T}$. A convex d-polytope T is called a *d-nontile* if it is not the combinatorial prototile of a monotypic *face-to-face* tiling $\mathcal{T}$ of E^d. A convex $(d+1)$-polytope P is *equifacetted* if all its facets are combinatorially isomorphic to a convex d-polytope T, the *facet type* of P. By a *d-nonfacet* we mean a convex d-polytope which is not the facet type of an equifacetted convex $(d+1)$-polytope. Each d-nontile is also a d-nonfacet (cf. Schulte 1984b).

For each $n \geqslant 3$ the plane admits a monotypic face-to-face tiling by convex n-gons. For 3-space many interesting examples of monotypic tilings were described in Danzer, Grünbaum and Shephard (1983). Schulte (1984b) discusses some projection methods for the construction of monotypic face-to-face tilings of E^d. These and similar techniques can be used to prove that the following types of convex 3-polytopes admit monotypic face-to-face tilings $\mathcal{T}$ of E^3: prisms, wedges, pyramids and bipyramids over n-gons ($n \geqslant 3$); for prisms and wedges there exist even normal tilings $\mathcal{T}$ (cf. Schulte 1988).

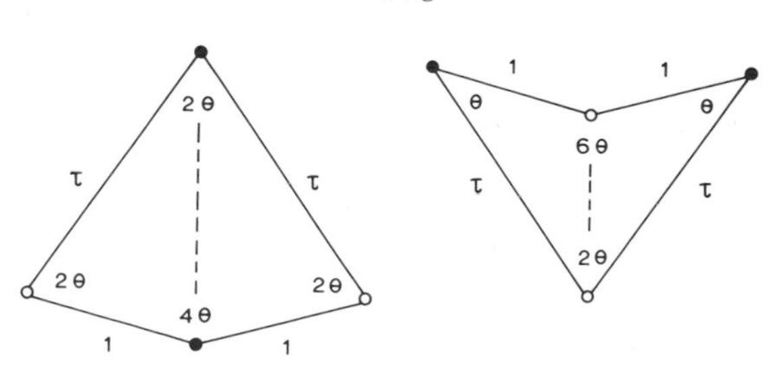

Figure 4. The Penrose set P2, kite and dart.

Our next two theorems show that 3-space plays a special role in the theory of monotypic tilings. Theorem 21 was proved in Grünbaum, Mani-Levitska and Shephard (1984), and Theorem 22 in Schulte (1984a). Recall that a convex d-polytope is called *simplicial* if all its faces are simplices.

Theorem 21. *Every simplicial* 3-*polytope is the combinatorial prototile of a monotypic face-to-face tiling of* E^3.

Theorem 22. *Every convex* 3-*polytope is the combinatorial prototile of a monotypic tiling of* E^3.

Theorem 23. *For each dimension* $d \geqslant 3$ *there are* d-*nontiles, simplicial if* $d \geqslant 5$.

Theorem 23 was proved in Schulte (1984b). In particular, it implies that Theorem 21 does not extend to $d \geqslant 5$ (and probably not to $d = 4$ either). The nontiles of Theorem 23 can be of various kinds; some even remain nontiles if the tiles are allowed to be topological polytopes and to tile manifolds. The easiest example in E^3 is the cuboctahedron; see also Schulte (1985). See also Danzer, Grünbaum and Shephard (1983) for a construction of 3-polytopes which are not combinatorial prototiles of normal monotypic face-to-face tilings of E^3. Furthermore, the constructions underlying Theorem 23 give many classes of nonfacets. Nonfacets were discovered earlier in Perles and Shephard (1967) and Barnette (1969, 1980).

Very little is known about tilings in which the tiles are not topologically d-cells. For examples of tilings of E^3 by handlebodies see Debrunner (1986).

5. Non-periodic tilings

One of the most fascinating discoveries in the theory of tilings is not much older than 20 years and concerns the phenomenon of aperiodicity. Recall that a tiling $\mathcal{T}$ in E^d is called *non-periodic* if its symmetry group $S(\mathcal{T})$ does not contain a non-trivial translation. A set $\mathcal{S}$ of prototiles in E^d is said to be *aperiodic* if $\mathcal{S}$ admits a tiling of E^d, yet all such tilings are non-periodic. Recall that in the planar case, if a finite set $\mathcal{S}$ of polygonal prototiles admits an edge-to-edge plane tiling with at least one non-trivial translational symmetry, then it also admits a periodic tiling; see Theorem 10.

In discussing non-periodic tilings $\mathcal{T}$ it is often convenient to work with sets $\mathcal{S}$ of *marked* prototiles. In a typical example the tiles in $\mathcal{S}$ are planar polygons with their corners and/or sides suitably colored (or marked); the condition is then that, in constructing a tiling, the colors of corners and/or sides of neighbouring tiles must match. The use of marked tiles together with certain matching conditions often avoids introducing (unmarked) tiles with more complicated metrical shapes.

We begin by reviewing some results on aperiodic planar prototiles. For a detailed exposition we refer to Grünbaum and Shephard (1986, chapters 10 and 11). See also Ammann, Grünbaum and Shephard (1992).

Historically, the first appearance of an aperiodic set is Berger's (1966) discovery of a set of 20426 so-called *Wang tiles*; these are square tiles with colored edges, which must be tiled edge-to-edge by translation only, such that colors of adjacent tiles match. This discovery refuted a conjecture of Wang (1961) according to which the planar "Tiling problem" should be decidable; here decidability means that there is an algorithm which, for any given set $\mathcal{S}$ of prototiles, decides in a finite number of steps whether $\mathcal{S}$ admits a tiling or not. See also Robinson (1971). By a result of Ammann, the number of tiles in an aperiodic set of Wang tiles can be reduced to 16; see Grünbaum and Shephard (1986, p. 594), and Robinson (1978).

Robinson (1971) discovered an aperiodic set of 6 tiles which are basically (unmarked) squares with modifications to their corners and sides. There is a similar aperiodic set of 6 tiles found by Ammann in 1977 (cf. Grünbaum and Shephard 1986, p. 529).

The next striking advance was Penrose's (1974, 1978) discovery in 1973 and 1974 of three sets of aperiodic planar prototiles, which are closely related to each other in that tilings with tiles from one set can be converted into tilings with tiles from another set. Tilings by these prototiles are now known as *Penrose tilings*. Below we will describe two of these sets and some of their properties. For more details see Penrose (1974, 1978), Gardner (1977) and Grünbaum and Shephard (1986, sections 10.3, 10.5 and 10.6). It must be mentioned here that many of the fascinating properties of the Penrose tilings were actually discovered by J.H. Conway.

The first Penrose aperiodic set P2 (notation as in Grünbaum and Shephard 1986) consists of two marked tiles known as *kite* and *dart*; see fig. 4. The sides of the tiles are of two lengths in the ratio $\tau : 1$ (with $\tau = (1 + \sqrt{5})/2$ the golden number) and the angle θ equals $\pi/5$. The corners are colored with two colors, black and white (say), as shown. The matching condition is that equal sides must be put together, so that the colors at the corners match. Figure 5 shows an example of a Penrose tiling by kites and darts; here, black and white vertices are denoted by solid and open circles, respectively.

As for many other aperiodic sets, the aperiodicity of the Penrose set P2 is based on two important transformations which apply uniquely to any Penrose tiling by kites and darts; these are *composition* and the corresponding inverse process of *decomposition*. By composition we mean the process of taking unions of tiles to form larger tiles of basically the same shapes as those of the original tiles, such that the markings of the original tiles specify a matching condition equivalent to the original one. Decomposition is the basis for the process of *inflation* which, when

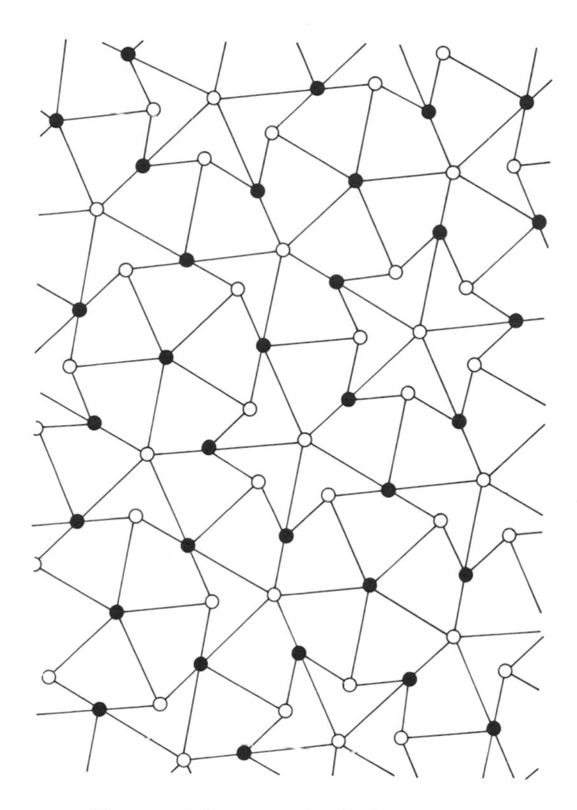

Figure 5. A Penrose tiling by kites and darts.

iterated, can be used to generate arbitrarily large patches or even global tilings by kites and darts. Here, inflation is the operation of homothetically enlarging a given patch of tiles (by some factor involving τ) and then decomposing it into tiles of the original size. Inflation and its inverse process, *deflation*, are characteristic features of many aperiodic sets and their tilings.

The inflation principle can be used to prove a property of the Penrose tilings known as *local isomorphism*: every patch of tiles in a tiling by Penrose kites and darts is congruent to infinitely many patches in every tiling by kites and darts. This property has remarkable consequences. For example, since there exist Penrose tilings with global D_5-symmetry, it follows that in each Penrose tiling there are arbitrarily large patches with D_5-symmetry. The Penrose tilings are the most prominent examples which illustrate the following remarkable fact recently proved by Radin and Wolff (1991). If a (suitable) finite set of tiles admits a tiling of Euclidean d-space, then it also admits a tiling with the local isomorphism property (in fact, a stronger version of it).

Figure 6 shows the second Penrose aperiodic set P3. It consists of two rhombs with colored corners and orientations given to some of their sides; again, $\theta = \pi/5$. The condition is that, in constructing a tiling, the colors of corners as well as lengths

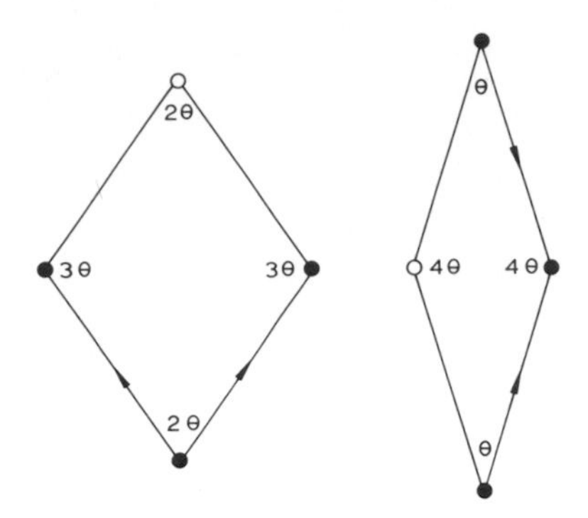

Figure 6. The aperiodic set P3 of Penrose rhombs.

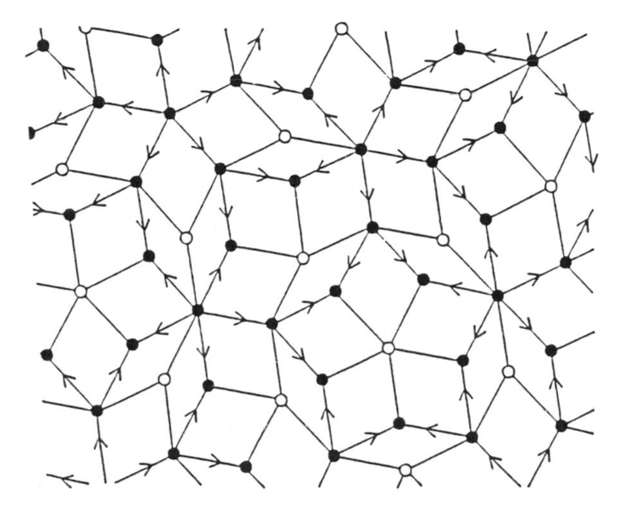

Figure 7. A Penrose tiling using the Penrose rhombs.

and orientations of sides must match. See fig. 7 for an example of a Penrose tiling using set P3. The Penrose set P3 has analogous properties as P2.

By starting from P3, Ammann has produced an aperiodic set of 3 unmarked convex polygons, one hexagon and two pentagons (cf. Grünbaum and Shephard 1986, p. 548); it is believed that there is no such example with only two unmarked tiles. It is an outstanding open problem whether or not there exists an aperiodic set consisting of only one marked or unmarked tile.

An important approach to the theory of Penrose tilings was de Bruijn's (1981) *pentagrid method* which relates to the duals of Penrose tilings. This approach also provided a new geometric interpretation of the Penrose tilings by recognizing these as projections of certain polyhedral surfaces made up from some of the square faces of the cubical tessellation in E^5. This was remarkable, since hereby the non-periodicity of the tilings was connected with the periodicity of the cubical tessellation in E^5. See also de Bruijn (1986).

In various papers this projection method was generalized to higher dimensions and applied to special situations; see, e.g., Levine and Steinhardt (1984), Kramer and Neri (1984), Katz and Duneau (1986), Katz (1988), Whittaker and Whittaker (1988), Korepin, Gähler and Rhyner (1988). See also Oguey, Duneau and Katz (1988) and Elser and Sloane (1987). Very remarkable is the discovery of non-periodic tilings of E^3 by two rhombohedra (known as *Ammann rhombohedra*) which have global icosahedral symmetry; this construction is related to a 6-dimensional representation of the icosahedral group and uses the projection method (cf. Kramer and Neri 1984). Until recently, for three and higher dimensions only global methods such as the projection method were known to construct non-periodic tilings. In a recent paper Katz (1988) succeeded in finding matching rules for the Ammann rhombohedra; more precisely, he discovered an aperiodic set of 22 marked prototiles whose underlying point sets are Ammann rhombohedra.

Traditionally crystallography studies structures which have a very definite form of periodicity and symmetry. Classical principles of crystallography were upset by the recent discovery in Shechtman et al. (1984) of certain aluminum–manganese alloys which solidify with a high degree of icosahedral symmetry. These *quasicrystal* have some, but not all, of the characteristic properties of ordinary crystals. To a large extent the recent interest in non-periodic tilings (with icosahedral symmetry) is motivated by the goal of finding an adaquate mathematical model explaining the structure of quasicrystals; the Penrose tilings and their analogues show a similar kind of quasiperiodicity and quasisymmetry as the quasicrystals (cf. de Bruijn 1986, Nelson 1986). For recent surveys on quasicrystals see Steurer (1990) and Senechal and Taylor (1990). See also chapter 3.7 in this Handbook.

There exist various other aperiodic sets of prototiles, which have several properties in common with the Penrose tilings and their analogues. Examples of such sets are the Ammann aperiodic tiles discussed in Grünbaum and Shephard (1986, section 10.4).

Very remarkable is Danzer's (1989) discovery of an aperiodic set of only 4 marked tetrahedral tiles in E^3 which admit tilings of E^3 with global icosahedral symmetry. Here, as in Katz's (1988) work, the use of matching conditions for the 3-dimensional tiles "localizes" the property of aperiodicity. Very recently McMullen (1990, 1991) discovered further new examples of non-periodic tilings in three and higher dimensions whose set of prototiles consists only of a small number of simplices. See also Coxeter (1991).

From Danzer (1989) also emerged the following suggestion (due to Danzer) for an appropriate definition of quasiperiodicity of tilings; in many publications this term is often used in an informal way. For $\rho > 0$ and $x \in E^d$ let $B^d(x, \rho)$ denote the closed d-dimensional ball of radius ρ centered at x. According to the proposed definition, a tiling $\mathcal{T}$ is *quasiperiodic* if there exists a constant $\gamma = \gamma(\mathcal{T})$ such that for all $\rho > 0$ and all $x, y \in E^d$ the ball $B(y, \gamma\rho)$ contains a "patch" congruent to

$$B(x, \rho) \cap \mathcal{T} \ (:= \{T \cap B(x, \rho) \mid T \in \mathcal{T}\}).$$

This condition is stronger than the local isomorphism property, but is indeed also satisfied by the Penrose tilings and their analogues. See also Danzer (1991).

It is likely that the study of non-periodic tilings will trigger the investigation of other non-periodic figures. For an example of a non-periodic sphere packing related to Penrose tilings see Wills (1990).

Concluding, for an interesting discussion of measures of "regularity" and "order" of geometric figures see also Radin (1991). Further, see Mozes (1989), Radin (1992), and Berend and Radin (1991) for an approach to aperiodicity of tilings in terms of dynamical systems.

References

Aleksandrov, A.D.
[1954] On filling of space by polytopes (in Russian), *Vestnik. Leningrad. Univ. (Ser. Mat. Fiz. Him.)* **9**, 33–43.

Ammann, R., B. Grünbaum and G.C. Shephard
[1992] Aperiodic tiles, *Discrete Comput. Geom.* **8**, 1–27.

Baranovskiĭ, E.P., and S.S. Ryškov
[1973] Primitive five-dimensional parallelohedra, *Dokl. Akad. Nauk. SSSR* **212**, 532–535 [*Soviet Math. Dokl.* **14**, 1391–1395].

Barnette, D.W.
[1969] A simple 4-dimensional nonfacet, *Israel J. Math.* **7**, 16–20.
[1980] Nonfacets for shellable spheres, *Israel J. Math.* **35**, 286–288.

Baumgartner, L.
[1971] Zerlegung des n-dimensionalen Raumes in kongruente Simplices, *Math. Nachr.* **48**, 213–224.

Berend, D., and C. Radin
[1991] Are there chaotic tilings?, preprint.

Berger, R.
[1966] The undecidability of the domino problem, *Mem. Amer. Math. Soc.*, Vol. 66.

Bezdek, A., and W. Kuperberg
[1990] Examples of space-tiling polyhedra related to Hilbert's Problem 18, Question 2, preprint.

Bieberbach, L.
[1910] Über die Bewegungsgruppen der Euklidischen Räume (Erste Abh.), *Math. Ann.* **70**, 297–336.
[1912] Über die Bewegungsgruppen der Euklidischen Räume (Zweite Abh.), *Math. Ann.* **72**, 400–412.

Bilinski, S.
[1949] Homogene Netze der Ebene, *Bull. Int. Acad. Yougoslave. Cl. Sci. Math. Phys. Tech. (N.S.)* **2**, 63–111.

Brown, H., R. Bülow, J. Neubüser, H. Wondratschek and H. Zassenhaus
[1978] *Crystallographic Groups of Four-dimensional Space* (Wiley, New York).

Burckhardt, J.J.
[1966] *Die Bewegungsgruppen der Kristallographie* (Birkhäuser, Basel, 2nd ed.).

Buser, P.
[1985] A geometric proof of Bieberbach's Theorems on crystallographic groups, *Enseign. Math.* **31**, 137–145.

Conway, J.H., and N.J.A. Sloane
[1988] *Sphere Packings, Lattices and Groups* (Springer, Berlin).

Coxeter, H.S.M.
[1962] The classification of zonohedra by means of projective diagrams, *J. Math. Pures Appl.* **41**, 137–156. Reprinted in: *Twelve Geometric Essays* (Southern Illinois Univ. Press, Carbondale, 1968) pp. 54–74.
[1973] *Regular Polytopes* (Dover, New York, 3rd ed.).
[1991] Cyclotomic integers, nondiscrete tessellations, and quasicrystals, preprint.

Coxeter, H.S.M., and W.O.J. Moser
[1980] *Generators and Relations for Discrete Groups* (Springer, Berlin, 4th ed.).

Danzer, L.
[1968] Zerlegbarkeit endlich-dimensionaler Räume in kongruente Simplices, *Math. Phys. Sem.-Ber., Neue Folge* **15**, 87.
[1989] Three-dimensional analogs of the planar Penrose tilings and quasicrystals, *Discrete Math.* **76**, 1–7.
[1991] Quasiperiodicity – Local and global aspects, in: *Lecture Notes in Physics*, Vol. 382, eds V.V. Dodonov et al. (Springer, Berlin) pp. 561–572.

Danzer, L., B. Grünbaum and G.C. Shephard
[1983] Does every type of polyhedron tile three-space?, *Structural Topology* **8**, 3–14.
[1987] Equitransitive tilings, or how to discover new mathematics, *Math. Mag.* **60**, 67–89.

De Bruijn, N.G.
[1981] Algebraic theory of Penrose's non-periodic tilings, *Nederl. Akad. Wetensch. Proc. Ser. A* **84**, 39–66.
[1986] Quasicrystals and their Fourier transforms, *Nederl. Akad. Wetensch. Proc. Ser. A* **89**, 123–152.

Debrunner, H.E.
[1985] Tiling Euclidean *d*-space with congruent simplices, in: *Discrete Geometry and Convexity*, eds J. Goodman, E. Lutwak, J. Malkevitch and R. Pollack, Ann. New York Acad. Sci., Vol. 440 (New York Acad. Sci., New York) pp. 230–261.
[1986] Tiling three-space with handlebodies, *Studia Sci. Math. Hungar.* **21**, 201–202.

Delone, B.N.
[1929] Sur la partition regulière de l'espace à 4 dimensions I, II, *Izv. Akad. Nauk SSSR, Ser. VII*, 79–110; 147–164.
[1959] The theory of planigons (in Russian), *Izv. Akad. Nauk SSSR Ser. Mat.* **23**, 365–386.
[1961] Proof of the fundamental theorem in the theory of stereohedra, *Dokl. Akad. Nauk SSSR* **138**, 1270–1272 [Soviet Math. Dokl. **2**, 812–815].

Delone, B.N., and N.N. Sandakova
[1961] Theory of stereohedra (in Russian), *Trudy Mat. Inst. Steklov.* **64**, 28–51.

Delone, B.N., N.P. Dolbilin and M.I. Štogrin
[1978] Combinatorial and metric theory of planigons (in Russian), *Trudy Mat. Inst. Steklov. Akad. Nauk SSSR* **148**, 109–140 [*Proc. Steklov. Inst. Math.* **4** (1980) 111–141].

Delone, B.N., R.V. Galiulin and M.I. Štogrin
[1979] The contemporary theory of regular decompositions of Euclidean space (in Russian), in: *Regular partitions of plane and space* (Russian translation of Fedorov, 1899), eds B.N. Delone et al. (Nauka, Leningrad) pp. 235–260.

Dress, A.W.M.
[1987] Presentations of discrete groups, acting on simply connected manifolds, in terms of parametrized systems of Coxeter matrices – A systematic approach, *Adv. in Math.* **63**, 196–212.

Dress, A.W.M., and R. Franz
[1992] Recursion formulas counting periodic tilings, to appear.

Dress, A.W.M., and D. Huson
[1987] On tilings of the plane, *Geom. Dedicata* **24**, 295–310.
[1992] Heaven and hell tilings, *Structural topology*, to appear.

Dress, A.W.M., and R. Scharlau
[1984] Zur Klassifikation äquivarianter Pflasterungen, *Mitt. Math. Sem. Giessen* **164**, 83–136.
[1986] The 37 combinatorial types of minimal, non-transitive, equivariant tilings, *Discrete Math.* **60**, 121–138.

Dress, A.W.M., D. Huson and E. Molnar
[1991] The classification of face-transitive 3D-tilings, preprint.

Elser, V., and N.J.A. Sloane
[1987] A highly symmetric four-dimensional quasicrystal, *J. Phys. A* **20**, 6161–6168.

Engel, P.
[1981] Über Wirkungsbereiche von kubischer Symmetrie, *Z. Krist.* **155**, 199–215.
[1986] *Geometric Crystallography – An Axiomatic Introduction to Crystallography* (Reidel, Boston).
[1989] New results on parallelotopes in five-dimensional space, in preparation.

Erdős, P., P.M. Gruber and J. Hammer
[1989] *Lattice Points*, Pitman Monograph (Longman, New York).

Fedorov, E.S.
[1885] *Elements of the Theory of Figures* (in Russian) (Imp. Acad. Sci., St. Petersburg). New edition: 1953 (Akad. Nauk SSSR).
[1892] Zusammenstellung der krystallographischen Resultate des Herrn Schoenflies und der meinigen, *Z. Kryst. Mineral.* **20**, 25–75.
[1899] Reguläre Plan- und Raumteilung, *Abh. Bayer. Akad. Wiss. (II. Klasse)* **20**, 465–588. Russian translation: Nauka, Leningrad, 1979.

Fejes Tóth, G.
[1983] New results in the theory of packing and covering, in: *Convexity and its Applications*, eds P.M. Gruber and J.M. Wills (Birkhäuser, Basel) pp. 318–359.

Fejes Tóth, L.
[1964] *Regular Figures* (Pergamon Press, New York).

Fischer, W.
[1979] Homogene Raumteilungen in konvexe Polyeder, in: *Collected Abstracts Symposium on Mathematical Crystallography, Riederalp, Wallis, Switzerland*, August, p. 5.

Föppl, L.
[1914] Der Fundamentalbereich des Diamantgitters, *Phys. Z.* **15**, 191–193.

Franz, R.
[1988] Zum rekursiven Aufbau der Familie der periodischen Plasterungen der euklidischen Ebene, Dissertation, Universität Bielefeld.

Frobenius, F.G.
[1911] Über die unzerlegbaren diskreten Bewegungsgruppen, *S.-Ber. Preuss. Akad. Wiss. Berlin, Phys. Math. Kl.* **29**, 654–665.

Gardner, M.
[1977] Extraordinary nonperiodic tilings that enriches the theory of tiles, *Sci. Amer.*, January, 110–121.

Goldberg, M.
[1978] On the space-filling heptahedra, *Geom. Dedicata* **7**, 175–184.

Groemer, H.
[1978/79] On multiple space subdivisions by zonotopes, *Monatsh. Math.* **86**, 185–188.

Gruber, P.M.
[1979] Geometry of numbers, in: *Contributions to Geometry*, eds J. Tölke and J.M. Wills (Birkhäuser, Basel) pp. 186–225.

Gruber, P.M., and C.G. Lekkerkerker
[1987] *Geometry of Numbers*, (North-Holland, Amsterdam, 2nd ed.).

Gruber, P.M., and S.S. Ryškov
[1989] Facet-to-facet implies face-to-face, *European J. Combin.* **10**, 83–84.

Grünbaum, B.
[1967] *Convex Polytopes* (Interscience, London).

Grünbaum, B., and G.C. Shephard
[1977] The eighty-one types of isohedral tilings in the plane, *Math. Proc. Cambridge Philos. Soc.* **82**, 177–196.
[1978] The ninety-one types of isogonal tilings in the plane, *Trans. Amer. Math. Soc.* **242**, 335–353; **249** (1979) 446.
[1980] Tilings with congruent tiles, *Bull. Amer. Math. Soc. N.S.* **3**, 951–973.
[1986] *Tilings and Patterns* (Freeman, San Francisco).

Grünbaum, B., P. Mani-Levitska and G.C. Shephard
[1984] Tiling three-space with polyhedral tiles of a given isomorphism type, *J. London Math. Soc. (2)* **29**, 181–191.

Hajós, G.
[1942] Über einfache und mehrfache Bedeckung des n-dimensionalen Raumes mit einem Würfelgitter, *Math. Z.* **47**, 427–467.

Heesch, H.
[1935] Aufbau der Ebene aus kongruenten Bereichen, *Nachr. Ges. Wiss. Gottingen, Neue Ser.* **1**, 115–117.
[1968] *Reguläres Parkettierungsproblem* (Westdeutscher Verlag, Köln).

Hilbert, D.
[1900] Mathematische Probleme, *Göttinger Nachr.*, 253–297 [1902, Mathematical problems, *Bull. Amer. Math. Soc.* **8**, 437–479].

Huson, D.
[1992] The generation and classification of tile-k-transitive tilings of the euclidean plane, the sphere and the hyperbolic plane, preprint.

Katz, A.
[1988] Theory of matching rules for the 3-dimensional Penrose tilings, *Commun. Math. Phys.* **118**, 263–288.

Katz, A., and M. Duneau
[1986] Quasiperiodic patterns and icosahedral symmetry, *J. Phys. (Paris)* **47**, 181–196.

Keller, O.H.
[1930] Über die lückenlose Erfüllung des Raumes mit Würfeln, *J. Reine Angew. Math.* **163**, 231–248.

Kepler, J.
[1619] *Harmonia Mundi* (Lincii). Also in: *Johannes Kepler Gesammelte Werke,* ed. M. Caspar, Band VI (Beck, München, 1940).

Kershner, R.B.
[1968] On paving the plane, *Amer. Math. Monthly* **75**, 839–844.

Klee, V.
[1986] Do infinite-dimensional Banach spaces admit nice tilings, *Studia Sci. Math. Hungar.* **21**, 415–427.

Koch, E., and W. Fischer
[1972] Wirkungsbereichstypen einer verzerrten Diamantkonfiguration mit Kugelpackungscharakter, *Z. Krist.* **135**, 73–92.

Korepin, V.E., F. Gähler and J. Rhyner
[1988] Quasi-periodic tilings – A generalized grid-projection method, *Acta Cryst. A* **44**, 667–672.

Kramer, P., and R. Neri
[1984] On periodic and non-periodic space fillings of E^m obtained by projection, *Acta Cryst. A* **40**, 580–587.

Lagarias, J.C., and P.W. Shor
[1992] Keller's cube-tiling conjecture is false in high dimensions, *Bull. Amer. Math. Soc. (N.S.)* **27**, 279–283.

Laves, F.
[1931] Ebenenteilung und Koordinationszahl, *Z. Krist.* **78**, 208–241.

Levine, D., and P.J. Steinhardt
[1984] Quasi-crystals, A new class of ordered structures, *Phys. Rev. Lett.* **53**, 2477–2480.

Löckenhoff, H.D., and E. Hellner
[1971] Die Wirkungsbereiche der invarianten kubischen Gitterkomplexe, *Neues Jahrb. Mineral. Monatsh.*, 155–174.

Mackay, A.L.
[1982] Crystallography and the Penrose pattern, *Phys. A* **114**, 609–613.

McMullen, P.
[1975] Space tiling zonotopes, *Mathematika* **22**, 202–211.
[1980] Convex bodies which tile space by translation, *Mathematika* **27**, 113–121; **28** (1981) 191.
[1990] Nondiscrete regular honeycombs, in: *Quasicrystals, Networks, and Molecules of Fivefold Symmetry,* ed. I. Hargittai (VCH Publishers, New York) ch. 12, pp. 201–221.
[1991] Quasi-periodic tilings of ordinary space, preprint.

Milnor, J.
[1976] Hilbert's Problem 18 – On crystallographic groups, fundamental domains, and on sphere packings, in: *Mathematical Developments arising from Hilbert Problems,* ed. F.E. Browder, Proc. Symposia in Pure Math., Vol. 28 (Amer. Math. Soc., Providence, RI) pp. 491–506.

Minkowski, H.
[1897] Allgemeine Lehrsätze über die konvexen Polyeder, *Nachr. Ges. Wiss. Göttingen, Math.-Phys. Kl.*, 198–219 [*Gesammelte Abhandlungen,* Vol. 2 (Chelsea, New York, 1967) pp. 103–121].
[1907] *Diophantische Approximationen* (Teubner, Leipzig).

Molnar, E.
[1983] Konvexe Fundamentalpolyeder und einfache D-V-Zellen für 29 Raumgruppen, die Coxetersche Spiegelungsgruppen enthalten, *Beiträge Algebra Geom.* **14**, 33–75.

Mozes, S.
[1989] Tilings, substitution systems and dynamical systems generated by them, *J. Analyse Math.* **53**, 139–186.

Nelson, D.R.
[1986] Quasicrystals, *Sci. Amer.* **255**, 32–41.

Nowacki, W.
[1935] Homogene Raumteilung und Kristallstruktur, Dissertation, E.T.H. Zürich.

Oguey, Ch., M. Duneau and A. Katz
[1988] A geometrical approach of quasiperiodic tilings, *Commun. Math. Phys.* **118**, 99–118.

Penrose, R.
[1974] The role of aesthetics in pure and applied mathematical research, *Bull. Inst. Math. Appl.* **10**, 266–271.
[1978] Pentaplexity, *Eureka* **39**, 16–22.

Perles, M.A., and G.C. Shephard
[1967] Facets and nonfacets of convex polytopes, *Acta Math.* **119**, 113–145.

Perron, O.
[1940a] Über lückenlose Ausfüllung des *n*-dimensionalen Raumes durch kongruente Würfel, *Math. Z.* **46**, 1–26; 161–180.

[1940b] Modulartige lückenlose Ausfüllung des R^n mit kongruenten Würfeln, *Math. Ann.* **117**, 415–447.

Radin, C.
[1991] Global order from local sources, *Bull. Amer. Math. Soc. (2)* **25**, 335–364.
[1992] Space tilings and substitutions, preprint.

Radin, C., and M. Wolff
[1991] Space tilings and local isomorphism, preprint.

Reinhardt, K.
[1918] Über die Zerlegung der Ebene in Polygone, Dissertation, Universität Frankfurt am Main (R. Noske, Leipzig).
[1928] Zur Zerlegung der euklidischen Räume in kongruente Polytope, *S.-Ber. Preuss. Akad. Wiss. Berlin*, 150–155.

Robinson, R.M.
[1971] Undecidability and nonperiodicity of tilings of the plane, *Invent. Math.* **12**, 177–209.
[1978] Undecidable tiling problems in the hyperbolic plane, *Invent. Math.* **44**, 259–264.

Ryškov, S.S., and E.P. Baranovskiĭ
[1976] The *C*-types of *n*-dimensional lattices and primitive five-dimensional parallelohedra (in Russian), *Trudy Mat. Inst. Akad. Nauk SSSR* **137**.

Schattschneider, D.
[1978] Tiling the plane with congruent pentagons, *Math. Mag.* **51**, 29–44.

Schneider, R., and W. Weil
[1983] Zonoids and related topics, in: *Convexity and its Applications,* eds P.M. Gruber and J.M. Wills (Birkhäuser, Basel) pp. 296–317.

Schoenflies, A.
[1891] *Krystallsysteme und Krystallstructur* (Teubner, Leipzig); reprinted: Springer, Berlin, 1984.

Schulte, E.
[1984a] Tiling three-space by combinatorially equivalent convex polytopes, *Proc. London Math. Soc. (3)* **49**, 128–140.
[1984b] Nontiles and nonfacets for the Euclidean space, spherical complexes and convex polytopes, *J. Reine Angew. Math.* **352**, 161–183.
[1985] The existence of non-tiles and non-facets in three dimensions, *J. Combin. Theory A* **38**, 75–81.
[1988] Combinatorial prototiles, in: *Shaping Space – A Polyhedral Approach,* eds M. Senechal and G. Fleck (Birkhäuser, Basel) pp. 198–204.

Senechal, M., and G. Fleck, eds
[1988] *Shaping Space – A Polyhedral Approach* (Birkhäuser, Basel).

Senechal, M., and J. Taylor
[1990] Quasicrystals, the view from Les Houches, *Math. Intelligencer* **12**(2), 54–64.

Shechtman, D., I. Blech, D. Gratias and J.W. Cahn
[1984] Metallic phase with long-range orientational order and no translational symmetry, *Phys. Rev. Lett.* **53**, 1951–1953.

Shephard, G.C.
[1974] Space filling zonotopes, *Mathematika* **21**, 261–269.

Smith, F.W.
[1965] The structure of aggregates – A class of 20-faced space-filling polyhedra, *Canad. J. Phys.* **43**, 2052–2055.

Stehling, T.
[1988] Über kombinatorische und graphentheoretische Eigenschaften normaler Pflasterungen, Dissertation, Universität Dortmund.

Stein, R.
[1985] A new pentagon tiler, *Math. Mag.* **58**, 308.

Stein, S.K.
[1972] A symmetric star-body that tiles but not as a lattice, *Proc. Amer. Math. Soc.* **36**, 543–548.
[1974] Algebraic tiling, *Amer. Math. Monthly* **81**, 445–462.
[1986] Tiling, packing, and covering by clusters, *Rocky Mountain J. Math.* **16**, 277–321.

Steurer, W.
[1990] The structure of quasicrystals, *Z. Krist.* **190**, 179–234.

Štogrin, M.I.
[1973] Regular Dirichlet–Voronoi partitions for the second triclinic group (in Russian), *Trudy Mat. Inst. Steklov.* **123** [*Proc. Steklov. Inst. Math.* **123** (1975)].

Šubnikov, A.V.
[1916] On the question of the structure in crystals (in Russian), *Bull. Acad. Imp. Sci., Ser. 6* **10**, 755–799.

Szabó, S.
[1985] A star polyhedron that tiles but not as a fundamental region, *Colloq. Math. Soc. János Bolyai* **48**, 531–544.

Venkov, B.A.
[1954] On a class of euclidean polytopes (in Russian), *Vestnik Leningrad. Univ., Ser. Mat. Fiz. Him.* **9**, 11–31.

Voderberg, H.
[1936] Zur Zerlegung der Umgebung eines ebenen Bereiches in kongruente, *Jber. Deutsch. Math.-Vereinig.* **46**, 229–231.

Voronoi, G.F.
[1908] Nouvelles applications des paramètres continus à la théorie des formes quadratiques, Deuxième Mémoire, Recherche sur les paralléloèdres primitifs, *J. Reine Angew. Math.* **134**, 198–287; **136** (1909) 67–181.

Wang, H.
[1961] Proving theorems by pattern recognition, II, *Bell System Tech. J.* **40**, 1–42.

Whittaker, E.J.W., and R.M. Whittaker
[1988] Some generalized Penrose patterns from projection of n-dimensional lattices, *Acta Cryst. A* **44**, 105–112.

Wills, J.M.
[1990] A quasi-crystalline sphere-packing with unexpected high density, *J. Phys. (Paris)* **51**, 1061–1064.

Žitomirskiĭ, O.K.
[1929] Verschärfung eines Satzes von Woronoi, *Ž. Leningrad Fiz.-Mat. Obšč.* **2**, 131–151.

CHAPTER 3.6

Valuations and Dissections

Peter McMULLEN

Department of Mathematics, University College, Gower Street, London WC1E 6BT, UK

Contents

HANDBOOK OF CONVEX GEOMETRY
Edited by P.M. Gruber and J.M. Wills

Introduction

The concept of a valuation lies at the very heart of geometry, as does the closely related concept of a dissection. Indeed, the word "geometry" means "measuring earth", and involves the notion of dividing up plots of ground and calculating their areas. The Greek definition of area was built on the idea of dissecting planar polygonal regions into triangles, whose areas are given by the familiar formula, although a rigorous proof that polygonal regions of the same area admit equidissections into congruent triangles had to wait until comparatively recently (see section 5.3).

The comparable 3-dimensional problem of comparing volumes proved much more difficult to the Greeks. The technique finally adopted, that of the method of exhaustion due to Eudoxus Archimedes (though first applied less formally in the plane by Antiphon), involves a limiting process. Whether an elementary dissection argument would also work here was asked by Gauss, if not earlier, but it was only with Dehn's negative solution of Hilbert's third problem (see section 4.5) that the problem was, at least partially, settled. What is now understood by Hilbert's third problem is that of finding necessary and sufficient conditions on polytopal regions for such equidissections (under various restrictions on the motions allowed) to be possible.

In this article, we shall discuss the current state of knowledge of valuations and dissections. An earlier survey article (McMullen and Schneider 1983) covered the same ground as this, and we shall draw on it extensively. There is, though, one striking difference. Recently, the concept of the polytope algebra, introduced by McMullen (1989), has done for general translation invariant valuations what the earlier algebra of polytopes of Jessen and Thorup (1978) did for translation invariant simple valuations. (The later work did, however, rely heavily on the earlier.) We shall therefore base our treatment of the abstract foundation of the theory of valuations on the polytope algebra (see section 3).

1. The basic theory

We begin with a discussion of the basic theory of valuations and dissections, including the background notions of Euclidean space which we employ. It is worth remarking, though, that much of this basic theory works in the more general context of a finite-dimensional linear space over an arbitrary ordered field, and we only choose the Euclidean context for simplicity of treatment (see, in particular, McMullen 1989).

1.1. Euclidean notions

We shall work, for the most part, in d-dimensional Euclidean space $\mathbb{E}^d$. A general vector in $\mathbb{E}^d$ is written $x = (\xi_1, \ldots, \xi_d)$, where $\xi_j \in \mathbb{R}$, the real numbers, for $j = 1, \ldots, d$. We endow $\mathbb{E}^d$ with the *inner product* $\langle x, y \rangle := \sum_{j=1}^{d} \xi_j \eta_j$, where

$y = (\eta_1, \dots, \eta_d)$, and corresponding *norm* $\|x\| := \sqrt{\langle x, x \rangle}$. Occasionally, though, we shall refer to a vector space V over an arbitrary (not necessarily Archimedean) ordered field $\mathbb{F}$. This too can be given an inner product, but there will not usually be a norm, because $\mathbb{F}$ will not be square-root-closed. However, the Gram–Schmidt orthogonalization process will enable us to define orthogonal projection onto subspaces of V.

We shall be as much concerned with affine as linear properties of $\mathbb{E}^d$; useful references here are Grünbaum (1967) or McMullen and Shephard (1971). In particular, the *vector* or *Minkowski sum* of $S, T \subseteq \mathbb{E}^d$ is defined by

$$S + T := \{x + y \mid x \in S, y \in T\},$$

the *translate* of S by $t \in \mathbb{E}^d$ is $S + t := S + \{t\}$, and the *scalar multiple* of S by $\lambda \in \mathbb{R}$ is

$$\lambda S := \{\lambda x \mid x \in S\}.$$

The *Hausdorff distance* $\rho(S, T)$ between two non-empty compact subsets S, T of $\mathbb{E}^d$ is defined by

$$\rho(S, T) := \min\{\rho \geqslant 0 \mid S \subseteq T + \rho B, T \subseteq S + \rho B\},$$

where $B = B^d := \{x \in \mathbb{E} \mid \|x\| \leqslant 1\}$ is the unit ball in $\mathbb{E}^d$. Continuity for functions on classes of compact subsets of $\mathbb{E}^d$ will usually be with respect to the Hausdorff metric.

1.2. Valuations

Let $\mathcal{S}$ be a family of sets in $\mathbb{E}^d$. We call a function φ on $\mathcal{S}$, taking values in some Abelian group, a *valuation*, or say that φ is *additive*, if $\varphi(S \cup T) + \varphi(S \cap T) = \varphi(S) + \varphi(T)$ whenever $S, T, S \cup T, S \cap T \in \mathcal{S}$. If $\emptyset \in \mathcal{S}$, we shall always suppose that $\varphi(\emptyset) = 0$. Our families $\mathcal{S}$ will usually be *intersectional*, which means that $S \cap T \in \mathcal{S}$ whenever $S, T \in \mathcal{S}$. For an intersectional family $\mathcal{S}$, we write $U\mathcal{S}$ for the family of finite unions of members of $\mathcal{S}$, and $\overline{U\mathcal{S}} := \{S \setminus T \mid S, T \in U\mathcal{S}\}$. Particular examples of intersectional families in $\mathbb{E}^d$ are the family $\mathcal{K}^d$ of compact convex sets or *convex bodies*, the family $\mathcal{P}^d$ of (convex) polytopes, and the family $\mathcal{C}^d$ of convex cones with apex the origin (zero vector) o. The family $U\mathcal{K}^d$ is called by Hadwiger (1957) the *convex ring*, although we shall avoid the term, and the members of $U\mathcal{P}^d$ are called *polyhedra*. In addition, a subscript $*$ on $\mathcal{K}^d$ or $\mathcal{P}^d$ will denote the subset of non-empty members of the appropriate family.

There are many examples of valuations, which we shall discuss in more detail in section 2. For now, let us note that the restriction of a measure to any family in $\mathbb{E}^d$ will yield a valuation on that family; in particular, volume is a valuation in every dimension. Surface area is also a valuation, as is the Euler characteristic on the three families just introduced. All these valuations φ are *translation invariant*, in that $\varphi(S + t) = \varphi(S)$ for each appropriate S and each $t \in \mathbb{E}^d$ (the definition has no force for the family $\mathcal{C}^d$). More generally, we shall say that a valuation φ

on a class $\mathcal{S}$ which is permuted by a group G of affinities of $\mathbb{E}^d$ is *G-invariant* if $\varphi(\Phi S) = \varphi(S)$ for each $S \in \mathcal{S}$ and $\Phi \in G$. Then these valuations just mentioned are actually D-invariant, where D is the group of all isometries of $\mathbb{E}^d$.

An important subclass of valuations consists of those that are simple. In general, if L is a linear subspace of $\mathbb{E}^d$, we say that a valuation φ on a family $\mathcal{S}$ is *L-simple* if $\varphi(S) = 0$ whenever $S \in \mathcal{S}(L) := \{S \in \mathcal{S} \mid S \subseteq L\}$ satisfies $\dim S < \dim L$ (our class here will consist of convex sets, for which the dimension can be defined). The term *simple* alone will mean $\mathbb{E}^d$-simple.

A function φ on a suitable family $\mathcal{S}$ is called *Minkowski additive* if $\varphi(S+T) = \varphi(S) + \varphi(T)$ for all $S, T \in \mathcal{S}$. Then there is a fundamental relation, due to Sallee (1966).

Lemma 1.1. *If $S, T, S \cup T \in \mathcal{K}^d_*$, then*

$$(S \cup T) + (S \cap T) = S + T.$$

Hence every Minkowski additive function on $\mathcal{K}^d_*$ is a valuation.

Since $\mathcal{K}^d_*$ is a semigroup under Minkowski addition which has a cancellation law, we can imbed $\mathcal{K}^d_*$ in an Abelian group; we can thus interpret this observation as saying that the identity map from $\mathcal{K}^d_*$ into itself is a valuation.

A closely related example of a valuation is the *support functional* $h(K, \cdot)$, defined by

$$h(K, u) = \max\{\langle x, u\rangle \mid x \in K\}$$

for $K \in \mathcal{K}^d_*$ and $u \in \mathbb{E}^d$, since Minkowski addition of convex bodies corresponds to addition of their support functionals. (By the way, we might remark here that the Hausdorff distance ρ is given by

$$\rho(K, L) = \max\{|h(K, u) - h(L, u)| \mid u \in \Omega\},$$

where $\Omega = \Omega^{d-1} = \{x \in \mathbb{E}^d \mid \|x\| = 1\}$ is the *unit sphere*.) If $C \in \mathcal{K}^d_*$ is fixed, then the map φ_C from $\mathcal{K}^d_*$ into itself defined by $\varphi_C(K) := K + C$ is a valuation, since

$$(S \cup T) + C = (S + C) \cup (T + C)$$

for all $S, T \in \mathcal{K}^d_*$, while

$$(S \cap T) + C = (S + C) \cap (T + C)$$

provided that $S \cup T \in \mathcal{K}^d_*$ also (see Hadwiger 1957, p. 144). This remark will play an important role later. This observation is also an immediate consequence of the following.

Lemma 1.2. *If $f : \mathcal{K}^d \to \mathcal{K}^d$ is a map such that $f(S \cup T) = f(S) \cup f(T)$ and $f(S \cap T) = f(S) \cap f(T)$ if $S, T, S \cup T \in \mathcal{K}^d$, then $\varphi \circ f$ is a valuation on $\mathcal{K}^d$ whenever φ is.*

For polytopes, a variant notion of valuation is useful. We call a map φ on $\mathcal{P}^d$ a *weak valuation* if $\varphi(P) + \varphi(P \cap H) = \varphi(P \cap H^+) + \varphi(P \cap H^-)$ whenever $P \in \mathcal{P}^d$ and

H is a hyperplane bounding the two closed half-spaces H^+ and H^-. Sallee (1968) (see also Groemer 1978) has shown:

Lemma 1.3. *Every weak valuation on $\mathcal{P}^d$ gives rise to a valuation.*

However, examples show that weak valuations on $\mathcal{K}^d$ need not arise from valuations in this way (see McMullen and Schneider 1983, p. 173).

1.3. Extensions

While it was not important for Jessen and Thorup (1978), it turns out to be vital for the approach of McMullen (1989) to be able to extend valuations on $\mathcal{S}$ to valuations on $\overline{U\mathcal{S}}$, for $\mathcal{S} = \mathcal{P}^d$ or $\mathcal{K}^d$. The quickest way to see that such extensions are possible follows Groemer (1977a) (we slightly modify his argument). If $S \in \mathcal{S}$, its *characteristic function* $S^\dagger$ is given by

$$S^\dagger(x) = \begin{cases} 1 & \text{if } x \in S, \\ 0 & \text{if } x \notin S. \end{cases}$$

The Abelian group of integer valued functions generated by the functions $S^\dagger$ with $S \in \mathcal{S}$ is denoted $X\mathcal{S}$. Then we have:

Lemma 1.4. *A valuation on $\mathcal{S}$ admits a unique extension to $\overline{U\mathcal{S}}$.*

With $\overline{U\mathcal{S}}$ replaced by $U\mathcal{S}$, the result is originally due to Volland (1957) (see also Perles and Sallee 1970). Since the basic idea of the proof is important, as well as simple, we shall outline it. First, note that, whenever $S, T \subseteq \mathbb{E}^d$, intersection is given in $X\mathcal{S}$ by

$$(S \cap T)^\dagger = S^\dagger T^\dagger$$

(the product is ordinary multiplication of functions), while complementation is given by

$$(\mathbb{E}^d \setminus S)^\dagger = 1 - S^\dagger.$$

(If necessary, we adjoin $\mathbb{E}^d$ to our intersectional family $\mathcal{S}$.) Thus union is given by

$$1 - (S_1 \cup \cdots \cup S_k)^\dagger = (1 - S_1^\dagger) \cdots (1 - S_k^\dagger)$$

(note that 1 occurs on both sides of the equation, so adjoining $\mathbb{E}^d$ to $\mathcal{S}$ is only a convenience), and this results in the *inclusion–exclusion principle*:

Lemma 1.5. *The extension of a valuation φ on $\mathcal{S}$ to $U\mathcal{S}$ is given by*

$$\varphi(S_1 \cup \cdots \cup S_k) = \sum_{j=1}^{k} (-1)^{j-1} \sum_{i(i)<\cdots<i(j)} \varphi(S_{i(1)} \cap \cdots \cap S_{i(j)}),$$

with $S_1, \ldots, S_k \in \mathcal{S}$, and further to $\overline{U\mathcal{S}}$ by

$$\varphi(A \backslash B) = \varphi(A) - \varphi(A \cap B),$$

with $A, B \in U\mathcal{S}$.

1.4. Dissections

Let $\mathcal{S}$ be a subfamily of $U\mathcal{Q}^d$, where $\mathcal{Q}^d$ consists of the *polyhedral sets*, or intersections of finitely many closed half-spaces, in $\mathbb{E}^d$. (We shall therefore consider the unions of polyhedral sets from the outset; the particular subfamilies we shall usually be concerned with are those obtained from the polytopes or polyhedral cones.) A *dissection* of $P \in \mathcal{S}$ is an expression of the form

$$P = P_1 \uplus \cdots \uplus P_k$$

with $P_1, \ldots, P_k \in \mathcal{S}$, which means that $P = P_1 \cup \cdots \cup P_k$, with $\operatorname{int}(P_i \cap P_j) = \emptyset$ whenever $i \neq j$; that is, a dissection is a union of sets with pairwise disjoint interiors.

Let G be a group of affinities of $\mathbb{E}^d$, which permutes the members of $\mathcal{S}$. We say that two members P, Q of $\mathcal{S}$ are *G-equidissectable*, written $P \approx_G Q$, if there are dissections $P = P_1 \uplus \cdots \uplus P_k$ and $Q = Q_1 \uplus \cdots \uplus Q_k$, and elements $\Phi_1, \ldots, \Phi_k \in G$, such that $Q_i = \Phi_i P_i$ for $i = 1, \ldots, k$. We say that $P, Q \in \mathcal{S}$ are *G-equicomplementable*, written $P \sim_G Q$, if there are $P', P'', Q', Q'' \in \mathcal{S}$ such that $P'' = P \uplus P'$, $Q'' = Q \uplus Q'$, and $P' \approx_G Q'$, $P'' \approx_G Q''$. We shall discuss equidissectability and equicomplementability, and the relationship between them, in section 4.4 and section 4.5 below. However, the reader is surely familiar with various games and puzzles, such as tangrams and pentominoes, which involve dissections. It is clear from the definitions that, if $\mathcal{S}$ is such a class as above, and φ is a G-invariant simple valuation on $\mathcal{S}$, then $\varphi(P) = \varphi(Q)$ whenever $P \approx_G Q$ (or $P \sim_G Q$); the core of Hilbert's third problem is to find sufficient conditions for G-equidissectability (or equicomplementability) in terms of suitable families of G-invariant simple valuations.

2. The classical examples

We shall now more formally introduce the classical examples of valuations. In the first two subsections, we consider volume, moment, and related valuations, while in the third, we treat the lattice-point enumerator, which differs in a number of important respects from the other examples.

2.1. Volume and derived valuations

Any measure on a ring of subsets of $\mathbb{E}^d$ which contains $\mathcal{K}^d$ which is finite on $\mathcal{K}^d$ will yield a valuation on $\mathcal{K}^d$. In particular, Lebesgue measure gives the ordinary volume V; this is a simple valuation. However, there is a quite different approach to volume, beginning with an elementary notion of volume of polytopes (which can be characterized axiomatically; see section 5.3 below), and extending this to general convex bodies by continuity arguments. For further details, see Hadwiger (1957), Boltyanskiĭ (1978) or Böhm and Hertel (1980).

We introduce a little notation. We shall write $\kappa_d := V(B^d)$ for the volume of the unit ball in $\mathbb{E}^d$, and σ for Lebesgue measure on Ω^{d-1}, so that $\sigma(\Omega^{d-1}) = d\kappa_d =:$ ω_{d-1} is the area of the unit sphere.

As we remarked in section 1.2, $\psi(K) := \varphi(K + C)$ is a valuation if φ is, and if $C \in \mathcal{K}^d_*$ is fixed. It is easy to show that, if $K_1, \ldots, K_k \in \mathcal{K}^d_*$ and $\lambda_1, \ldots, \lambda_k \geqslant 0$, then the volume of $\lambda_1 K_1 + \cdots + \lambda_k K_k$ is a homogeneous polynomial in $\lambda_1, \ldots, \lambda_k$ of degree d, say

$$V(\lambda_1 K_1 + \cdots + \lambda_k K_k) = \sum \lambda_{i(1)} \cdots \lambda_{i(d)} V(K_{i(1)}, \ldots, K_{i(k)}),$$

with $V(K_{i(1)}, \ldots, K_{i(k)})$ symmetric in the indices, and depending only on $K_{i(1)}, \ldots,$ $K_{i(k)}$. These coefficients are known as *mixed volumes*. We often write this expression in the form

$$V(\lambda_1 K_1 + \cdots + \lambda_k K_k) = \sum \binom{d}{r_1 \cdots r_k} \lambda_1^{r_1} \cdots \lambda_k^{r_k} V(K_1, r_1; \ldots; K_k, r_k),$$

where

$$\binom{d}{r_1 \cdots r_k} := \begin{cases} \dfrac{d!}{r_1! \cdots r_k!} & \text{if } \sum_{j=1}^k r_j = d \text{ and } r_j \geqslant 0 \; (j = 1, \ldots, k), \\ 0 & \text{otherwise,} \end{cases}$$

is the *multinomial coefficient*. Here, and elsewhere in such expressions, we use the abbreviation

$$\varphi(K_1, r_1; \ldots; K_k, r_k) = \varphi(\underbrace{K_1, \ldots, K_1}_{r_1 \text{ times}}, \ldots, \underbrace{K_k, \ldots, K_k}_{r_k \text{ times}}),$$

with $r_1 + \cdots + r_k = m$, for any function φ of m variables; we also write

$$\varphi(K_1, \ldots, K_p, \mathcal{C}) = \varphi(K_1, \ldots, K_p, L_{p+1}, \ldots, L_m),$$

when $\mathcal{C} = (L_{p+1}, \ldots, L_m)$ is a fixed $(m-p)$-tuple. With this convention, we then obtain

$$\begin{aligned} &V(\lambda_1 K_1 + \cdots + \lambda_k K_k, p; \mathcal{C}) \\ &\quad = \sum \binom{p}{r_1 \cdots r_k} \lambda_1^{r_1} \cdots \lambda_k^{r_k} V(K_1, r_1; \ldots; K_k, r_k; \mathcal{C}), \end{aligned}$$

with $m = d$ in the above.

Since the mapping

$$K \mapsto V(\lambda K + \lambda_{p+1} K_{p+1} + \cdots + \lambda_d K_d)$$

is a valuation on $\mathcal{K}^d$ for any $p \in \{1, \ldots, d\}$ and any $(d-p)$-tuple $\mathcal{C} = (K_{p+1}, \ldots, K_d)$, it follows by considering the coefficient of $(p!/d!)\lambda^p \lambda_{p+1} \cdots \lambda_d$ that the mapping φ, given by

$$\varphi(K) := V(K, p; \mathcal{C})$$

is also a valuation. In particular, the jth *quermassintegral* W_j, defined by

$$W_j(K) := V(K, d-j; B, j),$$

is a valuation, and corresponds to the *Steiner parallel formula*

$$V(K + \lambda B) = \sum_{j=0}^{d} \binom{d}{j} \lambda^j W_j(K).$$

We also define the normalized quermassintegral, or *intrinsic r-volume* V_r by

$$\kappa_{d-r} V_r := \binom{d}{r} W_{d-r};$$

since $W_d(K) = \kappa_d$ for all $K \in \mathcal{K}_*^d$, it follows that $V_0(K) = 1 (= \chi(K)$, the *Euler characteristic*, see below) for all such K. More generally, the normalization, due to McMullen (1975a), is such that $V_r(K)$ is the ordinary r-dimensional volume of K if $\dim K = r$; note that V_r is *homogeneous of degree* r, in that $V_r(\lambda K) = \lambda^r V_r(K)$ for all K and all $\lambda \geqslant 0$. Further, $S(K) := dW_1(K) = 2V_{d-1}(K)$ is the *surface area* of K.

While the valuation property of the quermassintegrals was pointed out by Blaschke (1937), and played an important role in the work of Hadwiger (see section 5), that of the general mixed volumes is, strangely, not mentioned in any standard textbooks to date.

Without going into details, let us also remark that the intrinsic volumes occur in various integral-geometric formulae, which average the intrinsic volumes of sections (projections) of a convex body by affine (on linear) subspaces of $\mathbb{E}^d$. [See Hadwiger (1957) or Santaló (1976, sections 13, 14) for the exact results and their proofs, and Hadwiger (1956, 1957) for generalizations.]

The extension property of valuations, discussed in section 1.3, enables the intrinsic volumes to be extended to $\overline{U\mathcal{K}^d}$. In particular, the extension of V_0 is the *Euler characteristic* χ. The Euler characteristic is, of course, of considerable importance in other branches of mathematics, and so a substantial literature has been devoted to it. Within convexity, we mention the inductive construction (in $U\mathcal{K}^d$, and on dimension) due to Hadwiger (1955a); this idea was further explored in Hadwiger (1957, 1959, 1968b, 1969c) and Hadwiger and Mani (1972) (see also Hadwiger 1974a, concerning planar polygons). For applications to combinatorial geometry, see (in addition) Hadwiger (1947), Klee (1963), and Rota (1964, 1971); the last paper provided a very general theoretical treatment. The extension of χ to relatively open polytopes was considered by Lenz (1970) and Groemer (1972); see also Hadwiger (1969c, 1973) for special cases, and Groemer (1973, 1974, 1975) for further generalizations.

The extension of the higher intrinsic volumes has received somewhat less attention. However, the Gauss–Bonnet theorem describes the Euler characteristic of surfaces in local terms. The Lipschitz–Killing curvatures, though initially defined in analytic (differential geometric) terms, turn out in the piecewise linear case just to be the extended intrinsic volumes, though this has not hitherto been recognized (see, for example, Cheeger, Müller and Schrader 1984 and Budach 1989; the desired connexion is most easily obtained from the latter).

Closely related to surface area are the *mixed area functions*. Using their properties of Minkowski additivity and uniform continuity in each argument, the Riesz representation theorem shows that, for given $K_1,\dots,K_{d-1}\in\mathcal{K}^d_*$, there is a unique (positive) measure $S(K_1,\dots,K_{d-1};\cdot)$ on the Borel subsets of the unit sphere Ω, such that

$$V(L,K_1,\dots,K_{d-1})=\frac{1}{d}\int_{\Omega}h(L,u)\,\mathrm{d}S(K_1,\dots,K_{d-1};u),$$

where $h(L,\cdot)$ is the support functional of $L\in\mathcal{K}^d_*$. This measure is due, independently, to Aleksandrov (1937) and Fenchel and Jessen (1938); see also Busemann (1958) (it is often called a *Fenchel–Jessen area measure*). In particular, one writes

$$S_p(K;\,\cdot\,):=S(K,p;B,d-p-1;\,\cdot\,)$$

for the *pth order area function* of K; we note that $S_p(K;\Omega)=dW_{d-p}(K)$.

If ω is a Borel set in Ω, then $S(K;\omega)$ is the (($d-1$)-dimensional Hausdorff) measure of that part of the boundary of K at which there is an outer normal vector (to a support hyperplane) lying in ω. The general mixed area function is then given by

$$\begin{aligned}&S_{d-1}(\lambda_1K_1+\cdots+\lambda_kK_k;\,\cdot\,)\\&\quad=\sum\lambda_{i(1)}\cdots\lambda_{i(d-1)}S(K_{i(1)},\dots,K_{i(d-1)};\,\cdot\,).\end{aligned}$$

It then follows that the mapping

$$K\mapsto S(K,p,\mathcal{C};\,\cdot\,)$$

is a valuation for each $p\in\{1,\dots,d-1\}$ and each $(d-p-1)$-tuple $\mathcal{C}=(L_{p+1},\dots,L_{d-1})$ in $\mathcal{K}^d_*$. In particular, each S_p is a valuation on $\mathcal{K}^d_*$ (with values in the vector space of signed Borel measures on Ω); this valuation property was first pointed out and used by Schneider (1975a).

The Fenchel–Jessen area measures can be obtained from a local version of the Steiner parallel formula. Variants on these measures are the *Federer measures*, which now depend additionally on a Borel set in $\mathbb{E}^d$ itself. As initially defined by Federer (1959), they applied to more general sets than convex surfaces. We shall not give any details here, but instead refer the reader to Federer's paper, and to Schneider (1978, 1979); the latter survey article contains other references. Subsequently, these ideas were further generalized by Wieacker (1982).

Schneider (1980) has shown how to extend the area functions and curvature measures to $U\mathcal{K}^d$; among other things, he obtains a variant of the Gauss–Bonnet theorem. Related notions (to this last) occur in Banchoff (1967, 1970) (see also Schneider 1977b).

2.2. *Moments and derived valuations*

The *moment vector* $z(K)$ of $K \in \mathcal{K}^d_*$ is defined by

$$z(K) := \int_K x \, \mathrm{d}V(x),$$

so that, if $\dim K = d$, then $g(K) := V(K)^{-1} z(K)$ is the centre of gravity of K. There is a polynomial expansion, exactly analoguous to that for volume, of the form

$$z(\lambda_1 K_1 + \cdots + \lambda_k K_k) = \sum \lambda_{i(0)} \cdots \lambda_{i(d)} \, z(K_{i(0)}, \ldots, K_{i(d)}),$$

whose coefficients are called *mixed moment vectors.* (In case $d = 3$, this expansion was already noticed by Minkowski (1911, section 23); the general theory was developed by Schneider 1972a,b.) As for mixed volumes, for each $p \in \{1, \ldots, d+1\}$ and $(d+1-p)$-tuple $\mathscr{C}$ of (fixed) convex bodies, the mapping

$$K \mapsto z(K, p; \mathscr{C})$$

is a valuation, which is homogeneous of degree p. Other properties of mixed volumes also carry over, but observe that, while they are invariant under translation of any of their arguments, the mixed moment vectors satisfy

$$z(K_0 + t, K_1, \ldots, K_d) = z(K_0, K_1, \ldots, K_d) + \frac{1}{d+1} V(K_1, \ldots, K_d) t,$$

which is a special case of translation covariance (see section 5.5 below).

2.3. *The lattice point enumerator*

We call P a *lattice polytope* if its vertices lie in the lattice $\mathbb{Z}^d$. Of great importance in a number of areas outside valuation theory (in, for example, the theory of numbers – see Gruber 1979) is the *lattice point enumerator* G, defined by $G(P) := \operatorname{card}(\mathbb{Z}^d \cap P)$. In fact, $G(S)$ can clearly be defined for any subset S of $\mathbb{E}^d$, and is a valuation which is invariant under *lattice translations*, that is, those in $\mathbb{Z}^d$. Various connexions with other functionals on convex bodies are discussed by Betke and Wills (1979), but we shall confine our attention here to those aspects strictly related to valuation theory.

We denote by $\mathscr{P}^d_L$ the family of lattice polytopes in $\mathbb{E}^d$. Ehrhart (1967a) showed that there is a polynomial expansion

$$G(nP) = \sum_{i=0}^{d} n^i G_i(P),$$

with n a non-negative integer, when $P \in \mathscr{P}^d_L$; the coefficients $G_i(P)$ do not depend on n. Besides making applications of this to various counting problems, Ehrhart (1967b) also discovered the *reciprocity law*

$$G(\operatorname{relint}(nP)) = (-1)^{\dim P} \sum_{i=0}^{d} (-n)^i G_i(P),$$

where again $P \in \mathcal{P}_L^d$ and (this time) n is a positive integer.

The polynomial expansions for mixed volumes have analogues for the lattice point enumerator. If $P_1,\ldots,P_k \in \mathcal{P}_L^d$ and $n_1,\ldots,n_k$ are non-negative integers, then

$$G(n_1P_1+\cdots+n_kP_k)=\sum n_1^{r_1}\cdots n_k^{r_k}G(P_1,r_1;\ldots;P_k,r_k),$$

where the sum extends over all non-negative integers $r_1,\ldots,r_k$ satisfying $r_1+\cdots+r_k \leqslant d$. This was independently found by Bernshtein (1976), at around the same time that more general results were discovered by McMullen (1975, 1977).

Various *weighted* lattice point numbers have been considered by Macdonald (1963, 1971); if

$$\alpha(x,P) := \lim_{\rho\searrow 0}\frac{V(P\cap(\rho B+x))}{V(\rho B)}$$

is that proportion of a sufficiently small ball centred at $x \in \mathbb{Z}^d$ which belongs to $P \in \mathcal{P}_L^d$, then

$$A(P) := \sum_x \alpha(x,P)$$

gives a simple valuation which is invariant under lattice translations. He proved an analogous polynomial expansion, and investigated the coefficients. Hadwiger (1957, p. 69), replaced the ball by the lattice-oriented cube, and obtained a criterion for equidissectability of lattice polytopes under lattice translations.

3. The polytope algebra

In section 4.1 below, we shall discuss the algebraic structure which underlies the theory of translation invariant simple valuations; this was described, independently, by Jessen and Thorup (1978) and Sah (1979). This structure is that of a vector space over $\mathbb{R}$ (or, more generally, over whichever ordered, but not necessarily Archimedean, field is the base field). It is therefore not surprising that the analogous structure underlying general translation invariant valuations behaves somewhat similarly. What is, perhaps, surprising is that the polytope algebra, introduced by McMullen (1989), has a far richer structure, and, indeed, fails to be a real graded (commutative) algebra in only one trivial respect.

In the following sections, we shall describe the polytope algebra; as usual, however, proofs of the results will be omitted (details can be found in McMullen 1989).

3.1. The algebra structure

The *polytope algebra* Π is, initially, the Abelian group with a generator $[P]$, called the *class* of P, for each $P \in \mathcal{P} := \mathcal{P}^d$ (with $[\emptyset] = 0$); these generators satisfy the relations (V) $[P \cup Q] + [P \cap Q] = [P] + [Q]$ whenever $P, Q, P \cup Q \in \mathcal{P}$ (the condition ensures that $P \cap Q \neq \emptyset$ if $P, Q \neq \emptyset$), and (T) $[P + t] = [P]$ whenever $P \in \mathcal{P}$ and $t \in \mathbb{E}^d$. Of course, (V) (which governs addition) and (T) just reflect the valuation property and translation invariance. It is clear that we have:

Lemma 3.1. *A translation invariant valuation on $\mathcal{P}^d$ (into some Abelian group) induces a homomorphism on Π, and conversely.*

We invariably employ the same symbol for the valuation and homomorphism.

We at once introduce the *multiplication*, which is induced by Minkowski addition, by (M) $[P] \cdot [Q] = [P + Q]$ for $P, Q \in \mathcal{P}^d$, and extend it by linearity to Π. In view of

$$P + (Q_1 \cup Q_2) = (P + Q_1) \cup (P + Q_2),$$

which always holds, and

$$P + (Q_1 \cap Q_2) = (P + Q_1) \cap (P + Q_2),$$

whenever $P \cup Q \in \mathcal{P}$ (compare Hadwiger 1957, section 1.2.2), the extension by linearity is compatible with addition on Π, so that Π becomes a commutative ring, with unity $1 := [o] = [t]$ for any $t \in \mathbb{E}^d$ (we write $[t]$ instead of $[\{t\}]$ for brevity).

An important role is played by *dilatation* (D) $\Delta(\lambda)[P] := [\lambda P]$ for $P \in \mathcal{P}^d$ and $\lambda \in \mathbb{R}$. It is clear that each $\Delta(\lambda)$ is a ring endomorphism of Π. We can now state the main structure theorem for Π.

Theorem 3.2. *The polytope algebra Π is almost a real graded (commutative) algebra, in the following sense:*

(a) *as an Abelian group, Π admits a direct sum decomposition*

$$\Pi = \bigoplus_{r=0}^{d} \Xi_r;$$

(b) *under multiplication,*

$$\Xi_r \cdot \Xi_s = \Xi_{r+s},$$

for $r, s = 0, \ldots, d$ (with $\Xi_r = \{0\}$ for $r > d$);

(c) *$\Xi_0 \cong \mathbb{Z}$, and for $r = 1, \ldots, d$, Ξ_r is a real vector space (with $\Xi_d \cong \mathbb{R}$);*

(d) *if $x, y \in Z_1 := \bigoplus_{r=1}^{d} \Xi_r$ and $\lambda \in \mathbb{R}$, then $(\lambda x)y = x(\lambda y) = \lambda(xy)$;*

(e) *the dilatations are algebra endomorphisms of Π, and, for $r = 0, \ldots, d$, if $x \in \Xi_r$ and $\lambda \geqslant 0$, then*

$$\Delta(\lambda)x = \lambda^r x,$$

with $\lambda^0 = 1$.

While we cannot prove Theorem 3.2 here, we can outline some of the ingredients of the proof. There are three stages. First, we establish the algebra structure over the rational numbers. Next, we introduce the real vector space structure on Ξ_1, and prove a special case of the algebra property (d). Last, after proving the separation Theorem 3.11, we extend the real vector space structure and the algebra property to the rest of Π.

The subgroup (actually subring) Ξ_0 of Π generated by the class 1 of a point clearly plays an anomalous role. The isomorphism $\Xi_0 \cong \mathbb{Z}$ is obvious. Writing Z_1 for the subgroup of Π generated by all elements of the form $[P]-1$ with $P \in \mathcal{P}^d_*$, we have:

Lemma 3.3. *As an Abelian group, $\Pi = \Xi_0 \oplus Z_1$. The projection from Π onto Ξ_0 is the dilatation $\Delta(0)$, and Z_1 is an ideal in Π, with $z \in Z_1$ if and only if $\Delta(0)z = 0$.*

If $a_0, a_1, \ldots, a_k \in \mathbb{E}^d$ are such that $\{a_1, \ldots, a_k\}$ is linearly independent, we write

$$T(a_1, \ldots, a_k) = \operatorname{conv}\{a_0, a_0 + a_1, \ldots, a_0 + \cdots + a_k\},$$

which is a k-simplex, and define

$$s(a_1, \ldots, a_k) := [T(a_1, \ldots, a_k)] - [T(a_1, \ldots, a_{k-1})],$$

with $s(\emptyset) = 1$. This is the class of a partly open simplex (lacking one facet), and these classes generate Π (an arbitrary polytope can be dissected into simplices – an easy proof is given by Tverberg 1974). The key tools are the analogues of the simplex dissection theorems of Hadwiger (1957).

Lemma 3.4. *For $\lambda, \mu \geqslant 0$,*

$$\Delta(\lambda + \mu)s(a_1, \ldots, a_k) = \sum_{j=0}^{k} (\Delta(\lambda)s(a_1, \ldots, a_j))(\Delta(\mu)s(a_{j+1}, \ldots, a_k)).$$

Lemma 3.5. *For $k \geqslant 1$ and integer $n \geqslant 0$,*

$$\Delta(n)s(a_1, \ldots, a_k) = \sum_{r=1}^{k} \binom{n}{r} z_r,$$

where

$$z_r = \sum_{0=j(0)<\cdots<j(r)=k} \prod_{i=1}^{r} s(a_{j(i-1)+1}, \ldots, a_{j(i)})$$

is independent of n.

Lemma 3.5 shows that, if $x \in \Pi$, there exist $y_0 \in \Xi_0$ and $y_1, \ldots, y_d \in Z_1$, such that

$$\Delta(n)x = \sum_{r=0}^{d} \binom{n}{r} y_r$$

for all integer $n \geqslant 0$; these y_r are unique, since, in fact,

$$y_r = \sum_{n=0}^{r} (-1)^{r-n} \binom{r}{n} \Delta(n)x.$$

If we compare

$$\Delta(n)[P] = [nP] = [P]^n = (([P]-1)+1)^n = \sum_{r=0}^{n} \binom{n}{r} ([P]-1)^r$$

with this, we see that $([P]-1)^r = 0$ whenever $P \in \mathcal{P}_*^d$ and $r > d$. We now let Z_r be the subgroup of Π generated by all elements of the form $([P]-1)^j$, with $P \in \mathcal{P}_*^d$ and $j \geqslant r$. Writing the relation above as

$$\Delta(n)([P]-1) = \sum_{k=1}^{d} \binom{n}{k} ([P]-1)^k,$$

taking the jth power (with $j \geqslant r$), and using the fact that $\Delta(n)$ is a ring endomorphism, shows that, if $x \in Z_r$, then $\Delta(n)x - n^r x \in Z_{r+1}$. It is now a short step to show that Z_1 is uniquely divisible, that is, for each $x \in Z_1$, there is a unique $y \in Z_1$, such that $x = ny$ (we recall here that $Z_{d+1} = \{0\}$).

At the next stage, we introduce the concepts of *logarithm* and *exponential*. These are defined by

$$\log(1+z) = \sum_{k \geqslant 1} \frac{(-1)^{k-1}}{k} z^{k-1},$$

$$\exp z = \sum_{k \geqslant 0} \frac{1}{k!} z^k$$

(with $z^0 = 1$), for every $z \in Z_1$. The nilpotence of Z_1 (which follows from its definition, and the nilpotence of elements $[P]-1$) shows that these are well-defined inverse functions on Z_1, with the usual properties of ordinary log and exp. If $P \in \mathcal{P}_*^d$, we write $\log P := \log[P]$; setting $z = [P]-1$, we recognize $\log P$ as the coefficient of n in the expansion of $[nP] = [P]^n$ given above. Indeed, since $\log[nP] = \log([P]^n) = n \log[P]$, and $\Delta(\lambda) \log P = \log(\lambda P)$ for rational λ, we deduce:

Lemma 3.6. *For $P \in \mathcal{P}_*^d$ and rational $\lambda \geqslant 0$, $\Delta(\lambda) \log P = \lambda \log P$.*

We now invert this relation. If $P \in \mathcal{P}_*^d$, $p = \log P$ and $\lambda \geqslant 0$ is rational, then, since $\Delta(\lambda)$ is a ring endomorphism, we have

$$[\lambda P] = \Delta(\lambda)[P] = \Delta(\lambda) \exp p = \exp(\Delta(\lambda)p) = \exp(\lambda p) = \sum_{r=0}^{d} \lambda^r \cdot \frac{1}{r!} p^r.$$

The sum terminates at $r = d$, because $p^{d+1} = 0$.

For $r = 1, \ldots, d$, we now define the *rth weight space* Ξ_r to be the subgroup of Π generated by all the elements p^r, with $p = \log P$ for some $P \in \mathcal{P}_*^d$. We then obtain (after a little more work) Theorem 3.2, except that the dilatations and scalar multiplications are, as yet, only by rationals. (Along the way, it is useful to characterize Ξ_r as the set of all $x \in \Pi$, such that $\Delta(\lambda)x = \lambda^r x$ for one single positive rational $\lambda \neq 1$.)

While it is not yet necessary, we now deal with Ξ_d. Comparison of the definition with Lemma 3.5 shows that Ξ_d is generated by all elements of the form $s(a_1)\cdots s(a_d)$, with $\{a_1, \ldots, a_d\}$ linearly independent. If $i \neq j$, it is easy to show that $s(a_i + \lambda a_j)s(a_j) = s(a_i)s(a_j)$ for any scalar λ, and, since $s(-a_j) = s(a_j)$, standard vector space theory shows that, if we choose a fixed basis $\{e_1, \ldots, e_d\}$ of $\mathbb{E}^d$, then $s(a_1)\cdots s(a_d) = s(\mu e_1)\cdots s(e_d)$, where $\mu = |\det(a_1, \ldots, a_d)|$, the determinant being with respect to the basis $\{e_1, \ldots, e_d\}$. Clearly also, $s((\mu + \nu)e_1) = s(\mu e_1) + s(\nu e_1)$ for $\mu, \nu \geqslant 0$. It now follows that the mapping

$$s(a_1)\cdots s(a_d) \mapsto |\det(a_1, \ldots, a_d)|$$

induces an isomorphism between the Abelian groups Ξ_d and $\mathbb{R}$. This isomorphism (together with the homomorphism it induces on Π, and the corresponding translation invariant valuation on $\mathcal{P}^d$) is called *volume*, and is denoted vol. More generally, each linear subspace L of $\mathbb{E}^d$ also admits a volume vol_L, which is unique up to positive scalar multiplication (see, for example, Hadwiger 1957, section 2.1.3).

The first weight space Ξ_1 admits an alternative characterization. Since Minkowski addition on $\mathcal{P}_*^d$ has a cancellation law, we can define an Abelian group $\mathcal{P}_T$, whose elements are the equivalence classes (P, Q) with $P, Q \in \mathcal{P}_*^d$ under the relation

$$(P, Q) \sim (P', Q') \Leftrightarrow P + Q' = P' + Q + t \quad \text{for some } t \in \mathbb{E}^d,$$

with addition $(P, Q) + (P', Q') = (P + P', Q + Q')$. Then, in fact:

Theorem 3.7. *The mapping* $\log : \mathcal{P}_*^d \to \Xi_1$ *induces an isomorphism between* $\mathcal{P}_T$ *and* Ξ_1.

Scalar multiplication on Ξ_1 can now be defined in one of two equivalent ways:

$$\lambda(P, Q) = \begin{cases} (\lambda P, \lambda Q) & \text{if } \lambda \geqslant 0, \\ (-\lambda Q, -\lambda P) & \text{if } \lambda < 0, \end{cases}$$

on $\mathcal{P}_T \cong \Xi_1$, or

$$\lambda x = \begin{cases} \Delta(\lambda)x & \text{if } \lambda \geqslant 0, \\ -\Delta(-\lambda)x & \text{if } \lambda < 0, \end{cases}$$

on Ξ_1 itself.

Checking that $(\lambda + \mu)x = \lambda x + \mu x$ is tedious rather than difficult; all the other vector space properties are obvious.

The next step involves proving a special case of Theorem 3.2(d). If we write $\Pi(L)$ for the subalgebra of Π generated by the polytopes in the linear subspace L of $\mathbb{E}^d$, and $\Xi_r(L)$ for its rth weight space, then we have:

Lemma 3.8. *If L and H are a complementary line and hyperplane in $\mathbb{E}^d$, E is a line segment in L and $e = \log E$, and $\lambda \in \mathbb{R}$, then $(\lambda e)x = e(\lambda x)$ for all $x \in \Xi_1(H)$.*

The method employed in McMullen (1989) to prove Lemma 3.8 closely followed that in Jessen and Thorup (1978) of the analogous result for the polytope group; for reasons of space, we shall not reproduce the details, even though the result is crucial for the discussion. We then appeal to the separation Theorem 3.11 (see section 3.3), to prove property (d) in full, after establishing it for $x, y \in \Xi_1$ when $d = 2$. The details are largely technical in nature, and do not merit much discussion. In the course of the proof, the remaining properties of Theorem 3.2 are also verified; once (d) is known, the rest follows relatively easily.

3.2. Negative dilatations and Euler-type relations

In section 3.1, the only dilatations considered were those by non-negative scalars λ. We now describe what happens when λ is allowed to be negative.

The *Euler map* * is defined on the generators $[P]$ of Π (with $P \in \mathcal{P}^d$) by (E) $[P]^* := \sum_F (-1)^{\dim F}[F]$, where the sum (here and elsewhere) extends over all faces F of P (including P itself).

Theorem 3.9. *The Euler map is an involutory algebra automorphism of Π. Moreover, for $r = 0, \ldots, d$, if $x \in \Xi_r$ and $\lambda < 0$, then*

$$\Delta(\lambda)x = \lambda^r x^*.$$

The proof of Theorem 3.9 uses the isomorphism Theorem 3.12, which, in turn, depends on the separation Theorem 3.11. In fact, the key result here is an abstract version of a theorem of Sommerville (1927) on polyhedral cones (we shall mention this in section 4.4 below); the idea of the proof is the same as that of the more concrete version proved for translation invariant valuations in McMullen (1977) (the result occurs without proof in McMullen 1975).

There is an amusing algebraic consequence of Theorem 3.9. It is clear that $x \in \Pi$ is invertible if and only if it is of the form $x = \pm(1+z)$ for some $z \in Z_1$. In particular, if $P \in \mathcal{P}^d_*$, then $[P]$ is invertible. Now, if $p = \log P$, then $[P]^{-1} = \exp(-p)$, and, in view of Theorem 3.9, $-p = (\Delta(-1)p)^*$. The algebra properties of the Euler map (which really follow from those of $\Delta(-1)$) now show that $[P]^{-1} = [-P]^*$; written in the form $[P] \cdot [-P]^* = 1 = [o]$, this has a nice interpretation as an equidecomposability result. (In fact, it can be shown not to depend on translation invariance – see section 3.5 below.)

Further properties of the Euler map, and its relations with the Euler characteristic, are given in McMullen (1989).

3.3. The separation theorem

If $o \neq u \in \mathbb{E}^d$, then the *face* of a polytope P *in direction* u is defined to be

$$P_u = \{v \in P \mid \langle v, u\rangle = h(P,u)\}.$$

Then we have:

Lemma 3.10. *The mapping $P \mapsto P_u$ induces an endomorphism $x \mapsto x_u$ of Π.*

Now let $U = (u_1, \ldots, u_{d-r})$ be a $(d-r)$-*frame*, that is, an ordered orthogonal set of non-zero vectors in $\mathbb{E}^d$. Defining recursively $P_U = (P_{u_1,\ldots,u_{d-r-1}})_{u_{d-r}}$, we see that we can define an induced endomorphism $x \mapsto x_U$. If

$$L = U^\perp := \{v \in \mathbb{E}^d \mid \langle v, u\rangle = 0 \text{ for all } u \in U\},$$

and we write $\mathrm{vol}_U := \mathrm{vol}_L$, then clearly the mapping f_U defined by

$$f_U(P) := \mathrm{vol}_U(P_U)$$

induces a (group) homomorphism $f_U : \Pi \to \mathbb{R}$, which we call a *frame functional of type* r (note that $\dim L = r$, so that f_U will be homogeneous of degree r). The frame functional of type d (with $U = \emptyset$) is just vol itself, and that of type 0 (which is, essentially, unique) is χ (or $\Delta(0)$). Then we have:

Theorem 3.11. *The frame functionals separate Π; that is, if $x \in \Pi$ is such that $f_U(x) = 0$ for all frames U, then $x = 0$.*

The key idea behind the proof, which may be found in McMullen (1989), is to use Lemma 3.8 to reduce the result to an inductive proof on the dimension. Again, this adapts the idea of Jessen and Thorup (1978).

The frame functionals are not independent, but although the family of relations or *syzygies* between them is conjectured in McMullen (1989), it is not yet completely established.

3.4. The cone group

Although we shall return to the question of spherical dissections in section 4.4 below, we need to begin the discussion of the cone group here. If L is a linear subspace of $\mathbb{E}^d$, the *cone group* $\widehat{\Sigma}(L)$ is defined to be the Abelian group generated by the *cone classes* $\langle K\rangle$ with $K \in \mathscr{C}(L) := \{K \in \mathscr{C}^d \mid K \subseteq L\}$, with the relations (V) (the valuation property) and (S) $\langle K\rangle = 0$ if $\dim K < \dim L$. Thus $\widehat{\Sigma}(L)$ is the abstract group underlying simple valuations on $\mathscr{C}(L)$. We further define the *full cone group* to be

$$\widehat{\Sigma} := \bigoplus_L \widehat{\Sigma}(L),$$

the sum, as usual, extending over all subspaces L of $\mathbb{E}^d$.

Let us write $P\|L$ to mean that the affine hull of the polyhedral set P is a translate of the linear subspace L, and say that P and L are *parallel*. If F is a face of a (non-empty) P, we define the *normal cone* $N(F,P)$ to P at F to be the set of outer normal vectors to hyperplanes which support P in F, and $n(F,P)$ to be its *intrinsic class*, that is, its class in $\widehat{\Sigma}(L)$, where $L\|N(F,P)$. We note that $\dim N(F,P) = d - \dim F$. We shall write $\operatorname{vol} P$ from now on to mean the volume of a polytope P measured relative to the subspace L parallel to P, with $\operatorname{vol} P = 1$ if P is a point. An important result, which is a consequence of the separation Theorem 3.11 is:

Theorem 3.12. *The mapping*

$$\sigma(P) := \sum_F \operatorname{vol} F \otimes n(F,P)$$

induces a monomorphism from Π into $\mathbb{R} \otimes \widehat{\Sigma}$.

If we define

$$\sigma_r(P) := \sum_{\dim F = r} \operatorname{vol} F \otimes n(F,P),$$

then, similarly, σ_r induces a monomorphism from Ξ_r into $\mathbb{R} \otimes \widehat{\Sigma}_{d-r}$, where

$$\widehat{\Sigma}_{d-r} := \bigoplus_{\dim L = d-r} \widehat{\Sigma}(L).$$

3.5. Translation covariance

Following McMullen (1983), we call a valuation φ on $\mathcal{P}^d$ taking values in the vector space $\mathcal{X}$ *translation covariant* if there exists a map $\Phi : \mathcal{P}^d \to \operatorname{Hom}_{\mathbb{Q}}(\mathbb{E}^d, \mathcal{X})$, such that $\varphi(P + t) = \varphi(P) + \Phi(P)t$ for all $P \in \mathcal{P}^d$ and $t \in \mathbb{E}^d$. With φ weakly continuous (see section 5.5 below), we can replace $\operatorname{Hom}_{\mathbb{Q}}$ by Hom.

The abstract theory underlying translation covariant valuations is, as yet, still in its infancy, and we outline the (unpublished) material only in the hope of stimulating further research. It is convenient, *for this subsection alone*, to change the notation somewhat. We now write Π_0 instead of Π, and define Π to be the Abelian group generated by the polytope classes under the relations (V) alone. As before, Π is actually a ring, with the multiplication given by (M); it still has a subring Ξ_0 generated by $1 := [o]$ which is isomorphic to $\mathbb{Z}$.

We let T be the ideal of Π generated by the elements of the form $[t] - 1$, with $t \in \mathbb{E}^d$. Then $\Pi_0 = \Pi/T$. The powers of T are also ideals of Π; we write $\Pi_k := \Pi/T^{k+1}$. Thus Π_1 is the abstract group underlying translation covariant valuations on $\mathcal{P}^d$. The Π_k for larger k correspond, in a similar way, to the theory of valuations which behave under translation like tensors.

It may be shown that each Π_k is almost (in the same sense as in Theorem 3.2) a graded commutative algebra over $\mathbb{Q}$, with graded terms $\Xi_{k,r}$ of degrees $r \leqslant d+k$. However, except in the relatively trivial case $d=1$ and $k=1$, we have not, so far, been able to extend the algebra properties to allow scalars in $\mathbb{R}$. The first weight space $\Xi_{k,1}$ is always a real vector space, and is, for $k \geqslant 1$, isomorphic to the space $\mathcal{P}_1$ defined like $\mathcal{P}_T$ in section 3.1, but with the equivalence relation not factoring out translations. The analogue of Lemma 3.8, though, has so far proved elusive.

It may be helpful to observe the following stability result. If we write Z for the ideal of Π generated by all the elements $[P]-1$, with $P \in \mathcal{P}^d_*$, then $T^j \subseteq Z^{d+j}$ for all $j \geqslant 0$. Define $\Xi_r := Z^r/Z^{r+1}$. Let Z_k be the corresponding ideal of Π_k, so that

$$Z_k = Z/(Z \cap T^{k+1}) \cong (Z + T^{k+1})/T^{k+1},$$

and hence, if $k \geqslant r$, then

$$\Xi_{k,r} = Z_k^r/Z_k^{r+1} \cong (Z^r + T^{k+1})/(Z^{r+1} + T^{k+1}) \cong Z^r/Z^{r+1} = \Xi_r.$$

What can be proved is that the Euler-type relation

$$\Delta(-1)x = (-1)^r x^*,$$

with $x \in \Xi_{k,r}$, remains valid for all k, and since

$$\bigcap_{k \geqslant 0} T^k = \emptyset,$$

it therefore follows that, if $P \in \mathcal{P}^d_*$, then $[P]^{-1} = [-P]^*$ in Π itself. Writing this out, and observing that we can identify $[P]$ with $P^\dagger$, we have the decomposability result:

Theorem 3.13. *If $P \in \mathcal{P}^d_*$, then*

$$\sum_F (-1)^{\dim F}(P-F)^\dagger = \{o\}^\dagger.$$

We end the section by remarking that Lawrence (unpublished) and Fischer and Shapiro (1992) have also investigated Π, which they call the *Minkowski ring*. In the former paper is shown the following. Let Y be an indeterminate, for $u \in \mathbb{E}^d$, define

$$\varepsilon_u(P) := Y^{h(P,u)},$$

with $P \in \mathcal{P}_*$, and extend to Π by linearity (it is clear from Lemma 1.1 that ε_u is a valuation). Then:

Theorem 3.14. *The valuations ε_u are multiplicative homomorphisms which separate Π.*

It is not clear how these valuations ε_u relate to the frame functionals (which are certainly not multiplicative).

The latter paper considers the subrings of Π generated by finitely many polytope classes, and their prime ideal structures.

3.6. Mixed polytopes

In section 2 we discussed mixed volumes and moment vectors. Not unnaturally, there are analogues in the polytope algebra Π itself (and, in fact, in the more general Π_k defined in section 3.5). The definition of a mixed polytope is straightforward: if $P_1,\dots,P_r \in \mathcal{P}^d_*$ and $p_i = \log P_i$ for $i = 1,\dots,r$, then the *mixed polytope* $m(p_1,\dots,p_r)$ is defined by

$$m(p_1,\dots,p_r) := \frac{1}{r!}\cdot p_1\cdots p_r.$$

Thus the rth weight space component p_r of a polytope class $[P]$ is just $p_r = m(p,\dots,p)$.

A theory of mixed polytopes was attempted by Meier (1977), but it appears that, at one point, his argument is flawed. Another theory, but only within the context of the polytope group (see section 4.1), was propounded in McMullen and Schneider (1983, section 6).

If φ is now a translation invariant valuation on $\mathcal{P}^d$ which is homogeneous of degree r, then the corresponding mixed valuation is

$$\varphi(P_1,\dots,P_r) = \varphi(m(p_1,\dots,p_r)).$$

Since

$$(\lambda_1 p_1)\cdots(\lambda_r p_r) = (\lambda_1\cdots\lambda_r)p_1\cdots p_r,$$

we have the curious consequence that, for fixed polytopes $P_1,\dots,P_r$, the value of the mixed valuation $\varphi(\lambda_1 p_1,\dots,\lambda_r p_r)$ depends only on the product $\lambda_1\cdots\lambda_r$, without any assumption on the continuity of φ.

We make one final remark in this subsection. It was observed by Groemer (1977a) that, if $K, L \in \mathcal{K}^d$ are such that $K\cup L$ is convex, then the mixed volume satisfies

$$V(K\cup L, K\cap L, \mathcal{C}) = V(K, L, \mathcal{C})$$

for any $(d-2)$-tuple $\mathcal{C}$ of convex bodies. An easier proof than the original was given in McMullen and Schneider (1983, section 3). However, the essence of the proof is algebraic, and in Π is almost trivial. Let $P, Q \in \mathcal{P}^d_*$ be such that $X := P\cup Q$ is convex, let $Y := P\cap Q$, and write $p := \log P$, and so on. Then the valuation property $[X]+[Y] = [P]+[Q]$ yields in Ξ_r:

$$x^r + y^r = p^r + q^r.$$

Hence

$$\begin{aligned} xy &= \tfrac{1}{2}((x+y)^2 - (x^2+y^2)) \\ &= \tfrac{1}{2}((p+q)^2 - (p^2+q^2)) \\ &= pq, \end{aligned}$$

which is the abstract version of the required property.

3.7. *Relatively open polytopes*

Schneider (1985) developed a theory of decomposition by translation, based on relatively open polytopes. We briefly describe this here, although we shall not set up the underlying abstract structure.

We know from section 3.1 that the class $[\operatorname{relint} P]$ of the relative interior of a polytope P exists in Π, and since

$$[P] = \sum_F [\operatorname{relint} F]$$

is obvious, Möbius inversion (see Rota 1964) yields

$$[\operatorname{relint} P] = \sum_F (-1)^{\dim P - \dim F} [F],$$

or:

Theorem 3.15. *The class of the relative interior of a polytope P is given by*

$$[\operatorname{relint} P] = (-1)^{\dim P} [P]^*.$$

In view of the fact that the mapping $[P] \mapsto [P]^*$ is an automorphism of Π, we see that $[\operatorname{relint} P] \mapsto (-1)^{\dim P}[P]$ also gives an automorphism – in retrospect this is transparent. Schneider (1985) gives a separation criterion based on relatively open polytopes (closely related to the isomorphisms of section 4.3); the simpler one given by McMullen (1989) is essentially the previous remark.

3.8. *Invariance under other groups*

When we impose invariance under bigger groups than translations on the polytope algebra, we not unnaturally lose some of its properties. In this subsection, we briefly discuss what is so far known to happen.

Let G be a group of affinities acting on $\mathbb{E}^d$, which contains the group $T \cong \mathbb{E}^d$ (as an additive group) of translations in $\mathbb{E}^d$. If we replace the translation invariance condition (T) in the definition of the polytope algebra Π by the stronger condition (G) $[\Phi P] = [P]$ for all $P \in \mathcal{P}^d$ and $\Phi \in G$, we obtain a new group Π_G. The polytope algebra Π itself is thus Π_T.

It is clear that Π_G is no longer a ring, unless $G = T$, because Minkowski addition is not compatible with affinities which are not translations. However, since Π_G is, as an Abelian group, a quotient of Π, a great deal of the structure of Π does survive.

Theorem 3.16. *For any group G of affinities of $\mathbb{E}^d$ which contains the translations, the polytope group Π_G has the following structure:*

(a) *Π_G has a direct sum decomposition*

$$\Pi_G = \bigoplus_{r=0}^{d} \Xi_r,$$

where $\Xi_0 \cong \mathbb{Z}$ *and, for* $r = 1, \ldots, d$, Ξ_r *is a real vector space;*

(b) *dilatation acts on* Ξ_r *by*

$$\Delta(\lambda)(x) = \begin{cases} \lambda^r x & \text{if } \lambda \geqslant 0, \\ \lambda^r x^* & \text{if } \lambda < 0 \end{cases}$$

(*with* $\lambda^0 = 1$) *if* $x \in \Xi_r$, *where* * *is the Euler map.*

We obtain the direct sum decomposition, because it is obvious that dilatations commute with the endomorphisms of Π induced by affinities.

For most groups G, we can say little more than this. However, there are some special cases.

Theorem 3.17. *If* G *contains a dilatation by some* $\lambda \neq \pm 1$, *then* $\Pi_G \cong \mathbb{Z}$.

Let A denote the group of all affinities of $\mathbb{E}^d$, and let EA denote the subgroup of *equiaffinities*, that is, the mappings of the form $v \mapsto \Phi v + t$, where Φ is a linear mapping with $\det \Phi = \pm 1$. Then we have:

Theorem 3.18. (a) $\Pi_A \cong \mathbb{Z}$;

(b) *For* $d \geqslant 1$, $\Pi_{\mathrm{EA}} \cong \mathbb{Z} \oplus \mathbb{R}$.

Part (a) is a consequence of the previous theorem, while in (b) the volume term additionally survives.

Finally, let TH be the group consisting of the translations and reflexions in points (mappings of the form $v \mapsto 2c - v$, where $c \in \mathbb{E}^d$). Since $\Delta(-1)$ acts as the identity on Π_{TH}, we have:

Theorem 3.19. *If* $G \supseteq \mathrm{TH}$, *then in* Π_G, *the rth weight space* Ξ_r *is generated by classes of polytopes of lower dimension than* d *if* $r \not\equiv d$ *modulo* 2.

4. Simple valuations and dissections

We now consider the groups which correspond to the simple valuations on $\mathcal{P}^d$. As in section 4.1, the translation invariant case is the fundamental one, about which most is known. However, since the rigid motion invariant case has provided much of the motivation for research in this area, we shall also devote a fair amount of space to it.

4.1. The algebra of polytopes

The *polytope group* $\widehat{\Pi}^d$ has a generator $\langle P \rangle$ for each $P \in \mathcal{P}^d$; these generators satisfy the relations (V) and (T) of the polytope algebra, together with (S) $\langle P \rangle = 0$ if $\dim P < d$, which corresponds to simple valuations. From Theorems 3.2 and 3.9, we deduce at once the main structure theorem of Jessen and Thorup (1978) and Sah (1979).

Theorem 4.1. (a) $\widehat{\Pi}^0 \cong \mathbb{Z}$.

(b) *If $d > 0$, then $\widehat{\Pi}^d$ has a direct sum decomposition*

$$\widehat{\Pi}^d = \bigoplus_{r=1}^{d} \widehat{\Xi}_r$$

into real vector spaces. Moreover, dilatation acts on $\widehat{\Xi}_r$ by

$$\Delta(\lambda)x = \begin{cases} \lambda^r x & \text{if } \lambda \geqslant 0, \\ (-1)^d \lambda^r x & \text{if } \lambda < 0. \end{cases}$$

Of course, dilatations are compatible with the relations (S). For the negative dilatations, we observe that the Euler map in $\widehat{\Pi}^d$ acts as

$$\langle P \rangle^* = \sum_F (-1)^{\dim F} \langle F \rangle = (-1)^d \langle P \rangle.$$

[In Jessen and Thorup (1978), this behaviour under negative dilatations was assumed; in Sah (1979), by contrast, it was proved.]

For the separation result, we need another concept. If $U = (u_1, \ldots, u_{d-r})$ is a frame, and $E = (\varepsilon_1, \ldots, \varepsilon_{d-r})$ is a vector with entries $\varepsilon_i = \pm 1$, we write $EU := (\varepsilon_1 u_1, \ldots, \varepsilon_{d-r} u_{d-r})$, and

$$\operatorname{sgn} E := \prod_{j=1}^{d-r} \varepsilon_j.$$

A *Hadwiger functional of type* r is then a map of the form

$$h_U := \sum_E \operatorname{sgn} E \, f_{EU},$$

where U is a $(d-r)$-frame. As with the frame functionals, $h_\emptyset = \mathrm{vol}$ is just ordinary volume. Then we have:

Theorem 4.2. *The Hadwiger functionals separate $\widehat{\Pi}^d$.*

This result, proved by Jessen and Thorup (1978) and Sah (1979), cannot be deduced from Theorem 3.11, but must be shown independently.

Just as there are syzygies between the frame functionals, so there are between the Hadwiger functionals. In Sah (1979), these syzygies were described, but the proof that they were the only ones was lacking. The proof was provided by Dupont (1982). In fact, there is a stronger result than Theorem 4.2.

Theorem 4.3. *Let $\mathscr{X}$ be a real vector space. Then all linear mappings $\varphi : \widehat{\Pi}^d \to \mathscr{X}$ are of the form*

$$\varphi = \sum_U f_U c_U,$$

where $U \mapsto c_U$ is an arbitrary function from frames into $\mathscr{X}$.

4.2. Cones and angles

We have already introduced the cone group $\widehat{\Sigma}$ in section 3.3. In order to proceed, we need to investigate it a little further.

First, we produce the analogue of the isomorphism Theorem 3.12. The subgroup of $\widehat{\Sigma}$ generated by the classes of cones which contain a line (and so have a face of apices of positive dimension) is denoted $\widehat{\Gamma}$. If $c \in \widehat{\Sigma}$, we write $\overline{c}$ for its image under the quotient map from $\widehat{\Sigma}$ onto $\widehat{\Sigma}/\widehat{\Gamma}$, and $\overline{n}(F,P) := \overline{n(F,P)}$ (with analogous notation employed subsequently). Then Theorem 4.2 implies:

Theorem 4.4. *The map $\overline{\sigma} : \mathcal{P}^d \to \mathbb{R} \otimes (\widehat{\Sigma}/\widehat{\Gamma})$, defined by*

$$\overline{\sigma}(P) = \sum_F \operatorname{vol}(F) \otimes \overline{n}(F,P)$$

induces a monomorphism from $\widehat{\Pi}^d$ into $\mathbb{R} \otimes (\widehat{\Sigma}/\widehat{\Gamma})$.

We shall see the suggestive role this result plays in section 4.5. For the moment, we just note that the classes of the normal cones to the faces of a lower-dimensional polytope lie in $\widehat{\Gamma}$.

An important notion is that of angle. In $\mathbb{E}^d$, we have a natural notion of rotation invariant angle, but over more general fields $\mathbb{F}$, where we do not usually have a full rotation group, an alternative approach is necessary. If L is a linear subspace, then an *angle* on $\mathscr{C}(L)$ is an L-simple valuation ω_L (into the base field), such that $\omega_L(L) = 1$. We can choose an angle on each subspace L simultaneously, as follows (the notion was suggested by U. Betke): pick any d-polytope Q with $o \in \operatorname{int}(Q)$, and define ω_L by

$$\omega_L(C) = \frac{\operatorname{vol}(C \cap Q)}{\operatorname{vol}(L \cap Q)}$$

for each $C \in \mathscr{C}(L)$, where $\operatorname{vol} = \operatorname{vol}_L$. (By the way, the same idea enables us to choose a particular scaling of all the volumes vol_L simultaneously – we just set $\operatorname{vol}_L(L \cap Q) = 1$ for all L.)

If we have an angle ω_L for each subspace L, then we can define a homomorphism ω on $\widehat{\Sigma}$ by $\omega = \omega_L(c)$ if $c \in SP(L)$. We shall also refer to such a homomorphism ω as an angle.

The *angle cone* of a polyhedral set P at its face F is the cone $A(F,P) := \operatorname{pos}(P - F)$ generated by P at (any relatively interior point of) F. We write $a(F,P)$ for the intrinsic class of $A(F,P)$ in $\widehat{\Sigma}$, and, if ω is an angle, we write $\alpha(F,P) := \omega(a(F,P))$, which we call an *inner angle*. We similarly write $\nu(F,P) := \omega(n(F,P))$, with $n(F,P)$ the class of the normal cone, which we call an *outer angle*. We observe that $\alpha(F,F) = 1 = \nu(F,F)$ for all (non-empty) faces F. Further, we call an inner angle α and outer angle ν *inverse* if

$$\sum_J (-1)^{\dim J - \dim F} \alpha(F,J)\nu(J,G) = \delta(F,G),$$

where

$$\delta(G,H) := \begin{cases} 1 & \text{if } F = G, \\ 0 & \text{if } F \neq G, \end{cases}$$

is the *delta function*.

The relationship between inner and outer angles is such that:

Lemma 4.5. *If α and ν are inverse inner and outer angles, then*

$$\sum_J (-1)^{\dim G - \dim J} \nu(F,J)\alpha(J,G) = \delta(F,G).$$

In fact, the best way to look at this is within the context of the *incidence algebra* of Rota (1964). This consists of the functions κ on ordered pairs of faces (taking values in the base field), such that $\kappa(F,G) = 0$ unless F is a face of G. Addition and multiplication of such functions are defined by

$$(\kappa + \lambda)(F,G) = \kappa(F,G) + \lambda(F,G),$$
$$(\kappa\lambda)(F,G) = \sum_J \kappa(F,J)\lambda(J,G).$$

These functions can be thought of as triangular matrices indexed by faces of polyhedral sets. The crucial result about angles is the following.

Lemma 4.6. *If ν is an outer angle, then there exists an inverse inner angle α, and conversely.*

When the angles are the ordinary normalized angles in $\mathbb{E}^d$, this result was first proved by McMullen (1975); the general result occurs in McMullen (1989).

Inner and outer angles can be used to find another relationship between the polytope algebra and the polytope groups (see section 4.3 below). The abstract results on which it depends are the following. In each case, $\mathcal{S}$ stands for $\mathcal{P}^d$ or $\mathcal{C}^d$, and in the latter case translation invariance is to be ignored.

Lemma 4.7. *Let $\mathcal{G}$ be an Abelian group, and for each subspace L, let $\psi_L : \mathcal{S}(L) \to \mathcal{G}$ be an L-simple translation invariant valuation. If $\psi : \mathcal{S} \to \mathcal{G}$ is defined by $\psi(P) := \psi_L(P)$ if $P \| L$, then the mapping $\varphi : \mathcal{S} \to \mathcal{G} \otimes \widehat{\Sigma}$ given by*

$$\varphi(P) := \sum_F \psi(F) \otimes n(F,P)$$

is a translation invariant valuation.

Lemma 4.8. *Let $\mathcal{G}$ be an Abelian group, and let $\varphi : \mathcal{S} \to \mathcal{G}$ be a translation invariant valuation. Then for each subspace L, the mapping $\psi_L : \mathcal{S}(L) \to \mathcal{G} \otimes \widehat{\Sigma}$ defined by*

$$\psi_L(P) := \begin{cases} \sum_F \varphi(F) \otimes (-1)^{\dim P - \dim F} a(F,P) & \textit{if } P \| L, \\ 0 & \textit{otherwise}, \end{cases}$$

is an L-simple translation invariant valuation.

The final lemma in this subsection is clear.

Lemma 4.9. *If $\mathscr{X}$ is a real vector space, and ω is an angle, then the mapping π : $\mathscr{X} \otimes \widehat{\Sigma} \to \mathscr{X}$ defined by $\pi(x \otimes c) := \omega(c)x$, with $x \in \mathscr{X}$ and $c \in \widehat{\Sigma}$, is a homomorphism.*

4.3. The polytope groups

Since a linear subspace L of $\mathbb{E}^d$ is itself a real vector space of the appropriate dimension, it also has associated with it a polytope group $\widehat{\Pi}(L)$; thus $\widehat{\Pi}^d = \widehat{\Pi}(\mathbb{E}^d)$. Before we discuss the connexion between the polytope groups and the polytope algebra, we shall mention a kind of multiplication, which is an analogue of the genuine multiplication in Π.

Theorem 4.10. *Let L and M be complementary linear subspaces of $\mathbb{E}^d$ of positive dimension. Then there is a natural embedding of $\widehat{\Pi}(L) \otimes \widehat{\Pi}(M)$ into $\widehat{\Pi}^d$. This embedding is compatible with the scalar multiplication in that, if $x \times y$ is the image of $x \otimes y$, with $x \in \widehat{\Pi}(L)$ and $y \in \widehat{\Pi}(M)$, then for each scalar λ,*

$$(\lambda x) \times y = x \times (\lambda y) = \lambda(x \times y).$$

The embedding is that which is obviously induced by the geometric direct sum. In fact, this theorem lies at the heart of the original proof of the structure Theorem 4.1 in Jessen and Thorup (1978) and Sah (1979).

We now write

$$\widehat{\Pi} := \bigoplus_L \widehat{\Pi}(L),$$

the sum, as usual, extending over all linear subspaces of $\mathbb{E}^d$. Our other isomorphism theorem for the polytope algebra is:

Theorem 4.11. $\Pi \cong \widehat{\Pi}$.

The isomorphism is easily described. Let α and ν be any pair of inverse inner and outer angles. First, we define the mapping $\varphi : \mathscr{P}^d \to \widehat{\Pi}$ by

$$\varphi(P) := \sum_F \nu(F,P)\langle F\rangle,$$

where $\langle F\rangle$ is the intrinsic class of F, that is, its class in $\widehat{\Pi}(L)$, where $L \| F$. There is no trouble with the vertices F^0 of P, even though $\widehat{\Pi}^0 \cong \mathbb{Z}$, since $\sum_{F^0} \nu(F^0,P) = 1$. Then φ induces a homomorphism from Π to $\widehat{\Pi}$ by Lemmas 4.7 and 4.9.

Next, for each subspace L of $\mathbb{E}^d$, we define a mapping $\psi_L : \mathscr{P}(L) \to \Pi$ by

$$\psi_L(P) = \begin{cases} \sum_F (-1)^{\dim P - \dim F} \alpha(F,P)[F] & \text{if } P \| L, \\ 0 & \text{otherwise.} \end{cases}$$

Again, we have no problems with the 0-components of the classes $[F]$, since

$$\sum_F (-1)^{\dim P - \dim F} \alpha(F,P) = \begin{cases} 1 & \text{if } \dim P = 0, \\ 0 & \text{if } \dim P > 0. \end{cases}$$

Then these ψ_L induce a homomorphism $\psi : \widehat{\Pi} \to \Pi$ by Lemmas 4.8 and 4.9.

Finally, the definition of inverse angles shows that φ and ψ are inverse homomorphisms, as required.

This proof closely parallels that in McMullen (1977) of the relationship between general and simple translation invariant valuations. However, this proof covers more than just the real-valued case considered there.

4.4. Spherical dissections

In preparation for the discussion of Hilbert's third problem in section 4.5, we must first discuss the analogous problem for spherical polytopes or cones (we can identify a spherical polytope on Ω^{d-1} with the cone in $\mathbb{E}^d$ which it spans, and so we shall usually consider the latter). In this section, we shall largely follow Sah (1979).

The group $\widehat{\Sigma}$ is, as a group, not of great interest. It is only when we impose additional relations on it that it begins to acquire some structure. As far as we are concerned, the most important case is the following. The group $\widehat{\Sigma}_O^d$ is $\widehat{\Sigma}(\mathbb{E}^d)$, with the additional relations (O) $\langle \Psi K \rangle = \langle K \rangle$ whenever $K \in \mathscr{C}^d$ and $\Psi \in O$, the orthogonal group. We define $\widehat{\Sigma}_O := \sum_{d \geqslant 0} \widehat{\Sigma}_O^d$, with $\widehat{\Sigma}_O^0 = \mathbb{Z}$. We then have a natural product $*$ on $\widehat{\Sigma}_O$, which is induced by orthogonal Cartesian product, and which is compatible with orthogonal transformations. Formally, if K_1 and K_2 are two cones, we define $\langle K_1 \rangle * \langle K_2 \rangle := \langle K_1 \times \Psi K_2 \rangle$, where Ψ is a suitable rotation taking K_2 into a subspace orthogonal to K_1.

Before we proceed further, let us make a remark. If SO is the subgroup consisting of the rotations in O, then we have (compare Lemma 4.31 below):

Lemma 4.12. *$\widehat{\Sigma}_O^d = \widehat{\Sigma}_{\mathrm{SO}}^d$ for every dimension $d \geqslant 2$.*

Clearly, $\widehat{\Sigma}_O^1 \cong \mathbb{Z}$, and is generated by the class p (which stands for "point") of a half-line. The same notion which proves Lemma 4.12 also lies at the heart of:

Lemma 4.13. *The group $\widehat{\Sigma}_O$ is 2-divisible.*

We also have:

Lemma 4.14. *For $d = 2$ or 3, $\widehat{\Sigma}_O^d \cong \mathbb{R}$.*

The case $d = 2$ is obvious; while the second is fairly familiar, we shall justify it in Theorem 4.17 below.

In order to investigate $\widehat{\Sigma}_O$, we need to introduce some further concepts. The *recession cone* $\operatorname{rec} K$ of a polyhedral set K is defined to be

$$\operatorname{rec} K := \{x \in \mathbb{E}^d \mid x + y \in K \text{ for all } y \in K\}.$$

If K is a polytope, then $\operatorname{rec} K = \{o\}$, while if K is a cone with apex a, then $\operatorname{rec} K = K - a$. A common generalization of results of Brianchon (1837) and Gram (1874) (see also Shephard 1967), and Sommerville (1927) is the following, due to McMullen (1983):

Theorem 4.15. *Let K be a polyhedral set in $\mathbb{E}^d$. Then*

$$\sum_F (-1)^{\dim F} a(F, K) = (-1)^d \langle \operatorname{rec}(-K) \rangle.$$

As in section 3.4, $a(F, K)$ denotes the class of the angle cone $A(F, K)$. We have kept $\operatorname{rec}(-K)$ instead of $\operatorname{rec} K$, to emphasize the geometric nature of the dissection result.

In case K is a pointed polyhedral cone in $\mathbb{E}^d$, we can apply Lemma 4.13 to Theorem 4.15, to deduce:

Theorem 4.16. *If d is odd, and K is a pointed polyhedral cone in $\mathbb{E}^d$, then*

$$\langle K \rangle = \frac{1}{2} \sum_{F \neq o} (-1)^{\dim F - 1} a(F, K).$$

An *r-fold join* is a just a product $k_1 * \cdots * k_r$, where $k_i = \langle K_i \rangle$ is the class of a cone of dimension at least 1 for $i = 1, \ldots, r$. We write $\widehat{\Sigma}^d_r$ for the subgroup of $\widehat{\Sigma}^d$ generated by the r-fold joins, and $\widehat{\Sigma}_r := \bigcup_{d \geqslant 0} \widehat{\Sigma}^d_r$. Then

$$\widehat{\Sigma} = \widehat{\Sigma}_1 \supseteq \widehat{\Sigma}_2 \supseteq \cdots,$$

and

$$\widehat{\Sigma}^{d_1}_{r_1} * \widehat{\Sigma}^{d_2}_{r_2} \subseteq \widehat{\Sigma}^{d_1+d_2}_{r_1+r_2},$$

for all r_1, r_2, d_1, d_2.

We may observe that each term in the sum of Theorem 4.16 is a join. Indeed, if we define the *intrinsic inner cone* of K at its face F to be $B(F, K) := A(F, K) \cap F^\perp$, where $F^\perp$ is the orthogonal complementary subspace to F in $\mathbb{E}^d$, then for $d \geqslant 1$, we can express $A(F, K)$ as a non-trivial product

$$A(F, K) = B(F, K) \times \operatorname{lin} F.$$

A closely related angle cone is

$$\widetilde{A}(F, K) := B(F, K) \times \mathbb{E}^{\dim F - 1}$$

whenever $\dim F \geqslant 1$, which corresponds to the angle cone at $F \cap \Omega$ of the spherical polytope $K \cap \Omega$. The mapping $e : \widehat{\Sigma}^d \to \widehat{\Sigma}^{d-1}$ given by

$$e(K) := \sum_{F \neq o} (-1)^{\dim F - 1} \widetilde{a}(F, K),$$

where $\widetilde{a}(F,K) := \langle \widetilde{A}(F,K) \rangle$, is called by Sah (1979) the *Gauss–Bonnet map* (the reasons for this name are not altogether clear). When d is even, Sah (1981) shows that $e(\langle K \rangle) = 0$, and Theorem 4.16 can be written in the form

$$\langle K \rangle = p * e(\langle K \rangle).$$

Extending the notation of section 3.4, let us write $\widehat{\Gamma}^d := p * \widehat{\Sigma}^{d-1}$. The crucial result of Sah (1981) is:

Theorem 4.17. *If $j \geqslant 0$, then $\widehat{\Sigma}^{2j+1} = \widehat{\Gamma}^{2j+1} = p * \widehat{\Sigma}^{2j}$, and the map $e; \widehat{\Sigma}^{2j+1} \to \widehat{\Sigma}^{2j}$ is an isomorphism inverse to $x \mapsto p * x$.*

It follows that, if $\widehat{\Gamma} := \bigoplus_{d \geqslant 1} \widehat{\Gamma}^d$, then $\widehat{\Sigma}/\widehat{\Gamma}$ is evenly graded by degree.

We now introduce the graded volume map. Our normalization of (spherical) volume gives the total volume of Ω^{d-1} as 1, in contrast to Sah (1979), who follows Schläfli in assigning it volume 2^d, and is just the rotation invariant angle. A natural way of defining this for a polyhedral cone K with apex o is by

$$\operatorname{vol} K := \int_K \exp(-\pi \|x\|^2)\, \mathrm{d}x,$$

where $\mathrm{d}x$ is ordinary Lebesgue measure in the subspace $\operatorname{lin} K$. Thus, the volume of a linear subspace is always 1. The *graded volume* of K is then defined by

$$\text{gr.vol}\langle K \rangle := \operatorname{vol} K \cdot T^{\dim K},$$

where T is an indeterminate. Further, Theorem 4.16 has the implication

$$\operatorname{vol} K = \tfrac{1}{2} \sum_{F \neq 0} \beta(F, K),$$

where $\beta(F,K) := \operatorname{vol} B(F,K) = \operatorname{vol} A(F,K)$, because of the normalization, and the elementary observation

$$\operatorname{vol}(K_1 \times K_2) = \operatorname{vol} K_1 \cdot \operatorname{vol} K_2$$

for orthogonal cones K_1 and K_2. Lemma 4.14 for $d = 3$ is now an immediate consequence of this and Theorem 4.17.

We now discuss various dissection results and their consequences. Let K be a polyhedral cone with apex o. First, when we note that each point $z \in \mathbb{E}^d$ admits a unique expression of the form $z = x + y$, where $x \in \operatorname{relint} F$ for some face F of K (possibly K itself) and $y \in N(F, K)$, we have:

Theorem 4.18. *Let K be a polyhedral cone in $\mathbb{E}^d$ with apex o. Then the cones F and $N(F,K)$, with F a face of K, are orthogonal, and $\mathbb{E}^d$ is dissected into the cones $F \times N(F,K)$.*

In $\widehat{\Sigma}$, this result leads to

$$\langle \mathbb{E}^d \rangle = \sum_F \langle F \rangle * n(F,K),$$

where $n(F,K) := \langle N(F,K) \rangle$.

The analogue for normal cones of $B(F,K)$ is the *intrinsic outer cone* $C(F,K) := N(F,K) \cap \operatorname{lin} K$, the normal cone to K at its face F in the subspace $\operatorname{lin} K$ which it spans; we write $c(F,K)$ for its class, and $\gamma(F,K)$ for its volume. Similarly, we write $Z(F,K)$ for the orthogonal complement of $\operatorname{lin} F$ in $\operatorname{lin} K$, and $z(F,K)$ for its class. Further, we define

$$m(F,K) := (-1)^{\dim Z(F,K)} z(F,K),$$

and the identity function i by

$$i(F,K) := \begin{cases} 1 \ (\in \mathbb{Z}) & \text{if } F = K, \\ 0 & \text{if } F \neq K. \end{cases}$$

We can extend the notation of the incidence algebra of Rota (1964) to the multiplication $*$, and write

$$f * g(F,G) := \sum_J f(F,J) g(J,G),$$

where F, G are faces of a polyhedral cone, and the summation is over all faces J, with the understanding that $f(F,J) = 0$ unless F is a face of J, and so on. The Euler relation for polyhedral cones implies:

Theorem 4.19. $m * z = i = z * m$.

If we define $\widetilde{b}$ by

$$\widetilde{b}(F,K) := (-1)^{\dim K - \dim F} b(F,K),$$

and $\widetilde{c}$ similarly, we can now rephrase Theorems 4.15 (for cones) and 4.18 as:

Theorem 4.20. (a) $m * b = \widetilde{b}$;
(b) $b * c = z$.

Combining these two results, we have consequences of McMullen (1983) (compare also McMullen 1975b):

Theorem 4.21. (a) $\widetilde{b} * c = i = b * \widetilde{c}$;
(b) $c * \widetilde{b} = i \ (= \widetilde{c} * b)$.

The basic result here is part (a), from which (b) follows by use of the incidence algebra, and polarity, which we shall talk about below (the second part of (b) is actually the same as the first). An important consequence of this (obtained by replacing the cone classes by their volumes or angles) is that β and γ are inverse inner and outer angles.

Now we introduce polarity. We define the *polar* K° of a cone K by

$$K^\circ := \{x \in \mathbb{E}^d \mid \langle x, y \rangle \leqslant 0 \text{ for all } y \in K\}.$$

Note that $K^{\circ\circ} = K$. As an example, $N(F,K) = A(F,K)^\circ$ for every face F of a polyhedral cone K. Our first remark is that polarity is compatible with the valuation property [this originates in Sah (1979), from which what follows is taken, but see also Lawrence (1988)].

Lemma 4.22. *Let K_1, K_2 be polyhedral cones with apex o. Then*

$$\begin{aligned}(K_1 \cap K_2)^\circ &= K_1^\circ + K_2^\circ \\ &= K_1^\circ \cup K_2^\circ,\end{aligned}$$

if $K_1^\circ \cup K_2^\circ$ is convex.

Further, if $\dim K < d$, then K° has $(\operatorname{lin} K)^\perp$ as its (non-trivial) face of apices, and so $\langle K^\circ \rangle \in \widehat{\Gamma}^d$; the converse is also obviously true. There then follows:

Theorem 4.23. *Polarity induces an involutory automorphism § of $\widehat{\Sigma}^d / \widehat{\Gamma}^d$, defined by*

$$\langle K \rangle^{\S} := \langle K^\circ \rangle.$$

This automorphism is called the *antipodal map*, and it extends to $\widehat{\Sigma}/\widehat{\Gamma}$ in the natural way, if § is now defined intrinsically; thus $b(F,K)^{\S} = c(F,K)$ in $\widehat{\Sigma}/\widehat{\Gamma}$. The algebra (ring) structure on $\widehat{\Sigma}$, with multiplication $*$, now induces an algebra structure on $\widehat{\Sigma}/\widehat{\Gamma}$. In fact, we also have a co-algebra structure.

Before we describe this, however, let us introduce something more general. If $x \in \widehat{\Sigma}$, we write $\overline{x}$ for its image in $\widehat{\Sigma}/\widehat{\Gamma}$; we also define $\overline{b}$ by $\overline{b}(F,K) := \overline{b(F,K)}$, and so on. The *total spherical Dehn invariant* of a pointed polyhedral cone K (or of the corresponding spherical polytope $K \cap \Omega$) is

$$\Psi_S := \sum_F^{*} \langle F \rangle \otimes \overline{b}(F,K) \in \widehat{\Sigma} \otimes (\widehat{\Sigma}/\widehat{\Gamma}),$$

where the sum extends over all faces $F \neq \{o\}$ with $\dim K - \dim F$ even. In fact, the terms with $\dim K - \dim F$ odd drop out anyway, and, when $\dim K$ is even, that for $F = \{o\}$ is not needed, because the information carried in the term $\langle K \rangle \otimes \overline{\langle o \rangle}$ contains that in $\langle o \rangle \otimes \overline{\langle K \rangle}$.

The map Ψ_S then induces a map $\overline{\Psi}_S : \widehat{\Sigma}/\widehat{\Gamma} \to (\widehat{\Sigma}/\widehat{\Gamma}) \otimes (\widehat{\Sigma}/\widehat{\Gamma})$, defined on the generators $\overline{\langle K \rangle}$ of $\widehat{\Sigma}/\widehat{\Gamma}$ by

$$\overline{\Psi}_S(\overline{\langle K \rangle}) := \sum_F^{*} \overline{\langle F \rangle} \otimes \overline{b}(F, K).$$

This is the *comultiplication* on $\widehat{\Sigma}/\widehat{\Gamma}$. The *co-unit* or *augmentation* (which is dual to the unit) is the natural mapping whose kernel is the set of elements of $\widehat{\Sigma}/\widehat{\Gamma}$ of positive degree. With these algebra and co-algebra structures and the antipodal map, $\widehat{\Sigma}/\widehat{\Gamma}$ then becomes a Hopf algebra (see Sah 1979).

In discussing equidissectability, it is natural to look for a suitable family of separating homomorphisms, as in section 3.3 for the polytope algebra. The map Ψ_S separates $\widehat{\Sigma}^d$, but in a rather trivial way, since it is obviously injective. More relevant is the *total classical Dehn invariant*

$$\Phi_S := (\text{gr.vol} \otimes \text{id}) \circ \Psi_S;$$

thus

$$\Psi_S(K) = \sum_F^{*} (\text{vol}\, F \cdot T^{\dim F}) \otimes \overline{b}(F, K) \in \mathbb{R}[T] \otimes (\widehat{\Sigma}/\widehat{\Gamma}),$$

with the same summation convention as above.

In the positive direction, we have:

Theorem 4.24. *Φ_S separates $\widehat{\Sigma}^d$ for $d = 2$ or 3.*

This is really just a restatement of Lemma 4.14. For $d \geqslant 4$, however, the situation is quite different, and the general equidissectability problem remains unsolved. For example, if $d = 4$, and the cone K has rational dihedral angles, then $\Phi_S(K) = \text{vol}\, K \cdot T^4$, so that, if Φ_S does separate $\widehat{\Sigma}^4$, then K should be equidissectable with a product cone. This is far from obviously true; indeed, such cones K are a possible source of torsion in $\widehat{\Sigma}^4$ (see Sah 1979).

A partial result in this direction by Dupont and Sah (1982) is the following.

Theorem 4.25. *A polyhedral cone which is the fundamental cone for a finite orthogonal group in $\mathbb{E}^d$ is equidissectable with a $(d-1)$-fold product cone.*

What is actually shown is that the fundamental polyhedral cones for two such orthogonal groups of the same order are equidissectable, and the core of the argument lies in proving it for p-groups (Sylow subgroups).

4.5. Hilbert's third problem

We now come to Hilbert's third problem and its variants. In section 1.4, we introduced the concepts of G-equidissectability and G-equicomplementability of polytopes (or polyhedra) under a group of affinities G of $\mathbb{E}^d$. As examples of typical such groups G, we have the groups T, TH, A and EA introduced in section 3.8, as well as the group D of all isometries of $\mathbb{E}^d$, and its subgroup SD of direct isometries or rigid motions. In what follows, we let G be a group of affinities. We begin with two important results of Hadwiger (1957).

Lemma 4.26. *Let $P, Q \in \mathcal{P}^d$. Then $P \approx_G Q$ if and only if $P \sim_G Q$.*

This result holds, in fact, whenever the base field $\mathbb{F}$ is Archimedean. We say that a simple valuation φ on $\mathcal{P}^d$ is *G-invariant* if $\varphi(\Phi P) = \varphi(P)$ whenever $\Phi \in G$. Then we have:

Theorem 4.27. *Let $P, Q \in \mathcal{P}^d$. Then $P \approx_G Q$ if and only if $\varphi(P) = \varphi(Q)$ for all G-invariant simple valuations φ on $\mathcal{P}^d$.*

Over a non-Archimedean field, $\approx_G$ has to be replaced by $\sim_G$.

Hadwiger's proof of Theorem 4.27 is highly non-constructive, because it uses the axiom of choice to pick a basis of the polytope group $\widehat{\Pi}_G$, which is obtained from $\widehat{\Pi}^d$ in the same way that Π_G is obtained from Π, namely by imposing on $\widehat{\Pi}^d$ the extra relations (G) $\langle \Phi P \rangle = \langle P \rangle$ for all $P \in \mathcal{P}^d$ and $\Phi \in G$.

In fact, the variants of *Hilbert's third problem* reduce to finding, for a given group G, a "nice" family of G-invariant simple valuations which separates $\widehat{\Pi}_G$. We have seen that the Hadwiger functionals provide such a family when $G = T$ (Theorem 4.2). Another group is easily dealt with.

Theorem 4.28. *Two d-polytopes P and Q are* TH*-equidissectable if and only if $h_U(P) = h_U(Q)$ for every Hadwiger functional whose type is congruent to d modulo 2.*

The reason is that, if h_U is of type r, then $h_U(-P) = h_U(P)$ for all P if and only if $r \equiv d$ modulo 2.

Another easy result, which is a consequence of Theorem 3.18(b), is the following.

Theorem 4.29. *Two d-polytopes P and Q are* EA*-equidissectable if and only if $V(P) = V(Q)$.*

As a last preliminary result, we have a result proved by Gerwien (1833a), which was also observed by F. Bólyai.

Theorem 4.30. *Two planar polygons are D-equidissectable if and only if they have the same area.*

This follows from the fact that a triangle is D- (or even SD- or TH-) equidissectable with a parallelogram; any two parallelograms of the same area are T-equidissectable. Hadwiger and Glur (1951) later showed that, if a group G of affinities is such that two planar polygons of the same area are always G-equidissectable, then $G \supseteq$ TH.

We now come to the classical version of Hilbert's third problem. In contrast to the planar situation, it was early recognized (by Gauss, among others) that the

3-dimensional case was likely to be more difficult. Hilbert (1900) formally posed the question of finding two 3-polytopes of the same volume (even pyramids with the same height on the same base) which were not D-equidissectable. Modifying an earlier attempt by Bricard (1896), Dehn (1900, 1902) found an example before the problem was published.

Before describing the example, we make some remarks about the polytope groups $\widehat{\Pi}^d_D$; we now distinguish the dimension. First, if two polytopes are symmetric in a hyperplane, then it may be shown that they are SD-equidissectable. Thus:

Lemma 4.31. $\widehat{\Pi}^d_D = \widehat{\Pi}^d_{\mathrm{SD}}$.

In view of the presence of scaling by -1, we also have (compare Theorem 4.28 above):

Lemma 4.32. $\widehat{\Pi}^d_D = \bigoplus_{r \equiv d(2), r>0} \widehat{\Xi}^d_r$.

That is, there are no graded terms $\widehat{\Xi}^d_r$ unless $r \equiv d$ modulo 2.

There is a natural product structure on $\widehat{\Pi}_D := \sum_{d \geqslant 0} \widehat{\Pi}^d_D$, induced by orthogonal Cartesian product, which, as in Theorem 4.10, we denote by $\times$. It is often helpful to work with $\widehat{\Pi}_D$, rather than with the individual terms $\widehat{\Pi}^d_D$. Observe that each term $\widehat{\Xi}_r$ with $r > 1$ is a sum of non-trivial products.

There are conjectures about separating functionals for $\widehat{\Pi}_D$, which are exactly analogous to those for separation of $\widehat{\Sigma}^d$ which were described in section 4.4. First, we have the *total Euclidean Dehn invariant* of a polytope P, defined by

$$\Psi_E(P) := \sum_F^{*} \langle F \rangle \otimes \overline{b}(F,P) \in \widehat{\Pi}_D \otimes (\widehat{\Sigma}/\widehat{\Gamma});$$

as in section 4.4, such sums are over all faces F of P with $\dim F > 0$ and $\dim P - \dim F$ even. Similarly, the *total classical Euclidean Dehn invariant* is

$$\Phi_E := (\mathrm{gr.vol} \otimes \mathrm{id}) \circ \Psi_E,$$

so that

$$\Phi_E(P) = \sum_F^{*} \mathrm{vol}\, F \cdot T^{\dim F} \otimes \overline{b}(F,P).$$

Theorem 4.30 shows that Φ_E separates $\widehat{\Pi}^d_D$ for $d = 2$ (and the case $d = 1$ is trivial). It was the considerable achievement of Sydler (1965) to extend this to the case $d = 3$. Jessen (1968) simplified Sydler's proof by using the language of the algebra of polytopes (see section 4.1), and then (Jessen 1972) extended it further to the case $d = 4$. Thus we have:

Theorem 4.33. *If $d \leqslant 4$, then Φ_E separates $\widehat{\Pi}^d_D$.*

We shall not give any of the details of the proof here; in the crucial case $d=3$, these involve clever dissection results for special tetrahedra. However, a disadvantage of the proof is that it involves an appeal to the axiom of choice at several stages. More recently, though, Dupont and Sah (1992) have produced a quite different approach, using the Eilenberg–MacLane homology of the classical groups, and Hochschild homology of the quaternions; this avoids both the special constructions of Sydler (which Jessen's proof retains) and the need for any appeal to the axiom of choice.

We should note that the case $d = 4$ follows directly from the case $d = 3$ and $\widehat{\Pi}_D^4 = \widehat{\Xi}_2^4 \oplus \widehat{\Xi}_4^4$, which says that every $x \in \widehat{\Pi}_D^4$ is equivalent to a prism $e \times y$, where e is the class of a line segment and $y \in \widehat{\Pi}_D^3$; Jessen (1972) gave a direct proof of this result.

We end this section with a few remarks. First, the antipodal map $\S S$ on the Hopf algebra $\widehat{\Sigma}/\widehat{\Gamma}$ (see section 4.4) enables us to replace $\overline{b}(F,P)$ by $\overline{c}(F,P)$ in the definitions of the two Euclidean Dehn invariants Ψ_E and Φ_E. In view of the isomorphism Theorems 3.12 and 4.4, this change is very natural, although Dupont and Sah (1992) offer contrary evidence in favour of retaining the present definitions.

Next, the Dehn invariants are compatible with the product structure. Indeed, since the angle cone of the orthogonal product $P \times Q$ at its face $F \times G$ is $B(F,P) \times B(G,Q)$, we have

$$\Psi_D(P \times Q) = \Psi_D(P)\Psi_D(Q).$$

A more general question along these lines, which is prompted as well by the proof of Theorem 4.33, is the following: *is every polytope equivalent to a direct sum of products of odd-dimensional components*? For example,

$$\widehat{\Pi}_D^2 = \widehat{\Xi}_2^2 \cong \widehat{\Xi}_1^1 \otimes \widehat{\Xi}_1^1,$$

$$\widehat{\Pi}_D^4 = \widehat{\Xi}_2^4 \oplus \widehat{\Xi}_4^4 \cong (\widehat{\Xi}_1^1 \otimes \widehat{\Xi}_1^3) \oplus (\otimes^4 \widehat{\Xi}_1^1),$$

since the other term $\widehat{\Xi}_1^2 \otimes \widehat{\Xi}_1^2$ in $\widehat{\Xi}_2^4$ vanishes. The first open question here concerns the case $d = 6$.

The space of indecomposable elements of $\widehat{\Pi}_D^d$ is certainly the sum of the $\widehat{\Xi}_1^{2s+1}$ for $s \geqslant 0$. We might also ask: *is $\widehat{\Pi}_D$ isomorphic to a symmetric algebra based on the space of indecomposable elements*? In particular: *is $\widehat{\Pi}_D$ an integral domain, or a Hopf algebra*?

A final question is: *what are the images of the Dehn Invariants*? This question is particularly interesting in case $d = 3$.

5. Characterization theorems

Certain valuations with invariance or covariance properties can be characterized in various ways. In this section, we shall survey the known results of this kind, as well as discussing several related open problems.

5.1. Continuity and monotonicity

We first consider some general relationships between valuations, involving continuity and similar notions. The natural metric on subfamilies of convex bodies is the Hausdorff metric, introduced in section 1.1, although other closely connected metrics have been used from time to time. In what follows, all functions will take values in some real vector space; a *functional* is then a real-valued function.

We shall call a functional φ on a subfamily $\mathcal{S}$ of $\mathcal{K}^d$ *monotone* if $\varphi(K) \leqslant \varphi(K')$ whenever $K, K' \in \mathcal{S}$ satisfy $K \subseteq K'$. The convention $\varphi(\emptyset) = 0$ for valuations φ ensures that a monotone valuation on $\mathcal{S}$ is non-negative.

A useful result of McMullen (1977) is the following (the case $d = 2$ was proved by Hadwiger 1951b).

Theorem 5.1. *A monotone translation invariant valuation is continuous.*

The theorem is initially proved for polytopes, and uses the polynomial expansion for translation invariant valuations of rational multiples of polytopes. The extension to general convex bodies is routine.

A different concept of continuity is often more appropriate for polytopes; it is due to Hadwiger (1952d). Let $U = (u_1, \ldots, u_n)$ be a (for the moment) fixed set of unit outer normal vectors, and write $\mathcal{P}^d(U)$ for the family of polytopes of the form

$$P(y) = \{x \in \mathbb{E}^d \mid \langle x, u_i \rangle \leqslant \eta_i \ (i = 1, \ldots, n)\},$$

where $y = (\eta_1, \ldots, \eta_n)$. We call a function φ on $\mathcal{P}^d_*$ *weakly continuous* if, for each such U, the function φ_U defined by $\varphi_U(y) = \varphi(P(y))$ is continuous. We clearly have:

Lemma 5.2. *A continuous function on $\mathcal{P}^d_*$ is weakly continuous.*

As we saw in sections 3.1 and 3.5, translation invariant (or even covariant) valuations admit polynomial expansions with rational coefficients. To extend these expansions to real coefficients, weak continuity, rather than continuity, will suffice. In fact, we have:

Theorem 5.3. *The following conditions on a translation invariant or covariant valuation φ on $\mathcal{P}^d_*$ are equivalent:*

(a) *φ is weakly continuous;*

(b) *$\varphi(\lambda_1 P_1 + \cdots + \lambda_k P_k)$ is a polynomial in $\lambda_1, \ldots, \lambda_k$ for all polytopes $P_1, \ldots, P_k$ and all real numbers $\lambda_1, \ldots, \lambda_k \geqslant 0$;*

(c) *for each U, the one-sided partial derivatives of φ_U exist.*

In addition, if φ is translation invariant, a further equivalent condition is

(d) *φ is continuous under dilatations; that is, the mapping Θ_P on $\mathbb{R}$ defined by $\Theta_P(\lambda) = \varphi(\lambda P)$ is continuous.*

The equivalence of conditions (a), (b) and (c) is due to McMullen (1977). The equivalence of (a) and (d), which was left open by Hadwiger (1952e), follows from the fact that $Z_1 \subset \Pi$ (the polytope algebra) is a real vector space. This latter equivalence for translation covariant valuations (which was inadvertently claimed in Theorem 11.2 of McMullen and Schneider 1983) would follow if it could be shown that Π_1 were a real algebra (see section 3.5); a stronger condition which is equivalent is that φ is continuous under translations as well as dilatations (compare McMullen 1983).

A final remark is often useful.

Theorem 5.4. *The mixed valuations derived from a (weakly) continuous translation invariant or covariant valuation are (weakly) continuous in each of their arguments.*

Whether this, or an analogous, result holds for monotone valuations is unknown.

5.2. Minkowski additive functions

A starting point for the characterization of more general valuations is often that of certain special types. We have already discussed the Euler characteristic in section 2.1, and we shall consider volume and moment in section 5.3 immediately following. A further important case is that of a *Minkowski additive* function φ, which means that $\varphi(K+K') = \varphi(K)+\varphi(K')$ for appropriate K and K'; because of the strong assumption, more specific results are available for such functions.

Lemma 1.1 says that a Minkowski additive function φ is a valuation. Further, for fixed K, we also have $\varphi(\lambda K) = \lambda\varphi(K)$ for rational $\lambda \geqslant 0$, and, if φ is continuous, this holds for real λ.

Since the *width* $w_u(K) = h(K,u)+h(K,-u)$ of K in the direction of the unit vector u obviously gives a translation invariant valuation, so does the *mean width* $\bar{b}$, given by

$$\bar{b}(K) = \frac{2}{\omega_{d-1}} \int_{\Omega} h(K,u)\,\mathrm{d}\sigma(u)$$

for $K \in \mathcal{K}_*^d$, which is a constant multiple of the intrinsic length V_1 (or quermass-integral W_{d-1}). In fact, V_1 admits the following characterization.

Theorem 5.5. *If $\varphi : \mathcal{K}_*^d \to \mathbb{R}$ is Minkowski additive, continuous and invariant under rigid motions, then $\varphi = \alpha V_1$, for some real constant α.*

The proof of Hadwiger (1957, p. 213), uses a rotation averaging process, and actually shows more: it is only necessary to assume that φ is continuous at the unit ball B.

However, it would be nice to have Theorem 5.5 for suitable subclasses of $\mathcal{K}_*^d$, in particular for $\mathcal{P}_*^d$. If φ is locally uniformly continuous on $\mathcal{P}_*^d$, then it can be extended uniquely to $\mathcal{K}_*^d$, and any additive and invariance properties carry over; by *local*, we mean that the uniform continuity (or other condition) holds for the elements in any fixed ball. It is natural to ask whether a continuous or uniformly bounded Minkowski additive functional on $\mathcal{P}^d_*$ must also be a constant multiple of V_1.

The vector-valued counterpart of mean width is the *Steiner point* s, which is defined on $\mathcal{K}^d_*$ by

$$s(K) = \frac{1}{\kappa_d} \int_{\Omega} h(K,u)\,\mathrm{d}\sigma(u);$$

this is clearly Minkowski additive. We call a mapping f on $\mathcal{K}^d_*$ or $\mathcal{P}^d_*$ taking values in $\mathbb{E}^d$ *rigid motion* (*translation*) *equivariant* if $f(\Phi K) = \Phi f(K)$ for every rigid motion Φ of $\mathbb{E}^d$ ($f(K+t) = f(K)+t$ for every translation $t \in \mathbb{E}^d$, respectively). Then we have the analogue of Theorem 5.5:

Theorem 5.6. *If $f : \mathcal{K}^d_* \to \mathbb{E}^d$ is Minkowski additive, continuous and rigid motion equivariant, then $f = s$, the Steiner point.*

Again, it need only be assumed that f is continuous at the unit ball. The first proof of Theorem 5.6 was by Schneider (1971); the case $d = 2$ was earlier shown by Shephard (1968b) using Fourier series, and Schneider's proof generalizes this (though not straightforwardly) by using spherical harmonics. A more elementary proof, which makes weaker assumptions, was given by Positel′skiĭ (1973). However, the application of spherical harmonics seems to be the proper tool in this context, since the method also treats other similar problems.

Earlier attempts to characterize the Steiner point, a problem which was first posed by Grünbaum (1963, p. 239), are worth mentioning. Grünbaum asked whether s can be characterized by Minkowski additivity and dilatation equivariance, but this was shown not to be the case by Sallee (1971) and (with an easier counter-example) by Schneider (1974a). Shephard (1968b) added the continuity assumption, and Meyer (1970) proved a weaker version of Theorem 5.6 by assuming uniform continuity; two other attempts to prove the general result (Schmitt 1968, Hadwiger 1969a) contained errors.

In case $d = 2$, elementary proofs were given by Hadwiger (1971) and Berg (1971), the latter obtaining additional results for polytopes. To describe them, let ν be any outer angle (see section 3.4), and define s_ν by

$$s_\nu(P) = \sum_{v \in \operatorname{vert} P} \nu(v,P)\,v,$$

where $\operatorname{vert} P$ denotes the set of vertices of the polytope P. If ν is the usual rotation invariant angle, then s_ν is the Steiner point. In general, s_ν is translation and dilatation covariant, and it may readily be shown that it is also Minkowski additive. Berg (1971) calls a Minkowski additive function $f : \mathcal{P}^d_* \to \mathbb{E}^d$ which is rigid motion and dilatation covariant an *abstract Steiner point*. If the angle ν is rotation invariant, then s_ν is an abstract Steiner point. Whether the converse holds when $d > 3$ is an open question; Berg (1971) established this for $d = 2$ and 3, showing additionally that (in these cases) a locally bounded abstract Steiner point is the usual Steiner point.

Another interesting family of Minkowski additive functions consists of those which take values in $\mathcal{K}_*^d$ itself. They should, perhaps, be called *endomorphisms*, but Schneider (1974a), answering certain questions raised by Grünbaum (1963), reserves this term for those which are also continuous and rigid motion equivariant, noting that such functions are compatible with the most natural geometric structures on $\mathcal{K}_*^d$. The cases $d = 2$ and $d \geqslant 3$ show different features, since the rotation group is only commutative if $d = 2$.

For $d = 2$, Schneider (1974a) characterizes endomorphisms as limits of functions of the form

$$\Phi(K) := \sum_{j=1}^{k} \lambda_j \Psi_j[K - s(K)] + s(K)$$

for $K \in \mathcal{K}_*^d$, where $\lambda_j \geqslant 0$ and $\Psi_j \in SO_2$, the rotation group, for $j = 1, \ldots, k$. More precisely, if we write $u(\alpha) = (\cos\alpha, \sin\alpha)$ for $\alpha \in [0, 2\pi)$ (with coordinate vectors relative to some orthonormal basis of $\mathbb{E}^2$), we have

Theorem 5.7. *Let Φ be an endomorphism of $\mathcal{K}_*^2$. Then there exists a (positive) measure ν on the Borel subsets of $[0, 2\pi)$, such that*

$$h(\Phi(K), u(\alpha)) = \int_0^{2\pi} h(K - s(K), u(\alpha + \beta)) \, \mathrm{d}\nu(\beta) + \langle s(K), u(\alpha) \rangle$$

for $\alpha \in [0, 2\pi)$ and all $K \in \mathcal{K}_*^2$.

The proof, by Schneider (1974b), uses a characterization by Hadwiger (1951b) of continuous translation invariant Minkowski additive functionals on $\mathcal{K}_*^2$; clearly, any Borel measure ν on $[0, 2\pi)$ defines an endomorphism Φ in this way. Schneider gives additional results about the uniqueness of ν for a given Φ, and about the nature of Φ when its image contains a polygon. In particular, if Φ maps $\mathcal{K}_*^2$ *onto* itself, it takes the form $\Phi(K) = \lambda \Psi(K - s(K)) + s(K)$ for some $\lambda > 0$ and $\Psi \in SO_2$. The space of all endomorphisms is also shown to have the structure of a convex cone, and certain properties proved by Inzinger (1949) for special endomorphisms are extended (after some normalization) to all of them.

For $d \geqslant 3$, as yet only partial results have been obtained, by Schneider (1974a). If $q : [0, \infty) \to [0, \infty)$ is a function for which the following integrals exist and are finite, then for $K \in \mathcal{K}_*^d$ there is a unique $\Phi_q(K) \in \mathcal{K}_*^d$ for which

$$h(\Phi_q(K), u) = \int_{\mathbb{E}^d} h(K - s(K), u - \|u\| z) q(\|z\|) \, \mathrm{d}z + \langle s(K), u \rangle$$

for $u \in \mathbb{E}^d$ (support functions are here defined for all vectors in $\mathbb{E}^d$); this Φ_q is then an endomorphism of $\mathcal{K}_*^d$. Such endomorphisms have proved useful in the study of certain approximation problems; see Berg (1969) and Weil (1975b). Schneider (1974a) obtained various results which characterized certain kinds of endomorphism Φ of $\mathcal{K}_*^d$ (for brevity, we tacitly assume this notation below, and any K will lie in $\mathcal{K}_*^d$).

Theorem 5.8. (a) *Every such Φ is uniquely determined by the image of some suitable convex body, such as a triangle with an irrational angle.*

(b) *If Φ takes a body which is not a point onto a point, then $\Phi(K) = s(K)$ for all K. If Φ takes some body onto a segment, then*

$$\Phi(K) = \lambda[K - s(K)] + \mu[-K + s(K)] + s(K)$$

for some $\lambda, \mu \geqslant 0$ with $\lambda + \mu > 0$.

(c) *If Φ is surjective, then*

$$\Phi(K) = \lambda[K - s(K)] + s(K)$$

for some $\lambda \neq 0$.

(d) *If Φ satisfies $V_r(\Phi(K)) = V_r(K)$ for some $r = 2, \ldots, d$, then*

$$\Phi(K) = \varepsilon[K - s(K)] + s(K)$$

for some $\varepsilon = \pm 1$.

Writing $\mathcal{K}_d^d$ for the family of full-dimensional convex bodies, we further have:

Theorem 5.9. *Let $\Phi : \mathcal{K}_d^d \to \mathcal{K}_*^d$ be a continuous Minkowski additive mapping, such that $\Phi(\Psi(K)) = \Psi(\Phi(K))$ for every affinity Ψ of $\mathbb{E}^d$. Then*

$$\Phi(K) = K + \lambda[K - K]$$

for some $\lambda \geqslant 0$.

If invariance or equivariance with respect to some group of affinities of $\mathbb{E}^d$ is not assumed, then other stronger conditions must be imposed. For example, Schneider (1974c) has shown:

Theorem 5.10. *Let $\Phi : \mathcal{K}_*^d \to \mathcal{K}_*^d$ be a Minkowski additive function such that $V(\Phi(K)) = V(K)$ for each K. Then there is an equiaffinity Ψ of $\mathbb{E}^d$ such that $\Phi(K)$ is a translate of $\Psi(K)$ for each K.*

Finally, we remark that Valette (1974) has studied the continuous maps $\Phi : \mathcal{K}_*^d \to \mathcal{K}_*^d$ (with $d \geqslant 2$) which commute with affine maps, and satisfy the weaker condition $\Phi(K_1 + K_2) \supseteq \Phi(K_1) + \Phi(K_2)$ for $K_1, K_2 \subset \mathcal{K}_*^d$.

5.3. Volume and moment

Until further notice, we shall take all valuations to be real-valued. We first deal with the characterizations of volume. From a geometric viewpoint, we should wish for something simpler than the fact that the essential uniqueness of Haar measure on $\mathbb{E}^d$ shows that Lebesgue measure (that is, volume) is the unique translation invariant (positive) measure φ on the Borel sets of $\mathbb{E}^d$ such that $\varphi(C) = 1$ for some fixed unit cube C. In particular, we should prefer to consider simple valuations on $\mathcal{P}^d$ or $\mathcal{K}^d$, rather than σ-additive functions on Borel sets. We shall, in fact, take $\mathcal{P}^d$ as our domain of definition, since the assumptions we have to impose will extend to uniqueness on $\mathcal{K}^d$ as well. Further, the extension of a simple valuation to $U\mathcal{P}^d$ will share any non-negativity or monotonicity property, or invariance under any group of affinities, of the original. Bearing these remarks in mind, we have:

Theorem 5.11. *If φ is a translation invariant, non-negative simple valuation on $\mathcal{P}^d$, then $\varphi = \alpha V$ for some $\alpha \geqslant 0$.*

Various proofs of this theorem, for example, that in Maak (1960) (see also Hadwiger 1955b), use techniques such as exhaustion or polyhedral approximation, and so are not strictly speaking elementary. Some continuity argument is necessary, since finite dissections alone will not suffice to compare volumes whose ratio is irrational. However, the limit process can be reduced to the essential uniqueness of monotone real-valued functions λ which satisfy Cauchy's equation $\lambda(\alpha+\beta) = \lambda(\alpha)+\lambda(\beta)$ (with $\alpha, \beta \in \mathbb{R}$), while the remainder of the proof does only use finite dissections. Such a proof was given by Hadwiger (1950d, 1957, section 2.1.3); and (1949a) for $d = 3$.

As shown by Schneider (1978), the analogous result holds in spherical spaces as well (for the notion of polytopes in these and hyperbolic spaces, see Böhm and Hertel 1980), when rotation invariance replaces translation invariance. An extension to compact homogeneous spaces was given by Schneider (1981), and a general approach which treats hyperbolic spaces also was outlined in McMullen and Schneider (1983, p. 226).

Returning to $\mathbb{E}^d$, we have the following alternative characterizations; the first is due to Hadwiger (1957, p. 79), the second to Hadwiger (1970), and the third to Hadwiger (1952e, 1957, p. 221).

Theorem 5.12. *A translation invariant valuation on $\mathcal{P}^d$ which is homogeneous of degree d is a constant multiple of volume.*

Theorem 5.13. *A non-negative simple valuation on $\mathcal{P}^d$ which is invariant under volume preserving linear mappings of $\mathbb{E}^d$ is a constant multiple of volume.*

Theorem 5.14. *A continuous rigid motion invariant simple valuation on $\mathcal{K}^d$ is a constant multiple of volume.*

For the last, one would like to replace $\mathcal{K}^d$ by $\mathcal{P}^d$, but it is so far unknown whether this is possible (the proof of Theorem 5.14 uses Theorem 5.5). Whether there is a characterization of the usual rotation invariant angle on convex cones (polyhedral or more general) analogous to Theorem 5.14 is also an open problem. We may observe that Theorem 4.17 gives a reduction from odd-dimensional polyhedral cones to products of lower-dimensional cones, but it is not clear that this remark is particularly helpful.

If $K \in \mathcal{K}_d^d$, then its *centroid* $c(K)$ is defined by $V(K)c(K) := z(K)$, the moment vector of K. We have the following counterparts to Theorems 5.11 and 5.14, which are due to Schneider (1973) and (1972b), respectively.

Theorem 5.15. *If $f : U\mathcal{P}_d^d \to \mathbb{E}^d$ is a translation equivariant function, such that Vf is a simple valuation and $f(P) \in \operatorname{conv} P$ for each P, then $f = c$.*

Theorem 5.16. *If $f : \mathcal{K}_d^d \to \mathbb{E}^d$ is a continuous rigid motion equivariant function such that Vf is a valuation, then $f = c$.*

5.4. Intrinsic volumes and moments

One of the central results in the theory of valuations is Hadwiger's famous characterization of linear combinations of quermassintegrals. Rephrasing this in the language of intrinsic volumes, it states:

Theorem 5.17. *If $\varphi : \mathcal{K}^d \to \mathbb{R}$ is a continuous rigid motion invariant valuation, then there are constants $\alpha_0, \ldots, \alpha_d$ such that*

$$\varphi(K) = \sum_{r=0}^{d} \alpha_r V_r(K)$$

for all $K \in \mathcal{K}^d$.

There is a variant on Theorem 5.17, with monotonicity replacing continuity.

Theorem 5.18. *If $\varphi : \mathcal{K}^d \to \mathbb{R}$ is a monotone rigid motion invariant valuation, then there are non-negative constants $\alpha_0, \ldots, \alpha_d$ such that*

$$\varphi(K) = \sum_{r=0}^{d} \alpha_r V_r(K)$$

for all $K \in \mathcal{K}^d$.

Blaschke (1937) was the first to produce a result of this kind (with $d = 3$), but he needed to make a somewhat artificial assumption about the "volume part" of a valuation. Hadwiger proved Theorem 5.17 for $d = 3$ in (1951a) (see also 1955b), and for general d in (1952e). Theorem 5.18 was proved in Hadwiger (1953a); it can also be deduced form Theorem 5.17 by means of Theorem 5.1. Both results appear in Hadwiger (1957, section 6.1.10) (see also Leichtweiß 1980).

Since Hadwiger's proof of Theorem 5.17 uses Theorem 5.14 (which in turn relies on Theorem 5.5), it is unclear whether $\mathcal{K}^d$ can be replaced by $\mathcal{P}^d$ in these theorems, possibly with local boundedness or non-negativity instead of continuity or monotonicity. It should be noted that these alternative conditions are inappropriate for general convex bodies; if $\varphi(K)$ is the sum of the $(d-1)$-dimensional volumes of the $(d-1)$-faces of the convex body K (or twice the $(d-1)$-dimensional volume of an at most $(d-1)$-dimensional body), then φ is a rigid motion invariant valuation which is locally bounded and non-negative, but which is clearly not a linear combination of intrinsic volumes. Theorem 5.17 has important applications to integral geometry. The basic idea, which is to show that certain integrals give continuous or monotone rigid motion invariant valuations, goes back to Blaschke (1937), but was systematically exploited by Hadwiger (1950e, 1955b, 1956, 1957) to derive both old and new integral geometric formulae. A different kind of application, to random sets, was made by Matheron (1975), and variants of the theorem are due to Groemer (1972) and Baddeley (1980).

It might be expected that analogues of Theorems 5.17 and 5.18 hold in spherical and hyperbolic space. For a polyhedral cone C with face of apices A, there are two analogues of the intrinsic volumes. Define $\beta(A,F)$ to be the ordinary normalized angle (in $\operatorname{lin} F$) of the face F of C, and let $\gamma(F,C)$ be the similarly normalized angle of the normal cone to C at F (these are the *intrinsic inner* and *outer angles*, see McMullen 1975). For $r = 0, \ldots, d$, define

$$\varphi_r(C) := \sum_{F^r} \beta(A,F^r)\gamma(F^r,C),$$

where the sum extends over all r-faces F^r of C, and

$$\psi_r := \sum_{m \geqslant 0} \varphi_{d+1-r+2m}.$$

Clearly, φ_d is increasing, it can be shown that φ_{d-1} is also (see Shephard 1968d), and by duality, φ_0 and φ_1 are decreasing; however, for $2 \leqslant r \leqslant d-2$, examples show that φ_r is neither increasing nor decreasing (see McMullen and Schneider 1983, section 3). Since $2\psi_r(C)$ is the normalized measure of the r-dimensional linear subspaces of $\mathbb{E}^d$ which do not meet C in the origin o alone, it is also increasing. Moreover, all these functions are continuous, and extend to general closed convex cones. It may be conjectured that continuous (monotone) rotation invariant valuations on $\mathscr{C}^d$ are linear combinations of the φ_r (ψ_r with non-negative coefficients, respectively); the first question has an affirmative answer if the corresponding characterization of spherical volume, mentioned in section 5.3, is valid.

5.5. Translation invariance and covariance

In a sense, the description of the polytope algebra in section 3, and particularly the isomorphism Theorem 3.12, tell us what a translation invariant valuation on $\mathscr{P}^d$ looks like; it is just the composition of σ with some group homomorphism. However, without some additional assumptions on the valuation, the resulting characterization will be too vague to be useful.

Natural conditions to impose include some form of continuity. We shall discuss the known results in this area. We begin with weak continuity. Throughout, $\mathscr{X}$ will denote a real vector space.

Theorem 5.19. *A function $\varphi : \mathscr{P}^d \to \mathscr{X}$ is a weakly continuous translation invariant valuation if and only if*

$$\varphi(P) = \sum_{F} \operatorname{vol} F \, \lambda(F,P),$$

where $\lambda : \widehat{\Sigma} \to \mathscr{X}$ is a simple valuation, and vol *is ordinary volume.*

This follows directly from Theorem 3.12, but was proved by McMullen (1983) using the following result of Hadwiger (1952e) (the extension from the original paper stated here is straightforward).

Theorem 5.20. *A function* $\varphi : \mathcal{P}^d \to \mathcal{X}$ *is a weakly continuous translation invariant simple valuation if and only if*

$$\varphi(P) = \sum_U \operatorname{vol} P_U \, \eta(U),$$

where η *is an odd function on frames, and* vol *is ordinary volume.*

In view of the isomorphism Theorem 4.4 for the polytope group $\widehat{\Pi}$, we could rephrase Theorem 5.20 in terms of mappings on $\mathbb{R} \otimes (\widehat{\Sigma}/\widehat{\Gamma})$. It should also be remarked that there are exactly analogous results for translation covariant valuations, which involve the moment vectors of faces as well as their volumes; see McMullen (1983) for details.

We now turn to continuity, and for simplicity confine our attention to real-valued valuations. The problem of characterizing continuous translation invariant valuations remains open; the supposed characterization of Betke and Goodey (1984) was unfortunately flawed. However, there are some partial results. For dimension $d = 2$ there is a complete solution by Hadwiger (1949, 1951b).

Theorem 5.21. (a) *If* φ *is a continuous translation invariant valuation on* $\mathcal{K}^2$, *then*

$$\varphi(K) = \alpha + \int_{\Omega^1} g(u) \, \mathrm{d}S_1(K; u) + \beta V_2(K)$$

for some constants α, β *and some continuous function* g.

(b) *If* φ *is a locally bounded translation invariant valuation on* $\mathcal{P}^2$, *then the same expression for* φ *holds, with* g *a bounded function.*

Actually, Hadwiger did not express his results in terms of the area function $S_1(K; \cdot)$. The function g is uniquely determined, up to a function of the form $\langle v, \cdot \rangle$, with v a constant vector. If φ is just Minkowski additive in part (b), then the same result holds with $\alpha = \beta = 0$.

For $d \geqslant 4$, no such explicit representations are known; the case $d = 3$ is covered by the results below. If we use the fact that a continuous translation invariant valuation φ can be written as a sum $\varphi = \sum_{r=0}^{d} \varphi_r$, with φ_r (continuous and) homogeneous of degree r, then we can obviously investigate the individual φ_r. By Hadwiger (1957, p. 79), φ_d is a constant multiple of volume, and φ_0 is constant. The only complete solutions for any of the remaining cases are those of McMullen (1980) for $r = d - 1$, and Goodey and Weil (1984) for $r = 1$.

Theorem 5.22. *Let* φ *be a continuous translation invariant valuation on* $\mathcal{K}^d$ *which is homogeneous of degree* $d - 1$. *Then there is a continuous function* g *on the unit sphere* Ω, *such that*

$$\varphi(K) = \int_{\Omega} g(u) \, \mathrm{d}S_{d-1}(K; u)$$

for each $K \in \mathcal{K}^d$.

As above, g is unique up to a function of the form $\langle v, \cdot \rangle$. The valuation can also be expressed as a limit

$$\varphi(K) = \lim_{i\to\infty} [V(K, d-1; L_i) - V(K, d-1; M_i)],$$

for suitable sequences $(L_i), (M_i)$ of convex bodies.

When $r = 1$, we have the following.

Theorem 5.23. *Let φ be a continuous translation invariant valuation on $\mathcal{K}^d$ which is homogeneous of degree 1. Then there are sequences $(L_i), (M_i)$ of convex bodies such that*

$$\varphi(K) = \lim_{i\to\infty} [V(K; L_i, d-1) - V(K; M_i, d-1)],$$

uniformly for all $K \subseteq mB$ and all $m > 0$.

For $r \in \{2, \ldots, d-2\}$, rather less is known. Clearly, suitable limits of mixed volumes will provide continuous translation invariant valuations, but it is open whether all such valuations can be obtained in this way. Goodey and Weil (1984) tried to relate such (mixed) valuations to distributions on tensor products of support functionals, of the form

$$\varphi(K_1, \ldots, K_r) = T(h(K_1, \cdot) \otimes \cdots \otimes h(K_r, \cdot)),$$

but an important part of their argument seems to be invalid.

With a stronger continuity assumption, McMullen (1980) (compare also Schneider 1974b) showed that a uniformly continuous translation invariant valuation on $\mathcal{K}^d$ which is homogeneous of degree 1 is of the form

$$\varphi(K) = V(K; L, d-1) - V(K; M, d-1)$$

for some convex bodies L, M; this can be deduced from the Riesz representation theorem.

A little more can be said about the case of monotone valuations.

Theorem 5.24. *Let $r = 1$ or $d-1$. If φ is a monotone translation invariant valuation on $\mathcal{K}^d$ which is homogeneous of degree r, then there exist convex bodies $L_{r+1}, \ldots, L_d$, such that*

$$\varphi(K) = V(K, r; L_{r+1}, \ldots, L_d).$$

The case $r = d$ is similar, with a non-negative multiple inserted (Theorem 5.11). For the theorem, the case $r = 1$ is due to Firey (1976), while the case $r = d-1$ was proved by McMullen (1990). It would be tempting to conjecture that the same result holds for all r, but the evidence to support this is meagre.

Finally, let us mention translation covariant valuations. McMullen (1983) proved the following analogue of Theorem 5.19. As before, $\mathcal{X}$ is a real vector space.

Theorem 5.25. *A function* $\varphi : \mathcal{P}^d \to \mathcal{X}$ *is a weakly continuous translation covariant valuation if and only if*

$$\varphi(P) = \sum_F (\operatorname{vol} F \, \lambda(F,P) + m(F) \, \Lambda(F,P)),$$

where vol *is volume, m is moment, and* $\lambda : \widehat{\Sigma} \to \mathcal{X}$ *and* $\Lambda : \widehat{\Sigma} \to \operatorname{Hom}(\mathbb{E}^d, \mathcal{X})$ *are simple valuations.*

There is a similar result for simple weakly continuous translation covariant valuations; compare Theorem 5.20.

5.6. Lattice invariant valuations

We call a function on subsets of $\mathbb{E}^d$ which is invariant under the translations of the integer lattice $\mathbb{Z}^d$ *lattice invariant*. A *unimodular* mapping of $\mathbb{E}^d$ is an affinity which leaves $\mathbb{Z}^d$ invariant; it is therefore the composition of a linear mapping whose matrix (with respect to the standard basis) has integer entries and determinant ± 1 with a lattice translation. The lattice point enumerator G, the derived functionals G_r and the weighted lattice point numbers A are examples of lattice invariant valuations, and the first two are also invariant under unimodular mappings. In this section, we consider various characterization theorems along the lines of, for example, Theorem 5.17, on the classes $\mathcal{P}_L^d$ of lattice polytopes and $\mathcal{P}_{\mathbb{Q}}^d$ of polytopes whose vertices have rational coordinate vectors. The first result is due to Betke (1979, unpublished a).

Theorem 5.26. *Let* $\varphi : \mathcal{P}_L^d \to \mathbb{R}$ *be a valuation which is invariant under unimodular mappings. Then there are constants* $\alpha_0, \ldots, \alpha_d$ *such that*

$$\varphi = \sum_{r=0}^{d} \alpha_r G_r.$$

Originally, Betke assumed that φ satisfied the stronger inclusion–exclusion principle, but Stein (1982) showed that this followed from the valuation property and lattice invariance; later, Betke (unpublished b) was able to remove this latter assumption.

A consequence of Theorem 5.26 and the method of its proof is the following description of the underlying abstract structure of valuations on $\mathcal{P}_L^d$ which are invariant under unimodular mappings.

Theorem 5.27. *The Abelian group* Π_L *generated by the equivalence classes of lattice polytopes under unimodular mappings, with addition defined by the valuation property* (V), *is isomorphic to* $\mathbb{Z}^{d+1}$.

In fact, the $d+1$ generators of the group are just the classes of the lattice polytopes $\operatorname{conv}\{o, e_1, \ldots, e_r\}$, for $r = 0, \ldots, d$. This result was proved by Betke and

Kneser (1985). Müller (1988) has extended these results to equidissectability with respect to more general crystallographic groups.

We finally discuss lattice invariant and covariant valuations on $\mathcal{P}^d_{\mathbb{Q}}$. McMullen (1983) proved the following analogue of Theorem 5.19.

Theorem 5.28. *A function* $\varphi : \mathcal{P}^d_{\mathbb{Q}} \to \mathbb{R}$ *is a lattice invariant valuation if and only if*

$$\varphi(P) = \sum_F \operatorname{vol} F \, \gamma(F, P),$$

where γ *is a real valued function on translates of normal cones, which depends only on the translation class of* $\operatorname{aff} F$ *modulo* $\mathbb{Z}^d$.

In McMullen (1978), the corresponding result for simple valuations was proved. There are results on covariant valuations analogous to Theorem 5.25; these are mentioned in McMullen (1983).

References

We have attempted to give a complete bibliography on valuations and dissections. As a consequence, some of the references listed here may not be mentioned in the text.

Aleksandrov, A.D.
[1937a] Zur Theorie der gemischte Volumina von konvexen Körpern, I: Verallgemeinerung einiger Begriffe der Theorie der konvexen Körper (in Russian), *Mat. Sb. N.S.* **2**, 947–972.
[1937b] Zur Theorie der gemischte Volumina von konvexen Körpern, II: Neue Ungleichungen zwischen den gemischten Volumina und ihre Anwendungen (in Russian), *Mat. Sb. N.S.* **2**, 1205–1238.

Allendoerfer, C.B.
[1948] Steiner's formulae on a general S^{n+1}, *Bull. Amer. Math. Soc.* **54**, 128–135.

Baddeley, A.
[1980] Absolute curvature in integral geometry, *Math. Proc. Cambridge Philos. Soc.* **88**, 45–58.

Banchoff, T.
[1967] Critical points and curvature for embedded polyhedra, *J. Discrete Geom.* **1**, 245–256.
[1970] Critical points and curvature for embedded polyhedral surfaces, *Amer. Math. Monthly* **77**, 475–485.

Bauer, H.
[1978] *Wahrscheinlichkeitstheorie und Grundzuge der Maßtheorie* (W. de Gruyter, Berlin, 3rd ed.).

Berg, C.
[1969] Corps convexes et potentiels sphériques, *Danske Vid. Selsk. Mat.-Fys. Medd.* **37**(6), 1–64.
[1971] Abstract Steiner points for convex polytopes, *J. London Math. Soc.(2)* **4**, 176–180.

Bernshtein, D.N.
[1976] The number of integral points in integral polyhedra (in Russian), *Funkcional. Anal. i Priložen.* **10**(3), 72–73 [*Functional Anal. Appl.* **10**, 293–294].

Betke, U.
[1979] Gitterpunkte und Gitterpunktfunktionale, Habilitionsschrift, Siegen.

Betke, U., and P.R. Goodey
[1984] Continuous translation invariant valuations on convex bodies, *Abh. Math. Sem. Univ. Hamburg* **54**, 95–105. Correction: ibid. **56** (1986) 253.

Betke, U., and P. Gritzmann
[1986] An application of valuation theory to two problems in discrete geometry, *Discrete Math.* **58**, 81–85.

Betke, U., and M. Kneser
[1985] Zerlegungen und Bewertungen von Gitterpolytopen, *J.Reine Angew. Math.* **358**, 202–208.

Betke, U., and J.M. Wills
[1979] Stetige und diskrete Funktionale konvexer Körper, in: *Contributions to Geometry*, eds J. Tölke and J.M. Wills (Birkhäuser, Basel).

Betke, U.
[unpublished, a] Ein Funktionalsatz für Gitterpolytope, Manuscript.
[unpublished, b] Das Einschließungs–Ausschließungsprinzip für Gitterpolytope, Manuscript.
[unpublished, c] Examples of continuous valuations on polytopes, Manuscript.

Blaschke, W.
[1937] *Vorlesungen über Integralgeometrie*, 1st ed. Third edition: VEB Deutsch. Verl. d. Wiss., Berlin, 1955.

Böhm, J., and E. Hertel
[1980] *Polyedergeometrie in* n*-dimensionalen Räumen konstanter Krümmung* (VEB Deutsch. Verl. d. Wiss., Berlin).

Boltyanskiĭ, V.G.
[1958] Zerlegungsgleichheit ebener Polygone, *Bul. Inst. Politehn. Jaşi* **4**(8), 33–38.
[1963] *Equivalent and Equidecomposable Figures* (Heath, Boston). Russian original: 1956.
[1966] Equidecomposability of polygons and polyhedra (in Russian), in: *Encyclopedia of Elementary Mathematics*, Vol. V (Moscow) pp. 142–180.
[1969] On Hilbert's Third Problem (in Russian), in: *Hilbert's Problems* (Izdat. Nauka, Moscow) pp. 92–94.
[1976] Decomposition equivalence of polyhedra and groups of motions (in Russian), *Dokl. Akad. Nauk. SSSR* **231**, 788–790 [*Soviet Math. Dokl.* **17**, 1628–1631].
[1978] *Hilbert's Third Problem* (Wiley, New York).

Brianchon, C.J.
[1837] Théorème nouveau sur les polyèdres, *J. École (Roy.) Polytech.* **15**, 317–319.

Bricard, R.
[1896] Sur une question de géométrie relative aux polyèdres, *Nouv. Ann. of Math.* **15**, 331–334.

Budach, L.
[1989] Lipschitz–Killing curvatures of angular partially ordered sets, *Adv. in Math.* **78**, 140–167.

Busemann, H.
[1958] *Convex Surfaces* (Wiley, New York).

Cerasoli, M., and G. Letta
[1988] On the Euler–Poincaré characteristic (in Italian), *Rend. Accad. Naz. Sci. XL Mem. Mat.(5)* **12**, 259–267.

Cheeger, J., W. Müller and R. Schrader
[1984] On the curvature of piecewise flat spaces, *Comm. Math. Phys.* **92**, 405–454.

Debrunner, H.
[1952] Translative Zerlegungsgleichheit von Würfeln, *Arch. Math.* **3**, 479–480.
[1969] Zerlegungsähnlichkeit von Polyedern, *Elem. Math.* **24**, 1–6.
[1978] Zerlegungsrelationen zwischen regulären Polyedern des E^d, *Arch. Math.* **30**, 656–660.
[1980] Über Zerlegungsgleichheit von Pilasterpolytopen mit Würfeln, *Arch. Math.* **35**, 583–587.

Dehn, M.
[1900] Über raumgleiche Polyeder, *Nachr. Akad. Wiss. Göttingen, Math.-Phys. Kl.*, 345–354.

[1902] Über den Rauminhalt, *Math. Ann.* **55**, 465–478.
[1905] Über den Inhalt sphärischer Dreiecke, *Math. Ann.* **60**, 166–174.

Dupont, J.L.
[1982] Algebra of polytopes and homology of flag complexes, *Osaka J. Math.* **19**, 599–611.

Dupont, J.L., and C.-H. Sah
[1982] Scissors congruence, II, *J. Pure Appl. Algebra* **25**, 159–195.
[1992] Homology of euclidean groups of motions made discrete and euclidean scissors congruence, to appear.

Eckhoff, J.
[1980] Die Euler-Charakteristik von Vereinigungen konvexer Mengen im R^d, *Abh. Math. Sem. Univ. Hamburg* **50**, 133–144.

Ehrhart, E.
[1967a] Sur un problème de géométrie diophantienne linéaire, I, *J. Reine Angew. Math.* **226**, 1–29.
[1967b] Démonstration de la loi de réciprocité pour un polyèdre entier, *C.R. Acad. Sci. Paris, Sér. A* **265**, 5–7.
[1967c] Démonstration de la loi de réciprocité du polyèdre rationnel, *C.R. Acad. Sci. Paris, Sér. A* **265**, 91–94.
[1968] Sur la loi de réciprocité des polyèdres rationnels, *C.R. Acad. Sci. Paris, Sér. A* **266**, 696–697.

Emch, A.
[1946] Endlichgleiche Zerschneidung von Parallelotopen in gewöhnlichen und höheren euklidischen Räumen, *Comment. Math. Helv.* **18**, 224–231.

Fáry, I.
[1961] Functionals related to mixed volumes, *Illinois J. Math.* **5**, 425–430.

Federer, H.
[1959] Curvature measures, *Trans. Amer. Math. Soc.* **93**, 418–491.

Fenchel, W., and B. Jessen
[1938] Mengenfunktionen und konvexe Körper, *Danske Vid. Selskab. Mat.-Fys. Medd.* **16**(3), 1–31.

Firey, W.J.
[1976] A functional characterization of certain mixed volumes, *Israel J. Math.* **24**, 274–281.

Fischer, K.G., and J. Shapiro
[1992] The prime ideal structure of the Minkowski ring of polytopes, *J. Pure Appl. Algebra* **78**, 239–251.

Gerwien, P.
[1833a] Zerschneidung jeder beliebigen Anzahl von gleichen geradlinigen Figuren in dieselben Stücke, *J. Reine Angew. Math.* **10**, 228–234.
[1833b] Zerschneidung jeder beliebigen Menge verschieden gestalteter Figuren von gleichem Inhalt auf der Kugelfläche in dieselben Stücke, *J. Reine Angew. Math.* **10**, 235–240.

Goodey, P.R., and W. Weil
[1984] Distributions and valuations, *Proc. London Math. Soc. (3)* **49**, 504–516.

Gram, J.P.
[1874] Om Rumvinklerne i et Polyeder, *Tidsskr. Math. (Copenhagen) (3)* **4**, 161–163.

Groemer, H.
[1972] Eulersche Characteristik, Projektionen und Quermaßintegrale, *Math. Ann.* **198**, 23–56.
[1973] Über einige Invarianzeigenschaften der Eulerschen Charakteristik, *Comment. Math. Helv.* **48**, 87–99.
[1973] The Euler chalacteristic and related functionals on convex surfaces, *Geom. Dedicata* **4**, 91–104.
[1974] On the Euler characteristic in spaces with a separability property, *Math. Ann.* **211**, 315–321.

[1977a] Minkowski addition and mixed volumes, *Geom. Dedicata* **6**, 141–163.
[1977b] On translative integral geometry, *Arch. Math.* **29**, 324–330.
[1978] On the existence of additive functionals on classes of convex sets, *Pacific J. Math.* **75**, 397–410.

Gruber, P.M.
[1979] Geometry of numbers, in: *Contributions to Geometry*, eds J. Tölke and J.M. Wills (Birkhäuser, Basel) pp. 186–225.

Gruber, P.M., and R. Schneider
[1979] Problems in geometric convexity, in: *Contributions to Geometry*, eds J. Tölke and J.M. Wills (Birkhäuser, Basel) pp. 255–278.

Grünbaum, B.
[1963] Measures of symmetry for convex bodies, *Proc. Symposia in Pure Math.* **7**, 223–270.
[1967] *Convex Polytopes* (Wiley, London).

Hadwiger, H.
[1947] Über eine symbolisch-topologische Formel, *Elem. Math.* **2**, 35–41. Portuguese translation: *Gazeta Mat.* **35** (1948) 6–9.
[1949a] Bemerkung zur elementaren Inhaltslehre des Räumes, *Elem. Math.* **4**, 3–7.
[1949b] Über beschränkte additive Funktionale konvexer Polygone, *Publ. Math. Debrecen* **1**, 104–108.
[1949c] Zerlegungsgleichheit und additive Polyederfunktionale, *Arch. Math.* **1**, 468–472.
[1950a] Zum Problem der Zerlegungsgleichheit der Polyeder, *Arch. Math.* **2**, 441–444.
[1950b] Zerlegungsgleichheit und additive Polyederfunktionale, *Comment. Math. Helv.* **24**, 204–218.
[1950c] Translative Zerlegungsgleichheit k-dimensionaler Parallelotope, *Collect. Math.* **3**, 11–23.
[1950d] Zur Inhaltstheorie der Polyeder, *Collect. Math.* **3**, 137–158.
[1950e] Einige Anwendungen eines Funktionalsatzes für konvexe Körper in der räumlichen Integralgeometrie, *Monatsh. Math.* **54**, 345–353.
[1951a] Beweis eines Funktionalsatzes für konvexe Körper, *Abh. Math. Sem. Univ. Hamburg* **17**, 69–76.
[1951b] Translationsinvariante, additive und stetige Eibereichfunktionale, *Publ. Math. Debrecen* **2**, 81–94.
[1952a] Ergänzungsgleichheit k-dimensionaler Polyeder, *Math. Z.* **55**, 292–298.
[1952b] Über addierbare Intervallfunktionale, *Tôhoku Math. J.* **4**, 32–37.
[1952c] Mittelpunktspolyeder und translative Zerlegungsgleichheit, *Math. Nachr.* **8**, 53–58.
[1952d] Translationsinvariante, additive und schwachstetige Polyederfunktionale, *Arch. Math.* **3**, 387–394.
[1952e] Additive Funktionale k-dimensionaler Eikörper, I, *Arch. Math.* **3**, 470–478.
[1953a] Additive Funktionale k-dimensionaler Eikörper, II, *Arch. Math.* **4**, 374–379.
[1953b] Über additive Funktionale k-dimensionaler Eipolyeder, *Publ. Math. Debrecen* **3**, 87–94.
[1953c] Lineare additive Polyederfunktionale und Zerlegungsgleichheit, *Math. Z.* **58**, 4–14.
[1953d] Über Gitter und Polyeder, *Monatsh. Math.* **57**, 246–254.
[1954a] Zum Problem der Zerlegungsgleichheit k-dimensionaler Polyeder, *Math. Ann.* **127**, 170–174.
[1954b] Zur Zerlegungstheorie euklidischer Polyeder, *Ann. Mat. Pura Appl. (IV)* **36**, 315–334.
[1955a] Eulers Charakteristik und kombinatorische Geometrie, *J. Reine Angew. Math.* **194**, 101–110.
[1955b] *Altes und Neues über konvexe Körper* (Birkhäuser, Basel).
[1956] Integralsätze im Konvexring, *Abh. Math. Sem. Univ. Hamburg* **20**, 136–154.
[1957] *Vorlesungen über Inhalt, Oberfläche und Isoperimetrie* (Springer, Berlin).
[1959] Normale Körper im euklidischen Raum und ihre topologischen und metrischen Eigenschaften, *Math. Z.* **71**, 124–140.
[1960] Zur Eulerschen Charakteristik euklidischer Polyeder, *Monatsh. Math.* **64**, 49–60.
[1963] Ungelöstes Problem Nr. 45, *Elem. Math.* **18**, 29–31.

[1968a] Translative Zerlegungsgleichheit der Polyeder des gewöhnlichen Raumes, *J. Reine Angew. Math.* **233**, 200–212.

[1968b] Eine Schnittrekursion für die Eulersche Charakteristik euklidischer Polyeder mit Anwendungen innerhalb der kombinatorischen Geometrie, *Elem. Math.* **23**, 121–132.

[1968c] Neuere Ergebnisse innerhalb der Zerlegungstheorie euklidischer Polyeder, *Jber. Deutsch. Math.-Vereinig.* **70**, 167–176.

[1969a] Zur axiomatischen Charakterisierung des Steinerpunktes konvexer Körper, *Israel J. Math.* **7**, 168–176.

[1969b] Eckenkrümmung beliebiger kompakter euklidischer Polyeder und Charakteristik von Euler–Poincaré, *Enseign. Math.* **15**, 147–151.

[1969c] Notiz zur Eulerschen Charakteristik offener und abgeschlossener Polyeder, *Studia Sci. Math. Hungar.* **4**, 385–387.

[1970] Zentralaffine Kennzeichnung des Jordanschen Inhaltes, *Elem. Math.* **25**, 25–27.

[1971] Zur axiomatischen Charakterisierung des Steinerpunktes konvexer Körper; Berichtigung und Nachtrag, *Israel J. Math.* **9**, 466–472.

[1972] Polytopes and translative equidecomposability, *Amer. Math. Monthly* **79**, 275–276.

[1973] Erweiterter Polyedersatz und Euler-Shephardsche Additionstheoreme, *Abh. Math. Sem. Univ. Hamburg* **39**, 120–129.

[1974a] Begründung der Eulerschen Charakteristik innerhalb der ebenen Elementargeometrie, *Enseign. Math.* **20**, 33–43.

[1974b] Homothetieinvariante und additive Polyederfunktionen, *Arch. Math.* **25**, 203–205.

[1975] Zerlegungsgleichheit euklidischer Polyeder bezüglich passender Abbildungsgruppen und invariante Funktionale, *Math.-Phys. Sem. Ber.* **22**, 125–133.

Hadwiger, H., and P. Glur

[1951] Zerlegungsgleichheit ebener Polygone, *Elem. Math.* **26**, 97–106.

Hadwiger, H., and P. Mani

[1972] On the Euler characteristic of spherical polyhedra and the Euler relation, *Mathematika* **19**, 139–143.

[1974] On polyhedra with extremal Euler characteristic, *J. Combin. Theory Ser. A* **17**, 345–349.

Hadwiger, H., and R. Schneider

[1971] Vektorielle Integralgeometrie, *Elem. Math.* **26**, 49–57.

Haraz̆švili, A.B.

[1977] Equidecomposition of polyhedra relative to the group of homotheties and translations (in Russian), *Dokl. Akad. Nauk. SSSR* **236**, 552–555 [*Soviet Math. Dokl.* **18**, 1246–1249].

[1978] Über Intervallpolygone, *Wiss. Z. Friedrich-Schiller-Univ. Jena, Math.-Nat. Reihe* **18**, 299–303.

Hertel, E.

[1969] Über Intervallpolygone, *Wiss. Z. Friedrich-Schiller-Univ. Jena, Math.-Nat. Reihe* **18**, 299–303.

[1971] Über Intervallpolyeder im R_n, *Beiträge Algebra Geom.* **1**, 77–83.

[1973] Zur translativen Zerlegungsgleichheit n-dimensionaler Polyeder, *Publ. Math. Debrecen* **20**, 133–140.

[1974a] Ein Subtraktionssatz der Polyederalgebra, *Beiträge Algebra Geom.* **2**, 83–86.

[1974b] Mittelpunktspolyeder im E^4, *Elem. Math.* **29**, 59–64.

[1974c] Polyederstrukturen, *Math. Nachr.* **62**, 57–63.

[1977] Neuere Ergebnisse und Richtigungen der Zerlegungstheorie von Polyedern, *Mitt. Math. Ges. DDR* **4**, 5–22.

Hertel, E., and H. Debrunner

[1980] Zur Rolle von Subtraktionssatz und Divisionssatz in Zerlegungsstrukturen, *Beiträge Algebra Geom.* **10**, 145–148.

Hilbert, D.

[1899] *Grundlagen der Geometrie* (Teubner, Leipzig).

[1900] Mathematische Probleme, *Nachr. Königl. Ges. Wiss. Göttingen, Math.-Phys. Kl.*, 253–297 [*Bull. Amer. Math. Soc.* **8** (1902) 437–479].

Inzinger, R.
[1949] Über eine lineare Transformation in den Mengen der konvexen und der stützbaren Bereiche einer Ebene, *Monatsh. Math.* **53**, 227–250.

Jessen, B.
[1939] Om Polyedres Rumfang, *Mat. Tidsskr. A*, 35–44.
[1941] En Bemaerkning om Polyedres Volumen, *Mat. Tidsskr. B*, 59–65.
[1946] Om Aekvivalens af Aggregater af regulaere Polyedre, *Mat. Tidsskr. B*, 145–148.
[1967] Orthogonal icosahedra, *Nordisk Mat. Tidsskr.* **15**, 90–96.
[1968] The algebra of polyhedra and the Dehn–Sydler theorem, *Math. Scand.* **22**, 241–256.
[1972] Zur Algebra der Polytope, *Nachr. Akad. Wiss. Göttingen, II. Math.-Phys. Kl.*, 47–53.
[1978] Einige Bemerkungen zur Algebra der Polyeder in nicht-euklidischen Räumen, *Comm. Math. Helv.* **53**, 525–528.

Jessen, B., and A. Thorup
[1978] The algebra of polytopes in affine spaces, *Math. Scand.* **43**, 211–240.

Jessen, B., J. Karpf and A. Thorup
[1968] Some functional equations in groups and rings, *Math. Scand.* **22**, 257–265.

Kagan, W.F.
[1903] Über die Transformation der Polyeder, *Math. Ann.* **57**, 421–424.

Kirsch, A.
[1978] Polyederfunktionale, die nicht translationsinvariant, aber injektiv sind, *Elem. Math.* **33**, 105–107.

Klee, V.L.
[1963] The Euler characteristic in combinatorial geometry, *Amer. Math. Monthly* **70**, 119–127.

Kuiper, N.H.
[1971] Morse relations for curvature and tightness, in: *Proc. Liverpool Singularities Symp.*, ed. C.T.C. Wall, Lecture Notes in Math., Vol. 209 (Springer, Berlin) pp. 77–89.

Kummer, H.
[1956] Translative Zerlegungsgleichheit k-dimensionaler Parallelotope, *Arch. Math.* **7**, 219–220.

Lawrence, J.
[1988] Valuations and polarity, *J. Discrete Comput. Geom.* **3**, 307–324.
[unpublished] Minkowski rings, to appear.

Lebesgue, H.
[1938] Sur l'équivalence des polyèdres, en particulier des polyèdres reguliers, et sur la dissection des polyèdres reguliers en polyèdres reguliers, *Ann. Soc. Math. Polon.* **17**, 193–226.
[1945] Sur l'équivalence des polyèdres, *Ann. Soc. Math. Polon.* **18**, 1–3.

Leichtweiss, K.
[1980] *Konvexe Mengen* (VEB Deutsch. Verl. d. Wiss., Berlin).

Lenz, H.
[1970] Mengenalgebra und Eulersche Charakteristik, *Abh. Math. Sem. Univ. Hamburg* **34**, 135–147.

Lindgren, H.
[1972] *Recreational Problems in Geometric Dissections and How to Solve Them,* revised and enlarged by Greg Frederickson (Dover, New York).

Maak, W.
[1960] *Differential- und Integralrechnung,* (Vandenhoeck and Ruprecht, Göttingen, 2nd ed.).

Macdonald, I.G.
[1963] The volume of a lattice polyhedron, *Proc. Cambridge Philos. Soc.* **59**, 719–726.
[1971] Polynomials associated with finite cell-complexes, *J. London Math. Soc. (2)* **4**, 181–192.

Mani, P.
[1971] On angle sums and Steiner points of polyhedra, *Israel J. Math.* **9**, 380–388.

Matheron, G.
[1975] *Random Sets and Integral Geometry* (Wiley, New York).

McMullen, P.
[1975a] Metrical and combinatorial properties of convex polytopes, in: *Proc. Int. Congr. Math., Vancouver, 1974*, pp. 491–495.
[1975b] Non-linear angle-sum relations for polyhedra cones and polytopes, *Math. Proc. Cambridge Philos. Soc.* **78**, 247–261.
[1977] Valuations and Euler-type relations on certian classes of convex polytopes, *Proc. London Math. Soc.* **35**, 113–135.
[1978] Lattice invariant valuations on rational polytopes, *Arch. Math.* **31**, 509–516.
[1980] Continuous translation invariant valuations on the space of compact convex sets, *Arch. Math.* **34**, 377–384.
[1983] Weakly continuous valuations on convex polytopes, *Arch. Math.* **41**, 555–564.
[1986] Angle-sum relations for polyhedral sets, *Mathematik* **33**, 175–188.
[1989] The polytope algebra, *Adv. in Math.* **78**, 76–130.
[1990] Monotone translation invariant valuations on convex bodies, *Arch. Math.* **55**, 595–598.

McMullen, P., and R. Schneider
[1983] Valuations on convex bodies, in: *Convexity and its Applications*, eds P.M. Gruber and J.M. Wills (Birkhäuser, Basel) pp. 170–247.

McMullen, P., and G.C. Shephard
[1971] *Convex Polytopes and the Upper-Bound Conjecture*, London Math. Soc. Lecture Notes Series, Vol. 3 (Cambridge).

Meier, C.
[1972] Zerlegungsähnlichkeit von Polyedern, *J. Reine Angew. Math.* **253**, 193–202.
[1977] Multilinearität bei Polyederaddition, *Arch. Math.* **29**, 210–217.
[unpublished] Ein Skalarprodukt für konvexe Körper, Manuscript.

Meyer, W.J.
[1970] Characterization of the Steiner point, *Pacific J. Math.* **35**, 717–725.

Minkowski, H.
[1911] Theorie der konvexen Körper, insbesondere Begründung ihres Oberflächenbegriffs, *Ges. Abh.*, Vol. II (Teubner, Leipzig) pp. 131–229.

Müller, C.
[1988] Equidecomposability of polyhedra with reference to crystallographic groups, *J. Discrete Comput. Geom.* **3**, 383–389.

Müller, H.R.
[1967] Zur axiomatischen Begründung der Eikörperfunktionale, *Monatsh. Math.* **71**, 338–343.

Munroe, M.E.
[1953] *Introduction to Measure and Integration* (Addison-Wesley, Reading, MA).

Mürner, P.
[1974] Zwei Beispiele zur Zerlegungsgleichheit 4-dimensionaler Polytope, *Elem. Math.* **29**, 132–135.
[1975] Translative Parkettierungspolyeder und Zerlegungsgleichheit, *Elem. Math.* **30**, 25–27.
[1977] Translative Zerlegungsgleichheit von Polytopen, *Arch. Math.* **2**, 218–224.

Nicoletti, O.
[1914] Sulla equivalenza dei poliedri, *Rend. Circ. Mat. Palermo* **37**, 47–75.
[1915] Sulla equivalenza dei poliedri, *Rend. Circ. Mat. Palermo* **40**, 194–210.

Perles, M.A., and G.T. Sallee
[1970] Cell complexes, valuations and the Euler relation, *Canad. J. Math.* **22**, 235–241.

Positselśkiĭ, E.D.
[1973] Characterization of Steiner points (in Russian), *Mat. Zametki* **14**, 243–247 [*Math. Notes* **14**, 698–700].

Rota, G.-C.
[1964] On the foundations of combinatorial theory, I: Theory of Möbius functions, *Z. Wahrschein. Verw. Geb.* **2**, 340–368.

[1971] On the combinatorics of the Euler characteristic, in: *Studies in Pure Mathematics (Papers Presented to Richard Rado)* (Academic Press, London) pp. 221–233.

Sah, C.-H.
[1979] *Hilbert's Third Problem: Scissors Congruence* (Pitman, San Francisco).
[1981] Scissors congruence, I: The Gauss–Bonnet map, *Math. Scand.* **49**, 181–210.

Sallee, G.T.
[1966] A valuation property of Steiner points, *Mathematika* **13**, 76–82.
[1968] Polytopes, valuations and the Euler relation, *Canad. J. Math.* **20**, 1412–1424.
[1971] A non-continuous "Steiner point", *Israel J. Math.* **10**, 1–5.
[1982] Euler's theorem and where it led, in: *Convexity and Related Combinatorial Geometry*, eds D.C. Kay and M. Breen (Marcel Dekker, New York) pp. 45–55.

Santaló, L.A.
[1976] *Integral Geometry and Geometric Probability* (Addison-Wesley, Reading, MA).

Scherk, P.
[1969] Über eine Klasse von Polyederfunktionalen, *Comment. Math. Helv.* **44**, 191–201.

Schmitt, K.A.
[1968] Kennzeichnung des Steinerpunktes konvexer Körper, *Math. Z.* **105**, 387–392.

Schneider, R.
[1971] On Steiner points of convex bodies, *Israel J. Math.* **9**, 241–249.
[1972a] Krümmungsschwerpunkte konvexer Körper, I, *Abh. Math. Sem. Univ. Hamburg* **37**, 112–132.
[1972b] Krümmungsschwerpunkte konvexer Körper, II, *Abh. Math. Sem. Univ. Hamburg* **37**, 204–217.
[1973] Volumen und Schwerpunkt von Polyedern, *Elem. Math.* **28**, 137–141.
[1974a] Equivariant endomorphisms of the space of convex bodies, *Trans. Amer. Math. Soc.* **194**, 53–78.
[1974b] Bewegungsäquivariante, additive und stetige Transformationen konvexe Bereiche, *Arch. Math.* **25**, 303–312.
[1974c] Additive Transformationen konvexe Körper, *Geom. Dedicata* **3**, 221–228.
[1975a] Kinematische Berürmaße für konvexe Körper, *Abh. Math. Sem. Univ. Hamburg* **44**, 12–23.
[1975b] Kinematische Berürmaße für konvexe Körper und Integralrelationen für Oberflächenmaße, *Math. Ann.* **218**, 253–267.
[1977a] Kritische Punkte und Krümmung für die Mengen des Konvexringes, *Enseig. Math.* **23**, 1–6.
[1977b] Ein kombinatorisches Analogon zum Satz von Gauss–Bonnet, *Elem. Math.* **32**, 105–108.
[1978] Curvature measures of convex bodies, *Ann. Mat. Pura Appl.* **11**, 101–134.
[1979] Boundary structure and curvature of convex bodies, in: *Contributions to Geometry*, eds J. Tölke and J.M. Wills (Birkhäuser, Basel).
[1980] Parallelmengen mit Vielfachheit und Steiner-Formeln, *Geom. Dedicata* **9**, 111–127.
[1981] A uniqueness theorem for finitely additive invariant measures on a compact homogeneous space, *Rend. Circ. Mat. Palermo, Ser. II* **30**, 341–344.
[1985] Equidecomposable polyhedra, *Colloq. Math. Soc. János Bolyai* **48**, 481–501.

Shephard, G.C.
[1966] The Steiner point of a convex polytope, *Canad. J. Math.* **18**, 1294–1300.
[1967] An elementary proof of Gram's theorem for convex polytopes, *Canad. J. Math.* **19**, 1214–1217.
[1968a] The mean width of a convex polytope, *J. London Math. Soc.* **43**, 207–209.
[1968b] Angle deficiencies of convex polytopes, *J. London Math. Soc.* **43**, 325–336.
[1968c] A uniqueness result for the Steiner point of a convex region, *J. London Math. Soc.* **43**, 439–444.
[1968d] Euler-type relations for convex polytopes, *Proc. London Math. Soc. (3)* **18**, 597–606.

Sommerville, D.M.Y.
[1927] The relations connecting the angle-sum and volume of a polytope in space of *n* dimensions, *Proc. Roy. Soc. London A* **115**, 103–119.

Speigel, W.
[1976a] Ein Zerlegungssatz für spezielle Eikörperabbildungen in den euklidischen Raum, *J. Reine Angew. Math.* **283/284**, 282–286.
[1976b] Zur Minkowski-Additivität bestimmter Eikörperabbildungen, *J. Reine Angew. Math.* **286/287**, 164–168.
[1978] Ein Beitrag über additive, translationsinvariante, stetige Eikörperfunktionale, *Geom. Dedicata* **7**, 9–19.
[1982] Non-negative, motion invariant valuations of convex polytopes, in: *Convexity and Related Combinatorial Geometry,* eds D.C. Kay and M. Breen (Marcel Dekker, New York) pp. 67–72.

Stachó, L.L.
[1979] On curvature measures, *Acta Sci. Math.* **41**, 191–207.

Stein, R.
[1982] Additivität und Einschließungs–Ausschließungsprinzip für Funktionale von Gitterpolytopen, Dissertation, Dortmund.

Strambach, K.
[1966] Über die Zerlegungsgleichheit von Polygonen bezüglich Untergruppen nichteuklidischer Bewegungsgruppen, *Math. Z.* **93**, 276–288.

Sydler, J.-P.
[1965] Conditions nécessaires et suffisantes pour l'équivalence des polyèdres de l'espace euclidien à trois dimensions, *Comment. Math. Helv.* **40**, 43–80.

Tverberg, H.
[1974] How to cut a convex polytope into simplices, *Geom. Dedicata* **3**, 239–240.

Valette, G.
[1974] Subadditive affine-variant transformations of convex bodies, *Geom. Dedicata* **2**, 461–465.

Volland, W.
[1957] Ein Fortsetzungssatz für additive Eipolyederfunktionale im euklidischen Raum, *Arch. Math.* **8**, 144–149.

Walkup, D.W., and R.J.-B. Wets
[1969] Lifting projections of convex polyhedra, *Pacific J. Math.* **28**, 465–475.

Weil, W.
[1974a] Über den Vektorraum der Differenzen von Stützfunktionen konvexer Körper, *Math. Nachr.* **59**, 353–369.
[1974b] Decomposition of convex bodies, *Mathematika* **21**, 19–25.
[1975a] On mixed volumes of nonconvex sets, *Proc. Amer. Math. Soc.* **53**, 191–194.
[1975b] Einschachtelung konvexer Körper, *Arch. Math.* **26**, 666–669.
[1981] Das gemischte Volumen als Distribution, *Manuscripta Math.* **3**, 1–18.

Wieacker, J.A.
[1982] Translative stochastische Geometrie der konvexen Körper, Dissertation, Freiburg.

Zähle, M.
[1984] Curvature measures and random sets, I, *Math. Nachr.* **119**, 327–339.
[1986] Curvature measures and random sets, II, *Probab. Theory Relat. Fields* **71**, 37–58.

Zylev, V.B.
[1965] Equidecomposability of equicomplementable pólyhedra (in Russian), *Dokl. Akad. Nauk. SSSR* **161**, 515–516 [*Soviet Math. Dokl.* **6**, 453–455].
[1968] G-decomposedness and G-complementability (in Russian), *Dokl. Akad. Nauk. SSSR* **179**, 529–530 [*Soviet Math. Dokl.* **9**, 403–404].

CHAPTER 3.7

Geometric Crystallography

Peter ENGEL

Laboratorium für Kristallographie, Universität Bern, Freiestrasse 3, CH-3012 Bern, Switzerland

Contents

HANDBOOK OF CONVEX GEOMETRY
Edited by P.M. Gruber and J.M. Wills

Introduction

Interest in the structure of crystals came up in the Renaissance, when the writings of the ancient Greek philosophers were studied again. Both the atomistic theory of Leucipos (approx 475 BC) and Democritus (approx 470–400 BC), and the space filling theory of Plato (427–347 BC), which was further developed by Aristotle (384–322 BC), were disputed up to our century. Scientists such as Johann Kepler (1571–1630), Robert Hook (1635–1703), and Christian Huygens (1629–1695) tried to explain the geometric forms of crystals by assuming small polyhedral or spherical particles arranged in a regular three-dimensional array.

The main achievements in geometric crystallography were obtained in the 19th century by René Just Haüy (1743–1822), Samuel Weiss (1780–1856), Ludwig August Seeber (1793–1855), Moritz Ludwig Frankenheim (1801–1872), August Bravais (1811–1863), and Leonhard Sohncke (1842–1897). It culminated in completing the list of the $219 + 11$ types of crystallographic groups by Arthur Schoenflies (1853–1928), and Evgraf Stepanovič Fedorov (1853–1919). But it was not until 1912, when Max von Laue (1879–1960) and his coworkers (see Friedrich, Knipping and von Laue 1912) discovered the diffraction of X-rays by crystals, that the elaborated theory of crystal symmetry could be verified experimentally. [See also Burke (1966), Burckhardt (1967, 1984, 1988), Šafranovskiĭ (1978, 1980), and Senechal (1990) for a more detailed historical survey.]

1. Regular systems of points

1.1. Basic definitions

The regular external forms of crystals suggest that within a crystal atomic building blocks, congruent to each other, are regularly arranged. Assuming the crystal to be infinite and the centres of the atoms to be represented by points, an infinite, discrete system of points, called a discontinuum, results which plays an essential role in crystallography. Moreover, such systems of points are of great importance in several branches of mathematics and physics. Whereas the existence of a continuum in nature cannot be shown, the discontinuum has an assured position in natural sciences.

We consider a set X of points in d-dimensional Euclidean space E^d which fulfils, after Hilbert and Cohn-Vossen (1931), the following conditions.

Conditions 1.1.

(i) the set X is discrete, i.e., around each point of X an open ball of fixed radius $r > 0$ can be drawn which contains no other point of X;

(ii) every interstitial ball, i.e., every ball which can be embedded into E^d such that its interior avoids all points of X, has a radius less than, or equal to, a fixed finite R;

(iii) the set X looks the same from every point of X.

The second condition ensures that the points are spread uniformly over the whole space. The points are said to be *relatively dense*. For relatively dense sets of points the number of points within any ball of radius $L \gg R$ increases with the dth power of L. A set X which fulfils the first two conditions only is called a *discontinuum* or, following Delone et al. (1976), an (r, R)-*system*. This more general kind of point set is important in the theory of amorphous matter, incommensurate solid state phases, and quasi-crystals which will be discussed later in section 7 on non-regular systems of points.

The third condition was proposed by Wiener (1863) in order to overcome the shortcomings of the arbitrary assumption that the atoms or molecules should be arranged by translations only. Following Sohncke (1874), this condition can be stated more precisely by considering the set of straight line segments drawn from any point of the set X to all the remaining points of X. The third condition requires that the line systems of any two points of X are directly or mirror congruent. It follows that for each pair of points there exists a rigid motion of Euclidean space which maps the two line systems and hence the whole set X onto itself. This condition also ensures that a pair of radii r, R exists (in an (r, R)-system the radii r and R represent the infimum, and the supreme of radii, respectively, and balls of radius r and R do not necessarily exist). A set X which fulfils all three conditions is called a *regular system of points* by Sohncke, or a *homogeneous discontinuum* by Niggli (1919). Regular systems of points have applications in the theory of ideal crystals. An *ideal crystal structure* is defined to be the union of one, or several regular systems of points. Each regular system of points corresponds to one atomic species.

Applying Wiener's condition, Sohncke (1874) determined in a purely geometric way, without using any symmetry arguments, thirteen construction types of regular systems of points in the plane. It was shown by Engel et al. (1984) that these construction types correspond to the thirteen types of stabilizers of regular point systems in the plane. Grünbaum and Shephard (1987) pointed out that Sohncke's construction almost completely coincides with the homeomeric classification of dot patterns in the plane. Sohncke's construction was a remarkable attempt to determine systematically regular systems of points without using the concept of symmetry. Strongly influenced by Jordan's paper on the groups of motions (1869), symmetry considerations for the classification of regular systems of points became important. At that time, symmetry groups were already used for the classification of crystal forms by Frankenheim (1826), and Hessel (1830). Indeed, the belief in symmetry became so strong that it was quite unexpected when quasi-crystals were first discovered.

1.2. Isometries

The regularity of a set X implies the existence of rigid motions which bring the set into self-coincidence. A rigid motion is called an *isometry*. We introduce a coordinate system with origin $\mathbf{0}$ and basis vectors $\hat{\boldsymbol{a}}_1, \ldots, \hat{\boldsymbol{a}}_d$, and *metric tensor* $\boldsymbol{C} := (c_{ij})$, where $c_{ij} := |\hat{\boldsymbol{a}}_i|\,|\hat{\boldsymbol{a}}_j| \cos \alpha_{ij}$. Synonymous with metric tensor also *Gram matrix* or *Hermite matrix* is used. An isometry σ in E^d can be represented by a non-singular $d \times d$ matrix $\mathbf{S} := (s_{ij})$, and a shift vector $\hat{s}$; they transform the coordinates of a point $\boldsymbol{x}$ into those of another point $\boldsymbol{x}'$:

$$\boldsymbol{x}' := \sigma \boldsymbol{x} = \mathbf{S}\boldsymbol{x} + \hat{s} .$$

Using Frobenius' symbol this equation is written, by abbreviation, as

$$\boldsymbol{x}' := (\mathbf{S}, \hat{s})\boldsymbol{x} .$$

The matrix $\mathbf{S} = \operatorname{rot} \sigma$ is called the *rotation part* and the shift vector $\hat{s} = \operatorname{trans} \sigma$ is called the *translation part* of the isometry σ. Every isometry that brings $\boldsymbol{x}$ into coincidence with $\boldsymbol{x}'$ also has to bring an arbitrary point $\boldsymbol{y}$ into coincidence with some point $\boldsymbol{y}'$. For an isometry it is required that the length of the vector $\hat{y} - \hat{x}$ is conserved under the motion.

$$|\hat{y} - \hat{x}|^2 = |\hat{y}' - \hat{x}'|^2 = |\mathbf{S}\hat{y} - \mathbf{S}\hat{x}|^2 = (\hat{y} - \hat{x})^{\mathrm{T}}\mathbf{S}^{\mathrm{T}}\boldsymbol{C}\mathbf{S}(\hat{y} - \hat{x}) .$$

This equation is valid for all $\hat{x}, \hat{y} \in \mathrm{E}^d$. Therefore, as a necessary and sufficient condition we have

$$\boldsymbol{C} \equiv \mathbf{S}^{\mathrm{T}}\boldsymbol{C}\mathbf{S} .$$

From $\det(\boldsymbol{C}) \equiv \det(\mathbf{S}^{\mathrm{T}}\boldsymbol{C}\mathbf{S}) = \det(\boldsymbol{C}) \det^2(\mathbf{S})$ it follows that $\det(\mathbf{S}) = \pm 1$. Thus $\mathbf{S}$ is an element of the *orthogonal group* $O(d)$, and $(\mathbf{S}, \hat{s})$ belongs to the *Euclidean group* $E(d)$. The value of the determinant $\det(\mathbf{S})$ establishes the *chirality character* of the isometry. Let T^d be a generic simplex in E^d such that it is not directly congruent to its mirror image. The simplex T^d and its mirror congruent copy are said to be *enantiomorph* to each other. If $\det(\mathbf{S}) = +1$ then the isometry σ carries the simplex T^d into a direct congruent copy. Such an isometry is called *proper*, or *of the first kind*. Otherwise, if $\det(\mathbf{S}) = -1$ then σ carries the simplex T^d into a mirror congruent copy. Such an isometry is called *improper*, or *of the second kind*.

There exist isometries which leave at least one point $\boldsymbol{p}$ fixed, $\sigma \boldsymbol{p} = \boldsymbol{p}$. If σ has a fixed point, then σ is called a *rotation* if $\det(\mathbf{S}) = +1$, or a *rotoreflection* if $\det(\mathbf{S}) = -1$. Otherwise, if σ has no fixed point, then it is called a *screw rotation*, or a *screw rotoreflection*, respectively. Special cases of rotoreflections are the reflection in a hyperplane and, in spaces of odd dimension, the inversion in a centre of symmetry. The translation is a special case of a screw rotation and a glide reflection is a special case of a screw rotoreflection. Instead of rotoreflection also *rotoinversion* is used. However, if we take the decomposition of space into symmetry-invariant subspaces, then rotoreflection seems to be more appropriate.

1.3. Groups of symmetry operations

Let M be any subset of E^d. We consider all the isometries which map M onto itself.

Definition 1.1. A *symmetry operation* acting on a set M is an isometry which maps M onto itself.

The symmetry operations of a set M have two important properties:

(i) A symmetry operation σ_1 followed by a second symmetry operation σ_2 is again a symmetry operation of M.

(ii) The symmetry operation σ^{-1} which reverses the symmetry operation σ is again a symmetry operation of M and the result is the identity operation.

It follows that the totality of symmetry operations of a set M generates a group in the mathematical sense. Symmetry groups correspond to linear representations of abstract groups in a vector space. Different symmetry groups may correspond to different representations of the same abstract groups. Thus, for classification of symmetry groups, not isomorphism of groups, but equivalence under the group of affine transformations will be used. More generally, we consider the action of a group on a certain set M, and classifications are made by operator isomorphisms.

Definition 1.2. Every group $\mathcal{K}[M]$ of symmetry operations acting on a set M, and which leaves at least one point $\boldsymbol{p} \in \mathrm{E}^d$ fixed, is called a *point group*.

Definition 1.3. The set of all symmetry operations of a group $\mathcal{G}$ which map a set M onto itself is called the *stabilizer* of M in $\mathcal{G}$.

Theorem 1.1. *Every group $\mathcal{K}[M]$ of symmetry operations acting on a bounded set M is a point group.*

We now consider the symmetry operations of a regular system of points X. By the regularity condition 1.1(iii), there exists, for every pair of points $\boldsymbol{x}, \boldsymbol{y} \in X$, a symmetry operation σ which carries $\boldsymbol{x}$ into $\boldsymbol{y}$ and thereby maps X onto itself. Thus, there exists a group of symmetry operations which acts transitively on X.

Definition 1.4. Every discrete group of symmetry operations which acts transitively on a regular system of points in E^d is a d-dimensional *space group*.

Synonymous with space group also *crystallographic group* is used. We will use equivalence under the affine group $A(d)$ to classify the space groups into isomorphism classes or space group types according to the following definition.

Definition 1.5. Two space groups $\mathcal{G}$ and $\tilde{\mathcal{G}}$ belong to the same *space group type* if there exists an affine transformation φ which maps $\mathcal{G}$ onto $\tilde{\mathcal{G}}$: $\tilde{\mathcal{G}} = \varphi\mathcal{G}\varphi^{-1}$.

Because of physical reasons, it is useful to consider only proper affine transformations φ^+ of the *special affine group* $A^+(d)$ that conserve the chirality of the coordinate system. For $d > 2$ the number of *special space-group types* is larger than the number of space-group types. We will add the number of enantiomorphic pairs to the number of space-group types, e.g., in E^3 the corresponding numbers will be stated as $219 + 11$ space-group types.

The hypothesis that the internal structure of a crystal is a three-dimensional lattice was proposed at the beginning of the 19th century by Seeber (1824). It was supported from earlier observations of the cleavage of calcite crystals. This important property of crystals to be periodic follows directly from the regularity condition 1.1(iii). This was proved by Schoenflies (1891) and Rohn (1900) for the three-dimensional case, and in general, for the d-dimensional case, by Bieberbach (1910) as part of his affirmative answer to the first part of Hilbert's 18th problem (1900).

Theorem 1.2. *Every discrete group of isometries acting on the d-dimensional Euclidean space* E^d *with compact fundamental domain contains d linearly independent translations.*

The second theorem of Bieberbach asserts the finiteness of isomorphism classes of space groups.

Theorem 1.3. *For each fixed d there are only finitely many isomorphism classes of d-dimensional crystallographic groups.*

Shortly after Bieberbach's proof came out, Frobenius (1911) gave a more accessible proof which, in one form or another, became standard in the contemporary literature. A geometric proof of Schoenflies' theorem was given by Delone and Štogrin (1974), and for the d-dimensional case by Buser (1985). As stated above, for symmetry groups equivalence under the affine group, as used in Definition 1.5, is more natural than isomorphism. Thus, in his proof of Bieberbach's theorems, Frobenius used the more restrictive affine equivalence. Finally, Bieberbach (1912) proved that, in contrast to point groups, space groups are affine equivalent if and only if they are isomorphic.

1.4. Regularity condition for an (r, R)-system

It has been shown by X-ray analysis that in most cases solidification of matter results in a periodic aggregation of an immense number of atoms or molecules. About 100 000 crystal structures have been determined up till now, and detailed information on their space groups are available. However, it must be emphasized that symmetry is not an intrinsic law of nature and we may not assume symmetry to be present a priori. All the more, it is surprising that such highly regular crystals grow in nature.

In order to better understand the crystalline state of matter, we have to investigate the local properties of an (r, R)-system X. From each point $\boldsymbol{x} \in X$ we draw the straight line segments to all the other points of X. The regularity condition 1.1(iii) requires that these infinite line systems are directly or mirror congruent. However, Delone et al. (1976) proved that the regularity condition can be weakened in the sense that for the infinite line systems, which they called spiders, congruence is only required within a given sphere of regularity. For a point $\boldsymbol{x} \in X$ let $B(\rho, \boldsymbol{x})$ be a ball of arbitrary radius ρ and centre $\boldsymbol{x}$, and denote by

$\mathcal{K}[B(\rho, \boldsymbol{x})]$ the stabilizer of $B(\rho, \boldsymbol{x})$. Because X is an (r, R)-system, there exists a radius ρ_0, $r \leqslant \rho_0 < 2R$, such that the order of $\mathcal{K}[B(\rho_0, \boldsymbol{x})]$ is finite.

Theorem 1.4. *An (r, R)-system X in d-dimensional space (Euclidean, spherical, or hyperbolic) is regular if for every point $\boldsymbol{x} \in X$ the system of straight line segments, drawn from $\boldsymbol{x}$ to all the other points $\boldsymbol{y} \in X$ within a ball of radius $\rho = 2R(\nu + 2)$ and centre in $\boldsymbol{x}$, are congruent, where ν is the number of prime factors in the order of $\mathcal{K}[B(\rho_0, \boldsymbol{x})]$.*

Thus periodicity, as observed in crystals, results from local conditions only. Using the Dirichlet domain partition, Engel (1986) could refine this result by showing that in E^2, $4R$, and in E^d, $d > 2$, $6R$ is an upper bound for a sphere of regularity. There exist (r, R)-systems that are locally regular in every point within a regularity sphere $4R$, but which are globally irregular. Remarkably, in E^3 such (r, R)-systems consist of infinite stacks of either trigonal, or else tetragonal nets. Always two adjacent nets compose a double layer. Depending on the symmetry of the nets, these double layers may be rotated one against the other by 60°, 120°, or 90°, respectively, such that the line system of every node remains congruent within the $4R$ regularity sphere. By rotating at random the double layers one against the other, an infinite number of non-regular stacks can be obtained, all being locally regular within the $4R$ regularity sphere. Similar variations in the stacking of layers are observed in nature. Compounds such as SiC, ZnS, or CdI which frequently form *polytypes* are examples of stacking faults [see Verma and Krishna (1966) for polytypism].

2. Dirichlet domains

The Dirichlet domain is an important mathematical tool for investigating the metrical and topological properties of point sets. It was introduced in 1850 by Peter Gustav Lejeune Dirichlet (1805–1859) in order to give a geometric interpretation of Seeber's reduction conditions (see also section 4.3). The same construction was used later by M. Georges Voronoï (1868–1908) in his investigation of higher-dimensional translation lattices (1908). In his topological structure analysis, Paul Niggli (1888–1953) introduced Dirichlet's construction in crystallography (1927). For a certain atom in a crystal structure, he constructed the Dirichlet domain with respect to all the remaining atoms. The domain thus obtained, he called the domain of influence. It has applications in the determination of atom coordination numbers and in finding maximal holes in crystal structures.

2.1. Definition and basic properties of Dirichlet domains

The metrical and topological properties of a set of points are revealed by the Dirichlet domain partition.

Definition 2.1. For a set of points X, the *Dirichlet domain* $D(\boldsymbol{x}) \subset \mathrm{E}^d$, $\boldsymbol{x} \in X$, is the part of space containing all points which are closer to $\boldsymbol{x}$ than to any other point $\boldsymbol{y} \in X$.

Synonymous with Dirichlet domain also *Voronoï region*, *Wigner–Seitz cell*, *Brillouin zone*, *domain of influence*, *honeycombe*, or *plesiohedron* is used. The Dirichlet domain of a point $\boldsymbol{x} \in X$ is obtained by the intersection of certain half spaces. For a pair of points $\boldsymbol{x}, \boldsymbol{y} \in X$, let H_y be the hyperplane which is normal to the straight line segment $\overline{\boldsymbol{xy}}$ and bisects it. The hyperplane H_y separates the space into two half spaces, H_y^+ and H_y^-, respectively. We assume H_y^+ to contain $\boldsymbol{x}$. By construction, all points $\boldsymbol{p} \in H_y^+$ lie closer to $\boldsymbol{x}$ then to $\boldsymbol{y}$. It follows that for a set of points X, the open Dirichlet domain $D^+(\boldsymbol{x})$ is the intersection of all possible half spaces H_y^+:

$$D^+(\boldsymbol{x}) := \bigcap_{\boldsymbol{y} \in X \setminus \boldsymbol{x}} H_y^+ .$$

By $D(\boldsymbol{x})$ we denote the closure of $D^+(\boldsymbol{x})$. The half spaces are convex and hence, $D(\boldsymbol{x})$ is convex. In general, not all points $\boldsymbol{y} \in X \setminus \boldsymbol{x}$ contribute, via their half spaces, to the Dirichlet domain. In order to show this, we consider a ball $B(|\hat{\boldsymbol{p}} - \hat{\boldsymbol{x}}|, \boldsymbol{p})$ with radius $|\hat{\boldsymbol{p}} - \hat{\boldsymbol{x}}|$, and centre at point $\boldsymbol{p}$. We define the region:

$$Q(\boldsymbol{x}) := \bigcup_{\boldsymbol{p} \in D(\boldsymbol{x})} B(|\hat{\boldsymbol{p}} - \hat{\boldsymbol{x}}|, \boldsymbol{p}) .$$

The following Theorems 2.1–2.4 reveal the relation of the Dirichlet domain to the radius R of an (r, R)-system. Proofs are given in Engel (1986).

Theorem 2.1. *In a system of points X, all points $\boldsymbol{y} \in X \setminus \boldsymbol{x}$ which generate facets of $D(\boldsymbol{x})$ lie on the boundary of the region $Q(\boldsymbol{x})$.*

Particularly, in an (r, R)-system, for every point $\boldsymbol{x} \in X$ the region $Q(\boldsymbol{x})$ is contained in a ball of radius $2R$. The following theorem holds.

Theorem 2.2. *In an (r, R)-system X, every Dirichlet domain $D(\boldsymbol{x})$, $\boldsymbol{x} \in X$, is determined through all points within a ball of radius $2R$ and centre $\boldsymbol{x}$. $D(\boldsymbol{x})$ is a bounded, convex polytope with a finite number of facets.*

For a bounded convex polytope, the vertices $\boldsymbol{v}_i$ are extreme points. Since the number of facets is finite, it follows that the number of vertices is finite too. The vertices of $D(\boldsymbol{x})$ are exactly those points of E^d whose distances from X are local maxima. The following theorems hold.

Theorem 2.3. *In an (r, R)-system X, the region $Q(\boldsymbol{x})$ for every Dirichlet domain $D(\boldsymbol{x})$, $\boldsymbol{x} \in X$, is determined through the union of all balls $B(|\hat{\boldsymbol{v}}_i - \hat{\boldsymbol{x}}|, \boldsymbol{v}_i)$ at the vertices $\boldsymbol{v}_i$ of $D(\boldsymbol{x})$.*

Theorem 2.4. *In an (r, R)-system X, the Dirichlet domain $D(\boldsymbol{x})$ of every point $\boldsymbol{x} \in X$, is contained within a ball $B(R, \boldsymbol{x})$ of radius R and centre $\boldsymbol{x}$.*

In an (r, R)-system X, the family of balls $B(R, \boldsymbol{x})$, $\boldsymbol{x} \in X$, *has the important covering property*:

Theorem 2.5. *In an (r, R)-system X, the set of closed balls $B(R, x)$, attached at every point $\boldsymbol{x} \in X$, covers space completely.*

2.2. Characterization of Dirichlet domains

A closed bounded Dirichlet domain is a convex polytope P consisting of 0-, 1-, . . . , d-dimensional elements, called the *k-faces* of P, where $0 \leq k \leq d$. The 0-faces F_i^0 are the *vertices* $\boldsymbol{v}_i$, the 1-faces F_j^1 are the *edges* E_j, and the $(d-1)$-faces F_h^{d-1} are the *facets* F_h. Sometimes the $(d-2)$-faces are called *ridges*. Among the k-faces of a polytope P exists a partial order with respect to the inclusion operation,

$$0\text{-}faces \subset 1\text{-}faces \subset \cdots \subset (d-1)\text{-}faces \subset P\,,$$

which determines the *hierarchical structure*, or the *face lattice* of the polytope P. Every k-face of P is again a k-dimensional convex polytope. In a face lattice every k-face is subordinated by at least $k+1$ $(k-1)$-faces. Particularly, every edge contains exactly two vertices. Inversely, every k-face has superordinated at least $d-k$ $(k+1)$-faces. Especially, in each $(d-2)$-face exactly two facets meet. Fast algorithms to calculate half-space intersections make extensive use of the face lattice as described by Engel (1986).

In the morphology of convex polytopes each polytope is completely characterized by its face lattice.

Definition 2.2. Two polytopes P and $\tilde{P}$ in d-dimensional space $\mathbb{R}^d$ are called *isomorphic* and belong to the same *combinatorial type* if P and $\tilde{P}$ have isomorphic face lattices.

The isomorphism of two polytopes is not easily recognized. In what follows some classification schemes are given. A rough classification of a polytope is obtained by the numbers N_k of k-faces, $0 \leq k < d$. A finer classification is obtained by the *k-subordination symbol*, where $2 \leq k < d$. The k-subordination symbol,

$$f_{1_{m_1}} f_{2_{m_2}} \cdots f_{r_{m_r}}\,,$$

gives the numbers m_i of those k-faces which have subordinated f_i $(k-1)$-faces. It follows that the number of k-faces is given by $N_k = \Sigma\, m_i$, $i = 1, \ldots, r$. In a similar way the *k-superordination symbol*, $0 \leq k \leq d-2$, is defined:

$$f_1^{m_1} f_2^{m_2} \cdots f_s^{m_s},$$

where m_i is the number of those k-faces which have superordinated f_i $(k+1)$-faces. A unique classification of a polytope is obtained by the *unified polytope scheme* in the following way. A subseries of mutually subordinated k-faces, $F^0 \subset F^1 \subset \cdots \subset F^{d-1} \subset P$, is called a *d-flag*. Given an arbitrary d-flag, Engel (1991) has shown that it is possible to number all k-faces of P in a unique way such that the face lattice of P becomes ordered. A polytope scheme is obtained by writing down for each facet, in turn, the numbers of its subordinated vertices in increasing order. This scheme depends only on the arbitrarily chosen initial d-flag. For each possible d-flag, a corresponding scheme is set up and only those schemes are retained wherein a smaller number first occurs. The unified polytope scheme is such a minimal scheme.

The number of identical unified polytope schemes corresponds to the order of the combinatorial automorphism group Aut$[P]$ of the face lattice. Let P be a polytope in Euclidean space E^d. By $\mathcal{K}[P]$ we denote the stabilizer of P in the orthogonal group $O(d)$. The symmetry group $\mathcal{K}[P]$ induces a group of automorphisms of the face lattice. Inversely, given a polytope P with combinatorial automorphism group Aut$[P]$, it is, for dimensions $d \geq 4$, in general not true that there also exists a polytope $\tilde{P}$ of the same combinatorial type with stabilizer $\mathcal{K}[P]$ which is isomorphic to Aut$[P]$. However, for dimensions $d \leq 3$, Mani (1971) proved the following theorem.

Theorem 2.6. *To every polytope P in E^d, with $d \leq 3$, there exists a polytope $\tilde{P}$ of the same combinatorial type such that its stabilizer $\mathcal{K}[\tilde{P}]$ is isomorphic to the combinatorial automorphism group* Aut$[P]$.

In crystallography the classification of polytopes with respect to their symmetries is important.

Definition 2.3. Two polytopes P and $\tilde{P}$ belong to the same *homeomeric type* if

(i) P and $\tilde{P}$ belong to the same combinatorial type;

(ii) P and $\tilde{P}$ have equivalent stabilizers $\mathcal{K}[P]$ and $\mathcal{K}[\tilde{P}]$ in $O(d)$, $\mathcal{K}[\tilde{P}] = \varphi\mathcal{K}[P]\varphi^{-1}$;

(iii) corresponding k-faces F^k and $\tilde{F}^k$ have equivalent stabilizers $\mathcal{K}[F^k]$ and $\mathcal{K}[\tilde{F}^k]$ in $\mathcal{K}[P]$ and $\mathcal{K}[\tilde{P}]$, respectively, $\mathcal{K}[\tilde{F}^k] = \varphi\mathcal{K}[F^k]\varphi^{-1}$.

2.3. Dirichlet domain partition

For an (r, R)-system X, the family of all Dirichlet domains $D(\boldsymbol{x})$, $\boldsymbol{x} \in X$, determines a space partition which has the following properties:

(i) the union of all closed Dirichlet domains $D(\boldsymbol{x})$, $\boldsymbol{x} \in X$, covers space completely;

(ii) the intersection of the interiors of any two Dirichlet domains is empty;

(iii) the intersection of two closed Dirichlet domains is either empty or a k-face of each.

The Dirichlet domain partition of an (r, R)-system is a special kind of a tiling. The tiles are called, following Fedorov (1885, 1896), *stereotopes*. Following Grünbaum and Shephard (1987), a tiling $\mathcal{T}$ is called *monohedral* if all of its tiles T_i are directly or mirror congruent to a *prototile* T_0. It is called *isohedral* if there exists a space group which acts transitively on the tiles. Synonymous to isohedral also *regular* or *homogeneous* is used. An isohedral tiling necessarily is monohedral. A tiling is called *face-to-face* if the intersection of any two tiles is either empty or a k-face of each. It is called *normal* if every tile of it is a topological ball, and if the intersection of every two tiles of it is a connected set, and if it is uniformly bounded. Uniformly bounded means that there exist for every tile T_i of $\mathcal{T}$ two balls, $B(u, \boldsymbol{v}_i)$ of radius $u > 0$ and $B(U, \boldsymbol{w}_i)$ of finite radius U, such that $B(u, \boldsymbol{v}_i) \subset T_i \subset B(U, \boldsymbol{w}_i)$. Clearly, the Dirichlet domain partition of a regular system of points is a normal, isohedral, face-to-face tiling. The classification of general space partitions is beyond our scope, and we restrict ourselves to the classification of Dirichlet domain partitions. The classification of space partitions is discussed in more detail in chapter 3.5 on tilings. In crystallography, we are mainly interested in classifying tilings with respect to topological properties and symmetries. Thus, the Dirichlet domain partitions will be assigned to topological and homeomeric types in the following way: let $\mathcal{T}$ be a Dirichlet domain partition with tiles T_i. The unified polytope scheme allows us to determine an adjacency symbol for a certain tile T_0 by assigning to each k-face $F_i^k \in T_0$ all the k-faces which meet T_0 in F_i^k.

Definition 2.4. Two isohedral face-to-face tilings $\mathcal{T}$ and $\tilde{\mathcal{T}}$ are said to be of the same *topological type* if $T_0 \in \mathcal{T}$ and $\tilde{\mathcal{T}}_0 \in \tilde{\mathcal{T}}$ have identical unified polytope-schemes and adjacency symbols.

The classification of Dirichlet domain partitions with respect to symmetries is done by affine mappings.

Definition 2.5. Two isohedral face-to-face tilings $\mathcal{T}$ and $\tilde{\mathcal{T}}$ are *isomorphic* and belong to the same homeomeric type if

(i) $\mathcal{T}$ and $\tilde{\mathcal{T}}$ belong to the same topological type;

(ii) $\mathcal{T}$ and $\tilde{\mathcal{T}}$ have affinely equivalent stabilizers $\mathcal{G}$ and $\tilde{\mathcal{G}}$, $\tilde{\mathcal{G}} = \varphi \mathcal{G} \varphi^{-1}$;

(iii) corresponding k-faces F^k and $\tilde{F}^k$ have equivalent stabilizers $\mathcal{K}[F^k]$ and $\mathcal{K}[\tilde{F}^k]$ in $\mathcal{G}$ and $\tilde{\mathcal{G}}$, respectively, such that $\mathcal{K}[\tilde{F}^k] = \varphi \mathcal{K}[F^k] \varphi^{-1}$.

In the plane E^2 many results are known which are described in chapter 3.5 on tilings. We only mention here Laves' 11 topological types (1931b), and Grünbaum and Shephard's 47 homeomeric types of isohedral face-to-face tilings. Laves (1931a) showed that his 11 topological types can be realized as Dirichlet domain partitions, but it is still unknown if such a realization also exists for the 47 homeomeric types.

Only rudimentary results are known in Euclidean space E^d, $d>2$. Delone and Sandakova (1961), considering isohedral space partitions in E^d, proved the following theorem (for a proof see also Štogrin 1973).

Theorem 2.7. *Let $\mathcal{T}$ be an isohedral tiling in E^d with stabilizer $\mathcal{G}$ and prototile T_0. The number of facets of T_0 is at most $N_{d-1} \leq 2^d(h+1)-2$, where h is the order of the factor group of $\mathcal{G}$ with respect to its translation subgroup.*

In three-dimensional space E^3 this upper bound is reached for small values of h but it becomes too high for the maximal value of $h=48$. The most complicated Dirichlet domains found in an isohedral space partition have 38 facets (Engel 1981). Detailed results on isohedral Dirichlet domain partitions are known for particular cases only. Löckenhoff and Hellner (1971) determined the Dirichlet domains of the special positions in the cubic space groups with no degree of freedom. These investigations were extended to the special positions with one and two degrees of freedom by Koch (1973) who determined 117 combinatorial types of Dirichlet domains. Štogrin (1973) determined the 165 combinatorial types of Dirichlet domains for the triclinic space group $P\bar{1}$, and Engel (1981) gave complete results for the symmorphic cubic space groups listing 91 combinatorial types of Dirichlet domains, 34 of which where already found by Koch. Dauter (1984) investigated the orthorhombic space group $P2_12_12_1$ assuming axes of equal length, and found 77 combinatorial types of Dirichlet domains with up to 22 facets.

2.4. Regularity condition for normal space partitions

In what follows we consider a normal space partition $\mathcal{T}$ with tiles T_i. By the normality condition every tile has a finite number of neighbours. By Theorem 2.2, this is true in particular for the Dirichlet domain partition of an (r, R)-system. From the regularity condition 1.1(iii), it follows that a normal space partition is regular if every tile $T_i \in \mathcal{T}$ is surrounded in exactly the same way by all the other tiles of $\mathcal{T}$. In a similar way as in section 1.4, the regularity condition can be weakened by considering the local properties of the tiling $\mathcal{T}$.

Definition 2.6. For any tile T_i of a tiling $\mathcal{T}$, the cell complex of all tiles $T_j \in \mathcal{T}$ which touch the tile T_i is called the *first corona* of T_i, $C^1(T_i) := \{T_j \mid T_j \cap T_i \neq \emptyset\}$.

Similarly, $C^k(T_i) := \{T_j \mid T_j \cap C^{k-1}(T_i) \neq \emptyset\}$ is called the *k-corona* of T_i. We define $C^0(T_i) := T_i$. We say $C^k(Ti)$ is congruent to $C^k(T_j)$ if there exists an isometry $\varphi_{ij} \in E(d)$ which maps $C^k(T_i)$ onto $C^k(T_j)$, $\varphi_{ij} : C^\nu(T_i) \rightarrow C^\nu(T_j)$, $\nu = k, k-1, \ldots, 0$. We denote by $\mathcal{K}[T_i]$ and $\mathcal{K}[C^k(T_i)]$ the stabilizers of T_i and $C^k(T_i)$, respectively, in the orthogonal group $O(d)$. By the normality condition, T_i has a finite number of facets and hence the order of $\mathcal{K}[T_i]$ is finite. Because every symmetry operation $\sigma \in \mathcal{K}[C^1(T_i)]$ maps T_i onto itself, it follows that $\mathcal{K}[C^1(T_i)] \leq \mathcal{K}[T_i]$. Thus the following theorem is a refinement of Theorem 1.4.

Theorem 2.8. *A normal space partition $\mathcal{T}$ in d-dimensional space (Euclidean, spherical, or hyperbolic) is regular if every tile has a congruent k-corona, with $k = \nu + 2$, where ν is the number of prime factors in the order of $\mathcal{K}[T_i]$.*

A further refinement of the regularity condition is obtained by investigating the possibilities of assembling an infinite set of tiles T_i, all of which are congruent to a prototile T_0. Let $\mathcal{T}_0$ be a normal space partition where all tiles $T_i \in \mathcal{T}_0$ have congruent first coronas. By this, clearly $\mathcal{T}_0$ is monohedral. The regularity of $\mathcal{T}_0$ depends on how the tiles T_i can be assembled in different ways and this, in turn, depends on the shape of the tiles. In the case of a tile which is shaped such that an infinite collection of congruent tiles can be assembled in only one unique way, then every tile is necessarily surrounded in exactly the same way by all the other tiles. It follows that in this case the tiling is regular. In the case of a tile with a more special shape such that different ways of assembling them are possible which result in non-congruent tilings $\mathcal{T}_i$, then the final tiling may not be regular. Well-known examples include bricks or dominoes which may be assembled in many different ways.

The possibilities of assembling the tiles become more restrictive if it is required that every tile keeps a congruent first corona. Introducing the concept of geometric extension, Engel (1986) has shown that in this case non-congruent tilings may be obtained if the tiles can be assembled such as to compose double layers which may be rotated one against the other. He conjectured:

Conjecture 2.9. A normal space partition $\mathcal{T}_0$ in E^d is isohedral if every tile of it has a congruent second corona.

The requirement of the second corona is an upper bound. In most cases the congruence of the first coronas or even the tiles themselves is sufficient to ensure regularity. This depends on the shape of the tiles and on their possibility to aggregate in different ways. For Dirichlet domain partitions such possibilities are more restricted. In E^2 the congruence of the first coronas is sufficient to prove the regularity. For E^3, however, Engel (1986) gave a list of 11 combinatorial types of Dirichlet domains which necessarily require the congruence of the second coronas in order to be regular.

3. Translation lattices

The periodicity of an ideal crystal structure is proved by Bieberbach's theorem 1.2. The regularity theorems 1.4 and 2.8, and Conjecture 2.9 reveal that periodicity depends on local properties only. Experimental evidence of periodicity in crystals comes from a very large number of X-ray investigations. The periodicity implies that every ideal crystal structure contains a translation lattice. Thus it is convenient to select a lattice basis as a crystal coordinate system with respect to which a crystal structure is described.

3.1. Lattice bases

Let $\Lambda^d := \{\hat{t} \mid \hat{t} = m_1\hat{a}_1 + \cdots + m_d\hat{a}_d,\ m_i \in \mathbb{Z}\}$ be a d-dimensional *translation lattice* in Euclidean space E^d, where $\hat{a}_1, \ldots, \hat{a}_d$ are d linearly independent lattice vectors. Λ^d is an additive group, which is also called a *translation group*.

Definition 3.1. A *lattice basis* of a translation lattice Λ^d consists of d linearly independent lattice vectors $\hat{a}_1, \ldots, \hat{a}_d$ with the property that every lattice vector $\hat{t} \in \Lambda^d$ has an integral representation $\hat{t} = m_1\hat{a}_1 + \cdots + m_d\hat{a}_d$.

The metric of a translation lattice Λ^d with lattice basis $\mathbf{B}^{\mathrm{T}} := (\hat{a}_1, \ldots, \hat{a}_d)$ is described, up to an element of the orthogonal group, through the metric tensor $\boldsymbol{C} := (c_{ij})$, where $c_{ij} := |\hat{a}_i|\,|\hat{a}_j| \cos \alpha_{ij}$ (see section 1.2). The squared volume of the parallelepiped P spanned by the basis vectors is given by the determinant of $\boldsymbol{C}$, $\mathrm{vol}^2(P) = \det(\boldsymbol{C})$. A new set of basis vectors, $\tilde{\mathbf{B}}^{\mathrm{T}} := (\hat{b}_1, \ldots, \hat{b}_d)$, is obtained by a linear transformation of the old basis vectors with transformation matrix $\mathbf{A} := (a_{ij})$ with integral coefficients, $\tilde{\mathbf{B}} = \mathbf{AB}$. Since the volume of the new parallelepiped $\tilde{P}$ has to be equal to the volume of P, it follows that $\mathbf{A}$ must be unimodular, i.e., $\det(\mathbf{A}) = \pm 1$. For $d > 1$ the number of such matrices $\mathbf{A}$ is unlimited, hence an infinite number of lattice bases exist for a translation lattice. Among these lattice bases a suitable one is chosen as a representative. Several reduction procedures to choose a lattice basis will be described in section 4. In reference to the lattice basis $\mathbf{B}$ a lattice vector $\hat{t}$ is written as $\hat{t} = m_1\hat{a}_1 + \cdots + m_d\hat{a}_d$, with integral components m_i. The squared length of $\hat{t}$ becomes

$$|\hat{t}|^2 = \hat{t}^{\mathrm{T}}\boldsymbol{C}\hat{t} = \sum_i \sum_j c_{ij} m_i m_j = f(m_1, \ldots, m_d)\,.$$

The function $f(m_1, \ldots, m_d)$ is called a *positive-definite d-nary quadratic form*. It is non-negative and assumes zero only for $m_1 = \cdots = m_d = 0$. Referred to a new lattice basis $\tilde{\mathbf{B}} = \mathbf{AB}$ the lattice vector $\hat{t}$ is written as $\hat{t}' = \tilde{m}_1\hat{b}_1 + \cdots + \tilde{m}_d\hat{b}_d$. The new components $\tilde{m}_1, \ldots, \tilde{m}_d$ transform with the contragredient matrix $\mathbf{A}^\circ = (\mathbf{A}^{-1})^{\mathrm{T}}$,

$$\hat{t}' = \mathbf{A}^\circ \hat{t}\,.$$

It is required that the length of a lattice vector $\hat{t}$ is conserved under a transformation of the basis vectors,

$$|\hat{t}|^2 = \hat{t}^{\mathrm{T}}\boldsymbol{C}\hat{t} = (\hat{t}')^{\mathrm{T}}\mathbf{A}\boldsymbol{C}\mathbf{A}^{\mathrm{T}}\hat{t}' = \hat{t}'^{\mathrm{T}}\tilde{\boldsymbol{C}}\hat{t}' = |\hat{t}'|^2\,.$$

It follows that the metric tensor $\tilde{\boldsymbol{C}}$, corresponding to the new lattice basis $\tilde{\mathbf{B}}$, becomes $\tilde{\boldsymbol{C}} = \mathbf{A}\boldsymbol{C}\mathbf{A}^{\mathrm{T}}$. This is an equivalence relation and $\tilde{\boldsymbol{C}}$ is said to be equivalent to $\boldsymbol{C}$. The set of all lattice bases of a translation lattice Λ^d corresponds to a set of equivalent metric tensors.

Let $(\mathbf{S}, \hat{s})$ be a symmetry operation of a regular system of points. Under a change of basis vectors with transformation matrix $\mathbf{A}$, the symmetry operation transforms according to

$$(\tilde{\mathbf{S}}, \hat{s}') = (\mathbf{A}^{\circ}\mathbf{S}\mathbf{A}^{\mathrm{T}}, \mathbf{A}^{\circ}\hat{s}) .$$

For a translation lattice Λ^d with lattice basis $\mathbf{B}$, the *reciprocal* or *dual basis* $(\mathbf{B}^*)^{\mathrm{T}} := (a_1^*, \ldots, a_d^*)$ is defined by $\mathbf{B}^* = \boldsymbol{C}^{-1}\mathbf{B}$. The metric tensor $\boldsymbol{C}^*$ of the dual basis is given by $\boldsymbol{C}^* = \boldsymbol{C}^{-1}\boldsymbol{C}\boldsymbol{C}^{\circ} = \boldsymbol{C}^{-1}$. The translation lattice is called *integral* if $\boldsymbol{C}$ has integral coefficients. For an integral lattice with unimodular metric tensor $\boldsymbol{C}$ it holds that $\mathbf{B}^*$ is equivalent to $\mathbf{B}$, and then the lattice is called *self-dual*.

In many cases it is convenient to work with a Cartesian coordinate system instead of using a lattice basis. A set of mutual orthogonal basis vectors $\hat{\mathbf{B}}^{\mathrm{T}} := (\hat{e}_1, \ldots, \hat{e}_d)$ is obtained from Schmidt's orthogonalization procedure as $\hat{\mathbf{B}} = \mathbf{U}\mathbf{B}$, where $\mathbf{U}$ is a lower triangular matrix. It follows that $\boldsymbol{C} = \mathbf{U}^{-1}\mathbf{U}^{0}$. The matrix $\mathbf{U}$ can directly be calculated from the metric tensor $\boldsymbol{C}$ by transforming $\boldsymbol{C}$ into diagonal form.

3.2. Lattice planes

Crystals grow by addition of atoms, or clusters of atoms, to its surface. The correspondence rule predicts that the new material is added in layers along lattice planes having large lattice plane spacings. Thus, the polyhedral shape of crystals is mainly determined by such lattice planes. [See Hartman (1973) for an introduction to crystal growth.]

Let Λ^k be a sublattice of Λ^d with $k < d$, and denote by F^k the affine hull of Λ^k. In what follows we assume that $\Lambda^k = F^k \cap \Lambda^d$. The coset decomposition becomes $\Lambda^d = \Lambda^k \cup \hat{t}_1 + \Lambda^k \cup \cdots$. A coset $\hat{t}_i + \Lambda^k$ is called a *lattice plane*. Particularly, if $k = 1$ then the coset is called a *lattice line*, or if $k = d - 1$ then it is called a *lattice hyperplane*. A lattice hyperplane F is characterized by the facet vector $\hat{f}^*$ perpendicular to it. One of the fundamental laws in crystallography, obtained by Haüy and by Weiss from goniometric measurements of crystals, states that a lattice hyperplane has rational intercepts. The following theorem is known as the law of rationality.

Theorem 3.1. *Every lattice hyperplane F has a facet vector $\hat{f}^*$ perpendicular to it which has integral components if referred to the dual lattice basis, $\hat{f}^* := h_1\hat{\boldsymbol{a}}_1^* + \cdots + h_d\hat{\boldsymbol{a}}_d^*$.*

A proof of this theorem is given in Engel (1986). The integers $(h_1, \ldots, h_d)$, put into parentheses, are called the *Miller indices* of the lattice hyperplane. For a lattice hyperplane F, the cosets $\hat{t}_i + \Lambda^{d-1}$ define an infinite stack of lattice hyperplanes F_i which can be labelled in the following way:

$$\ldots, F_{-2}, F_{-1}, F_0, F_1, F_2, \ldots .$$

For each F_i, the index i is calculated by its representative $\hat{t}_i$, $i = (\hat{t}_i, \hat{f}) = m_1h_1 + \cdots + m_dh_d$. The lattice hyperplane spacing δ between two adjacent lattice hyperplanes is given by

$$\delta := \frac{1}{|\hat{f}^*|} = \frac{1}{\sqrt{(\hat{f}^*)^{\mathrm{T}}\boldsymbol{C}^{-1}\hat{f}^*}} .$$

3.3. Parallelotopes

The metrical and topological properties of a translation lattice Λ^d are best revealed by its parallelotope. If we take the point $\boldsymbol{x}_0$ as an origin, then the action of the translation lattice Λ^d on $\boldsymbol{x}_0$ generates a *point lattice* L^d. With respect to L^d, the Dirichlet domain can be constructed as described in section 2.1. It will be called the *Dirichlet parallelotope* $P(\boldsymbol{x}_0)$, and it has the property that translates of $P(\boldsymbol{x}_0)$ cover the Euclidean space E^d completely.

Definition 3.2. A convex polytope P, congruent copies of which tile the Euclidean space E^d by translation, is called a *parallelotope*.

The Dirichlet parallelotope $P(\boldsymbol{x}_0)$ is a special kind of a parallelotope. Although in this section only Dirichlet parallelotopes will be considered, most results remain valid for general parallelotopes. Voronoï (1908) implicitly conjectured that every parallelotope is combinatorially equivalent to a Dirichlet parallelotope but he proved it for primitive parallelotopes only. Delone (1929) proved the conjecture for $n \leqslant 4$. Following Voronoï, a parallelotope P in a face-to-face tiling of E^d is called *primitive* if in every k-face of P exactly $d-k+1$ adjacent parallelotopes meet. The following theorem was proved by Voronoï (1908).

Theorem 3.2. *A parallelotope P in a face-to-face tiling of E^d is primitive if and only if in every vertex of P $d+1$ contiguous parallelotopes meet.*

The following classification of lattice vectors is useful. If the lattice vector $\hat{t}$ carries $P(\boldsymbol{x}_0)$ onto $P(\boldsymbol{x}_i)$ such that $P(\boldsymbol{x}_0) \cap P(\boldsymbol{x}_i) = F$ is a facet of $P(\boldsymbol{x}_0)$, then $\hat{t}$ is called a *facet vector*. If $P(\boldsymbol{x}_0) \cap P(\boldsymbol{x}_i) \neq \emptyset$, then $\hat{t}$ is called a *corona vector*. We denote by S, F, and C the set of shortest vectors, facet vectors, and corona vectors, respectively. It holds that $\mathrm{S} \subseteq \mathrm{F} \subseteq \mathrm{C}$. In order to specify the facet vectors we introduce, following Voronoï, the sublattice

$$2\Lambda^d := \{\hat{t} \mid \hat{t} = 2m_1\hat{\boldsymbol{a}}_1 + \cdots + 2m_d\hat{\boldsymbol{a}}_d,\ m_i \in \mathbb{Z}\}$$

of index 2^d. The following theorem of Voronoï (1908) completely characterizes the facet vectors of $P(\boldsymbol{x}_0)$.

Theorem 3.3. *The lattice vectors $\pm\hat{t}$ are facet vectors of the Dirichlet parallelotope $P(\boldsymbol{x}_0)$ if and only if $\pm\hat{t}$ are exactly the two shortest vectors in their coset $\hat{t} + 2\Lambda^d$.*

By Theorem 2.1, the number of facets of $P(\boldsymbol{x}_0)$ is bounded. The following theorem of Minkowski holds.

Theorem 3.4. *The number of facet vectors of a parallelotope is at least* $2d$ *and at most* $2(2^d - 1)$.

The minimum is attained for the orthogonal lattice (orthogonal direct sums of 1-dimensional lattices). In this case the number of corona vectors is maximal, $|\mathrm{C}| = 3^d - 1$. The upper bound $2(2^d - 1)$ is reached for the primitive parallelotopes. In this case $\mathrm{F} = \mathrm{C}$, but, beginning with dimensions $d \geqslant 4$, there exist also non-primitive parallelotopes for which $\mathrm{F} = \mathrm{C}$. An upper bound for the number of vertices of primitive parallelotopes was given by Voronoï (1908):

Theorem 3.5. *The number of vertices of a primitive parallelotope is* $N_0 \leqslant (d+1)!$.

Beginning with $d \leqslant 5$ there exist primitive parallelotopes with less than $(d+1)!$ vertices. Of particular importance are the zones and the belts of a parallelotope.

Definition 3.3. A *zone* of a parallelotope P in E^d is a class of parallel 1-faces of P.

Definition 3.4. A *belt* of a parallelotope P in E^d is a class of parallel $(d-2)$-faces of P.

From the properties of a translation lattice it immediately follows that every Dirichlet parallelotope $P(\boldsymbol{x}_0)$ is centrosymmetric and has centrosymmetric facets. Moreover, every belt contains four or six $(d-2)$-faces, and hence, four or six facets. Venkov (1954) and McMullen (1980) independently proved the more general result:

Theorem 3.6. *A convex polytope* P *in* E^d *is a parallelotope if and only if it is centrosymmetric, and every facet of it is centrosymmetry, and if every belt consists of four or six facets.*

Engel (1986) gave an upper bound for the number of belts:

Theorem 3.7. *The number of belts of a parallelotope* P *in* E^d, $d \geqslant 2$ *is at most*

$$N = \sum_{k=2}^{d} \left[K \binom{d}{k} + \sum_{l=2}^{d} \varepsilon \binom{d}{k} \binom{d-k}{l} \right] - \binom{d}{2},$$

where $\varepsilon = 1$ *if* $l < k$, *or* $\varepsilon = \frac{1}{2}$ *if* $l = k$ *and* $\binom{d-k}{l} = 0$ *if* $d - k < l$.

The upper bound is realized for the primitive parallelotopes. The numbers N for dimensions 2 to 9 are given in table 1.

Let Z be a zone of a parallelotope P. A zone Z_i is called *closed* if every 2-face of P that contains an edge $E_j \in Z_i$ contains exactly a second edge $E_k \in Z_i$.

Table 1

d	2	3	4	5	6	7	8	9
N	1	6	25	90	301	966	3025	9330

Otherwise Z_i is called *open*. Starting from P, a reduced parallelotope P' is obtained by reducing a closed zone Z_i in the following way. Let E_1 be a shortest edge of Z_i. The zone Z_i is reduced by shortening every edge $E_j \in Z_i$ by E_1. Through the reduction process the zone Z_i becomes open, or disappears completely, and it cannot be reduced further. It follows that P and P' are of different combinatorial type. However, all properties according to Theorem 3.4 that distinguish a parallelotope are conserved under a zone reduction. In general, P' is no longer a Dirichlet parallelotope, and it is an open problem if there exists a Dirichlet parallelotope $\tilde{P}'$ of the same combinatorial type as P'. If P collapses under a zone reduction, then P' is a $(d-1)$-dimensional parallelotope. In d-dimensional space E^d we call a parallelotope *maximal* if it cannot be obtained through a zone reduction in E^d. Clearly, the primitive parallelotopes are maximal. A parallelotope is called *totally reduced* if all of its zones are open, or it is called *relatively reduced* when no reduction can be made without a collapse of the parallelotope. All parallelotopes that can be reduced to the same totally-reduced parallelotope belong to the same *reduction lattice*. If all zones in a parallelotope P are closed, and if within every single zone the edges are of the same length, then P is a *zonotope*. Since a zone reduction includes metric properties, it occurs that, for dimensions $d \geqslant 5$, parallelotopes of the same combinatorial type may belong to different reduction lattices. Parallelotopes which are zonotopes reduce to a point. Thus they all belong to a single reduction lattice.

3.4. Results on parallelotopes

The combinatorial types of parallelogons in the Euclidean plane E^2, the hexagon and quadrilateral, both being centrosymmetric, are known for pavings since the antique. The hexagon is primitive and thus maximal. From the hexagon the quadrilateral is obtained through a zone reduction, it is relatively reduced. The hexagon and the quadrilateral are both zonotopes.

In E^3 there exist five combinatorial types of parallelohedra, which were determined by Fedorov (1885, 1896). The unique primitive type is the cuboctahedron. The four non-primitive types, the rhomb-dodecahedron, the elongated rhomb-dodecahedron, the hexagonal prism, and the cube, are obtained through successive zone reductions from the cuboctahedron. The cube is relatively reduced. All of them are zonotopes and belong to the same reduction lattice. Fedorov (1896) and Delone (1932) gave a finer classification of the five parallelohedra into 24 homeomeric types. A detailed description of these homeomeric types is also given by Burzlaff and Zimmermann (1977).

In E^4 there exist three types of primitive parallelotopes which were determined by Voronoï (1908). Delone (1929) found another 48 non-primitive parallelotopes and the missing one was discovered by Štogrin (1973). Altogether there exist 52 combinatorial types of parallelotopes. Remarkably, among these exists one non-primitive type with the maximal number of facets given by Minkowski's bound, and with F = C. It is also maximal. Engel (1986) has shown that starting from the four maximal types, exactly 48 further combinatorial types of parallelotopes are obtained by successive zone reductions. They belong to two reduction lattices. The first one contains 17 types which all are zonotopes, with the hypercube being relatively reduced. The second one is determined by the totally reduced type 24-1 having 24 facets and stabilizer of order 1124. It is uniquely characterized by its 3-subordination symbol 8_{24}. For all combinatorial types of parallelotopes P_i, $1 \leqslant i \leqslant 52$, Engel (1992) could find a realization by Dirichlet parallelotope $\tilde{P}_i$ such that the stabilizer $\mathcal{K}[\tilde{P}_i]$ is isomorphic to $\mathrm{Aut}[P_i]$. Štogrin (1974) gave a finer classification of the combinatorial types of parallelotopes into homeomeric types. He found 295 homeomeric types, but this result has not yet been verified.

Only few results are known in E^5. According to Baranovskiĭ and Ryškov (1973) there exist 221 combinatorial types of primitive parallelotopes. However, Engel (1989) has found 223 types, of which 22 have only 708 vertices, as well as 225 non-primitive types having the maximal number of facets given by Minkowski's bound. Clearly these are maximal types. But there exist other non-primitive types having less facets which are maximal. Altogether 79 different reduction lattices were discovered. One of them contains the parallelotopes which are zonotopes (see Engel 1988). There exist 2 combinatorial types of relatively reduced, and 77 types of totally reduced parallelotopes. It is not known if all of them can be realized as Dirichlet parallelotopes.

In higher dimensions Dirichlet parallelotopes were determined by Conway and Sloane (1982) for the self dual-lattices.

3.5. Determination of a lattice basis

Let Λ^d be a translation lattice, and let the lattice vectors be given referred to a Cartesian coordinate system. In spaces of dimension $d \leqslant 3$, a lattice basis of Λ^d is given through d shortest, linearly independent lattice vectors of Λ^d (see section 4). However, in spaces of dimension $d \geqslant 4$ this is no longer true, and the determination of a lattice basis becomes more complicated. Since an infinite family of congruent parallelotopes covers space completely by translation it follows that the set of facet vectors F is a generating set of the translation lattice Λ^d. Moreover, the following theorem holds.

Theorem 3.8. *Every translation lattice contains a basis of facet vectors.*

A proof is given in Engel, Michel and Senechal (1993). For primitive parallelotopes a lattice basis can directly be obtained.

Theorem 3.9. *Let Λ^d be a translation lattice having a primitive Dirichlet parallelotope $P(\boldsymbol{x}_0)$ which has the maximal number of vertices, then for every vertex*

$\boldsymbol{v}_h \in P(\boldsymbol{x}_0)$ *the d facet vectors whose facets are incident to* $\boldsymbol{v}_h$ *determine a lattice basis of* Λ^d.

In the general case, a lattice basis is obtained in the following way. Suppose a translation lattice Λ^d is generated by the set of its facet vectors $\mathrm{F} := \{\hat{\boldsymbol{f}}_1, \ldots, \hat{\boldsymbol{f}}_s\}$, where the lattice vectors $\hat{\boldsymbol{f}}_i \in \mathrm{F}$ are given referred to a Cartesian coordinate system $\hat{\mathbf{B}}$ in E^n, with $n \geq d$. In order to determine a lattice basis of Λ^d, we select from F, as a first step, d linearly independent lattice vectors as basis vectors and denote them by $\tilde{\mathbf{B}}^{\mathrm{T}} := (\hat{\boldsymbol{a}}_1, \ldots, \hat{\boldsymbol{a}}_d)$, with $\tilde{\mathbf{B}} = \mathbf{A}\hat{\mathbf{B}}$. The metric tensor becomes $\boldsymbol{C} = \mathbf{A}\mathbf{A}^{\mathrm{T}}$. Since $\boldsymbol{C}$ is a $d \times d$ matrix, $\mathbf{A}^{\circ} = \boldsymbol{C}^{-1}\mathbf{A}$ can be calculated. Suppose $\tilde{\mathbf{B}}$ generates a sublattice $\tilde{\Lambda}^d \subset \Lambda^d$ of index k. The next step is to replace one basis vector of $\tilde{\mathbf{B}}$ by a suitable lattice vector in a coset of Λ^d with respect to $\tilde{\Lambda}^d$ in order to obtain a new basis which generates a sublattice of lower index. Referred to the basis $\tilde{\mathbf{B}}$, the generating facet vectors $\hat{\boldsymbol{f}}_i$ are given by $\hat{\boldsymbol{f}}'_i = \mathbf{A}^{\circ}\hat{\boldsymbol{f}}_i$, $i = 1, \ldots, s$. Since $\tilde{\Lambda}^d$ is a sublattice of Λ^d, it follows that there exists a lattice vector $\hat{\boldsymbol{t}}$ with a non-integral component τ_j. By adding a suitable multiple of $\hat{\boldsymbol{a}}_j$ to $\hat{\boldsymbol{t}}$, it is always possible to obtain $\tilde{\tau}_j = \tau_j + m$, such that $0 < |\tilde{\tau}_j| \leq \frac{1}{2}$. A new basis $\tilde{\mathbf{B}}$ is obtained by replacing $\hat{\boldsymbol{a}}_j$ by $\hat{\boldsymbol{t}} + m\hat{\boldsymbol{a}}_j$. The new basis $\tilde{\mathbf{B}}$ generates another sublattice of index k', where $k' = k|\tilde{\tau}_j|$. By repeating the second step, a lattice basis $\mathbf{B}$ of Λ^d can be constructed in a finite number of steps. The lattice basis $\mathbf{B}$ thus obtained is not unique, and a reduction procedure as described in section 4 will be used to obtain a reduced form within the infinite class of equivalent forms.

4. Reduction of quadratic forms

Crystals are classified according to their symmetry and their lattice parameters. In order to get a unified description of the known crystal structures, reduction theory is of importance. There is another practical application of reduction theory: in X-ray diffraction of crystals, usually an arbitrary lattice basis is determined by standard procedures. In general, such a lattice basis does not show the symmetry of the translation lattice, and only the reduced basis reveals symmetry completely.

4.1. Definition of the $\mathbb{Z}$*-reduced form*

In section 3.1, the positive-definite d-nary quadratic form was defined to be

$$|\hat{\boldsymbol{t}}|^2 = \hat{\boldsymbol{t}}^{\mathrm{T}}\boldsymbol{C}\hat{\boldsymbol{t}} = \sum_i \sum_j c_{ij} m_i m_j = f(m_1, \ldots, m_d)\,, \quad m_i \in \mathbb{Z}\,,$$

where $\boldsymbol{C} = (c_{ij})$ is the metric tensor of a translation lattice Λ^d. Every quadratic form determines by its coefficients c_{ij}, $i \leq j$, a point in Euclidean space E^N, $N = \frac{1}{2}d(d+1)$. In this space the set of all positive-definite forms defines a convex cone with apex at the origin. Under a change of the lattice basis, $\boldsymbol{C}$ transforms according to $\tilde{\mathbf{C}} = \mathbf{A}\mathbf{C}\mathbf{A}^{\mathrm{T}}$. This is an equivalence relation and $\tilde{\mathbf{C}}$ is said to be equivalent to $\mathbf{C}$. The transformation matrix $\mathbf{A} := (a_{ij})$ is unimodular with integral coefficients. Particularly, the matrix $\mathbf{A}$ is called a *transposition* if for fixed k, l, with $1 \leqslant k \neq l \leqslant d$, it holds that $a_{ij} = \delta_{ij}$ for $ij \neq kl$, and $a_{kl} = \pm 1$. According to Jacobi (1846), a symmetric positive definite matrix $\mathbf{C}$ can always be brought into diagonal form by successive transformations of the form $\tilde{\mathbf{C}} = \mathbf{S}^{\mathrm{T}}\mathbf{C}\mathbf{S}$, where the rotation matrix $\mathbf{S} \in O(d)$ has real coefficients. It corresponds to a reduction of $\mathbf{C}$ over the real numbers $\mathbb{R}$. Similarly, the reduction of a quadratic form over the integers $\mathbb{Z}$ means to bring the metric tensor $\mathbf{C}$ by successive transpositions into a suitable form still to be defined. In general, reduction over $\mathbb{Z}$ is more intriguing than reduction over $\mathbb{R}$.

Definition 4.1. In a class of equivalent positive-definite d-nary quadratic forms, a form is called *$\mathbb{Z}$-reduced* if it assumes successive minima for the d-tuples $(1, 0, \ldots, 0), \ldots, (0, \ldots, 0, 1)$, and, if some of the minima are equal, it fulfils a system of selection rules.

The following theorem is due to Lagrange (1773) for $d = 2$, to Seeber (1831) for $d = 3$, and, in the general case, to Minkowski (1905).

Theorem 4.1. *In each class of equivalent positive-definite d-nary quadratic forms there exists a unique $\mathbb{Z}$-reduction form.*

In the following subsections various reduction schemes, as used in crystallography will be discussed. Alternative $\mathbb{Z}$-reduced forms will be given in sections 4.4 and 4.6.

4.2. The reduction scheme of Lagrange

Lagrange investigated positive-definite binary quadratic forms. His reduction scheme is based on the following theorem.

Theorem 4.2. *In a translation lattice Λ^2, two shortest linearly independent lattice vectors form a lattice basis.*

According to Lagrange (1773), the following reduction scheme determines a complete system of reduction conditions.

Definition 4.2. A positive-definite binary quadratic form is called *$\mathbb{Z}$-reduced according to Lagrange* if the following conditions are fulfilled:

$$0 \leqslant 2c_{12} \leqslant c_{11} \leqslant c_{22} .$$

These reduction conditions determine in each class of equivalent binary quadratic forms a unique $\mathbb{Z}$-reduced form.

4.3. *The reduction scheme of Seeber*

Seeber (1824) was the first to introduce a modern lattice theory of crystals. With his investigation of the positive-definite ternary quadratic forms he hoped to find all possible types of translation lattices in Λ^3 and thus, to describe all crystal structures. Although his aim was too eager at that time, his method of classifying translation lattices has become important today. The reduction scheme of Seeber is based on the following theorem.

Theorem 4.3. *In a translation lattice Λ^3, three shortest linearly independent lattice vectors form a lattice basis.*

For a proof see Engel (1986). The following reduction scheme, first set up by Seeber (1831) in terms of the dual lattice, and put into the present form by Eisenstein (1854), determines a complete system of reduction conditions. The proof that the conditions are complete was worked out in detail by Seeber. A geometric interpretation of Seeber's reduction conditions was given by Dirichlet (1850). Only in 1928, Niggli introduced the reduction theory of Seeber–Eisenstein into crystallography. We note that the angles between the basis vectors can be chosen either all acute or else all obtuse or right. Correspondingly a $\mathbb{Z}$-reduced form is called here acute or obtuse, respectively. (Seeber called it positive or negative, respectively, which however is confusing with the expression positive-definite).

Definition 4.3. A positive-definite ternary quadratic form is called *$\mathbb{Z}$-reduced according to Seeber* (also *Niggli reduced*) if the following conditions are fulfilled:

(1) For acute $\mathbb{Z}$-reduced forms with $c_{12}, c_{13}, c_{23} > 0$.

Main conditions:

$c_{11} \leq c_{22} \leq c_{33}$,
$2c_{12} \leq c_{11}$, $2c_{13} \leq c_{11}$, $2c_{23} \leq c_{22}$.

Auxillary conditions:

$c_{23} \leq c_{13}$ if $c_{11} = c_{22}$,
$c_{13} \leq c_{12}$ if $c_{22} = c_{33}$,
$c_{12} \leq 2c_{13}$ if $2c_{23} = c_{22}$,
$c_{12} \leq 2c_{23}$ if $2c_{13} = c_{11}$,
$c_{13} \leq 2c_{23}$ if $2c_{12} = c_{11}$.

(2) For obtuse $\mathbb{Z}$-reduced forms with $c_{12}, c_{13}, c_{23} \leq 0$.

Main conditions:

$c_{11} \leq c_{22} \leq c_{33}$,
$2|c_{12}| \leq c_{11}$, $2|c_{13}| \leq c_{11}$, $2|c_{23}| \leq c_{22}$,
$2|c_{12} + c_{13} + c_{23}| \leq c_{11} + c_{22}$.

Auxillary conditions:

$|c_{23}| \leq |c_{13}|$ if $c_{11} = c_{22}$,
$|c_{13}| \leq |c_{12}|$ if $c_{22} = c_{33}$,
$|c_{12}| = 0$ if $2|c_{23}| = c_{22}$,

$$|c_{12}| = 0 \text{ if } 2|c_{13}| = c_{11},$$
$$|c_{13}| = 0 \text{ if } 2|c_{12}| = c_{11},$$
$$c_{11} \leqslant |c_{12}| + 2|c_{13}| \text{ if } 2|c_{12} + c_{13} + c_{23}| = c_{11} + c_{22}.$$

This system of reduction conditions determines in each class of equivalent ternary quadratic forms a unique $\mathbb{Z}$-reduced form and hence, up to a rotation, a unique translation lattice. The corresponding lattice basis is called a *reduced lattice basis*. In E^N, $N = \frac{1}{2}n(n+1)$, the reduction conditions determine a number of half spaces which cut out a convex cone.

Seeber described an algorithm to determine a reduced lattice basis using successive transpositions. A similar algorithm was described independently by Křivý and Gruber (1976).

A finer classification of the reduced forms, with respect to the symmetry of the translation lattice, was performed by Niggli (1928). By $\mathscr{K}[\Lambda^d]$ we denote the stabilizer of the translation lattice Λ^d in the orthogonal group $O(d)$. Every symmetry operation $\mathbf{S} \in \mathscr{K}[\Lambda^d]$ leaves the metric tensor $\boldsymbol{C}$ of Λ^d invariant, $\boldsymbol{C} \equiv \mathbf{S}^{\mathrm{T}}\boldsymbol{C}\mathbf{S}$. We say that two reduced quadratic forms, with metric tensor $\boldsymbol{C}$ and $\tilde{\boldsymbol{C}}$, respectively, belong to the same *type of $\mathbb{Z}$-reduced forms* if the corresponding coefficients c_{ij} and $\tilde{c}_{ij}$ fulfil the same reduction conditions. According to Niggli, a homeomeric classification of the $\mathbb{Z}$-reduced forms is obtained as follows.

Definition 4.4. Two positive-definite d-nary quadratic forms, with metric tensors $\boldsymbol{C}$ and $\tilde{\boldsymbol{C}}$, respectively, belong to the same *homeomeric type of $\mathbb{Z}$-reduced form* if

(i) they belong to the same type of $\mathbb{Z}$-reduced form;

(ii) Λ^d and $\tilde{\Lambda}^d$ have equivalent stabilizers $\mathscr{K}[\Lambda^d]$ and $\mathscr{K}[\tilde{\Lambda}^d]$ in $O(d)$, $\mathscr{K}[\tilde{\Lambda}^d] = \varphi\mathscr{K}[\Lambda^d]\varphi^{-1}$;

(iii) corresponding lattice vectors $\hat{\boldsymbol{t}}$ and $\hat{\boldsymbol{t}}'$ have equivalent stabilizers $\mathscr{K}[\hat{\boldsymbol{t}}]$ and $\mathscr{K}[\hat{\boldsymbol{t}}]$ in $\mathscr{K}[\Lambda^d]$ and $\mathscr{K}[\tilde{\Lambda}^d]$, respectively, $\mathscr{K}[\hat{\boldsymbol{t}}'] = \varphi\mathscr{K}[\hat{\boldsymbol{t}}]\varphi^{-1}$.

Niggli (1928) distinguished 44 homeomeric types of $\mathbb{Z}$-reduced forms. A complete list of these 44 types is also given by Burzlaff, Zimmermann, and de Wolff in the International Tables for Crystallography, Hahn (1987), by Engel (1986), and by de Wolff and Gruber (1991).

4.4. *The reduction scheme of Selling*

An alternative reduction scheme in E^3, which however is not conform to our Definition 4.1, was described by Selling (1874). Let $\hat{\boldsymbol{a}}_1, \hat{\boldsymbol{a}}_2, \hat{\boldsymbol{a}}_3$ be a lattice basis of a translation lattice Λ^3. Selling introduced a fourth lattice vector $\hat{\boldsymbol{a}}_4 := -\hat{\boldsymbol{a}}_1 - \hat{\boldsymbol{a}}_2 - \hat{\boldsymbol{a}}_3$. The four lattice vectors define, following Selling, six *homogeneous coefficients* $c_{12}, c_{13}, c_{14}, c_{23}, c_{24}$, and c_{34}.

Definition 4.5. The six homogeneous coefficients are *$\mathbb{Z}$-reduced according to Selling* (also *Delone reduced*) if $\sigma := |\hat{\boldsymbol{a}}_1|^2 + |\hat{\boldsymbol{a}}_2|^2 + |\hat{\boldsymbol{a}}_3|^2 + |\hat{\boldsymbol{a}}_4|^2$ is minimal.

Selling proved that the minimum of σ is obtained if all six homogeneous coefficients are negative and he asserts the existence of such a minimum for each translation lattice. Delone (1932) introduced Selling's reduction into crystallography, and he gave a reduction algorithm, where he identified the six homogeneous coefficients with the edges of a tetrahedron in $\mathbb{R}^3$. This algorithm is also described by Burzlaff and Zimmermann (1977), and Engel (1986). Although the $\mathbb{Z}$-reduced homogeneous coefficients are unique within a class of equivalent forms, it was shown by Patterson and Love (1957) that it is not always possible to obtain a unique lattice basis.

4.5. *The reduction scheme of Minkowski*

In dimensions $d \geq 4$ d shortest linearly independent lattice vectors do not necessarily form a lattice basis, as it was shown by Bieberbach and Schur (1928). Let $\hat{\boldsymbol{a}}_1, \ldots, \hat{\boldsymbol{a}}_{k-1}$ be $k-1$ shortest linearly independent lattice vectors of a translation lattice Λ^d. In general, for $k > 3$, a shortest lattice vector $\hat{\boldsymbol{a}}_k$ which is linearly independent of $\hat{\boldsymbol{a}}_1, \ldots, \hat{\boldsymbol{a}}_{k-1}$ does not necessarily belong to a lattice basis of Λ^d. Minkowski proved the following theorem.

Theorem 4.4. *Let $\hat{\boldsymbol{a}}_1, \ldots, \hat{\boldsymbol{a}}_d$ be a lattice basis of Λ^d. The lattice vector $\hat{\boldsymbol{t}} := m_1\hat{\boldsymbol{a}}_1 + \cdots + m_d\hat{\boldsymbol{a}}_d$ forms together with the basis vectors $\hat{\boldsymbol{a}}_1, \ldots, \hat{\boldsymbol{a}}_{k-1}$, $1 < k \leq d$, a lattice basis of a sublattice Λ^k if the greatest common divisor of $m_k, \ldots, m_d$ is equal to* 1 $[\gcd(m_k, \ldots, m_d) = 1]$.

Based on the above theorem Minkowski gave the following reduction conditions.

Definition 4.6. A positive-definite d-nary quadratic form is *$\mathbb{Z}$-reduced according to Minkowski* if the following conditions are fulfilled:

Main condition:

$c_{kk} \leq f(m_1, \ldots, m_d)$ with $\gcd(m_k, \ldots, m_d) = 1$, $m_i \in \mathbb{Z}$.

Normalizing condition:

$c_{k,k+1} \geq 0$.

This definition contains an unlimited set of reduction conditions. Therefore, the first finiteness theorem of Minkowski is decisive:

Theorem 4.5. *The reduction conditions according to Minkowski all arise from a finite number amongst them.*

Proofs are given by Minkowski (1905), and Bieberbach and Schur (1928). Indeed, suppose that the lattice basis $\mathbf{B}$ is $\mathbb{Z}$-reduced. For $1 < k \leq d$, the first $k-1$ basis vectors $\hat{\boldsymbol{a}}_1, \ldots, \hat{\boldsymbol{a}}_{k-1}$ of $\mathbf{B}$ determine a sublattice Λ^{k-1}. Let P' be its Dirichlet parallelotope. The basis vector $\hat{\boldsymbol{a}}_k$ has an orthogonal projection onto Λ^{k-1} which has to fall into, or on the boundary of P', because otherwise $\hat{\boldsymbol{a}}_k$ would not be relatively shortest. Let $\hat{f} = m_1\hat{a}_1 + \cdots + m_{k-1}\hat{a}_{k-1}$ be a facet vector of P', then the inequality $\hat{f}^{\mathrm{T}}\boldsymbol{C}\hat{a}_k \leqslant \frac{1}{2}\hat{f}^{\mathrm{T}}\boldsymbol{C}\hat{f}$ holds. Since P' has a finite number of facets, a finite set of inequalities determines $\hat{a}_k$. Explicit expressions of the main conditions for $d = 4$ were given by Weber (1962), and van der Waerden and Gross (1968). However, we should note that Minkowski's conditions do not specify a unique $\mathbb{Z}$-reduced form in the special case when some of the minima are equal. In order to make Minkowski's reduction scheme complete a set of selection rules has to be added. A more detailed account of Minkowski's reduction theory is given by Gruber and Lekkerkerker (1987).

4.6. *Optimal bases*

In two- and three-dimensional space two and three, respectively, shortest linearly independent lattice vectors form a lattice basis. They are always facet vectors of the corresponding Dirichlet parallelotope. In higher dimensions, however, not necessarily all basis vectors of a Minkowski reduced form are facet vectors. A better lattice basis is obtained if we look for one with respect to which all lattice vectors of a given length have minimal components.

Definition 4.7. For a translation lattice Λ^d, and a given real value ρ, a lattice basis $\breve{\mathbf{B}}$ with metric tensor $\breve{\boldsymbol{C}} := (\breve{c}_{ij})$ is called *optimal* if the maximum of the sum of the squared components of the lattice vectors $\hat{t}$ of norm $|\hat{t}| \leqslant \rho$ is minimal.

For a translation lattice Λ^d, let $\boldsymbol{C} = (c_{ij})$ be the metric tensor of a Minkowski reduced form, and let $\mathbf{B}$ be the corresponding lattice basis. The metric tensor of the dual basis $\mathbf{B}^*$ is given by $\boldsymbol{C}^{-1}$. We denote by $M(\rho)$ the set of all lattice vectors $\hat{t} \in \Lambda^d$ with $|\hat{t}| \leqslant \rho$. In reference to the lattice basis $\mathbf{B}$, a lattice vector $\hat{t} \in M(\rho)$ has components $m_1, \ldots, m_d$. By $m(\mathbf{B}, \rho)$ we denote the maximum of the $|m_i|$ for all $\hat{t} \in M(\rho)$. In what follows we determine an upper bound for $m(\mathbf{B}, \rho)$. By $\hat{\mathbf{B}} = \mathbf{UB}$ we introduce a Cartesian coordinate system. A lattice vector $\hat{t}$ transforms according to the contragredient matrix, $\hat{u} = \mathbf{U}^{\circ}\hat{t}$. It follows that $\hat{t}^{\mathrm{T}}\hat{t} = \hat{u}^{\mathrm{T}}\mathbf{U}\mathbf{U}^{\mathrm{T}}\hat{u}$. The matrix $\mathbf{Q} := \mathbf{U}\mathbf{U}^{\mathrm{T}}$ is symmetric and positive definite. Let λ_m be a maximal eigenvalue of $\mathbf{Q}$, and let $\hat{x}$ be the corresponding eigenvector of norm $|\hat{x}| = 1$. From $\hat{x}^{\mathrm{T}}\mathbf{Q}\hat{x} = \lambda_m\hat{x}^{\mathrm{T}}\hat{x}$, it follows that $\hat{t}^{\mathrm{T}}\hat{t} \leqslant \lambda_m\rho^2$ for all $\hat{t} \in M(\rho)$. Since $\mathbf{Q} = \mathbf{U}\mathbf{U}^{\mathrm{T}}$ has the same eigenvalues as $\boldsymbol{C}^{-1} = \mathbf{U}^{\mathrm{T}}\mathbf{U}$, the following theorem holds.

Theorem 4.6. *A lattice basis $\breve{\mathbf{B}}$ of translation lattice Λ^d is optimal, if the maximal eigenvalues of the metric tensor $\breve{\boldsymbol{C}}^{-1}$ of the dual $\breve{\mathbf{B}}^*$ is minimal.*

An optimal basis $\mathbf{B}$ can be obtained, starting from a Minkowski reduced basis, by reducing the metric tensor $\boldsymbol{C}^{-1}$ of the dual basis. In general, it is not possible to have $\breve{\boldsymbol{C}}$ and $\breve{\boldsymbol{C}}^{-1}$ simultaneously Minkowski reduced, thus the weaker condition is applied that $\max(c_{ii})$ is of the same order as $\max(c_{ii}^{-1})$. Still, the optimal basis is not yet unique in the special case that several of the coefficients $\breve{c}_{ii}$ are equal, but the possibilities are strongly restricted. From the d eigenvectors $\hat{x}_i$ we get an upper bound of $m(\breve{\mathbf{B}}, \bar{\rho})$ for the set S of shortest vectors, having length $\bar{\rho}$.

Referred to the optimal length basis $\breve{\mathbf{B}}$, the eigenvectors $\hat{x}_i$ become $\hat{v}_i = \mathbf{U}\hat{x}_i$. We denote by μ_m the maximal component of all eigenvectors $\hat{v}_i$. For the shortest vectors it holds that

$$m(\breve{\mathbf{B}}, \bar{\rho}) \leqslant \text{floor}(\bar{\rho}\mu_m + 1) .$$

In Engel, Michel and Senechal (1993) it is proved that $m(\breve{\mathbf{B}}, \sqrt{c_{dd}}) = 1$ for dimensions $d \leqslant 7$, and it is conjectured that $m(\breve{\mathbf{B}}, \sqrt{c_{dd}}) \leqslant \frac{1}{4}d$ for $d \geqslant 8$. These authors also give optimal bases for the self-dual lattices, having no vectors of norm 1, for dimension $d = 24$.

5. Finite groups of symmetry operations

The concept of symmetry was introduced in crystallography at the same time by Weiss (1815) and, in a more definite way, by Haüy (1815). Point groups were the first symmetry laws studied on well shaped crystals. Frankenheim (1826) was pioneering by his systematic derivation of the 32 crystallographic point groups in E^3. Independently, they were redetermined by Hessel (1830), Bravais (1849), who however considered the rotoreflection not to be a crystallographic symmetry operation, and later once more by Gadolin (1869), who, for the first time, gave stereographic projections of the crystallographic point groups. Symmetry groups have become important in crystallography for describing and classifying crystal structures.

5.1. *Crystallographic symmetry operations*

Let $\mathcal{G}$ be a space group in E^d which acts transitively on a regular system of points X. Because of the periodicity of X, it follows that any symmetry operation $(\mathbf{S}, \hat{s}) \in \mathcal{G}$ has to map the translation lattice Λ^d onto itself.

Definition 5.1. A *crystallographic symmetry operation* in E^d is an isometry which brings some translation lattice Λ^d into self-coincidence.

Every translation $(\mathbf{I}, \hat{t})$ of Λ^d is a crystallographic symmetry operation. By conjugation of $(\mathbf{I}, \hat{t})$ in $\mathcal{G}$ we obtain $(\mathbf{S}, \hat{s})(\mathbf{I}, \hat{t})(\mathbf{S}, \hat{s})^{-1} = (\mathbf{I}, \mathbf{S}\hat{t})$. This has to be again a translation of Λ^d, thus the rotation part $\mathbf{S}$ maps the translation lattice onto itself and thereby leaves the origin fixed. A necessary condition for a symmetry operation $(\mathbf{S}, \hat{s})$ to induce an automorphism of Λ^d is that the rotation part $\mathbf{S}$ has an integral representation by a $d \times d$ matrix. Such an integral representation is obtained with respect to any lattice basis. Every rotation $\mathbf{S}$ maps any lattice vector $\hat{t}$ of finite length $|\hat{t}|$ onto another lattice vector of equal length. By condition 1.1(i), only a finite number of lattice vectors of length $|\hat{t}|$ exist within a ball of radius $|\hat{t}|$, hence the order of $\mathbf{S}$ has to be finite. Let k be the smallest integer such that $\mathbf{S}^k = \mathbf{I}$. We calculate

$$(\mathbf{S}, \hat{s})^k = (\mathbf{I}, \mathbf{S}^{k-1}\hat{s} + \cdots + \mathbf{S}\hat{s} + \hat{s}) .$$

The vector $\hat{t} := \mathbf{S}^{k-1}\hat{s} + \cdots + \mathbf{S}\hat{s} + \hat{s}$ is called the *screw vector*. It is again a lattice vector of Λ^d, which poses restrictions on the translation part $\hat{s}$.

Theorem 5.1. *Let* $(\mathbf{S}, \hat{s})$ *be a crystallographic symmetry operation, and let k be the order of* $\mathbf{S}$. *There exists a representation of* $(\mathbf{S}, \hat{s})$ *in* E^d *such that* $\hat{s}$ *has rational components with denominator k.*

The following theorem of Schläfli (1866) helps to understand rotations in E^d.

Theorem 5.2. *Every rotation* $\mathbf{S}$ *in* E^d *can be represented through rotations in a set of mutually orthogonal one- and two-dimensional invariant subspaces.*

A geometric proof of this theorem is given by Buser (1982). The two-dimensional invariant subspaces were called *principal planes* by Schläfli. Referred to a Cartesian coordinate system $\hat{\mathbf{B}}$ assumed within these invariant subspaces, the matrix representation of $\mathbf{S}$ has a block-diagonal structure:

$$\mathbf{S} := \begin{pmatrix} \mathbf{I}_h & & & & \\ & -\mathbf{I}_k & & & \\ & & \mathbf{S}_1 & & \\ & & & \ddots & \\ & & & & \mathbf{S}_m \end{pmatrix},$$

where $\mathbf{I}_h, \mathbf{I}_k$ are identity matrices of dimensions h and k, respectively, and $\mathbf{S}_i$, $i = 1, \ldots, m$, is a proper rotation of the form

$$\mathbf{S}_i := \begin{pmatrix} \cos\varphi_i & -\sin\varphi_i \\ \sin\varphi_i & \cos\varphi_j \end{pmatrix}.$$

Every $\mathbf{S}_i$ represents a rotation in a principal plane π_i through an angle φ_i. If some of the angles φ_i are equal, then the decomposition of E^d into principal planes is not unique. Since the early beginnings of crystallography, it was observed that rotations of order $k = 5$ and $k \geqslant 7$ do not occur in crystals. This phenomenon was known as the *crystallographic restriction*. For the determination of those rotations which leave a translation lattice invariant we use the concept of *$\mathbb{Z}$-irreducibility*. An integral $\mathbb{Z}$-irreducible representation of $\mathbf{S}$ in E^d maps a translation lattice Λ^d onto itself and leaves no proper sublattice of lower rank invariant. The allowed orders of $\mathbb{Z}$-irreducible crystallographic symmetry operations are given by a theorem of Hermann (1949):

Theorem 5.3. *A rotation* $\mathbf{S}$ *of order k with* $\mathbb{Z}$*-irreducible integral representation exists only in a translation lattice* Λ^d *of dimension* $d = \psi(k)$.

The number-theoretic Euler totient function $\psi(k)$ gives the number of integers less than and relatively prime to k. For $k > 2$ Euler's totient function assumes even values only. Let the rotation $\mathbf{S}$ of order k be $\mathbb{Z}$-irreducible in E^{2m}. By this, the rotation angles φ_i in each principal plane π_i are uniquely determined. Since the characteristic polynomial $\det(\mathbf{S} - \lambda \mathbf{I})$ is invariant under a transformation of the basis vectors, it follows that the trace of $\mathbf{S}$, with respect to the Cartesian coordinate system $\hat{\mathbf{B}}$, must be integral:

$$\mathrm{tr}(\mathbf{S}) = 2\cos 2\pi l_1/k + \cdots + 2\cos 2\pi l_m/k = n\,.$$

This is fulfilled if the l_i are less than $\frac{1}{2}k$ and are relatively prime to k. There exist exactly m such integers l_i, which are uniquely determined.

The chirality character of a rotation in E^d is revealed by the following theorem due to Hermann (1949).

Theorem 5.4. *Let the matrix representation of a crystallographic rotation be fully $\mathbb{Z}$-reduced. Then, if the number of one-dimensional submatrices of order* 2 *is even, the rotation is a proper rotation, otherwise it is an improper rotation.*

This allows us to find the allowed crystallographic rotations in E^d. In table 2 each crystallographic rotation is given with respect to its invariant sublattice decomposition in the form $(k_1/\cdots/k_s)$. The numbers k_i give the orders of rotation in every invariant sublattice whose dimension is given by Euler's totient function. Improper rotations are indicated with an "i".

In E^3, as already mentioned, the rotation of order 5 is not a crystallographic rotation. The symbol (4/1) represents a rotation around a fixed axis with rotation angle $\varphi = 90°$. The symbol (4/2)i represents a rotoreflection which is composed of a rotation around a fixed axis with rotation angle $\varphi = 90°$, followed by a reflection in the mirror plane perpendicular to the rotation axis.

It is common in crystallography to visualize symmetry operations by geometric

Table 2
Allowed crystallographic rotations of order k in E^3 and E^4

d	k	Crystallographic rotations
3	1	(1/1/1)
	2	(2/1/1)i (2/2/1) (2/2/2)i
	3	(3/1)
	4	(4/1) (4/2)i
	6	(6/1) (6/2)i (3.2)
4	1	(1/1/1/1)
	2	(2/1/1/1)i (2/2/1/1) (2/2/2/1)i (2/2/2/2)
	3	(3/1/1) (3/3)
	4	(4/1/1) (4/2/1)i (4/2/2) (4/4)
	5	(5)
	6	(3/2/1)i (3/2/2) (6/1/1) (6/2/1)i (6/2/2) (6/3) (6/6)
	8	(8)
	10	(10)
	12	(12) (4/3) (6/4)

models. We can do this in the following way. A rotation $\mathbf{S} \in O(d)$ leaves at least the origin $\mathbf{o}$ fixed. We get the set of all fixed points from the equation $\mathbf{S}\boldsymbol{x} = \boldsymbol{x}$.

Definition 5.2. The subspace $U \subset \mathrm{E}^d$ which remains fixed under the action of the rotation $\mathbf{S}$ is called the *symmetry support* of $\mathbf{S}$.

In E^3, a centre of inversion, a rotation axis, or a mirror plane are examples of symmetry supports. We now consider a general symmetry operation $(\mathbf{S}, \hat{s})$ in E^d. Let $U^{\|}$ be the symmetry support of the rotation part $\mathbf{S}$ and let $U^{\perp}$ be its orthogonal complement in E^d. By $\hat{s}^{\|}$ and $\hat{s}^{\perp}$ we denote the orthogonal projections of $\hat{s}$ into $U^{\|}$ and $U^{\perp}$, respectively. With respect to $U^{\|}$ and $U^{\perp}$ we have the following decomposition of $(\mathbf{S}, \hat{s})$:

$$\begin{pmatrix} \mathbf{I}_h & 0 \\ 0 & \mathbf{S}^{\perp} \end{pmatrix}, \quad \begin{pmatrix} \hat{s}^{\|} \\ \hat{s}^{\perp} \end{pmatrix}.$$

With $\boldsymbol{p}^{\perp}$ we denote the fixed point of $(\mathbf{S}^{\perp}, \hat{s}^{\perp})$ in $U^{\perp}$. As a generalization of Definition 5.2, the coset $\boldsymbol{p}^{\perp} + U^{\|}$ is called the symmetry support of the symmetry operation $(\mathbf{S}, \hat{s})$. If $(\mathbf{S}, \hat{s})$ represents a screw rotation, or a glide reflection, the coset $\boldsymbol{p}^{\perp} + U^{\|}$ remains invariant, whereas for a rotation, or a reflection, it remains pointwise fixed.

5.2. *Crystallographic point groups*

We now use the important concept of a symmetry group as defined in section 1.3.

Definition 5.3. A *crystallographic point group* in E^d is a group of symmetry operations which brings a translation lattice Λ^d into self-coincidence and leaves at least one point fixed.

Symmetry groups are classified into equivalence classes with the following definition.

Definition 5.4. Two point groups $\mathcal{K}$ and $\tilde{\mathcal{K}}$ are *equivalent* and belong to the same *geometric class* if there exists a transformation φ that maps $\mathcal{K}$ onto $\tilde{\mathcal{K}}$, $\tilde{\mathcal{K}} = \varphi\mathcal{K}\varphi^{-1}$.

Synonymous to geometric class also $\mathbb{Q}$-*class* is used. Burckhardt (1966) proved that there exists an orthogonal transformation φ:

Theorem 5.5. *If two point groups $\mathcal{K}$ and $\tilde{\mathcal{K}}$ are equivalent, then they are orthogonal equivalent.*

Point groups which contain only proper symmetry operations are called *rotation groups*, or *point groups of the first kind*. Non-cyclic point groups of the first kind are obtained by combining rotations around different rotation axes. Point groups of the *second kind* contain improper symmetry operations. Every point group of the second kind has a rotation group as a normal subgroup of index 2. They are obtained by adding an improper symmetry operation to the rotation group. Point groups can also be generated by reflections on the walls of a fundamental simplex. A reflection in a hyperplane determines a one-dimensional invariant subspace. It is generated by the *root vector* perpendicular to the hyperplane. For a point group the system of all root vectors is called the *root system*. A natural classification of point groups is obtained by assigning them to root classes.

Definition 5.5. Two point groups belong to the same *root class* if they have congruent root systems.

A point group of maximal order in a root class is called the *holohedry* of that root class. We can visualize point groups in the following way. For each rotation $\mathbf{S}_i$ of a point group $\mathcal{K}$ we determine the symmetry support U_i. Different rotations may have identical symmetry supports, thus we assign a weight to each symmetry support which counts the number k of its occurrence, and we say that U is a *k-fold symmetry support*.

Definition 5.6. The *symmetry scaffolding* of a point group $\mathcal{K}$ is the weighted union of the symmetry supports of all symmetry operations $\mathbf{S} \in \mathcal{K}$.

In crystallography the term *symmetry element* is used for the symmetry scaffolding of a cyclic group. We avoid this term because of its confusion with group elements. We can define the symmetry support of a point group as the intersection of the symmetry supports of all its symmetry operations.

In E^3, the point groups are easily derived. The possibilities to intersect different rotation axes are completely determined by Euler's construction (see, e.g., Buerger 1971). They exactly correspond to the arrangement of the rotation axes in the icosahedron, the octahedron, the tetrahedron, and the k-fold prism. The tetrahedron, the octahedron, and the icosahedron are three among the five convex *regular polyhedra*, or *Platonic solids*. The point groups of the second kind are obtained by adding the reflection in a mirror plane in such a way that the rotation group remains invariant. Geometrically this means that the reflection has to map the symmetry scaffolding of the rotation group onto itself. Thus only a few non-equivalent possibilities exist to add reflections.

In E^4, point groups were first investigated by Goursat (1889) by forming direct products of two- and three-dimensional point groups, and by generating groups by reflections on the walls of a simplex on the sphere S^3. Point groups generated by reflections were investigated by Dyck (1882), Klein (1884), and then mainly by Coxeter (1973). Point groups were also studied using quaternions (see, e.g., Du Val 1964). Engel (1986) derived rotation groups by extending Euler's construction to higher dimensions.

Among all point groups in E^d we have to determine the crystallographic point groups and to classify them into crystal classes. The crystallographic point groups are exactly those which have an integral representation in E^d. By the Jordan–Minkowski–Zassenhaus theorem (1880, 1905, 1938) their number is finite for every d. The actual derivation of the crystal classes in E^4 however, started from the maximal irreducible integral matrix groups determined by Dade (1965) and Ryškov (1972). Plesken and Pohst (1977) determined the maximal irreducible integral matrix groups up to $d = 9$. The known results are summarized in the following theorem.

Theorem 5.6. *The number of geometric crystal classes is* 2 *in* E^1, 10 *in* E^2, 32 *in* E^3, *and* 277 *in* E^4.

The classification of crystal classes is based on the equivalence of their matrix representations in E^d. The point groups can also be classified according to their isomorphism type.

Definition 5.7. Two groups $\mathcal{K}$ and $\tilde{\mathcal{K}}$ are *isomorphic* and belong to the same *isomorphism type* if there is a bijective mapping of $\mathcal{K}$ onto $\tilde{\mathcal{K}}$ which conserves the group composition.

The results of this classification are summarized in the following theorem.

Theorem 5.7. *The number of isomorphism types of crystal classes is* 2 *in* E^1, 9 *in* E^2, 18 *in* E^3 *and* 118 *in* E^4.

Lists of geometric crystal classes and isomorphism types of crystal classes are given in Brown et al. (1978).

5.3. Symmetry of translation lattices

Let Λ^d be a translation lattice in E^d. The full symmetry of Λ^d, which fixes the origin, is completely determined by the stabilizer of its Dirichlet parallelotope.

Definition 5.8. A Bravais point group $\mathcal{B}$ is a *crystallographic point group* which contains all the symmetry operations of a translation lattice that leave the origin fixed.

Similarly as with Definition 5.4, Bravais point groups may be classified into *geometric Bravais classes*. There is a natural classification of the set of all translation lattices according to their Bravais point groups.

Definition 5.9. Two lattices Λ^d and $\tilde{\Lambda}^d$ belong to the same *Bravais system* if they have equivalent geometric Bravais point groups.

Among the crystallographic point groups the Bravais point groups are distinguished to have the following properties given in Theorems 5.8 and 5.9.

Theorem 5.8. *A Bravais point group* $\mathcal{B}$ *contains the central inversion* $-\mathbf{I}$.

Let $\mathbf{S}$ be a symmetry operation of a Bravais point group $\mathcal{B}$ of order $k > 1$. We denote by $U^{\parallel}$ the symmetry support of $\mathbf{S}$ and by $U^{\perp}$ the orthogonal complement of $U^{\parallel}$ in E^d.

Theorem 5.9. *If the Bravais point group* $\mathcal{B}$ *of* Λ^d *has a symmetry operation* $\mathbf{S}$ *of order* $k > 1$, *and if* $U^{\perp}$ *has a decomposition into m mutually orthogonal* $\mathbb{R}$-*irreducible subspaces, then* $\mathcal{B}$ *contains an involution* $\mathbf{Z} := \mathbf{R}_1\mathbf{R}_2 \cdots \mathbf{R}_m$ *as a generating symmetry operation, where* $\mathbf{R}_i$, $i = 1, \ldots, m$, *is a reflection in a hyperplane with corresponding root vector in the principal plane* π_i *which maps* Λ^d *onto itself.*

A proof is given in Engel (1986). The number of geometric Bravais classes is given in the following theorem.

Theorem 5.10. *The number of geometric Bravais classes is* 1 *in* E^1, 4 *in* E^2, 7 *in* E^3, *and* 33 *in* E^4.

Following Burckhardt (1934), a finer classification of the crystallographic point groups is obtained if we take into account the action of $\mathcal{K}$ on Λ^d.

Definition 5.10. Two crystallographic point groups $\mathcal{K}$ and $\tilde{\mathcal{K}}$ with lattices Λ^d and $\tilde{\Lambda}^d$ are *arithmetically equivalent* and belong to the same *arithmetic crystal class* if there exists an affine mapping φ which maps $\mathcal{K}$ onto $\tilde{\mathcal{K}}$ and thereby maps Λ^d onto $\tilde{\Lambda}^d$, $\tilde{\mathcal{K}} = \varphi\mathcal{K}\varphi^{-1}$ and $\tilde{\Lambda}^d = \varphi\Lambda^d$.

Geometrically this means that two point groups which belong to the same geometric crystal class are arithmetically non-equivalent if their symmetry scaffoldings have different orientations with respect to the translation lattice Λ^d. Referred to a lattice basis, $\mathcal{K}$ has an integral representation in E^d. Therefore, if two crystallographic point groups are arithmetically equivalent, then they have $\mathbb{Z}$-conjugate matrix representations in $\mathrm{GL}(d, \mathbb{Z})$, the group of all integral $d \times d$ matrices with determinant ± 1. Thus, synonymous with arithmetic crystal class, $\mathbb{Z}$-*class* is used.

Theorem 5.11. *The number of arithmetic crystal classes is* 2 *in* E^1, 13 *in* E^2, 73 *in* E^3 *and* 710 *in* E^4.

In a similar way as the crystallographic point groups, the Bravais point groups may be classified into *geometric* and *arithmetic Bravais classes*. The arithmetic Bravais classes classify the translation lattices into *Bravais types of lattices*. There exists no straightforward procedure to determine all possible lattices which are invariant under a given geometric Bravais point group $\mathcal{B}$, although for low dimensions they can easily be determined. Let $M := \{\mathbf{S}_1, \ldots, \mathbf{S}_n\}$ be a set of generating symmetry operations of $\mathcal{B}$. By $\tau(\mathbf{S})$ we denote the set of all lattices invariant under $\mathbf{S}$. The set of all lattices invariant under $\mathscr{B}$ is given by

$$\tau(\mathscr{B}) = \bigcap_{\mathbf{S}_i \in M} \tau(\mathbf{S}_i) .$$

Thus it is sufficient to determine for each $\mathbf{S}_i$ the set of all invariant lattices Λ^d. Let $U_1, \ldots, U_m$ be the set of $\mathbb{Z}$-irreducible subspaces of $\mathbf{S}_i$. By $\Lambda_i = \Lambda^d \cap U_i$ we denote the sublattice in U_i. For dimensions up to $d = 21$, every $\mathbb{Z}$-irreducible rotation determines exactly one Bravais type of lattice. This is no longer true in higher dimensions. For the rotation of order 21 which has a $\mathbb{Z}$-irreducible representation in dimension 22 non-equivalent Bravais type of lattices exist. For each sublattice Λ_i a lattice basis is chosen. If these basis vectors, taken for all sublattices, generate the whole lattice Λ^d, then the lattice is called *primitive*, otherwise the lattice Λ^d is called *centred*.

The Bravais types of lattices in E^3 were determined by Frankenheim (1842), who however counted one twice, and Bravais (1850). In E^4 they were determined by Bülow (see Brown et al. 1978) by calculating the subgroups of the maximal integral 4×4 matrix groups. In E^5 Plesken (1981), and in E^6 Plesken and Hanrath (1984) determined the types of Bravais lattices by splitting up the geometric Bravais classes into $\mathbb{Z}$-classes. The results are summarized in the following theorem.

Theorem 5.12. *The number of arithmetic Bravais classes is* 1 *in* E^1, 5 *in* E^2, 14 *in* E^3, 64 *in* E^4, 189 *in* E^5, *and* 826 *in* E^6.

There is a one-to-one relation between the arithmetic Bravais classes and the Bravais types of lattices. By Definitions 5.5 and 5.9, there is a natural classification of point groups into root classes and of translation lattices into Bravais systems [see also Michel and Mozrzymas (1989) for classification of crystallographic groups]. In E^3 there exist 7 root classes which contain crystallographic point groups. By chance the holohedries of these root classes exactly correspond to the Bravais point groups. Because of this correspondence, holohedries are defined in the International Table for Crystallography to be Bravais point groups. As originally proposed by Weiss (1815), in E^3, there is a natural classification of the crystallographic point groups into 7 crystal systems, each being determined by a root class. However, in dimensions $d \geq 4$ there exist root classes which contain crystallographic point groups, but which have a non-crystallographic holohedry. An extension to higher dimensions of the concepts of crystal systems and crystal families was proposed by Neubüser, Plesken and Wondratschek (1981) (see also the International Table for Crystallography, Hahn 1987).

5.4. Crystal forms

Crystallography has its origins in the 17th and 18th century when crystals of well-defined external shape were described. However, the complete understanding of the external shape of crystals was only achieved after the theories of translation lattices and symmetry groups had been developed. According to the correspondence rule the facets of a crystal are parallel to lattice planes (cf. section 3.2). In general, the growth of a facet with facet vectors $\hat{f}^*$ is not the same in the positive or negative direction of $\hat{f}^*$. A facet vector $\hat{f}_0^*$, together with a finite positive real number δ, determines a rational half space H_0. By Theorem 3.1, the facet vector has integral components, if referred to the dual basis $\mathbf{B}^*$, and therefore, the half space is called *rational*. We determine the orbit of $\hat{f}_0^*$ under a crystallographic point group $\mathcal{K}$, and thus obtain a set of symmetry related rational half spaces.

Definition 5.11. A *crystal form* P is the intersection of all rational half spaces H_i which are equivalent to H_0 under a crystallographic point group $\mathcal{K}$.

A form is called *open* or *closed* if P is an unbounded or a bounded set, respectively. By $\mathcal{K}[P]$ we denote the stabilizer of P in $O(d)$. It may be of higher order than the generating crystallographic point group $\mathcal{K}$, $\mathcal{K}[P] \geq \mathcal{K}$. A form is called *characteristic* if $\mathcal{K}[P] = \mathcal{K}$, and *non-characteristic* if $\mathcal{K}[P] > \mathcal{K}$. Let F be a facet (bounded or unbounded) of P and let $\mathcal{K}[F]$ be the stabilizer of F in $\mathcal{K}[P]$. The form is called *general* if $\mathcal{K}[F] = \mathcal{C}_1$ (cyclic group of order 1), or *special* if $\mathcal{K}[F] > \mathcal{C}_1$. We use Definition 2.3 to classify forms into *types of forms*. For a given generating point group $\mathcal{K}$, the sphere S^{d-1} of radius $|\hat{f}^*|$ is divided into fields of existence each corresponding to the same type of crystal form. The dimension of an existence field is equal to the *degree of freedom* for that form.

Theorem 5.13. *The number of types of crystal forms is* 2 *in* E^1, 9 *in* E^2, *and* 47 *in* E^3.

The crystal forms are given in the International Tables for Crystallography, Hahn (1987). Clearly, the number of types of forms increases if we neglect the law of rationality. It was shown by Galiulin (1980) that there exist 56 types of forms for the crystallographic point groups in E^3. The crystal forms are usually not perfectly developed on a crystal, thus the symmetry has to be determined from the distribution of the facet vectors measured on a goniometer. In general, the external shape of a crystal is the union of several crystal forms belonging to the same crystallographic point group $\mathcal{K}$. For a given crystal species, its characteristic combination of crystal forms is called the *habit* of a crystal, it depends on the internal structure of the crystal and on its growth conditions. As a general rule, a crystal will be bounded by its faces of slowest growth. According to Bravais (1851), they correspond to those lattice planes which are most densely packed. Wulff (1908) showed that the most probable habit of a crystal would correspond to the Dirichlet parallelotope of the dual lattice Λ^*.

6. Infinite groups of symmetry operations

Space groups are needed for the description of internal crystal structures. Space groups with proper motions only were determined by Leonhard Sohncke (1879).

But soon it was realized by Pierre Curie (1885) and Leonhard Wulff (1887) that certain crystals can only be described by assuming improper motions. Thus the mathematician Arthur Schoenflies, prompted by Felix Klein, determined the types of space groups in E^3 including improper motions. Independently, they were also determined by the crystallographer Evgraf Stepanovič Fedorov. But only after they had compared their results, the correct number of space group types was established in 1890/91. Space groups are now indispensable in crystallography and solid state physics.

6.1. *Space groups*

Space groups in E^d were defined in Definition 1.4. By Theorem 1.2, they contain d linearly independent translations which generate a free Abelian translation group Λ^d. Let $\mathcal{K}$ be a crystallographic point group which maps Λ^d onto itself. Following Ascher and Janner (1965, 1968), a space group $\mathcal{G}$ is an extension of Λ^d by $\mathcal{K}$. The following two theorems reveal important properties of space groups.

Theorem 6.1. *The translation group Λ^d is normal in $\mathcal{G}$ and is a maximal free Abelian subgroup of $\mathcal{G}$.*

Theorem 6.2. *The quotient group $\mathcal{G}/\Lambda^d$ is isomorphic to a crystallographic point group $\mathcal{K}$.*

Proofs of these two theorems are given by Schwarzenberger (1980). These properties are expressed as a short exact sequence of group homomorphisms:

$$0 \rightarrow \Lambda^d \xrightarrow{\alpha} \mathcal{G} \xrightarrow{\beta} \mathcal{K} \xrightarrow{\gamma} 1 .$$

Let h be the order of $\mathcal{K}$. The coset decomposition of $\mathcal{G}$ with respect to Λ^d becomes

$$\mathcal{G} = \Lambda^d \cup \Lambda^d(\mathbf{S}_2, \hat{s}_2) \cup \cdots \cup \Lambda^d(\mathbf{S}_h, \hat{s}_h) .$$

Since $\mathcal{G}/\Lambda^d \simeq \mathcal{K}$, it follows that the coset multiplication is closed. From this the *Frobenius congruence* is obtained:

$$\mathbf{S}_i\hat{s}_j + \hat{s}_i - \hat{s}_k = \hat{\boldsymbol{t}}_{ij} = f(\mathbf{S}_i, \mathbf{S}_j) .$$

The function $f(\mathbf{S}_i, \mathbf{S}_j)$ is called a *factor* and has values in Λ^d. It is independent of the choice of origin. The normalizing condition is $f(\mathbf{I}, \mathbf{S}_j) = f(\mathbf{S}_i, \mathbf{I}) = 0$. A set of factors $f(\mathbf{S}_i, \mathbf{S}_j)$, for $i, j = 1, \ldots, h$, is called a *normalized 2-cocycle*. It is one solution of the Frobenius congruence. The space group corresponding to the zero solution which has shift vectors $\hat{s}_i = 0$, is called a *symmorphic space group*. There is a one-to-one correspondence between the arithmetic crystal classes and the symmorphic types of space groups. The set of all 2-cocycles is denoted by

$Z^2(\mathcal{K}, \Lambda^d)$. They form an additive Abelian group. The 2-cocycles depend on the choice of coset representatives. If we change the representatives, then a 2-cocycle changes by a 2-coboundary $\mathbf{S}_i\hat{t}_j + \hat{t}_i - \hat{t}_k$. The set of all 2-coboundaries is denoted by $B^2(\mathcal{K}, \Lambda^d)$. It is a subgroup of $Z^2(\mathcal{K}, \Lambda^d)$. Each element of the second cohomology group

$$H^2(\mathcal{K}, \Lambda^d) := Z^2(\mathcal{K}, \Lambda^d)/B^2(\mathcal{K}, \Lambda^d)$$

gives an extension of Λ^d by $\mathcal{K}$. The number of isomorphism classes of space groups belonging to an arithmetic crystal class $(\mathcal{K}, \Lambda^d)$ is equal to the number of orbits of the action of the normalizer $N_{O(d)}(\mathcal{K})$ on $H^2(\mathcal{K}, \Lambda^d)$. Algorithms to derive the space groups were devised by Burckhardt (1934), and Zassenhaus (1948). Schwarzenberger (1980) used the isomorphism between the space-group types of the orthogonal system and certain classes of graphs with d vertices to derive the corresponding space-group types in any dimension d. A related method was applied by Niggli (1949), by using characters. Engel (1986) derived the space groups by orbit splitting. He proved the following theorem.

Theorem 6.3. *If the crystallographic point group $\mathcal{K}$ has a symmetry operation $\mathbf{S}$ which leaves only the origin fixed, then every space group $\mathcal{G}$ derived from any arithmetic crystal class $(\mathcal{K}, \Lambda^d)$ has a finite subgroup which is equivalent to the point group generated by $\mathbf{S}$.*

Burckhardt (1966) proved the following theorem.

Theorem 6.4. *From an arithmetic crystal class $(\mathcal{K}, \Lambda^d)$, where $\mathcal{K}$ is a cyclic, $\mathbb{Z}$-irreducible crystallographic point group, arises only the symmorphic space group.*

Schwarzenberger (1980), and Michel and Mozrzymas (1984) proved the next theorem.

Theorem 6.5. *If the crystallographic point group $\mathcal{K}$ has the central inversion $-\mathbf{I}$, then for any space group $\mathcal{G}$ derived from $(\mathcal{K}, \Lambda^d)$ the shift vectors $\hat{s}$ can be chosen such that $\hat{s} = \frac{1}{2}\hat{t}$, $\hat{t} \in \Lambda^d$.*

The plane-group types in E^2 were determined by Fedorov (1891), and independently again by Pólya (1924). The space-group types in E^3 were determined by Schoenflies (1891), and Fedorov (1891). A detailed description of the space-group types in E^2 and E^3 is given in the International Tables for Crystallography, Hahn (1987). Finally, the space-group types in E^4 were calculated by Brown et al. (1978). The results are summarized in the following theorem.

Theorem 6.6. *The number of space-group types is* 2 *in* E^1, 17 *in* E^2, 219 + 11 *in* E^3, *and* 4783 + 112 *in* E^4.

For many theoretical investigations of symmetry groups, and for a unified description of crystal structures, the normalizer $N_\Gamma(\mathcal{G})$ of a group $\mathcal{G}$ in a group Γ is important:

$$N_\Gamma(\mathcal{G}) := \{\varphi \in \Gamma \mid \mathcal{G} = \varphi \mathcal{G} \varphi^{-1}\} .$$

For most applications the normalizer is taken in the orthogonal group $O(d)$, or in the Euclidean group $E(d)$, or in the affine group $A(d)$. The normalizers $N_{O(3)}(\mathcal{K})$ of the crystallographic point groups were determined by Galiulin (1980). The Euclidean normalizers of the space groups depend on the symmetry of the translation lattice. In order to assign Euclidean normalizers to the space-group types, we need the concept of a *minimal group*, which was introduced by Schwarzenberger (1980). In each space-group type there exists a minimal space group $\mathcal{G}$ which has a lattice of minimal Bravais point group. The normalizers $N_{E(3)}(\mathcal{G})$ of the minimal space groups in E^3 were determined by Hirshfeld (1968). The corresponding affine normalizers $N_{A(3)}(\mathcal{G})$ were determined by Burzlaff and Zimmermann (1980), Billiet, Burzlaff and Zimmermann (1982), and Gubler (1982).

6.2. Crystallographic orbits

An orbit,

$$\mathcal{O}(\mathcal{G}, \boldsymbol{x}_0) := \{\boldsymbol{x}_i \mid \boldsymbol{x}_i = (\mathbf{S}_i, \hat{s}_i)\boldsymbol{x}_0, (\mathbf{S}_i, \hat{s}_i) \in \mathcal{G}\} ,$$

which is generated by a space group $\mathcal{G}$ is called a *crystallographic orbit* because of its discrete nature. A systematology of crystal structures requires a classification of the crystallographic orbits as it was done by Niggli (1919), and Wyckoff (1922).

Definition 6.1. The set of all points $\boldsymbol{p} \in \mathrm{E}^d$ that have conjugate stabilizers $\mathcal{K}[\boldsymbol{p}]$ in $\mathcal{G}$ is called a *Wyckoff position* if conjugation is taken in $\mathcal{G}$, and it is called a *Wyckoff set* if conjugation is taken in the affine normalizer $N_{A(d)}(\mathcal{G})$.

Wyckoff positions are the *strata* of $\mathcal{G}$. They correspond to the symmetry supports of the finite subgroups of $\mathcal{G}$. Wyckoff positions or Wyckoff sets only classify crystallographic orbits of one particular space group. We now consider ordered pairs $(\mathcal{G}, \mathcal{W})$ of a space group $\mathcal{G}$ and a Wyckoff set $\mathcal{W}$.

Definition 6.2. Two ordered pairs $(\mathcal{G}, \mathcal{W})$ and $(\tilde{\mathcal{G}}, \tilde{\mathcal{W}})$ are *equivalent* and belong to the same *type of Wyckoff sets*, if there exists an affine mapping $\varphi \in A(d)$ such that $\tilde{\mathcal{G}} = \varphi \mathcal{G} \varphi^{-1}$ and $\tilde{\mathcal{W}} = \varphi \mathcal{W}$.

The types of Wyckoff sets in E^3 were determined by Koch and Fischer (1975). Every crystallographic orbit corresponds to a regular system of points X. However, the types of Wyckoff sets do not classify regular systems of points. Such a classification is obtained through the stabilizer $\mathcal{S}[X]$ of X in $E(d)$. By Definition

1.4, $\mathcal{S}[X]$ is a space group. A crystallographic orbit $\mathcal{O}(\mathcal{G}, x_0) \equiv X$ is called characteristic or non-characteristic with respect to $\mathcal{G}$ if $\mathcal{S}[X] = \mathcal{G}$ or $\mathcal{S}[X] > \mathcal{G}$, respectively. In E^2 the non-characteristic orbits were determined by Matsumoto and Wondratschek (1987). In E^3 the non-characteristic orbits of the minimal space groups were determined by Engel et al. (1984). Every regular system of points X can be assigned to a stabilizer $\mathcal{S}[X]$ and a unique Wyckoff set $\mathcal{W}$ in $\mathcal{S}[X]$. We obtain a homeomeric classification of all regular systems of points in the following way:

Definition 6.3. Two regular systems of points, X and $\tilde{X}$, belong to the same *orbit type* if:

(i) X and $\tilde{X}$ have affinely equivalent stabilizers $\mathcal{S}[X]$ and $\mathcal{S}[\tilde{X}]$ in $E(d)$, $\mathcal{S}[\tilde{X}] = \varphi\mathcal{S}[X]\varphi^{-1}$;

(ii) corresponding points $\boldsymbol{x}$ and $\tilde{\boldsymbol{x}}$ have equivalent stabilizers $\mathcal{K}[\boldsymbol{x}]$ and $\mathcal{K}[\tilde{\boldsymbol{x}}]$ in $\mathcal{S}[X]$ and $\mathcal{S}[\tilde{X}]$, respectively, $\mathcal{K}[\tilde{\boldsymbol{x}}] = \varphi\mathcal{K}[\boldsymbol{x}]\varphi^{-1}$.

Synonymous with orbit type, *homeomeric type of dot patterns* is used (compare section 1.1). In E^2 only 13 out of 17 plane-groups types, and in E^3 166 + 11 out of 219 + 11 space-group types occur as stabilizers of regular systems of points. In a different way *lattice complexes* were *introduced* by Fischer et al. (1973). Every orbit type corresponds exactly to a lattice complex, but not vice-versa.

Theorem 6.7. *The number of orbit types is* 2 *in* E^1, 30 *in* E^2, *and* 402 + 25 *in* E^3.

A finer classification of crystallographic orbits is obtained by considering their topological properties which are given through their Dirichlet domain partition. In analogy to Definition 2.5 we can define topological types of orbits. Only partial results on this classification are known (see section 2.3). Koch and Fischer (1974) used Dirichlet domains to define fundamental regions for the cubic space groups.

6.3. Packing of balls

Atoms may be considered as small rigid balls. The packing of balls is an important geometrical concept in crystallography to describe the arrangements of atoms within a crystal. Many crystal structures are based on densest packings of balls or ellipsoids.

In E^d, we consider a family of d-dimensional closed balls. If two balls touch each other, then we join their centres with a straight line segment. We define a packing of balls as follows:

Definition 6.4. A family of balls κ is said to form a *packing* into E^d if:

(i) the intersection of the interiors of any two balls of κ is empty;

(ii) there exists for any two balls of κ a chain of joins connecting them.

A packing of balls is called *monospherical* if all of its balls have the same finite radius μ. It is called *regular* or *homogeneous* if there exists a space group $\mathcal{G}$ which acts transitively on the balls. Thus the centres of the balls form an orbit of $\mathcal{G}$. It is called a *lattice packing* if a translation group Λ^d acts transitively on the balls. The set of all centres together with the joins form a *packing graph*.

Definition 6.5. Two packings of balls, κ and $\tilde{\kappa}$, are *isomorphic* and belong to the same *topological type of packing* if they have isomorphic packing graphs.

In dimensions $d > 2$ it may become very difficult to verify the isomorphism of two packing graphs. Even for regular packings of balls, there exists no general algorithm that allows a unique classification of packing graphs. A finer classification of regular packings of balls is obtained if we consider the stabilizer $\mathcal{S}[\kappa]$ of the packing and the way $\mathcal{S}[\kappa]$ acts on the balls. We obtain a homeomeric classification of all regular packings of balls in the following way:

Definition 6.6. Two regular packings of balls, κ and $\tilde{\kappa}$, belong to the same *type of regular packing* if:

(i) κ and $\tilde{\kappa}$ belong to the same topological type of packing;

(ii) κ and $\tilde{\kappa}$ have affinely equivalent stabilizers $\mathcal{S}[\kappa]$ and $\mathcal{S}[\tilde{\kappa}]$ in $E(d)$, $\mathcal{S}[\tilde{\kappa}] = \varphi\mathcal{S}[\kappa]\varphi^{-1}$;

(iii) corresponding balls B and $\tilde{B}$ have equivalent stabilizers $\mathcal{K}[B]$ and $\mathcal{K}[\tilde{B}]$ in $\mathcal{S}[\kappa]$ and $\mathcal{S}[\tilde{\kappa}]$, respectively, $\mathcal{K}[\tilde{B}] = \varphi\mathcal{K}[B]\varphi^{-1}$.

In order to determine the regular packings of balls for a given space group $\mathcal{G}$, Niggli (1927) introduced the symmetry domain of a cyclic subgroup $\mathcal{C} < \mathcal{G}$. Each crystallographic orbit $\mathcal{O}(\mathcal{G}, \boldsymbol{x}_0)$ can be separated into two disjoint subsets, $\mathcal{O}(\mathcal{G}, \boldsymbol{x}_0) = \mathcal{O}(\mathcal{C}, \boldsymbol{x}_0) \cup \mathcal{O}'$. For any $\boldsymbol{x}_0$, there exist two points $\boldsymbol{x}_1 \in \mathcal{O}(\mathcal{C}, \boldsymbol{x}_0)\backslash\boldsymbol{x}_0$ and $\boldsymbol{x}_2 \in \mathcal{O}'$ such that $|\hat{\boldsymbol{x}}_1 - \hat{\boldsymbol{x}}_0|$ and $|\hat{\boldsymbol{x}}_2 - \hat{\boldsymbol{x}}_0|$ are minimal.

Definition 6.7. The *symmetry domain* $\Delta[\mathcal{C}] \subset \mathrm{E}^d$ of a cyclic subgroup $\mathcal{C} < \mathcal{G}$ is the set of all points $\boldsymbol{x}_0$ for which

$$\inf_{\boldsymbol{x}_1 \in \mathcal{O}(\mathcal{C}, \boldsymbol{x}_0)\backslash \boldsymbol{x}_0} |\hat{\boldsymbol{x}}_1 - \hat{\boldsymbol{x}}_0| < \inf_{\boldsymbol{x}_2 \in \mathcal{O}'} |\hat{\boldsymbol{x}}_2 - \hat{\boldsymbol{x}}_0| .$$

For a given space group $\mathcal{G}$, the space E^d is uniquely partitioned into symmetry domains. The boundary region between two adjacent symmetry domains is given by the equal sign in the above inequality. They correspond to surfaces of the second degree. Necessarily, the centres of balls of a regular packing lie on boundary surfaces. In order to obtain a regular packing of balls into E^d, at least d shortest distances $|\hat{\boldsymbol{x}}_1 - \hat{\boldsymbol{x}}_0|$ of equal length must exist. Following Nowacki (1935), it is necessary and sufficient that the corresponding d symmetry operations $(\mathbf{S}_i, \hat{s}_i)$, $i = 1, \ldots, d$, define a generating set of $\mathcal{G}$.

The types of regular packings of circular disks into the plane E^2 were determined by Niggli (1927). His results are summarized in the following theorem.

Theorem 6.8. *There exist* 31 *types of regular packings of disks into the plane* E^2.

Nowacki (1948) investigated regular packings of ellipses into the plane. His list was corrected and extended by Grünbaum and Shephard (1987). The results are summarized in the following theorem.

Theorem 6.9. *There exist* 57 *types of regular packings of ellipses into the plane* E^2.

Densest packings of balls into E^3 are generated by stacking densest packed layers. The numbers of different packing sequences having a given repeat period were determined by Iglesias (1981), and McLarnan (1981). Of particular importance are the *cubic* and the *hexagonal closest* packings. Sinogowitz (1943) found 681 types of regular packings of balls in E^3 generated by space groups of the *triclinic*, *monoclinic*, and *orthorhombic* crystal systems. Fischer (1971, 1973, 1991) determined 593 types of regular packings generated by space groups of the *tetragonal* and *cubic* crystal systems. Densest lattice packings in higher dimensions were investigated by Conway and Sloane (1988) (see also chapter 3.3).

6.4. Colour groups

Since ancient times, colour groups can be observed in decorative art. Speiser (1956) pointed out that in most coloured patterns each colour was fixed by a subgroup of the symmetry group of the uncoloured pattern. In crystallography, the observation of spin moments in ferromagnetic crystals made it necessary to enlarge the concept of symmetry. Together with an isometry we also consider a change of some material property. Symbolically, this is expressed by a change of colours.

In E^d we consider a set M of objects $\mu_i \subset E^d$ such that for every pair $\mu_i, \mu_j \in M$ it holds that $\operatorname{int} \mu_i \cap \operatorname{int} \mu_j = \emptyset$. Let $\mathcal{G}$ be a symmetry group which acts transitively on M. To each symmetry operation $\sigma \in \mathcal{G}$ we associate a permutation $P(\sigma)$ which depends on σ only. We can represent $P(\sigma)$ through a change of colours of the objects μ_i. The ordered pairs $(\sigma, P(\sigma))$ form a group by the direct product with composition law:

$$(\sigma_1, P(\sigma_1)) \circ (\sigma_2, P(\sigma_2)) = (\sigma_1 \sigma_2, P(\sigma_1 \sigma_2)) .$$

Following Speiser (1956) and van der Waerden and Burckhardt (1961), we define a colour group as follows:

Definition 6.8. A group of ordered pairs $(\sigma_i, P(\sigma_i))$, of a symmetry operation $\sigma_i \in \mathcal{G}$ and permutation $P(\sigma_i)$, which acts on a set M is called a *colour group* if there exists a subgroup $\mathcal{G}' < \mathcal{G}$ which acts transitively on the objects of the same colour.

Suppose the subgroup $\mathcal{G}'$ is of index k in $\mathcal{G}$. The left multiplication of the cosets $\rho_i \mathcal{G}'$ with some $\sigma \in \mathcal{G}$ gives a permutation $P(\sigma)$ of the colours of M,

$$P(\sigma) := \begin{pmatrix} \mathcal{G}' & \cdots & \rho_k \mathcal{G}' \\ \sigma \mathcal{G}' & \cdots & \sigma \rho_k \mathcal{G}' \end{pmatrix} .$$

The $P(\sigma)$ give a representation of $\mathscr{G}$ by permutations. The group of permutations is a homeomorphic image of $\mathscr{G}$. Thus, each colour group is determined by an ordered pair $(\mathscr{G}, \mathscr{G}')$ of a group and a subgroup. Following Jarrat and Schwarzenberger (1980), a classification of colour groups is obtained as follows:

Definition 6.9. Two colour groups $(\mathscr{G}, \mathscr{G}')$ and $(\tilde{\mathscr{G}}, \tilde{\mathscr{G}}')$ are *equivalent* and belong to the same *colour-group type* if there exists an affine transformation $\varphi \in A(d)$ which maps $\mathscr{G}$ onto $\tilde{\mathscr{G}}$ and simultaneously maps $\mathscr{G}'$ onto $\tilde{\mathscr{G}}'$, $\tilde{\mathscr{G}} = \varphi\mathscr{G}\varphi^{-1}$ and $\tilde{\mathscr{G}}' = \varphi\mathscr{G}'\varphi^{-1}$.

Harker (1976) determined the number of colour-group types of the crystallographic point groups in E^3. The number of colour-group types of the plane groups were determined, up to subgroups of index 15, by Jarrat and Schwarzenberger (1980), and up to subgroups of index 60, by Wieting (1982). In E^3 the 1191 colour-group types having subgroups of index 2 were determined by Zamorzaev (1953), and Harker (1981) determined 423 colour-group types having subgroups of index 3. Kotzev and Alexandrova (1988) determined the 2571 colour-group types having equitranslational subgroups. Colour groups and colourings of patterns were also investigated by Senechal (1979, 1988) [see also Schwarzenberger (1984) and Grünbaum and Shephard (1987) for a detailed review on colour groups].

Koptsik (1975, 1988) proposed generalized symmetry groups by taking the wreath product of $\mathscr{G}$ and P. He thus abandons the requirement that a colour remains fixed under a subgroup of $\mathscr{G}$.

6.5. Subperiodic groups

In what follows we leave the concept of a discontinuum and consider more general symmetry groups. A symmetry group in E^d is called a *subperiodic group* if it contains less than d linearly independent translations. Following Bohm (1979), we denote subperiodic groups by the symbol $\mathscr{G}^t_{p\cdots v}$, where t is the dimension of a subspace U which is spanned by t linearly independent translations and $p, \ldots, v$ denote the dimensions of symmetry invariant subspaces $U_p, \ldots, U_v$ which have no translations. The subspaces are mutually orthogonal and it holds that $d = t + p + \cdots + v$. The fundamental region of a subperiodic group is unbounded. If any of the integers $p, \ldots, v$ is larger than 1, then rotations of any order may occur. In crystallography, it is common to consider only subperiodic groups which are subgroups of space groups.

The 7 types of border groups $\mathscr{G}^1_1$ are important in ornamental art. They were determined by Pólya (1924), and again by Niggli (1926). Speiser (1956) determined the 31 types of two-sided border groups $\mathscr{G}^1_{1,1}$. Layer groups $\mathscr{G}^2_1$ have applications in ornamental art and in crystallography. The 80 types of layer groups were determined almost at the same time and independently by Hermann (1929), Weber (1929), and Alexander and Herrmann (1929). The 75 types of rod groups $\mathscr{G}^1_2$ were determined by Alexander (1929), and Hermann (1929). Kopský

(1989) constructed three-dimensional reducible space groups from rod and layer groups. Koch and Fischer (1978) determined the lattice complexes of rod and layer groups.

7. Non-regular systems of points

In this section we will consider general (r, R)-systems which fulfil conditions 1.1(i) and (ii) only. Although the set of all (r, R)-systems is very large, rather few is known on its structure. Since research in this field is very recent, this section will have a provisional character.

Solidification of matter not necessarily results in well-built crystals. By its absence of facets, glasses have very early been recognized to be of non-crystalline nature. Glasses are *amorphous*. Between the amorphous and the crystalline state, there exist many intermediate states. Crystal twins have been described since the beginnings of crystallography. Other examples of non-periodic solids are some polytypes (see section 2.4), incommensurately modulated structures (see Janner and Janssen 1980), and helical structures, as they are observed in biological tissues (see Cochran, Crick and Vand 1952). But it was a great surprise when a solid state phase of an alloy of aluminium and manganese was discovered (see Shechtman et al. 1984) which gives a diffraction pattern with bright spots showing icosahedral symmetry, which is forbidden for three-dimensional periodic crystals. For some of these alloys dodecahedrally shaped quasicrystals could be grown. They clearly violate the classical law of rationality [see Steurer (1990) for a recent survey on quasicrystals].

7.1. *Definitions*

Guided by diffraction experiments, the Fourier transform has become important in order to delimit substructures in the set of all (r, R)-systems. In this, however, the main difficulty is to define what does "bright spots" mean. The Fourier transform $T[K]$ of an ideal crystal K, with translation lattice Λ^d, is an infinite sum of weighted Dirac delta functions situated at the nodes of the dual lattice $(\Lambda^*)^d$. In a diffraction pattern of a real crystal with habit H, which we assume to be sufficiently small in size such that absorption is negligible, every Dirac delta function is convoluted with the Fourier transform of the shape function $f(H)$, and it is spread out by some small amount because of the mosaic structure of the crystal. If a weighted Dirac delta function is above a certain threshold, then the result is called a *Bragg peak*. Let $\hat{\boldsymbol{a}}_1, \ldots, \hat{\boldsymbol{a}}_k$ be $k > d$ rationally independent vectors which span E^d. The set $\{\hat{\boldsymbol{t}} \mid \hat{\boldsymbol{t}} = m_1\hat{\boldsymbol{a}}_1 + \cdots + m_k\hat{\boldsymbol{a}}_k,\ m_i \in \mathbb{Z}\}$ is called a $\mathbb{Z}$-module of rank k and dimension d. Following Janner and Janssen (1980) and Katz and Duneau (1987) we define:

Definition 7.1. A solid state phase is called a *quasicrystal* if its Fourier transform shows a finite set of Bragg peaks which are situated at the nodes of a $\mathbb{Z}$-module of rank $k > d$. It is called a *crystal* if $k = d$.

Another definition of quasicrystals is based on properties in direct space. For a set of points X, we denote by $X - X$ its *difference set*, or *Patterson vector set*. It contains for every pair of points $\boldsymbol{x}_i, \boldsymbol{x}_j \in X$ the difference vector $\hat{\boldsymbol{x}}_j - \hat{\boldsymbol{x}}_i$. Following Galiulin (1989) we define:

Definition 7.2. An (r, R)-system X is *quasiregular* if the difference set $X - X$ is also an (r', R')-system.

It seems that Definition 7.2 is more general than 7.1. Tilings with a finite number of different prototiles also have a finite number of different first and second coronas. We can get a rough classification of tilings by assigning to every tile a tiling symbol (N_0, N_1, N_2), where N_i, $i = 0, 1, 2$, gives the number of different i-coronas. Clearly, a regular tiling with only one prototile will have the symbol $(1, 1, 1)$.

7.2. *The projection method*

Much interest has concentrated on the projection method introduced by de Bruijn (1981), and worked out by Duneau and Katz (1985). Let $\mathcal{G}$ be a space group in E^k, and let $U^\perp$ be a d-dimensional subspace with $d < k$. By $U^\|$ we denote the orthogonal complement of $U^\perp$ in E^k. In $U^\|$ we choose a bounded region $W \subset U^\|$ which is called a *window*. By $\mathscr{C} := W \times U^\perp$ we denote the cylinder over W with cylinder axis $U^\perp$. Let $\mathcal{O}(\mathcal{G}, \boldsymbol{x}_0)$ be a crystallographic orbit of $\mathcal{G}$. Every point $\boldsymbol{x}_i \in \mathcal{O}(\mathcal{G}, \boldsymbol{x}_0)$ which lies in $\mathscr{C}$ is projected along $U^\|$ onto the subspace $U^\perp$. The projected points $\tilde{\boldsymbol{x}}_i$ form an (r, R)-system $\tilde{X}$ in $U^\perp$. It is periodic if $U^\perp$ is a rational plane with respect to the translation lattice Λ^k of $\mathcal{G}$, and it is non-periodic if $U^\perp$ is an irrational plane, i.e., $U^\perp$ intersects Λ^k only in the origin $\mathbf{o}$. In order to determine the Fourier transform of $\tilde{X}$, we define a mask $M(\boldsymbol{x})$, with $M(\boldsymbol{x}) = 1$ if $\boldsymbol{x} \in \mathscr{C}$, and $M(\boldsymbol{x}) = 0$ if $\boldsymbol{x} \not\in \mathscr{C}$. To every point of $\mathcal{O}(\mathcal{G}, \boldsymbol{x}_0)$ we assign a weight $\mu \in \mathbb{R}$. The product $\mathcal{O}' := M(\boldsymbol{x}) \cdot \mathcal{O}(\mathcal{G}, \boldsymbol{x}_0)$ determines the part of $\mathcal{O}(\mathcal{G}, \boldsymbol{x}_0)$ within $\mathscr{C}$. By $T[M]$ and $T[\mathcal{O}]$ we denote the Fourier transforms of $M(\boldsymbol{x})$ and weighted $\mathcal{O}(\mathcal{G}, \boldsymbol{x}_0)$, respectively. $T[\mathcal{O}]$ has values in Fourier space $(\mathrm{E}^*)^k$, the space dual to E^k, at discrete points situated at the nodes of the dual translation lattice $(\Lambda^*)^k$. By the convolution theorem, it holds that $T[\mathcal{O}'] = T[M] * T[\mathcal{O}]$. The projection along $U^\|$ corresponds to an intersection of $(\mathrm{E}^*)^k$ with $(U^\perp)^*$. Since $T[M]$ has values in $(U^\|)^*$ it follows that $(U^\perp)^* \cap T[\mathcal{O}']$ results in a dense set of Dirac delta functions with weights $\mu^* \in \mathbb{C}$, which are positioned at the nodes of a $\mathbb{Z}$-module in $(U^\perp)^*$ of rank k and dimension d. Since the window W is bounded, it follows that within any bounded region of $(U^\perp)^*$ there exist only a finite number of Dirac delta functions with an amplitude larger than a given threshold δ.

It was demonstrated by Duneau and Katz (1985), de Bruijn (1986), Elser (1986), Gähler and Rhyner (1986), Conway and Knowles (1986), and Whittaker and Whittaker (1988) that quasiperiodic tilings may be obtained by the projection method.

7.3. *Quasiregular tilings*

During the solidification process, we may assume that single atoms, or small aggregates of atoms, are added to the growing nucleus. As mentioned in section 2.4, solidification of matter may be considered as a process of assembling an infinite collection of tiles. Up to now, no quantitative structure analysis of a quasicrystal has been carried out. However, very impressive images of quasicrystalline alloys were obtained by high resolution transmission electron microscopy by Hiraga et al. (1987), and Nissen (1990). Coordination polyhedra with high coordination numbers in the range 12–16 are frequently observed in alloys. These coordination polyhedra were described by Frank and Kasper (1958). The images provided by Nissen support a random aggregation of different Frank–Kasper polyhedra which are only determined in orientation by rigid links between the polyhedra.

Stimulated by Penrose's non-periodic pattern (1974), much effort was taken to find three-dimensional analogues of it. Mackay (1981) described a tiling due to Ammann which has an acute and an obtuse rhombohedra as prototiles. Kramer (1982) constructed a non-periodic tiling, which preserves the icosahedral symmetry group, using seven different prototiles. Haase et al. (1987) used the projection method to construct zonotopes which are candidates for prototiles of non-periodic tilings. A more complicated tiling having four tetrahedral prototiles was constructed by Danzer (1989).

In view of the electron microscope images, random packings of polyhedra with well-defined rigid links, together with their Fourier transform, should be more investigated. In the plane, a random packing of bricks or dominoes would be an example. Guyot, Audier and Lequette (1986) have investigated a random packing of icosahedra.

Also another model of quasicrystals has been proposed. Sadoc and Mosseri (1985) considered ordered structures in three-dimensional curved space, which is then flattened out by a hierarchical network of disclinations.

Dense non-periodic packings of balls were also investigated. Bagley (1970) described a packing of balls in E^3 with five-fold pseudosymmetry having a packing density of 0.73406. Wills (1990a) described a packing of balls in a cylindrical three-dimensional Penrose tiling with packing density of 0.67798 and in (1990b) he constructed a sequence of sphere packings with densities approaching 0.7341 . . . , which is only 0.85% less than the densest lattice packing.

References

Alexander, E.
[1929] Systematik der eindimensionalen Raumgruppen, *Z. Krist.* **70**, 367–382.

Alexander, E., and K. Herrmann
[1929] Die 80 zweidimensionalen Raumgruppen, *Z. Krist.* **70**, 309–327.

Ascher, E., and A. Janner
[1965] Algebraic aspects of crystallography. I. Space groups as extensions, *Helv. Phys. Acta* **38**, 551–572.
[1968] Algebraic aspects of crystallography. II. Non-primitive translations in space groups, *Commun. Math. Phys.* **11**, 138–167.

Bagley, B.G.
[1970] Five-fold pseudosymmetry, *Nature* **225**, 1040–1041.

Baranovskiĭ, E.P., and S.S. Ryškov
[1973] Primitive five-dimensional parallelohedra, *Soviet Math. Dokl.* **14**, 1391–1395.

Bieberbach, L.
[1910] Über die Bewegungsgruppen der euklidischen Räume, *Math. Ann.* **70**, 297–336.
[1912] Über die Bewegungsgruppen der euklidischen Räume. Zweite Abhandlung, *Math. Ann.* **72**, 400–412.

Bieberbach, L., and I. Schur
[1928] Über die Minkowskische Reduktionstheorie, *Sitzungsber. Kgl. Preuss. Akad. Wiss. Berlin. Phys.-Math. Kl.*, 510–535.

Billiet, Y., H. Burzlaff and H. Zimmermann
[1982] Comment on the paper of H. Burzlaff and H. Zimmermann, 'On the choice of origin in the description of space groups', *Z. Krist.* **160**, 155–157.

Bohm, J.
[1979] Zur Anzahl kristallographischer Symmetriegruppenarten, *Z. Krist.* **150**, 115–123.

Bravais, A.
[1849] Mémoire sur les polyèdres de forme symétriques, *Liouville J. Math. Pures Appl.* **14**, 141–180.
[1850] Mémoire sur les systèmes formés par des points distribués régulièrement sur un plan ou dans l'espace, *J. École Polytechn.* **19**, 1–128.
[1851] Etudes cristallographiques, *J. École Polytechn.* **20**, 101–278.

Brown, H., J. Bülow, R. Neubüser, H. Wondratschek and H. Zassenhaus
[1978] *Crystallographic Groups of Four-dimensional Space* (Wiley, New York).

Buerger, M.J.
[1971] *Introduction to Crystal Geometry* (McGraw-Hill, New York).

Burckhardt, J.J.
[1934] Zur Theorie der Bewegungsgruppen, *Comment. Math. Helv.* **6**, 159–184.
[1966] *Die Bewegungsgruppen der Kristallographie* (Birkhäuser, Basel, 2. Ausgabe).
[1967] Zur Geschichte der Entdeckung der 230 Raumgruppen, *Arch. Hist. Exact. Sci.* **4**, 235–246.
[1984] Die Entdeckung der 32 Kristallklassen durch M.L. Frankenheim im Jahre 1826, *N. Jb. Miner. Mh.* **11**, 481–482.
[1988] *Die Symmetrie der Kristalle* (Birkhäuser, Basel).

Burke, J.G.
[1966] *Origins of the Science of Crystals* (Univ. of California Press, Berkeley).

Burzlaff, H., and H. Zimmermann
[1977] *Symmetrielehre* (Georg Thieme Verlag, Stuttgart).
[1980] On the choice of origin in the description of space groups, *Z. Krist.* **153**, 151–179.

Buser, P.
[1982] A geometric approach to invariant subspaces of orthogonal matrices, *Amer. Math. Monthly* **89**, 751.
[1985] A geometric proof of Bieberbach's theorems on crystallographic groups, *Enseign. Math.* **31**, 137–145.

Cochran, W., F.H.C. Crick and V. Vand
[1952] The structure of synthetic polypeptides. I. The transform of atoms on a helix, *Acta Cryst.* **5**, 581–586.

Conway, J.H., and K.M. Knowles
[1986] Quasiperiodic tiling in two and three dimensions, *J. Phys. A* **19**, 3645–3653.
Conway, J.H., and N.J.A. Sloane
[1982] Voronoi regions of lattices, second moments of polytopes, and quantization, *IEEE Trans. Inform. Theory* **28**, 211–226.
[1988] *Sphere Packings. Lattices and Groups* (Springer, Berlin).
Coxeter, H.S.M.
[1973] *Regular Polytopes* (Collier–Macmillan, Dover, New York, 3rd ed.).
Curie, P.
[1885] Sur les répétitions et la symétrie, *C.R. Acad. Sci. Paris* **100**, 1393–1396.
Dade, E.C.
[1965] The maximal finite groups of 4×4 matrices, *Illinois J. Math.* **9**, 99–122.
Danzer, L.
[1989] Three dimensional analog of the planar Penrose tiling and quasicrystal, *Discrete Math.* **76**, 1–7.
Dauter, Z.
[1984] Domains in pseudocubic $P2_12_12_1$ space group, *Abstracts Paul Niggli Symp., Zürich*, pp. 23–24.
De Bruijn, N.G.
[1981] Algebraic theory of Penrose's non-periodic tilings of the plane, *Nederl. Akad. Wetensch. Proc. Ser. A* **84**, 38–66.
[1986] Quasicrystals and their Fourier transform, *Math. Proc. A* **89**, 123–151.
De Wolff, P.M., and B. Gruber
[1991] Niggli lattice characters: Definition and graphical representations, *Acta Cryst.* **A47**, 29–36.
Delone, B.N.
[1932] Neue Darstellung der geometrischen Kristallographie, *Z. Krist.* **84**, 109–149.
Delone, B.N., and N.N. Sandakova
[1961] Theory of stereohedra (in Russian), *Trudy. Mat. Inst. Akad. Nauk. SSSR* **64**, 28–51.
Delone, B.N., and M.I. Štogrin
[1974] A simplified proof of Schoenflies' theorem (in Russian), *Dokl. Akad. Nauk. SSSR* **219**, 95–98.
Delone, B.N., N.P. Dolbilin, M.I. Štogrin and R.V. Galiulin
[1976] A local criterion for regularity of a system of points, *Soviet Math. Dokl.* **17**, 319–322.
Delone, B.N. (Delaunay)
[1929] Sur la partition régulière de l'espace à 4 dimensions, *Izv. Akad. Nauk. SSSR Otdel. Fiz.-Mat. Nauk*, 79–110; 145–164.
Dirichlet, P.G.L.
[1850] Über die Reduction der positiven quadratischen Formen mit drei unbestimmten ganzen Zahlen, *J. Reine Angew. Math.* **40**, 209–227.
Du Val, P.
[1964] *Homographies, Quaternions and Rotations* (Oxford Univ. Press, London).
Duneau, M., and A. Katz
[1985] Quasiperiodic patterns, *Phys. Rev. Lett.* **54**, 2688–2691.
Dyck, W.
[1882] Gruppentheoretische Studien, *Math. Ann.* **20**, 1–44.
Eisenstein, G.
[1854] Tabelle der reducierten positiven quadratischen Formen nebst den Resultaten neuer Forschungen über diese Formen, insbesondere Berücksichtigung auf ihre tabellarische Berechnung, *J. Reine Angew. Math.* **41**, 141–190.

Elser, V.
[1986] The diffraction pattern of projected structures, *Acta Cryst. A* **42**, 36–43.

Engel, P.
[1981] Über Wirkungsbereichsteilungen von kubischer Symmetrie, *Z. Krist.* **154**, 199–215; II. Die Typen von Wirkungsbereichspolyedern in den symmorphen kubischen Raumgruppen, **157**, 259–275.
[1986] *Geometric Crystallography. An Axiomatic Introduction to Crystallography* (Reidel, Dordrecht).
[1988] Mathematical problems in modern crystallography, *Comput. Math. Appl.* **16**, 425–436.
[1989] New results on parallelotopes in five-dimensional space, to appear.
[1991] On the enumeration of four-dimensional polytopes, *Discrete Math.* **91**, 9–31.
[1992] On the symmetry classification of the four-dimensional parallelohedra, *Z. Krist.* **200**, 199–213.

Engel, P., T. Matsumoto, G. Steinmann and H. Wondratschek
[1984] The non-characteristic orbits of the space groups, *Z. Krist.*, Suppl. Issue **1**, p. 217.

Engel, P., L. Michel and M. Senechal
[1993] *Lattice Geometry*, to appear.

Fedorov, E.S.
[1885] An introduction to the theory of figures (in Russian), *Zap. Imper. St. Peterb. Mineral. Ova* **21**, 1–279.
[1891] The symmetry of regular systems of figures (in Russian), *Zap. Imper. St. Peterb. Mineral. Ova* **28**, 1–146 [ACA Monograph, Vol. 7, Amer. Cryst. Assoc., 1971].
[1896] Theorie der Krystallstructur. I. Mögliche Structurarten, *Z. Krist.* **25**, 113–224 [ACA Monograph, Vol. 7, Amer. Cryst. Assoc., 1971].

Fischer, W.
[1971] Existenzbedingungen homogener Kugelpackungen in Raumgruppen tetragonaler Symmetrie, *Z. Krist.* **133**, 18–42.
[1973] Existenzbedingungen homogener Kugelpackungen zu kubischen Gitterkomplexen mit weniger als drei Freiheitsgraden, *Z. Krist.* **138**, 129–146; **140**, 50–74.
[1991] Tetragonal sphere packings. I. Lattice complexes with zero or one degree of freedom, *Z. Krist.* **194**, 67–85; II. Lattice complexes with two degrees of freedom, **194**, 87–110.

Fischer, W., H. Burzlaff, E. Hellner and J.D.H. Donnay
[1973] *Space Groups and Lattice Complexes,* National Bur. Standards Monograph, Washington DC, Vol. 134.

Frank, F.C., and J.S. Kasper
[1958] Complex alloy structures regarded as sphere packings. I. Definitions and basic principles, *Acta Cryst.* **11**, 184–190; II. Analysis and classification of representative structures, **12**, 483–499.

Frankenheim, M.L.
[1826] Crystallonomische Aufsätze, *ISIS Enzyklopädische Zeitung von Onken* **5**, 497–515.
[1842] System der Crystalle, *Nova Acta Leopoldina* **19**, 471–660.

Friedrich, W., P. Knipping and M. von Laue
[1912] Interferenzerscheinungen bei Röntgenstrahlen, *Sitzungsber. Math.-Phys. Kl. Kgl. Akad. Wiss. München*, 303–322.

Frobenius, F.G.
[1911] Über die unzerlegbaren diskreten Bewegungsgruppen, *Sitzungsber. Kgl. Preuss. Akad. Wiss. Berlin. Phys.-Math. Kl.*, 654–665.

Gadolin, A.V.
[1869] Deduction of all crystallographic systems and their subdivisions by means of a single principle (in Russian), *Zap. Imper. St. Peterb. Mineral. Ova* **4**, 112–200. German translation: 1896, *Ostwald's Klassiker der exakten Wissenschaften,* Vol. 75 (Engelmann, Leipzig).

Gähler, F., and J. Rhyner
[1986] Equivalence of the generalized grid and projection methods for the construction of quasiperiodic tilings, *J. Phys. A* **19**, 267–277.

Galiulin, R.V.
[1980] Classification of directions in crystallographic point groups according to the symmetry principle, *Acta Cryst. A* **36**, 864–869.
[1989] Zonohedral Delone systems, in: *Collected Abstracts. XII Europ. Crystallogr. Meeting, Moscow,* Vol. 1, p. 21.

Goursat, E.
[1889] Sur les substitutions orthogonal et les divisions régulières de l'espace, *Ann. Sci. École Norm. Sup.* **6**, 9–102.

Gruber, P.M., and C.G. Lekkerkerker
[1987] *Geometry of Numbers* (North-Holland, Amsterdam).

Grünbaum, B., and G.C. Shephard
[1987] *Tilings and Patterns* (Freeman, New York).

Gubler, M.
[1982] Normalizer groups and automorphism groups of symmetry groups, *Z. Krist.* **158**, 1–26.

Guyot, P., M. Audier and R. Lequette
[1986] Quasi-crystal and crystal in AlMn and AlMnSi. Model structure of the icosahedral phase, *J. Phys.* **47**, C3-389–404.

Haase, R.W., L. Kramer, P. Kramer and H. Lalvani
[1987] Polyhedra of three quasilattices associated with the icosahedral group, *Acta Cryst. A* **43**, 574–587.

Hahn, T.
[1987] *International Tables for Crystallography,* Vol. A (Reidel, Dordrecht, 2nd ed.).

Harker, D.
[1976] A table of the colored crystallographic and icosahedral point groups, *Acta Cryst. A* **23**, 133–139.
[1981] The three-colored three-dimensional space groups, *Acta Cryst. A* **37**, 286–292.

Hartman, P.
[1973] *Crystal Growth: An Introduction* (North-Holland, Amsterdam).

Haüy, R.-J.
[1815] Mémoire sur une loi de cristallisation, appelée loi de symétrie, *J. Mines* **37**, 215–235, 347–369; **38**, 5–34, 161–174.

Hermann, C.
[1929] Zur systematischen Strukturtheorie III. Ketten-und Netzgruppen, *Z. Krist.* **69**, 250–270.
[1949] Kristallographie in Räumen beliebiger Dimensionszahl. I. Die Symmetrie operationen, *Acta Cryst.* **2**, 139–145.

Hessel, J.F.C.
[1830] Krystallometrie, in: *Gehler's Phys. Wörterbuch,* Bd. 8. (Leipzig).

Hilbert, D.
[1900] Mathematische Probleme, *Nachr. Kgl. Ges. Wiss. Göttingen. Math.-Phys. Kl.*, 253–297.

Hilbert, D., and S. Cohn-Vossen
[1931] *Anschauliche Geometrie* (Wiss. Buchgesellschaft, Darmstadt). English translation: Chelsea, New York, 1953.

Hiraga, K., M. Hirabayashi, A. Inoue and T. Matsumoto
[1987] High-resolution electron microscopy, *Japan J. Appl. Phys.* **27**, L951–L953.

Hirshfeld, F.L.
[1968] Symmetry in the generation of trial structures, *Acta Cryst. A* **24**, 301–311.

Iglesias, J.E.
[1981] A formula for the numbers of closest packings of equal spheres having a given repeat period, *Z. Krist.* **155**, 121–127.

Jacobi, C.G.J.
[1846] Über ein leichtes Verfahren, die in der Theorie der Säkularströmungen vorkommenden Gleichungen numerisch aufzulösen, *J. Reine Angew. Math.* **30**, 51–94.

Janner, A., and T. Janssen
[1980] Symmetry of incommensurate crystal phases. II. Incommensurate basic structures, *Acta Cryst. A* **36**, 408–415.

Jarrat, J.D., and R.L.E. Schwarzenberger
[1980] Coloured plane groups, *Acta Cryst. A* **36**, 884–888.

Jordan, C.
[1869] Mémoire sur les groupes de mouvement, *Ann. Mat. Pura Appl.* **2**, 167–215; 322–345; *Oeuvres* **4**, 231–302.
[1880] Sur la détermination des groupes d'ordre fini contenus dans le group linéaire, *Atti Accad. Nap.* **8** No. 11; *Oeuvres* **2**, 177–218.

Katz, A., and M. Duneau
[1987] Quasiperiodic structures obtained by the projection method, *J. Phys. (Paris) Suppl.* **47**, 103–112.

Klein, F.
[1884] *Vorlesung über das Ikosaeder und die Auflösung der Gleichungen fünften Grades* (Teubner, Leipzig).

Koch, E.
[1973] Die Wirkungsbereichspolyeder und Wirkungsbereichsteilungen zu kubischen Gitterkomplexen mit weniger als drei Freiheitsgraden, *Z. Krist.* **138**, 196–215.

Koch, E., and W. Fischer
[1974] Zur Bestimmung asymmetrischer Einheiten kubischer Raumgruppen mit Hilfe von Wirkungsbereichen, *Acta Cryst. A* **30**, 490–496.
[1975] Automorphismengruppen von Raumgruppen und die Zuordnung von Punktlagen zu Konfigurationslagen, *Acta Cryst. A* **31**, 88–95.

Kopský, V.
[1989] Subperiodic groups as factor groups of reducible space groups, *Acta Cryst. A* **45**, 805–815; Subperiodic classes of reducible space groups, 815–823.

Koptsik, V.A.
[1975] Advances in theoretical crystallography. Color symmetry of defect crystals, *Kristall und Technik* **10**, 231–245.
[1988] Generalized symmetry in crystal physics, *Comput. Math. Appl.* **16**, 407–424.

Kotzev, J.N., and D.A. Alexandrova
[1988] Full tables of colour space groups with colour-reserving translations, *Acta Cryst. A* **44**, 1082–1096.

Kramer, P.
[1982] Non-periodic central space filling with icosahedral symmetry using copies of seven elementary cells, *Acta Cryst. A* **38**, 257–264.

Křivý, S., and B. Gruber
[1976] A unified algorithm for determining the reduced (Niggli) cell, *Acta Cryst. A* **32**, 297–298.

Lagrange, J.L.
[1773] Recherches d'arithmetique, *Oeuvre III* (Gauthier–Villars, Paris) pp. 695–795.

Laves, F.
[1931a] Ebenenteilung in Wirkungsbereiche, *Z. Krist.* **76**, 277–284.
[1931b] Ebenenteilung und Koordinationszahl, *Z. Krist.* **78**, 208–241.

Löckenhoff, H.D., and E. Hellner
[1971] Die Wirkungsbereiche der invarianten kubischen Gitterkomplexe, *N. Jb. Miner. Mh.*, 155–174.

Mackay, A.L.
[1981] De nive quinquangula: on the pentagonal snowflake, *Soviet Phys. Cryst.* **26**, 517–522.

Mani, P.
[1971] Automorphismen von polyedrischen Graphen, *Math. Ann.* **192**, 279–303.

Matsumoto, T., and H. Wondratschek
[1987] The non-characteristic G-orbits of the plane group G, *Z. Krist.* **179**, 7–30.

McLarnan, T.J.
[1981] The numbers of polytopes in close-packings and related structures, *Z. Krist.* **155**, 269–291.

McMullen, P.
[1980] Convex bodies which tile space by translation, *Mathematica* **27**, 113–121; **28**, 191.

Michel, L., and J. Mozrzymas
[1984] Structure des classes de conjugaison d'un group crystallographique, *C.R. Acad. Sci. Paris, Sér. II* **299**, 387–390.
[1989] Les concepts fondamentaux de la crystallographie, *C.R. Acad. Sci. Paris, Sér. II* **308**, 151–158.

Minkowski, H.
[1897] Allgemeine Lehrsätze über die konvexen Polyeder, *Nachr. Akad. Wiss. Göttingen. Math.-Phys. Kl.*, 198–219 [Gesammelte Abhandlungen, Vol. II (Chelsea, New York, 1967) pp. 103–121].
[1905] Diskontinuitätsbereich für arithmetische Äquivalenz, *J. Reine Angew. Math.* **129**, 220–274 [Gesammelte Abhandlungen, Vol. II (Chelsea, New York, 1967) pp. 53–100].

Neubüser, J., W. Plesken and H. Wondratschek
[1981] An emendatory discursion on defining crystal systems, *Match* **10**, 77–96.

Niggli, P.
[1919] *Geometrische Kristallographie des Diskontinuums* (Bornträger, Leipzig).
[1926] Die regelmässige Punktverteilung längs einer Geraden in einer Ebene, *Z. Krist.* **63**, 255–274.
[1927] Die topologische Strukturanalyse. I, *Z. Krist.* **65**, 391–415; II, **68**, 404–466.
[1928] *Handbuch der Experimentalphysik: Kristallographische und theoretische Grundbegriffe* (Akad. Verlagsgemeinschaft, Leipzig).
[1949] Die vollständige und eindeutige Kennzeichnung der Raumsysteme durch Charakterentafeln. I, *Acta Cryst.* **2**, 263–270; II, **3**, 429–433.

Nissen, H.-U.
[1990] in: *Quasicrystals, Networks and Molecules of Fivefold Symmetry*, ed. I. Hargittai (VCH Publishers, New York).

Nowacki, W.
[1935] *Homogene Raumteilungen und Kristallstruktur,* Doctoral thesis, ETH Zürich.
[1948] Über Ellipsenpackungen in der Kristallebene, *Schweiz. Miner. Petr. Mitt.* **28**, 502–508.

Patterson, A.L., and W.E. Love
[1957] Remarks on the Delaunay reduction, *Acta Cryst.* **10**, 111–116.

Penrose, R.
[1974] The role of aesthetics in pure and applied mathematical research, *Bull. Inst. Math. Appl.* **10**, 266–271.

Plesken, W.
[1981] Bravais groups in low dimensions, *Match* **10**, 97–119.

Plesken, W., and W. Hanrath
[1984] The lattices of six-dimensional euclidean space, *Math. Comp.* **43**, 573–587.

Plesken, W., and M. Pohst

[1977] On maximal finite irreducible subgroups of GL(n, Z). I. The five and the seven dimensional case, *Math. Comp.* **31**, 536–551; II. The six dimensional case, **31**, 552–573; III. The nine dimensional case, **34**, 245–258; IV. Remarks on even dimension with applications to $n = 8$, **34**, 259–275; V. The eight dimensional case and complete description of dimensions less than 10, **34**, 277–301.

Pólya, G.

[1924] Über die Analogie der Kristallsymmetrie in der Kristallebene, *Z. Krist.* **60**, 278–282.

Rohn, K.

[1900] Einige Sätze über regelmässige Punktgruppen, *Math. Ann.* **53**, 440–449.

Ryškov, S.S.

[1972] On maximal finite groups of integer $n \times n$ matrices, *Soviet Math. Dokl.* **13**, 720–724.

Sadoc, J.F., and R. Mosseri

[1985] Hierarchical interlaced networks of disclination lines in non-periodic structures, *J. Phys. (Paris)* **46**, 1809–1826.

Šafranovskiĭ, I.I.

[1978] *History of Crystallography: From Ancient Times to the Beginning of the 19th Century* (in Russian) (Nauka, Leningrad).

[1980] *History of Crystallography. 19th Century* (in Russian) (Nauka, Leningrad).

Schläfli, J.

[1866] Über invariante Elemente einer Orthogonalen Substitution, wenn dieselbe als Ausdruck einer Bewegung jeder Gruppe von Werthen der Variablen aus dem identischen Zustande in den transformierten gefasst wird, *J. Reine Angew. Math.* **65**, 185–187.

Schoenflies, A.

[1891] *Krystallsysteme und Krystallstructur* (Teubner, Leipzig). Reprinted: Springer, Berlin, 1984.

Schwarzenberger, R.L.E.

[1980] *N-dimensional Crystallography,* Research Notes in Mathematics, Vol. 41 (Pitman, San Francisco).

[1984] Colour symmetry, *Bull. London Math. Soc.* **16**, 209–240.

Seeber, L.A.

[1824] Versuch einer Erklärung des innern Baues der festen Körper, *Gilbert's Ann. Phys.* **76**, 229–248, 349–372.

[1831] *Untersuchungen über die Eigenschaften der positiven ternären quadratischen Formen* (Freiburg).

Selling, E.

[1874] Über die binären und ternären quadratischen Formen, *J. Reine Angew. Math.* **77**, 143–229.

Senechal, M.

[1979] Color groups, *Discrete Appl. Math.* **1**, 51–73.

[1988] Color Symmetry, *Comput. Math. Appl.* **16**, 545–553.

[1990] Brief history of geometric crystallography, in: *Historical Atlas of Crystallography,* ed. J. Lima-de-Faria (Kluwer Academic Publisher, Dordrecht, Boston, London).

Shechtman, D., I. Blech, D. Gratias and J.W. Cahn

[1984] Metallic phase with long-range orientational order and no translational symmetry, *Phys. Rev. Lett.* **53**, 1951–1953.

Sinogowitz, U.

[1943] Herleitung aller homogenen nicht kubischen Kugelpackungen, *Z. Krist.* **105**, 23–52.

Sohncke, L.

[1874] Die regelmässigen ebenen Punktsysteme von unbegrenzter Ausdehnung, *J. Reine Angew. Math.* **77**, 47–101.

[1879] *Entwicklung einer Theorie der Krystallstruktur* (Teubner, Leipzig).

Speiser, A.

[1956] *Die Theorie der Gruppen endlicher Ordnung,* (Birkhäuser, Basel, 4th. ed.).

Steurer, W.

[1990] The structure of quasicrystals, *Z. Krist.* **190**, 179–234.

Štogrin, M.I.

[1973] Regular Dirichlet–Voronoï partitions for the second triclinic group, *Proc. Steklov. Inst. Math.* **123**.

[1974] On Bravais, Voronoï and Delone classification of four dimensional lattices, *Soviet Math. Dokl.* **15**, 1386–1390.

Van der Waerden, B.L., and J.J. Burckhardt

[1961] Farbgruppen, *Z. Krist.* **115**, 231–234.

Van der Waerden, B.L., and H. Gross

[1968] *Studien zur Theorie der quadratischen Formen* (Birkhäuser, Basel).

Venkov, B.A.

[1954] On a class of euclidean polytopes (in Russian), *Vestnik Leningr. Univ. Ser. Mat. Fiz. Him.* **9**, 11–31.

Verma, A.R., and P. Krishna

[1966] *Polymorphism and Polytypism in crystals* (Wiley, New York).

Voronoï, G.M.

[1908] Nouvelles applications des paramètres continus à la théorie des formes quadratiques. Deuxième Mémoire. Recherches sur les paralléloèdres primitifs, *J. Reine Angew. Math.* **134**, 198–287; **136**, 67–178.

Weber, L.

[1929] Die Symmetrie homogener ebener Punktsysteme, *Z. Krist.* **70**, 309–327.

Weber, O.

[1962] Über die Reduktion und die Darstellung positiver quarternärer quadratischer Formen, *Comment. Math. Helv.* **36**, 181–213.

Weiss, C.S.

[1815] Übersichtliche Darstellung der verschiedenen natürlichen Abteilungen der Krystallsysteme, *Abh. Kgl. Akad. Wiss. Berlin*, 289–336.

Whittaker, E.J.W., and R.M. Whittaker

[1988] Some generalized Penrose patterns from projections of n-dimensional lattices, *Acta Cryst. A* **44**, 105–112.

Wiener, Chr.

[1863] *Grundzüge der Weltordnung* (Leipzig/Heidelberg).

Wieting, T.W.

[1982] *The Mathematical Theory of Chromatic Plane Ornaments* (Marcel Dekker, New York).

Wills, J.M.

[1990a] A quasi-crystalline sphere-packing with unexpected high density, *J. Phys. (France)* **51**, 1061–1064.

[1990b] Dense sphere packings in cylindrical tilings, *Phys. Rev. B* **42**, 4610–4612.

Wulff, G.

[1908] Zur Theorie des Kristallhabitus, *Z. Krist.* **45**, 433–772.

Wulff, L.

[1887] Über die regelmässigen Punktsysteme, *Z. Krist.* **13**, 503–566.

Wyckoff, R.

[1922] *The Analytical Expression of the Results of the Theory of Space Groups* (Washington).

Zamorzaev, A.M.

[1953] Generalization of the space groups (in Russian), Doctoral thesis, Leningrad.

Zassenhaus, H.

[1938] Neuer Beweis der Endlichkeit der Klassenzahl bei unimodularer Äquivalenz endlicher ganzzahliger Substitutionsgruppen, *Abh. Math. Sem. Univ. Hamburg* **12**, 276–288.

[1948] Über einen Algorithmus zur Bestimmung der Raumgruppen, *Comment. Math. Helv.* **21**, 117–141.

Part 4
Analytic Aspects of Convexity

CHAPTER 4.1

Convexity and Differential Geometry

Kurt LEICHTWEIß

Mathematisches Institut B, Universität Stuttgart, Pfaffenwaldring 57, D-70569 Stuttgart, Germany

Contents

HANDBOOK OF CONVEX GEOMETRY
Edited by P.M. Gruber and J.M. Wills

Introduction

In differential geometry intrinsic geometric properties of a differentiable m-dimensional manifold M as well as extrinsic properties with respect to an immersion of M into the d-dimensional Euclidean space $\mathbb{R}^d$ $(m < d)$ are studied. By such an immersion we mean a differentiable map $x : M \to \mathbb{R}^d$, given by $x = x(u^1, \ldots, u^m)$ in the local coordinates $u^1, \ldots, u^m$ of M, where the induced linear map $x_* : T_pM \to \mathbb{R}^d$ is of maximal rank or, equivalently, the partial derivatives $x_i := \partial x / \partial u^i$, $i = 1, \ldots, m$, are linearly independent at each point $p \in M$. The map x defines a differentiable submanifold F of $\mathbb{R}^d$ which is called hypersurface in the special case $m = d - 1$. A differentiable hypersurface F may be oriented if M is covered by coordinate systems with coordinate changes of positive functional determinant everywhere. In this case the normalized vector product of $x_1, \ldots, x_{d-1}$ represents the so-called "unit normal vector" n of F which is defined independently from the chosen coordinate system of M.

The importance of convexity in differential geometry consists in the fact that certain differentiability assumptions easily may be removed if a differentiable hypersurface F of $\mathbb{R}^d$ is "convex", i.e., the point set $x(M)$ lies in the boundary of some suitable closed convex set K of $\mathbb{R}^d$. By this way one obtains convex geometric generalizations of notions and theorems in differential geometry. Conversely, if the boundary of a closed convex set K of $\mathbb{R}^d$ is the point set of some differentiable hypersurface F, then it is not hard to compute convex geometric entities and to prove convex geometric theorems in a differential geometric manner.

In order to get an impression of the influence of the convexity of a differentiable hypersurface F with $x : M \to \mathbb{R}^d$ of differentiability class C_2 we choose the first $d-1$ coordinates $x^1, \ldots, x^{d-1}$ of a suitable Cartesian coordinate system $\{x^1, \ldots, x^d\}$ in $\mathbb{R}^d$ as local parameters of F such that

$$x^d = f(x^1, \ldots, x^{d-1}), \tag{1}$$

with f of C_2 is a local representation of F. Without loss of generality we may assume that $H^-: x^d \geq 0$ is a supporting halfspace of $x(M)$ with the additional property that H: $x^d = 0$ touches $x(M)$ at the origin. hence

$$f(0, \ldots, 0) = f_i(0, \ldots, 0) = 0 \quad \text{and} \quad f(x^1, \ldots, x^{d-1}) \geq 0 \tag{2}$$

($f_i := \partial f / \partial x^i$, $i = 1, \ldots, d-1$). Moreover, f must be a convex function, and it is well known that this is equivalent to the fact that the quadratic form

$$f_{ij}\xi^i\xi^j \quad \left(f_{ij} := \frac{\partial^2 f}{\partial x^i \, \partial x^j}, \quad i, j = 1, \ldots, d-1\right) \tag{3}$$

is positive semidefinite everywhere.

This may be expressed in a more geometrical manner: the curvature k of a plane normal section of F is given as the quotient of the "second fundamental form" of F

$$h_{ij}\xi^i\xi^j \quad \left(h_{ij} := \left\langle \frac{\partial^2 x}{\partial x^i\, \partial x^j}, n \right\rangle = \frac{f_{ij}}{(1+\Sigma_{i=1}^{d-1}(f_i)^2)^{1/2}}\,, \quad i, j = 1, \ldots, d-1\right) \tag{4}$$

and the (positive definite) "first fundamental form" of F

$$g_{ij}\xi^i\xi^j \quad (g_{ij} := \langle x_i, x_j \rangle\,, \quad i, j = 1, \ldots, d-1)\,. \tag{5}$$

If the position of the cutting plane (given by $\xi^1, \ldots, \xi^{d-1}$) varies, k attains $d-1$ stationary values $k_1, \ldots, k_{d-1}$, the so-called "principal curvatures" of F. According to Lagrange's multiplicator method they are the roots of the characteristic equation

$$\det(h_{ij}g^{jk} - z\delta_i^k) = 0\,. \tag{6}$$

Their normalized elementary symmetric functions,

$$\begin{aligned} &H_1 := \frac{1}{d-1}(k_1 + \cdots + k_{d-1})\,, \\ &\vdots \\ &H_{d-1} := k_1 \cdots k_{d-1}\,, \end{aligned} \tag{7}$$

are defined as the "mean curvatures" of F (especially H_{d-1} is called the "Gaussian curvature" of F).

Now the positive semidefiniteness of $f_{ij}\xi^i\xi^j$ is equivalent to the positive semidefiniteness of the second fundamental form of F, eq. (4), or to the nonnegativity of all principal curvatures of F. We comprehend: *locally, a hypersurface F in $\mathbb{R}^d$ is convex iff (after changing the orientation of F eventually) all principal curvatures $k_1, \ldots, k_{d-1}$ of F are nonnegative. In this case also the Gaussian curvature H_{d-1} of F is nonnegative.*

1. Differential geometric characterization of convexity

A basic result on the global characterization of the convexity of a compact hypersurface in $\mathbb{R}^d$ is:

Theorem 1.1 (Hadamard 1897, pp. 352–353). *Let $x: M \to \mathbb{R}^d$ $(d \geq 3)$ be a compact oriented hypersurface F of class C_2 with positive Gaussian curvature*:

$$H_{d-1} > 0\,. \tag{8}$$

Then x is an embedding and $x(M)$ is equal to the boundary of a suitable compact convex body K in $\mathbb{R}^d$.

Proof. We consider the spherical map ν of F, defined by $x(p) \in x(M) \mapsto -n(p) \in S^{d-1}$. Taking into account, besides eqs. (8) and (7), a special point $p_0 \in M$ where a hyperplane of $\mathbb{R}^d$ touches the compact set $x(M)$ in $\mathbb{R}^d$, we can conclude that the principal curvatures $k_1, \ldots, k_{d-1}$ of F at $x(p_0)$ are all of the same sign, say $+1$, after changing the orientation of F eventually. This remains true for all points $p \in M$ [as no principal curvature of F changes its sign because of (8)] and justifies us to say that F is "one-sided curved". Moreover, the induced linear map $\nu_* : x_*(T_pM) \to T_{-n(p)}S^{d-1}$, defined by the Weingarten equations:

$$-n_i(p) = h_{ij}(p)g^{jk}(p)x_k(p) \quad \left(n_i := \frac{\partial n}{\partial u^i}\,, \quad i = 1, \ldots, d-1\right) \tag{9}$$

[compare (4) and (5)], has the eigenvalues $k_1, \ldots, k_{d-1}$ and a positive determinant

$$\det \nu_* = \det(h_{ij}g^{jk}) = k_1 \cdots k_{d-1} > 0 \tag{10}$$

everywhere, see (6). As a consequence the image of the spherical map ν itself is a covering of S^{d-1} without boundary or branch points. Therefore, S^{d-1} being simply connected for $d \geq 3$, this covering must be onefold. So each tangential hyperplane $H(p)$ of F with the "outer" unit normal $-n(p)$ is a supporting hyperplane of F. Namely otherwise, by a compactness argument of $x(M)$, there would exist a supporting hyperplane of F with the same outer normal vector $-n(p)$ and different from $H(p)$, which is impossible. This yields the fact that $x(M)$ coincides with the boundary of the compact convex intersection K of all closed supporting halfspaces $H^-(p)$ of F and that x is injective. □

Remark 1.2. Theorem 1.1 becomes wrong if we have $d = 2$ (regard the slightly disturbed k times traversed unit circle in the plane!).

The assumptions in Theorem 1.1 may be essentially weakened in the sense that the positivity of the principal curvatures k_i of F is replaced by their nonnegativity (i.e., the local convexity of F), with the exception of one point, and the compactness of F by its completeness. This is a trivial consequence of:

Theorem 1.3 (van Heijenoort 1952, p. 241). *If $x : M \to \mathbb{R}^d$ $(d \geq 3)$ is a complete, locally homeomorphic mapping F of the $(d-1)$-dimensional topological manifold M which is*

(i) *locally convex, and*

(ii) *"absolutely convex" at a point p of M (i.e., there is locally a supporting hyperplane of $x(M)$ which intersects $x(M)$ only at p),*

then x is homeomorphic and $x(M)$ is the boundary of a suitable convex body in $\mathbb{R}^d$.

We omit the proof of this theorem – where differential geometry is not involved – but make the following remark.

Remark 1.4. Theorem 1.3 becomes wrong if assumption (ii) is dropped. Counterexample: Cartesian product of $\mathbb{R}^{d-2}$ and the curve in $\mathbb{R}^2$ of Remark 1.2.

Remark 1.5. Before van Heijenoort, Stoker (1936) had just proven that Theorem 1.1 remains valid if $d = 3$ and F is complete instead of being compact (with K not necessarily compact).

The last step of extensions of Theorem 1.1 was done in:

Theorem 1.6 (Sacksteder 1960). *Let $x: M \to \mathbb{R}^d$ $(d \geq 3)$ be a complete hypersurface F of class C_{d+1} with*

(i) *a (positive or negative) semidefinite second fundamental form* (4) *everywhere,*

or equivalently,

(ii) *nonnegative sectional curvature*

$$K(\xi, \eta) := \frac{R_{ijkl}\xi^i\eta^j\xi^k\eta^l}{(g_{ik}g_{jl} - g_{il}g_{jk})\xi^i\eta^j\xi^k\eta^l} \quad (\xi, \eta = \textit{linearly independent}) \tag{11}$$

of the induced Riemannian metric (5) *on M everywhere $[R_{ijkl} = h_{ik}h_{jl} - h_{il}h_{jk}$ = components of the Riemannian curvature tensor of the metric* (5)].

Then either

(a) *$x(M)$ is the boundary of a convex body K in $\mathbb{R}^d$ (if $K(\xi, \eta) \not\equiv 0$) or*

(b) *$x(M)$ is a "hypercylinder" consisting of ∞^1 parallel $(d-2)$-flats in $\mathbb{R}^d$ (if $K(\xi, \eta) \equiv 0$).*

Proof (*Outline*). M is divided into the set M_0 of flat points $(h_{ij}\xi^i\xi^j \equiv 0)$ and its complement M_1 $(h_{ij}\xi^i\xi^j \not\equiv 0)$. Then for each (connected) component C of M_0 the unit normal vector $n|_C$ is constant. Namely, a theorem of Sard (1942, p. 888) says that $\nu(C) = -n(C)$ is a one-dimensional zero set of S^{d-1} and therefore totally disconnected as image of flat points of M where the rank of the induced linear map ν_*, given by (9), is 0. But simultaneously, $\nu(C)$ must be connected as continuous map of the connected set C. Thus $n(C) = n_0$ and $\langle n_0, x\rangle|_C = \text{const}$, whence

$$x(C) \subset H \tag{12}$$

for a suitable hyperplane H in $\mathbb{R}^d$. But eq. (12) and simple convexity arguments, applied to the convex hypersurfaces $x(D_\alpha)$ (D_α = component of M_1) imply that *$x(C)$ is a convex set in $\mathbb{R}^d$*. Conversely, a k-flat $x(L)$ on $x(M)$ has a constant unit normal vector. These two facts, together with the completeness of F and rather complicated topological considerations permit to reduce the proof of Theorem 1.6 to Theorem 1.3 by factoring out eventually a suitable k-flat of $x(M)$ in case (a). □

Remark 1.7. The assumption of completeness of F is essential in Theorem 1.6 as seen in the following example: $d = 3$ and F: $x^3 = (x^1)^3(1 + (x^2)^2)$ with $(x^2)^2 < \frac{1}{2}$. A simple calculation shows that the second fundamental form of F is positive definite for $x^1 > 0$, zero for $x^1 = 0$ and negative definite for $x^1 < 0$, i.e., F cannot be locally convex at the origin. Indeed F is not complete because of the restriction for x^2.

Remark 1.8. Moreover, Theorem 1.6 fails to be true if the assumption of the nonnegativity of the sectional curvature (11) is replaced by the nonnegativity of the Gaussian curvature H_{d-1}, see (7)! Namely, Chern and Lashof (1958, pp. 10–12) gave the following counterexamples: (1) $d \geq 4$ even, $M = S^1 \times S^{d-2}$, $x(M)$: $(\sqrt{(x^1)^2 + \cdots + (x^{d-1})^2} - 2)^2 + (x^d)^2 = 1$; (2) $d \geq 5$ odd, $M = S^2 \times S^{d-3}$, $x(M)$: $(\sqrt{(x^1)^2 + \cdots + (x^{d-2})^2} - 2)^2 + (x^{d-1})^2 + (x^d)^2 = 1$.

Remark 1.9. Before Sacksteder, Chern and Lashof (1958, p.6) (see also Voss 1960, p. 125) gave proofs of Theorem 1.6 in the special case $d = 3$ and $F =$ compact.

At the end of this section we would like to mention that there is also a possibility to characterize the convexity of an m-dimensional submanifold F with $x : M \to \mathbb{R}^d$ $(0 < m < d)$, regarded as hypersurface in its affine hull aff$(x(M))$, in a differential geometric manner. This was first done by Fenchel (1929) for $m = 1$ and $d = 3$, and, for general m and d, in the following theorem.

Theorem 1.10 (Chern and Lashof 1957, p. 307). *We suppose that $x : M \to \mathbb{R}^d$ is a compact oriented m-dimensional submanifold F of class C_∞ ith the generalized spherical map*

$$\nu : (x(p), n(p)) \mapsto -n(p) := -\sum_{\alpha=m+1}^{d} \lambda_\alpha \overset{\alpha}{n}(p)$$

$$\left(\overset{m+1}{n}, \ldots, \overset{d}{n} = \textit{orthonormal normal vector fields of } F, \sum_{\alpha=m+1}^{d} (\lambda_\alpha)^2 = 1\right)$$

of the unit normal bundle B_ν of F into the unit sphere S^{d-1} about the origin. Then

(a) *The induced linear map $\nu_* : x_*(T_pM) \oplus T_{-n(p)}S^{d-m-1} \to T_{-n(p)}S^{d-1}$, given by*

$$-\overset{\alpha}{n}_i(p) = \overset{\alpha}{h}_{ij}(p) g^{jk}(p) x_k(p) + \sum_{\beta=m+1}^{d} \overset{\alpha\beta}{t}_i(p) \overset{\beta}{n}(p),$$

$$i = 1, \ldots, m, \ \alpha = m+1, \ldots, d,$$

$$\overset{\alpha}{h}_{ij} := \left\langle \frac{\partial^2 x}{\partial u^i \, \partial u^j}, \overset{\alpha}{n} \right\rangle, \quad \overset{\alpha\beta}{t}_i := -\left\langle \overset{\alpha}{n}_i, \overset{\beta}{n} \right\rangle = -\overset{\beta\alpha}{t}_i, \tag{13}$$

[*compare eq.* (9)] *has the determinant*

$$det\ \nu_* := \det\Big(\sum_{\alpha=m+1}^{d} \lambda_\alpha \overset{\alpha}{h}_{ij} g^{jk} \Big) \tag{14}$$

[*compare eq.* (10)], *which is the ratio of the volume element* $\mathrm{d}\omega^{d-1}$ *of* S^{d-1} *and the volume element* $\mathrm{d}\omega^{d-m-1} \wedge \mathrm{d}F$ *of* B_ν *relative to the map* ν_* ($\mathrm{d}\omega^{d-m-1}$ = *volume element of the sphere* S^{d-m-1} *of the unit normal vectors of* F *at a fixed point,* $\mathrm{d}F$ = *volume element of* F),

(b) *If* $G(p, n(p)) := \det \nu_*(p)$ *is the so-called "Lipschitz–Killing curvature" of* F *at* $(p, n(p))$, *and*

$$K^*(p) := \int_{S^{d-m-1}(p)} |G(p, n(P))| \, \mathrm{d}\omega^{d-m-1}, \tag{15}$$

then the "total absolute curvature"

$$\int_M K^* \, \mathrm{d}F \tag{16}$$

of F *attains its minimal value* $2d\kappa_d$ (κ_d = *volume of the d-dimensional unit ball*) *iff* $x(M)$ *is the boundary of a compact convex body* K *in a subspace* $\mathbb{R}^{m+1}$ *of* $\mathbb{R}^d$.

The proof of Theorem 1.10(a) is obvious, and the proof of 1.10(b) needs a detailed study of the covering of S^{d-1} by the endpoints of the unit normal vectors of F. It is highly connected with integral geometry and with Morse theory for the nondegenerate critical points of differentiable functions. Finally it is noteworthy that:

Corollary 1.11 (of Theorem 1.10). *A compact and (suitably) oriented hypersurface* F *of class* C_∞ *in* $\mathbb{R}^d$ *bounds a compact convex body in* $\mathbb{R}^d$ *iff*

$$H_{d-1} \geq 0 \tag{17}$$

and

$$\deg \nu = +1 \,. \tag{18}$$

This follows easily from the fact that (17) and (18) imply $K^* = 2H_{d-1}$ [see eqs. (15), (14), (10) and (17)] and therefore

$$\int_M K^* \, \mathrm{d}F = 2 \int_M H_{d-1} \, \mathrm{d}F = 2(\deg \nu) d\kappa_d = 2d\kappa_d \,.$$

Corollary 1.11 fails to be true if the assumption (18) is omitted (see the counterexample in Remark 1.8).

2. Elementary symmetric functions of principal curvatures respectively principal radii of curvature at Euler points

We return now to the mean curvatures (7) of a convex hypersurface $F: M \to \mathbb{R}^d$ of class C_2. They may be geometrically characterized in the following manner: Steiner's formula of the volume for the shell U_λ, bounded by F: $x = x(u^1, \ldots, u^{d-1})$ and its outer parallel hypersurface F_λ: $x = x(u^1, \ldots, u^{d-1}) - \lambda n(u^1, \ldots, u^{d-1})$ ($\lambda = \text{const} > 0$) indicates that we have by means of eqs. (9), (6) and (7)

$$
\begin{aligned}
(d-1)!V(U_\lambda) &= \int_0^\lambda \Big[\int_M (\mathrm{d}x - \mu\, \mathrm{d}n, \underbrace{\ldots}_{d-1}, \mathrm{d}x - \mu\, \mathrm{d}n, n) \Big] \mathrm{d}\mu \\
&= \int_0^\lambda \Big[\int_M \sum_{\nu=0}^{d-1} \mu^{d-1-\nu} \binom{d-1}{\nu} (-\mathrm{d}n, \underbrace{\ldots}_{d-1-\nu}, -\mathrm{d}n, \mathrm{d}x, \underbrace{\ldots}_{\nu}, \mathrm{d}x, n) \Big] \mathrm{d}\mu \\
&= \sum_{\nu=0}^{d-1} \binom{d-1}{\nu} \frac{1}{d-\nu} \lambda^{d-\nu} \int_M H_{d-1-\nu}(\mathrm{d}x, \ldots, \mathrm{d}x, n) \\
&= (d-1)! \frac{1}{d} \sum_{\nu=0}^{d-1} \binom{d}{\nu} \lambda^{d-\nu} \int_M H_{d-1-\nu}\, \mathrm{d}F\,. \qquad (19)
\end{aligned}
$$

Here $\mathrm{d}x := \Sigma_{i=1}^{d-1} (\partial x / \partial u^i)\, \mathrm{d}u^i$, $\mathrm{d}n := \Sigma_{i=1}^{d-1} (\partial n / \partial u^i)\, \mathrm{d}u^i$, and the determinants with differential form valued column vectors have to be computed as in Leichtweiss (1973). Equation (19) establishes that *the mean curvature $H_{d-1-\nu}$ is the density of the νth "curvature measure" in the sense of Federer of a convex body bounded by F* ($\nu = 0, \ldots, d-1$, $H_0 := 1$; see chapter 1.8). In the case of an oriented and compact M with F of class C_3 these mean curvatures are related by Minkowski's integral formulas

$$
\begin{aligned}
\int_M H_{\nu-1}\, \mathrm{d}F &= \frac{1}{(d-1)!} \int_M (-\mathrm{d}n, \underbrace{\ldots}_{\nu-1}, -\mathrm{d}n, \mathrm{d}x, \underbrace{\ldots}_{d-\nu}, \mathrm{d}x, n) \\
&= \frac{1}{(d-1)!} \int_M -(-\mathrm{d}n, \underbrace{\ldots}_{\nu}, -\mathrm{d}n, \mathrm{d}x, \underbrace{\ldots}_{d-1-\nu}, \mathrm{d}x, x) \\
&= \int_M H_\nu h\, \mathrm{d}F\,, \quad \nu = 1, \ldots, d-1\,, \qquad (20)
\end{aligned}
$$

where

$$
h := -\langle x, n \rangle \qquad (21)
$$

is the support function of F, a direct consequence of Stokes' theorem. The eqs. (20) are an important tool for proving results in the global differential geometry of hypersurfaces.

If our convex hypersurface F has positive Gaussian curvature H_{d-1} everywhere,

$$H_{d-1} > 0 , \tag{22}$$

then all the principal curvatures $k_1, \ldots, k_{d-1}$ of F must be positive [see eq. (7)], and we may define their reciprocal values

$$\begin{aligned} R_1 &:= \frac{1}{k_1} > 0 , \\ &\vdots \\ R_{d-1} &:= \frac{1}{k_{d-1}} > 0 , \end{aligned} \tag{23}$$

as the "principal radii of curvature" of F. In this case we can locally use the spherical image $\nu(x(M)) \subset S^{d-1}$ of F as parameter manifold for F and represent the points of F by their unit normal vector n in the following manner:

$$x(-n) = \left(\frac{\partial H}{\partial y^1}(-n), \ldots, \frac{\partial H}{\partial y^d}(-n) \right) . \tag{24}$$

Herein, H is the support function of a convex body, bounded by F:

$$\begin{aligned} &H(y^1, \ldots, y^d) := \|y\| h\left(\frac{y^1}{\|y\|}, \ldots, \frac{y^d}{\|y\|} \right) , \\ &\quad y := (y^1, \ldots, y^d) \in \mathbb{R}^d \backslash \{0\} \end{aligned} \tag{25}$$

[see eq. (21)], convex, positively homogeneous of degree 1 and of class C_2 (see chapter 1.2). The homogeneity of H yields

$$\sum_{b=1}^{d} \frac{\partial H}{\partial y^b} y^b = H , \tag{26}$$

whence after partial differentiation

$$\sum_{b=1}^{d} \frac{\partial^2 H}{\partial y^a \, \partial y^b} (-n) n^b = 0 , \quad a = 1, \ldots, d . \tag{27}$$

Therefore, the Hessian of H at $-n$ has the eigenvector n to the eigenvalue 0. Because of the relations

$$\begin{aligned} &x_i(-n) = \left(-\sum_{b=1}^{d} \frac{\partial^2 H}{\partial y^1 \, \partial y^b} (-n)(n^b)_i, \ldots, -\sum_{b=1}^{d} \frac{\partial^2 H}{\partial y^d \, \partial y^b} (-n)(n^b)_i \right) , \\ &\quad i = 1, \ldots, d-1 \end{aligned} \tag{28}$$

– resulting from (24) by differentiation with respect to u^i $(i = 1, \ldots, d-1)$ – the other $d-1$ eigenvalues of $(\operatorname{Hess} H)(-n)$ coincide with the eigenvalues $R_1(-n), \ldots, R_{d-1}(-n)$ of the linear map $(\nu^{-1})_* : T_{-n}S^{d-1} \to x_*(T_{p(-n)}M)$, induced by the (locally existing) inverse ν^{-1} of the spherical map ν [see eq. (23)].

From these facts we deduce that we have for the normalized elementary symmetric functions

$$\begin{aligned} P_1 &:= \frac{1}{d-1}(R_1 + \cdots + R_{d-1}), \\ &\vdots \\ P_{d-1} &:= R_1 \cdots R_{d-1}, \end{aligned} \tag{29}$$

defined as the "mean radii of curvature" of F, the equation:

$$\binom{d-1}{\nu} P_\nu(-n) = D_\nu(H)(-n), \quad \nu = 1, \ldots, d-1, \ -n \in \nu(x(M)), \tag{30}$$

where $D_\nu(H)$ denotes the sum of all principal minors with ν rows of the matrix Hess H. Now eqs. (7), (29) and (23) prove

$$P_\nu = \frac{H_{d-1-\nu}}{H_{d-1}}, \quad \nu = 0, \ldots, d-1, \ P_0 := 1 \tag{31}$$

[see eq. (22)], whence (19) respectively (20) may be transformed into the "dual" formulas

$$V(U_\lambda) = \frac{1}{d} \sum_{\nu=0}^{d-1} \binom{d}{\nu} \lambda^{d-\nu} \int_{\nu(x(M))} P_\nu \, d\omega \tag{32}$$

($d\omega$ = volume element of S^{d-1}) and, for $d > 2$ (see Theorem 1.1) and F of class C_3,

$$\int_{S^{d-1}} P_{d-\nu} \, d\omega = \int_{S^{d-1}} P_{d-\nu-1} h \, d\omega, \quad \nu = 1, \ldots, d-1. \tag{33}$$

Equation (32) shows that the *mean radius of curvature P_ν is the density of the νth "surface area measure" in the sense of Aleksandrov of a convex body bounded by F* ($\nu = 0, \ldots, d-1$; see chapter 1.8).

A very important theorem about the influence of the infinitesimal behaviour of a convex hypersurface to its global behaviour is the following:

Theorem 2.1. (Blaschke 1956 (1916), p. 118). *A sphere S_R of radius $R > 0$ in $\mathbb{R}^d$ is rolling freely in the interior K of an "ovaloid" F, i.e., a hypersurface $x : M \to \mathbb{R}^d$ of class C_2 with positive Gaussian curvature bounding a compact convex body K, if the condition*

$$R \leq \min_{-n \in S^{d-1}} \{R_1(-n), \ldots, R_{d-1}(-n)\} \tag{34}$$

holds.

Proof. In the first step we prove Theorem 2.1 in the case $d=2$. Here, eq. (30) becomes

$$R_1(-n)=P_1(-n)=\left(\frac{\partial^2 H}{(\partial y^1)^2}+\frac{\partial^2 H}{(\partial y^2)^2}\right)(-n)=g(\varphi)+\frac{\mathrm{d}^2 g}{\mathrm{d}\varphi^2}(\varphi)\,, \qquad (35)$$

after the introduction of the angle φ by

$$-n := (\cos\varphi, \sin\varphi) \qquad (36)$$

and of the auxiliary function g by

$$g(\varphi) := h(\cos\varphi, \sin\varphi)\,, \quad 0 \leqslant \varphi \leqslant 2\pi\,, \qquad (37)$$

see eq. (25). If we denote $R_1(-n)=R_1(\cos\varphi, \sin\varphi)$ by $r(\varphi)$, we find by the method of variation of constants the following solution of the inhomogeneous linear ordinary differential equation of second order (35):

$$g(\varphi)=\int_{\varphi_0}^{\varphi} r(\psi)\sin(\varphi-\psi)\,\mathrm{d}\psi + C_1\cos\varphi + C_2\sin\varphi\,, \qquad (38)$$

with given initial conditions at φ_0 implying

$$C_1=\cos\varphi_0 g(\varphi_0)-\sin\varphi_0 g'(\varphi_0)\,, \qquad C_2=\sin\varphi_0 g(\varphi_0)+\cos\varphi_0 g'(\varphi_0)\,. \qquad (39)$$

Now we compare g with the support function g_R of the circle $S_R(\varphi_0)$ of radius R touching the curve F from the inner side at an arbitrary point with the outer unit normal vector $(\cos\varphi_0, \sin\varphi_0)$. Clearly, g_R has the representation

$$g_R(\varphi)=\int_{\varphi_0}^{\varphi} R\sin(\varphi-\psi)\,\mathrm{d}\psi + C_1\cos\varphi + C_2\sin\varphi \qquad (40)$$

[analogous to eq. (38)], with the same integration constants C_1 and C_2 because of (39). Therefore, subtraction of (40) from (38) gives

$$g(\varphi)-g_R(\varphi)=\int_{\varphi_0}^{\varphi}(r(\psi)-R)\sin(\varphi-\psi)\,\mathrm{d}\psi\,, \quad \varphi_0-\pi\leqslant\varphi\leqslant\varphi_0+\pi\,, \qquad (41)$$

and the condition (34) or $R\leqslant r(\psi)$ for all ψ yields

$$g(\varphi)\geqslant g_R(\varphi)\,, \quad \varphi_0-\pi\leqslant\varphi\leqslant\varphi_0+\pi\,. \qquad (42)$$

Equation (37) together with eq. (42) proves that the circle $S_R(\varphi_0)$ totally lies in the interior of F so that S_R "is rolling freely" there as the contact point of S_R and F may be chosen arbitrarily.

The second step in the proof of Theorem 2.1 consists in the reduction of dimension by an orthogonal projection π of $\mathbb{R}^d$ onto a suitable plane $\mathbb{R}^2$ in $\mathbb{R}^d$ containing the origin. After (21) and (25) the support function $H^{(\pi)}$ of the curve $F^{(\pi)} := \pi \circ F$ in the plane $\mathbb{R}^2$ equals the restriction of the support function H of F to $\mathbb{R}^2$. Therefore, the radius of curvature $R_1^{(\pi)}(-n)$ of $F^{(\pi)}$ at a point with the unit normal vector $-n \in \mathbb{R}^2$ has the value

$$\begin{aligned} R_1^{(\pi)}(-n) &= \min_{\substack{z\in\mathbb{R}^2 \\ \|z\|=1,\ \langle z,n\rangle=0}} \frac{\partial^2 H}{\partial y^a\, \partial y^b}(-n) z^a z^b \\ &\geqslant \min_{\substack{z\in\mathbb{R}^d \\ \|z\|=1,\ \langle z,n\rangle=0}} \frac{\partial^2 H}{\partial y^a\, \partial y^b}(-n) z^a z^b \\ &= \min\{R_1(-n), \ldots, R_{d-1}(-n)\}, \end{aligned} \tag{43}$$

as $R_1(-n), \ldots, R_{d-1}(-n)$ are the eigenvalues of $(\operatorname{Hess} H)(-n)$ different from the eigenvalue 0 for its eigenvector n. Now the assumption (34) implies $R \leqslant R_1^{(\pi)}(-n)$ so that $\pi(S_R)$ is rolling freely in the interior $\pi(K)$ of $F^{(\pi)}$. But consequently S_R is rolling freely in the interior of F because otherwise there would exist a sphere S_R touching F from the inner side at $x(p)$ and with a point $x(q)$ of F in the proper interior of S_R; and the projection π along the $(d-2)$-dimensional direction simultaneously orthogonal to the normals of F at $x(p)$ and $x(q)$ would produce a contradiction to $\pi(S_R) \subset \pi(K)$. □

Remark 2.2. Using the definition $R_i = 1/k_i$ of the principal radii of curvature where the k_i are the extremal values of curvature of the normal sections of F by planes $(i = 1, \ldots, d-1)$ it can easily be seen that the condition (34) is also necessary for the assertion of Theorem 2.1.

Remark 2.3. By the same method of proof it follows that an ovaloid $\tilde{F}$ is rolling freely in the interior of another ovaloid F, iff

$$\max_{-n\in S^{d-1}} \{\tilde{R}_1(-n), \ldots, \tilde{R}_{d-1}(-n)\} \leqslant \min_{-n\in S^{d-1}} \{R_1(-n), \ldots, R_{d-1}(-n)\}$$

($\tilde{R}_1, \ldots, \tilde{R}_{d-1}$ = principal radii of curvature of $\tilde{F}$) holds. This fact also possesses a local version. For further information see Schneider (1988, Theorem 1), and Brooks and Strantzen (1989).

At the end of this section we shall explain how the notion of principal curvatures or principal radii of curvature of a convex hypersurface F of class C_2 in $\mathbb{R}^d$, given by (1), can be extended to the case of an arbitrary convex hypersurface. Then f is no more continuously twice differentiable but only convex. However, a famous theorem of Aleksandrov (1939) says that *f must be nevertheless twice differentiable in a certain abstract sense* (*due to Frechet*) *almost everywhere* (see chapter 4.2). This implies the validity of Taylor's formula at the points $(u_{(0)}, f(u_{(0)}))$ of twice differentiability of f in the form

$$
\begin{aligned}
&|f(u) - f(u_{(0)}) - \mathrm{d}f_{(u_{(0)})}(u - u_{(0)}) - \tfrac{1}{2}\,\mathrm{d}^2 f_{(u_{(0)})}(u - u_{(0)}, u - u_{(0)})| \\
&\quad \leqslant R(\|u - u_{(0)}\|)\|u - u_{(0)}\|^2 , \qquad (44) \\
&(u := x^1, \ldots, x^{d-1} , \quad u_{(0)} := x^1_{(0)}, \ldots, x^{d-1}_{(0)} , \\
&R : \mathbb{R}^+ \to \mathbb{R}^+ \text{a monotone increasing function with } \lim_{t \to +0} R(t) = 0) ,
\end{aligned}
$$

where $\mathrm{d}f_{(u_{(0)})}$ is a linear and $\mathrm{d}^2 f_{(u_{(0)})}$ is a positive semidefinite quadratic function on $\mathbb{R}^{d-1}$ (see Bangert 1979, Lemma 4.8).

In order to understand the geometrical meaning of (44) it is convenient to choose a coordinate system in $\mathbb{R}^d$ with $u_{(0)} = 0$, $f(u_{(0)}) = 0$, $\mathrm{d}f_{(u_{(0)})} = 0$ and $f(u) \geqslant 0$, and to consider the intersection D_h of a convex body K, bounded by F, with the hyperplane $x^d = h = \text{const} > 0$, expanded by the factor $1/\sqrt{2h}$ and projected orthogonally onto the supporting hyperplane $x^d = 0$. The boundary bd D_h of the convex body D_h is given in polar coordinates by

$$
r = \frac{1}{\sqrt{2h}}\, \rho_h(v) , \tag{45}
$$

where $\rho_h(v)$ fulfils

$$
f(\rho_h(v)v) = h , \quad v \in \mathbb{R}^{d-1} , \ \|v\| = 1 . \tag{46}
$$

By this we conclude from (44) that $\lim_{h \to +0} D_h$ exists (in Hausdorff sense) with the representation of its boundary in the form

$$
\begin{aligned}
r &= \lim_{h \to +0} \frac{1}{\sqrt{2h}}\, \rho_h(v) = \lim_{h \to +0} \frac{\rho_h(v)}{\sqrt{2f(\rho_h(v)v)}} \\
&= \lim_{t \to \lim_{h \to +0} \rho_h(v) + 0} \frac{1}{\sqrt{\dfrac{2f(tv)}{t^2}}} = \frac{1}{\sqrt{\mathrm{d}^2 f_{(0)}(v, v)}}
\end{aligned} \tag{47}
$$

or, equivalently,

$$
\mathrm{d}^2 f_{(0)}(u, u) = 1 . \tag{48}
$$

Therefore, the boundary is a quadric in $\mathbb{R}^{d-1}$ with the origin as midpoint, namely an ellipsoid if $\mathrm{d}^2 f_{(0)}$ is positive definite, or an elliptic cylinder if $\mathrm{d}^2 f_{(0)}$ is positive semidefinite but not positive definite or zero, or the empty set if $\mathrm{d}^2 f_{(0)}$ is zero. This leads to:

Definition 2.4 (Aleksandrov 1939). A point $x(p)$ of a general convex hypersurface F in $\mathbb{R}^d$ is called *normal* iff $\lim_{h\to+0} D_h$ exists and its boundary $I(p)$ is a quadric in the tangent hyperplane of F at $x(p)$ with $x(p)$ as its midpoint, the so-called "indicatrix of Dupin".

We have just seen that $x(p)$ is a normal point of a convex hypersurface F if the function f, representing F, is twice differentiable at this point. However, also the converse is true as shown by Aleksandrov (1939). It is convenient to define the *inverse of the squared length of the semiaxes of* $I(p)$ as the (generalized) "principal curvatures" of F at a normal point $x(p)$ [compare (47) in the C_2-case!]. Their product may be regarded as a (generalized) Gaussian curvature $H_{d-1}(x(p))$ of F which turns out to be the appropriately defined derivative at $x(p)$ of the (generalized) Gauss map ν as a set function (see Aleksandrov 1939). Finally, if in addition the (convex) support function H of a convex body, bounded by F, is twice differentiable at $-n(p)$, then the product of the "principal radii of curvature" (inverse to the principal curvatures) $1/H_{d-1}(x(p))$ is finite and equals to the entity $D_{d-1}(H)(-n(p))$ of (30) (see also Aleksandrov 1939). For further information we refer to a paper of Busemann and Feller (1936) and a paper of Schneider (1979).

3. Mixed discriminants and mixed volumes

In this section we want to express the volume of a compact convex body K respectively the mixed volume of such bodies $K_1, \ldots, K_m$ in $\mathbb{R}^d$, provided that they are "regular", i.e., they have C_2-boundaries F, respectively $F_1, \ldots, F_m$, with positive Gaussian curvatures, by suitable integrals over F, respectively $F_1, \ldots, F_m$, and also over the unit sphere S^{d-1}. This will produce certain "counterparts" of important theorems on mixed volumes (see chapter 1.2). We begin with the representation

$$\begin{aligned} V(K) &= \frac{1}{d!}\int_K (\underbrace{\mathrm{d}x, \ldots, \mathrm{d}x}_{d}) = \frac{1}{d!}\int_F (\underbrace{\mathrm{d}x, \ldots, \mathrm{d}x}_{d-1}, -x) = \frac{1}{d}\int_F h\,\mathrm{d}F \\ &= \frac{1}{d}\int_{S^{d-1}} HP_{d-1}\,\mathrm{d}\omega = \frac{1}{d}\int_{S^{d-1}} HD_{d-1}(H)\,\mathrm{d}\omega \end{aligned} \tag{49}$$

of the volume of K, a consequence of Stokes' theorem after a suitable orientation of F [see also eqs. (21), (31) and (30)]. If we insert in (49) for x a linear combination $x = \Sigma_{l=1}^m \lambda_l x^{(l)}$ ($\lambda_1 \geq 0, \ldots, \lambda_m \geq 0$) for corresponding points $x^{(1)}, \ldots, x^{(m)}$ (with the same unit normal vector n), the comparison of the coefficients of $\lambda_{l_1} \cdots \lambda_{l_d}$ yields

$$\begin{aligned} &V(K_{l_1}, \ldots, K_{l_d}) = \frac{1}{d!}\int_F (\mathrm{d}x^{(l_1)}, \ldots, \mathrm{d}x^{(l_{d-1})}, -x^{(l_d)}), \\ &1 \leq l_1 \leq \cdots \leq l_d \leq m. \end{aligned} \tag{50}$$

In order to get this formula we used the symmetry of the integral in (50) with respect to the indices $l_1, \ldots, l_d$ arising from Stokes' theorem and the compactness of the oriented hypersurface F. Furthermore, the insertion of (28) into (50) gives

$$V(K_{l_1}, \ldots, K_{l_d}) = \frac{1}{d!} \int_F (\mathrm{d}x^{(l_1)}, \ldots, \mathrm{d}x^{(l_{d-1})}, n) h^{(l_d)} = \frac{1}{d} \int_{S^{d-1}} D_{d-1}(H^{(l_1)}, \ldots, H^{(l_{d-1})}) H^{(l_d)} \, \mathrm{d}\omega \,, \tag{51}$$

where $(d-1)! D_{d-1}(H^{(l_1)}, \ldots, H^{(l_{d-1})})$ denotes the sum of all "mixed" principal minors with $d-1$ rows of the Hessians of the support functions $H^{(l_1)}, \ldots, H^{(l_{d-1})}$ of $K_{l_1}, \ldots, K_{l_{d-1}}$. We can easily realize (51) if we use the fact that this term equals the coefficient of $\lambda_0 \cdot \lambda_{l_1} \cdots \lambda_{l_{d-1}}$ in the determinant

$$\det\left[\lambda_0 \delta_{ab} + \lambda_{l_1} \frac{\partial^2 H^{(l_1)}}{\partial y^a \, \partial y^b} + \cdots + \lambda_{l_{d-1}} \frac{\partial^2 H^{(l_{d-1})}}{\partial y^a \, \partial y^b} \right],$$

and that this determinant is invariant against orthogonal transformations of $\mathbb{R}^d$, so that it suffices to prove (51) under the additional assumption $n = (0, \ldots, 0, 1)$ and therefore $\mathrm{d}n^d = 0$ as well $\partial^2 H / \partial y^a \, \partial y^d = 0$ [$a = 1, \ldots, d$; see eq. (27)] at the corresponding point. For all these reasons it is convenient to define:

Definition 3.1. Let $Q_1, \ldots, Q_t$ be arbitrary symmetric $(d \times d)$-matrices. Then the coefficients $D(Q_{s_1}, \ldots, Q_{s_d})$ of the expansion

$$\det(w_1 Q_1 + \cdots + w_t Q_t) = \sum_{s_1=1}^{t} \cdots \sum_{s_d=1}^{t} w_{s_1} \cdots w_{s_d} D(Q_{s_1}, \ldots, Q_{s_d}) \tag{52}$$

which are required to be symmetric in $s_1, \ldots, s_d$ are called *mixed discriminants* of $Q_1, \ldots, Q_t$.

These mixed discriminants $D(Q_{s_1}, \ldots, Q_{s_d})$ are depending only on the matrices $Q_{s_1}, \ldots, Q_{s_d}$, and we see easily that

$$D(Q, \ldots, Q) = \det Q \tag{53}$$

in the case $Q_1 = \cdots = Q_t =: Q$. Moreover, the mixed discriminants are linear in each argument:

$$D(\alpha_1 Q_{s_1}^{(1)} + \alpha_2 Q_{s_1}^{(2)}, Q_{s_2}, \ldots, Q_{s_d}) = \alpha_1 D(Q_{s_1}^{(1)}, Q_{s_2}, \ldots, Q_{s_d}) + \alpha_2 D(Q_{s_1}^{(2)}, Q_{s_2}, \ldots, Q_{s_d}) \,. \tag{54}$$

As we have just shown, there exists the relation

$$D_{d-1}(H^{(l_1)}, \ldots, H^{(l_{d-1})}) = d \cdot D(\text{Hess } H^{(l_1)}, \ldots, \text{Hess } H^{(l_{d-1})}, E) \tag{55}$$

[$E = (d \times d)$ – unit matrix] for the integrand in (51). In the same way the equation

$$D_\nu(H^{(l_1)}, \ldots, H^{(l_\nu)}) = \binom{d}{\nu} D(\text{Hess } H^{(l_1)}, \ldots, \text{Hess } H^{(l_\nu)}, \underbrace{E, \ldots, E}_{d-\nu}) \tag{56}$$

[$1 \leq \nu \leq d-1$; see eq. (30)!] becomes obvious.

A first important property for mixed discriminants – a counterpart to the nonnegativity of the mixed volume – is expressed in:

Proposition 3.2. *If all the (symmetric) matrices $Q_1, \ldots, Q_t$ in Definition 3.1 are positive semidefinite (respectively positive definite), then all mixed discriminants $D(Q_{s_1}, \ldots, Q_{s_d})$ are nonnegative (positive) ($1 \leq s_1 \leq t, \ldots, 1 \leq s_d \leq t$).*

Proof. After suitable approximation of positive semidefinite matrices $Q_1, \ldots, Q_t$ by positive definite ones we can see that it suffices to prove this proposition for positive definite matrices. This will be done by induction with respect to d. The case $d = 1$ is trivial. We suppose Proposition 3.2 to be valid for $d-1$. Then at first the application of Sylvesters' law of inertia for the positive definite matrix Q_{s_d} together with the simultaneous multiplication of all matrices $Q_1, \ldots, Q_t$ by T^* on the left and T on the right implies

$$D(Q_{s_1}, \ldots, Q_{s_d}) = (\det T)^{-2} D(Q'_{s_1}, \ldots, Q'_{s_{d-1}}, E), \tag{57}$$

where $\det T \neq 0$ and $Q'_{s_1} := T^* Q_{s_1} T, \ldots, Q'_{s_{d-1}} := T^* Q_{s_{d-1}} T$, $E = T^* Q_{s_d} T$ [see eq. (52)]. However, $d! \cdot D(Q'_{s_1}, \ldots, Q'_{s_{d-1}}, E)$ is the sum of all "mixed" principal minors with $d-1$ rows of the positive definite matrices $Q'_{s_1}, \ldots, Q'_{s_{d-1}}$ and therefore positive by the inductive assumption so that also $D(Q_{s_1}, \ldots, Q_{s_d})$ must be positive because of (57). □

Remark 3.3. Proposition 3.2 and eqs. (51), (55) together with the relation

$$\begin{aligned} &d \cdot D(\text{Hess } H^{(l_1)}, \ldots, \text{Hess } H^{(l_{d-1})}, E) \\ &\quad = D'(\text{Hess}' \, H^{(l_1)}, \ldots, \text{Hess}' \, H^{(l_{d-1})}) > 0, \end{aligned} \tag{58}$$

where D' denotes the mixed discriminant of the (positive definite) restrictions of $\text{Hess } H^{(l_1)}(-n), \ldots, \text{Hess } H^{(l_{d-1})}(-n)$ to the $(d-1)$-dimensional sum of their eigenspaces corresponding to their positive eigenvalues (the radii of curvature of $F_{l_1}, \ldots, F_{l_{d-1}}$), imply the well-known property for regular compact convex bodies

$$V(K_{l_1}, \ldots, K_{l_d}) > 0 . \tag{59}$$

Hereby the positivity of $H^{(l_d)}$ was used coming from the choice of the origin in the interior of the convex body K_{l_d}. The same argument yields the monotonicity property

$$V(K_{l_1}, \ldots, K_{l_{d-1}}, K_{l_d}) \leq V(K_{l_1}, \ldots, K_{l_{d-1}}, \tilde{K}_{l_d}) \quad \text{if } K_{l_d} \subset \tilde{K}_{l_d} , \tag{60}$$

with equality iff $K_{l_d} = \tilde{K}_{l_d}$.

We now proceed to the following counterpart of the Aleksandrov–Fenchel–Jessen inequalities for mixed volumes (see chapter 1.2).

Theorem 3.4 (Aleksandrov 1938). *If the matrices $Q_{s_1}, \ldots, Q_{s_{d-1}}$ are positive definite and Q_{s_d} is any (symmetric) matrix, then*

$$\begin{aligned} &[D(Q_{s_1}, \ldots, Q_{s_{d-2}}, Q_{s_{d-1}}, Q_{s_d})]^2 \\ &\quad \geq D(Q_{s_1}, \ldots, Q_{s_{d-2}}, Q_{s_{d-1}}, Q_{s_{d-1}}) \cdot D(Q_{s_1}, \ldots, Q_{s_{d-2}}, Q_{s_d}, Q_{s_d}) , \end{aligned} \tag{61}$$

where the equality holds only if Q_{s_d} is proportional to $Q_{s_{d-1}}$.

Proof (*See also* Busemann 1958, pp. 53–56)**.** After introduction of the quadratic form $g_{s_1 \cdots s_{d-2}}$ in the essential elements ξ_{ij} $(1 \leq i \leq j \leq d)$ of an arbitrary symmetric matrix X by:

$$\begin{aligned} g_{s_1 \cdots s_{d-2}}(X, X) &:= D(Q_{s_1}, \ldots, Q_{s_{d-2}}, X, X) \\ &= \sum_{i=1}^{d} \sum_{j=1}^{d} (D(Q_{s_1}, \ldots, Q_{s_{d-2}}, X))_{ij} \xi_{ij} \\ &= \sum_{i=1}^{d} \sum_{j=1}^{d} \sum_{k=1}^{d} \sum_{l=1}^{d} (D(Q_{s_1}, \ldots, Q_{s_{d-2}}))_{ij,kl} \xi_{ij} \xi_{kl} , \end{aligned} \tag{62}$$

the inequality (61) takes the form

$$[g_{s_1 \cdots s_{d-2}}(Q_{s_{d-1}}, Q_{s_d})]^2 \geq g_{s_1 \cdots s_{d-2}}(Q_{s_{d-1}}, Q_{s_{d-1}}) \cdot g_{s_1 \cdots s_{d-2}}(Q_{s_d}, Q_{s_d}) , \tag{63}$$

with equality iff

$$Q_{s_d} = \gamma \cdot Q_{s_{d-1}} , \quad \gamma \in \mathbb{R} . \tag{64}$$

This is equivalent to the implication

$$g_{s_1 \cdots s_{d-2}}(Q_{s_{d-1}}, Q) = 0 \Rightarrow g_{s_1 \cdots s_{d-2}}(Q, Q) \leq 0 \tag{65}$$

for a symmetric matrix Q, with equality iff

$$Q = 0 . \tag{66}$$

Indeed, eqs. (63) and (62) together with Proposition 3.2 imply (65) and (66). Conversely, after applying (65) and (66) for

$$Q := Q_{s_d} - \frac{g_{s_1 \cdots s_{d-2}}(Q_{s_{d-1}}, Q_{s_d})}{g_{s_1 \cdots s_{d-2}}(Q_{s_{d-1}}, Q_{s_{d-1}})} \cdot Q_{s_{d-1}}$$

we get (63) and (64).

In the following, (65) and (66) will be proven by induction with respect to d. (65) and (66) may be easily shown in the case $d = 2$ if the positive definite (respectively symmetric) matrices Q_{s_1} (Q) have been simultaneously transformed into diagonal forms

$$\begin{bmatrix} a_{11} & 0 \\ 0 & a_{22} \end{bmatrix}, \quad a_{11} > 0 , \; a_{22} > 0 \qquad \left(\text{respectively} \begin{bmatrix} b_{11} & 0 \\ 0 & b_{22} \end{bmatrix}\right)$$

by a unimodular transformation, not changing the mixed discriminants. Indeed, here the implication

$$a_{11} b_{22} + a_{22} b_{11} = 0 \Rightarrow b_{11} b_{22} \leqslant 0 \tag{67}$$

is trivial, with equality iff

$$b_{11} = b_{22} = 0 . \tag{68}$$

Now we assume (65) and (66) to be true for $d - 1$ and show their validity for d by consideration of the nonnegative eigenvalues of $g_{s_1 \cdots s_{d-2}}$ with respect to the quadratic form

$$\sum_{i=1}^{d} \sum_{j=1}^{d} (\xi_{ij})^2 . \tag{69}$$

At first this will be done for the special quadratic form

$$g^{(0)}(X, X) := D(E, \ldots, E, X, X) = \binom{d}{2}^{-1} \sum_{1 \leqslant i < j \leqslant d} (\xi_{ii} \xi_{jj} - (\xi_{ij})^2) . \tag{70}$$

It results from an elementary calculation involving Lagrange's multiplicator method that $g^{(0)}$ has $1/d$ as single positive, and $-1/d(d-1)$ as $(\frac{1}{2}d(d+1) - 1)$-fold negative eigenvalue, so that 0 cannot be an eigenvalue of $g^{(0)}$. *More general, 0 is not an eigenvalue for* $g_{s_1 \cdots s_{d-2}}$. Otherwise we would have

$$g_{s_1 \cdots s_{d-2}}(Q_0, X) = D(Q_{s_1}, \ldots, Q_{s_{d-2}}, Q_0, X) = 0 \tag{71}$$

for a symmetric matrix $Q_0 \neq 0$ and an arbitrary symmetric matrix X [see eq. (62)], and this relation remains unchanged after a simultaneous unimodular matrix transformation of $Q_{s_1}, \ldots, Q_{s_{d-2}}, Q_0, X$ such that the positive definite $Q_{s_{d-2}}$ and the symmetric Q_0 are transformed into diagonal matrices with elements

$$a_{ij}^{(s_{d-2})} = a_{ii}^{(d-2)} \delta_{ij} \quad (a_{ii}^{(d-2)} > 0) \quad \text{and} \quad b_{ij}^{(0)} = b_{ii}\delta_{ij}\,, \qquad 1 \leq i \leq j \leq d\,. \tag{72}$$

So we may suppose the validity of (72) without loss of generality and deduce from (62) and (71) the equations $(D(Q_{s_1}, \ldots, Q_{s_{d-2}}, Q_0))_{ii} = 0$, $i = 1, \ldots, d$, where $d(D(Q_{s_1}, \ldots, Q_{s_{d-2}}, Q_0))_{ii}$ equals the mixed discriminant of $Q_{s_1}, \ldots, Q_{s_{d-2}}, Q_0$ with all elements with index i having been deleted. For this reason we have by the inductive assumption

$$(D(Q_{s_1}, \ldots, Q_{s_{d-3}}, Q_0, Q_0))_{ii} \leq 0\,, \quad 1 \leq i \leq d\,, \tag{73}$$

and by (71)

$$\sum_{i=1}^{d} (D(Q_{s_1}, \ldots, Q_{s_{d-3}}, Q_0, Q_0))_{ii} a_{ii}^{(d-2)} = D(Q_{s_1}, \ldots, Q_{s_{d-2}}, Q_0, Q_0) = 0\,.$$

But this and eq. (72) imply equality in all the inequalities (73), and therefore $Q_0 = 0$ by the inductive assumption which contradicts $Q_0 \neq 0$. Now a continuity argument, applied to the eigenvalues of the forms

$$g^{(\tau)}(X, X) := D((1-\tau)E + \tau Q_{s_1}, \ldots, (1-\tau)E + \tau Q_{s_{d-2}}, X, X)\,,$$
$$0 \leq \tau \leq 1\,,$$

connecting $g^{(0)}$ with $g_{s_1 \cdots s_{d-2}}$, shows that $g_{s_1 \cdots s_{d-2}}$ *has exactly one positive and no zero eigenvalue* as well as $g^{(0)}$ because no eigenvalue of $g^{(\tau)}$ passes through 0 when τ runs from 0 to 1.

At the end of the proof we interpret the essential elements ξ_{ij} $(1 \leq i \leq j \leq d)$ of a (nonvanishing) symmetric matrix X as projective coordinates of the points of a $(\frac{1}{2}d(d+1) - 1)$ – dimensional projective space P. Then, by the previously mentioned property of $g_{s_1 \cdots s_{d-2}}$,

$$g_{s_1 \cdots s_{d-2}}(X, X) = 0 \tag{74}$$

is the equation of a hyperellipsoid in P. Because of (62) and Proposition 3.2 we have $g_{s_1 \cdots s_{d-2}}(Q_{s_{d-1}}, Q_{s_{d-1}}) > 0$, which means that the point $Q_{s_{d-1}}$ lies in the component of the hyperellipsoid (74) which does not contain a full line of P. But then each point $Q \neq 0$ with $g_{s_1 \cdots s_{d-2}}(Q_{s_{d-1}}, Q) = 0$, i.e., each point Q conjugate to $Q_{s_{d-1}}$ with respect to (74) lies in the other component, whence $g_{s_1 \cdots s_{d-2}}(Q, Q) < 0$. Therefore, the implication (65) is true with equality iff (66) holds which completes the proof of Theorem 3.4. □

4. Differential geometric proof of the Aleksandrov–Fenchel–Jessen inequalities

In this section we shall apply Theorem 3.4 for a proof, formally analogous to that of Theorem 3.4 for mixed discriminants, of the following *Aleksandrov–Fenchel–Jessen inequalities for mixed volumes*:

$$\begin{aligned}&[V(K_{l_1},\dots,K_{l_{d-2}},K_{l_{d-1}},K_{l_d})]^2\\ &\quad \geqslant V(K_{l_1},\dots,K_{l_{d-2}},K_{l_{d-1}},K_{l_{d-1}})\cdot V(K_{l_1},\dots,K_{l_{d-2}},K_{l_d},K_{l_d}),\\ &\quad 1\leqslant l_1,\dots,l_d\leqslant m,\end{aligned} \tag{75}$$

where $K_1,\dots,K_m$ are assumed to be regular compact convex bodies in $\mathbb{R}^d$ with the origin in the interior and equality occurs iff $K_{l_{d-1}}$ and K_{l_d} are homothetic:

$$K_{l_d}=\gamma\cdot K_{l_{d-1}}+a,\quad \gamma>0,\ a\in\mathbb{R}^d, \tag{76}$$

see Aleksandrov (1938, §6) and Busemann (1958, pp. 56–59).

For this purpose Aleksandrov extends [similar to his first proof of eq. (75)] the notion of the mixed volume (51) as a functional over the d-fold Cartesian product of the space $C_2^{\text{conv}}(S^{d-1})$ of all those positive C_2-functions on the sphere S^{d-1} with convex positively homogeneous extensions H of degree 1 to $\mathbb{R}^d\setminus\{0\}$ whose Hessians have $d-1$ positive eigenvalues (orthogonal to the position vector of the argument) in the following manner: let Z be the positively homogeneous extension of degree 1 of an *arbitrary C_2-function on S^{d-1}* to $\mathbb{R}^d\setminus\{0\}$, and let $C_2(S^{d-1})$ be the space of all such functions. Then, by adding to Z a suitable positive multiple αH_0 of the special function $H_0(y):=(\Sigma_{a=1}^d (y^a)^2)^{1/2}$ with a Hessian of $d-1$ eigenvalues 1 on S^{d-1}, we get another function $H_1\in C_2^{\text{conv}}(S^{d-1})$ (because of the compactness of S^{d-1}!) so that the representation

$$Z=H_1-\alpha H_0,\quad H_1\in C_2^{\text{conv}}(S^{d-1}),\ H_0\in C_2^{\text{conv}}(S^{d-1}) \tag{77}$$

holds for Z. This fact permits us to define a bilinear function $q_{l_1\cdots l_{d-2}}$ on the space $C_2(S^{d-1})\times C_2(S^{d-1})$ by a well-defined bilinear extension of the mixed volume (51) considered as a bilinear form on $C_2^{\text{conv}}(S^{d-1})\times C_2^{\text{conv}}(S^{d-1})$ with the fixed support functions $H^{(l_1)},\dots,H^{(l_{d-2})}\in C_2^{\text{conv}}(S^{d-1})$ of the fixed regular compact convex bodies $K_{l_1},\dots,K_{l_{d-2}}$. This extension has the property

$$q_{l_1\cdots l_{d-2}}(H^{(l_{d-1})},H^{(l_d)})=V(K_{l_1},\dots,K_{l_d}), \tag{78}$$

and we shall prove, instead of (75) with the equality condition (76), the following stronger version:

$$\begin{aligned}&[q_{l_1\cdots l_{d-2}}(H^{(l_{d-1})},Z)]^2\geqslant q_{l_1\cdots l_{d-2}}(H^{(l_{d-1})},H^{(l_{d-1})})\cdot q_{l_1\cdots l_{d-2}}(Z,Z),\\ &\quad Z\in C_2(S^{d-1}),\end{aligned} \tag{79}$$

with equality iff

$$Z = \gamma \cdot H^{(l_{d-1})} + \langle a, \cdot \rangle \,, \quad \gamma \in \mathbb{R} \,,\ a \in \mathbb{R}^d \,, \tag{80}$$

compare (78).

Now a first step of the proof is to show that (79) *and* (80) *are equivalent to the fact that*

$$q_{l_1 \cdots l_{d-2}}(H^{(l_{d-1})}, Z) = 0 \,, \quad Z \in C_2(S^{d-1}) \,, \tag{81}$$

implies

$$q_{l_1 \cdots l_{d-2}}(Z, Z) \leq 0 \,, \tag{82}$$

with equality iff

$$Z = \langle a, \cdot \rangle \,. \tag{83}$$

Clearly, from (79) and (81) we may deduce (82) considering (78) and (59), and the insertion of (80) into (81) yields $\gamma = 0$, i.e., (83) because of the translation invariance of the mixed volume. Conversely,

$$\hat{Z} := Z - \frac{q_{l_1 \cdots l_{d-2}}(H^{(l_{d-1})}, Z)}{q_{l_1 \cdots l_{d-2}}(H^{(l_{d-1})}, H^{(l_{d-1})})} \cdot H^{(l_{d-1})} \tag{84}$$

fulfils (81) implying (82) for $\hat{Z}$, which is equivalent to (79). There equality occurs iff this is true for (82) with $\hat{Z}$, i.e., $\hat{Z} = \langle a, \cdot \rangle$ or (80) by definition (84).

Then a next step is the application of Hilbert's "parametrix method" (see Hilbert 1912, chapter 18 in the case $d = 3$), the essential idea of Aleksandrov's proof. This will be done by defining a "weighted" inner product on the space $C_2(S^{d-1})$ by:

$$\langle Z, W \rangle := \int_{S^{d-1}} Z \cdot W \, w \, \mathrm{d}\omega \,, \tag{85}$$

with the weight

$$w := \frac{D_{d-1}(H^{(l_1)}, \ldots, H^{(l_{d-1})})}{H^{(l_{d-1})}} > 0 \,; \tag{86}$$

and a linear differential operator L on $C_2(S^{d-1})$ by:

$$L(Z) := D_{d-1}(H^{(l_1)}, \ldots, H^{(l_{d-2})}, Z) \,. \tag{87}$$

This operator is symmetric because of the relation

$$\int_{S^{d-1}} L(Z)\cdot W\,\mathrm{d}\omega = \int_{S^{d-1}} D_{d-1}(H^{(l_1)},\ldots,H^{(l_{d-2})},Z)\cdot W\,\mathrm{d}\omega$$

$$= \int_{S^{d-1}} D_{d-1}(H^{(l_1)},\ldots,H^{(l_{d-2})},W)\cdot Z\,\mathrm{d}\omega = \int_{S^{d-1}} Z\cdot L(W)\,\mathrm{d}\omega\,, \quad (88)$$

arising from (51) and the symmetry of the mixed volume with respect to $K_{l_1},\ldots,K_{l_d}$ after extension of $C_2^{\mathrm{conv}}(S^{d-1})$ to $C_2(S^{d-1})$. Moreover, L is of elliptic type what means that the (symmetric) matrix C' of the cofactors of the elements $\partial^2 Z/\partial u'^i\,\partial u'^j$ $(i,j=1,\ldots,d-1)$ in the expansion

$$L(Z) = D'(\mathrm{Hess}'\,H^{(l_1)},\ldots,\mathrm{Hess}'\,H^{(l_{d-2})},\mathrm{Hess}'\,Z)$$

$$= \sum_{i=1}^{d-1}\sum_{j=1}^{d-1}(D'(\mathrm{Hess}'\,H^{(l_1)},\ldots,\mathrm{Hess}'\,H^{(l_{d-2})}))_{ij}\cdot\frac{\partial^2 Z}{\partial u'^i\,\partial u'^j} \quad (89)$$

[see eqs. (87), (55) and (58)], is positive definite. This can be seen as follows. By a suitable orthogonal transformation of the Cartesian coordinates $u'^1,\ldots,u'^{d-1}$ of the tangent hyperplane $T_{-n}S^{d-1}$ of S^{d-1} which acts on the Hessians of $H^{(l_1)},\ldots,H^{(l_{d-2})}$ simultaneously by:

$$\mathrm{Hess}''\,H^{(l_1)} = T'^*\,\mathrm{Hess}'\,H^{(l_1)}\,T'\,,$$
$$\vdots$$
$$\mathrm{Hess}''\,H^{(l_{d-2})} = T'^*\,\mathrm{Hess}'\,H^{(l_{d-2})}\,T'\,,$$

and for this reason on the matrix C' of cofactors (in the same manner as on the matrix of the cofactors of a determinant) by:

$$C'' = (\det T')^2(T')^{-1}C'(T'^*)^{-1} = T'^*C'T'$$

we can make C'' a diagonal matrix. Therefore, C'' and thus C' are positive definite if the diagonal elements of C'' are positive:

$$(D''(\mathrm{Hess}''\,H^{(l_1)},\ldots,\mathrm{Hess}''\,H^{(l_{d-2})}))_{ii} > 0\,,\quad i=1,\ldots,d-1\,. \quad (90)$$

But these cofactors are, up to the factor $1/(d-1)$, the mixed discriminants of the matrices $\mathrm{Hess}''\,H^{(l_1)},\ldots,\mathrm{Hess}''\,H^{(l_{d-2})}$ with suppressed elements with the index i, being all positive definite, so that (90) holds because of Proposition 3.2.

Now the principal result of Hilbert's parametrix method is contained in:

Proposition 4.1 [Hilbert (1912, p. 241) for $d=3$ and the C_∞-case].

(a) *If L is any linear symmetric differential operator of second order and elliptic type with continuous coefficients on the space $C_2(S^{d-1})$, and if w is any continuous positive weight function on S^{d-1}, then the differential equation*

$$L(z) + \lambda wz = 0 \quad (91)$$

for $z := Z|_{S^{d-1}}$, *has a countable number of eigenvalues* $\lambda_1 \leqslant \lambda_2 \leqslant \cdots$ *with a* [*relatively to* $C_2(S^{d-1})$] *closed system of eigenfunctions* $z_1 = Z_1|_{S^{d-1}}$, $z_2 = Z_2|_{S^{d-1}}$, ..., *mutually orthogonal with respect to the inner product* (85):

$$\langle Z_\mu, Z_\nu \rangle = 0 \quad \text{if } \mu \neq \nu\,, \tag{92}$$

and normalized by

$$\langle Z_\nu, Z_\nu \rangle = 1\,. \tag{93}$$

(b) *Hereby,* λ_ν *is characterized by the property*:

$$\lambda_\nu = \min_{\langle Z, Z\rangle = 1,\, \langle Z_1, Z\rangle = \cdots = \langle Z_{\nu-1}, Z\rangle = 0} \left(-\int_{S^{d-1}} Z \cdot L(Z)\, \mathrm{d}\omega \right), \tag{94}$$

attained for $Z = Z_\nu$.

(c) *If* L *and* w *depend analytically on a parameter* t, *then* λ_ν *depends continuously on* t.

Applying this result to our differential operator (87), together with the weight function (86) and the inner product (85), the implication (81) $\Rightarrow$ (82), with equality iff (83) holds, or equivalently,

$$\langle H^{(l_{d-1})}, Z \rangle = 0 \Rightarrow -\int_{S^{d-1}} Z \cdot L(Z)\, \mathrm{d}\omega \geqslant 0\,, \tag{95}$$

with equality iff

$$Z(y) = \sum_{a=1}^{d} a_a y_a\,, \quad a_a = \text{const}\,, \tag{96}$$

will be proved if we can show:

$$\lambda_1 = -1 \quad \text{with eigenfunction } Z_1 = H^{(l_{d-1})}\,, \tag{97}$$

$$\lambda_2 = \cdots = \lambda_{d+1} = 0 \quad \text{with eigenfunctions } Z_2 = y_1, \ldots, Z_{d+1} = y_d \tag{98}$$

and

$$\lambda_{d+2} > 0\,. \tag{99}$$

That $H^{(l_{d-1})}$ is eigenfunction of L for the eigenvalue -1 is a direct consequence of (87) and (86). Moreover, the linearly independent linear functions $y_1, \ldots, y_d$ with vanishing Hessians are eigenfunctions of L for the eigenvalue 0 because of (87). But this is not enough to prove (97), (98) and (99) because there might exist further nonpositive eigenvalues for L. Therefore, Aleksandrov considers at first the special case $K_{l_1} = \cdots = K_{l_{d-1}} = K_0 := B^d$, with $H^{(l_1)}(y) = \cdots = H^{(l_{d-1})}(y) = H_0(y) = (\Sigma_{a=1}^d (y^a)^2)^{1/2}$, and thus

$$\text{Hess}' H^{(l_1)}(-n) = \cdots = \text{Hess}' H^{(l_{d-1})}(-n) = E' . \tag{100}$$

The insertion of (100) into (89) leads to

$$L_0(Z) = \frac{1}{d-1} \sum_{i=1}^{d-1} \frac{\partial^2 Z}{(\partial u'^i)^2} := \frac{1}{d-1} \Delta' Z \tag{101}$$

as well as

$$w = 1 . \tag{102}$$

Now it is well known that the so-called "spherical harmonics", defined by:

$$s_n(y) := \frac{p_n(y)}{(\Sigma_{a=1}^d (y^a)^2)^{n/2}} , \quad y \neq 0 , \tag{103}$$

with suitably normalized harmonic homogeneous polynomials $p_n(y)$ of degree n in the coordinates $y^1, \ldots, y^d$,

$$\Delta p_n = 0 , \quad n = 0, 1, 2, \ldots , \tag{104}$$

form a closed system for the continuous functions on S^{d-1} (see Müller 1966). These spherical harmonics are mutually orthogonal with respect to the inner product (85) with $w = 1$. Their positively homogeneous extensions

$$Z_n^{(0)}(y) := \frac{p_n(y)}{(\Sigma_{a=1}^d (y^a)^2)^{(n-1)/2}} , \quad y \neq 0 , \tag{105}$$

of degree 1 to $\mathbb{R}^d \backslash \{0\}$ provide *all* (normalized) eigenfunctions of our differential operator L_0 because (101), (105), (104) and Euler's homogeneity relation yield

$$L_0(Z_n^{(0)}) := \frac{1}{d-1} \Delta' Z_n^{(0)} = \frac{1}{d-1} \Delta Z_n^{(0)} = \frac{(n-1)(-n-d+1)}{d-1} \cdot Z_n^{(0)} \tag{106}$$

on S^{d-1}. Therefore, we see that in fact $(n-1)(n+d-1)/(d-1)$, $n = 0, 1, 2, \ldots$, are all the possible eigenvalues for L_0. *Especially* $\lambda_1^{(0)} = -1$ *is the smallest eigenvalue of* L_0 *with multiplicity* 1 *and* $\lambda_2^{(0)} = \cdots = \lambda_{d+1}^{(0)} = 0$ *is the next one with multiplicity* d; *there exist no other nonpositive eigenvalues for* L_0. Finally, Aleksandrov applies the "continuity method" in order to complete his proof by showing (97), (98) and (99). As we have just seen this is true in the special case $L = L_0$. It remains to prove this fact for a general differential operator L. For this reason we consider the array of differential operators L_τ respectively of weights

w_τ, defined by (87) respectively (86) with respect to the regular compact convex bodies $K_\tau^{(l_1)} := (1-\tau)K_0 + \tau K_{l_1}, \ldots, K_\tau^{(l_{d-1})} = (1-\tau)K_0 + \tau K_{l_{d-1}}$ $(0 \leqslant \tau \leqslant 1)$. Proposition 4.1(c) says that the eigenvalues $\lambda_\nu^{(\tau)}$ of L_τ together with w_τ $(\nu = 1, 2, \ldots)$ depend continuously on τ. Therefore, it suffices to prove that 0 is eigenvalue for all L_τ, w_τ *with multiplicity d* because then no higher eigenvalue $\lambda_\nu^{(0)}$ of L_0 with $\nu \geqslant d+2$ can move into the interval $(-\infty, 0]$ when τ runs from 0 to 1.

This will be done as follows. Let Z be an arbitrary solution of

$$L_\tau(Z) := D'(\text{Hess}'\, H_\tau^{(l_1)}, \ldots, \text{Hess}'\, H_\tau^{(l_{d-2})}, \text{Hess}'\, Z) = 0\,, \tag{107}$$

see (55) and (58). Then (61) implies

$$D'(\text{Hess}'\, H_\tau^{(l_1)}, \ldots, \text{Hess}'\, H_\tau^{(l_{d-3})}, \text{Hess}'\, Z, \text{Hess}'\, Z) \leqslant 0 \quad \text{on } S^{d-1} \tag{108}$$

Now

$$\int_{S^{d-1}} D'(\text{Hess}'\, H_\tau^{(l_1)}, \ldots, \text{Hess}'\, H_\tau^{(l_{d-3})}, \text{Hess}'\, Z, \text{Hess}'\, Z) \cdot H_\tau^{(l_{d-2})}\, \mathrm{d}\omega$$

$$= \int_{S^{d-1}} D'(\text{Hess}'\, H_\tau^{(l_1)}, \ldots, \text{Hess}'\, H_\tau^{(l_{d-2})}, \text{Hess}'\, Z) \cdot Z\, \mathrm{d}\omega = 0\,, \tag{109}$$

because of (88) and (107), so that we may deduce the equality in (108) from (109) and $H_\tau^{(l_{d-2})} > 0$. But Theorem 3.4 says that this occurs only if the relation

$$\text{Hess}'\, Z = q \cdot \text{Hess}'\, H_\tau^{(l_{d-2})} \tag{110}$$

holds on S^{d-1} with $q \equiv 0$ as it can be seen by insertion of (110) into (107). Thus $\text{Hess}'\, Z = \text{Hess}\, Z \equiv 0$, whence $Z(y) = \langle a, y \rangle = \Sigma_{b=1}^d a_b y^b$, with $a_1, \ldots, a_d =$ const, so that $Z_2^{(\tau)} = y^1, \ldots, Z_{d+1}^{(\tau)} = y^d$ are all linearly independent eigenfunctions of L_τ for the eigenvalue 0. This completes Aleksandrov's proof. □

Remark 4.2. By an approximation argument it may be seen that the Aleksandrov–Fenchel–Jessen inequalities (75) hold for arbitrary compact convex bodies. But this argument gives no information for the equality case which has not yet been completely solved. For this topic we refer to chapter 1.2.

5. Uniqueness theorems for convex hypersurfaces

In literature numerous theorems about the uniqueness of a regular compact convex body in $\mathbb{R}^d$ with prescribed infinitesimal behaviour of its boundary (up to certain congruence) exist. Most typical result in this direction is the following theorem.

Theorem 5.1. [Aleksandrov (1937, Section 7) and Fenchel and Jessen (1938), both without differentiability assumptions]. *If the boundaries of two regular compact convex bodies* K_0 and K_1 *in* $\mathbb{R}^d$ *have the same* ν*th mean radius of curvature at corresponding points with equal unit normal,*

$$P_\nu^{(0)}(-n) = P_\nu^{(1)}(-n)\,, \quad -n \in S^{d-1} \tag{111}$$

for a fixed ν $(1 \leq \nu \leq d-1)$, *then* K_0 *and* K_1 *only differ by a translation.*

Proof for $\nu > 1$ (*Outline*) [Chern (1959) *in case of differentiability* C_∞]. In the same manner as the Aleksandrov–Fenchel–Jessen inequalities (75) imply the concavity of the Brunn–Minkowski-function

$$\Psi(t) := \Big(V(\underbrace{K_t, \ldots, K_t}_{\nu+1}, \underbrace{B^d, \ldots, B^d}_{d-1-\nu})\Big)^{1/\nu+1}\,,$$
$$K_t := (1-t)K_0 + tK_1\,, \quad 0 \leq t \leq 1$$

(see Leichtweiss 1980, Satz 24.1), the corresponding inequalities (61) for the (multilinear) mixed discriminants imply the concavity of the function

$$\begin{aligned}\Phi(t) &:= \left[\binom{d-1}{\nu} D'(\underbrace{\mathrm{Hess}'\, H^{(t)}, \ldots, \mathrm{Hess}'\, H^{(t)}}_{\nu}, \underbrace{E', \ldots, E'}_{d-1-\nu})(-n)\right]^{1/\nu} \\ &= \left[\binom{d}{\nu} D(\underbrace{\mathrm{Hess}\, H^{(t)}, \ldots, \mathrm{Hess}\, H^{(t)}}_{\nu}, \underbrace{E, \ldots, E}_{d-\nu})(-n)\right]^{1/\nu} \\ &= [D_\nu(H^{(t)})(-n)]^{1/\nu}\end{aligned} \tag{112}$$

[compare eq. (56)], where $H^{(t)} := (1-t)H^{(0)} + tH^{(1)}$ is the support function of K_t. Therefore, after applying (30) and (111), the inequalities

$$D_\nu(\underbrace{H^{(0)}, \ldots, H^{(0)}}_{\nu-1}, H^{(1)}) \geq [D_\nu(H^{(0)})]^{(\nu-1)/\nu} \cdot [D_\nu(H^{(1)})]^{1/\nu} = D_\nu(H^{(1)}) \tag{113}$$

and (changing the part of K_0 and K_1)

$$D_\nu(H^{(0)}, \underbrace{H^{(1)}, \ldots, H^{(1)}}_{\nu-1}) \geq [D_\nu(H^{(0)})]^{1/\nu} \cdot [D_\nu(H^{(1)})]^{(\nu-1)/\nu} = D_\nu(H^{(0)}) \tag{114}$$

hold (see Leichtweiss 1980, Hilfssatz 22.2). Hereby equality only occurs if $\Phi(t)$ is linear in t, i.e., if we have equality in the underlying inequalities (61), whence (because of $\nu > 1$), after using (111), $\mathrm{Hess}'\, H^{(0)} = \mathrm{Hess}'\, H^{(1)}$. For this reason, in order to prove Theorem 5.1, it suffices to show equality in (113) and (114). This may be done by subtracting the generalized Minkowski's integral formulas

$$\int_{S^{d-1}} h^{(0)}\binom{d-1}{\nu}^{-1} D_\nu^{(\nu-1,1)}\,\mathrm{d}\omega = \int_{S^{d-1}} \binom{d-1}{\nu+1}^{-1} D_{\nu+1}^{(\nu,1)}\,\mathrm{d}\omega$$
$$= \int_{S^{d-1}} h^{(1)}\binom{d-1}{\nu}^{-1} D_\nu^{(\nu,0)}\,\mathrm{d}\omega \qquad (115)$$

and vice versa

$$\int_{S^{d-1}} h^{(1)}\binom{d-1}{\nu}^{-1} D_\nu^{(1,\nu-1)}\,\mathrm{d}\omega = \int_{S^{d-1}} \binom{d-1}{\nu+1}^{-1} D_{\nu+1}^{(1,\nu)}\,\mathrm{d}\omega$$
$$= \int_{S^{d-1}} h^{(0)}\binom{d-1}{\nu}^{-1} D_\nu^{(0,\nu)}\,\mathrm{d}\omega\,, \qquad (116)$$

with

$$D_\nu^{(\lambda,\nu-\lambda)} = D_\nu(\underbrace{H^{(0)},\ldots,H^{(0)}}_{\lambda},\underbrace{H^{(1)},\ldots,H^{(1)}}_{\nu-\lambda})\,,\quad 0\leq\lambda\leq\nu \qquad (117)$$

[compare eqs. (30) and (33)!], from each other. Namely, this yields

$$\int_{S^{d-1}} [h^{(0)}(D_\nu^{(\nu-1,1)} - D_\nu^{(0,\nu)}) + h^{(1)}(D_\nu^{(1,\nu-1)} - D_\nu^{(\nu,0)})]\,\mathrm{d}\omega = 0\,, \qquad (118)$$

which indeed implies equality in (113) and (114) if we assume (without loss of generality) $h^{(0)}>0$ and $h^{(1)}>0$ on S^{d-1}. □

In the case $\nu=1$, Chern gave a proof of Theorem 5.1 [which does not make use of the profound inequalities (113) and (114)] by another integral formula. This theorem was at first proved for $d=3$ and $\nu=1$ by Christoffel (1865), and for $d=3$ and $\nu=2$ by Minkowski (1903). Some important generalizations of it, replacing P_ν by a suitable function $\Phi(R_1,\ldots,R_{d-1},n,x)$ with $\partial\Phi/\partial R_i>0$ for $i=1,\ldots,d-1$, and using a maximum principle for the solutions of elliptic differential equations, are due to Aleksandrov (1962). For other generalizations with regard to parts of convex hypersurfaces, see Oliker (1979). We want to mention further that Theorem 5.1 is obviously equivalent to the following fact: *Two regular convex hypersurfaces F_0 and F_1 in $\mathbb{R}^d$ are congruent if they have equal "third fundamental form"*

$$e_{ij}\xi^i\xi^j\,,\quad e_{ij} = \left\langle \frac{\partial n}{\partial u^i}, \frac{\partial n}{\partial u^j}\right\rangle,\ i,j=1,\ldots,d-1\,, \qquad (119)$$

and equal νth mean radius of curvature at corresponding points, because the first condition induces the congruence of the spherical maps of F_0 and F_1.

Corollary 5.2 (Süss 1929, Satz 2). *A regular convex hypersurface F in $\mathbb{R}^d$ with the property*

$$P_\nu = \text{const}\,, \quad \nu \textit{ fixed}\,,\ 1 \leqslant \nu \leqslant d-1\,, \tag{120}$$

is a sphere.

"Dual" to Corollary 5.2 is the following theorem.

Theorem 5.3 [Liebmann (1990, pp. 107 and 109) in the case $d = 3$]. *A regular convex hypersurface F in $\mathbb{R}^d$ with the property*

$$H_\nu = \text{const}\,, \quad \nu \textit{ fixed}\,,\ 1 \leqslant \nu \leqslant d-1\,, \tag{121}$$

is a sphere.

Proof (*in the C_3-case for $\nu < d-1$*). By Newton's formulas (see Hardy, Littlewood and Pòlya 1934, p. 104), there is

$$\frac{H_1}{H_0} \geqslant \frac{H_2}{H_1} \geqslant \cdots \geqslant \frac{H_{\nu+1}}{H_\nu}\,, \quad H_0 := 1\,, \tag{122}$$

with equality only in the case $k_1 = \cdots = k_{d-1}$ or $F = \text{sphere}$. But (121) and the Minkowski-formulas (20) imply

$$\int_M (H_1 H_\nu - H_{\nu+1}) h \,\mathrm{d}F = \int_M (H_\nu - H_{\nu+1} h)\,\mathrm{d}F - H_\nu \int_M (H_0 - H_1 h)\,\mathrm{d}F$$
$$= 0\,, \tag{123}$$

whence indeed follows equality in (122) or $H_1 H_\nu - H_{\nu+1} \geqslant 0$ if we assume, without loss of generality, $h > 0$. □

For a proof of Theorem 5.3 in the case $\nu = d-1$ by a slightly different integral formula, see Walter (1989, pp. 186–187). Furthermore, we want to draw attention to a paper of Walter (1985) with far reaching extensions of Theorem 5.3 involving special "isoparametric hypersurfaces" (whose principal curvatures are all constant).

Theorem 5.3 is a consequence of a reflection theorem and – more general – of a uniqueness theorem for hypersurfaces of another type as theorem 5.1:

Theorem 5.4 (Voss 1956, Satz VI). *If the boundaries F_0 and F_1 of two regular compact convex bodies K_0 and K_1 in $\mathbb{R}^d$ of differentiability class C_3 have the same "lower" and "upper" orthogonal projection along the direction $e \in \mathbb{R}^d$ on a hyperplane in $\mathbb{R}^d$ as well as equal νth mean curvature at corresponding points with respect to this projection,*

$$H_\nu^{(0)}(x^{(0)}) = H_\nu^{(1)}(x^{(1)})\,, \quad \nu \textit{ fixed}\,,\ 1 \leqslant \nu \leqslant d-1\,, \tag{124}$$

where

$$x^{(1)} = x^{(0)} + \chi e \,, \tag{125}$$

then K_0 *and* K_1 *only differ by a translation along* (*the constant vector*) e.

The idea of the proof of this theorem consists in a modification of Steiner's continuous symmetrization: F_0 and F_1 are joined by a linear array F_t, given by

$$x^{(t)} := x^{(0)} + t\chi e \,, \quad 0 \leq t \leq 1 \,, \tag{126}$$

compare eq. (125)! Then Theorem 5.4 turns out to be a direct consequence of the integral formula

$$\begin{aligned}
0 &= \int_M (H_\nu^{(1)} - H_\nu^{(0)}) \langle \chi e, n^{(0)} \rangle \, \mathrm{d}F_0 \\
&= \int_0^1 \left[\int_M \frac{\partial H_\nu^{(t)}}{\partial t} \langle \chi e, n^{(t)} \rangle \, \mathrm{d}F_t \right] \mathrm{d}t \\
&= \frac{\nu}{(d-1)!} \int_0^1 \left[\int_M \Big(-\mathrm{d} \frac{\partial n^{(t)}}{\partial t}, \underbrace{-\mathrm{d}n^{(t)}, \ldots, -\mathrm{d}n^{(t)}}_{\nu-1}, \underbrace{\mathrm{d}x^{(t)}, \ldots, \mathrm{d}x^{(t)}}_{d-1-\nu}, \chi e \Big) \right] \mathrm{d}t \\
&= \frac{\nu}{(d-1)!} \int_0^1 \left[\int_M \Big(\underbrace{-\mathrm{d}n^{(t)}, \ldots, -\mathrm{d}n^{(t)}}_{\nu-1}, \underbrace{\mathrm{d}x^{(t)}, \ldots, \mathrm{d}x^{(t)}}_{d-1-\nu}, \mathrm{d}\chi e, -\frac{\partial n^{(t)}}{\partial t} \Big) \right] \mathrm{d}t \\
&= -\frac{\nu}{d-1} \int_M C_{(\nu)}^{ij} (\langle e, n^{(0)} \rangle)^2 \frac{\partial x}{\partial u^i} \frac{\partial \chi}{\partial u^j} \, \mathrm{d}F_0
\end{aligned} \tag{127}$$

[compare eq. (20)!], because hereby the matrix $(C_{(\nu)}^{ij})$ is positive definite and $\langle e, n^{(0)} \rangle$ vanishes only on the "shadow boundary" of $F_{(0)}$ with respect to projection along e.

Remark 5.5 (Aeppli 1959, *Satz* 10). Theorem 5.4 remains valid if we replace there the "parallel map" (125) by a "radial map"

$$x^{(1)} = \chi x^{(0)} \,, \quad \chi > 0 \,, \tag{128}$$

the equality (124) by the equality of the (dilatation invariant) "reduced νth mean curvature" at corresponding points

$$\|x^{(0)}\|^\nu H_\nu^{(0)}(x^{(0)}) = \|x^{(1)}\|^\nu H_\nu^{(1)}(x^{(1)}) \,, \quad \nu \text{ fixed} \,, \ 1 \leq \nu \leq d-1 \,, \tag{129}$$

and the translation along e by a dilatation with the origin of $\mathbb{R}^d$ as the center – under the additional assumption that all "joining hypersurfaces"

$$x^{(t)} := (1 + t(\chi - 1))x^{(0)}, \quad 0 \leq t \leq 1 \tag{130}$$

[compare eq. (126)], be regular convex hypersurfaces.

It should be mentioned that two different common generalizations of Theorem 5.4 and Remark 5.5, involving 1-parameter transformation groups of a Riemannian target space $\mathbb{R}^d$, have been made by Hopf and Katsurada (1968a,b), based on Stokes' theorem and on a maximum principle. For further information on the topic of this section, especially for $d = 3$, see Huck et al. (1973).

6. Convexity and relative geometry

In the so-called "relative differential geometry" of convex hypersurfaces, a regular convex hypersurface $F: M \to \mathbb{R}^d$ of differentiability class C_3, given by $x = x(u^1, \ldots, u^{d-1})$, is referred to a regular convex "gauge hypersurface" $N: M \to \mathbb{R}^d$ of differentiability class C_2, containing the origin in the "interior" and given by $-y = -y(u^1, \ldots, u^{d-1})$, by means of the so-called "Peterson map" between (equally oriented) points of F and N with parallel tangent hyperplanes:

$$-y_i := -\frac{\partial y}{\partial u^i} = B_i^k x_k, \quad i = 1, \ldots, d-1 \tag{131}$$

[compare eq. (9)], with

$$(x_1, \ldots, x_{d-1}, y) > 0. \tag{132}$$

(Thus Euclidean differential geometry is the special case $y = n$ of relative differential geometry!) Besides the "normalization vector" y of F we consider also its "conormal vector" X, defined by:

$$\langle X, x_i \rangle = \langle X, y_i \rangle = 0, \quad i = 1, \ldots, d-1, \tag{133}$$

and

$$\langle X, y \rangle = 1. \tag{134}$$

Because of the obvious symmetry relations

$$B_{ij} = B_i^k G_{kj} = \left\langle \frac{\partial X}{\partial u^j}, y_i \right\rangle = -\left\langle X, \frac{\partial^2 y}{\partial u^i \, \partial u^j} \right\rangle = \left\langle \frac{\partial X}{\partial u^i}, y_j \right\rangle = B_j^k G_{ki} = B_{ji} \tag{135}$$

for the coefficients of the (quadratic) "third fundamental form"

$$B_{ij} \xi^i \xi^j \tag{136}$$

of (F, N), compare eq. (119), with respect to the coefficients

$$G_{ij} := \langle X, \partial^2 x / \partial u^i \, \partial u^j \rangle = G_{ji}$$

of its *positive definite* (quadratic) "second fundamental form"

$$G_{ij} \xi^i \xi^j \tag{137}$$

[compare eq. (4)], the roots of the characteristic equation $\det(B_i^k - z\delta_i^k) = 0$ are all real. They are called the "*relative principal curvatures*" ${}_{\mathrm{r}}k_1, \ldots, {}_{\mathrm{r}}k_{d-1}$ of (F, N), and their inverses, ${}_{\mathrm{r}}R_1, \ldots, {}_{\mathrm{r}}R_{d-1}$, the "*relative principal radii of curvature*" of (F, N) with the normalized elementary symmetric functions ${}_{\mathrm{r}}H_\nu$ respectively ${}_{\mathrm{r}}P_\nu$ $(\nu = 1, \ldots, d-1)$. As we have assumed also the gauge hypersurface N to be regular convex, all the entities ${}_{\mathrm{r}}k_\nu$, ${}_{\mathrm{r}}R_\nu$, ${}_{\mathrm{r}}H_\nu$, ${}_{\mathrm{r}}P_\nu$ $(\nu = 1, \ldots, d-1)$ are positive in view of the positive definiteness of the matrix $(B_{ij}) = (\langle X, -\partial^2 y / \partial u^i \, \partial u^j \rangle)$, see eq. (135). They are related by the "relative geometric Minkowski's integral formulas"

$$\int_M {}_{\mathrm{r}}H_{\nu-1} \, \mathrm{d}_{\mathrm{r}}F = \int_M {}_{\mathrm{r}}H_\nu \, {}_{\mathrm{r}}h \, \mathrm{d}_{\mathrm{r}}F \,, \quad \nu = 1, \ldots, d-1 \,, \tag{138}$$

$$\int_M {}_{\mathrm{r}}P_{d-\nu} \, \mathrm{d}_{\mathrm{r}}N = \int_M {}_{\mathrm{r}}P_{d-\nu-1} \, {}_{\mathrm{r}}h \, \mathrm{d}_{\mathrm{r}}N \,, \quad \nu = 1, \ldots, d-1 \,, \tag{139}$$

involving the "relative geometric support function"

$${}_{\mathrm{r}}h := -\langle X, x \rangle \tag{140}$$

[compare eq. (21)] and the "relative surface area"

$$\int_M \mathrm{d}_{\mathrm{r}}F = \frac{1}{(d-1)!} \int_M (\underbrace{\mathrm{d}x, \ldots, \mathrm{d}x}_{d-1}, y) \tag{141}$$

of F, respectively the "relative surface area"

$$\int_M \mathrm{d}_{\mathrm{r}}N = \frac{1}{(d-1)!} \int_M (\underbrace{-\mathrm{d}y, \ldots, -\mathrm{d}y}_{d-1}, y) \tag{142}$$

of N [compare eqs. (20) and (33)!]. These Minkowski's formulas may be proved exactly as in the Euclidean case after replacing the Euclidean unit normal vector n by the normalization vector y. As an important consequence we note:

Theorem 6.1 [Süss (1927, p. 69) for $d = 3$]. *A pair (F, N) of regular compact convex hypersurfaces of differentiability class (C_3, C_2) with constant relative νth mean curvature ${}_{\mathrm{r}}H_\nu$ for a fixed ν with $1 \leq \nu \leq d-1$, is a "relative sphere", i.e., a pair of homothetic hypersurfaces F and N (compare Theorem* 5.3!).

The proof of this theorem is totally analogous to the proof of Theorem 5.3; it uses the Minkowski's formulas (138). Theorem 6.1 remains valid with ${}_{\mathrm{r}}H_{\nu}$ having been replaced by ${}_{\mathrm{r}}P_{\nu}$ ($1 \leq \nu \leq d-1$) [compare Corollary 5.2 and use Minkowski's formulas (139)!]. We remark that this fact is a trivial consequence of the following relative geometric generalization of Theorem 5.1.

Theorem 6.2. *The hypersurface F of the pair (F, N) of regular compact convex hypersurfaces of class (C_3, C_2) is uniquely determined up to a translation by N and the relative νth mean radius of curvature ${}_{\mathrm{r}}P_{\nu}$ for an arbitrarily fixed ν with $1 \leq \nu \leq d-1$.*

This theorem may be proved totally analogous to Theorem 5.1 for $\nu > 1$; for $\nu = 1$ there exists (in the C_{∞}-case) a proof, applying a suitable integral formula, due to Oliker and Simon (1984, Theorem 3.1). Finally, we cite in this context:

Theorem 6.3 [Schneider 1967, Satz 4.4 (4.1)]. *A pair (F, N) of regular compact convex hypersurfaces of class (C_3, C_3) is uniquely determined up to a nondegenerate affinity by its third (second) fundamental form* (136) (*eq.* (137)), *the relative $(d-1)$th (first) mean curvature ${}_{\mathrm{r}}H_{d-1}$ (${}_{\mathrm{r}}H_1$) and the "Tschebycheff-vector"*

$$V_i := G^{jk} A_{ijk}\,, \quad i = 1, \ldots, d-1\,, \tag{143}$$

where (G^{jk}) is inverse to (G_{ij}) and $A_{ijk} := \langle X, x_{i;j;k} \rangle$ are the (symmetric) coefficients of the "cubic fundamental form"

$$A_{ijk} \xi^i \xi^j \xi^k \tag{144}$$

of (F, N) ($x_{i;j;k}$ denote covariant derivatives with respect to the Riemannian metric $\mathrm{d}s^2 = G_{ij}\, \mathrm{d}u^i\, \mathrm{d}u^j$!).

7. Convexity and affine differential geometry

In the so-called "equiaffine differential geometry" the special conormal vector

$$X := \frac{[x_1, \ldots, x_{d-1}]}{[\det((x_1, \ldots, x_{d-1}, x_{ij}))]^{1/(d+1)}} \tag{145}$$

is assigned to a regular convex hypersurface F in $\mathbb{R}^d$ of differentiability class C_3, invariant against orientation preserving parameter transformations of F. Then the envelope of the hyperplanes $\langle X, z \rangle = 1$, reflected at the origin, may be used as a gauge hypersurface $-C$ (reflected "curvature image") for F (we assume for the moment that $-C$ must not be regular convex). Its position vector $-y$ [compare eqs. (134) and (133)] is called the negative "affine normal vector" of F, as $-C$ transforms itself in a homogeneous equivariant manner when F is transformed by an equiaffine (i.e., volume preserving affine) map of $\mathbb{R}^d$. From eqs. (145) and (134) we conclude for the "affine surface area" of F [compare eq. (141)]

$$\int_M \mathrm{d}_a F = \frac{1}{(d-1)!} \int_M (\mathrm{d}x, \underbrace{\ldots}_{d-1}, \mathrm{d}x, y)$$

$$= \int_M (\det(G_{ij}))^{1/2}\, \mathrm{d}u^1 \cdots \mathrm{d}u^{d-1}\,, \tag{146}$$

whence after logarithmic partial differentiation of the integrands in (146),

$$V_i = 0\,, \quad i = 1, \ldots, d-1 \tag{147}$$

("apolarity conditions"). So we have natural specializations of Theorem 6.3 in the equiaffine differential geometry (e.g., see Schneider 1967, Satz 4.3). Now the notions of relative mean curvatures (respectively relative mean radii of curvature) immediately translate to equiaffine differential geometry and we have the following theorem.

Theorem 7.1. *A regular compact convex hypersurface F of class C_5 with constant affine νth mean curvature ${}_aH_\nu$ for a fixed ν with $1 \leqslant \nu \leqslant d-1$, is an ellipsoid* (*compare Theorem* 5.3!).

This follows from the fact that such a hypersurface must be a "proper affine sphere", i.e., a hypersurface with homothetic F and $-C$ (compare Theorem 6.1), and therefore an ellipsoid by a famous theorem of Blaschke (1923, §74 and §77) for $d = 3$ and Deicke (1953) for general d. For a short proof, see Schneider (1967, pp. 395–396), for further information, Simon (1985).

There are special results for regular compact convex hypersurfaces F *with regular convex reflected curvature image* $-C$ which we will denote as being "*of elliptic type*". For an example, we cite the following theorem.

Theorem 7.2 (Leichtweiss 1990, Satz 1). *A regular compact convex hypersurface F_0 of class C_4 and elliptic type, containing another one F_1* (*not necessarily of elliptic type*) *in its interior, has a bigger affine surface area* (146) *than F_1 unless F_0 and F_1 coincide.*

Finally, we mention affine geometrical interpretations of the notions "affine normal vector" and "affine surface area", given by Blaschke (1923, §43 and §47) for $d = 3$, and by Leichtweiss (1989, 1986) for general d, that are connected with each other in a certain sense.

References

Aeppli, A.
[1959] Einige Ähnlichkeits- und Symmetriesätze für differenzierbare Flächen im Raum, *Comment. Math. Helv.* **33**, 174–195.

Aleksandrov, A.D.
[1937] Neue Ungleichungen zwischen den gemischten Volumina und ihre Anwendungen (in Russian), *Mat. Sb. N.S.* **2**, 1205–1238.

[1938] Mixed discriminants and mixed volumes (in Russian), *Mat. Sb. N.S.* **3**, 227–251.
[1939] Almost everywhere existence of the second differential of a convex function and some properties of convex surfaces connected with it (in Russian), *Učn. Zap. Leningrad Gos. Univ. Math. Ser.* **6**, 3–35.
[1962] Uniqueness theorems for surfaces in the large I–V, *Amer. Math. Soc. Transl. Ser. 2* **21**, 341–416.

Bangert, V.
[1979] Analytische Eigenschaften konvexer Funktionen auf Riemannschen Mannigfaltigkeiten, *J. Reine Angew. Math.* **307**, 309–327.

Blaschke, W.
[1923] *Vorlesungen über Differentialgeometrie II* (Springer, Berlin).
[1956] *Kreis und Kugel* (Walter de Gruyter, Berlin, 2. Aufl.).

Brooks, J.N., and J.B. Strantzen
[1989] Blaschkes rolling theorem in R^n, *Mem. Amer. Math. Soc.* **405**, 101 pp.

Busemann, H.
[1958] *Convex surfaces* (Interscience, New York).

Busemann, H., and W. Feller
[1936] Krümmungseigenschaften konvexer Flächen, *Acta Math.* **66**, 1–47.

Chern, S.S.
[1959] Integral formulas for hypersurfaces in euclidean spaces and their applications to uniqueness theorems, *J. Math. Mech.* **8**, 5–12.

Chern, S.S., and R. Lashof
[1957] On the total curvature of immersed manifolds, *Amer. J. Math.* **79**, 306–318.
[1958] On the total curvature of immersed manifolds II, *Michigan Math. J.* **5**, 5–12.

Christoffel, E.B.
[1865] Über die Bestimmung einer krummen Fläche durch lokale Messungen auf derselben, *J. Reine Angew. Math.* **64**, 193–209.

Deicke, A.
[1953] Über die Finsler-Räume mit $A_i = 0$, *Arch. Math.* **4**, 45–51.

Fenchel, W.
[1929] Über die Krümmung und Windung geschlossener Raumkurven, *Math. Ann.* **101**, 238–252.

Fenchel, W., and B. Jessen
[1938] Mengenfunktionen und konvexe Körper, *Danske Vidensk. Selsk. Math.-Fys. Medd.* **16**, 1–31.

Hadamard, J.
[1897] Sur certaines propriétés des trajectoires en dynamique, *J. Math. Pures Appl.* **3**, 331–387.

Hardy, G.H., J.E. Littlewood and G. Pòlya
[1934] *Inequalities* (Cambridge Univ. Press).

Hilbert, D.
[1912] *Grundzüge einer allgemeinen Theorie linearer Integralgleichungen* (Teubner, Leipzig).

Hopf, H., and Y. Katsurada
[1968a] Some congruence theorems for closed hypersurfaces in Riemann spaces I, *Comment. Math. Helv.* **43**, 176–194.
[1968b] Some congruence theorems for closed hypersurfaces in Riemann spaces II, *Comment. Math. Helv.* **43**, 217–223.

Huck, H., et al.
[1973] *Beweismethoden der Differentialgeometrie im Großen*, Lecture Notes in Mathematics, Vol. 335 (Springer, Berlin).

Leichtweiss, K.
[1973] Über den Kalkül von K. Voss zur Herleitung von Integralformeln, *Abh. Math. Sem. Univ. Hamburg* **39**, 15–20.

[1980] *Konvexe Mengen* (Deuts. Verl. Wiss., Berlin).
[1986] Über eine Formel Blaschkes zur Affinoberfläche, *Studia Sci. Math. Hungar.* **21**, 453–474.
[1989] Über eine Deutung des Affinnormalenvektors einseitig gekrümmter Hyperflächen, *Arch. Math.* **53**, 613–621.
[1990] Bemerkungen zur Monotonie der Affinoberfläche von Eihyperflächen, *Math. Nachr.* **147**, 47–60.

Liebmann, H.
[1900] Über die Verbiegung geschlossener Flächen positiver Krümmung, *Math. Ann.* **53**, 91–112.

Minkowski, H.
[1903] Volumen und Oberfläche, *Math. Ann.* **57**, 447–495.

Müller, C.
[1966] *Spherical harmonics,* Lecture Notes in Mathematics, Vol. 17 (Springer, Berlin).

Oliker, V.I.
[1979] On certain elliptic differential equations on a hypersphere and their geometric applications, *Indiana Univ. Math. J.* **28**, 35–51.

Oliker, V.I., and U. Simon
[1984] The Christoffel problem in relative differential geometry, *Colloq. Math. Soc. János Bolyai* **46**, 973–1000.

Sacksteder, R.
[1960] On hypersurfaces with no negative sectional curvatures, *Amer. J. Math.* **82**, 609–630.

Sard, A.
[1942] The measure of the critical values of differentiable maps, *Bull. Amer. Math. Soc.* **48**, 883–890.

Schneider, R.
[1967] Zur affinen Differentialgeometrie im Großen I, *Math. Z.* **101**, 375–406.
[1979] Boundary structure and curvature of convex bodies, *Proc. Geom. Symp., Siegen, 1978*, pp. 13–59.
[1988] Closed convex hypersurfaces with curvature restrictions, *Proc. Amer. Math. Soc.* **103**, 1201–1204.

Simon, U.
[1985] Zur Entwicklung der affinen Differentialgeometrie nach Blaschke, in: *Wilhelm Blaschke Gesammelte Werke,* Bd. 4 (Thales Verlag, Essen).

Stoker, J.J.
[1936] Über die Gestalt positiv gekrümmter offener Flächen im dreidimensionalen Raum, *Compositio Math.* **3**, 55–88.

Süss, W.
[1927] Zur relativen Differentialgeometrie I: Über Eilinien und Eiflächen in der elementaren und affinen Differentialgeometrie, *Japan J. Math.* **4**, 57–75.
[1929] Zur relativen Differentialgeometrie V: Über Eiflächen im R^{n+1}, *Tôhoku Math. J.* **31**, 202–209.

van Heijenoort, J.
[1952] On locally convex manifolds, *Comm. Pure Appl. Math.* **5**, 223–242.

Voss, K.
[1956] Einige differentialgeometrische Kongruenzsätze für geschlossene Flächen und Hyperflächen, *Math. Ann.* **131**, 180–218.
[1960] Differentialgeometrie geschlossener Flächen im Euklidischen Raum I, *Jber. Deutsch. Math.-Vereinig.* **63**, 117–135.

Walter, R.
[1985] Compact hypersurfaces with a constant higher mean curvature function, *Math. Ann.* **270**, 125–145.
[1989] *Differentialgeometrie* (Bibliographisches Institut, Mannheim, 2. Aufl.).

CHAPTER 4.2

Convex Functions

A.W. ROBERTS

Mathematics Department, Macalester College, St. Paul, MN 55105, USA

Contents

HANDBOOK OF CONVEX GEOMETRY
Edited by P.M. Gruber and J.M. Wills

A real valued function f defined on an interval I of the real line is said to be convex if for all x, $y \in I$ and $\lambda \in [0, 1]$,

$$f[\lambda x + (1-\lambda)y] \leq \lambda f(x) + (1-\lambda) f(y) .$$

We say f is strictly convex if the inequality is strict for all x and y, $x \neq y$. In terms of a graph, the definition requires that if P, Q, and R are any three points on the graph of f with Q between P and R, then Q is on or below the chord PR (fig. 1). The definition can be taken as a statement about the slopes of the segments pictured in fig. 1,

$$\text{slope } PQ \leq \text{slope } PR \leq \text{slope } QR .$$

The papers of Jensen (1905, 1906) are generally cited as the first systematic study of the class of convex functions, but earlier work that noted properties of such functions is summarized in Roberts and Varberg (1973, p. 8).

It is easily seen that f is convex if and only if the set of points above its graph, its *epigraph*, (fig. 1) is a convex set in the plane. The study of convex functions is therefore subsumed by the study of convex sets, so that most studies of convexity (including this handbook) focus on convex sets. Convex functions do arise naturally, however, in optimization, analytic inequalities, functional analysis, and applied mathematics, so that there has arisen a vast literature that treats convexity in language familiar to analysts, that of functions. Our purpose is to survey that literature, according to the following outline:

1. Basic notions: Mid-convexity and continuity; Lower semi-continuity and closure of convex functions; Conjugate convex functions.

2. Differentiability of convex functions: Functions defined on $\mathbb{R}$; A function defined on $\mathbb{R}^2$; Functions defined on a linear space $\mathscr{L}$; Differentiable convex functions.

3. Inequalities: Classical inequalities obtained from convex functions; Matrix inequalities.

We conclude this introductory section with a briefly annotated bibliography of surveys of our topic, the complete citations of which can be found in the references listed at the end of the article.

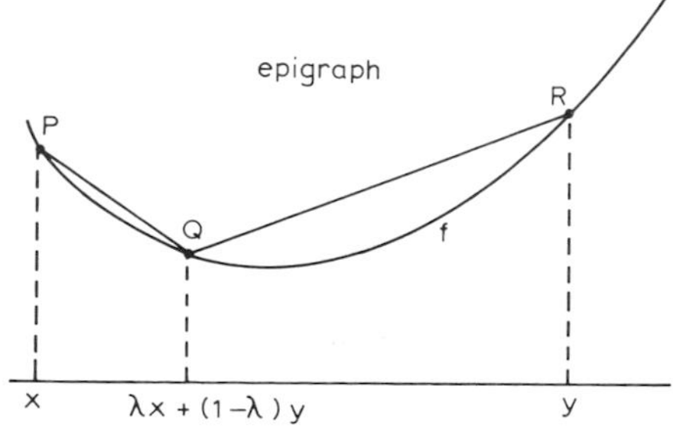

Figure 1.

(Beckenbach 1948). This article from the Bulletin of the AMS gives the flavor of some later, more extensive, frequently cited, but unpublished notes by the author.

(Ekeland and Temam 1976). Translated from the French, this was one of the first general works to introduce convex analysis into the calculus of variations.

(Fenchel 1953). Another set of frequently quoted but unpublished notes used as the basis of lectures at Princeton, these notes are given special mention for their influence on Rockafellar's *Convex Analysis*.

(Giles 1982). This text, acknowledging its debt to the unpublished 1978 notes of Phelps, focused on making accessible to graduate students the research on differentiability of convex functions.

(Moreau 1966). These are again lecture notes that were influential in the development of what is now called convex analysis.

(Phelps 1989). Perhaps prompted by the Giles book, Phelps finally brought up to date and then published the 1978 notes he had used at the University of London.

(Roberts and Varberg 1973). This text, written for an undergraduate audience, gives accessible proofs to most of the fundamental properties of convex functions.

(Rockafellar 1970b). Restricted to convex functions on $\mathbb{R}^n$, this carefully written book stands as the most complete reference on the topic. It makes extensive use of the notion of conjugate functions.

(Van Tiel 1984). Here is another book aimed at the undergraduate which provides a handy place to look for proofs of the basic properties of convex functions.

1. Basic notions

1.1. Midconvexity and continuity

Jensen, in his classic papers, said a function f was convex if it satisfied an inequality we shall take as the definition of midconvexity; f is *midconvex* on an interval I if for every $x, y \in I$,

$$f\left(\frac{x+y}{2}\right) \leqslant \tfrac{1}{2}[f(x) + f(y)] .$$

Examples of discontinuous functions satisfying this inequality were known to Jensen, so he naturally addressed the question of what minimal additional conditions would guarantee the continuity of a midconvex function. He first established what is now known as Jensen's inequality.

Theorem 1. *f is midconvex on I if and only if*

$$f\left(\sum_{i=1}^{n} \alpha_i x_i\right) \leqslant \sum_{i=1}^{n} \alpha_i f(x_i)$$

for any n points $x_i \in I$ and any n nonnegative rational α_i such that $\Sigma_{i=1}^{n} \alpha_i = 1$.

He then showed that a midconvex function defined and bounded on an open interval would be continuous there.

This set in motion a series of papers that still continue, in which people strive for minimal additional conditions to imply continuity. It is known, for example, that a midconvex function defined on $[a, b]$ will be continuous on (a, b) if it is bounded above on a set M of Lebesgue measure $m(M) > 0$, or if it is bounded above on a second category Baire set. Roberts and Varberg (1973, chapter 7) summarize and give references to papers giving conditions that, along with midconvexity, imply continuity.

Results of this kind also occur for functions defined on spaces other than the real line. If U is a convex set in a linear space $\mathscr{L}$, and if $f : U \to \mathbb{R}$ is a real valued function defined on U, then the definitions of both convexity and midconvexity still make sense. Of course, to talk about the continuity of f, we need some sort of topology on $\mathscr{L}$, and most convex analysis is carried out with the understanding that $\mathscr{L}$ is a normed linear space. That seems the right context in which to end our discussion of the continuity of midconvex functions. A midconvex function defined on an open set U in a normed linear space is continuous on U if it is bounded above in a neighborhood of a single point of U.

1.2. Lower semi-continuity and closure of convex functions

If a convex function is defined on a nonempty open set in a locally convex linear topological space $\mathscr{L}$, then it is quite easy to describe the continuity properties of f.

Theorem 2. *Let $f : U \to \mathbb{R}$ be convex on a nonempty open set $U \subset \mathscr{L}$. If f is bounded above in a neighborhood of just one point p of U, then f is continuous on U* (Roberts and Varberg 1973, p. 67).

It is possible, of course, that a convex function may fail to be bounded above in a neighborhood of even a single point; well-known examples of discontinuous linear functionals defined on infinite dimensional linear spaces all fail to be bounded above in a neighborhood of any point. One might summarize with the observation that a convex function defined on an open set U is either continuous on U, or wildly discontinuous there. In particular, if $U \subset \mathscr{L} = \mathbb{R}^n$, then f is bounded in a neighborhood of every point $p \in U$, and is continuous on U (Roberts and Varberg 1973, p. 93).

Two difficulties that enter into this otherwise tidy situation can be illustrated by letting U be the half plane subset of $\mathbb{R}^3$

$$U = \{(x, y, 0): x > 0\}$$

and defining the convex function $f : U \to \mathbb{R}$ by:

$$f(x, y, z) = y^2/x .$$

There is a natural sense in which U has interior points, but since no point of U

has, in the topology of $\mathbb{R}^3$, a neighborhood in U, it is not meaningful to talk about the continuity of f until we embed U in its affine hull (the entire xy plane in our example). Then we can say that f is continuous on U in the relative topology. Though easily done, this procedure introduces the complications of phrasing everything in terms of the relevant topology, and it is commonly avoided by the expedient described below of taking any convex function to be defined on all $\mathscr{L}$.

There are, in fact, far more compelling reasons than the one just mentioned for extending to all of $\mathscr{L}$ the definition of a convex function originally defined only on a subset of $\mathscr{L}$. It is very useful in any setting, such as optimization, where we are working with a large set of functions, to have them all defined on a common domain; in convex programming problems, it enables us to build constraints into the objective function to be minimized by defining the function to be infinity outside of its feasible set; and it is absolutely essential to have our functions globally defined if we are to take full advantage of the duality we shall meet when we introduce conjugate convex functions in the next section.

To properly set the stage for this program, we need to consider functions $f: U \to \bar{\mathbb{R}}$, where $\bar{\mathbb{R}} = \{\mathbb{R} \cup (+\infty) \cup (-\infty)\}$, possibly taking the values $\pm\infty$. This requires careful, but common sense arithmetic rules involving $+\infty$ and $-\infty$, and we must modify our definition to say that f is convex if and only if for any x and y in U for which there are real numbers α and β with $f(x) < \alpha$, $f(y) < \beta$,

$$f[\lambda x + (1-\lambda)y] \leq \lambda\alpha + (1-\lambda)\beta$$

when $0 < \lambda < 1$. In this setting, we say that the *effective domain* of f is $\mathrm{dom}(f) = \{x \in \mathscr{L}: f(x) < +\infty\}$, and a convex function is called *proper* if $\mathrm{dom}(f) \neq \emptyset$. We shall avoid some technical difficulties with the understanding that all the convex functions we discuss are proper. Proper convex functions never take the value of $-\infty$.

Now given a function $f: U \to \mathbb{R}$ convex on $U \subset \mathscr{L}$, it would be possible to extend f to all of $\mathscr{L}$ and preserve convexity by simply defining $f(x) = \infty$ for any $x \notin U$. Quite obviously, however, we would sacrifice whatever continuity properties f might have had. It is natural to wonder, in the case of a convex function continuous on U (perhaps with an appropriately chosen relative topology), whether we might do better, and it is here that our example above illustrates another difficulty. Since an approach to $(0, 0)$ along the parabolic path $x = y^2/m$ results in a limit of m for any $m > 0$, no definition of $f(0, 0)$ will make f continuous at $(0, 0)$. We face the fact that even in the relative topology that enables us to talk about the continuity of f on U, we may still be unable to extend f to the closure of U so as to retain the continuity of f.

It is easy to show (Fenchel 1949) that if a convex function f is defined at a limit point p of its domain U, then

$$\liminf_{u \to p} f(u) \leq f(p) .$$

This suggests that lower semi-continuity is the right goal to have in mind when extending the definition of a convex function to include the limit points of its domain.

This works particularly well when $U \subset \mathbb{R}^n$. Then given a convex function $f : U \to \bar{\mathbb{R}}$ that never takes the value $-\infty$, we define clf to be identical with f on the interior of U, and we define it at the limit points p of U by:

$$\mathrm{clf}(p) = \liminf_{u \to p} f(u)$$

when this limit is finite; otherwise we set $f(p) = \infty$. To complete the definition, let clf be the constant $-\infty$ for the case in which f assumes the value $-\infty$ at some point. This function is lower semi-continuous.

Theorem 3. *Let $f : U \to \bar{\mathbb{R}}$ be a convex function that is not identically $+\infty$ on $U \subset \mathbb{R}^n$. Then* clf *is a lower semi-continuous convex function that agrees with f except possibly at relative boundary points of the* dom(f) (Rockafellar 1970b, p. 56).

Apart from necessary fussiness over details, we may conclude that any convex function can, with some possible redefinitions on the boundary for which we have a constructive method, be taken to be lower semi-continuous; and since all this will be true if we extend clf to all of $\mathbb{R}^n$ by setting $\mathrm{clf}(x) = +\infty$ for all x not in the closure of U, we may take any convex function to be globally defined.

For functions $f : \mathscr{L} \to \bar{\mathbb{R}}$, we define the epigraph of f to be the set $\mathrm{epi}(f) = \{(x, \alpha) \in \mathbb{R}^n \times \mathbb{R}: \alpha \geqslant f(x)\}$. It is still true that f is convex if and only if its epigraph, epi f is a convex subset $\mathscr{L} \times \bar{\mathbb{R}}$. It can also be shown that f is lower semi-continuous if and only if epi f is a closed subset of $\mathscr{L} \times \bar{\mathbb{R}}$. For this reason, clf is called the *closure* of f.

At this point we see an argument for subordinating the study of convex functions to an epigraphical viewpoint that immerses the study of functions in a study of their epigraphs. In their argument for this approach, Rockafellar and Wets (1984) point out that the graph of $f : \mathbb{R}^n \to \bar{\mathbb{R}}$ is not well defined as a subset of $\mathbb{R}^{n+1}$ because $f(x)$ may be $+\infty$; the graph is really a subset of $\mathbb{R}^n \times \bar{\mathbb{R}}$. The epigraph, however, does lie entirely in $\mathbb{R}^{n+1}$.

A basic reference for the epigraphical perspective is the monograph of Attouch (1984), which considers various convergence notions for convex functions in terms of their epigraphs. One important concept here for the case when the underlying space is reflexive is Mosco convergence (Mosco 1969). A promising approach for general normed linear spaces is the Attouch–Wets convergence, where convergence of epigraphs means uniform convergence on bounded subsets of $\mathscr{L} \times \mathbb{R}$ of the distance functions for the epigraphs. This reduces to ordinary Hausdorff metric convergence when restricted to closed and bounded convex sets and is well suited for estimation and approximation. Also, this notion of convergence is stable with respect to duality without reflexivity (Beer 1990).

Corresponding to a nonempty convex set U in a normed linear space $\mathscr{L}$, there are four globally defined convex functions that play a prominent role in the literature:

The *indicator function*

$$I(x|U) = \begin{cases} 0 & \text{if } x \in U\,, \\ +\infty & \text{if } x \notin U\,. \end{cases}$$

The *gauge* function

$$G(x|U) = \inf\{\lambda \geqslant 0 \mid x \in \lambda U\}\,.$$

The *support* function

$$S(\ell, U) = \sup\{\langle \ell, x\rangle \mid x \in U\} \quad \text{for } U \subset \mathbb{R}^n,\ \ell \in \mathcal{L}^*\,.$$

The *distance* function

$$D(x|U) = \inf\{\|x - y\| \mid y \in U\}\,.$$

1.3. Conjugate convex functions

The two-variable version of the geometric mean–arithmetic mean inequality as stated in Theorem 18 below can be written, for $x > 0$, $y > 0$, in the form

$$xy \leqslant f(x) + g(y)\,, \tag{1}$$

where f and g are convex functions defined, with $p > 0$, $q > 0$ and $1/p + 1/q = 1$, by:

$$f(x) = \frac{1}{p}\, x^p\,, \qquad g(y) = \frac{1}{q}\, y^q\,.$$

Our interest is in inequalities of the form (1).

Another inequality of this form from classical analysis involves integrals. Consider a function $h : [0, \infty) \to [0, \infty)$ that is strictly increasing and continuous with $h(0) = 0$ and $\lim_{t\to\infty} h(t) = \infty$; in such a case, h^{-1} exists and has the same property as h. This allows us to define two convex functions

$$f(x) = \int_0^x h(t)\,\mathrm{d}t\,, \qquad g(y) = \int_0^y h^{-1}(t)\,\mathrm{d}t$$

for which Young's inequality then says

$$xy \leqslant f(x) + g(y)\,.$$

Fenchel began a seminal paper (1949) by calling attention to Young's inequality and to one more example. If we set the support function S and the gauge function G defined at the end of the previous section for $U \subset \mathbb{R}^n$ equal to f and g, respectively, they satisfy the inequality (1). Generalizing from these observations, he then proved a theorem that has turned out to be a fundamental tool in the study of convex functions.

Theorem 4. *Let f be a convex function defined on a convex point set $\mathcal{F} \subset \mathbb{R}^n$ so as to be lower semi-continuous and such that* $\lim_{x \to p} f(x) = \infty$ *for each boundary point p of $\mathcal{F}$ that does not belong to $\mathcal{F}$. Then there exists a unique convex function g defined on a convex set $\mathcal{G}$ with exactly the same properties and such that*

$$x_1 y_1 + \cdots + x_n y_n \leq f(x_1, \ldots, x_n) + g(y_1, \ldots, y_n) .$$

Moreover, the relationship is dual in the sense that if we begin with g defined on the convex set $\mathcal{G}$, we obtain f and $\mathcal{F}$.

The functions f and g are called *conjugate* convex functions.

Beginning with a convex function f having the required continuity properties, the unique conjugate function guaranteed by Theorem 4 is denoted by f^*. The usual procedure is simply to define, for $y \in \mathcal{L}$,

$$f^*(y) = \sup_{x \in \mathcal{F}} \{\langle x, y \rangle - f(x)\}$$

and then show that it has the desired properties. The notation $\langle x, y \rangle$ is used for $y(x)$ to emphasize the duality.

If one is not so careful about the lower semi-continuity of f, it is still possible to define the conjugate f^*, but it can then be proved that f^* is a closed convex function; that is, its epigraph will be a closed set in $\mathcal{L} \times \mathbb{R}$. There is no hope, then, of achieving the complete duality of having $f^{**} = f$ unless f is a closed function to begin with. It is for this reason that we find the conventions mentioned at the end of the last section to be convenient.

Though pairs of functions satisfying (1) entered into an earlier paper by Birnbaum and Orlicz (1931), Fenchel (1949) gave the first general treatment of conjugate convex functions, and his work has had far reaching ramifications to which we can only allude here.

We shall see when we ask about the existence of the derivatives of convex functions that one of the most satisfying answers comes in the form of the relationship between the subdifferential of f^* and the inverse of the subdifferential of f; $\partial(f^*) = (\partial f)^{-1}$.

The convex programming problem, which is to minimize a convex function f over a constraint set K, can be replaced, according to the Rockafellar–Fenchel Theorem (Holmes 1972, p. 68) with the dual problem of maximizing the sum of the conjugate of f and the indicator function of the set K. Rockafellar (1970b, 1974) has developed these ideals very fully.

After exploring the role of the duality of conjugate convex functions in the calculus of variations and in minimax theory, Ekeland and Temam (1976) take up applications to numerical analysis, control theory, mechanics, and economics.

This list does not exhaust the applications that are made of conjugate convex functions. In addition to the authors already mentioned, Ioffe and Tikhomirov (1968) provide a good survey of the applications of conjugacy.

2. Differentiability

We begin our consideration of the differentiability of convex functions with a careful look at functions of a single variable, first because easily drawn graphs often expose the heart of proofs that can be carried to more abstract settings, and secondly because the properties we shall discover help us anticipate what is true in general. In section 2.2, we use a particular function of two variables to introduce related concepts central to our survey in section 2.3 of differentiability of convex functions defined on Banach spaces. In section 2.4 we turn from questions of existence to look at what can be proved about convex functions known to be differentiable throughout an open set.

2.1. Functions defined on $\mathbb{R}$

The characterization of convexity in terms of slopes of secant chords says for the four points shown in fig. 2 that

$$\text{slope } PQ \leqslant \text{slope } PR \leqslant \text{slope } QR \leqslant \text{slope } QS \leqslant \text{slope } RS\ .$$

In particular,

$$\frac{f(x)-f(y)}{x-y} \leqslant \frac{f(z)-f(y)}{z-y}\ . \tag{2}$$

Since slope $PR \leqslant$ slope QR, it is clear that slope QR increases as $x \uparrow y$. Similarly, slope RS decreases as $z \downarrow y$. Thus the quotient on the left side of (2) increases as

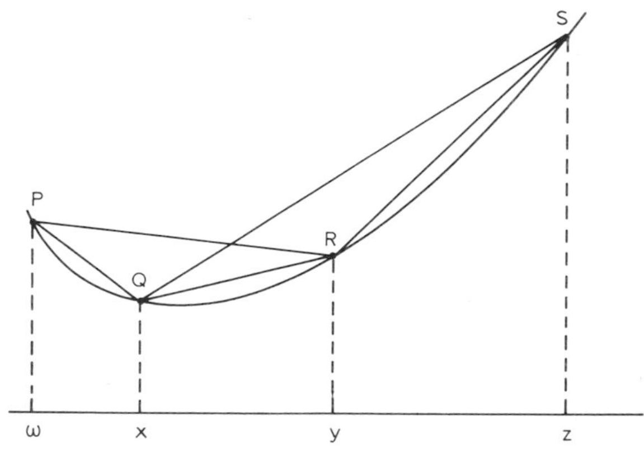

Figure 2.

$x \uparrow y$, the one on the right decreases as $z \downarrow y$, and we have established our first theorem.

Theorem 5. *At an arbitrary point y interior to its interval of definition, a convex function f has both a left derivative* $f'_-(y)$ *and a right derivative* $f'_+(y)$; *moreover,* $f'_-(y) \leqslant f'_+(y)$.

When $f'_-(y) = f'_+(y) = m$, then f is differentiable at y, and a line with slope m is tangent to the graph at $R(y, f(y))$. Otherwise we may choose any m satisfying $f'_-(y) < m < f'_+(y)$ and draw a line with slope m through R that lies entirely under the graph of f. Such a line is said to be a *line of support* for f at y.

Referring once again to our statement comparing slopes of secant lines, we see that

$$f'_+(w) \leqslant \frac{f(x) - f(w)}{x - w} \leqslant \frac{f(y) - f(x)}{y - x} \leqslant f'_-(y)$$

with all inequalities strict if f is strictly convex. Thus, drawing on Theorem 5, we can write

$$f'_-(w) \leqslant f'_+(w) \leqslant f'_-(y) \leqslant f'_+(y) .$$

Theorem 6. *If* $f : I \to \mathbb{R}$ *is convex (strictly convex), then* $f'_-(x)$ *and* $f'_+(x)$ *exist and are increasing (strictly increasing) on their respective domains.*

Further analysis of fig. 2 establishes quickly that

$$\lim_{x \uparrow y} f'_+(x) = f'_-(y) \quad \text{and} \quad \lim_{z \downarrow y} f'_+(z) = f'_+(y) .$$

From these two facts, we conclude that $f'_-(y) = f'_+(y)$ if and only if f'_+ is continuous at y. Stated another way, the derivative fails to exist at precisely those points where the increasing function f'_+ is discontinuous. Since an increasing function can be discontinuous on at most a countable set, we have proved the following theorem.

Theorem 7. *If* $f : I \to \mathbb{R}$ *is convex on an interval I, the set E where f' fails to exist is countable. Moreover, f' is continuous on* $I - E$.

Finally, let us note that if the convex function f is differentiable at x_0, then for every $x > x_0$ in the domain of f,

$$f[(1 - \alpha)x_0 + \alpha x] \leqslant (1 - \alpha)f(x_0) + \alpha f(x) .$$

If we set $h = \alpha(x - x_0)$, we see that

$$\frac{f(x_0 + h) - f(x_0)}{h} \leqslant \frac{f(x) - f(x_0)}{x - x_0} .$$

Consideration of the case $x < x_0$ leads to the same inequality, and taking limits as $h \to 0$ gives us our last theorem of this section.

Theorem 8. *If f is convex on an interval I and differentiable at x_0, then for any $x \in I$,*

$$f(x) - f(x_0) \geq f'(x_0)(x - x_0) .$$

2.2. Related concepts

Many of the very deep results about the differentiation of convex functions are couched in terms that can be nicely illustrated with a function of two variables. Consider the graph (fig. 3) of the function

$$f(x, y) = \begin{cases} x^2 + y^2 , & x^2 + y^2 \geq 2 , \\ 2 , & x^2 + y^2 \leq 2 . \end{cases}$$

If $P(x_0, y_0, f(x_0, y_0))$ is a point on the graph, then

$$z = A(x, y) = f(x_0, y_0) + a(x - x_0) + b(y - y_0)$$

is a plane that meets the graph of f at P. It is instructive to look at two such planes:

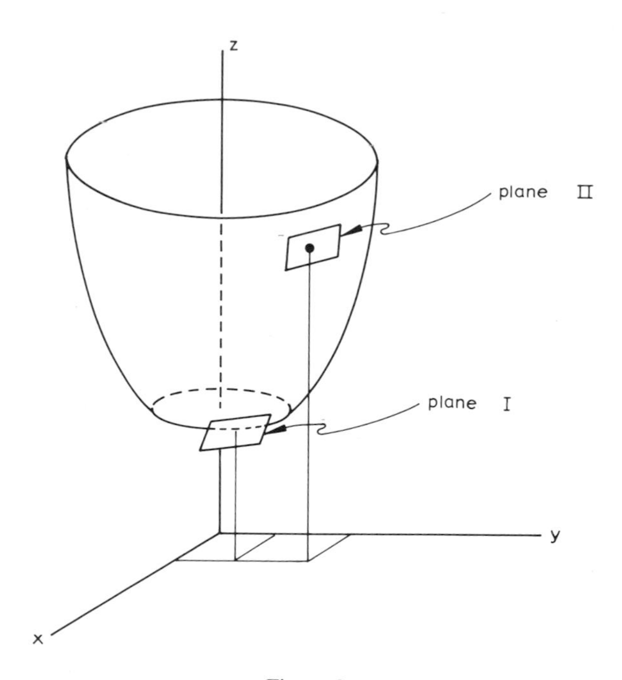

Figure 3.

$$\text{Plane I, through } (1,1,2)\,, \quad z = 2 + \tfrac{3}{2}(x-1) + \tfrac{3}{2}(y-1)\,,$$

$$\text{Plane II, through } (1,2,5)\,, \quad z = 5 + 2(x-1) + 4(y-2)\,.$$

Both meet the graph of f in exactly one point, and otherwise lie below it. A plane that meets the graph of $z = f(x, y)$ in at least one point and never rises above it is called a *support plane*. Our example illustrates the following facts about support planes.

(1) A plane of support may meet the graph of $z = f(x, y)$ in many points; $z = 2$ is a support plane that meets the graph in fig. 3 in infinitely many points.

(2) There may be many planes of support passing through the same point on the graph of $z = f(x, y)$; besides Plane I and the plane $z = 2$, there are obviously many other planes of support to the graph of fig. 3 at $(1, 1, 2)$.

(3) The plane of support at a point (x_0, y_0, z_0) might be unique, as in Plane II, the plane tangent to the graph at $(1, 2, 5)$ in our example.

(4) A function $f(x, y)$ is convex if and only if it has at least one plane of support of each point (x_0, y_0, z_0) on its graph.

The most obvious generalization of the derivative of a function of one variable to a function of two variables (not necessarily convex) is the so-called directional derivative. Starting at $\boldsymbol{p}_0(x_0, y_0)$ and moving in the direction of the unit vector $\boldsymbol{v} = [r\ s]$, we define:

$$f'(\boldsymbol{p}_0)(\boldsymbol{v}) = \lim_{t \to 0} \frac{f(\boldsymbol{p}_0 + t\boldsymbol{v}) - f(\boldsymbol{p}_0)}{t}\,.$$

In our example, with $\boldsymbol{p}_0(1, 2)$ and $\boldsymbol{v} = [\frac{4}{5}\ \frac{3}{5}]$, it may be shown that $f'(\boldsymbol{p}_0)(\boldsymbol{v}) = 4$. This is the slope of the line formed by the intersection of the tangent Plane II and the plane through $(1, 2)$ that contains $\boldsymbol{v}$ and is perpendicular to the xy plane.

The directional derivatives for the choice $\boldsymbol{v} = [1\ 0]$ and $\boldsymbol{v} = [0\ 1]$, when they exist, are called the partial derivatives $\partial f/\partial x(x_0, y_0)$ and $\partial f/\partial y(x_0, y_0)$. When the directional derivative exists at $\boldsymbol{p}_0$ for every choice of $\boldsymbol{v}$, we say f is *Gateaux* differentiable at $\boldsymbol{p}_0$.

In general, Gateaux differentiability does not guarantee all that we would like to be true of a differentiable function. The Gateaux derivative of f may exist at a point even if f is discontinuous there; and $f'(\boldsymbol{p}_0)(\boldsymbol{v})$ need not be linear in $\boldsymbol{v}$. All this changes if it is known that f is convex; convexity imposes orderlines. If a convex function is Gateaux differentiable at a point, it is continuous there, and $f'(\boldsymbol{p}_0)(\boldsymbol{v})$ will be linear in $\boldsymbol{v}$.

If the convex function f has continuous partial derivatives in a neighborhood of (x_0, y_0), then the support plane, $A(x, y)$ as defined above, is unique. It is the plane tangent to the graph at (x_0, y_0, z_0), and the coefficients a and b that determine the plane are the components of the gradient vector ∇f of elementary calculus which, with an eye to the future we shall call

$$\mathrm{d}f = \left[\frac{\partial f}{\partial x}(x_0, y_0) \quad \frac{\partial f}{\partial y}(x_0, y_0) \right].$$

We turn now to a second generalization of the derivative which, though it brings us to the same place for convex functions defined on $U \subset \mathbb{R}^n$, will prove to be a stronger concept for functions defined on a normed linear space.

Consider again the function $A(x, y)$ defined above. It is said to be *affine*. It is the sum of a constant $f(x_0, y_0)$ and a linear function, a fact that can be emphasized by writing it in the form

$$A(x, y) = f(x_0, y_0) + [a\ b]\begin{bmatrix} x - x_0 \\ y - y_0 \end{bmatrix}.$$

We now adopt the viewpoint that the derivative of f at (x_0, y_0) is the linear transformation from $\mathbb{R}^2$ to $\mathbb{R}$, represented by the matrix $[a\ b]$. The special thing about this linear transformation is, of course that it is a linear transformation that closely approximates $A(x, y) - f(x_0, y_0)$. That is the idea that gives rise to our definition.

Using $\Delta x = x - x_0$ and $\Delta y = y - y_0$, we say that f is Frechet differentiable at (x_0, y_0) if there exists a linear transformation L such that

$$f(x, y) = f(x_0, y_0) + L(\Delta x, \Delta y) + |(\Delta x, \Delta y)|\varepsilon(x_0, y_0, \Delta x, \Delta y)\ ,$$

where $\varepsilon(x_0, y_0, \Delta x, \Delta y) \to 0$ as $(\Delta x, \Delta y) \to 0$. The linear transformation L, if it exists, is easily shown to be unique; it is called the *Frechet derivative* $f'(x_0, y_0)$ of f at (x_0, y_0).

As we have already said, for a convex function f defined on $U \subset \mathbb{R}^n$, the Frechet derivative is identical to the Gateaux derivative:

$$\mathrm{d}f = \left[\frac{\partial f}{\partial x}(x_0, y_0)\quad \frac{\partial f}{\partial y}(x_0, y_0)\right] = f'(x_0, y_0)\ .$$

There is yet a third way to look at this expression. We may regard d as a special case of a set valued operator ∂ that maps a point (x_0, y_0) into a set of linear transformations, any one of which will define a plane of support. Thus, when f is differentiable at (x_0, y_0), $\partial f(x_0, y_0)$ is the single transformation $\mathrm{d}f(x_0, y_0)$, but when f is not differentiable, $\partial f(x_0, y_0)$ is many valued. In our example,

$$\partial(1, 1) = [t\ t]\ ,\quad \text{where } t \in [0, 2]\ .$$

The set valued operator ∂ is called the *subdifferential* of f at (x_0, y_0), and a particular member of the set, such as $[\frac{3}{4}\ \frac{3}{4}]$ is called a subgradient. Our example makes it clear that a function will have a subdifferential at points where it may not have a derivative in the sense of either Gateaux or Frechet. It is this fact that makes the subdifferential such a useful tool in convex analysis.

We have seen that a convex function of a single real variable has, by virtue of its convexity, numerous differentiability properties. It is already clear from the example in this section, which is nondifferentiable on $x^2 + y^2 = 2$, that we cannot hope to establish differentiability on all but a countable set. Under what conditions will a convex function on $\mathscr{L}$ have a Gateaux differential? a Frechet derivative? When will these derivatives be equal? How can we characterize sets of nondifferentiability? These questions have led to a rich, deep, and largely satisfying literature to which we now turn.

2.3. *Functions defined on a linear space $\mathscr{L}$*

A function $f: U \to \mathbb{R}$ defined on an open set U in a linear topological space $\mathscr{L}$ is said to be Gateaux differentiable at $x_0 \in U$ provided that

$$\lim_{t \to 0} \frac{f(x_0 + tv) - f(x_0)}{t}$$

exists for every $v \in \mathscr{L}$. The limit, when it exists, is called the Gateaux differential $\mathrm{d}f(x_0)$.

Though not true of the Gateaux differential in general, $\mathrm{d}f(x_0)$ turns out to be a linear functional on $\mathscr{L}$ if f is a convex function; and while $\mathrm{d}f(x_0)$ may be a discontinuous linear functional (as, for example, when f itself is a discontinuous linear functional) it can be shown to be continuous if the convex function f is continuous at x_0.

Under what conditions will f have a Gateaux differential at x_0? For convex functions defined on $U \subset \mathbb{R}^n$, the answer is easy: it is sufficient that the n partial derivatives exist and are finite. To answer this in greater generality, we need to extend to $\mathscr{L}$ some of the concepts first met in $\mathbb{R}^2$. We say $f: U \to \mathbb{R}$ has *support* at $x_0 \in U$ if there exists an affine function $A: \mathscr{L} \to \mathbb{R}$ such that $A(x_0) = f(x_0)$ and $A(v) \leq f(v)$ for every $v \in \mathscr{L}$. With what we know about convex functions of a single real variable and a standard application of the Hahn–Banach Theorem, it can be shown that f is convex on U if and only if f has support at each $x \in U$. If the convex function f is continuous, then the support functions will be continuous; the converse is more tricky to prove, but it will be true if $\mathscr{L}$ is a normed linear space.

Theorem 9. *The convex function $f: U \to \mathbb{R}$ has a Gateaux differential at $x_0 \in U$ if and only if f has unique support at x_0.*

The existence of a convex function on ℓ^∞ that is everywhere continuous and nowhere Gateaux differentiable (Phelps 1989, p. 13) shows that we will not extend to an arbitrary space $\mathscr{L}$ the fact that a convex function of a single real variable is differentiable except on at most a countable set. The usual response is to put conditions on $\mathscr{L}$ as well as on f, and the most familiar theorem of this sort is due to Mazur (1933).

Theorem 10. *If $\mathscr{L}$ is a separable Banach space and f is a continuous convex function defined on an open set $U \subset \mathscr{L}$, then the set of points x where the Gateaux differential* $\mathrm{d}f(x)$ *exists is a dense G_δ set in U.*

Another not quite equivalent definition of a derivative gets a lot of attention in the literature. We say that $f : U \to \mathbb{R}$ is Frechet differentiable at $x_0 \in U$ if there exists a linear functional $L : \mathcal{L} \to \mathbb{R}$ such that for sufficiently small $v \in \mathcal{L}$,

$$f(x_0 + v) = f(x_0) + L(v) + \|v\| \varepsilon(x_0, v) ,$$

where $\varepsilon(x_0, v) \to 0$ as $v \to 0$. When such a functional L exists, it is, as was noted above, unique; it is called the *Frechet derivative* and is designated by $f'(x_0)$.

If the Frechet derivative $f'(x_0)$ exists, then so does the Gateaux differential $\mathrm{d}f(x_0)$ and $f'(x_0) = \mathrm{d}f(x_0)$, but not conversely. There are examples of convex functions that are Gateaux differentiable at every nonzero point but nowhere Frechet differentiable. Consequently, Frechet and Gateaux derivatives are sometimes referred to as strong and weak derivatives respectively.

A word about continuity is in order. A linear transformation $f : \mathcal{L} \to \mathbb{R}$ is both Gateaux and Frechet differentiable at any x_0 in $\mathcal{L}$, and $f = \mathrm{d}f(x_0) = f'(x_0)$. Since there are, on infinite dimensional spaces $\mathcal{L}$, well-known examples of discontinuous linear (and so convex) functions, it follows that neither the Gateaux nor the Frechet derivative of a convex function needs to be continuous.

If, however, the convex function is continuous at x_0, then the existence of $\mathrm{d}f(x_0)$ is enough to also guarantee that $\mathrm{d}f(x_0)$ is a continuous linear functional; similarly with $f'(x_0)$. Moreover, for the Frechet derivative, it is a two way street. The derivative $f'(x_0)$ of a convex function is a continuous linear transformation if and only if the function f is continuous at x_0.

It is enough for our purposes to say that caution must be exercised, even when $\mathcal{L} = \mathbb{R}^n$, where the existence of the Gateaux differential $\mathrm{d}f(x_0)$ does not by itself guarantee the continuity of f. We shall simplify our own exposition by assuming henceforth that our function $f : U \to \mathbb{R}$ is convex and continuous on $U \subset \mathcal{L}$. This will guarantee that when either $\mathrm{d}f(x_0)$ or $f'(x_0)$ exists, it will be a member of the dual space $\mathcal{L}^*$ of continuous linear functionals on $\mathcal{L}$.

If $U \subset \mathbb{R}^n$, it can be proved that a continuous convex function $f : U \to \mathbb{R}$ has a Frechet derivative almost everywhere (in the sense of Lebesgue) (Roberts and Varberg 1973, p. 116). In the context of general linear spaces, attempts to characterize sets on which $f : U \to \mathbb{R}$ is differentiable, Gateaux or Frechet, have led to deep and revealing connections between convex functions, the geometry of Banach spaces, measure theory, and the study of monotone operators. Progress on the general question is generally dated from the work of Asplund (1968) who took a two pronged approach. He examined spaces more general than separable Banach spaces in which the results of Mazur's theorem still held, and he also examined a sub-class of Banach spaces in which the conclusion of Mazur's Theorem held for Frechet differentiability. He called these latter strong differentiability spaces, but common terminology now associates his name with both classes of spaces.

A Banach space $\mathcal{L}$ is an *Asplund space* if every continuous convex function defined on a nonempty open convex set $U \subset \mathcal{L}$ is Frechet differentiable at each point of a dense G_δ subset of D; and it is a *weak Asplund space* if Frechet is replaced by Gateaux in the definition.

In this terminology, Mazur's Theorem says that if a Banach space $\mathscr{L}$ is separable, then $\mathscr{L}$ is a weak Asplund space. Asplund proved that if the dual $\mathscr{L}^*$ of a Banach space $\mathscr{L}$ is separable, then $\mathscr{L}$ is an Asplund space. The converse was proved later (Namioka and Phelps 1975), and provides a very satisfying characterization.

Theorem 11. *A separable Banach space is an Asplund space if and only if its dual space $\mathscr{L}^*$ is separable.*

In fact, more can be said. A Banach space is an Asplund space if and only if every separable closed subspace has a separable dual.

The rich interconnections mentioned above have made it possible to characterize Asplund spaces in other ways. It turns out that a Banach space $\mathscr{L}$ is an Asplund space if and only if $\mathscr{L}^*$ has the Raydon–Nikodym Property; that is, roughly speaking, the classical Raydon–Nikodym Theorem holds for suitably restricted vector-valued measures with values in $\mathscr{L}^*$. This idea is developed very fully by Diestral and Uhl (1977) and by Bourgin (1983).

Another characterization figures into what we shall say below, so we shall state it as a theorem here (Giles 1982, p. 202; Phelps 1989, p. 80).

Theorem 12. *A Banach space $\mathscr{L}$ is an Asplund space if and only if every nonempty weak* compact convex subset of $\mathscr{L}^*$ is the weak* closed convex hull of its weak* strongly exposed points.*

Characterization of weak Asplund spaces turns out to be more difficult. Asplund showed that if the norm of a given Banach space can be replaced by an equivalent norm that has a strictly convex dual norm, then $\mathscr{L}$ is a weak Asplund space; and Preiss, Phelps and Namioka (1990) showed very recently that the existence of an equivalent Gateaux differentiable norm on a Banach space $\mathscr{L}$ implies that $\mathscr{L}$ is a weak Asplund space. Complete characterization, however, remains a research problem.

A case can be made (Phelps 1989, chapter 6) for trying instead to characterize *Gateaux differentiability spaces*. These are defined to be Banach spaces in which any convex function defined on a nonempty open set U is Gateaux differentiable on a set G that is dense in U. This differs from a weak Asplund space only in that the set G does not need to be a G_δ set, but the difference is enough to allow an analogue to Theorem 12. The Banach space will be a Gateaux differentiability space under exactly the same conditions specified in Theorem 12 except that one speaks about exposed points rather than strongly exposed points.

We turn finally to the topic of the subdifferential of a convex function $f : U \to \mathbb{R}$ defined on a set U of a linear space $\mathscr{L}$. Since the concept finds its application not only in the treatment of nondifferentiable functions, but in the theory of optimization where discontinuous convex functions arise in natural ways, we abandon in this context the understanding we have had that the functions under consideration are continuous.

Roughly speaking, the subdifferential is an operator that maps each $x \in U$ into a subset $\partial f(x)$ of $\mathscr{L}^*$ that contains all the possible candidates for $f'(x)$. More formally,

$$\partial f(x_0) = \{L \in \mathscr{L}^*\colon A(x) = f(x_0) + L(x - x_0) \leq f(x) \text{ for all } x \in U\}\ .$$

There can be points for which $\partial f(x)$ is empty, even when f is convex (Rockafellar 1970a, p. 215), but the important fact is that $\partial f(x)$ is most frequently nonempty, in which case we say that f is *subdifferentiable* at x. When it is recalled that any convex function is, or can by re-definitions at certain boundary points be made lower semi-continuous, one sees the importance of the Bronsted–Rockafellar Theorem which guarantees that if $\mathscr{L}$ is a Banach space, then ∂f exists on a dense subset of the effective domain of f. Of course much more can be said when f is continuous at $x_0 \in U$. Then $\partial f(x_0)$ is nonempty, convex, and weak* compact in $\mathscr{L}^*$ (Giles 1982, p. 132).

Let x and y be two points in the interior of an open set $U \subset \mathscr{L}$ on which f is convex, and choose two subgradients $L_x \in \partial f(x)$ and $L_y \in \partial f(y)$. Then

$$L_x(y - x) \leq f(y) - f(x)\ ,$$
$$L_y(x - y) \leq f(x) - f(y)$$

and addition gives

$$(L_y - L_x)(y - x) \geq 0\ .$$

A set valued function $T : \mathscr{L} \to \mathscr{L}^*$ is called a *monotone* operator if for all x and y whenever $L_x \in T(x)$ and $L_y \in T(y)$, $(L_y - L_x)(y - x) \geq 0$, so the calculation above shows that a subdifferential is a monotone operator. Subdifferentials may therefore be studied in the context of what is known about monotone operators, a program that has been carried out by Phelps (1989) for functions defined on a normed linear space $\mathscr{L}$, and in great detail for the case in which $\mathscr{L} = \mathbb{R}^n$ by Rockafellar (1970b). Rockafellar (1970a) has in fact been able to characterize those monotone operators which are subdifferentials of convex functions, and recent work by Simons (1991) offers a much shorter proof.

The convex function f is Gateaux differentiable at x_0 if and only if $\partial f(x_0)$ consists of exactly one $L \in \mathscr{L}^*$; and it is Frechet differentiable at x_0 if and only if in addition to the uniqueness of $L \in \partial f(x_0)$, the operator ∂f is norm to norm upper-semi-continuous at x_0 (Giles 1982, p. 122). The great utility of the subdifferential comes, of course, in those situations where it enables one to say something about a nondifferentiable function. This occurs in optimization, for instance (Phelps 1989, p. 44).

Theorem 13. *A proper lower semi-continuous convex function f has a global minimum at x_0 if and only if $\mathscr{O} \in \partial f(x_0)$.*

2.4. *Differentiable convex functions*

When a function $f: U \to \mathbb{R}$ is convex on $U \subset \mathcal{L}$ and Frechet differentiable at $x_0 \in U$, then $\partial f(x_0)$ consists of the single linear functional $f'(x_0)$, and for any other $x \in U$,

$$f(x) - f(x_0) \geq f'(x_0)(x - x_0).$$

This inequality characterizes convex functions that are differentiable throughout U (Roberts and Varberg 1973, p. 98).

Theorem 14. *If if is differentiable throughout $U \subset \mathcal{L}$, then f is convex if and only if*

$$f(x) - f(x_0) \geq f'(x_0)(x - x_0)$$

for all x and $x_0 \in U$. Moreover, f is strictly convex if and only if the inequality is strict whenever $x \neq x_0$.

If f is continuous and differentiable throughout an open set $U \subset \mathcal{L}$, the computations performed above for differentials show that f' is a monotone operator. This also characterizes convex functions that are differentiable throughout U.

Theorem 15. *Let $f: U \to \mathbb{R}$ be continuous and differentiable on the open convex set $U \subset \mathcal{L}$. Then f is convex if and only if*

$$[f'(x) - f'(y)](x - y) \geq 0$$

and f is strictly convex if and only if the inequality is strict whenever $x \neq y$.

The convexity of functions $f: U \to \mathbb{R}$ that are twice differentiable on $U \subset \mathcal{L}$ can also be characterized in terms of their second derivatives. The Frechet derivative is a mapping $f': U \to \mathcal{L}^*$ from one normed space to another, and as such may itself be differentiable, in which case $f''(x)$ is a bilinear transformation on $\mathcal{L}$, commonly written $f''(x)(h, k)$. Under reasonably general conditions (Dieudonne 1960, p. 175), $f''(x)$ is symmetric, enabling us to ask whether $f''(x)$ is nonnegative definite.

Theorem 16. *Let f be twice differentiable throughout $U \subset \mathcal{L}$. Then f is convex if and only if $f''(x)$ is nonnegative definite for every $x \in U$; and if $f''(x)$ is positive definite on U, then f is strictly convex.*

This is of course what one would expect when thinking intuitively about the graph of a convex function of a single variable. The tangent line will rotate counterclockwise as it moves along the curve.

3. Inequalities

3.1. Classical inequalities

A rich study at which we can only hint in the available space uses convex functions to derive well-known and not so well-known inequalities with great elegance. One chain of inequalities with their proofs will give the idea.

Theorem 17 (Jensen's Inequality). *Let f be a convex function defined on the (possibly infinite) open interval (a, b), and let $x_i \in (a, b)$. If $\alpha_i > 0$ and $\Sigma_1^n \alpha_i = 1$, then*

$$f\Big(\sum_1^n \alpha_i x_i\Big) \leqslant \sum_1^n \alpha_i f(x_i) .$$

Proof. Let $x_0 = \Sigma_1^n \alpha_i x_i$; the convex function f has support at x_0, so there is a support function $L(x) = f(x_0) + m(x - x_0)$ that satisfies $L(x_i) \leqslant f(x_i)$ for each i. Multiply each side by α_i and sum, remembering that $\Sigma_1^n a_i = 1$, to obtain the inequality. □

We will now use Jensen's inequality with the choice of $f(t) = \exp(t)$.

Theorem 18 (The Geometric Mean–Arithmetic Mean Inequality). *If $x_i \geqslant 0$, $\alpha_i > 0$, and $\Sigma_1^n \alpha_i = 1$, then*

$$x_1^{\alpha_1} x_2^{\alpha_2} \cdots x_n^{\alpha_n} \leqslant \alpha_1 x_1 + \alpha_2 x_2 + \cdots + \alpha_n x_n .$$

Proof. Since the inequality is obvious if any $x_i = 0$, we need only consider the case where all $x_i > 0$, enabling us to define $t_i = \ln x_i$. Then

$$x_i^{\alpha_i} = \exp(\alpha_i \ln x_i) = \exp(\alpha_i t_i) ,$$

$$\begin{aligned} x_1^{\alpha_1} x_2^{\alpha_2} \cdots x_n^{\alpha_n} &= \exp(\alpha_1 t_1 + \cdots + \alpha_n t_n) \\ &\leqslant \alpha_1 \exp(t_1) + \cdots + \alpha_n \exp(t_n) \\ &= \alpha_1 x_1 + \cdots + \alpha_n x_n . \end{aligned}$$ □

The choice of $\alpha_i = 1/n$ for each i gives us the GM–AM inequality in its classic form and with this inequality available, it is but a short jump to the inequalities of Hölder, Cauchy–Bunyakovski–Schwarz, and Minkowski. We shall move on to a less well-known inequality in the chain we are pursuing.

Theorem 19. *If $x_i \geqslant 0$, $y_i \geqslant 0$, and n is a positive integer, then*

$$\Big[\prod_1^n (x_i + y_i)\Big]^{1/n} \geqslant \Big(\prod_1^n x_i\Big)^{1/n} + \Big(\prod_1^n y_i\Big)^{1/n} .$$

Proof. Again we need only concern ourselves with the case where $x_i + y_i > 0$ for all i. Some algebraic manipulation and the classic form of the GM–AM inequality gives us

$$\frac{\left(\prod_1^n x_i\right)^{1/n} + \left(\prod_1^n y_i\right)^{1/n}}{\left[\prod_1^n (x_i + y_i)\right]^{1/n}} = \left(\prod_1^n \frac{x_i}{x_i + y_i}\right)^{1/n} + \left(\prod_1^n \frac{y_i}{x_i + y_i}\right)^{1/n}$$

$$\leqslant \frac{1}{n} \sum_1^n \frac{x_i}{x_i + y_i} + \frac{1}{n} \sum_1^n \frac{y_i}{x_i + y_i} = 1 . \qquad \square$$

3.2. Matrix inequalities

We shall conclude this section by deriving some matrix inequalities. Let Ω_n be the class of doubly stochastic matrices, that is matrices with nonnegative entries in which the sum of elements in any row or any column is 1. Ω_n is easily shown to be a convex set in the space of all $n \times n$ matrices, and it can be shown that the extreme points of Ω_n are the permutation matrices, that is matrices obtained by some permutation of the rows of the identity matrix. If, for a fixed vector $\boldsymbol{v} \in \mathbb{R}^n$, we let

$$K_v = \{\boldsymbol{x} \in \mathbb{R}^n : \boldsymbol{x} = S\boldsymbol{v},\ S \in \Omega_n\}$$

and observe that K_v is a convex set in $\mathbb{R}^n$, we can obtain an optimization theorem for matrices.

Theorem 20. *Given a vector $\boldsymbol{v} \in \mathbb{R}^n$ and a convex function $f : U \to \mathbb{R}$ where U contains K_v, define $g : \Omega_n \to \mathbb{R}$ by:*

$$g(S) = f(S\boldsymbol{v}) = f(\langle \boldsymbol{s}_1, \boldsymbol{v} \rangle, \ldots, \langle \boldsymbol{s}_n, \boldsymbol{v} \rangle) ,$$

where s_i is the ith row vector of S. Then g must be convex, and it assumes its maximum value at a permutation matrix P.

Proof. Verification of the convexity of g follows in a straightforward way from the definition. Now let $P_1 \cdots P_m$ be the $m = n!$ permutation matrices in Ω_n. Since they are the extreme points of Ω_n, any $S \in \Omega_n$ may be written in the form $S = \Sigma_1^m \alpha_i P_i$ for some choice of $\alpha_i \geqslant 0$ where $\Sigma_1^m \alpha_i = 1$. The convexity of g then enables us to write

$$g(S) \leqslant \sum_1^m \alpha_i g(P_i) \leqslant \sum_1^m \alpha_i g(P_v) = g(P_v) ,$$

where P_v is one of the P_i, chosen so that $g(P_v) = \max\{g(P_i)\}$. $\square$

We are now in a position to prove a general inequality for convex functions of symmetric matrices, from which numerous matrix inequalities may be easily derived.

Theorem 21. *Let A be a real symmetric $n \times n$ matrix with eigenvalues $r_1, \ldots, r_n$. Set $\boldsymbol{v} = (r_1, \ldots, r_n)$ and let $f : U \to \mathbb{R}$ be convex on a set U that contains K_v. Then for any orthonormal set $\{\boldsymbol{p}_1, \ldots, \boldsymbol{p}_n\}$,*

$$f(\langle A\boldsymbol{p}_1, \boldsymbol{p}_1\rangle, \ldots, \langle A\boldsymbol{p}_n, \boldsymbol{p}_n\rangle) \leqslant f(P_v \boldsymbol{v}) ,$$

where P_v is the permutation matrix whose existence is asserted by Theorem 20.

Proof. Let $\{\boldsymbol{u}_1, \ldots, \boldsymbol{u}_n\}$ be the normalized eigenvectors of A corresponding respectively to $r_1, \ldots, r_n$. Define S to be the matrix with entries $s_{ij} = \langle \boldsymbol{u}_i, \boldsymbol{p}_j \rangle^2$. S is clearly symmetric and it is an exercise to show it doubly stochastic. Next we note that for any j,

$$\begin{aligned} A\boldsymbol{p}_j &= A[\langle \boldsymbol{u}_1, \boldsymbol{p}_j\rangle \boldsymbol{u}_1 + \cdots + \langle \boldsymbol{u}_n, \boldsymbol{p}_j\rangle \boldsymbol{u}_n] \\ &= \langle \boldsymbol{u}_1, \boldsymbol{p}_j\rangle r_1 \boldsymbol{u}_1 + \cdots + \langle \boldsymbol{u}_n, \boldsymbol{p}_j\rangle r_n \boldsymbol{u}_n \end{aligned}$$

so

$$\begin{aligned} \langle A\boldsymbol{p}_j, \boldsymbol{p}_j\rangle &= \langle \boldsymbol{u}_1, \boldsymbol{p}_j\rangle^2 r_1 + \cdots + \langle \boldsymbol{u}_n, \boldsymbol{p}_j\rangle^2 r_n \\ &= s_{1j} r_1 + \cdots + s_{nj} r_n = \langle \boldsymbol{s}_j, \boldsymbol{v}\rangle , \end{aligned}$$

where $\boldsymbol{s}_j$ is the jth row vector of S.

$$f(\langle A\boldsymbol{p}_1, \boldsymbol{p}_1\rangle, \ldots, \langle A\boldsymbol{p}_n, \boldsymbol{p}_n\rangle) = f(\langle \boldsymbol{s}_1, \boldsymbol{v}\rangle, \ldots, \langle \boldsymbol{s}_n, \boldsymbol{v}\rangle) = f(S\boldsymbol{v}) .$$

Since f is convex on a set that contains K_v, we know from Theorem 20 that $f(S\boldsymbol{v}) \leqslant f(P_v \boldsymbol{v})$.

To prove our last result, it will be convenient to know something about the function f defined for $\boldsymbol{x} = (x_1, \ldots, x_n) \in \mathbb{R}^n_+$, the positive orthant where $x_i \geqslant 0$ for all i, by:

$$f(\boldsymbol{x}) = f(x_1, \ldots, x_n) = (x_1 \cdot x_2 \cdots x_n)^{1/n} .$$

According to Theorem 19, for $\alpha + \beta = 1$ and $\boldsymbol{x}, \boldsymbol{y} \in \mathbb{R}^n_+$,

$$\begin{aligned} f(\alpha \boldsymbol{x} + \beta \boldsymbol{y}) = \left[\prod_1^n (\alpha x_i + \beta y_i)\right]^{1/n} &\geqslant \left(\prod_1^n \alpha x_i\right)^{1/n} + \left(\prod_1^n \beta y_i\right)^{1/n} \\ &\geqslant \alpha f(\boldsymbol{x}) + \beta f(\boldsymbol{y}) \end{aligned}$$

so $-f$ is convex on $\mathbb{R}^n_+$. □

Theorem 22 (Hadamard's Determinant Theorem). *If A is a real nonnegative definite matrix with entries a_{ij}, then* $\det A \leqslant a_{11} \cdot a_{22} \cdots a_{nn}$.

Proof. Let the eigenvalues of A be $r_1, \ldots, r_n$ and form $\boldsymbol{v} = (r_1, \ldots, r_n)$. The fact that the determinant of a matrix is equal to the product of its eigenvalues can be expressed in terms of the function f above by writing $\det A = [f(\boldsymbol{v})]^n$; indeed, $\det A = f(P\boldsymbol{v})$ for any permutation matrix P. Our appeal is to Theorem 18. The domain of f surely contains K_v; we take the orthonormal set $\{\boldsymbol{p}_1, \ldots, \boldsymbol{p}_n\}$ to be the standard basis, meaning that $\boldsymbol{p}_j$ has a 1 in the jth position as its only nonzero entry, and that $\langle A\boldsymbol{p}_j, \boldsymbol{p}_j \rangle = a_{jj}$. We conclude for the convex $-f$ that

$$-f(\langle A\boldsymbol{p}_1, \boldsymbol{p}_1 \rangle, \ldots, \langle A\boldsymbol{p}_n, \boldsymbol{p}_n \rangle) \leqslant -f(P_v \boldsymbol{v})$$

which is equivalent to $a_{11} \cdot a_{22} \cdots a_{nn} \geqslant \det A$. □

References

Asplund, E.
[1968] Frechet differentiability of convex functions, *Acta Math.* **121**, 31–47.

Attouch, H.
[1984] *Variational Convergence for Functions and Operators* (Pitman, Boston).

Beckenbach, E.F.
[1948] Convex functions, *Bull. Amer. Math. Soc.* **54**, 439–460.

Beer, G.
[1990] Conjugate convex functions and the epi-distance topology, *Proc. Amer. Math. Soc.* **108**, 117–126.

Birnbaum, Z., and W. Orlicz
[1931] Über die Verallgemeinerung des Begriffes der zueinander konjugierte Potenzen, *Studia Math.* **3**, 1–67.

Bourgin, R.D.
[1983] *Geometric Aspects of Convex Sets with the Radon–Nikodym Property*, Lecture Notes in Mathematics, Vol. 993 (Springer, Berlin).

Diestel, J., and J.J. Uhl
[1977] *Vector Measures*, Math. Surveys, Vol. 15 (Amer. Math. Soc., Providence, RI).

Dieudonne, J.
[1960] *Foundations of Modern Analysis* (Academic Press, New York).

Ekeland, I., and R. Temam
[1976] *Convex Analysis and Variational Problems* (North-Holland, Amsterdam).

Fenchel, W.
[1949] On conjugate convex functions, *Canad. J. Math.* **1**, 73–77.
[1953] *Convex cones, sets, and functions*, Mimeographed lecture notes (Princeton University Press, Princeton, NJ).

Giles, J.R.
[1982] *Convex Analysis with Application in Differentiation of Convex Functions* (Pitman, Boston).

Holmes, R.B.
[1972] *A Course on Optimization and Best Approximation*, Lecture Notes in Mathematics, Vol. 257 (Springer, Berlin).

Ioffe, A.D., and V.M. Tikhomirov
[1968] Duality of convex functions and extremal problems, *Russian Math. Surveys* **23**(6), 53–124.
Jensen, J.L.W.V.
[1905] Om konvexe Funktioner og Uligheder mellem Middelvaerdier, *Nyt Tidsskr. Math.B* **16**, 49–69.
[1906] Sur les fonctions convexes et les inégalites entre les valeurs moyennes, *Acta Math.* **30**, 175–193.
Mazur, S.
[1933] Über konvexe Mengen in linearen normierten Raumen, *Studia Math.* **4**, 70–84.
Moreau, J.-J.
[1966] *Fonctionelles convexes, Séminaire sur les équations aux dérivees partielles II* (Collège de France, Paris).
Mosco, U.
[1969] Convergence of convex sets and of solutions of variational inequalities, *Adv. in Math.* **3**, 510–585.
Namioka, I., and R.R. Phelps
[1975] Banach spaces which are Asplund spaces, *Duke Math. J.* **42**, 735–750.
Phelps, R.R.
[1989] *Convex Functions, Monotone Operators, and Differentiability*, Lecture Notes in Mathematics, Vol. 1364 (Springer, Berlin).
Preiss, D., R.R. Phelps and I. Namioka
[1990] Smooth Banach spaces, weak Asplund spaces and monotone or usco mappings, *Israel J. Math.* **71**, 1–23.
Roberts, A.W., and D.E. Varberg
[1973] *Convex Functions* (Academic Press, New York).
Rockafellar, R.T.
[1970a] On the maximal monotonicity of subdifferential mappings, *Pacific J. Math.* **33**, 209–216.
[1970b] *Convex Analysis* (Princeton University Press, Princeton, NJ).
[1974] *Conjugate Duality and Optimization* (Society for Industrial and Applied Mathematics, Philadelphia, PA).
Rockafellar, R.T., and R.J.-B. Wets
[1984] Variational systems, an introduction, in: *Multifunctions and Integrands*, Lecture Notes in Mathematics, Vol. 1091 (Springer, Berlin).
Simons, S.
[1991] The least slope of a convex function and the maximal monotonicity of its subdifferential, *J. Optim. Theory Appl.* **71**, 127–136.
Van Tiel, J.
[1984] *Convex Analysis* (Wiley, New York).

CHAPTER 4.3

Convexity and Calculus of Variations

Ursula BRECHTKEN-MANDERSCHEID

Mathematisches Institut der Universität Würzburg, D-97074 Würzburg am Hubland, Germany

Erhard HEIL

Fachbereich Mathematik, TH Darmstadt, Schlossgartenstrasse 7, D-64289 Darmstadt, Germany

Contents

HANDBOOK OF CONVEX GEOMETRY
Edited by P.M. Gruber and J.M. Wills

Introduction

Before we start with the calculus of variations we recall some relevant facts about extrema of functions on $\mathbb{R}$ (section 1). Then in sections 2.1, 2.2 we describe the basic problem of minimizing an integral

$$I[y] = \int_a^b f(x, y(x), y'(x))\, \mathrm{d}x \tag{0.1}$$

under given boundary conditions. (Replacing f by $-f$ converts a maximum problem into a minimum problem.) A sufficiently smooth minimizer satisfies the Euler–Lagrange differential equation. Usually solutions of this equation are called extremals, but this does not mean that they minimize or maximize the integral. In fact, one fundamental problem is to find sufficient conditions which guarantee that extremals are minimizers. Other fundamental problems are the existence and regularity (i.e., smoothness) of minimizers.

In section 2.3 we consider (0.1) for a function f which is convex with respect to (y, y'). Then I is convex and any extremal satisfying the boundary conditions is an absolute minimizer. Next we consider the case where f is convex with respect to y' and show what impact this has on the three fundamental questions mentioned above: in section 2.4 we discuss sufficient conditions, in sections 2.5, 2.6 and 2.7 existence and regularity.

Some aspects become clearer in the more geometric form

$$I[x, y] = \int_\alpha^\beta F(x(t), y(t), \dot{x}(t), \dot{y}(t))\, \mathrm{d}t, \tag{0.2}$$

where the function graphs are described as curves with parameter representation. If $F \geqslant 0$ then (0.2) may be considered as a generalized arc length of curves. The spaces with such a measurement (here 2-dimensional) are called Finsler spaces; special cases are Riemannian spaces. The convexity of F with respect to $(\dot{x}, \dot{y})$ is equivalent to the convexity of the unit balls in the tangent spaces, and it is not surprising that this is a far reaching property. We describe this in section 2.8.

In section 3 we consider multiple integrals, for instance

$$I[z] = \int_\Omega f(x, y, z(x, y), z_x(x, y), z_y(x, y))\, \mathrm{d}x\, \mathrm{d}y, \tag{0.3}$$

where the values of z are prescribed on the boundary of the compact region $\Omega \subset \mathbb{R}^2$ and z is to minimize I. In section 3.1 we present the analogue of section 2.3: if f is convex with respect to z and its derivatives z_x, z_y, then I is convex and any extremal with the prescribed boundary values is an absolute minimizer. Apart from this case, the sufficiency theory is much more complicated for multiple integrals than for simple integrals. In sections 3.2 and 3.3 we present several concepts of convexity with respect to the derivational arguments which are involved in necessary and sufficient conditions. The existence theory for (0.3) (and its higher-dimensional analogues) rests on the concept of quasiconvexity which we present shortly in section 3.4 without going into the real subtleties of existence and regularity theory.

We will not give an introduction to the calculus of variations, but sections 1 and 2 should be comprehensible to everybody who knows what the Euler–Lagrange equations are. As a first introduction we recommend Elsgolc (1961, 1970). Bolza (1909) and Cesari (1983) are books of encyclopedic character concerning simple integrals. For multiple integrals see Morrey (1966), Klötzler (1970), Giaquinta (1983) and Dacorogna (1989).

1. Extremum problems for functions

Even functions $\varphi : [a,b] \to \mathbb{R}$ show which natural role convexity plays when dealing with extremum problems. If φ is differentiable and attains a minimum at an inner point x_0 of $[a,b]$, then necessarily

$$\varphi'(x_0) = 0. \tag{1.1}$$

1.1. *Convexity as a sufficient condition*

A real valued function is convex if its epigraph is convex (see chapter 4.2). The epigraph is the set lying above the graph, that is here

$$\{(x,y) \mid a \leqslant x \leqslant b,\ \varphi(x) \leqslant y\}.$$

Theorem 1.1. *If φ is convex, then* (1.1) *implies that φ has an absolute minimum at x_0.*

On the other hand (1.1) is not necessary for a minimum since the minimum may be attained at a or b. In the framework of convex analysis a subdifferential $\partial\varphi(x_0)$ is defined such that a convex function φ (not necessarily differentiable) has an absolute minimum at x_0 if and only if $0 \in \partial\varphi(x_0)$. Neither here nor in the following presentation of the calculus of variations we will use the subdifferential. For more information see for instance Clarke (1983).

1.2. *The second derivative test*

If φ is twice continuously differentiable, then (1.1) together with

$$\varphi''(x_0) > 0 \tag{1.2}$$

implies that φ has a local minimum at x_0. This is because (1.2) implies that there is a neighborhood of x_0 in which φ is convex, and the assertion follows from Theorem 1.1.

1.3. Weierstrass' theorem on the existence of minima

If $\varphi:[a,b] \to \mathbb{R}$ is continuous, then there is $x_0 \in [a,b]$ such that φ assumes a minimum at x_0 $(a,b \in \mathbb{R})$. This is Weierstrass' classical theorem. In order to see how it gains relation to convexity in the calculus of variations we consider its generalization by Baire: let K be a compact set in a metric space and $\Phi: K \to \mathbb{R}$. Continuity can be split into two conditions. For the existence of minima lower semicontinuity it is sufficient that

$$\Phi(x) - \Phi(x_0) > -\varepsilon$$

if x is close to x_0. Equivalently

$$\liminf_{x_n \to x_0} \Phi(x_n) \geqslant \Phi(x_0)$$

for all sequences $x_n \to x_0$.

Theorem 1.2. *If $\Phi : K \to \mathbb{R}$ is lower semicontinuous, then there is a point x_0 in the compact metric space K at which Φ assumes its minimum.*

The relation to convexity arises in the following way. The lower semicontinuity of the functionals considered in the calculus of variations is implied by a convexity condition. For instance, for the integral (0.1) it is the convexity of the integrand f in its last argument.

2. The basic problem of the calculus of variations

2.1. Examples

Which is the shortest plane curve joining two points $(x_1, y_1), (x_2, y_2)$? If $x_1 < x_2$ and if we restrict our search to curves representable as graphs of functions which moreover are continuously differentiable, then we have to minimize

$$\int_{x_1}^{x_2} \sqrt{1 + y'(x)^2}\, \mathrm{d}x \tag{2.1}$$

among the functions y with $y(x_1) = y_1$, $y(x_2) = y_2$. Everybody "knows" that the solution is given by the function y which represents the segment joining the two given points.

The problem of rotational minimal surfaces leads to a variational problem of the same kind. Let the meridian of a rotational surface in Euclidian 3-space be given as the graph of a function y. Then its surface area is given by

$$2\pi \int_{x_1}^{x_2} y(x)\sqrt{1+y'(x)^2}\, \mathrm{d}x. \tag{2.2}$$

This has to be minimal subject to the conditions $y(x_1) = y_1,\ y(x_2) = y_2$ which correspond to bounding circular wires. The solution is not as obvious as in the preceding example though it can be found in many text books, see, e.g., Elsgolc (1961, 1970). The surface may physically be realized as a soap film. So it is plausible that there is no solution if the two wires are too far apart (or too small).

2.2. *A necessary condition: The Euler–Lagrange equation*

How can one find the solution of such a minimal problem, for instance the shape of a rotational minimal surface? Consider the problem of minimizing (0.1)

$$I[y] = \int_a^b f(x, y(x), y'(x))\, \mathrm{d}x$$

within the class of C^1-functions (i.e., continuously differentiable functions) with $y(a) = y_a,\ y(b) = y_b$. A necessary condition for a minimizer y_0, corresponding to (1.1), is the Euler–Lagrange equation

$$\frac{\mathrm{d}}{\mathrm{d}x} f_{y'} = f_y. \tag{2.3}$$

Here $f_y, f_{y'}$ denote partial derivatives of f with respect to its second and third argument. (Let us assume $f \in C^2$.) After inserting the function y_0 and its derivative y_0' for the second and third variable this has to be an identity in x. For instance, for (2.2) equation (2.3) means

$$\frac{\mathrm{d}}{\mathrm{d}x}\left(\frac{y\,y'}{\sqrt{1+y'^{\,2}}}\right) = \sqrt{1+y'^{\,2}}.$$

For a proof of the necessary condition (2.3) we refer to Elsgolc (1961, 1970). As a first step in solving the minimum problem we may look for C^2-functions obeying (2.3), i.e., satisfying

$$f_{y'y'}\, y'' + f_{y'y}\, y' + f_{y'x} = f_y. \tag{2.4}$$

This is an ordinary differential equation of the second order. If there is a solution with the prescribed boundary conditions, then it may be a relative or absolute minimizer. But this will not always be the case, and we already know from example (2.2) that a minimizer does not have to exist. We also have to expect difficulties with (2.4) if the regularity condition

$$f_{y'y'} \neq 0 \tag{2.5}$$

is not fulfilled. Moreover, we will see that the differentiability properties of y_0 which we assumed above are not as natural as one might think (see section 2.5).

2.3. Convex integrals

Theorem 1.1 has an analogue for the variational problems considered in section 2.2. We describe it first informally. If f is convex in (y, y'), then I is convex, i.e., the epigraph of I is convex. Therefore a function y_0 at which the epigraph has a horizontal tangent hyperplane is a global minimizer. That means: a solution of the Euler–Lagrange equation (2.3) with the prescribed boundary values is a global minimizer. The exact formulation of what we just said involves convex sets and other concepts in infinite-dimensional spaces. Though this provides no great difficulty there is no need to use them.

Theorem 2.1. *For $f \in C^2$ consider* (0.1)

$$I[y] = \int_a^b f(x, y, y')\, \mathrm{d}x$$

on the set $\mathfrak{D} = \{y \in C^1 |\ y(a) = y_a,\ y(b) = y_b\}$. If f is convex in (y, y') and $y_0 \in \mathfrak{D}$ is a solution of the Euler–Lagrange equation (2.3), *then y_0 is an absolute minimizer of I in $\mathfrak{D}$.*

Proof. For the moment we consider f as a function on $\mathbb{R}^2$, namely as a function of its second and third argument, which we denote (η, ζ). Since f is convex by assumption, its graph always lies above its tangent plane, i.e., we have

$$f(x, \eta + \delta, \zeta + \varepsilon) \geqslant f(x, \eta, \zeta) + f_\eta(x, \eta, \zeta)\delta + f_\zeta(x, \eta, \zeta)\varepsilon. \tag{2.6}$$

If $y \in \mathfrak{D}$ and $v \in C^1$ with $v(a) = v(b) = 0$, then $(y + v) \in \mathfrak{D}$. From (2.6) we get

$$\begin{aligned}
&\int_a^b f(x, y(x) + v(x), y'(x) + v'(x))\, \mathrm{d}x \\
&\quad \geqslant \int_a^b f(x, y(x), y'(x))\, \mathrm{d}x \\
&\qquad + \int_a^b [f_\eta(x, y(x), y'(x))v(x) + f_\zeta(x, y(x), y'(x))v'(x)]\, \mathrm{d}x.
\end{aligned}$$

With (2.6) and our former notation for the partial derivatives of f we may write

$$I[y+v] \geqslant I[y] + \int_a^b [f_y(x, y(x), y'(x))v(x) + f_{y'}(\ldots)v'(x)] \, \mathrm{d}x.$$

Now y_0 is a solution of (2.3):

$$f_y(x, y_0(x), y_0'(x)) = \frac{\mathrm{d}}{\mathrm{d}x} [f_{y'}(x, y_0(x), y_0'(x))].$$

Therefore

$$\begin{aligned}
&\int_a^b [f_y(x, y_0(x), y_0'(x))v(x) + f_{y'}(\ldots)v'(x)] \, \mathrm{d}x \\
&= \int_a^b \{\frac{\mathrm{d}}{\mathrm{d}x}[f_{y'}(x, y_0(x), y_0'(x))]v(x) + f_{y'}(\ldots)v'(x)\} \, \mathrm{d}x \\
&= \int_a^b \frac{\mathrm{d}}{\mathrm{d}x}[f_{y'}(x, y_0(x), y_0'(x))v(x)] \, \mathrm{d}x \\
&= f_{y'}(x, y_0(x), y_0'(x))v(x)|_a^b = 0,
\end{aligned}$$

since $v(a) = v(b) = 0$. Thus we have

$$I[y_0 + v] \geqslant I[y_0],$$

i.e., y_0 is an absolute minimizer in $\mathcal{D}$. □

Remark 2.1. We do not claim the existence of a minimizer. Indeed, as in the case of a convex function φ, where no point with $\varphi' = 0$ need exist, the Euler–Lagrange equation need have no solution with the prescribed boundary conditions. The convexity hypothesis of the theorem is very strong. For instance, it is not fullfilled by (2.2). But if f does not depend on y, then it reduces to the assumption $f_{y'y'} \geqslant 0$, which is quite natural (see sections 2.4–2.6). If moreover f does not depend on x and $f_{y'y'} > 0$ holds, then (2.3) reduces to $y'' = 0$. Extremals are the functions $y(x) = a_1 x + a_2$ representing straight lines. The theorem shows that these are minimizers. An example is given by (2.1).

Remark 2.2. In contrast to the sufficient conditions described in section 2.4, which were developed in the last century, Theorem 2.1 found its way into textbooks only recently: there is a short hint by Courant and Hilbert (1930, p. 186); Ewing (1969, p. 78) shows that for convex I every weak local minimum is a global minimum; Theorem 2.1 is given by Hadley and Kemp (1971, p. 102), Brechtken-Manderscheid (1983, p. 48), Troutman (1983, pp. 57, 208); Troutman also discusses the case of uniqueness.

2.4. *Further sufficient conditions*

We again ask when an extremal is a minimizer, but this time for those f which are convex only with respect to y'. In order to see which extra condition then has to be demanded we consider two examples.

Example 2.1. Consider an arc of a great circle on the sphere S^2. Let λ denote its length. Apart from the exceptional case that the two endpoints are antipodal ($\lambda = \pi$) there are two cases:
(α) $\lambda < \pi$: the arc is a unique absolute minimizer,
(β) $\lambda > \pi$: the arc is not even a relative minimizer.
The length of arcs on S^2 may be written in the form (0.1), but since this is pretty complicated we give a simpler example.

Example 2.2. Minimize

$$I[y] = \int_0^b (y'^2 - y^2)\, dx$$

subject to $y(0) = 0,\ y(b) = 0,\ 0 < b < 2\pi$.
The Euler–Lagrange equation is

$$y'' + y = 0,$$

and the extremals with $y(0) = 0$ are $y(x) = \alpha \sin x,\ \alpha \in \mathbb{R}$. As we shall see (Theorem 2.2), for $b < \pi$ there is the unique absolute minimizer $y(x) = 0$. For $\pi < b < 2\pi$ the only extremal with the prescribed boundary values is again $y(x) = 0$, and $I[0] = 0$. But there are non-extremals with a smaller value of I. For instance take

$$y_b(x) = \sin(\pi/b)x.$$

A short calculation gives

$$I[\gamma y_b] = \gamma^2 \frac{\pi}{2}\left(\frac{\pi}{b} - \frac{b}{\pi}\right) < 0,$$

so that $\inf I = -\infty$, since $\gamma \in \mathbb{R}$ is arbitrary. This example shows that at $x = \pi$ the extremal $y(x) = 0$ ceases to be a minimizer. For $b < \pi$ the extremal which attains the boundary values is embedded in a field of non-intersecting extremals which cover the strip $0 < x < \pi$ of $\mathbb{R}^2$.

Generally, we call a continuous 1-parameter family F of extremals of (0.1) a *field* in the simply connected plane region R if through every point of R passes a unique extremal of the family. Then we have a continuous function $p: R \to \mathbb{R}$, the *slope function* which assigns to each point $P \in R$ the value of the derivative of the extremal passing through P.

We also need *Weierstrass' $\mathscr{E}$ function.* For fixed x, y we consider f as a function of its last argument. Figure 1 shows its graph and depending on p, q (and x, y), the value of $\mathscr{E}$. Thus, writing as before $f_{y'}$ for the derivative with respect to the last variable we have

$$\mathscr{E}(x,y,p,q) = f(x,y,q) - f(x,y,p) - f_{y'}(x,y,p)(q-p).$$

Since a differentiable function is convex if and only if its graph lies above all tangent lines, we have

$$\mathscr{E} \geqslant 0 \quad \text{if and only if} \quad f_{y'y'} \geqslant 0. \tag{2.7}$$

Now let y_0 be an extremal which for the interval $[a,b]$ belongs to a field F of extremals. Let y be a function with $y(a) = y_0(a)$, $y(b) = y_0(b)$, the graph of which lies in the region of F (cf. fig. 2).

Then with some effort it is posssible to deduce

$$I[y] = I[y_0] + \int_a^b \mathscr{E}(x, y(x), p(x, y(x)), y'(x))\, \mathrm{d}x. \tag{2.8}$$

"This is the famous formula of Weierstrass, which revolutionized the calculus of variations" (Young 1969, p. 27). From (2.7) and (2.8) we get the following theorem.

Theorem 2.2. *Consider the problem of Theorem* 2.1, *but assume that f is convex only with respect to y'. Let $y_0 \in \mathscr{D}$ be a solution of the Euler–Lagrange equation*

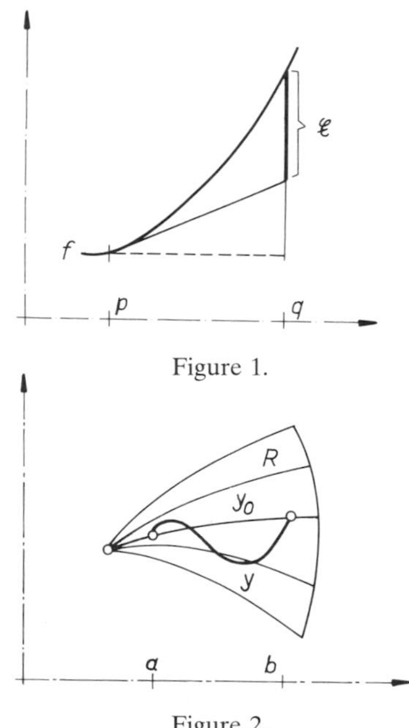

Figure 1.

Figure 2.

(2.3) *which can be embedded in a field F. Then y_0 is a minimizer with respect to all other functions of $\mathfrak{D}$ whose graph lies within the region R covered by F.*

As an example consider again Example 2.2 with $b < \pi$. Theorem 2.2 implies that $y(x) = 0$ is the unique absolute minimizer. A suitable field of extremals is for instance given by $y = \alpha \sin(x + (\pi - b)/2)$ where α is the parameter of the family.

The problem with Theorem 2.2 is of course that in general it is difficult to embed y_0 in a field. It is therefore useful to have a criterion which says whether y_0 is embeddable or not, but which does not afford the knowledge of a 1-parameter family of extremals. For in general the complete solution of (2.3) is not known. Without further explanation we here remark that there is a condition, the *Jacobi condition*, which facilitates this task. One has to show that the next conjugate point of y_0 (the antipodal point in Example 2.1) lies beyond b. Then y_0 can be embedded over $(a, b]$.

Several sufficient conditions can be derived from Theorem 2.2. Here we give one of them.

An extremal y_0 satisfying the boundary conditions is a local minimizer for (0.1) if

(1) the Jacobi condition is fulfilled,

(2) $\mathscr{E}(x, y, p, q) \geqslant 0$ for all points (x, y) in a neighborhood N of the curve represented by y_0 and for all $p, q \in \mathbb{R}$.

This is the form in which criteria like this are usually given in text books. As we have seen, (2) just means that for $(x, y) \in N$ the function f is convex in y' or, since we assume $f \in C^2$, that $f_{y'y'}(x, y, p) \geqslant 0$ for $(x, y) \in N$ and $p \in \mathbb{R}$.

2.5. Broken extremals

The following example shows that minimizers may have corners. As we will see this is not possible if f in (0.1) is strictly convex with respect to y'. Let

$$I[y] = \int_0^2 y^2(y' - 1)^2 \, \mathrm{d}x \quad \text{with } y(0) = 0, \; y(2) = 1.$$

Obviously $I[y] \geqslant 0$ and $I[y_0] = 0$ for the function

$$y_0(x) = \begin{cases} 0, & 0 \leqslant x \leqslant 1, \\ x - 1, & 1 \leqslant x \leqslant 2. \end{cases}$$

Thus y_0 is a global minimizer. It is composed of two extremal arcs and has a corner at $x = 1$. Also it is not difficult to see that it is the only global minimizer.

Already in 1779 Euler found an example of this kind. In the last century the calculus of variations was developed for piecewise continuously differentiable functions y. These functions are continuously differentiable except at finitely many points where the one-sided derivatives y'_- and y'_+ exist and are equal to the one-sided limits of the derivatives. If a piecewise differentiable function y_0 which satisfies the Euler–Lagrange equation (2.3) at the non-exceptional points is a minimizer, then at an exceptional point x_1 the Weierstrass–Erdmann corner condition is fulfilled:

$$f_{y'}(x_1, y_0(x_1), y'_{0^-}(x_1)) = f_{y'}(x_1, y_0(x_1), y'_{0^+}(x_1)). \tag{2.9}$$

Such a curve is called a *broken extremal.* In the above example both values in (2.9) are 0 at $x_1 = 1$.

Now assume that f is *strictly convex* in y'. That means: for fixed (x, y) $f(x, y, \cdot)$ is convex and $f_{y'}(x, y, p_1) = f_{y'}(x, y, p_2)$ implies $p_1 = p_2$. For such f (2.9) implies $y'_{0^-}(x_1) = y'_{0^+}(x_1) = y'_0(x_1)$. Therefore, in this case a minimizer y_0 which is a broken extremal has no corners, i.e., is an extremal. Note that we do not claim that this convexity condition implies that a minimizer y_0 is an extremal. It is indeed possible that there are points where y_0 is not differentiable, even infinitely many such points may exist.

2.6. *Existence and regularity. Lower semicontinuity*

We consider two examples with non-convex integrands in order to see what difficulties we have to expect when looking for conditions which assure existence of minimizers.

Example 2.3. Consider $I[y] = \int_0^2 (y'^2 - 1)^2 \, dx$ with $y(0) = y(2) = 0$. Obviously $I[y] \geqslant 0$. Let

$$y_1 = \begin{cases} x, & 0 \leqslant x \leqslant 1, \\ 2 - x, & 1 \leqslant x \leqslant 2. \end{cases}$$

Then $I[y_1] = 0$. Also, for y_n as indicated in fig. 3b, $I[y_n] = 0$. So there are minimizers with arbitrarily many corners, and it is easy to construct a minimizer with infinitely many corners in a similar way. It is also not difficult to see that $I[y] > 0$ for $y \in C^1$, and by rounding the corner of y_1 by small circles that $\inf I[y] = 0$ over C^1. This indicates that in considering the existence problem it is not appropriate to restrict oneself to functions which are C^1 or piecewise C^1.

Example 2.4. By a small alteration of Example 2.3 we see that we cannot always expect the existence of a minimizer. Let

$$J[y] = \int_0^2 [(y'^2 - 1)^2 + y^2] \, dx \quad \text{with } y(0) = y(2) = 0.$$

Again $J[y] \geqslant 0$ and for y_n as in Example 2.3

$$J[y_n] = \int_0^2 y_n^2 \, dx = \frac{2}{3n^2} \to 0.$$

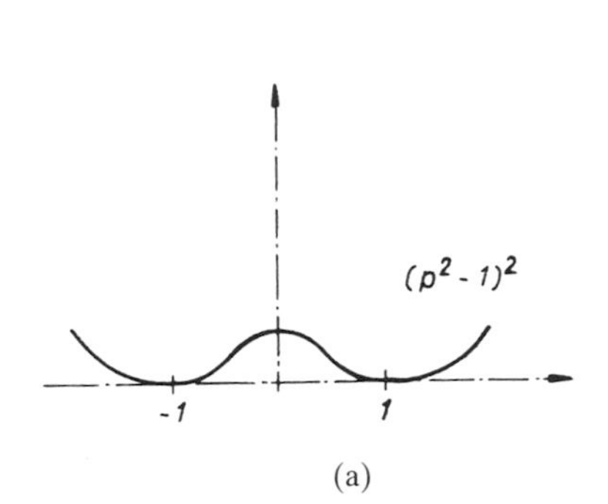

(a)

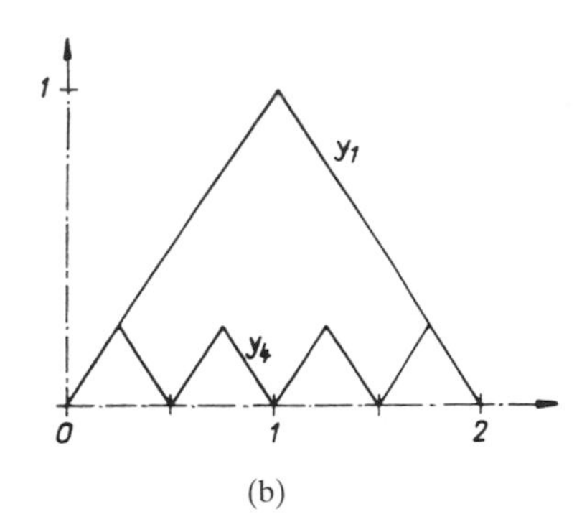

(b)

Figure 3.

Thus $y_n \to 0$ uniformly and $J[y_n] \to \inf J[y] = 0$, and one should expect that $y = 0$ is a minimizer, but

$$J[0] = 2.$$

The general procedure in proving existence theorems via the direct method of the calculus of variations is as follows. For an integral (0.1) we consider a class A of functions with certain differentiability properties and attaining given boundary conditions. We assume

$$\inf_{y \in A} I[y] > -\infty.$$

Then there is a minimizing sequence y_n :

$$I[y_n] \to \inf_{y \in A} I[y].$$

In order to show the existence of a minimizer y_0 we need two properties,

(1) *compactness:* there is a subsequence (call it again y_n) converging to $y_0 \in A$,
(2) *lower semicontinuity* of I: this implies

$$\lim_{n \to \infty} I[y_n] = I[y_0] = \inf_{y \in A} I[y].$$

Then y_0 is an absolute minimizer (compare section 1.3). For further explanation we begin with (2). Of course continuity of I would suffice, but it is given only in exceptional cases or for too strong modes of convergence. (Consider for instance

(2.1) or (2.2), which are not continuous with respect to uniform convergence, but are continuous if in addition y' converges uniformly.) For our purpose lower semicontinuity in the following sense suffices: if y_n converges to y_0 in the chosen mode of convergence, then

$$\liminf_{n\to\infty} I[y_n] \geqslant I[y_0].$$

For then we have, because of $I[y_n] \to \inf I$ and $I[y_0] \geqslant \inf I$,

$$\inf I = \lim I[y_n] \geqslant I[y_0] \geqslant \inf I$$

so that indeed

$$I[y_0] = \inf I.$$

The crucial condition assuring *lower semicontinuity* of I is *convexity* of the integrand f with respect to its last argument. For instance we have the following theorem of Tonelli (1921, p. 400, 1934, p. 205).

Theorem 2.3. *Assume that f in* (0.1) *is C^2. Then I is lower semicontinuous with respect to uniform convergence if and only if for all (x, y) the function $f(x, y, \cdot)$ is convex.*

Examples. (2.1) and Example 2.2 are lower semicontinuous. This is true for (2.2) if $y > 0$. Examples 2.3, 2.4 are not lower semicontinuous: $y_n \to 0$ uniformly, $I[y_n] = 0$, $I[0] = 2$, $J[y_n] \to 0$, $J[0] = 2$.

A minimizing sequence of C_1 functions may uniformly converge to a function which is not even piecewise C_1. Therefore the compactness argument (1) requires that these two classes of functions, which laid the ground for the calculus of variations in the last century, have to be replaced by a larger class. Tonelli (1921, 1923) was the first to choose the class of absolutely continuous functions together with Lebesgue's concept of integral. A function f is *absolutely continuous* in $[a, b]$ if and only if f' exists almost everywhere, f' is Lebesgue integrable, and $\int_a^x f'(\xi)\, \mathrm{d}\xi = f(x)$, see, e.g., Weir (1973, p. 67).

The compactness argument (1) also requires that f satisfies a *growth condition* with respect to y', sometimes called *coercivity condition.* We then have for instance the following theorem of Tonelli (1934, p. 208), see also Cesari (1983, p. 112).

Theorem 2.4. *Consider $I[y] = \int_a^b f(x, y, y')\, \mathrm{d}x$ on the set $A = \{y \mid y$ absolutely continuous, $y(a) = y_a,\ y(b) = y_b\}$. Assume that f is C^2, convex in y', and that there is a function Φ bounded below for $t > 0$ with $\lim_{t\to\infty} \Phi(t)/t = \infty$ such that*

$$f(x, y, p) \geqslant \Phi(|p|). \tag{2.10}$$

Then there exists $y_0 \in A$ such that $I[y_0] = \inf_{y\in A} I[y]$.

Examples. The theorem can be applied to $\int (y'^2 + y^4)\,dx$. The growth condition (2.10) is not satisfied for (2.1) and (2.2), and indeed (2.2) does not have a minimizer for all boundary conditions. (2.10) is also not satisfied for Example 2.2. But the theorem can be applied after a small change: if (x,y) is restricted to a compact region B, then (2.10) can be satisfied, and there will be a minimizer y_0, but its graph may partially coincide with ∂B. Find the minimizer in Example 2.2, e.g. for $y(0) = y(4) = 0$, $|y| \leqslant 1$. In Example 2.4 there is no minimizer. Here the convexity assumption is violated.

The last task is to show that a minimizer, if it exists, is in C^2 and a solution of the Euler–Lagrange equation (2.4) because usually this is the only way to find it.

This is a more delicate problem since it may happen that a minimizer has an infinite derivative on a set of measure zero. It will be smooth and obey (2.4) if it is Lipschitz continuous and if f fulfills (2.10) and some natural assumptions. For details we refer to Cesari (1983), especially to pp. 58, 60. But the situation is not really satisfying since only in very special cases one can assure in advance, i.e. directly from the function f, that a minimizer will be Lipschitz continuous.

Remark. Uniform convergence is a simple concept. But it has some advantages to use a more sophisticated concept. In the language of Sobolew spaces the class of absolutely continuous functions on $[a,b]$ is the underlying set of the space $W^{1,1}(a,b)$. *Weak convergence in* $W^{1,1}(a,b)$ is equivalent to uniform convergence and weak convergence of the derivatives with respect to $L^1(a,b)$, cf. Dacorogna (1989, pp. 15, 17), Cesari (1983, p. 99). If we use this mode of convergence it is enough in Theorem 2.4 to assume that f is continuous (instead of C^2). But what is more important, with respect to weak convergence in $W^{1,1}$ this theorem remains valid for $\mathbb{R}^n$-valued y. (Compare Cesari 1983, pp. 104, 112, 115, and 1974, p. 468.)

We close this subsection with the remark that the direct method of the calculus of variations has great significance in the numerical treatment of variational problems and partial differential equations (methods of Ritz and Galerkin, see, e.g., Mikhlin 1965).

2.7. Relaxation and convexification

In section 2.6 we have seen the great role played by lower semicontinuity and convexity in the existence problem. If a functional I, as for instance (0.1), is not lower semicontinuous, then it may be replaced by the *relaxed functional* $\tilde{I}$, which is the greatest lower semicontinuous functional less than I (on the given set of functions). Under certain conditions $\tilde{I}$ has the same infimum as I and possesses a minimizer, which is approximated by the minimizing sequences of I. For more details see Ekeland and Temam (1972, ch. IX, X), Dacorogna (1989, ch. 5), and the articles of Buttazzo and Mascolo in Hildebrandt, Kinderlehrer and Miranda (1988). In general an explicit expression for $\tilde{I}$ may be complicated, but for (0.1) it is very easy: convexify $f(x,y,\cdot)$ for every (x,y) :

$$\tilde{f}(x,y,p) = \sup\{g(p) \mid g \text{ convex},\ g(p) \leqslant f(x,y,p)\},$$

see Alekseev, Tikhomirov and Fomin (1987, p. 41). Consider for instance Example 2.4. From fig. 3a we see that

$$\tilde{f}(x,y,p) = \begin{cases} (p^2-1)^2 + y^2 & \text{if } |p| \geqslant 1, \\ y^2 & \text{if } |p| \leqslant 1. \end{cases}$$

As for J we have $\tilde{J} \geqslant 0$, $\tilde{J}[y_n] \to 0$, y_n as in fig. 3b. But whereas $J[0] = 2$ we now have $\tilde{J}[0] = 0$ so that a minimizer exists.

2.8. Problems in parametric form. Geometric theory

The theory presented in sections 2.1–2.6 may be extended to problems involving several unknown functions, i.e., y may be $\mathbb{R}^n$-valued. No great changes are needed. Cesari (1983) treats this case from the beginning. The Euler–Lagrange equation becomes a system, the regularity condition (2.5) is now a determinant condition and convexity is not just monotonicity of the derivative.

We obtain a problem for two unknown functions from the ordinary problem (0.1) if we drop the distinction between the variables x, y. This certainly makes sense for the geometric problem of arc length, which then becomes

$$\int_{t_1}^{t_2} \sqrt{\dot{x}_1^2 + \dot{x}_2^2}\ \mathrm{d}t. \tag{2.11}$$

Also some of the difficulties with existence and regularity vanish since jumps of functions are then represented by vertical segments and vertical tangents play no exceptional role.

Any integral (0.1) can be brought into the form

$$\int_{t_1}^{t_2} F(x(t), y(t), \dot{x}(t), \dot{y}(t))\ \mathrm{d}t \tag{2.12}$$

by a substitution $x = x(t)$, $y = y(t)$, which means that the graph of the function y is described by a parameter representation. We have

$$\int_a^b f(x,y,y')\ \mathrm{d}x = \int_{t_1}^{t_2} f(x,y,\dot{y}/\dot{x})\dot{x}\ \mathrm{d}t. \tag{2.13}$$

But minimum problems for (0.1) and (2.12) are not equivalent since not every curve in the plane is a graph of a function. Also, as (2.13) shows, continuity or differentiability may be lost. For the relations between (0.1) and (2.12) see also Bolza (1902).

We now write x instead of (x,y) and X for $(\dot{x},\dot{y})$. We may then interpret the formulas also in higher dimensions. From (2.13) we see that F is homogeneous:

$$F(x,\lambda X)=\lambda F(x,X),\quad \lambda\geqslant 0, \tag{2.14}$$

and this is equivalent to the invariance of (2.12) against parameter transformation. Usually (2.14) is not demanded for all $\lambda\in\mathbb{R}$, which means that (2.12) may depend on the orientation of the traversed curve. For instance the quickest path in crossing a river will in general not be the same for both directions. If (2.14) holds and $F(x,X)>0$ for $X\neq 0$, then

$$\int_{t_1}^{t_2} F(x,\dot{x})\,\mathrm{d}t$$

may be considered as a generalized arc length. Indeed, under convexity and regularity conditions, a manifold on which F is invariantly defined is called a *Finsler space*, see Rund (1959), Matsumoto (1986). Special cases are *Riemannian spaces*, where

$$F(x,X)^2=\sum_{i,j} g_{ij}(x)X_iX_j,$$

and among them *surfaces in Euclidian space.*

For simplicity we now restrict ourselves to such F which are independent of x and try to illustrate the part played by convexity. So assume that F is a homogeneous (2.14), convex function with $F(X)>0$ for $X\neq 0$, defined on $\mathbb{R}^2$ (or $\mathbb{R}^n$). Then the graph of F is the boundary of a convex cone with vertex 0 lying above the X-plane. Apart from symmetry F has the properties of a *norm*, indeed, in the presence of (2.14) F is convex if and only if the triangle inequality

$$F(X+Y)\leqslant F(X)+F(Y)$$

is valid. Strict convexity may be defined by the additional requirement

$$\operatorname{grad} F(X)=\operatorname{grad} F(Y)\Rightarrow X=\lambda Y,\ \lambda>0,$$

for $X,Y\neq 0$. The regularity condition corresponding to (2.5), $\det F_{X_iX_j}\neq 0$, is never fulfilled in consequence of (2.14), but may be replaced by

$$(F^2)_{X_iX_j}\ \text{is positive definite.} \tag{2.15}$$

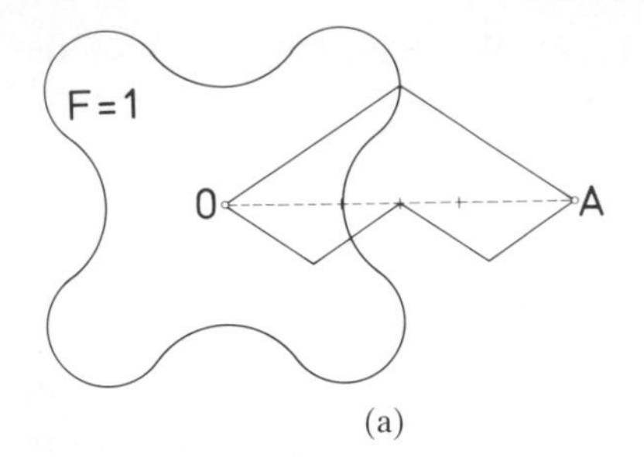

(a)

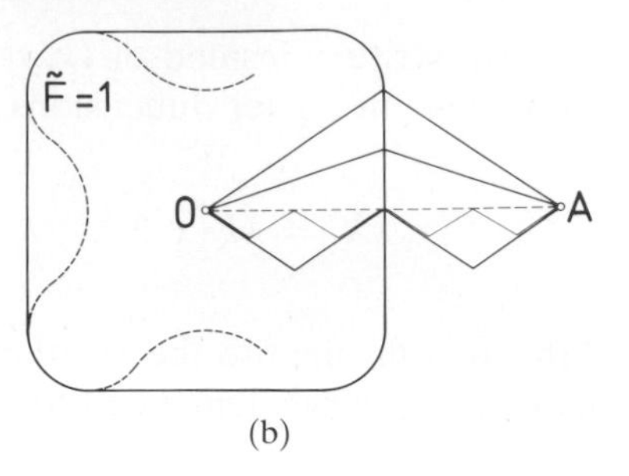

(b)

Figure 4.

To see what this means we consider the curve (or hypersurface) given by

$$F(X) = 1.$$

This curve, called *indicatrix*, encloses the origin since $F(X) > 0$ for $X \neq 0$. It is convex if F is convex. It generalizes the unit circle (or sphere) to which it reduces for (2.11). Condition (2.15) means that the indicatrix has no flat points. If the indicatrix is given it is easy to measure the "length" $\int F(\dot{x})\,\mathrm{d}t$ of polygons. We use this to illustrate the role of convexity by means of an example with non-convex indicatrix. (Compare section 2.6.) In fig. 4a we show two paths from O to A of length 2 whereas the segment joining O and A has length 3. For the convexified function $\tilde{F}$ (cf. section 2.7) all polygonal paths drawn in fig. 4b have length 2. One easily constructs a sequence of zigzag curves with length 2 converging uniformly to the segment OA. This illustrates that the "length" is not lower semicontinuous for F, but is so for $\tilde{F}$. Moreover, fig. 4b shows that "shortest" lines are not unique even locally if the indicatrix is not strictly convex.

3. Multiple integrals in the calculus of variations

We consider multiple variational integrals

$$I[y] = \int_{\Omega} f(x, y(x), \mathrm{D}y(x))\,\mathrm{d}x. \tag{3.1}$$

Here Ω is a bounded open set of $\mathbb{R}^n$ with Lipschitz boundary; $y : \bar{\Omega} \to \mathbb{R}^m$ is a function in a given class A of continuous functions on $\bar{\Omega}$ with given values on the boundary of Ω and certain differentiability properties; $\mathrm{D}y = \left(\partial y^i / \partial x^\alpha\right)$, $i = 1, \ldots, m$, $\alpha = 1, \ldots, n$, is the Jacobian of y; the integrand $f : (x, y, u) \in \Omega \times \mathbb{R}^m \times \mathbb{R}^{nm} \to f(x, y, u) \in \mathbb{R}$ is of class C^2 in its arguments.

A $C^2(\Omega)$-solution y_0 of the variational problem

$$\inf\{I[y] \mid y \in A\}$$

satisfies the Euler–Lagrange equations:

$$E_i[y_0](x) := f_{y^i}(x, y_0(x), \mathrm{D}y_0(x)) - \sum_{\alpha=1}^{n} \frac{\partial}{\partial x^\alpha}(f_{u^i_\alpha}(\cdot, y_0(\cdot), \mathrm{D}y_0(\cdot))(x) = 0, \qquad (3.2)$$
$$x \in \Omega,\ i = 1, \ldots, m.$$

Solutions of (3.2) are called *extremals* of (3.1).

Example 3.1. The *Dirichlet integral* $(n = 2, m = 1)$

$$\int_\Omega \left[\left(\frac{\partial y}{\partial x^1}\right)^2(x) + \left(\frac{\partial y}{\partial x^2}\right)^2(x)\right] \mathrm{d}x.$$

Its Euler–Lagrange equation is the Laplace equation.

Example 3.2. The *Plateau problem* $(n = 2, m = 1)$. For a given boundary curve in $\mathbb{R}^3$, given as the graph of a function $g: \operatorname{bd}\Omega \to \mathbb{R}$, we look, in the class A of $C^1(\Omega)$-functions y with $y|_{\operatorname{bd}\Omega} = g$, for functions whose graphs have minimal area

$$\inf_{y \in A} \int_\Omega \left[1 + \left(\frac{\partial y}{\partial x^1}\right)^2(x) + \left(\frac{\partial y}{\partial x^2}\right)^2(x)\right]^{1/2} \mathrm{d}x.$$

This problem can be generalized:

Example 3.3. In the class A of all differentiable parametrized surfaces (representation $y : \Omega \to \mathbb{R}^3$) with a given curve as boundary, we look for the surface with minimal area. Then we have $n = 2$ parameters x^α and $m = 3$ coordinate functions y^i; the surface area is given by

$$\int_\Omega \left[\left(\frac{\partial(y^1, y^2)}{\partial(x^1, x^2)}\right)^2(x) + \left(\frac{\partial(y^1, y^3)}{\partial(x^1, x^2)}\right)^2(x) + \left(\frac{\partial(y^2, y^3)}{\partial(x^1, x^2)}\right)^2(x)\right]^{1/2} \mathrm{d}x. \qquad (3.3)$$

This integral is in parametric form (cf. 3.5).

3.1. Convex integrals

The following theorem essentially depends on the fact that I is a convex functional if $f(x, \cdot, \cdot)$ is convex for all $x \in \Omega$.

Theorem 3.1. *Let $f(x, \cdot, \cdot)$ be convex for all $x \in \Omega$; then I is convex and an extremal $y_0 \in A$ is a solution of the variational problem* $\inf_{y \in A} I[y]$.

Proof. From the convexity inequality corresponding to (2.6) we have

$$I[y] - I[y_0] \geqslant \int_\Omega \Big[\sum_{1 \leqslant i \leqslant m} (f_{y^i}(x, y_0(x), \mathrm{D}y_0(x))(y^i(x) - y_0^i(x)) + \sum_{\substack{1 \leqslant i \leqslant m \\ 1 \leqslant \alpha \leqslant n}} f_{u^i_\alpha}(x, y_0(x), \mathrm{D}y_0(x))(\frac{\partial}{\partial x^\alpha}(y^i - y_0^i)(x))\Big] \, \mathrm{d}x.$$

Application of the Gauss–Green theorem yields

$$I[y] - I[y_0] \geqslant \int_\Omega \sum_{1 \leqslant i \leqslant m} E_i[y_0](x)(y^i(x) - y_0^i(x)) \, \mathrm{d}x = 0,$$

since $E_i[y_0] = 0$. This proves the theorem. Moreover it is easily seen that I is convex. □

In Example 3.1 we have this convexity property, but not in Example 3.2.

Convexity of f with respect to the second and the third group of variables is very restrictive. There is a hierarchy of convexity conditions for f with respect to the third group of (the directional) variables, which are in part necessary and in part sufficient for solutions. These conditions are discussed in the following sections. There is a remarkable difference between the cases $m = 1$ or $n = 1$ and the cases where m and n both are > 1.

3.2. The necessary conditions of Weierstrass and Legendre

Let y_0 be a differentiable solution of the variational problem. The necessary Weierstrass condition asserts that for all $x \in \Omega$

$$\begin{aligned} &\mathscr{E}(x, y_0(x), \mathrm{D}y_0(x), v) \\ &:= f(x, y_0(x), v) - f(x, y_0(x), \mathrm{D}y_0(x)) \\ &\quad - \sum_{\substack{1 \leqslant \alpha \leqslant n \\ 1 \leqslant i \leqslant m}} f_{u^i_\alpha}(x, y_0(x), \mathrm{D}y_0(x)) \cdot (v^i_\alpha - \frac{\partial y_0^i}{\partial x^\alpha}(x)) \geqslant 0 \end{aligned}$$

for all $v \in \mathbb{R}^{m \cdot n}$ with rank $(v - \mathrm{D}y_0(x)) \leqslant 1$. For $m = n = 1$ this means that $f(x, y_0(x), \cdot)$ is supported by the tangent in $y_0'(x)$ (cf. fig. 1). When n and m both are > 1 then, because of the rank qualification, $\mathscr{E} \geqslant 0$ is much weaker than the analogous support property.

The necessary *condition of Legendre* says that along the solution y_0 of (3.1)

$$\sum_{\substack{1\leqslant i,j\leqslant m\\1\leqslant \alpha,\beta\leqslant n}} \frac{\partial^2 f}{\partial u^i_\alpha \partial u^j_\beta}(x, y_0(x), \mathrm{D}y_0(x))\left(v^i_\alpha - \frac{\partial y^i_0}{\partial x^\alpha}(x)\right)\left(v^j_\beta - \frac{\partial y^j_0}{\partial x^\beta}(x)\right) \geqslant 0$$

for every $(x,v) \in \Omega \times \mathbb{R}^{mn}$ with rank $(v - \mathrm{D}y_0(x)) \leqslant 1$. For details and further literature on these classical conditions see Klötzler (1970, p. 112).

In connection with these necessary conditions the following property is introduced. A function $\underline{f}: \mathbb{R}^{nm} \to \mathbb{R}$ is said to be *rank one convex* at $u \in \mathbb{R}^{nm}$, if $\underline{f}(tu+(1-t)v)$ as a function of t is convex for every $v \in \mathbb{R}^{nm}$ with rank $(v-u) \leqslant 1$.

3.3. *Sufficient conditions with invariant integrals*

One way to get sufficient conditions for an extremal $y_0 \in A$ to be a solution of the problem $\inf I[y]$ is to use Hilbert invariant integrals. By Stokes' theorem the integral of an exact differential form $\psi = \mathrm{d}\chi$ over a surface S, here the graph of a function $y \in A$, equals the integral of χ over the boundary $\operatorname{bd} S$ of S. As the functions $y \in A$ have fixed boundary values

$$\int_S \mathrm{d}\chi = \int_{\operatorname{bd} S} \chi = \int_{\operatorname{bd} S_0} \chi = \int_{S_0} \mathrm{d}\chi,$$

where S, S_0 are the graphs of y, y_0, respectively. There is a differentiable function f^* such that

$$\int_S \mathrm{d}\chi = \int_\Omega f^*(x, y(x), \mathrm{D}y(x))\, \mathrm{d}x. \tag{3.4}$$

If we now have a differential $(n-1)$-form χ such that

$$f^*(x, y_0(x), \mathrm{D}y_0(x)) = f(x, y_0(x), \mathrm{D}y_0(x)) \quad \text{for all } x \in \Omega \tag{3.5}$$

and

$$f^*(x,y,u) \leqslant f(x,y,u) \quad \text{for all } x, y \text{ and } u, \tag{3.6}$$

then

$$I[y] - I[y_0] = I[y] - \int_{S_0} \mathrm{d}\chi = I[y] - \int_S \mathrm{d}\chi$$

$$= \int_\Omega [f(x, y(x), \mathrm{D}y(x)) - f^*(x, y(x), \mathrm{D}y(x))]\ \mathrm{d}x \geqslant 0.$$

The difference $f - f^*$ is the generalization of the Weierstrass excess-function $\mathscr{E}$. We do not want to discuss here the existence of a function f^* with (3.4) (see

Klötzler 1970, p. 158 f, and for historical remarks Funk 1962, p. 410 ff). Conditions (3.5), (3.6) express a support or convexity condition (see fig. 1). Whereas for $n = 1$ or $m = 1$ there is at most one support element of $f(x, y, \cdot)$ at u (for differentiable f), the support elements will in general form a manifold of dimension $\binom{m+n}{n} - 1 - m \cdot n$ in the cases where m and n are both > 1. Approaches to fulfilling (3.4) were made by de Donder and Weyl and by Carathéodory. They used support elements of special structures, and as a consequence they got different types of convexity conditions, which are best understood by transforming (3.1) to the homogeneous form (cf. section 3.5 below). The de Donder–Weyl convexity condition means convexity of $f(x, y, \cdot)$ in the usual sense. The generalized convexity with Carathéodory type invariant integrals (see Carathéodory 1929) is *total convexity* of $f(x, y, \cdot)$, as described in section 3.5 below. The weakest form of such a convexity condition is *polyconvexity* of $f(x, y, \cdot)$. A function $\underline{f}: \mathbb{R}^{nm} \to \mathbb{R}$ is called *polyconvex* if a convex function g exists, such that $\underline{f}(u) = g(\underline{T}(u))$, where $T(u)$ is a vector whose components are the determinants of all $r \times r$ submatrices of $u, 1 \leqslant r \leqslant \min(n, m)$.

For example the function

$$f(u^i_\alpha) := (u^1_1)^2 + (u^1_1 u^2_2 - u^2_1 u^1_2), \quad i = 1, 2, \quad \alpha = 1, 2, \quad (n = m = 2)$$

is polyconvex (with $g(u^i_\alpha, u^{12}_{12}) := (u^1_1)^2 + u^{12}_{12}$) but not convex.

3.4. Lower semicontinuity and existence

The lower semicontinuity of I is the essential tool for existence theorems; compare section 2.6. A Borel measurable and locally integrable function $\underline{f}: \mathbb{R}^{nm} \to \mathbb{R}$ is said to be *quasiconvex* if for every $u_0 \in \mathbb{R}^{nm}$, for every domain $G \subset \Omega$ and for every φ of the Sobolev space $W^{1,\infty}_0(G, \mathbb{R}^m)$

$$\underline{f}(u_0) \cdot \text{meas}\, G \leqslant \int_G \underline{f}(u_0 + \mathrm{D}\varphi)\, \mathrm{d}x. \tag{3.7}$$

When $\underline{f} > 0$, the geometrical meaning of $\underline{f}(u_0) \cdot \text{meas}\, G$ in (3.7) is the $\underline{f}$-area of the graph of the linear mapping $l: x \in G \to u \cdot x_0 \in \mathbb{R}^m$. If $\varphi \in W^{1,\infty}_0(G, \mathbb{R}^m)$ is a piecewise affine mapping, the graph of $l + \varphi$ is a polyhedron with the same boundary as the graph of l. In this case (3.7) says, that the $\underline{f}$-area of one face of a closed polyhedron is less than or equal to the sum of the $\underline{f}$-areas of the other faces *(polyhedron inequality)*.

The functional I, defined on the Sobolev space $W^{1,p}(\Omega, \mathbb{R}^m), 1 < p \leqslant \infty$, is weakly lower semicontinuous for $1 < p < \infty$ (weak * lower semicontinuous for $p = \infty$) if and only if, for every (x, y), $f(x, y, \cdot)$ is quasiconvex, provided f satisfies some growth conditions (see Morrey 1966, Dacorogna 1989, p. 156 ff).

The *existence* of a solution of

$$\inf\{I[y] = \int_\Omega f(x,y(x),\mathrm{D}y(x))\,\mathrm{d}x \mid y \in y_0 + W_0^{1,p}(\Omega,\mathbb{R}^m)\} \tag{3.8}$$

can be proved if $f(x,y,\cdot)$ is quasiconvex for every $(x,y) \in \Omega \times \mathbb{R}^m$ and satisfies some growth and coercivity conditions.

In the case where f is not quasiconvex, one considers the corresponding relaxed problem $\inf I_\varphi[y]$, where I_φ is the variational integral with the integrand

$$\varphi := \sup\{g \leqslant f \mid g(x,y,\cdot) \text{ is quasiconvex for all } x \text{ and } y\}$$

(the quasiconvex envelope of f). The solutions of the relaxed problem are related to generalized surfaces (in the sense of L.C. Young) but are in general not found in a Sobolev space.

3.5. Problems in parametric form

A variational problem is called homogeneous or in parametric form if the variational integral I_{hom} is defined for a curve, surface or n-dimensional submanifold of $\mathbb{R}^{m+n}$ but not for a function. Locally these submanifolds are described by parametric representations $z: x \in \Omega \subset \mathbb{R}^n \to z(x) \in \mathbb{R}^{m+n}$ and the integral I_{hom} is formulated with these z. As the value of the homogeneous variational integral must be independent of the choice of the special parametric representations, the integrand F of I_{hom} has certain properties: it is independent of the parameters x of the manifold and it is positively homogeneous in the directional variables: for any z and w

$$F(z,wC) = \det C \cdot F(z,w) \text{ for every } n \times n\text{-matrix } C \text{ with } \det C > 0.$$

Thus there exists a function $F^*: D \times G_n^{m+n} \to \mathbb{R}$, where D is open in $\mathbb{R}^{m+n}$ and G_n^{m+n} is the Grassmann cone of the decomposable n-vectors $w_1 \wedge \cdots \wedge w_n$ ($w_\rho =$ the ρth column of the $(m+n) \times n$-matrix w) such that

$$F(z,w) = F^*(z, w_1 \wedge \cdots \wedge w_n).$$

We remark that G_n^{m+n} is a convex subset of the linear space $\bigwedge^n \mathbb{R}^{n+m}$ (of all n-vectors) *only* in the cases $n = 1$ or $m = 1$ of curves or hypersurfaces. Therefore convexity of $F^*(z,\cdot)$ must be redefined when $n > 1$ and $m > 1$.

Let us consider the special case where $z(x) = (x^1,\ldots,x^n,y^1(x),\ldots,y^m(x))$ is a parametric representation of the submanifold. Then $\mathrm{D}z = \binom{I}{\mathrm{D}y}$, where I is the $n \times n$ identity matrix. Now we are able to compare the integral in parametric form with the former integral $I[y]$:

$$\int_\Omega F^*(z(x), \frac{\partial z}{\partial x^1}(x) \wedge \cdots \wedge \frac{\partial z}{\partial x^n}(x))\,\mathrm{d}x$$

$$= \int_\Omega F(x,y(x),\mathrm{D}z(x))\,\mathrm{d}x = I[y] = \int_\Omega f(x,y(x),\mathrm{D}y(x))\,\mathrm{d}x$$

if

$$f(x,y(x),\mathrm{D}y(x)) := F\left(z(x), \binom{I}{\mathrm{D}y(x)}\right)$$

$$= F^*(z(x), \frac{\partial z}{\partial x^1}(x)\wedge\cdots\wedge\frac{\partial z}{\partial x^n}(x)). \tag{3.9}$$

We call f, defined by (3.9), the associated integrand. An integrand f of (3.1) does not fully determine the values of F^* (or F) such that f is the associated integrand of F^*. But by (3.9) the values $F^*(z, w_1\wedge\cdots\wedge w_n)$ are determined for those n-tuples $(w_1,\ldots,w_n)$ of vectors $\in \mathbb{R}^{n+m}$ for which the matrix $(w^i_\alpha), 1\leqslant i\leqslant m+n, 1\leqslant\alpha\leqslant n$, has the form $\binom{I}{u}$ and moreover, because of the positive homogeneity of $F^*(z,\cdot)$ for all $(w_1,\ldots,w_n)$, such that $\det(w^i_\alpha)_{1\leqslant i,\alpha\leqslant n} > 0$. We call D_f the set of $w_1\wedge\cdots\wedge w_n \in G^{m+n}_n$, such that $\det(w^i_\alpha)_{1\leqslant i,\alpha\leqslant n} > 0$.

In sections 3.2 and 3.3 we have seen the importance of the convexity properties of $f(x,y,\cdot)$ for the variational problem. In terms of the integrand F^* of the parametric problem these properties read as follows:

Let D^* be an open subset of G^{m+n}_n and the function $F^*(z,\cdot): D^* \to \mathbb{R}$ be positively homogeneous. Then $F^*(z,\cdot)$ is said to be *weakly convex* if the restrictions of $F^*(z,\cdot)$ to the convex subsets of D^* are convex. $F^*(z,\cdot)$, defined on $D^* = D_f$, is weakly convex if and only if the associated $f(x,y,\cdot)$ is rank one convex.

$F^*(z,\cdot)$ is called *convex* if there exists an open convex set M in $\bigwedge^n \mathbb{R}^{n+m}, D^* \subset M$ and a convex function $G^*: M \to \mathbb{R}$ which is a continuation of $F^*(z,\cdot)$. Convexity of $F^*(z,\cdot)$, defined on D_f, is equivalent to the polyconvexity of the associated function $f(x,y,\cdot)$; in order to see this, let w_α be the αth column of $\binom{I}{u}$; the coordinates of $w_1\wedge\cdots\wedge w_n$ with regard to the canonical base of $\bigwedge^n \mathbb{R}^{n+m}$ are the determinants $T(u)$ of the submatrices of u (see section 3.3).

$F^*(z,\cdot)$ is called *totally convex* if $F^*(z,\cdot)$ is convex and if for every $X \in D^*$ there exists a decomposable $X^* \in G^{m+n}_n$, such that

$$F^*(z,Z) \geqslant F^*(z,X) + \langle X^*, Z-X\rangle \quad \text{for all } Z \in D^*$$

where $\langle\cdot,\cdot\rangle$ denotes the canonical scalar product in $\bigwedge^n \mathbb{R}^{m+n}$ (see Busemann–Shephard (1967) and the series of papers "Convex bodies and convexity on Grassmann cones" by Busemann, Ewald and Shephard quoted there).

A function f is called totally convex if f is the associated integrand of a totally convex F^*.

For the case $m,n > 1$ we have the following hierarchy of convexity properties (in terms of f):

$$f \text{ totally convex} \underset{\not\leftarrow}{\to} f \text{ convex} \underset{\not\leftarrow}{\to} f \text{ polyconvex}$$

$$\underset{\not\leftarrow}{\to} f \text{ quasiconvex} \underset{\not\leftarrow}{\to} f \text{ rank one convex}.$$

All these properties are equivalent when $m = 1$ or $n = 1$. For details see Ball

(1977). That for $m \geqslant 3, n \geqslant 2$ rank one convexity does not imply quasiconvexity was recently shown by Šverák (1992).

References

Alekseev, V.M., V.M. Tikhomirov and S.V. Fomin
[1987] *Optimal Control* (Consultants Bureau, New York).

Ball, J.M.
[1977] Convexity conditions and existence theorems in nonlinear elasticity, *Arch. Rational Mech. Anal.* **63**, 337–403.

Bolza, O.
[1902] Some instructive examples in the calculus of variations, *Bull. Amer. Math. Soc.* **9**, 1–10.
[1909] *Vorlesungen über Variationsrechnung* (Teubner, Leipzig).

Brechtken-Manderscheid, U.
[1983] *Einführung in die Variationsrechnung* (Wissensch. Buchgesellschaft, Darmstadt).

Busemann, H., and G.C. Shephard
[1967] Convexity on nonconvex sets, *Proc. Coll. Convexity, Copenhagen, 1965*, pp.20–33.

Carathéodory, C.
[1929] Über die Variationsrechnung bei mehrfachen Integralen, *Acta Litt. Sci. Sect. Sci. Math. (Szeged)* **4**, 193–216 [*Gesammelte mathematische Schriften I* (Beck, München, 1954) pp. 401–426].

Cesari, L.
[1974] A necessary and sufficient condition for lower semicontinuity, *Bull. Amer. Math. Soc.* **80**, 467–472.
[1983] *Optimization – Theory and Applications. Problems with Ordinary Differential Equations* (Springer, New York).

Clarke, F.H.
[1983] *Optimization and Nonsmooth Analysis* (Wiley, New York).

Courant, R., and D. Hilbert
[1930] *Methoden der Mathematischen Physik I* (Springer, Berlin, 2. Aufl.).

Dacorogna, B.
[1989] *Direct Methods in the Calculus of Variations* (Springer, New York).

Ekeland, I., and R. Temam
[1972] *Analyse Convexe et Problèmes Variationnels* (Dunod, Paris).

Elsgolc, L.E.
[1961] *Calculus of variations* (Pergamon Press, Oxford). *Variationsrechnung* (Bibl. Inst., Mannheim, 1970).

Ewing, G.M.
[1969] *Calculus of Variations with Applications* (Norton, New York).

Funk, P.
[1962] *Variationsrechnung und ihre Anwendungen in Physik und Technik* (Springer, Berlin).

Giaquinta, M.
[1983] *Multiple Integrals in the Calculus of Variations and Nonlinear Elliptic Systems* (Princeton University Press, Princeton, NJ).

Hadley, G., and M.C. Kemp
[1971] *Variational Methods in Economics* (North-Holland, Amsterdam).

Hildebrandt, S., D. Kinderlehrer and M. Miranda, eds
[1988] *Calculus of Variations and Partial Differential Equations*, Lecture Notes in Mathematics, Vol. 1340 (Springer, Berlin).

Klötzler, R.
[1970] *Mehrdimensionale Variationsrechnung* (Birkhäuser, Basel).

Matsumoto, M.
[1986] *Foundations of Finsler Geometry and Special Finsler Spaces* (Kaiseisha Press, Otsu-Shi, Japan).

Mikhlin, S.G.
[1965] *The Problem of the Minimum of a Quadratic Functional* (Holden-Day, San Francisco).

Morrey Jr, Ch.B.
[1966] *Multiple Integrals in the Calculus of Variations* (Springer, Berlin).

Rund, H.
[1959] *The Differential Geometry of Finsler Spaces* (Springer, Berlin).

Šverák, V.
[1992] Rank-one convexity does not imply quasiconvexity, *Proc. Roy. Soc. Edinburgh Sect. A* **120**, 185–189.

Tonelli, L.
[1921] *Fondamenti di Calcolo delle Variazioni I* (Zanichelli, Bologna).
[1923] *Fondamenti di Calcolo delle Variazioni II* (Zanichelli, Bologna).
[1934] Su gli integrali del calcolo delle variazioni in forma ordinaria, *Ann. Scuola Norm. Sup. Pisa* **3**, 401–450 [*Opere scelte* III (Cremonese, Roma, 1962) pp. 192–254].

Troutman, J.L.
[1983] *Variational Calculus with Elementary Convexity* (Springer, New York).

Weir, A.J.
[1973] *Lebesgue Integration and Measure* (Cambridge University Press).

Young, L.C.
[1969] *Lectures on the Calculus of Variations and Optimal Control Theory* (Saunders, Philadelphia).

CHAPTER 4.4

On Isoperimetric Theorems of Mathematical Physics

Giorgio TALENTI

Istituto Matematico dell' Università, Viale Morgagni 67/A, I-50134, Firenze, Italy

Contents

HANDBOOK OF CONVEX GEOMETRY
Edited by P.M. Gruber and J.M. Wills

1. Introduction

The following proposition holds: *Let G be a 3-dimensional body (i.e., a compact subset of Euclidean space* $\mathbb{R}^3$ *having a non-empty interior), and let* $J(G)$ *be the energy of the gravitational field generated by G. Suppose the density, or specific gravity, is a given constant, and the volume of G is fixed. Then* $J(G)$ *is a maximum if G is a ball.*

This proposition, whose proof is sketched in section 3, may be considered as an example of the so-called *isoperimetric theorems of mathematical physics*. Typically, an isoperimetric theorem of mathematical physics portrays the maximum or the minimum, subject to possible side conditions, of a function – such as J above – whose domain is a collection of sets and that has a special physical significance. All isoperimetric theorems of mathematical physics fall in the category of isoperimetric theorems, of course. Some are close relatives of the standard isoperimetric theorem, which says that among all subsets of Euclidean n-dimensional space $\mathbb{R}^n$ having a prescribed measure the balls take the least perimeter. Some are related to boundary value problems for differential equations, thus involve such quantities as eigenvalues or eigensolutions.

In this chapter we discuss some representative isoperimetric theorems of mathematical physics and try to explain key ideals behind them. A closely related topic – rearrangements of functions à la Hardy and Littlewood – is briefly discussed.

2. Historical and bibliographical comments

Important prototypes were conjectured by de Saint Venant (1856), Lord Rayleigh (1894), and Poincaré (1903). De Saint Venant conjectured that among all cylindrical beams having a prescribed cross-sectional area the circular beam has the highest torsional rigidity. Lord Rayleigh conjectured that among all vibrating clamped membranes having a constant density and a prescribed area the circular membrane has the lowest principal frequency. Poincaré conjectured that among all bodies having a given volume the ball has the least electrostatic capacity. The conjectures by de Saint Venant and Poincaré were demonstrated by Pòlya (1948) and Szegö (1930), respectively; a proof of Rayleigh's conjecture was devised by Faber (1923) and Krahn (1925). Modern statements and proofs are outlined in next sections of this chapter.

A wealth of beautiful isoperimetric theorems of mathematical physics is due to Pòlya and Szegö, and the numerous people attracted by them. Collaborators included Hersch, Payne, Protter, Schiffer, and Weinberger. The activity of Pòlya and Szegö resulted in a source book on this matter (1951). Later developments are exhaustively reported in Payne (1967, 1991). More detailed information is in Bandle (1980), Mossino (1984), and Hersch (1988).

3. Rearrangements

Central to several isoperimetric theorems of mathematical physics is a *rearrangement process*. Loosely speaking, rearrangements are intended for enhancing special qualities of a function or a set without modifying specific traits. The Steiner symmetrization – which converts a subset of $\mathbb{R}^n$ into a subset of $\mathbb{R}^n$ having the *same* measure, but a *smaller* perimeter and an *extra* symmetry – is an example of rearrangement. Rearrangements of functions were introduced by Hardy and Littlewood, see Hardy, Littlewood and Pòlya (1964). Anticipations exist – motivated by a problem of fluid dynamics, Somigliana and Volterra considered the subject, see Somigliana (1899) – as well as generalizations and variants. A catalog of rearrangements is in Kawohl (1985). Rearrangements of functions are used in real and harmonic analysis and in investigations about function spaces; a good many isoperimetric theorems can be demonstrated via rearrangements; rearrangements fit well in the theory of elliptic second-order partial differential equations – see, e.g., Talenti (1991), and the references quoted therein. Rearranging a function à la Hardy and Littlewood basically amounts to subjecting its graph to a Schwarz symmetrization. More formal definitions will follow.

Let G be a measurable subset of Euclidean space $\mathbb{R}^n$ and let u be a real-valued function defined in G. Suppose $m(G)$, the Lebesgue measure of G, is not zero. Suppose u is measurable and μ is the distribution function of u, i.e., for every nonnegative t

$$\{x \in \mathbb{R}^n \colon |u(x)| > t\}\,, \tag{3.1}$$

a typical level set of u, is measurable and $\mu(t)$ is the Lebesgue measure of such a level set.

Definition 1. The *decreasing rearrangement* of u, u^*, is the restriction to $[0, \infty[$ of the distribution function of μ.

As μ decreases and is right-continuous,

$$\begin{aligned} u^*(s) &= \sup\{t \geq 0 \colon \mu(t) > s\} \\ &= \min\{t \geq 0 \colon \mu(t) \leq s\} \end{aligned} \tag{3.2}$$

for every nonnegative s. Clearly:

(i) u^* is a decreasing map from $[0, \infty[$ into $[0, \infty]$;

(ii) u^* is right-continuous;

(iii) sprt u^*, the support of u^*, is the interval $[0, m(\text{sprt } u)]$;

(iv) $u^*(0) = \text{ess sup } |u|$.

The following proposition is crucial:

(v) The distribution function of u^* is exactly μ; in other words, u and u^* are *equimeasurable* or *equidistributed*.

Properties (i), (ii) and (v) *characterize* u^*. Property (v) has important consequences, e.g.,

$$\text{(vi)} \quad \int_G A(|u(x)|)\,\mathrm{d}x = \int_0^\infty A(u^*(s))\,\mathrm{d}s\,,$$

where A is any continuous map from $[0, \infty]$ into $[0, \infty]$.

Definition 2. The *spherically symmetric rearrangement* of u, $u^\star$, is the nonnegative function defined in $\mathbb{R}^n$ by:

$$u^\star(x) = u^*(\kappa_n |x|^n)\,, \tag{3.3}$$

where $|x| = (x_1^2 + \cdots + x_n^2)^{1/2}$, the length of x, and κ_n is the measure of the unit n-dimensional ball.

Clearly: (i) $u^\star$ is a *radial* function – i.e., invariant under rotations about the origin – that *decreases* as the distance from the origin decreases; (ii) u and $u^\star$ are *equidistributed*; (iii) sprt $u^\star$ is included in $G^\star$, the ball whose center is the origin and whose measure equals $m(G)$. In other words, $\{x \in \mathbb{R}^n\colon u^\star(x) > t\}$, a level set of $u^\star$, is the *ball* whose center is the origin and whose measure *equals* the measure of $\{x \in G\colon |u(x)| > t\}$, the allied level set of u – here t is any nonnegative number. Thus $u^\star$ is a function whose graph results from a *Schwarz symmetrization* of the graph of u.

In most applications the main theorems on rearrangements à la Hardy and Littlewood are Theorems A, B, and C below.

Theorem A. *Suppose G is a measurable subset of $\mathbb{R}^n$, u and v are nonnegative measurable functions defined in G. Then*

$$\int_G u(x)v(x)\,\mathrm{d}x \leqslant \int_0^{m(G)} u^*(s)v^*(s)\,\mathrm{d}s\,. \tag{3.4}$$

Theorem B. *Suppose f, g, h are nonnegative measurable functions defined in $\mathbb{R}^n$. Then*

$$\int_{\mathbb{R}^n} \mathrm{d}x \int_{\mathbb{R}^n} f(x)g(y)h(x-y)\,\mathrm{d}y \leqslant \int_{\mathbb{R}^n} \mathrm{d}x \int_{\mathbb{R}^n} f^\star(x)g^\star(y)h^\star(x-y)\,\mathrm{d}y\,. \tag{3.5}$$

Theorem C. *Suppose u is a real-valued locally integrable function defined in $\mathbb{R}^n$ and the first-order (weak, or distributional) derivatives of u are p-integrable over the whole of $\mathbb{R}^n$. Then the (weak, or distributional) derivatives of $u^\star$ are p-integrable in $\mathbb{R}^n$ and the following inequality holds*:

$$\int_{\mathbb{R}^n} |\mathrm{grad}\, u|^p\,\mathrm{d}x \geqslant \int_{\mathbb{R}^n} |\mathrm{grad}\, u^\star|^p\,\mathrm{d}x\,. \tag{3.6}$$

Here $p \geqslant 1$, and grad *is a shorthand for gradient.*

Theorem A is by Hardy and Littlewood, a proof is, e.g., in Hardy, Littlewood and Pòlya (1964). Theorem A is simple, but decisive: one may claim that most theorems from real and harmonic analysis, which can be demonstrated via rearrangements, are consequences of Theorem A.

Theorem B is due to Riesz (1930). A proof is in Hardy, Littlewood and Pòlya (1964), generalizations and improvements are in Brascamp, Lieb and Luttinger (1974), Friedman and McLeod (1986), and Lieb (1983a).

Theorem C implies that the total variation and Dirichlet type integrals of sufficiently smooth functions, which vanish on the boundary of their domain, decrease under spherically symmetric rearrangement. An easy generalization of Theorem C yields the following inequality:

$$\int_{\mathbb{R}^n} A(|\text{grad } u|)\, \mathrm{d}x \geq \int_{\mathbb{R}^n} A(|\text{grad } u^\star|)\, \mathrm{d}x\,, \tag{3.7}$$

where A is any nonnegative nondecreasing *convex* function vanishing at 0. A proof of Theorem C is based on ideas that Pòlya and Szegö mastered – see Pòlya and Szegö (1951) – and ultimately can be found in the papers by Faber (1923) and Krahn (1925). However, implementing these ideas needs tools from geometric measure theory that have been set up in recent times only. Exhaustive proofs of Theorem C, and generalizations or variants of it, are offered in Brothers and Ziemer (1988), Friedman and McLeod (1986), Garsia and Rodemich (1974), Hilden (1976), Sperner (1973, 1974), Spiegel (1973), and Talenti (1976a). An ingenious proof of the $p = 2$ case is due to Lieb (1977). It should be pointed out that Theorem C is basically a consequence of the standard isoperimetric inequality in $\mathbb{R}^n$, i.e.,

$$\begin{aligned} &n\kappa_n^{1/n} \cdot (n\text{-dim. measure of } G)^{1-1/n} \\ &\quad \leq (n-1)\text{-dim. measure of the boundary of } G\,, \end{aligned} \tag{3.8}$$

where G is any sufficiently smooth subset of $\mathbb{R}^n$ having finite measure: this inequality is precisely what allows one, when applied to level sets of the relevant functions, to compare the left- and right-hand side of (3.6).

As an application of Theorem B let us give a proof of the proposition opening the introduction.

Proof. We have

$$J(G) = \int_{\mathbb{R}^3} |\text{grad } u|^2\, \mathrm{d}x\,,$$

where u – a potential of the relevant gravitational field – is given (apart from a numerical factor) by

$$u(x) = \int_{\mathbb{R}^3} \chi_G(y)\, \frac{\mathrm{d}y}{|x-y|}$$

and satisfies

$$-\Delta u = 4\pi\chi_G ,$$
$$u(x) = \mathrm{O}(|x|^{-1}) \quad \text{as } |x| \to \infty .$$

Here

$$\Delta = \sum_{i=1}^{3} \frac{\partial^2}{\partial x_i^2} ,$$

the *Laplace operator*, and χ_G stands for the *characteristic function* of G. Integrations by parts show

$$J(G) = \int_{\mathbb{R}^3} u \cdot (-\Delta u)\, \mathrm{d}x ,$$

hence

$$J(G) = 4\pi \int_{\mathbb{R}^3} \mathrm{d}x \int_{\mathbb{R}^3} \chi_G(x)\chi_G(y)|x-y|^{-1}\, \mathrm{d}y .$$

Note that $(\chi_G)^\star$, the spherically symmetric rearrangement of χ_G, is exactly the characteristic function of $G^\star$, the ball having its center at the origin and the same volume as G. Thus Theorem B gives

$$J(G) \leqslant J(G^\star) .$$

This completes the proof. □

4. Capacity

Let G be a 3-dimensional body, and consider a condenser made up by G and a ball of infinitely large radius. The *capacity* of G may be defined as the capacitance of such a condenser. Thus cap(G), the capacity of G, obeys (up to a multiplicative constant)

$$\operatorname{cap}(G) = \int_{\mathbb{R}^3 \setminus G} |\operatorname{grad} u|^2\, \mathrm{d}x , \tag{4.1}$$

provided u is a potential of the relevant electric field, i.e., satisfies

$$\Delta u = 0 \quad \text{out of } G , \tag{4.2a}$$
$$u = 1 \quad \text{on the boundary of } G , \tag{4.2b}$$
$$u = 0 \quad \text{at infinity} , \tag{4.2c}$$

and belongs to an appropriate function space. A standard apparatus from the calculus of variations – see, e.g., Morrey (1966) – tells us that the appropriate solution to (4.2) can be recovered via a Dirichlet type minimum principle. In other words, the capacity of G may be alternatively defined as follows:

$$\mathrm{cap}(G) = \text{minimum of} \int_{\mathbb{R}^3} |\mathrm{grad}\, u|^2 \, \mathrm{d}x \,, \tag{4.3}$$

as u is constrained by the following conditions: (i) u is locally integrable in $\mathbb{R}^3$ and the first-order (weak) derivatives of u are square integrable over $\mathbb{R}^3$; (ii) u decays at infinity in the proper way; (iii) the level set $\{x \in \mathbb{R}^3: u(x) \geq 1\}$ contains G. For more information about capacity see Frehse (1982).

Theorem. *If G is any 3-dimensional body, then*

$$\mathrm{cap}(G) \geq 48^{1/3} \pi^{2/3} \cdot (\text{volume of } G)^{1/3} \,. \tag{4.4}$$

The right-hand side of (4.4) *is exactly the capacity of G, if G is a ball. Thus, among all 3-dimensional bodies having a prescribed volume the balls take the smallest capacity.*

Proof. The main ingredients of the present proof are formula (4.3) and Theorem C, section 3. Let u be a minimizer in (4.3). Consider $u^\star$, the spherically symmetric rearrangement of u, and $G^\star$, the ball having its center at the origin and the same volume as G. The equimeasurability of u and $u^\star$ implies $\{x \in \mathbb{R}^3: u^\star(x) \geq 1\} \supseteq G^\star$. Consequently,

$$\mathrm{cap}(G^\star) \leq \int_{\mathbb{R}^3} |\mathrm{grad}\, u^\star|^2 \, \mathrm{d}x \,,$$

thanks to the very definition of capacity. Theorem C implies

$$\int_{\mathbb{R}^3} |\mathrm{grad}\, u^\star|^2 \, \mathrm{d}x \leq \int_{\mathbb{R}^3} |\mathrm{grad}\, u|^2 \, \mathrm{d}x \,.$$

Thus

$$\mathrm{cap}(G^\star) \leq \mathrm{cap}(G) \,.$$

The capacity of a ball can be easily evaluated via eqs. (4.1) and (4.2). Thus

$$\mathrm{cap}(G^\star) = 48^{1/3} \pi^{2/3} \cdot (\text{volume of } G^\star)^{1/3} \,.$$

This completes the proof. □

5. Torsional rigidity

Elastic torsion

Let G be a 2-dimensional open bounded region. Consider a cylindrical beam, made of an *elastic homogeneous* material, whose cross-section is G, and suppose the beam is twisted. The *torsional rigidity* of G – call it $S(G)$ – may be defined as the resistance of the beam against the twisting force. The theory of elasticity tells us that $S(G)$ obeys (up to a multiplicative factor)

$$S(G) = \int_G |\text{grad } u|^2 \, dx \,, \tag{5.1}$$

provided u belongs to an appropriate function space and satisfies

$$\Delta u = -1 \quad \text{in } G \,, \tag{5.2a}$$

$$u = 0 \quad \text{on the boundary of } G \,, \tag{5.2b}$$

a standard Dirichlet boundary value problem for a nonhomogeneous Laplace equation. As we learn from the calculus of variations, the functional whose values are

$$\int_G |\text{grad } u|^2 \, dx - 2 \int_G u \, dx \tag{5.3}$$

has a minimum and a unique minimizer within a special space of functions that vanish on the boundary of G – the Sobolev space $W_0^{1,2}(G)$. The *convexity* of such a functional is decisive in this context. Moreover, the minimizer is a solution of (5.2) and the minimum is exactly the negative of $S(G)$. Thus the definition of torsional rigidity can be recast in the following form:

$$-S(G) = \text{minimum of} \int_G |\text{grad } u|^2 \, dx - 2 \int_G u \, dx \,, \tag{5.4}$$

as u belongs to $W_0^{1,2}(G)$. Recall that, if G is a subset of $\mathbb{R}^n$, $W_0^{1,2}(G)$ is the collection of real-valued functions u such that (i) u is square integrable in $\mathbb{R}^n$ together with its first-order (weak) derivatives, and (ii) u vanishes out of G (see, e.g., Adams 1975).

Theorem. *If G is any open bounded subset of $\mathbb{R}^2$, then*

$$S(G) \leq \frac{1}{8\pi} (\text{area of } G)^2 \,. \tag{5.5}$$

The right-hand side of (5.5) *is precisely the torsional rigidity of G, if G is a disk.*

Thus among all 2-dimensional open bounded regions the disks take the smallest torsional rigidity.

Proof. Formula (5.4), the equimeasurability of rearrangements and Theorem C, section 3, immediately give

$$S(G) \leqslant (S(G^\star) ,$$

where $G^\star$ is a disk having the same area as G. The torsional rigidity of a disk is easily computed via (5.1) and (5.2). One finds

$$S(G^\star) = \frac{1}{8\pi} (\text{area of } G^\star)^2 .$$

This completes the proof. □

Remarks. Inequality (5.5) can be viewed as an a priori bound for the solution of the boundary value problem (5.2). The question arises whether similar a priori bounds may be derived – for the solution of the same problem – that have the form of isoperimetric inequalities. Let

$$S_p(G) = \int_G |\text{grad } u|^p \, dx , \tag{5.6}$$

where G is any open bounded subset of $\mathbb{R}^2$ and u satisfies (5.2). Results from Talenti (1976b) ensure that if $0 < p \leqslant 2$, then actually

$$S_p(G) \times (\text{area of } G)^{-1-p/2} \leqslant \frac{2^{1-p}\pi^{-p/2}}{2+p} \tag{5.7}$$

and equality holds if G is a disk. We stress that *inequality* (5.7) *fails to be true if* p *exceeds* 2. For instance, the left-hand side of (5.7) equals

$$\frac{4\pi^{-1-p/2}}{2+p} \, \frac{t^{p/2}}{(1+t^2)^p} \int_0^{\pi/2} (\cos^2\phi + t^2 \sin^2\phi)^{p/2} \, d\phi \tag{5.8}$$

if $t = a_1/a_2$ and G is the *ellipse* where

$$(x_1/a_1)^2 + (x_2/a_2)^2 < 1 ; \tag{5.9}$$

as is easy to check, (5.8) is *not* a maximum when $t = 1$ if p is larger than 6. Thus, if p is large enough, the left-hand side of (5.7) *does not* take its maximum value when G is a disk. An open problem would be to give an exponent p larger than 2 and detect those members G from an appropriate collection of 2-dimensional regions which render the left-hand side of (5.7) a maximum.

Plastic torsion

Consider a cylindrical rod made up of *distinct plastic* materials, and subject to torsion. Suppose the rod has a given length; suppose the number, the quantities and the plastic yield limits of the materials are given. The following isoperimetric theorem holds: the capacity of the rod to withstand a twisting moment is largest *if and only if* the cross-section is a disk, and the materials are arranged in concentric annuli – the hardest material outermost, the softest material innermost, and the others orderly in between. Recall from the theory of plasticity that the torsional moment in question is proportional to the integral of a stress function u, and the relevant data are stored precisely in the distribution function of $|\text{grad}\, u|$, see Aronsson (1979) for details. The above theorem can be found in Aronsson (1979) and derived from the following theorem by Aronsson and Talenti (1981).

Theorem. *Let u be a real-valued function defined on Euclidean space $\mathbb{R}^n$. Suppose u is integrable over $\mathbb{R}^n$ together with its first-order (weak) derivatives, and the support of u has finite measure. Let E be this support and $F(t) = m\{x \in E\colon |\text{grad}\, u(x)| \leq t\}$. Then*

$$\left|\int_{\mathbb{R}^n} u(x)\,\mathrm{d}x\right| \leqslant \frac{[\Gamma(1+n/2)]^{1/n}}{(n+1)\sqrt{\pi}} \int_0^\infty [m(E)^{1+1/n} - F(t)^{1+1/n}]\,\mathrm{d}t\,. \tag{5.10}$$

Equality holds in (5.10) *if and only if either u or $-u$ equals almost everywhere a* dome *function* (*i.e., a nonnegative continuous function with* spherical support, *whose restriction to that support is* spherically symmetric *and* concave).

6. Clamped membranes

Let

$$\lambda_0(G),\ \lambda_1(G),\ldots,\ \lambda_k(G),\ldots \tag{6.1}$$

be the *eigenvalues*, arranged in increasing order, of the following eigenvalue problem:

$$\Delta u + \lambda u = 0 \quad \text{in } G\,, \tag{6.2a}$$

$$u = 0 \quad \text{on the boundary of } G\,. \tag{6.2b}$$

Here Δ is the Laplace operator, G is an open bounded subset of Euclidean space $\mathbb{R}^n$. In the case where the dimension n is 2, eigenvalues (6.1) represent the frequencies of an elastic homogeneous vibrating *membrane* in the shape of G, clamped at the boundary. An updated version of Rayleigh's conjecture and the Faber and Krahn theorem may sound as follows.

Theorem. *If G is any open bounded subset of* $\mathbb{R}^n$, *then the first eigenvalue* $\lambda_0(G)$ *of problem* (6.2) *obeys*

$$\lambda_0(G) \geqslant j^2_{n/2-1,1} \cdot \kappa_n^{2/n} \times (n\text{-dim. measure of } G)^{-2/n} . \tag{6.3}$$

The right-hand side of (6.3) *is the first eigenvalue of problem* (6.2), *if G is a ball. Thus among all open bounded subsets of* $\mathbb{R}^n$ *having a given measure the balls render the first eigenvalue of problem* (6.2) *a minimum.*

As usual, we denote by $j_{\nu,k}$ the kth positive zero of Bessel function J_ν, see Abramowitz and Stegun (1964). Recall that κ_n denotes the measure of the n-dimensional unit ball.

Proof. Recall from the calculus of variations – see, e.g., Weinberger (1974) – that

$$\lambda_0(G) = \text{minimum of} \int_G |\text{grad } u|^2 \, dx , \tag{6.4a}$$

as u is constrained by the following conditions:

$$u \text{ belongs to Sobolev space } W_0^{1,2}(G) , \tag{6.4b}$$

$$\int_G u^2 \, dx = 1 . \tag{6.4c}$$

Let u be a minimizer in (6.4). Incidentally, both u and $-u$ are eigenfunctions associated with $\lambda_0(G)$ – the only eigenfunctions of problem (6.2) that do not change their sign. Consider $G^\star$, the ball having its center at the origin and the same measure as G, and $u^\star$, the spherically symmetric rearrangement of u. Theorem C, section 3, gives

$$u^\star \text{ belong to } W_0^{1,2}(G^\star) ,$$

$$\int_{G^\star} |\text{grad } u^\star|^2 \, dx \leqslant \lambda_0(G) .$$

The equimeasurability of rearrangements implies

$$\int_{G^\star} (u^\star)^2 \, dx = 1 .$$

Consequently,

$$\lambda_0(G^\star) \leqslant \lambda_0(G) ,$$

thanks to the variational characterization of the first eigenvalue. An inspection shows

$$\lambda_0(G^\star) = j^2_{n/2-1,1} \cdot \kappa_n^{2/n} \times (n\text{-dim. measure of } G^\star)^{-2/n},$$

which completes the proof. □

Protter (1981) showed

$$\lambda_0(G) \geq (\tfrac{1}{2}\pi)^2[\rho^{-2} + (n-1)d^{-2}], \tag{6.5}$$

provided G is *convex*, ρ is the *inradius* of G – i.e., the radius of the largest ball included in G – and d is the *diameter* of G. Inequality (6.5) is isoperimetric, in the sense that equality holds in the limit as G approaches a slab.

Lieb (1983b) proved for any open bounded subset G of $\mathbb{R}^n$

$$\lambda_0(G) \geq j^2_{n/2-1,1}(\varepsilon^{-2/n} - 1)R^{-2}, \tag{6.6}$$

where ε is any number between 0 and 1 and R is the radius of the largest n-dimensional ball B such that

$$(\text{measure of } G \cap B) > \varepsilon \times (\text{measure of } B).$$

Payne, Pòlya and Weinberger (1956) proved a number of inequalities for successive eigenvalues of problem (6.2). Developments deriving from these inequalities are reviewed in Protter (1988). Payne et al. conjectured that $\lambda_1(G)/\lambda_0(G)$, the ratio of the first two eigenvalues, is a maximum when G is a ball. Contributions towards a proof of this conjecture were given by Brands (1964), de Vries (1967), Chiti (1983), and also by Hile and Protter (1980), and Marcellini (1980). The conjecture has recently been settled by Ashbaugh and Benguria (1991, 1992), who have proved the following theorem.

Theorem. *If G is any open bounded subset of $\mathbb{R}^n$, then the ratio of the first two eigenvalues of problem* (6.2) *satisfies*

$$\frac{\lambda_1(G)}{\lambda_0(G)} \leq \frac{j^2_{n/2,1}}{j^2_{n/2-1,1}}. \tag{6.7}$$

Equality holds in (6.7) *if and only if G is a ball. Thus, among all open bounded subsets G of $\mathbb{R}^n$ the balls render the ratio $\lambda_1(G)/\lambda_0(G)$ a maximum.*

7. Clamped plates

Consider the following eigenvalue problem:

$$\Delta^2 u - \lambda u = 0 \qquad \text{in } G, \tag{7.1a}$$

$$u = |\text{grad } u| = 0 \quad \text{on the boundary of } G. \tag{7.1b}$$

Here Δ is the Laplace operator, $\Delta^2 = \Delta\Delta$, and G is an open bounded subset of Euclidean space $\mathbb{R}^n$. Problem (7.1) arises in the theory of small vibrations of an elastic thin *plate*, clamped at the boundary. Problem (7.1) has a countable set of eigenvalues of finite multiplicity and a complete system of eigenfunctions. Let

$$\lambda(G) \tag{7.2a}$$

be the *smallest eigenvalue* of problem (7.1). The calculus of variations tells us that

$$\lambda(G) = \text{minimum of} \int_G (\Delta u)^2 \, dx , \tag{7.2b}$$

as u is constrained by the following conditions: (i) u belongs to Sobolev space $W_0^{2,2}(G)$, i.e., u is square integrable over the whole of $\mathbb{R}^n$ together with its first-order and second-order (weak) derivatives and vanishes out of G; (ii) $\int_G u^2 \, dx = 1$.

The question arises whether $\lambda(G)$ enjoys the *isoperimetric property*:

$$\lambda(G) \geqslant \lambda(G^\star) , \tag{7.3a}$$

where

$$G^\star = \text{a ball having the same measure as } G . \tag{7.3b}$$

By the way, $\lambda(G^\star)$ is explicitly computed by making

$$[\lambda(G^\star)]^{1/4} \times \kappa_n^{-1/n} \times (\text{measure of } G)^{1/n} \tag{7.4}$$

equal to the smallest positive root of the transcendental equation

$$I_{n/2-1}(z) J_{n/2}(z) + I_{n/2}(z) J_{n/2-1}(z) = 0 .$$

Here κ_n is the measure of the n-dimensional unit ball, I and J stand for Bessel functions.

Szegö proved in 1950 and 1958 that (7.3) holds if G fulfils the following hypothesis: a *principal* eigenfunction, i.e., one of the eigenfunctions associated with $\lambda(G)$, exists which *does not* change its sign. Szegö's proof was originally written in dimension 2, but can be easily carried out in any dimension n greater than 1. Unfortunately, the absence of nodal lines seems to be a crucial hypothesis for Szegö's arguments, on the other hand no criterion is available for deciding whether a given domain fulfils such hypothesis or not. Domains G such that *all* eigenfunctions of problem (7.1) *do change* their sign actually exist. For instance, the first eigenvalue has multiplicity 2 and all principal eigenfunctions have a diametral nodal line if G is a 2-dimensional annulus, the outer radius is 1 and the inner radius <0.001311774 – see Coffman, Duffin and Shaffer (1979). Any eigenfunction oscillates infinitely many times on any ray issuing from a corner if G is a square (see Coffman 1982).

A numerical analysis gives evidence to inequality (7.3) in selected cases not covered by Szegö's theorem, e.g., if G is an annulus, a square, a triangle, a rhombus, a trapezium – see Coffman, Duffin and Shaffer (1979), Hackbush and Hofman (1980), and Kuttler and Sigillito (1980, 1981). Incquality (7.3) has been neither proved in general nor disproved yet. The following result is in Talenti (1981): *A constant c_n exists such that $\lambda(G) \geqslant c_n \lambda(G^\star)$ for any open bounded subset G of $\mathbb{R}^n$. Values of c_n are displayed in table* 1.

Table 1

n	c_n
2	0.97768
3	0.73910
4	0.65242
5	0.60925

References

Abramowitz, M., and I. Stegun, eds
[1964] *Handbook of Mathematical Functions* (U.S. Government Printing Office).

Adams, R.A.
[1975] *Sobolev Spaces* (Academic Press, New York).

Aronsson, G.
[1979] An integral inequality and plastic torsion, *Arch. Rational Mech. Anal.* **72**, 23–39.

Aronsson, G., and G. Talenti
[1981] Estimating the integral of a function in terms of a distribution function of its gradient, *Boll. Un. Mat. Ital. B(5)* **18**, 885–894.

Ashbaugh, M.S., and R.D. Benguria
[1991] Proof of the Payne–Pòlya–Weinberger conjecture, *Bull. Amer. Math. Soc.* **25**, 19–29.
[1992] Sharp bound for the ratio of the first two eigenvalues of Dirichlet Laplacians and extensions, *Ann. of Math.*, to appear.

Bandle, C.
[1980] *Isoperimetric Inequalities and Applications* (Pitman, Boston).

Brands, J.J.A.M.
[1964] Bounds for the ratios of the first three membrane eigenvalues, *Arch. Rational Mech. Anal.* **16**, 265–268.

Brascamp, H.J., E.H. Lieb and J.M. Luttinger
[1974] A general rearrangement inequality for multiple integrals, *J. Funct. Anal.* **17**, 227–237.

Brothers, J.E., and W.P. Ziemer
[1988] Minimal rearrangements of Sobolev functions, *J. Reine Angew. Math.* **384**, 153–179.

Chiti, G.
[1983] A bound for the ratio of the first two eigenvalues of a membrane, *SIAM J. Math. Anal.* **14**, 1163–1167.

Coffman, C.V.
[1982] On the structure of solutions to $\Delta^2 u = \lambda u$ which satisfy the clamped plate conditions on a right angle, *SIAM J. Math. Anal.* **13**, 746–757.

Coffman, C.V., R.J. Duffin and D.H. Shaffer
[1979] The fundamental mode of vibration of a clamped annular plate is not of one sign, in: *Constructive Approaches to Mathematical Models* (Academic Press, New York) pp. 267–277.

de Saint Venant, B.
[1856] Mémoire sur la torsion des prismes, *Acad. Sci.* **14**, 233–560.

de Vries, H.L.
[1967] On the upper bound for the ratio of the first two membrane eigenvalues, *Z. Naturforsch. A* **22**, 152–153.

Faber, G.
[1923] Beweis, dass unter allen homogenen Membranen von gleicher Fläche und gleicher Spannung die kreisförmige den tiefsten Grundton gibt, *Sitzungsber. Math.-Phys. Kl. Bayer. Akad. Wiss. München*, 169–172.

Frehse, J.
[1982] Capacity methods in the theory of partial differential equations, *Jber. Deutsch. Math.-Vereinig.* **84**, 1–44.

Friedman, A., and B. McLeod
[1986] Strict inequalities for integrals of decreasingly rearranged functions, *Proc. Roy. Soc. Edinburgh Sect. A* **102**, 277–289.

Garsia, A.M., and E. Rodemich
[1974] Monotonicity of certain functionals under rearrangement, *Ann. Inst. Fourier Grenoble* **24**, 69–116.

Hackbusch, W., and G. Hofman
[1980] Results of the eigenvalue problem for the plate equation, *Z. Angew. Math. Phys.* **31**, 730–739.

Hardy, G.H., J.E. Littlewood and G. Pòlya
[1964] *Inequalities* (Cambridge Univ. Press).

Hersch, J.
[1988] Isoperimetric monotonicity: some properties and conjectures (connections between isoperimetric inequalities), *SIAM Rev.* **30**, 551–577.

Hilden, K.
[1976] Symmetrization of functions in Sobolev spaces and the isoperimetric inequality, *Manuscripta Math.* **18**, 215–235.

Hile, G.N., and M.H. Protter
[1980] Inequalities for eigenvalues of the Laplacian, *Indiana Univ. Math. J.* **29**, 523–538.

Kawohl, B.
[1985] *Rearrangements and Convexity of Level Sets in PDE* (Springer, Berlin).

Krahn, E.
[1925] Über eine von Rayleigh formulierte Minimaleigenschaft des Kreises, *Math. Ann.* **94**, 97–100.

Kuttler, J.R., and V.G. Sigillito
[1980] Upper and lower bounds for frequencies of clamped rhombical plates, *J. Sound Vibration (4)* **68**, 597–590.
[1981] Upper and lower bounds for frequencies of trapezoidal and triangular plates, *J. Sound Vibration (4)* **78**, 585–590.

Lieb, E.H.
[1977] Existence and uniqueness of the minimizing solution of Choquard's nonlinear equation, *Studia Appl. Math.* **57**, 93–105.
[1983a] Sharp Constants in the Hardy–Littlewood–Sobolev and related inequalities, *Ann. of Math.* **118**, 349–374.
[1983b] On the lowest eigenvalue of the Laplacian for the intersection of two domains, *Invent. Math.* **74**, 441–448.

Marcellini, P.
[1980] Bounds for the third membrane eigenvalue, *J. Differential Equations* **37**, 438–443.

Morrey Jr, Ch.B.
[1966] *Multiple Integrals in the Calculus of Variations* (Springer, Berlin).

Mossino, J.
[1984] *Inégalités Isopérimetriques et Applications en Physique* (Herman, Paris).

Payne, L.E.
[1967] Isoperimetric inequalities and their applications, *SIAM Rev.* **9**, 453–488.
[1991] Some comments on the past fifty years of isoperimetric inequalities, in: *Inequalities*, ed. W.N. Everitt (Marcel Dekker, New York) pp. 143–161.

Payne, L.E., G. Pòlya and H.F. Weinberger
[1956] On the ratio of consecutive eigenvalues, *J. Math. Phys.* **35**, 289–298.

Poincaré, H.
[1903] *Figures d'Equilibre d'une Masse Fluide* (Naud, Paris).

Pòlya, G.
[1948] Torsional rigidity, principal frequency, electrostatic capacity and symmetrization, *Quart. Appl. Math.* **6**, 267–277.

Pòlya, G., and G. Szegö
[1951] *Isoperimetric Inequalities in Mathematical Physics* (Princeton University Press, Princeton, NJ).

Protter, M.H.
[1981] A lower bound for the fundamental frequency of a convex region, *Proc. Amer. Math. Soc.* **81**, 65–70.
[1988] Universal inequalities for eigenvalues, in: *Maximum Principles and Eigenvalue Problems in PDE*, ed. P.W. Schaefer (Longman, New York) pp. 111–120.

Rayleigh, Lord
[1894/96] *The Theory of Sound* (Macmillan, New York).

Riesz, F.
[1930] Sur une inégalité intégrale, *J. London Math. Soc.* **5**, 162–168.

Somigliana, C.
[1899] Sulle funzioni reali di una variabile, Considerazioni sulle funzioni ordinate, *Rend. R. Accad. Lincei* **8**, 173–198.

Sperner Jr, E.
[1973] Zur symmetrisierung für Funktionen auf Sphären, *Math. Z.* **134**, 317–327.
[1974] Symmetrisierung für Funktionen mehrerer reeller Variablen, *Manuscripta Math.* **11**, 159–170.

Spiegel, W.
[1973] Über die symmetrisierung stetiger Funktionen im euklidischen Raum, *Arch. Math.* **24**, 545–551.

Szegö, G.
[1930] Über einige neue Extremalaufgaben der Potentialtheorie, *Math. Z.* **31**, 583–593.
[1950] On membranes and plates, *Proc. Nat. Acad. Sci. U.S.A.* **36**, 210–216.
[1958] Note to my paper "On membranes and plates", *Proc. Nat. Acad. Sci. U.S.A.* **44**, 314–316.

Talenti, G.
[1976a] Best constant in Sobolev inequality, *Ann. Mat. Pura Appl. (4)* **110**, 353–372.
[1976b] Elliptic equations and rearrangements, *Ann. Scuola Norm. Sup. Pisa* **3**, 697–718.
[1981] On the first eigenvalue of the clamped plate, *Ann. Mat. Pura Appl. (4)* **129**, 265–280.
[1991] Rearrangements and partial differential equations, in: *Inequalities*, ed. W.N. Everitt (Marcel Dekker, New York) pp. 211–230.

Weinberger, H.F.
[1974] *Variational Methods for Eigenvalue Approximation*, CBNS Reg. Conf. Ser. Appl. Math., Vol. 15 (SIAM, Philadelphia, PA).

CHAPTER 4.5

The Local Theory of Normed Spaces and its Applications to Convexity*

J. LINDENSTRAUSS

Department of Mathematics, The Hebrew University, Jerusalem, Israel

V.D. MILMAN

Department of Mathematics, Tel-Aviv University, Ramat Aviv, 69978 Tel-Aviv, Israel

Contents

* Supported in part by a grant from the U.S.-Israel BSF.

HANDBOOK OF CONVEX GEOMETRY
Edited by P.M. Gruber and J.M. Wills

1. Introduction

The local theory of Banach spaces deals with convex bodies in $\mathbb{R}^n$ where n is finite but large. The main theme of the theory is a quantitative study of the structure of such sets and asymptotic estimates of various parameters associated with them as $n \to \infty$.

The theory grew out of, and in a sense can still be considered as a central part of, functional analysis. The main idea in functional analysis is to consider complex elements (say a function or the state of a physical system) as points in a linear space and to investigate the relation between such objects in analogy to the study of points, lines, planes, etc., in geometry using also an appropriate norm for quantitative estimates. This approach leads naturally in many examples to the study of infinite-dimensional Banach spaces, i.e., geometrically to the study of convex sets in infinite-dimensional spaces. The definition of convexity involves only 2-dimensional subspaces of a linear space. Many fundamental facts concerning convex sets in the plane or 3-space generalize in a neat and very useful way to infinite dimensions (e.g., the Hahn–Banach or Krein–Milman theorems). These results are responsible for the great initial success of functional analysis and continue to yield new interesting applications to the present. The arguments used in connection with these theorems were either 1-, 2- or 3-dimensional geometrical arguments or, on the other hand, arguments of a purely infinite-dimensional nature. Only much later was it realized that for a deeper understanding of functional analysis it is imperative to study the quantitative theory of n-dimensional spaces with n finite but large. This theory relies on inequalities which for say $n = 3$ are too crude to be interesting but on the other hand cannot be formulated at all in infinite-dimensional spaces. This study led to a deep theory which had many very surprising consequences in analysis and geometry. This theory is now commonly called the local theory of Banach spaces.

Actually the name "local theory" is applied to two somewhat different topics:

(a) The quantitative study of n-dimensional normed spaces as $n \to \infty$.

(b) The relation between the structure of an infinite-dimensional space and its finite-dimensional subspaces.

It is topic (a) above which has by now an elaborate and deep theory. This survey intends to outline the main points in topic (a) from a functional analytic and geometric point of view. The last section of the survey deals specifically with results of a purely geometric nature which were obtained in the framework of the local theory of Banach spaces.

We state now two theorems which can be considered as the starting points of the local theory. We first recall the definition of the finite-dimensional $L_p(\mu)$ spaces. For $1 \leqslant p \leqslant \infty$ and n an integer we denote by ℓ_p^n the space of n-tuples of scalars $x = (x_1, \ldots, x_n)$ with $\|x\| = \left(\sum_{i=1}^n |x_i|^p\right)^{1/p}$ $(= \max_i |x_i|$ if $p = \infty)$. These spaces play a role in local theory which is even more central and natural than the role of $L_p(\mu)$ spaces in the theory of normed spaces in general.

Theorem 1.1 (Dvoretzky 1959, 1961). *Let X be an infinite-dimensional Banach space. Then for every integer n and every $\varepsilon > 0$ there is an operator $T: \ell_2^n \to X$ so that $\|u\| \leqslant \|Tu\| \leqslant (1+\varepsilon)\|u\|$ for every $u \in \ell_2^n$.*

Theorem 1.2 (James 1964). *Let X be a non-reflexive infinite-dimensional Banach space. Then for every $\varepsilon > 0$ there is an operator $T: \ell_1^2 \to X$ so that $\|u\| \leqslant \|Tu\| \leqslant (1+\varepsilon)\|u\|$ for every $u \in \ell_1^2$.*

Geometrically, Theorem 1.1 (respectively Theorem 1.2) means that the unit ball of every infinite-dimensional space X has, for every n, n-dimensional sections which are arbitrarily close to ellipsoids (respectively, if X is non-reflexive, there are also 2-dimensional sections which are arbitrarily close to a parallelogram).

As stated both theorems belong to topic (b) above rather than topic (a). However, both theorems have natural formulations also in the context of topic (a). (This will be explained below.) This situation is quite common. Also in the sequel we shall encounter results concerning topic (a) which are most conveniently formulated in the context of infinite-dimensional Banach spaces.

The statement that Theorems 1.1 and 1.2 were the starting points in the development of local theory does not mean that there are no earlier theorems which play an important role in the subject. We shall present several such theorems in the sequel.

We end this introduction by mentioning some books which are mainly devoted to some topics in local theory: Pisier (1986), Milman and Schechtman (1986), Tomczak-Jaegermann (1989) and Pisier (1989b) and to a large extent also Ledoux and Talagrand (1991). There are also some expository articles devoted to local theory or some aspects of it: Pelczynski (1980), Figiel (1983), Maurey (1983), Gluskin (1986), Milman (1986a, 1988a) and Lindenstrauss (1992).

2. Basic concepts

2.1. A few classic facts and definitions

The subject of our study are finite-dimensional normed spaces over the reals $X = (\mathbb{R}^n, \|\cdot\|)$. Most of the results we shall mention will be valid also in the complex case but unless stated so explicitly we shall consider only real spaces. There is a one to one correspondence between norms in $\mathbb{R}^n$ and convex bodies (= compact convex sets with non-empty interior) K in $\mathbb{R}^n$ which are symmetric with respect to the origin. Indeed, the unit ball K_X of $X = (\mathbb{R}^n, \|\cdot\|)$ is such a body, and conversely, given a symmetric convex body K in $\mathbb{R}^n$ then its Minkowski functional $\|x\|_K = \inf\{\lambda^{-1}\colon \lambda x \in K, \lambda > 0\}$ defines a norm in $\mathbb{R}^n$ whose unit ball is K.

A basic notion for the quantitative study of normed spaces is that of the Banach–Mazur (multiplicative) distance, defined as follows:

$$d(X,Y) = \inf\{\|T\|\,\|T^{-1}\|\colon\ T: X \to Y \text{ a linear isomorphism}\}. \tag{2.1.1}$$

Notice that $d(X,Y)$ is finite if and only if there is a (bounded linear) isomorphism from X onto Y. In the setting of finite-dimensional spaces this clearly is the case if and only if $\dim X = \dim Y$. It is trivial to verify by a compactness argument that if X and Y are of the same finite dimension then the inf in (2.1.1) is always attained and in particular $d(X,Y) = 1$ if and only if there is an isometry T from X onto Y (such an isometry is necessarily linear as shown by Mazur and Ulam 1932). Also, for finite-dimensional spaces obviously $d(X,Y) = d(X^*,Y^*)$ where X^* denotes the dual of X. Note that for all X,Y and Z

$$d(X,Y)d(Y,Z) \geqslant d(X,Z), \qquad d(X,Y) = d(Y,X) \geqslant 1\,. \tag{2.1.2}$$

In terms of convex symmetric bodies in $\mathbb{R}^n$ the Banach–Mazur distance gives a (multiplicative) distance between equivalence classes of such bodies. Two bodies K_1 and K_2 belong to the same equivalence class if there is a $T \in \mathrm{GL}_n$ such that $TK_1 = K_2$. We have that if $\dim X = n$ then $d(X,Y) \leqslant d$ iff there is a $T \in \mathrm{GL}_n$ so that

$$K_Y \subset TK_X \subset d \cdot K_Y. \tag{2.1.3}$$

For many purposes it will be convenient to assume that the normed space $X = (\mathbb{R}^n, \|\cdot\|)$ is also equipped with a specific Euclidean norm $|\cdot|$. Geometrically this means that we designate a specific ellipsoid in $\mathbb{R}^n$ to serve as the unit ball of $|\cdot|$. In some cases we take simply the standard Euclidean norm $|x| = \left(\sum_{i=1}^n |x_i|^2\right)^{1/2}$ without any relation to the given norm in X. A more canonical choice of an ellipsoid is the ellipsoid $\mathscr{E}_K$ which is contained in $K = K_X$ and has the maximal volume among all such ellipsoids. Löwner was the first to consider this ellipsoid and he proved that it is uniquely determined by K. John (1948) proved that for every convex symmetric K we have that $K \subset \sqrt{n}\mathscr{E}_K$. This fact clearly implies that

$$d_X = d(X, \ell_2^n) \leqslant \sqrt{n}, \qquad \dim X = n. \tag{2.1.4}$$

Note that in (2.1.4) we also introduced a special notation d_X for $d(X, \ell_2^n)$ since these special distances have a central role in the theory and appear frequently in various formulas. The result of John is sharp, it is easy to check that $d(\ell_\infty^n, \ell_2^n) = \sqrt{n}$ for every n.

In the same paper John also gave a sharp estimate on the "distance" between a general convex body K in $\mathbb{R}^n$ and ellipsoids. He proved that if we denote again by $\mathscr{E}_K$ the ellipsoid of maximal volume contained in K then $K \subset n\mathscr{E}_K$ (if K is a simplex this estimate cannot be improved).

Remarks.

(1) In this survey we shall concentrate our attention on symmetric convex bodies which enter naturally in functional analysis as unit balls of norms. Some of the results presented below can be (and have been) generalized to the setting of general convex bodies with only minor changes in the proof and appropriate changes

in the result. (In the case of John's theorem the change required in the proof is minor but the constant in the result has to be changed from $\sqrt{n}$ to n which is quite substantial). In some other cases the passage from symmetric convex bodies to general convex bodies is more substantial and actually much remains to be done in this direction. For example the whole nice theory of bodies of constant width has not be investigated till now from the point of view of the local theory of Banach spaces.

(2) Let K be a convex symmetric body in $\mathbb{R}^n$. Its polar K° is defined as usual by

$$K^\circ = \{y \in \mathbb{R}^n\colon\ |\langle x, y\rangle| \leqslant 1 \text{ for all } x \in K\}. \tag{2.1.5}$$

Note that if $K = K_X$ then $K^\circ = K_{X^*}$. If $\mathscr{E}_K$ is the ellipsoid of maximal volume contained in K then $\mathscr{E}_K^\circ$ is the ellipsoid of minimal volume containing K° (which is again, by Löwner's result, determined uniquely). The dual formulation of the result of John is thus the following. If $\mathscr{E}'_K$ denotes the ellipsoid of minimal volume containing a symmetric convex body K in $\mathbb{R}^n$ then $K \subset \mathscr{E}'_K \subset \sqrt{n}K$.

(3) On the subject of choosing suitable ellipsoids we will have much more to say in section 4 below.

Besides ellipsoids the most important bodies in the theory are the unit balls of ℓ_∞^n and ℓ_1^n (i.e., parallelopipeds and their polars-crosspolytopes). Let $X = (\mathbb{R}^n, \|\cdot\|)$ be a normed space, $K = K_X$ its unit ball and $\mathscr{P}$ a parallelopiped of minimal volume containing K (note that $\mathscr{P}$ is in general not uniquely determined by K). By definition $\mathscr{P}$ is given as $\bigcap_{i=1}^n \{x\colon |\langle x, x_i^*\rangle| \leqslant 1\}$ where $x_i^* \in X^*$ and $\|x_i^*\| = 1$ for all i. A minute of reflection shows that for every i there is an $x_i \in X$ with $\|x_i\| = 1$ so that $\langle x_i, x_j^*\rangle = \delta_{i,j}$ for all i and j. In other words, we get the result first proved by Auerbach that for every Banach space $X = (\mathbb{R}^n, \|\cdot\|)$ there exists a system $\{x_i\}_{i=1}^n \subset X$, $\{x_i^*\}_{i=1}^n \subset X^*$ (called an Auerbach system) and satisfying

$$\|x_i\| = \|x_i^*\| = 1, \quad 1 \leqslant i \leqslant n; \qquad \langle x_i, x_j^*\rangle = \delta_{i,j}, \quad 1 \leqslant i, j \leqslant n. \tag{2.1.6}$$

Note that for every $x \in X$, $x = \sum_{i=1}^n \langle x, x_i^*\rangle x_i$ and thus in particular $K_X \subset \mathscr{P} \subset nK_X$.

The Auerbach system provides a possible choice of a basis. There are many other possible choices. For bodies which have a rich group of symmetries (or "almost symmetries") one naturally prefers to use bases which are related to these symmetries. A basis $\{x_i\}_{i=1}^n$ of X is said to be 1-unconditional if

$$\left\|\sum_{i=1}^n \alpha_i x_i\right\| = \left\|\sum_{i=1}^n \alpha_i \theta_i x_i\right\|, \tag{2.1.7}$$

for every choice of scalars $\{\alpha_i\}_{i=1}^n$ and $\theta_i = \pm 1$, $1 \leqslant i \leqslant n$. If the space X has a 1-unconditional basis then the reflections with respect to the co-ordinate hyperplanes map K_X onto itself. For a general basis $\{x_i\}_{i=1}^n$ in X we call its unconditional constant the smallest λ satisfying

$$\left\| \sum_{i=1}^{n} \alpha_i \theta_i x_i \right\| \leqslant \lambda \left\| \sum_{i=1}^{n} \alpha_i x_i \right\|; \quad \text{all scalars } \alpha_i, \theta_i = \pm 1 \,. \tag{2.1.8}$$

A basis satisfying (2.1.8) is called λ-unconditional. If λ is small then the reflections with respect to the coordinate hyperplanes are in a sense "almost isometries". If we consider besides the operation of changing signs also the permutation of the coordinates we get the notion of a λ-symmetric basis. A basis $\{x_i\}_{i=1}^n$ in X is called λ-symmetric if

$$\left\| \sum_{i=1}^{n} \alpha_{\pi(i)} \theta_i x_i \right\| \leqslant \lambda \left\| \sum_{i=1}^{n} \alpha_i x_i \right\|, \tag{2.1.9}$$

for every choice of scalars $\{\alpha_i\}_{i=1}^n$, signs $\theta_i = \pm 1$ and permutation π of $\{1, 2, \ldots, n\}$. For example, the unit vector basis in ℓ_p^n is 1-symmetric for every choice of n and $1 \leqslant p \leqslant \infty$. The word "symmetry" as used in this paragraph (and in the literature in general) has different meanings in different contexts. When we say simply that a convex body K is symmetric we mean that $K = -K$. For such a K to be the unit ball of a space with a 1-symmetric basis it has to have much more symmetries. We trust that the use of the word symmetric in the sequel will be clear from the context.

If $\{x_i\}_{i=1}^n$ is a 1-symmetric basis of X and $\{x_i^*\}_{i=1}^n$ are the biorthogonal functionals in X^*, then it is trivial to check that

$$\left\| \sum_{i=1}^{n} x_i \right\| \cdot \left\| \sum_{i=1}^{n} x_i^* \right\| = n. \tag{2.1.10}$$

An interesting and often useful variant of (2.1.10) holds if we just assume that the $\{x_i\}_{i=1}^n$ are 1-unconditional. There are positive $\{\lambda_i\}_{i=1}^n$ so that

$$\left\| \sum_{i=1}^{n} \lambda_i x_i \right\| \cdot \left\| \sum_{i=1}^{n} \lambda_i^{-1} x_i^* \right\| = n. \tag{2.1.11}$$

This fact is due to Lozanovskii (1969) (see Pisier 1989b, p. 30, for a simple proof).

2.2. *Type, cotype and related notions*

Much of the early development of the local theory was motivated by the theory of infinite series in infinite-dimensional Banach spaces. This holds in connection with Dvoretzky's theorem (Theorem 1.1) as well as for the notions of type and cotype which are the main subject of the present subsection. Nordlander (1961) introduced these notions (without giving them names) in the context of $L_p(\mu)$ spaces for the study of convergence of series of the form $\sum_{n=1}^{\infty} \theta_n x_n$ where $\{x_n\}_{n=1}^{\infty}$ are vectors in an infinite-dimensional Banach space and $\{\theta_n\}_{n=1}^{\infty}$ are signs chosen randomly and independently from each other. Hoffmann-Jørgensen (1974) extended the study to arbitrary Banach spaces and introduced the notions of type and cotype explicitly.

The notions of type and cotype (in the context of $L_p(\mu)$ spaces) arose also in the work of Stein (1961) concerning weak type inequalities for maximal operators. This direction of research was continued by several authors, in particular Nikishin (1970) and Maurey (1974b). (See also the expository paper Gilbert (1979) on this topic.) These notions became a fundamental tool in local theory mainly in view of the extensive study and deep results obtained by Maurey and Pisier (e.g., Maurey and Pisier 1976).

Let X be a normed space, $1 \leqslant p < \infty$ and n an integer. We denote by $T_p(X,n)$ (= the type p constant of X determined by n vectors) the smallest constant λ such that

$$\left(\text{Average}_{\theta_i=\pm 1}\left\|\sum_{i=1}^{n}\theta_i x_i\right\|^2\right)^{1/2} \leqslant \lambda\left(\sum_{i=1}^{n}\|x_i\|^p\right)^{1/p} \tag{2.2.1}$$

holds for every choice of $\{x_i\}_{i=1}^n$ in X. Similarly for $1 \leqslant q < \infty$ the cotype q constant of X determined by n vectors $C_q(X,n)$ is the smallest positive λ for which

$$\left(\text{Average}_{\theta_i=\pm 1}\left\|\sum_{i=1}^{n}\theta_i x_i\right\|^2\right)^{1/2} \geqslant \lambda^{-1}\left(\sum_{i=1}^{n}\|x_i\|^q\right)^{1/q} \tag{2.2.2}$$

holds for all choices of $\{x_i\}_{i=1}^n$ in X. The averages in (2.2.1) and (2.2.2) are with respect to all 2^n possible choices of signs. A convenient way to write the average is via the Rademacher functions defined by

$$r_0(t) \equiv 1, \qquad r_n(t) = \operatorname{sign}\sin(2^n\pi t), \quad n \geqslant 1; \qquad t \in [0,1]. \tag{2.2.3}$$

One has

$$\text{Average}\left\|\sum_{i=1}^{n}\theta_i x_i\right\|^2 = \int_0^1\left\|\sum_{i=1}^{n}r_i(t)x_i\right\|^2 \mathrm{d}t. \tag{2.2.4}$$

The type p constant (respectively the cotype q constant) of a Banach space X is defined by

$$T_p(X) = \sup_n T_n(X,n), \qquad C_q(X) = \sup_n C_q(X,n)\ . \tag{2.2.5}$$

If $T_p(X) < \infty$ (respectively $C_q(X) < \infty$) we say that the Banach space X has type p (respectively cotype q). Note that by definition $T_p(X,n)$ (and therefore also $T_p(X)$) is for every X and n a non-decreasing function of p. Similarly the cotype constants are non-increasing functions of q. The parallelogram identity for inner product spaces shows that in any inner product spaces X (in particular in a 1-dimensional space)

$$\int_0^1 \left\| \sum_{i=1}^n r_i(t)x_i \right\|^2 \mathrm{d}t = \sum_{i=1}^n \|x_i\|^2, \qquad \{x_i\}_{i=1}^n \in X.$$

Hence $T_2(\ell_2) = C_2(\ell_2) = 1$ and moreover for any space X, $T_p(X)$ (respectively $C_q(X)$) can be finite only if $p \leqslant 2$ (respectively $q \geqslant 2$). Also, trivially, $T_1(X) = 1$ for every Banach space. It follows from the above that for every finite-dimensional space X, $T_p(X) < \infty$ iff $p \leqslant 2$ and $C_q(X) < \infty$ iff $q \geqslant 2$. Somewhat less trivial, but still easy, is the fact that if X is an infinite-dimensional $L_p(\mu)$ space, $1 \leqslant p < \infty$, then

$$X \text{ is of type } \max(p,2) \text{ and cotype } \min(p,2)\ . \tag{2.2.6}$$

(This fact is the main point in the paper of Nordlander mentioned above.)

We mention now some useful facts concerning type and cotype whose proofs require more work.

The spaces isomorphic to Hilbert space are the only spaces which have the best type and cotype (i.e., are of type 2 and cotype 2). This result is due to Kwapien (1972) who proved actually the following quantitative version of this result. For every Banach space X there is a Hilbert space H so that

$$d_X = d(X,H) \leqslant C_2(X)T_2(X). \tag{2.2.7}$$

The proof of (2.2.7) reduces eventually to an application of the separation (i.e., Hahn–Banach) theorem.

In spaces which have type $p > 1$, a dimension-free variant of Caratheodory's theorem holds. The statement is the following: Let $\{x_i\}_{i-1}^m$ be elements in the unit ball K_X of X. Let $u \in \operatorname{conv}\{x_i\}_{i=1}^m$ and let $k \leqslant m$. Then there is a v which is in the convex hull of k vectors out of the $\{x_i\}_{i=1}^m$ so that

$$\|v - u\| \leqslant cT_p(X,k) \cdot k^{p^{-1}-1}, \tag{2.2.8}$$

for some absolute constant c. This fact is due to Maurey (see Pisier 1981) and the estimate is sharp (consider $x_i = i$th unit vector in ℓ_p, $1 < p \leqslant 2$).

In the inequalities defining the type and cotype constants we used only the L_2 norms of $\sum_{i=1}^n r_i(t)x_i$. A result of Kahane (1968) (which generalizes the classical Khintchine inequality) states that for expressions of the form $\sum_{i=1}^n r_i(t)x_i$ all the L_p norms $1 \leqslant p < \infty$ are equivalent. More precisely one has for every Banach space X, any integer n and vectors $\{x_i\}_{i=1}^n \in X$ and any $1 \leqslant p < \infty$

$$\int_0^1 \left\| \sum_{i=1}^n r_i(t)x_i \right\| \mathrm{d}t \leqslant \left(\int_0^1 \left\| \sum_{i=1}^n r_i(t)x_i \right\|^p dt \right)^{1/p} \leqslant K_p \int_0^1 \left\| \sum_{i=1}^n r_i(t)x_i \right\| \mathrm{d}t, \tag{2.2.9}$$

where $K_p \leqslant Cp^{1/2}$ and C an absolute constant [this estimate of K_p is due to Kwapien (1976), see also Borell (1979) and Lindenstrauss and Tzafriri (1979)]. It follows of course from (2.2.9) that one could have used in the definition of type and cotype, instead of $\left(\text{Average}\,\|\sum_{i=1}^n \theta_i x_i\|^2\right)^{1/2}$ also $\left(\text{Average}\,\|\sum_{i=1}^n \theta_i x_i\|^r\right)^{1/r}$ and the constants thus obtained would have differed from those given in (2.2.1) or (2.2.2) by at most a multiple depending only on r.

Another possible variant in the definition of type and cotype is obtained by using instead of the Rademacher functions $\{r_i(t)\}_{i=1}^\infty$, which form a sequence of equally distributed independent random variables, another sequence of such variables, namely a sequence of independent normalized (i.e., with mean 0 and variance 1) Gaussian variables $g_i(\omega)$ on some probability space (Ω, μ). Using these variables one defines the Gaussian type p (respectively cotype q) constants $\alpha_p(X,n)$ (respectively $\beta_q(X,n)$) of a Banach space X to be the smallest numbers for which

$$\beta_q(X,n)^{-1}\left(\sum_{i=1}^n \|x_i\|^q\right)^{1/q} \leqslant \left(\int_\Omega \left\|\sum_{i=1}^n g_i(\omega)x_i\right\|^2 \mathrm{d}\mu(\omega)\right)^{1/2} \leqslant \alpha_p(X,n)\left(\sum_{i=1}^n \|x_i\|^p\right)^{1/p} \tag{2.2.10}$$

holds for all choices of $\{x_i\}_{i=1}^n$ in X. The numbers $\alpha_p(X), \beta_q(X)$ and Gaussian type or cotype of X are defined in an obvious way. The importance of the Gaussian type and cotype in the geometry of Banach spaces stems from the following obvious identity which relates the integral appearing in (2.2.10) with integration on the sphere S^{n-1} with respect to the normalized rotation invariant measure σ_n on it

$$\int_\Omega \left\|\sum_{i=1}^n g_i(\omega)x_i\right\|^2 \mathrm{d}\mu(\omega) = n\int_{S^{n-1}} \left\|\sum_{i=1}^n t_i x_i\right\|^2 \mathrm{d}\sigma_n(t), \tag{2.2.11}$$

where $t = (t_1, \ldots, t_n)$ are points on S^{n-1}. The Gaussian type and cotype constants are very close to the usual (Rademacher) type and cotype constants. This follows from the following facts (due to Maurey and Pisier 1976). One has

$$\int_0^1 \left\|\sum_{i=1}^n r_i(t)x_i\right\|^2 \mathrm{d}t \leqslant \lambda_1 \int_\Omega \left\|\sum_{i=1}^n g_i(\omega)x_i\right\|^2 \mathrm{d}\mu(\omega)\,, \tag{2.2.12}$$

where λ_1 is an absolute constant (not depending on X, n or $\{x_i\}_{i=1}^n$) and on the other hand (and this is the deep part) if X has Rademacher cotype q for some $q < \infty$ then

$$\int_\Omega \left\|\sum_{i=1}^n g_i(\omega)x_i\right\|^2 \mathrm{d}\mu(\omega) \leqslant \lambda_2 \cdot q \cdot C_q(X)^2 \int_0^1 \left\|\sum_{i=1}^n r_i(t)x_i\right\|^2 \mathrm{d}t\,, \tag{2.2.13}$$

for some absolute constant λ_2. Another inequality in the same direction where the dimension of X enters instead of its cotype constant is

$$\int_\Omega \left\| \sum_{i=1}^n g_i(\omega)x_i \right\|^2 d\mu(\omega) \leqslant \lambda_3 \log(m+1) \int_0^1 \left\| \sum_{i=1}^n r_i(t)x_i \right\|^2 dt\ , \tag{2.2.14}$$

whenever $\dim X = m$ and where λ_3 is an absolute constant. We also mention that the obvious analog of Kahane's inequality (2.2.9) holds for the expressions $\sum_{i=1}^n g_i(\omega)x_i$.

For the computation of the type and cotype constants the following facts are sometimes useful. Tomczak-Jaegermann (1979a) showed that there is an absolute constant c ($\sqrt{2\pi}$ will do) so that if $\dim X = n$ then

$$\alpha_2(X) \leqslant c\alpha_2(X,n), \qquad \beta_2(X) \leqslant c\beta_2(X,n), \qquad T_2(X) \leqslant cT_2(X,n). \tag{2.2.15}$$

It seems to be unknown if a similar statement is true for $C_2(X)$. Of course, in a weak form such an inequality follows immediately from (2.2.15), (2.2.12) and (2.2.14). König (1980) has generalized (2.2.15) to Gaussian type p and cotype q constants. Szarek (1991a) improved König's result by showing that (2.2.15) holds if 2 is replaced by p, respectively q, with a constant c independent of p, respectively q. Junge (1991) gave a simple proof of the König–Szarek result, reducing it directly to the case $p = q = 2$ (i.e., to (2.2.15)).

Also, for $\alpha_2(X,n), \beta_2(X,n)$ and $T_2(X,n)$ (here n is any number, not related to $\dim X$) it is known (cf. Bourgain, Kalton and Tzafriri 1989) that one can restrict oneself in the computation of them to vectors $\{x_i\}_{i=1}^n$ all having norm 1. This again will affect the constants involved only by a multiplicative absolute constant c. Again it seems to be unknown whether this applies also to $C_2(X,n)$. It is known that this is no longer true for type p constants $p < 2$ or cotype q constants $q > 2$.

There is a duality relation between the notions of type and cotype. It follows directly from the definitions that for every Banach space X, every n and dual indices p and q ($p^{-1} + q^{-1} = 1$, $1 \leqslant p \leqslant 2$)

$$C_q(X,n) \leqslant T_p(X^*,n), \tag{2.2.16}$$

and similarly for the Gaussian type and cotype. There is in general no reverse inequality to (2.2.16). This can be seen from the fact that $L_1(\mu)$ has cotype 2 while its dual $L_\infty(\mu)$ (if it is infinite-dimensional) fails to have type 2 (or any type > 1). The notion which enters naturally into the study of the reverse inequality to (2.2.16) is that of the Rademacher projection. Let X be a Banach space and n an integer. The subspace of $L_2([0,1],X) = \{f\colon [0,1] \to X$, measurable with $\|f\| = \left(\int_0^1 \|f(t)\|^2\, dt\right)^{1/2} < \infty\}$ consisting of functions of the form $\sum_{i=1}^n r_i(t)x_i$ is denoted by $\mathrm{Rad}_n X$. There is an obvious projection from $L_2([0,1],X)$ onto $\mathrm{Rad}_n X$ which is also denoted by Rad_n,

$$\mathrm{Rad}_n f = \sum_{i=1}^{n} r_i(t) \cdot \int_0^1 r_i(s) f(s)\,\mathrm{d}s\ . \tag{2.2.17}$$

By $\|\mathrm{Rad}_n X\|$ we mean the norm of this projection as an operator in $L_2([0,1],X)$. It is easy to verify that the following inverse inequality to (2.2.16) holds

$$T_p(X^*,n) \leqslant \|\mathrm{Rad}_n X\| C_q(X,n). \tag{2.2.18}$$

Observe that if X is a Hilbert space then Rad_n is an orthogonal projection in the Hilbert space $L_2([0,1],X)$ and hence in this case $\|\mathrm{Rad}_n X\| = 1$. It follows that for every Banach space isomorphic to Hilbert space $\|\mathrm{Rad}_n X\| \leqslant d_X$. A remarkable fact due to Pisier (1980) is that $\|\mathrm{Rad}_n X\|$ is actually much smaller. He proved that

$$K(X) = \sup_n \|\mathrm{Rad}_n X\| \leqslant (\mathrm{e}+1)\log(1+d_X). \tag{2.2.19}$$

It was proved by Bourgain (1984a) that the estimate (2.2.19) is optimal (up to the value of the constant e+1). Another result of Pisier (1982) (which is most conveniently formulated in terms of infinite-dimensional Banach spaces) characterizes spaces X for which $K(X)$ (defined in (2.2.19)) is finite.

Theorem 2.2.1. *Let X be a Banach space. The following four assertions are equivalent:*

(i) $K(X) < \infty$.

(ii) *X has type p for some $p > 1$.*

(iii) *For every constant λ there is an integer n so that $d(X_n, \ell_1^n) \geqslant \lambda$ for every n-dimensional subspace X_n of X.*

(iv) *The same as* (iii) *but with "for some $\lambda > 1$" instead of "every λ".*

Remarks.

(1) The deep part of the theorem is the implication that (ii) (or (iii)) imply (i). An interesting fact concerning the proof of this implication is that it uses a result of Beurling concerning analytic functions.

(2) The negation of (iii) means that X contains subspaces uniformly isomorphic to ℓ_1^n for every n, while the negation of (iv) means that X contains almost isometric copies of ℓ_1^n for all n. The equivalence of (iii) and (iv) actually means the following. For every $\varepsilon > 0, \lambda > 1$ and integer k there is an $n(\varepsilon,\lambda,k)$ so that if $n \geqslant n(\varepsilon,\lambda,k)$ and $d(Y,\ell_1^n) < \lambda$ then Y has a k-dimensional subspace Y_0 so that $d(Y_0,\ell_1^k) \leqslant 1+\varepsilon$. This fact was proved first by Giesy (1966). It is observed by Amir and Milman (1980) that the estimate for $n(\varepsilon,\lambda,k)$ obtained in that paper ($\approx k^{\log\lambda/\varepsilon}$) cannot be improved.

(3) It was noted by Maurey and Pisier (1976) that the assertion $K(X) < \infty$ is equivalent also to the assertion that the norms of the natural projections from

$L_2(\Omega, X)$ onto the subspaces consisting of the functions $\sum_{i=1}^{n} g_i(\omega)x_i$, $n = 1, 2, \ldots$, are uniformly bounded.

(4) A space X for which $K(X) < \infty$ is called K-convex. In the literature such spaces are also called B-convex. This terminology arose in view of a paper of Beck (1962) in which he proved that condition (iii) of the theorem is equivalent to the assertion that the law of large numbers holds for X-valued random variables.

A notion which is close to type, cotype and especially K-convexity is uniform convexifiability. Recall that a normed space X is called uniformly convex if for every $\varepsilon > 0$ there is a $\delta(\varepsilon) > 0$ so that

$$\|x\| = \|y\| = 1, \qquad \|x - y\| \geqslant \varepsilon \Rightarrow \|x + y\| \leqslant 2 - \delta(\varepsilon). \tag{2.2.20}$$

The main result concerning the existence of such norms is the following theorem combining work of James (1972) and Enflo (1972).

Theorem 2.2.2. *The following assertions concerning a Banach space X are equivalent:*

(i) *X has an equivalent uniformly convex norm.*

(ii) *For every $\lambda > 0$ there is an $n = n(\lambda)$ so that the unit ball of X does not contain a λ separated binary tree of size n. In other words there does not exist a system of vectors $x_{\theta_1,\ldots,\theta_k}$ in the unit ball of X with $\theta_i = \pm 1$, $1 \leqslant k \leqslant n$, so that for all $\{\theta_i\}_{i=1}^{k}$ and $k < n$*

$$\begin{aligned} x_{\theta_1,\theta_2,\ldots,\theta_k} &= (x_{\theta_1,\theta_2,\ldots,\theta_k,1} + x_{\theta_1,\ldots,\theta_k,-1})/2, \\ \|x_{\theta_1,\ldots,\theta_k,1} &- x_{\theta_1,\theta_2,\ldots,\theta_k,-1}\| \geqslant \lambda. \end{aligned} \tag{2.2.21}$$

(iii) *For every $\varepsilon > 0$ there is an $n = n(\varepsilon)$ so that whenever $\{x_i\}_{i=1}^{n}$ are of norm 1 in X then for some $0 \leqslant k \leqslant n$*

$$\|x_1 + x_2 + \cdots + x_k - x_{k+1} - x_{k+1} - \cdots - x_n\| \leqslant n - \varepsilon. \tag{2.2.22}$$

Remarks.

(1) Assertion (ii) is an isomorphic condition in analogy to assertion (iii) of Theorem 2.2.1. On the other hand assertion (iii) is an "almost isometric" condition which is evidently a stronger form of assertion (iv) of Theorem 2.2.1.

(2) Pisier (1975) gave a new proof of the implication (ii) $\Rightarrow$ (i) in which he obtained also a good, power-type, estimate for the modulus of convexity $\delta(\varepsilon)$ of the norm appearing in (i).

(3) It was observed by James that condition (ii) is selfdual. Thus, in view of the theorem, the dual of a uniformly convex space has an equivalent uniformly convex norm.

(4) If $\|x\| = \|y\| = 1$ and $\|x \pm y\| \geqslant 2 - \varepsilon$ then $d(Z, \ell_1^2) \leqslant 1 + 2\varepsilon$ where $Z = \operatorname{span}\{x, y\}$. It follows that the negation of (iii) implies that X contains almost isometric copies of ℓ_1^2. This is the way James proved the earlier result quoted

as Theorem 1.2 above. On the other hand James (1974) (cf. also James and Lindenstrauss 1974) proved that the negation of (iii) does not imply the existence of almost isometric copies of ℓ_1^3. Hence the assertions in Theorem 2.2.2 are strictly stronger than those of Theorem 2.2.1. James (1978) (cf. also Pisier and Xu 1987) showed that there is even a Banach space of type 2 which fails to have an equivalent uniformly convex norm.

(5) V. Kadec (1982) showed that the negation of (iii) implies that X contains an almost isometric copy of any 2-dimensional space. This fact and the work of James show that Theorem 1.2 can be strengthened: If X is non-reflexive and $\dim Y = 2$ then for every $\varepsilon > 0$ there is a linear map $T: Y \to X$ so that $\|y\| \leqslant \|Ty\| \leqslant (1+\varepsilon)\|y\|$ for all $y \in Y$.

2.3. *Ideal norms and p-absolutely summing operators*

As in any mathematical theory the study of certain objects naturally requires also a serious study of the morphisms acting on these objects, in our case the space $L(X,Y)$ of linear operators from X to Y. Since our interest is the local theory we restrict ourselves from the outset mainly to finite-dimensional X and Y. (This will save us from some complications related to non-reflexive spaces and mainly from problems related to the approximation property.) Besides the usual operator norm on $L(X,Y)$ denoted as usual by $\|\cdot\|$ we shall have to consider also other norms. Many of these norms will be what is called ideal norms. An ideal norm i is a norm defined on $L(X,Y)$ for all finite-dimensional normed spaces X and Y which satisfies

$$i(T) = \|T\| \quad \text{if } \operatorname{rank} T = 1, \tag{2.3.1}$$

$$i(UTV) \leqslant \|U\| i(T) \|V\|, \tag{2.3.2}$$

whenever $V \in L(Z,X)$, $T \in L(X,Y)$, $U \in L(Y,W)$. Here are some examples of ideal norms (besides $\|T\|$ itself). The nuclear norm of $T, n(T)$ defined by

$$n(T) = \inf\left\{ \sum_{i=1}^{m} \|x_i^*\|\,\|y_i\| \colon\ Tx = \sum_{i=1}^{m} \langle x, x_i^* \rangle\, y_i \right\} \tag{2.3.3}$$

(the infimum is with respect to all possible representations of T, m is arbitrary and may be ∞) is easily seen to be an ideal norm. For $1 \leqslant p \leqslant \infty$ we let $\gamma_p(T)$ be defined as

$$\gamma_p(T) = \inf\left\{ \|U\|\,\|V\| \colon\ T = UV,\ V: X \to L_p(\mu),\ U: L_p(\mu) \to Y \right\}. \tag{2.3.4}$$

In other words $\gamma_p(T)$ is the infimum of $\|U\|\,\|V\|$ taken over all possible factorizations of T through an $L_p(\mu)$ space for an arbitrary measure μ. (The space $L_p(\mu)$ may be infinite-dimensional but it is easy to verify that for finite rank operators we may as well restrict ourselves to factorizations through ℓ_p^n-spaces and get the same inf.) It is a nice exercise to verify that γ_p is a norm (this requires a little trick) but it is easy to see that it satisfies (2.3.1) and (2.3.2).

An important ideal norm is the p-absolutely summing norm denoted by $\pi_p(T)$. An operator $T : X \to Y$ is called p-absolutely summing if there is a constant λ so that

$$\left(\sum_{i=1}^{m} \|Tx_i\|^p\right)^{1/p} \leqslant \lambda \sup\left\{\left(\sum_{i=1}^{m} |\langle x_i, x^*\rangle|^p\right)^{1/p} : x^* \in X^*,\ \|x^*\| \leqslant 1\right\}, \tag{2.3.5}$$

for every m and every choice of $\{x_i\}_{i=1}^m \in X$. The smallest λ satisfying (2.3.5) is denoted by $\pi_p(T)$ and it is easily checked that it is an ideal norm. The reason for this terminology is the fact that the 1-absolutely summing operators T are exactly those for which $\sum_{i=1}^{\infty} \|Tx_i\| < \infty$ whenever $\sum_{i=1}^{\infty} x_i$ is unconditionally convergent in X. The identity operator I_p from $L_\infty(\mu)$ to $L_p(\mu)$ where μ is a probability measure is easily seen to satisfy $\pi_p(I_p) = 1$. A very useful result of Pietsch (1967) shows that in a sense I_p is the canonical p-absolutely summing operator. If $T : X \to Y$ is p-absolutely summing then there is a probability measure μ on the unit ball of X^* so that

$$\|Tx\| \leqslant \pi_p(T)\left(\int_{K_{X^*}} |\langle x, x^*\rangle|^p \,\mathrm{d}\mu(x^*)\right)^{1/p}. \tag{2.3.6}$$

In other words (2.3.6) means that T can be factored as follows:

$$\begin{array}{ccc}
C(K) & \xrightarrow{j} & L_p(\mu) \\
\cup & & \cup \\
W & \longrightarrow & V \\
{\scriptstyle i}\uparrow & & \downarrow{\scriptstyle \tilde{T}} \\
X & \xrightarrow{T} & Y
\end{array}$$

$T = \widetilde{T} \cdot j_{|W} \cdot i$ where K is K_{X^*} (which is compact in its w^*-topology), $i : X \to C(K)$ is the natural injection $(ix)(x^*) = \langle x, x^*\rangle$, $W = iX$, $j : C(K) \to L_p(\mu)$ is the natural identity mapping, $V = jW$ and $\widetilde{T} : V \to Y$ is a linear operator of norm at most $\pi_p(T)$. The proof of (2.3.6) is, like that of (2.2.7), done by reducing it to a separation argument. From (2.3.6) we get immediately that for $1 \leqslant r < s \leqslant \infty$

$$\|T\| = \pi_\infty(T) \leqslant \pi_s(T) \leqslant \pi_r(T) \leqslant \pi_1(T)\,. \tag{2.3.7}$$

Also, since for $p = 2$ the operator $\widetilde{T}$ can be extended without change of norm to $L_2(\mu)$ it follows that

$$\gamma_2(T) \leqslant \pi_2(T). \tag{2.3.8}$$

If $X = Y = \ell_2$ then it is easily checked that $T\colon X \to Y$ is 2-absolutely summing iff T is a Hilbert–Schmidt operator and $\pi_2(T)$ is exactly equal to the Hilbert–Schmidt norm $\mathrm{HS}(T)$. As a matter of fact Pelczynski (1967) verified that for every $1 \leqslant p < \infty$ there are constants $c_p > 0$, $d_p < \infty$ so that

$$c_p \mathrm{HS}(T) \leqslant \pi_p(T) \leqslant d_p \mathrm{HS}(T), \quad T\colon \ell_2 \to \ell_2. \tag{2.3.9}$$

The best values of the constants c_p and d_p for which (2.3.9) holds were computed by Garling (1970). There are several results which connect the notions of p-summing operators to type and cotype theory. For example, the following result, due to Maurey, which is not hard to deduce from the classical Khintchine inequality. Let $T : X \to Y$ and let $2 < q < \infty$. Then

$$\pi_2(T) \leqslant c q^{1/2} C_2(Y) \pi_q(T), \tag{2.3.10}$$

where c is an absolute constant.

For operators T of rank n it is possible to give estimates on the number $m = m(n,p)$ of vectors $\{x_i\}_{i=1}^m$ needed in (2.3.5) in order to evaluate $\pi_p(T)$ up to a multiplicative constant. These estimates are closely related to results on computing type and cotype constants, like (2.2.15) and the results mentioned just after this formula. The estimates are also connected to the question of embedding n-dimensional subspaces of $L_p(\mu)$ in ℓ_p^m to be discussed in section 3.2. For the strongest known results in this direction see Johnson and Schechtman (1992).

An important fact concerning absolutely summing operators is a result of Grothendieck (1956a). This result has many formulations and has found applications in various directions (local theory, structure theory of infinite-dimensional Banach spaces and applications of Banach space theory to harmonic and complex analysis). Here we just bring the following two consequences of the Grothendieck result (also called Grothendieck's inequality, cf. also Lindenstrauss and Pelczynski 1968)

$$\pi_1(T) \leqslant K_G \|T\| \quad \text{for } T\colon \ell_1^m \to \ell_2^m, \tag{2.3.11}$$

$$\pi_2(T) \leqslant K_G \|T\| \quad \text{for } T\colon \ell_\infty^m \to \ell_1^m, \tag{2.3.12}$$

where K_G is an absolute constant called Grothendieck's constant (it depends on the scalar field used).

There are many more useful ideal norms. A detailed study of those can be found in the book of Pietsch (1978). Also, a comprehensive discussion of the material presented above is given in Pisier (1986).

We pass next to duality of ideal norms, a topic first introduced in the context of Banach space theory by Grothendieck and later studied in detail by Persson and Pietsch (1969). Here the assumption that $\dim X, \dim Y$ are both finite will be crucial. Assume that i is an ideal norm. The dual ideal norm i^* is defined as follows for $T \in L(X,Y)$:

$$i^*(T) = \sup \{ \mathrm{trace}(ST)\colon S \in L(Y,X),\ i(S) \leqslant 1 \}.$$

Recall that trace(ST) = trace(TS). It is easy to see that i^* is again an ideal norm and that $i^{**} = i$. In some cases the dual norm can be quite easily computed explicitly. One has for example

$$\pi_2^* = \pi_2, \qquad \pi_1^* = \gamma_\infty , \qquad \|\cdot\|^* = n(\cdot) \tag{2.3.14}$$

(the equalities are exact, not only equivalences). The duality relations appearing in (2.3.14) are not only the simplest but actually the most important ones in applications.

A very useful result relating i with i^* is the following result of Lewis (1979) (cf. also Pelczynski 1980). Let $\dim X = \dim Y = n$ and let i be an ideal norm. Then there exists a $W \in L(X, Y)$ such that

$$i(W)i^*(W^{-1}) = n. \tag{2.3.15}$$

Such a W (in general not unique even up to scalar multiples) is obtained by taking among all $W \in L(X,Y)$ with $i(W) = 1$ one for which the volume of WK_X (or $\det W$) is maximal. The proof is a simple variational argument and originates from John's proof of (2.1.4). The result (2.3.15) actually implies John's result and also Auerbach's result (2.1.6) (obtained by taking $X = \ell_1^n$ and $i = \|\cdot\|$), as well as other results (e.g., (2.1.11)). Observe that, by the definition of the dual norm, we have for every invertible U in $L(X,Y)$, $i(U)i^*(U^{-1}) \geqslant \mathrm{trace}(I_X) = n$.

Besides ideal norms defined on $L(X,Y)$ for every X and Y there are some important norms in local theory which are defined only for special X or Y. One such norm is the ℓ norm defined on $L(\ell_2^n, X)$ $(n = 1, 2, \ldots)$ by

$$\ell(T) = n^{1/2}\left(\int_{S^{n-1}} \|Ty\|^2 \,\mathrm{d}\sigma_n(y)\right)^{1/2}, \tag{2.3.16}$$

where σ_n is the normalized rotation invariant measure on S^{n-1}. Like any integral on S^{n-1} with respect to σ_n we can write (2.3.16) also in terms of independent normalized Gaussian variables $\{g_i\}_{i=1}^n$, namely

$$\ell(T) = \left(E\left\|\sum_{i=1}^n g_i T e_i\right\|\right)^{1/2}, \tag{2.3.17}$$

where $\{e_i\}_{i=1}^n$ is an orthonormal basis in ℓ_2^n and E means, as usual, expectation, i.e., integration on the probability space on which the g_i are defined. The norm ℓ satisfies (2.3.1) and (2.3.2) as long as it makes sense (i.e., only for $V \in L(\ell_2^n, \ell_2^n)$). The dual norm ℓ^* to ℓ is defined by (2.3.13) for $T \in L(X, \ell_2^n)$. Lewis' result (2.3.15) remains valid with the same proof for $i = \ell$ if $X = \ell_2^n$ and $\dim Y = n$.

3. ℓ_p^n Subspaces of Banach spaces

3.1. Hilbertian subspaces

We start with some notations. We say that ℓ_p^n is λ embedded in a normed space X and write $\ell_p^n \overset{\lambda}{\hookrightarrow} X$ iff there is a subspace E of X with $d(\ell_p^n, E) \leqslant \lambda$. In most cases we will be interested in λ near 1. For finite-dimensional spaces X we denote

$$k_p(X,\varepsilon) = \sup\{k : \ell_p^k \overset{1+\varepsilon}{\hookrightarrow} X\}. \tag{3.1.1}$$

We shall first treat the case $p = 2$ which turns out to be the most important case from many points of view, in particular from the geometric point of view. The basic result in this direction is Dvoretzky's Theorem 1.1 which we have already formulated in the introduction. This theorem was conjectured by Grothendieck (1956b). An obviously equivalent form of stating Dvoretzky's theorem is the following: For every $\varepsilon > 0$

$$\lim_{n\to\infty} k_2(n,\varepsilon) = \infty, \tag{3.1.2}$$

where

$$k_2(n,\varepsilon) = \inf\left\{k_2(X,\varepsilon)\colon \dim X = n\right\}. \tag{3.1.3}$$

Obviously, it is of great interest to get good estimates on $k_2(n,\varepsilon)$ as well as related constants which are obtained by taking the inf in (3.1.3) only with respect to certain distinguished subclasses of spaces X of dimension n. The estimate obtained by Dvoretzky (1961) was $k_2(n,\varepsilon) \geqslant c(\varepsilon)(\log n/\log\log n)^{1/2}$. This estimate was improved by Milman (1971) who proved that $k_2(n,\varepsilon) \geqslant c\varepsilon^2|\log\varepsilon|^{-1}\log n$. As shown by Gordon (1988) (cf. also Schechtman 1989) this estimate can be still slightly improved and one gets

$$k_2(n,\varepsilon) \geqslant c\varepsilon^2 \log n, \tag{3.1.4}$$

where of course as usual c denotes an absolute constant. We shall outline briefly the approach of Milman below. By now there are very many different proofs of Dvoretzky's theorem and its variants (see, e.g., the expository paper, Lindenstrauss 1992). Let us note that the proof of Milman (1971) works also for complex scalars (this is not obvious for the original proof of Dvoretzky).

For some range of ε and n the estimate (3.1.4) is sharp. In particular, as far as the dependence of $k_2(n,\varepsilon)$ on n is concerned (for fixed ε) (3.1.4) is sharp. It is easy to check that

$$k_2(\ell_n^\infty, 1) \leqslant c_1 \log n. \tag{3.1.5}$$

In other words, up to a possible absolute constant factor $k_2(X,1)$ achieves its minimum for spaces of dimension n at $X = \ell_n^\infty$. This is of interest since for $X = \ell_n^\infty$, Dvoretzky's theorem is obvious, and actually ℓ_n^∞ contains subspaces of dimension $k = c_2 \log n$ which are 2-isomorphic to an arbitrary space of this dimension. In fact, if $\dim Y = k$ then the unit ball of Y^* contains a δ net of size n, $\{y_i^*\}_{i=1}^n$ (δ depends on c_2). The map $T : Y \to \ell_n^\infty$ defined by

$$Ty = \left(\langle y, y_1^*\rangle, \ldots, \langle y, y_n^*\rangle\right)$$

satisfies $\|y\|/2 \leqslant \|Ty\| \leqslant \|y\|$ if δ is small enough.

The proof of Milman also gives the following estimate

$$k_2(X, \varepsilon) \geqslant c(\varepsilon)n/d_X^2. \tag{3.1.6}$$

Thus if $d(X, \ell_2^n) \leqslant 2$, say, then by (3.1.6) $k_2(X, \varepsilon)$ is proportional to n for fixed ε. It follows that for investigating the dependence of $k_2(X, \varepsilon)$ on $\dim X$ for fixed ε, the value of ε itself does not matter much and therefore it makes sense to introduce further the following notation

$$k(X) = k_2(X, 1). \tag{3.1.7}$$

As for the dependence of $k_2(n, \varepsilon)$ on ε for fixed n (3.1.4) seems to give a poor estimate and a satisfactory estimate is unknown. In order to formulate the problem in a precise way it makes more sense to talk of $n(k, \varepsilon)$ which is the smallest n so that for every X with $\dim X = n$, $\ell_2^k \overset{1+\varepsilon}{\hookrightarrow} X$. By taking $X = \ell_n^\infty$ one easily finds that $\ell_2^k \overset{1+\varepsilon}{\hookrightarrow} X$ only if $n \geqslant c(k)\varepsilon^{-(k-1)/2}$. Again, it seems that ℓ_n^∞ is the worst example (in the sense that if $\ell_2^k \overset{1+\varepsilon}{\hookrightarrow} \ell_n^\infty$ then $\ell_2^k \overset{1+\varepsilon}{\hookrightarrow} X$ for every X with $\dim X = c(k)n$). Milman (1988c) has noted that such a statement would follow from a well-known conjecture (Knaster's hypothesis) in topology. In fact, Gromov observed that a topological argument proves this statement for $k = 2$. By using a different approach (via the theory of irregularity of distribution for points on a sphere) Bourgain and Lindenstrauss (1989) verified the statement (in a slightly weaker form) for every k provided X has a 1-symmetric basis.

We outline now the approach of Milman (1971) which was later greatly extended by Figiel, Lindenstrauss and Milman (1977). Let f be a real-valued continuous function on S^{n-1} (with its usual metric and normalized rotation invariant measure σ_n). The median μ of f is, as usual, a value for which a set of half the measure of S^{n-1} has value $\geqslant \mu$ and a set of half the measure has value $\leqslant \mu$. Let $A = \{x \colon x \in S^{n-1}$ with $f(x) = \mu\}$ and let A_δ be $\{x \colon d(x, A) \leqslant \delta\}$. Then as observed by P. Lévy in 1919 (see Lévy 1951)

$$\sigma_n(A_\delta) \geqslant 1 - (\pi/2)^{1/2}\, \mathrm{e}^{-\delta^2 n/2}. \tag{3.1.8}$$

For f a linear function (3.1.8) follows by a direct computation (A is then an equator) and the general case reduces to the linear one by the isoperimetric inequality for subsets of S^{n-1}. Hence, if f has Lipschitz constant α we get from (3.1.8)

$$\sigma_n\{x\colon |f(x) - \mu| \geqslant \lambda\} \leqslant (\pi/2)^{1/2}\, \mathrm{e}^{-\lambda^2 n/2\alpha^2}. \tag{3.1.9}$$

It follows from this that if we are given m points $\{x_i\}_{i=1}^m$ on S^{n-1} with

$$m \cdot (\pi/2)^{1/2}\, \mathrm{e}^{-\lambda^2 n/2\alpha^2} < 1, \tag{3.1.10}$$

then there is a rotation U of S^{n-1} so that $|f(Ux_i) - \mu| < \lambda$ for every i. We apply this fact to $X = (\mathbb{R}^n, \|\cdot\|)$ where the coordinate system is chosen so that the Löwner ellipsoid is the Euclidean unit ball. By John's result (2.1.4) we have

$$n^{-1/2}|x| \leqslant \|x\| \leqslant |x|, \quad x \in \mathbb{R}^n,$$

where $|\cdot|$ is the Euclidean norm. Note that the Lipschitz constant of $f(x) = \|x\|$ is $\leqslant 1$. By another result concerning the Löwner ellipsoid (the Dvoretzky–Rogers lemma; to be discussed in section 4) and a standard fact concerning the expectation of the maximum of n normalized Gaussians, the median μ of f cannot be as low as $n^{-1/2}$; it has to be at least $(c \log n/n)^{1/2}$. We use now the preceding argument with $\lambda = \varepsilon\mu$ and $\{x_i\}_{i=1}^m$ an ε net on $S^{k-1} \subset S^{n-1}$ (m has to be of size $\approx \varepsilon^{-k}$, and thus (3.1.10) gives a bound on k) and get that there is a rotation U on $\mathbb{R}^n$ so that

$$\mu(1-\varepsilon) \leqslant \|Ux_i\| \leqslant \mu(1+\varepsilon), \quad 1 \leqslant i \leqslant m\,. \tag{3.1.11}$$

It follows easily from (3.1.11) that a slightly weaker inequality holds for every $x \in S^{k-1}$, i.e., that $U\mathbb{R}^k$ is an almost Hilbertian subspace of $\mathbb{R}^n$ ($\mathbb{R}^k$ is the subspace so that $S^{k-1} = \mathbb{R}^k \cap S^{n-1}$).

Remarks.

(1) The main point here was the fact that most of the mass of S^{n-1} is highly concentrated near an equator and therefore every Lipschitz function is concentrated highly around the median. Because of this concentration we may also replace the median by the mean which is much more easy to handle. In particular we introduce the following notation which will be much used below

$$M_X = \int_{S^{n-1}} \|x\| \, \mathrm{d}\sigma_n(x). \tag{3.1.12}$$

Note that M_X depends not only on the normed space X but also on the coordinate system, or equivalently the Euclidean structure we choose in $\mathbb{R}^n$. Note also that

$$M_{X^*} = \int_{S^{n-1}} \|x\|^* \, \mathrm{d}\sigma_n(x) \tag{3.1.13}$$

is up to a possible normalization constant the mean width of the body K_X. Since for $x \in S^{n-1}$, $1 = \langle x, x\rangle \leqslant \|x\| \, \|x\|^*$ it follows that always

$$1 \leqslant M_X M_{X^*}. \tag{3.1.14}$$

For obtaining a good upper bound of $M_X M_{X^*}$ one has to choose properly the Euclidean structure. This will be discussed in section 4.

(2) The argument above is probabilistic in nature. In many cases it is unknown how to construct explicitly almost Euclidean sections of the large dimension given by this approach. On the other hand because of the probabilistic nature of the proof we get not only the existence of almost Euclidean sections but that most subspaces of a given dimension k (in the sense of the natural measure on the Grassmanian $G_{n,k}$) are almost Euclidean. Again, it is important to note that the measure on the Grassmannian depends on the choice of the Euclidean structure in X.

We mention now several results obtained by this method in the paper of Figiel, Lindenstrauss and Milman mentioned above. One involves duality

$$k(X)k(X^*) \geqslant cn^2/d_X^2, \quad n = \dim X \tag{3.1.15}$$

and, in particular, since $d_X \leqslant n^{1/2}$

$$k(X)k(X^*) \geqslant cn. \tag{3.1.16}$$

A fact in combinatorial geometry which follows immediately from (3.1.15) and (3.1.16) is that if P is symmetric compact polytope in $\mathbb{R}^n$ with interior having $v(P)$ vertices and $f(P)$ maximal faces then

$$\log v(P) \cdot \log f(P) \geqslant cn. \tag{3.1.17}$$

This result as well as (3.1.15) and (3.1.16), are sharp up to the value of the constant. There exist, e.g., a symmetric polytope P in $\mathbb{R}^n$ so that $\log v(P) \approx \log f(P) \approx n^{1/2}$.

The proof of (3.1.15) also gives information on the existence of good projections on almost Euclidean subspaces of X or X^*. Assume, e.g., that $k(X) \leqslant k(X^*)$, then X has a subspace Y with $d(Y, \ell_2^k) \leqslant 2$ where $k = k(X)$ on which there is a projection of norm at most $cM_X M_{X^*}$ (if $k(X^*) \leqslant k(X)$ there is a suitable subspace in X^*).

Another result in the paper just quoted connects $k(X)$ with the cotype of X. One has ($\dim X = n$, $2 \leqslant q < \infty$)

$$k(X) \geqslant cn^{2/q}/C_q(X)^2. \tag{3.1.18}$$

In particular if $C_2(X)$ is bounded by a constant independent of n then $k(X)$ is proportional to n, for example

$$k(\ell_p^n) \geqslant cn, \quad 1 \leqslant p \leqslant 2, \tag{3.1.19}$$

where c is an absolute positive constant.

For $p = 1$, (3.1.19) was proved independently by Kashin (1977) by a different method. He showed that for any $\alpha < 1$ there is a $\lambda = \lambda(\alpha)$ so that $\ell_2^{[\alpha n]} \overset{\lambda}{\hookrightarrow} \ell_1^n$. For $\alpha = \frac{1}{2}$ (assuming for simplicity that n is even) his approach shows that one can find a pair of orthogonal (with respect to the usual Euclidean structure) subspaces V_1 and V_2 of dimension $n/2$ of $\mathbb{R}^n$ so that

$$\begin{aligned} &\mathbb{R}^n = V_1 \oplus V_2 \quad \text{(obviously)}; \\ &\lambda^{-1}\|x\|_1 \leqslant \|x\|_2 \leqslant \|x\|_1, \quad x \in V_i,\ i = 1,2, \end{aligned} \tag{3.1.20}$$

where λ is independent of n.

The proof of Kashin was simplified by Szarek (1978). He showed that the result follows from the fact that for $X = \ell_1^n$ the "volume-ratio" of X defined by

$$\mathrm{vr}(X) = \left(\frac{\mathrm{Vol}\, K_X}{\mathrm{Vol}\, \mathscr{E}_K}\right)^{1/n}, \quad \dim X = n, \tag{3.1.21}$$

where $\mathscr{E}_K$ is the Löwner ellipsoid of K_X, is bounded by a constant independent of n (the notion (3.1.21) itself was formally introduced somewhat later in Szarek and Tomczak-Jaegermann 1980). It is worthwhile to recall briefly the argument of Szarek, and see how assumptions concerning notions like volume are used in the theory we survey here.

We may assume that $\mathscr{E}_K$ is the Euclidean ball of $\mathbb{R}^n$. Then clearly, by assumption,

$$\operatorname{vr}(X) = \left(\int_{S^{n-1}} \|x\|^{-n} \, \mathrm{d}\sigma_n(x) \right)^{1/n} \leqslant C \tag{3.1.22}$$

($\|\cdot\|$ is the norm in X, $|\cdot|$ will denote the Euclidean norm and C is independent of n). By the usual Fubini like argument (the integral on S^{n-1} can be considered as the average with respect to the measure on the Grassmanian $G_{n,k}$ of integrals on the spheres of k-dimensional sections of S^{n-1}) there is for every $k < n$ an $E \subset \mathbb{R}^n$ with $\dim E = k$ so that

$$\int_{E \cap S^{n-1}} \|x\|^{-n} \, \mathrm{d}\sigma_k(x) \leqslant \operatorname{vr}(X)^n \leqslant C^n. \tag{3.1.23}$$

Hence for any $r > 0$, $\sigma_k\{x \in E \cap S^{n-1},\ \|x\| < r\} \leqslant (rC)^n$. The $r/2$ neighbourhood (with respect to $|\cdot|$) of a point in S^{k-1} has σ_k measure $\geqslant (\beta r)^k$ where β is an absolute constant. Hence if

$$(rC)^n < (\beta r)^k, \tag{3.1.24}$$

the set $\{x\colon \|x\| \geqslant r\}$ is an $r/2$ net on $E \cap S^{k-1}$ and hence (by the triangle inequality, and the fact that $\|x\| \leqslant |x|$) one gets $\|x\| > r|x|/2$ for every $x \in E$. Clearly, for every $\alpha < 1$ if $k = [\alpha n]$ there is an $r(\alpha) > 0$ satisfying (3.1.24) as desired.

Remarks.

(1) The argument we have just recalled clearly shows also the existence of decompositions in X of the type described (in ℓ_1^n) in (3.1.20).

(2) Our interest here is with bodies having a small volume ratio. Nevertheless it is worth mentioning that the body with the largest volume ratio is the unit cube (see Ball 1989).

Bourgain and Milman (1985, 1987) showed that $\operatorname{vr}(X)$ can be estimated by $C_2(X)$. Their result, as slightly improved later in Milman and Pisier (1986), is

$$\operatorname{vr}(X) \leqslant c(\varepsilon) C_2(X) \left[\log\left(1 + C_2(X)\right) \right]^{1+\varepsilon}, \quad (\varepsilon > 0). \tag{3.1.25}$$

From the nature of the constants $\operatorname{vr}(X)$ and $C_2(X)$ it follows that an estimate in the reverse direction cannot exist. In fact, $\operatorname{vr}(X)$ depends just on the global structure of X while $C_2(X)$ may be affected considerably by a "bad" small-dimensional subspace. There are reasonable ways to ask about existence of inequalities in the reverse direction to (3.1.25) but examples show that also the answer to the "reasonable questions" is negative (this involves the so-called "weak Hilbert spaces"; see Pisier (1989b) for a detailed discussion).

The previous results show that under special conditions on X (with $\dim X = n$) one can find subspaces not far from Euclidean of dimension proportional to n. In general not more than dimension $c \log n$ is ensured. If we allow also the dual operation to that of taking subspaces, i.e., passing to quotient spaces we can always get $c\sqrt{n}$ but no more. However, if we perform both operations, of taking subspaces and quotients simultaneously, we can get cn in the general case. This result is due to Milman (1985a) and called the QS (quotient of subspace) theorem.

Theorem 3.1.1. *Let $\frac{1}{2} \leqslant \alpha < 1$ and let X be a Banach space of dimension n. Then there exist subspaces $E \supset F$ of X with*

$$k = \dim E/F \geqslant \alpha n, \qquad d(E/F, \ell_2^k) \leqslant c(1-\alpha)^{-1}\big|\log(1-\alpha)\big|. \tag{3.1.26}$$

Geometrically, the QS theorem means the following. Let $\frac{1}{2} \leqslant \alpha < 1$ and let K be a symmetric convex body in $\mathbb{R}^n$. Then there are subspaces $E \supset G$ in $\mathbb{R}^n$ with $\dim G \geqslant \alpha n$ and an ellipsoid $\mathscr{E}$ in G so that

$$\mathscr{E} \subset P_G(E \cap K) \subset c(1-\alpha)^{-1}\big|\log(1-\alpha)\big|\mathscr{E}, \tag{3.1.27}$$

where P_G is the orthogonal projection from $\mathbb{R}^n$ onto G.

Several consequences of the QS theorem will be discussed in section 4.

3.2. Embedding of and into ℓ_p^n

Every subspace of a Hilbert space is again a Hilbert space, hence for no spaces besides the ℓ_2^n spaces one can get as general embedding results as those considered in the previous section. Krivine (1976) was able however to obtain an embedding result of general ℓ_p^n spaces into Banach spaces (cf. also Lemberg 1981). The p which one gets in Krivine's theorem is often different from 2 and this of course is the source of the interest in Krivine's theorem. We shall state now a special case of Krivine's result where the p is given a priori (another part of Krivine's result enters implicitly in Theorem 3.2.2).

Theorem 3.2.1. *For every integer k, $1 \leqslant p \leqslant \infty$, $\lambda < \infty$ and $\varepsilon > 0$ there is an $n_0(p,k,\lambda,\varepsilon)$ so that if $n \geqslant n_0$ and $d(X, \ell_p^n) \leqslant \lambda$, then $\ell_p^k \overset{1+\varepsilon}{\hookrightarrow} X$.*

For $p = 1$ (and dually for $p = \infty$) this result is much simpler than the general case. As mentioned already in section 2 this special case is due to Giesy (1966) and in this case there is available a sharp estimate namely

$$n_0(p,k,\lambda,\varepsilon) = k^{c\varepsilon^{-1}\log\lambda}, \quad p = 1, \infty. \tag{3.2.1}$$

For $p=2$ we are in the setting of the previous subsection (and (3.1.6) means that $n_0(2,k,\lambda,\varepsilon)=k\lambda^2/c(\varepsilon)$). Note however that even for $p=2$ the proof of Theorem 3.2.1 provides new information. Krivine produces in his proof almost Euclidean subspaces of a very special form. They are generated by "block bases" of a given basis. The really interesting part of Theorem 3.2.1 involves $1<p<\infty$, $p\neq 2$. An estimate for $n_0(p,k,\lambda,\varepsilon)$ (which is probably not sharp) in this case was given by Amir and Milman (1980, 1985). The first part of their proof consists of showing that if $d(X,\ell_p^n)\leqslant\lambda$ then X contains a sequence $\{y_j\}_{j=1}^m$ which is $1+\varepsilon$ symmetric and satisfies

$$\Big(\sum_{j=1}^m |a_j|^p\Big)^{1/p}\leqslant\Big\|\sum_{j=1}^m a_jy_j\Big\|\leqslant\lambda\Big(\sum_{j=1}^m |a_j|^p\Big)^{1/p}\quad\text{for all }\{a_j\}_{j=1}^m,$$

where $m\approx n^{1/3}$ (the equivalence depends on p,λ and ε of course). This is proved by means of arguments involving concentration of measure, similar to the one outlined in the previous paragraph (the sphere S^{n-1} is however replaced by a suitable finite space with a natural measure and distance function). For this step a sharper result was obtained by Gowers (1989). He showed that one can actually get $m\approx n/\log n$ and that this is essentially optimal. This first step reduced the proof to the case where X has a 1-symmetric basis. This is the setting in which one can use Krivine's original argument and obtain an estimate for n_0 namely $n_0(p,k,\lambda,\varepsilon)\leqslant c_1(p)k^{c_2(\lambda/\varepsilon)^p}$. The argument in the second step is deterministic in contrast to the probabilistic nature of the proofs mentioned in the previous subsection.

Based on the work of Krivine, Maurey and Pisier (1976) obtained the following remarkable result (see also Milman and Sharir 1979).

Theorem 3.2.2. *Let X be an infinite-dimensional Banach space and put*

$$p_X=\sup\{p\colon\ T_p(X)<\infty\},\qquad q_X=\inf\{q\colon\ C_q(X)<\infty\}.\tag{3.2.2}$$

Then for every $\varepsilon>0$ and integer n

$$\ell_{p_X}^n\overset{1+\varepsilon}{\hookrightarrow}X,\qquad \ell_{q_X}^n\overset{1+\varepsilon}{\hookrightarrow}X.\tag{3.2.3}$$

As mentioned in the previous subsection (see (3.1.18)) a result of Figiel, Lindenstrauss and Milman shows that if $\dim X=n$ and the cotype q constant of X is bounded by a number independent of n for some $q<\infty$ then X has almost Hilbertian subspaces of dimension a power of n. If no such q exists it follows from Theorem 3.2.2 (but actually this special case is much simpler) that $\ell_\infty^k\overset{1+\varepsilon}{\hookrightarrow}X$ for large k (tending to ∞ as $n\to\infty$). This observation was put into quantitative form by Alon and Milman (1983) using combinatorial methods. They showed that

$$k_2(X,1)k_\infty(X,1)\geqslant e^{c(\log n)^{1/2}},\quad n=\dim X.\tag{3.2.4}$$

A quantitative version of the type part of Theorem 3.2.2 was obtained by Pisier (1983). He proved that if $1 < p < 2$ and if q is the conjugate exponent of p $(p^{-1} + q^{-1} = 1)$ then

$$k_p(X, \varepsilon) \geqslant c(p, \varepsilon) T_p(X)^q. \tag{3.2.5}$$

Remarks.

(1) Note that (3.2.5) shows that if $\dim X = \infty$ then $\ell_p^n \overset{1+\varepsilon}{\hookrightarrow} X$ for every n and ε also for $p_X < p \leqslant 2$. An analogue statement concerning the cotype does not hold.

(2) Pisier's result is actually stronger than (3.2.5). In his result $T_p(X)$ is replaced by a possibly larger constant, namely the stable type p constant defined in analogy to the Rademacher and Gaussian types by using p-stable random variables. We do not discuss these generalized types here.

Since clearly $T_p(\ell_1^n) \geqslant n^{1/q}$ it follows from (3.2.5) that

$$\ell_p^k \overset{1+\varepsilon}{\hookrightarrow} \ell_1^n \quad \text{for } k \geqslant \alpha(p, \varepsilon)n, \quad 1 \leqslant p \leqslant 2. \tag{3.2.6}$$

In fact, (3.2.6) was proved first by Johnson and Schechtman (1982) who also used p-stable random variables in their proof. Geometrically (3.2.6) has the following interesting meaning. The crosspolytope of dimension n has for every $1 < p \leqslant 2$ sections of dimension proportional to n which are (almost) affinely equivalent to the unit ball of ℓ_p spaces of the appropriate dimension. We shall come back to the geometric interpretation in section 6, and continue here with the discussion from the functional analytic point of view.

Instead of ℓ_p^k, $1 < p \leqslant 2$, one might want to consider a general k-dimensional subspace X of $L_1(0,1)$. (For this discussion it is worthwhile to recall that for $1 \leqslant p < r \leqslant 2$, $L_r(0,1)$ is isometric to a subspace of $L_p(0,1)$.) A simple but very effective approach to this question was suggested by Schechtman (1987) – the so-called empirical distribution method. One maps X into ℓ_1^n by considering the map

$$L_1(0,1) \supset X \ni f \to \big(f(t_1), \ldots, f(t_n)\big) \in \ell_1^n.$$

Then if X is properly situated in $L_1(0,1)$ and n is large enough one proves that for most choices of $\{t_i\}_{i=1}^n$ (chosen independently and uniformly distributed on $(0,1)$) the map above is almost an isometry. By using this approach of Schechtman it was proved by Bourgain, Lindenstrauss and Milman (1989a) that (3.2.6) holds if ℓ_p^k is replaced by any k-dimensional subspaces of $L_p(0,1)$, $1 < p \leqslant 2$, and "almost" holds for a general k-dimensional subspace of $L_1(0,1)$. A slightly stronger form of this result was obtained by Talagrand (1990) (with a considerably simpler proof). The result states that if $X \subset L_1(0,1)$ with $\dim X = k > 1$ then $X \overset{1+\varepsilon}{\hookrightarrow} \ell_1^n$ with

$$n \leqslant cK(X)^2 \varepsilon^{-2} k \leqslant c_1 \varepsilon^{-2} k \log k, \tag{3.2.7}$$

where $K(X)$ is the Rademacher projection constant of X (see (2.2.19)). It is not known if the term $K(X)^2$ (or for that matter $\log k$) are really needed in (3.2.7).

A similar result was proved in Bourgain, Lindenstrauss and Milman (1989a) also for embedding into ℓ_p^n. For $1 < p \leqslant 2$ it is shown that if $X \subset L_p(0,1)$ with $\dim X = k > 1$ then $X \xrightarrow{1+\varepsilon} \ell_p^n$ for $n \leqslant c(p,\varepsilon)k \cdot \log^3 k$ and the $\log^3 k$ factor can be removed if $X \subset L_r(0,1)$ with $p < r \leqslant 2$ (provided $c(p,\varepsilon)$ is replaced by an appropriate $c(p,r,\varepsilon)$). For $2 < p < \infty$ it is shown in that paper that $X \subset L_p(0,1)$ with $\dim X = k$, $1+\varepsilon$ embeds in ℓ_p^n for $n \geqslant c(p,\varepsilon)k^{p/2} \log k$. In conclusion, we get for arbitrary subspaces of $L_p(0,1)$ (for any $1 \leqslant p \leqslant \infty$) the same results for embedding into ℓ_p^n as those obtained in the previous paragraph concerning Hilbertian sections (up to possible logarithmic factors). Recall that by (3.1.6) and (3.1.19) $k(\ell_p^n) \geqslant c(p)n^{2/\max(2,p)}$, $1 \leqslant p < \infty$, and that this estimate is sharp also for $p > 2$ as shown by Bennett et al. (1977).

In the above results we have emphasized mainly the dependence of n on k. Also the dependence of n on ε for fixed k is sometimes of geometric interest. Some quite precise results in this direction are contained in Bourgain, Lindenstrauss and Milman (1989a), and Bourgain and Lindenstrauss (1988, 1989) (see section 6.2 below).

In contrast to the spaces ℓ_2^n, the spaces ℓ_p^n have a quite complicated internal structure which has been studied extensively. Some of these studies are related to the results mentioned above. One may ask for example the following. Let X be a subspace of ℓ_p^n of dimension m. For what k are we ensured that ℓ_p^k embeds (up to 2, say) in X?

The first results in this direction were obtained by Figiel and Johnson (1980). They treated the case $p = \infty$ and showed that for $m = cn$ one gets $k \approx m^{1/2}$ and that this result is sharp. Their result was extended to smaller m (more precisely if $m \geqslant cn^\delta$ for some $\delta > 0$) by Bourgain (1984b). Geometrically this means that if a centrally symmetric polytope P in R^m has at most m^r faces (r independent of m) then P has a central k-dimensional section close to the k-cube for k of the order $m^{1/2}$.

Assume that $2 < p < \infty$. By the results mentioned above (e.g., about Hilbertian sections) one can say nothing as long as $m \leqslant n^{2/p}$. For $m > n^{2/p}$ a sharp result is available. Every such X contains a good copy of ℓ_p^k with

$$k = c \min\left(m^{q/2}, (m/n^{2/p})^{p/(p-2)}\right), \tag{3.2.8}$$

where q is the adjoint index of p. This is proved in Bourgain and Tzafriri (1990). A dual result (concerning quotient spaces of such X) is proved in Bourgain, Lindenstrauss and Milman (1989). For $1 \leqslant p < 2$, because of the large Euclidean sections, nothing can be said on the question above unless m is very near to n, i.e., $m = n - o(n)$. This case is treated in Gluskin, Tomczak-Jaegermann and Tzafriri (1992).

Many theorems concerning the existence of nice embeddings into or from ℓ_p^n spaces have analogues concerning projections. We restrict ourselves here to one result on projections in ℓ_p^n. This result is due to Johnson and Schechtman (1991) (and based on results in Bourgain and Tzafriri (1987b) and the above-mentioned paper of Bourgain, Lindenstrauss and Milman) Let P be a projection in ℓ_p^n and put $X = P\ell_p^n$, $Y = \ker P$, $k = \dim X$. Then

$$d(Y, \ell_p^{n-k}) \leqslant f(\|P\|, p, d_X), \tag{3.2.9}$$

for a suitable function f. In other words if the norm of P is controlled and its range is not far from ℓ_2^k then the kernel of P is not far from ℓ_p^{n-k}.

4. Ellipsoids in local theory

4.1. John's theorem and ellipsoids related to it

In this section we discuss in more detail the question of introducing "good" Euclidean structures in a given n-dimensional normed space X. Geometrically the situation is the following. We are given a symmetric convex body K_X (the unit ball of X) in $\mathbb{R}^n$ and are interested in choosing a suitable Euclidean norm (or inner product) in $\mathbb{R}^n$ related in a specific way to K_X. It will be sometimes convenient to look at this situation in the following equivalent way. We are given $K = K_X$ and the standard Euclidean norm $\|x\|_2 = \left(\sum_{i=1}^n |x_i|^2\right)^{1/2}$ in $\mathbb{R}^n$ and look for a suitable "position" UK of K where $U \in \mathrm{GL}_n$. Of course, by definition, for every $U \in \mathrm{GL}_n$, UK induces in $\mathbb{R}^n$ a norm with respect to which $\mathbb{R}^n$ is isometric to X. Thus instead of leaving K fixed and choosing a suitable ellipsoid as a unit ball for the Euclidean structure we may fix the ellipsoid and move K.

We start the discussion by recalling the (mutually dual) notions of maximal volume (respectively minimal volume) ellipsoids contained in (respectively containing) K, which were already mentioned in section 2. The main fact concerning these (uniquely defined) ellipsoids is a result of John (1948) on the contact points between the boundaries of K and these ellipsoids.

Theorem 4.1.1. *Let $X = (\mathbb{R}^n, \|\cdot\|)$ be a normed space and let $|\cdot|$ be the Euclidean norm on $\mathbb{R}^n$ induced by the ellipsoid $\mathscr{E}_K$ of maximal volume in K_X. Then there exist $\{x_j\}_{j=1}^s$ in X with $n \leqslant s \leqslant n(n+1)/2$ and positive scalars $\{\lambda_j\}_{j=1}^s$ so that*

$$\|x_j\| = |x_j| = 1, \quad 1 \leqslant j \leqslant s, \tag{4.1.1}$$

$$x = \sum_{j=1}^s \lambda_j \langle x, x_j \rangle x_j, \quad x \in X. \tag{4.1.2}$$

Remarks.

(1) It follows from (4.1.1) that the only support functional to K_X at x_j is given by x_j itself, $1 \leqslant j \leqslant s$, and hence also $\|x_j\|^*$ (the norm of x_j in X^*) is equal to 1. Hence, since $\mathscr{E}_K^\circ$ is the ellipsoid of minimal volume of $K_{X^*} = K_X^\circ$, it follows that Theorem 4.1.1 holds verbatum also for the Euclidean norm induced by the minimal volume ellipsoid.

(2) By considering the trace of the identity operator of X it follows from (4.1.2) that

$$\sum_{j=1}^{s} \lambda_j = n. \tag{4.1.3}$$

From (4.1.3) and the Cauchy–Schwartz inequality we get that $\|x\| \leqslant |x| \leqslant n^{1/2}\|x\|$ for every $x \in X$, and this is the part of the assertion of the theorem which was quoted in section 2. However, more can be deduced from (4.1.2) and this is best stated in terms of 2-absolutely summing norms. Let T denote the identity map from $(X, |\cdot|)$ to $(X, \|\cdot\|)$. Then it follows from (4.1.2) and (4.1.3) that $\pi_2(T) \leqslant n^{1/2}$ and $\pi_2(T^{-1}) \leqslant n^{1/2}$. Hence if I is the identity map of $(X, \|\cdot\|)$, $I = TT^{-1}$ and $\pi_2(I) \leqslant \|T\|\pi_2(T^{-1}) \leqslant n^{1/2}$. This fact, the facts that $\pi_2 = \pi_2^*$ (see (2.3.14)) and $\operatorname{trace} I = n$ imply that all the preceding inequalities are actually equalities. In particular

$$\pi_2(I) = n^{1/2}, \quad I \text{ the identity of } X, \dim X = n. \tag{4.1.4}$$

(The first explicit statement of this consequence of Theorem 4.1.1 is in Garling and Gordon 1971).

(3) It was noted by John that if $|||\cdot|||$ is any norm in $\mathbb{R}^n$ so that $|||x_j||| = 1$ for every j and $|||x||| \leqslant |x|$ for every $x \in X$ then $\mathscr{E}_K$ (the unit ball of $|\cdot|$) is also the ellipsoid of maximal volume in the unit ball of $|||\cdot|||$.

Dvoretzky and Rogers (1950) also analyzed the points of contact between the boundaries of $K = K_X$ and $\mathscr{E}_K$, the maximal volume ellipsoid of K. They proved that this set contains a large subset which is almost orthogonal with respect to the norm $|\cdot|$ induced by $\mathscr{E}_K$. It follows from their result that there is an orthonormal basis $\{e_i\}_{i=1}^n$ in $(\mathbb{R}^n, |\cdot|)$ so that

$$1 \geqslant \|e_i\| \geqslant \frac{\sqrt{n-i+1}}{\sqrt{n}+\sqrt{i-1}}, \quad 1 \leqslant i \leqslant n. \tag{4.1.5}$$

In particular $\|e_i\| \geqslant \frac{1}{3}$ for $i \leqslant n/2$. As pointed out to us by Johnson, this fact trivially implies the existence of (a different) orthonormal basis $\{f_i\}_{i=1}^n$ of $(\mathbb{R}^n, |\cdot|)$ so that $\|f_i\| \geqslant \frac{1}{8}$ for all $1 \leqslant i \leqslant n$.

An interesting geometric consequence of Theorem 4.1.1 (or equivalently of the Dvoretzky–Rogers result) was pointed out by Ball (1989). He introduced the "cubical ratio" of an n-dimensional normed space X by putting

$$\operatorname{cr}(X) = \inf\{\operatorname{Vol} Q / \operatorname{Vol} K_X\colon\ Q \supset K_X,\ Q \text{ a parallelopiped}\}^{1/n}. \tag{4.1.6}$$

Ball proved that

$$\operatorname{vr}(\ell_\infty^n) \leqslant \operatorname{cr}(X) \cdot \operatorname{vr}(X) \leqslant \sqrt{e}\ \operatorname{vr}(\ell_\infty^n). \tag{4.1.7}$$

The left-hand inequality of (4.1.7) is just a trivial consequence of the definition; the right-hand side inequality is the one which was deduced from Theorem 4.1.1 (see also Pelczynski and Szarek (1991) for a very slight improvement). By applying (4.1.7) to X and X^* and using the geometric meaning of determinants one gets

$$\begin{aligned} \operatorname{Vol}\mathscr{E}_{\min}/\operatorname{Vol}\mathscr{E}_{\max} &\leqslant \max\left\{ \det\left\{ \langle x_i, x_j^* \rangle \right\}_{i,j=1}^n : \|x_i\| \leqslant 1,\ \|x_j^*\| \leqslant 1 \right\} \\ &\leqslant e^n \operatorname{Vol}\mathscr{E}_{\min}/\operatorname{Vol}\mathscr{E}_{\max}, \end{aligned} \tag{4.1.8}$$

where $\mathscr{E}_{\min}$ is the ellipsoid of minimal volume containing K_X and $\mathscr{E}_{\max}$ the ellipsoid of maximal volume contained in K_X (denoted above also by $\mathscr{E}_K$). This consequence of (4.1.7) was stated explicitely first by Geiss (1992). The spaces X for which $(\operatorname{Vol}\mathscr{E}_{\min}/\operatorname{Vol}\mathscr{E}_{\max})^{1/n}$ are bounded by a constant independent of the dimension are those which are studied in the theory of weak Hilbert spaces, to which we hinted after (3.1.25) (see Pisier (1989b) for details).

As we mentioned in section 2 the proof of John of Theorem 4.1.1 was generalized by Lewis to the setting of general ideal norms (in Theorem 4.1.1, itself the relevant norm is the π_2 norm). By applying this generalization (see (2.3.15)) to the setting of the ℓ norm (see (2.3.17)) Figiel and Tomczak-Jaegermann (1979) obtain the following. In every finite-dimensional space $X = (\mathbb{R}^n, \|\cdot\|)$ there is an Euclidean norm $|\cdot|$ (called the ℓ-structure) such that if T is the identity map from $(\mathbb{R}^n, |\cdot|)$ into X then

$$\ell(T)\ell((T^{-1})^*) \leqslant nK(X), \tag{4.1.9}$$

where $K(X)$ is the Rademacher projection constant defined in (2.2.19). It follows from (4.1.9) the definition of M_X (see (3.1.12), the Euclidean structure being the ℓ-structure) and (2.2.19) that for some absolute constants c_1 and c_2

$$M_X M_{X^*} \leqslant c_1 K(X) \leqslant c_2 \log(1 + d_X). \tag{4.1.10}$$

Another natural ellipsoid associated with $X = (\mathbb{R}^n, \|\cdot\|)$ is the so-called distance ellipsoid $\mathscr{E}_d$ for which the Banach–Mazur distance $d_X = d(\ell_2^n, X)$ is attained. In other words $\mathscr{E}_d$ is an ellipsoid which satisfies

$$\mathscr{E}_d \subset \alpha K_X, \quad K_X \subset \beta \mathscr{E}_d, \alpha \cdot \beta = d_X. \tag{4.1.11}$$

Easy examples show that $\mathscr{E}_d$ is not uniquely defined even if we normalize (4.1.11) by requiring say that $\alpha = 1$. It is also easy to see that $\mathscr{E}_d$ may be very far from the ellipsoids considered above. There is however an analogue to the results of John or Dvoretzky and Rogers concerning contact points for the distance ellipsoid(s). Larman and Mani (1975) proved the following. Let $X = (\mathbb{R}^n, \|\cdot\|)$ and let $|\cdot|$ be an Euclidean norm in $\mathbb{R}^n$ so that $|x| \leqslant \|x\| \leqslant d_X|x|$ for every $x \in X$. Let $\delta > 0$ and let $m = m(\delta) = \min([d_X/4], [\delta^2 d_X^2/4])$. Then there exist m vectors $\{x_i\}_{i=1}^m$ in X and m orthogonal vectors $\{e_i\}_{i=1}^m$ so that

$$\|x_i\| = 1, \quad |x_i| = |e_i| = 1/d_X, \quad |x_i - e_i| \leqslant \delta/d_X, \qquad 1 \leqslant i \leqslant m. \tag{4.1.12}$$

The information contained in this result is more useful for X with large d_X.

The convex hull of two ellipsoids is of course not an ellipsoid but it is near to an ellipsoid; its Banach–Mazur distance from an ellipsoid is at most $\sqrt{2}$. This fact often allows us to "combine" good properties of several different Euclidean structures. For instance, using this observation Bourgain and Milman (1986) combined the properties of the maximal volume ellipsoid with the distance ellipsoid to obtain for every $X = (\mathbb{R}^n, \|\cdot\|)$ an Euclidean structure, or equivalently an operator $T : \ell_2^n \to X$ so that

$$|y| \geqslant \|Ty\| \geqslant |y|/\sqrt{2}d_X, \quad y \in \ell_2^n; \qquad \pi_2(T^{-1}) \leqslant \sqrt{2n}. \tag{4.1.13}$$

They also combined this structure with the ℓ-structure. These structures turned out to be useful in some results concerning the Banach–Mazur distance.

A much more delicate result of this nature was obtained by Bourgain and Szarek (1988). They proved the following stronger form of the Dvoretzky–Rogers lemma. For every $1 > \delta > 0$ there is a $b(\delta) > 0$ so that if $X = (\mathbb{R}^n, \|\cdot\|)$ then there is a Euclidean norm $|\cdot|$ on X and vectors $\{x_j\}_{j=1}^m$ in X with $m \geqslant (1-\delta)n$ and $\|x_j\| \leqslant 1$ for all j so that

$$|x| \leqslant \|x\| \leqslant \sqrt{2}d_X|x|, \quad x \in X; \qquad b(\delta)\Big(\sum_{j=1}^m |\alpha_j|^2\Big)^{1/2} \leqslant \Big|\sum_{j=1}^m \alpha_j x_j\Big|, \tag{4.1.14}$$

for all scalars $\{\alpha_j\}_{j=1}^m$. They also obtained an estimate on $b(\delta)$ which was later considerably improved by Szarek and Talagrand (1989) to $b(\delta) = c\delta^2$.

4.2. *Inertia ellipsoids, isotropic position and generalized Khintchine inequality*

A natural ellipsoid which is associated to any symmetric convex body K (actually any compact set with positive measure) in $\mathbb{R}^n$ is the Binet ellipsoid $\mathcal{E}_{\mathrm{B}}$ of K. It is defined by the inner product norm it induces on $\mathbb{R}^n$

$$|x|_{\mathcal{E}_{\mathrm{B}}}^2 = (\operatorname{Vol} K)^{-1} \int_K |\langle x, y\rangle|^2 \, \mathrm{d}y \, , \tag{4.2.1}$$

the integration being with respect to the Lebesgue measure. It is easy to check that if $\mathcal{E}$ is an ellipsoid in $\mathbb{R}^n$ and $\mathcal{E}^\circ$ denotes, as usual, its polar then

$$\int_{\mathcal{E}} |\langle x, y\rangle|^2 \, \mathrm{d}y = (n+2)^{-1} \operatorname{Vol}\mathcal{E} \cdot |x|_{\mathcal{E}^\circ}^2. \tag{4.2.2}$$

It follows from (4.2.1) and (4.2.2) that

$$\int_{\mathcal{E}_{\mathrm{L}}} |\langle x, y\rangle|^2 \, \mathrm{d}y = \int_K |\langle x, y\rangle|^2 \, \mathrm{d}y, \quad x \in \mathbb{R}^n, \tag{4.2.3}$$

where $\mathscr{E}_{\rm L}$ is the ellipsoid defined by

$$\mathscr{E}_{\rm B} = (n+2)^{1/2}(\mathrm{Vol}\,\mathscr{E}_{\rm L})^{-1/2}\mathscr{E}_{\rm L}^{\circ}\,. \tag{4.2.4}$$

The ellipsoid $\mathscr{E}_{\rm L}$ defined by (4.2.3) or (4.2.4) is called the Legendre ellipsoid of K. By definition, it is the unique ellipsoid which has the same moments of inertia as K with respect to the axes and is therefore also called the Inertia ellipsoid. The Binet and Legendre ellipsoids are notions from classical mechanics. The duality relation between them is classical.

The Legendre ellipsoid arises naturally also from probabilistic considerations. Let ξ be a random variable uniformly distributed on the symmetric convex body K. Let $\{\xi_i\}_{i=1}^{\infty}$ be mutually independent copies of ξ. Then by the central limit theorem the sequence $(\xi_1 + \xi_2 + \cdots + \xi_m)/\sqrt{m}$ converges to a Gaussian variable G which satisfies

$$E|\langle G,y\rangle|^2 = (\mathrm{Vol}\,K)^{-1}\int_K |\langle x,y\rangle|^2\,\mathrm{d}x. \tag{4.2.5}$$

Another context (this time geometric) in which $\mathscr{E}_{\rm L}$ arises will be explained later on in the present subsection.

A symmetric convex body K is said to be in isotropic position in $\mathbb{R}^n$ if $\mathrm{Vol}\,K = 1$ and its Binet (or for that matter its Legendre) ellipsoid is homothetic to the Euclidean ball in $\mathbb{R}^n$, i.e., if

$$\int_K |\langle x,y\rangle|^2\,\mathrm{d}x = L_K^2, \quad y \in S^{n-1}. \tag{4.2.6}$$

Clearly, for every symmetric K there is a $T \in \mathrm{GL}_n$ so that $TK = \widehat{K}$ is in isotropic position. The transformation T is determined uniquely up to an orthogonal transformation. The constant L_K for a general convex symmetric K is defined to be the constant $L_{\widehat{K}}$ appearing in (4.2.6) for $\widehat{K}$. The isotropic position TK of K of volume 1 can be also characterized by the fact that $\int_{UK} |x|^2\,\mathrm{d}x$ attains its minimum as U varies over SL_n for $U = T$. (See Milman and Pajor (1989) which contains a detailed survey of most of the material in the present subsection).

It was shown by Blaschke (1918) (for $n = 3$) and John (1937) (for general n) that $L_K \geqslant L_{B_n}$ where B_n is the Euclidean ball in $\mathbb{R}^n$. It is easy to check that L_{B_n} tends to a positive limit as $n \to \infty$ and thus for some positive constant c

$$L_K \geqslant c > 0. \tag{4.2.7}$$

An interesting open problem (stated first explicitly in this form by Bourgain 1986a) is:

$$\text{is } \sup\{L_K\colon\ K \subset \mathbb{R}^n,\ n = 1,2,\ldots\} < \infty? \tag{4.2.8}$$

It is a consequence of (2.1.4) that for a symmetric convex body K in $\mathbb{R}^n$ we have $L_K \leqslant c_1\sqrt{n}$. If K is the unit ball of a space with a 1-unconditional basis (or for that matter λ-unconditional basis with a control on λ) or if K is a zonoid (for this concept see chapter 4.9 in this Handbook or section 6 below) then indeed L_K is bounded by a constant independent of K or the dimension n. This is also true for K a polar of a zonoid (cf. Ball 1991a). For a general $K \subset \mathbb{R}^n$ the best upper estimate known at present is $L_K \leqslant c_2 n^{1/4} \log n$ (cf. Bourgain 1991).

The question (4.2.8) has many equivalent formulations some going back to problems formulated by Busemann. A detailed discussion of this is given in the Milman and Pajor paper. Here we mention just one formulation which involves the ($(n-1)$-dimensional) volume of central sections of K. Assume that K is a symmetric convex body in $\mathbb{R}^n$ in isotropic position. Hensley (1980) proved that if H is any hyperplane in $\mathbb{R}^n$ (through the origin) then

$$c_1 \operatorname{Vol}(K \cap H) \leqslant L_K^{-1} \leqslant c_2 \operatorname{Vol}(K \cap H) \tag{4.2.9}$$

for some universal constants c_1 and c_2. It follows, of course, from (4.2.9) that, with K as above, for every pair of hyperplanes H_1 and H_2, $\operatorname{Vol}(K \cap H_1)/\operatorname{Vol}(K \cap H_2)$ is bounded by a universal constant. Problem (4.2.8) is equivalent to the question whether for a symmetric convex K of volume 1 one can always find a hyperplane H so that $\operatorname{Vol}(K \cap H) \geqslant \delta$ where δ is a positive absolute constant.

The result of Hensley mentioned above is a special case of a general phenomenon of concentration of measure in convex sets which we would like to explain next in some detail. Let K be a symmetric convex body in $\mathbb{R}^n$ and let B be another convex body (not necessarily symmetric) in $\mathbb{R}^n$. The following result was deduced by Borell (1975) from the classical Brunn–Minkowski inequality

$$\operatorname{Vol}(B \cap K) \geqslant \tfrac{2}{3} \operatorname{Vol} K \Rightarrow \operatorname{Vol}(tB \cap K) \geqslant \left(1 - \frac{\sqrt{2}}{3} 2^{-t/2}\right) \operatorname{Vol} K \quad \text{for } t > 1. \tag{4.2.10}$$

The choice of the parameter $\frac{2}{3}$ in (4.2.10) is of course arbitrary and done just for being specific. The point in (4.2.10) is that if $B \cap K$ contains a large part of K then the proportion of K outside tB decreases exponentially in t as $t \to \infty$ in a rate which is independent of the sets or the dimension. For a function $f(x)$ defined on K we denote by $\|f\|_p$ its L_p norm with respect to the probability measure $\mathrm{d}x/\operatorname{Vol} K$, i.e.,

$$\|f\|_p = \left((\operatorname{Vol} K)^{-1} \int_K |f(x)|^p \, \mathrm{d}x\right)^{1/p}, \quad 0 < p < \infty.$$

It follows easily from (4.2.10) that if $f(x)$ is a linear function, i.e., $f(x) = \langle x, y \rangle$ for some $y \in \mathbb{R}^n$ then

$$\|f\|_q \leqslant \|f\|_p \leqslant c_{p,q} \|f\|_q, \quad 0 < q < p \tag{4.2.11}$$

for some universal constants $c_{p,q}$ (the inequality in the left is trivial while the right-hand side inequality follows by taking in (4.2.10) $B = \{x\colon |f(x)| \leqslant \lambda\}$ for a suitable λ, see Gromov and Milman 1984). Inequality (4.2.11) is very similar to the classical Khintchine inequality. In fact, for $K =$ the unit cube (i.e., the unit ball of ℓ_∞^n) this is Khintchine's inequality. By using (4.2.10) for K the unit cube, it is also easy to deduce Kahane's generalized Khintchine inequality (2.2.9) (cf. Milman and Schechtman 1986).

From (4.2.11) it is easy to deduce another ("isomorphic") description of the Legendre ellipsoid to which we hinted above. Let K again be a symmetric convex body in $\mathbb{R}^n$. The centroid body $Z(K)$ of K which was introduced by Petty (1961) is defined as follows,

$$Z(K) = (\operatorname{Vol} K)^{-1} \int_K [0, x]\, \mathrm{d}x. \tag{4.2.12}$$

The integral in (4.2.12) is an integral of sets and can be defined as a limit (in the Hausdorff distance) of Minkowski sums of segments $\sum_{j=1}^m \lambda_j [0, x_j]$. Thus $Z(K)$ is a zonoid. An equivalent way to define $Z(K)$ is to describe its support functional (i.e., the norm induced by $Z(K)^\circ$ in $\mathbb{R}^n$), since this is described by an ordinary integral

$$\|y\|_{Z(K)^\circ} = \tfrac{1}{2} (\operatorname{Vol} K)^{-1} \int_K |\langle x, y\rangle|\, \mathrm{d}x. \tag{4.2.13}$$

From (4.2.1), (4.2.4) and the Khintchine type inequality (4.2.11) we deduce that

$$c_1 \mathscr{E}_L \subset (n \operatorname{Vol} K / \operatorname{Vol} \mathscr{E}_L)^{1/2} Z(K) \subset c_2 \mathscr{E}_L, \tag{4.2.14}$$

where c_1 and c_2 are universal constants and $\mathscr{E}_L$ the Legendre ellipsoid of K.

Coming back to (4.2.11) it can be checked that the constant $c_{p,q}$ appearing in it can be chosen to depend just on p alone (see Ullrich (1988), for the case K the unit cube, and Milman and Pajor (1989) for general K). Hence, by letting $q \to 0$ in (4.2.11) one gets

$$\left((\operatorname{Vol} K)^{-1} \int_K |\langle x, y\rangle|^p\, \mathrm{d}x \right)^{1/p} \leqslant c_p \exp\left[(\operatorname{Vol} K)^{-1} \int_K \log |\langle x, y\rangle|\, \mathrm{d}x \right]. \tag{4.2.15}$$

By using this inequality it was shown by Milman and Pajor that for $y \in S^{n-1}$ and $H = \{x\colon\ \langle x, y\rangle = 0\}$

$$\frac{\mathrm{e}^{-1}}{2} \frac{\operatorname{Vol} K}{\operatorname{Vol}(K \cap H)} \leqslant \exp\left[(\operatorname{Vol} K)^{-1} \int_K \log |\langle x, y\rangle|\, \mathrm{d}x \right] \leqslant \frac{\operatorname{Vol} K}{\operatorname{Vol}(K \cap H)} \frac{\mathrm{e}^{-\gamma}}{2}. \tag{4.2.16}$$

The constant γ is the Euler constant. We wrote in (4.2.16) explicit constants since they can be checked to be sharp (as $n \to \infty$). Note that by (4.2.15) it follows that if K is in isotropic position then up to a multiplicative constant the middle term in (4.2.16) is independent of the choice of $y \in S^{n-1}$ and thus equivalent to L_K. It follows that (4.2.9) is a direct consequence of (4.2.16).

Inequality (4.2.11) can be generalized to some non-linear functions as well. Bourgain (1991) proved that if $f(x)$ is a polynomial of degree d in $\mathbb{R}^n$ then

$$\|f\|_p \leqslant c_{p,q,d}\|f\|_q, \quad 0<q<p, \tag{4.2.17}$$

where $c_{p,q,d}$ depend just on the indicated parameters (and not on K or the dimension n). This fact lies deeper than (4.2.11) since we have no convexity condition on the level sets of f. The convexity is introduced in the proof by applying the Knothe (1957) map.

4.3. M ellipsoids and the inverse Brunn–Minkowski inequality

The basic result of this subsection is the following theorem of Milman (1986b) concerning what now are called M ellipsoids.

Theorem 4.3.1. *There is an absolute constant C so that the following holds. For every symmetric convex body K in $\mathbb{R}^n$, $n = 2, \ldots$ there is an ellipsoid $\mathscr{E}_M$ in $\mathbb{R}^n$ so that*

$$\big(\mathrm{Vol}\,\big(\mathrm{Conv}(\mathscr{E}_M \cup K)\big)/\,\mathrm{Vol}(\mathscr{E}_M \cap K)\big)^{1/n} \leqslant C, \tag{4.3.1}$$

$$\big(\mathrm{Vol}\,\big(\mathrm{Conv}(\mathscr{E}_M^\circ \cup K^\circ)\big)/\,\mathrm{Vol}(\mathscr{E}_M^\circ \cap K^\circ)\big)^{1/n} \leqslant C. \tag{4.3.2}$$

Obviously $\mathscr{E}_M$ is not uniquely determined by K and in particular it depends on the choice of C. The smallest possible C for which the theorem holds is not known and not really meaningful since there are various ways to state the theorem and different versions will have different best constants. There are by now several proofs of Theorem 4.3.1 in the literature. Two proofs are presented in the book of Pisier (1989b), which contains a nice exposition of the material of this subsection (including also more historical background). Still another proof is contained in Pisier (1989a). A geometric proof of Milman (1988b) will be outlined in section 6 below.

As is evident from the statement of the theorem and will be clarified further below $\mathscr{E}_M$ may replace K in many volume computations provided we are not interested in precise results but only "isomorphic computations" which are valid up to an absolute multiplicative constant.

As a first application of Theorem 4.3.1 we present the so-called Blaschke–Santaló inequality and in particular its inverse which is due to Bourgain and Milman (1985, 1987) (which was proved however prior to Theorem 4.3.1). Note that (4.3.1) implies in particular that $(\mathrm{Vol}\,K)^{1/n}$ is up to an absolute constant the same as $(\mathrm{Vol}\,\mathscr{E}_M)^{1/n}$ and (4.3.2) shows that the same is true concerning the pair $(\mathrm{Vol}\,K^\circ)^{1/n}$

and $(\mathrm{Vol}\,\mathscr{E}_M^\circ)^{1/n}$. Observe also that the product $\mathrm{Vol}\,K \cdot \mathrm{Vol}\,K^\circ$ is an affine invariant, i.e.,

$$\mathrm{Vol}\,TK \cdot \mathrm{Vol}(TK)^\circ = \mathrm{Vol}\,K \cdot \mathrm{Vol}\,K^\circ, \quad T \in \mathrm{GL}_n. \tag{4.3.3}$$

Hence, $\mathrm{Vol}\,\mathscr{E}_M \cdot \mathrm{Vol}\,\mathscr{E}_M^\circ$ is equal to $(\mathrm{Vol}\,B_n)^2$ where B_n is the Euclidean ball in $\mathbb{R}^n$. It is trivial to check that $n \cdot (\mathrm{Vol}\,B_n)^{2/n}$ tends to a positive limit as $n \to \infty$ and hence we get for some universal constant C_1 and all symmetric convex bodies K in $\mathbb{R}^n$

$$C_1^{-1}/n \leqslant (\mathrm{Vol}\,K \cdot \mathrm{Vol}\,K^\circ)^{1/n} \leqslant C_1/n\,. \tag{4.3.4}$$

The estimate from above is classic and known to hold in a sharper form. It was proved by Blaschke (for $n = 3$) and Santaló (for general n) that for every $K \subset \mathbb{R}^n$

$$\mathrm{Vol}\,K \cdot \mathrm{Vol}\,K^\circ \leqslant (\mathrm{Vol}\,B_n)^2. \tag{4.3.5}$$

(An elegant proof of this, using Steiner symmetrization, was found by Ball (1986a). The same proof was found independently by Meyer and Pajor (1989) who also treated the more delicate non-symmetric case in Meyer and Pajor (1990).) For the lower estimate it was conjectured by Mahler that $\mathrm{Vol}\,K \cdot \mathrm{Vol}\,K^\circ$ attains its minimum for K the cube (i.e., unit ball of ℓ_∞^n). This conjecture has been proved in some special cases, by Saint-Raymond, Meyer, Reisner and others (e.g., K the unit ball of a space with a 1-unconditional basis or K a zonoid) but the general case is still open. The lower estimate in (4.3.4) proves Mahler's conjecture for general K in a weak sense. For a more complete discussion of this topic and the relevant references we refer to Pisier's book.

In order to describe further consequences of Theorem 4.3.1 we introduce the notion of covering numbers. Let K_1 and K_2 be two symmetric convex bodies in $\mathbb{R}^n$. The covering number $N(K_1, K_2)$ is defined by

$$N(K_1, K_2) = \min\left\{N\colon\ \exists\,\{x_j\}_{j=1}^N \in \mathbb{R}^n,\ K_1 \subset \bigcup_{j=1}^N (x_j + K_2)\right\}. \tag{4.3.6}$$

The notion of covering number was introduced and studied first by von Neumann (1942). He was mainly interested in the case where $K_1, K_2 \subset \mathbb{R}^{n^2}$ are the unit balls of the space of operators from ℓ_2^n to itself with respect to the usual operator norm and the Hilbert–Schmidt norm. The systematic study of this notion in general was started by Kolmogoroff (see, in particular, Kolmogoroff and Tichomirov 1959). By now the literature on this subject is vast (see, for example, the recent book Carl and Stephani 1990).

A fundamental property of the covering numbers which follows easily from their definition is

$$\mathrm{Vol}\,K_1/\,\mathrm{Vol}\,K_2 \leqslant N(K_1, K_2) \leqslant 3^n\ \mathrm{Vol}\,K_1/\,\mathrm{Vol}\,K_2 \quad \text{if } K_1 \supset K_2. \tag{4.3.7}$$

The left-hand side in (4.3.7) trivially holds without the assumption that $K_1 \supset K_2$ while for the right-hand side the assumption is essential. A consequence of (4.3.4) and (4.3.7) (observed in König and Milman 1987) is the following. For all symmetric convex bodies K_1, K_2 in $\mathbb{R}^n$

$$C^{-1} \leqslant \left[N(K_1,K_2)/N(K_2^\circ,K_1^\circ)\right]^{1/n} \leqslant C, \tag{4.3.8}$$

where C is an absolute constant. Indeed, if $K_1 \supset K_2$ this is obvious. The general case can be reduced to this situation by using the following easily checked observations

$$N\big(\operatorname{conv}(K_1 \cup K_2), K_2\big) \leqslant 3^n N\big(\operatorname{conv}(K_1 \cup K_2), 2K_2\big) \leqslant 3^n N(K_1,K_2)$$

and similarly

$$N(K_1, K_1 \cap K_2) \leqslant 3^n N\big(K_1, 2(K_1 \cap K_2)\big) \leqslant 3^n N(K_1,K_2).$$

Inequality (4.3.8) proves a special case of a still open general problem concerning the relation between so-called entropy numbers of an operator and its adjoint. For details about this problem, including references and the best known results at present we refer to Bourgain et al. (1989).

It follows from (4.3.1) that the nth root of $N(K,\mathscr{E}_M)$ and $N(\mathscr{E}_M,K)$ (as well as the covering numbers of their polars) are uniformly bounded. Hence if K and $\mathscr{E}_M$ are as above and A any compact set in $\mathbb{R}^n$ then

$$\operatorname{Vol}(K+A) \leqslant N(K,\mathscr{E}_M)\operatorname{Vol}(\mathscr{E}_M+A) \leqslant C^n \operatorname{Vol}(\mathscr{E}_M+A). \tag{4.3.9}$$

We now turn to the Brunn–Minkowski inequality and its inverse. Recall that the Brunn–Minkowski inequality (valid for any compact A_1, A_2 in $\mathbb{R}^n$) states that

$$\operatorname{Vol}(A_1+A_2)^{1/n} \geqslant (\operatorname{Vol}A_1)^{1/n} + (\operatorname{Vol}A_2)^{1/n} . \tag{4.3.10}$$

It is trivial that if A_1 and A_2 are homothetic convex bodies then equality holds in (4.3.10) (it is well known that if A_1 and A_2 are convex bodies this is, up to translation, the only case of equality). It follows from (4.3.9) that if K_1 and K_2 are convex symmetric bodies which have homethetic M ellipsoids then the reverse inequality to (4.3.10) holds up to a universal multiplicative factor. It is worthwhile to put this result in the following form. For every convex symmetric body K, let $T_K \in \mathrm{SL}_n$ be such that it maps the M ellipsoid of K to a suitable multiple of the Euclidean ball. Then for the "position" $T_K K$ of K (which evidently has the same volume as K), we have the following "reverse Brunn–Minkowski inequality" (cf. Milman 1986b)

$$\operatorname{Vol}(T_{K_1}K_1 + T_{K_2}K_2)^{1/n} \leqslant C\big((\operatorname{Vol}T_{K_1}K_1)^{1/n} + (\operatorname{Vol}T_{K_2}K_2)^{1/n}\big) \tag{4.3.11}$$

holds for all $K_1, K_2 \subset \mathbb{R}^n$, $n = 2, 3, \ldots$ and a universal constant C. Note that in view of the duality part of Theorem 4.3.1, T_K may be chosen so that for every symmetric convex body K and any scalar $\lambda \neq 0$

$$T_{K^\circ} = (T_K^*)^{-1}, \qquad T_{\lambda K} = \boldsymbol{T}_K\ . \tag{4.3.12}$$

In some cases it is not hard to identify an M ellipsoid of K. If the Banach space X has a bounded cotype 2 constant $C_2(X)$ (e.g., if $X = \ell_1^n$) then $\mathscr{E}_M$ of K_X may be taken to be the ellipsoid of maximal volume in K_X (i.e., the Löwner ellipsoid). In this case the constant C for which (4.3.1) and (4.3.2) are valid for K_X depends on $C_2(X)$ (one has $C \approx C_2(X) \log\left(1 + C_2(X)\right)$, this is related to (3.1.25)). Similarly if X^* has a bounded cotype 2 constant (e.g., if $X = \ell_\infty^n$) then $\mathscr{E}_M$ of K_X can be taken to be the ellipsoid of minimal volume containing K_X. Thus, for example, in the case where K_1 is the unit cube in $\mathbb{R}^n$ and K_2 is the standard cross polytope of the same dimension, then (4.3.11) (and (4.3.10)) mean that

$$\mathrm{Vol}(t_1 K_1 + t_2 K_2)^{1/n} \approx t_1 + t_2/n, \quad t_1, t_2 > 0. \tag{4.3.13}$$

To conclude this paragraph let us note that Theorem 4.3.1 is closely related to the QS theorem (Theorem 3.1.1). The original proofs of (4.3.4) and of Theorem 4.3.1 relied on the QS theorem. Conversely, it is possible to deduce the QS theorem from Theorem 4.3.1. This is due to the fact that by (4.3.1) respectively (4.3.2) the volume ratios of the norms defined by $\mathrm{Conv}(K \cup \mathscr{E}_M)$ (respectively $\mathrm{Conv}(K^\circ \cup \mathscr{E}_M^\circ)$) are bounded (where K is the unit ball of a given space X). A large subspace E of $\mathbb{R}^n$ so that $K \cap E$ is almost spherical is obtained by using the argument sketched in (3.1.22)–(3.1.24). If P_E denotes the orthogonal projection on E one verifies (essentially by a mixed volume argument) that $P_E K^\circ$ has a small volume ratio (in E). Using again the argument sketched in section 3.1 the desired quotient of subspace is obtained (cf. Milman 1985b).

5. Distances and projections

5.1. Distances from and projections on classical spaces, especially ℓ_2^n

The computation of the Banach–Mazur distance between given Banach spaces of the same dimension is usually quite tricky. Already the computation of the distance between the simplest non-Euclidean spaces, namely ℓ_p^n spaces leads to unexpected results. For p and q on the same side of 2 we have as expected

$$d(\ell_p^n, \ell_q^n) = n^{|1/p - 1/q|}, \quad (p-2)(q-2) > 0. \tag{5.1.1}$$

Indeed, this is trivial for $p = 2$ and $q = \infty$ and the general case follows from this and the "triangle inequality" (2.1.2). In this case the operator $T : \ell_p^n \to \ell_q^n$ for which $\|T\|\,\|T^{-1}\| = d(\ell_p^n, \ell_q^n)$ is simply the formal identity operator. However, if p

and q are on different sides of 2 (e.g., $p = 1$, $q = \infty$) then there are operators T which are much better than the formal identity. Gurari, Kadec and Mačaev (1966) showed by using as T the Walsh-matrix (for $n = 2^m$) that

$$d(\ell_p^n, \ell_q^n) \approx \max(n^{1/p-1/2}, n^{1/2-1/q}), \quad \infty \geqslant q > 2 > p \geqslant 1. \tag{5.1.2}$$

As another example consider the n^2-dimensional spaces C_p^n of operator $U : \ell_2^n \to \ell_2^n$ with $\|U\|_p = \left(\operatorname{trace}(UU^*)^{p/2}\right)^{1/p}$. It was proved by Tomczak-Jaegermann (1978) that for all p and q

$$d(C_p^n, C_q^n) \approx d(\ell_p^n, \ell_q^n). \tag{5.1.3}$$

Again, the distance is smaller than expected if we consider only "obvious" isomorphisms.

A more general result on distances was obtained by Lewis (1979) who applied (2.3.15) to the ideal norm π_p to obtain

$$d_X \leqslant n^{|1/p-1/2|} \quad \text{if } X \subset L_p(0,1),\ \dim X = n,\ 1 \leqslant p \leqslant \infty. \tag{5.1.4}$$

Thus the subspaces of $L_p(0,1)$ which are farthest from Euclidean are the spaces ℓ_p^n. A result which contains (5.1.2) and generalizes (5.1.4) in somewhat less precise form (because of the constant involved) was obtained by Bourgain and Milman (1986)

$$\begin{gathered} d(X,Y) \leqslant C(p,q)\max(n^{1/p-1/2}, n^{1/2-1/q}), \\ X \subset L_p(0,1),\ Y \subset L_q(0,1),\ 1 \leqslant p \leqslant 2 \leqslant q \leqslant \infty. \end{gathered} \tag{5.1.5}$$

In connection with the result of John (2.1.4), it is of interest to study spaces X for which the largest distance from Hilbert space is achieved. It was proved by Milman and Wolfson (1978) that $d_X = \sqrt{n}$ ($n = \dim X$) implies that X contains isometrically copies of ℓ_1^k with $k \geqslant c \log n$ (the example $X = \ell_n^\infty$ shows that this result is optimal). It is obvious that there are many examples of spaces X of dimension n (besides ℓ_n^∞ and ℓ_n^1) so that $d_X = \sqrt{n}$. In view of this fact it is somewhat suprising that, in the non-symmetric case, the simplex is the unique convex set whose Banach–Mazur distance from the Euclidean ball is equal to the dimension n. This was shown by Palmon (1992).

The isomorphic version of the Milman–Wolfson result is more delicate. If $d_X \geqslant \alpha\sqrt{n}$ for some $\alpha \geqslant 0$ then X contains 2 isomorphic copies of ℓ_1^k with k a certain power of $\log n$. This result, with a weaker estimate, appears in the above mentioned paper of Milman and Wolfson. Their result was improved by several mathematicians (Kashin, Pisier, Bourgain) and the strongest known estimate appears in the book of Tomczak-Jaegermann (1989). It is still not clear what the situation is if we assume $d_X = o(\sqrt{n})$ but not too far from $\sqrt{n}$ (say $\sqrt{n}/\log n$). For instance the following problem is open. Do there exist $c(p,\lambda) < \infty$ and $\alpha(p,\lambda) > 0$ defined for $p > 1$ and $0 < \lambda < \infty$ so that

$$d_X \leqslant c\big(p, T_p(X)\big) \cdot n^{1/2-\alpha(p,T_p(X))}$$

for every n-dimensional Banach space X? Recall that the absence in X of 2-isomorphic copies of ℓ_1^k with large k is equivalent to the boundedness of $T_p(X)$ for some $p > 1$ (see Theorem 2.2.1, in view of it we could also formulate the problem using $K(X)$ instead of $T_p(X)$). The best known result in this direction seems to be

$$d_X \leqslant cT_p(X)C_q(X)n^{1/p-1/q}, \quad \dim X = n, \ 1 < p \leqslant 2 \leqslant q < \infty \tag{5.1.6}$$

for a suitable constant c. The result (5.1.6) is a combination of a result in Figiel, Lindenstrauss and Milman (1977) with (2.2.15). It was proved by Bourgain (1984b) that (5.1.6) gives a sharp estimate provided, of course, that $p^{-1} - q^{-1} < \frac{1}{2}$ (see also Tomczak-Jaegermann 1989, Proposition 27.5).

We saw in section 2 that if X is an n-dimensional Banach space then for a suitable k "most" k-dimensional subspaces of X are almost Euclidean. If however, all k-dimensional subspaces of X are, say, 2-Euclidean and k is sufficiently large X itself must be not far from a Hilbert space. More precisely it was proved in the Figiel, Lindenstrauss and Milman paper that for $2 \leqslant k < n < \infty$

$$d_X \leqslant C^{2(1+\log n/\log k)}, \quad \dim X = n, \ d_Y \leqslant C \text{ for all } Y \subset X \text{ with } \dim Y = k. \tag{5.1.7}$$

In particular if $k = [n^s]$ for some $0 < s < 1$ then $d_X \leqslant f(s, c)$ for a suitable function f.

The preceding result is related to a question which was formulated already in Banach's classical book. Given $1 < k < n$ and assume that X is a Banach space of dimension n so that all k-dimensional subspaces of X are mutually isometric. Is X isometric to a Hilbert space? This question was treated by Gromov (1967) by topological methods. He gave a positive answer to this question for most pairs k and n. The only case which is still open is n an even number and $k = n - 1$.

An isomorphic version of this result was investigated by Bourgain (1988). He proved that there is an $0 < \alpha < 1$ so that if dim $X = n$ and $d(Y, Z) \leqslant C$ whenever $Y, Z \subset X$ with dim Y = dim $Z = [\alpha n]$ then $d_X \leqslant \varphi(C)$ for suitable function φ. This result of Bourgain was strengthened by Mankiewicz and Tomczak-Jaegermann (1991) who proved that instead of saying above "there is an α" it is possible to say for "every $0 < \alpha < 1$" (of course in this case φ has to depend on C and α). The papers of Bourgain and of Mankiewicz and Tomczak-Jaegermann use much of the machinery of local Banach space theory which we outlined in sections 3 and 4 as well as the Gluskin technique which will be explained in section 5.2 below.

We turn now to the discussion of projections. There is a close analogy between some problems concerning the estimation of projection constants and problems concerning evaluation of Banach–Mazur distances. We define first the projection constant $\lambda(X)$ of a finite-dimensional Banach space X,

$$\lambda(X) = \min\{\lambda\colon \text{ for all } Y \supset X \text{ there is a projection of norm} \leqslant \lambda \text{ from } Y \text{ on } X\}. \tag{5.1.8}$$

Recall that by applying the Hahn–Banach theorem coordinatewise we get that for all pairs of Banach spaces $Y \supset X$ and every operator $T: X \to \ell_\infty^m$ there is an extension $\widehat{T}$ of T from Y into ℓ_∞^m with $\|\widehat{T}\| = \|T\|$. From this remark it follows trivially that $\lambda(\ell_\infty^n) = 1$ for all n and that whenever $X \subset \ell_\infty^m$ for some m (i.e., the unit ball of X is polyhedral) then $\lambda(X)$ is the minimal norm of a projection from ℓ_∞^m onto X. If X is a general finite-dimensional space one gets similarly that $\lambda(X)$ is the minimal norm of a projection from any $C(K)$ space containing X isometrically onto X. This simplifies the computation of $\lambda(X)$ since we do not have to consider all possible embeddings of X into general spaces Y.

The spaces ℓ_∞^n are the only finite-dimensional spaces whose projection constant is 1. This was proved, independently, by Goodner (1950) and Nachbin (1950) as a special case of an infinite-dimensional result. A very simple direct proof of the finite-dimensional assertion is given by Zippin (1981a). It is an interesting open problem whether the isomorphic version of this result is true. More precisely:

$$\text{does there exist a function } f(t) \text{ on } [1,\infty) \text{ so that } d(X,\ell_\infty^n) \leqslant f\big(\lambda(X)\big)? \tag{5.1.9}$$

There are some known partial results concerning this question. It was observed by Lindenstrauss and Pelczynski (1968), by using (2.3.11), that the answer is yes if one assumes that X has a γ-unconditional basis with bounded γ. In other words there is a function g so that $d(X,\ell_\infty^n) \leqslant g\big(\lambda(X),\gamma\big)$. It was proved by Bourgain (1981a) that there is a function $\varphi(\lambda)$ so that if $\dim X = n$ then X has a $k = \big[\varphi\big(\lambda(X)\big)^{-1} n\big]$-dimensional subspace Y with $d(Y,\ell_\infty^k) \leqslant \varphi\big(\lambda(X)\big)$. The deepest known result here is the fact that there is an $\varepsilon_0 > 0$ so that f satisfying (5.1.9) exists on the interval $[1, 1+\varepsilon_0]$. This is due to Zippin (1981a,b).

The result of John on maximal volume ellipsoids has also a consequence concerning projection constants. We recall that by John's result $\pi_2(I_X) = \sqrt{n}$ where I_X is the identity operator on X ($\dim X = n$, cf. (4.1.4)). By the Pietsch factorization theorem (cf. (2.3.6)) there is an operator T from $C(K_{X^*})$ into X so that $T \cdot i = I_X$ where $i: X \to C(K_{X^*})$ is the canonical isometry and $\|T\| \leqslant \pi_2(I_X) = \sqrt{n}$. Hence it follows from the remarks made above that

$$\lambda(X) \leqslant \sqrt{n}, \quad \dim X = n. \tag{5.1.10}$$

This inequality was first pointed out by Kadec and Snobar (1971) who deduced it directly from John's theorem 4.1.1 (without using the π_2 norm). It was noted in König and Lewis (1987) that the inequality in (5.1.10) is always strict. Later Lewis (1988) showed that $\lambda(X) \leqslant \sqrt{n} - c \cdot 5^{-n}$ for some positive constant c. Finally, it was shown by König and Tomczak-Jaegermann (1990) that for some $c > 0$

$$\lambda(X) \leqslant \sqrt{n} - \frac{c}{\sqrt{n}}, \quad \dim X = n. \tag{5.1.11}$$

Up to the value of c (5.1.11) is sharp; there are spaces X with $\dim X = n$ so that $\lambda(X) \geqslant \sqrt{n} - (1/2\sqrt{n})$ (this is the case for complex scalars, for real scalars known examples yield only $\lambda(X) \geqslant \sqrt{n}-2$). The proofs in the paper of König and Tomczak-Jaegermann involve computations with spherical functions. It also involves estimations of π_1 summing norms. That the notion of $\lambda(X)$ is closely related to π_1 summing norms of operators follows from the relation $\pi_1^* = \gamma_\infty$ (see the end of section 5.2 below).

The constants $\lambda(\ell_1^n)$ were calculated precisely by Grünbaum (1960). He also estimated from above $\lambda(\ell_2^n)$ and his estimate was shown to be precise by Rutovitz (1965). From these computations, the trivial fact that $\lambda(X) \leqslant d(X,Z)\lambda(Z)$ (in particular $\lambda(X) \leqslant d(X,\ell_n^\infty)$, $\dim X = n$) (5.1.1), (5.1.2), (5.1.10) and a result of Gordon (1968), it follows that

$$\lambda(\ell_p^n) \approx d(\ell_p^n, \ell_\infty^n), \quad 1 \leqslant p \leqslant \infty. \tag{5.1.12}$$

Note that, in particular, $\lambda(\ell_p^n) \approx \sqrt{n}$ for $1 \leqslant p \leqslant 2$, and thus $\lambda(X)$ may be asymptotically of the maximal order even for X which do not contain copies of ℓ_1^k spaces. Nevertheless, there is a version of the Milman and Wolfson result for projections. It was shown independently by Milman and Pisier (see Pisier 1978) that if $\dim X = n$ and if $Y \supset X$ is such that the smallest norm of a projection from Y onto X is $\geqslant \alpha\sqrt{n}$ then Y has to contain 2-isomorphic copies of ℓ_1^k for large k.

Just as (5.1.10) follows from John's theorem also the proof of the result (5.1.4) of Lewis on distances gives information on projections as well. On any n-dimensional subspace X of $L_p(0,1)$ there is a projection P with $\|P\| \leqslant n^{|(1/p)-(1/2)|}$, $1 \leqslant p \leqslant \infty$ (for $p = \infty$ this is just (5.1.10)).

We turn next to the question of existence of good projections from a Banach space on its subspaces. In a Hilbert space X there is a projection of norm 1 on any closed subspace and the same is true (by the Hahn–Banach theorem) for every 2-dimensional space X. If X is a Banach space of dimension $\geqslant 3$ such that there is a projection of norm 1 on any closed subspace, then X is isometric to a Hilbert space. This theorem, which reduces trivially to the case $\dim X = 3$, was stated first by Kakutani (1939). Kakutani translated the assumption on X to a geometric property of K_X and noted that it was shown earlier by Blaschke that this property characterizes ellipsoids. However, Blaschke made a smoothness assumption which cannot be trivially eliminated. The first complete proof of the theorem was given a year later by Phillips (1940). For a detailed discussion of this isometric characterization of Hilbert space as well as further references we refer to the book of Amir (1986).

An isomorphic version of the preceding characterization of Hilbert space was obtained by Lindenstrauss and Tzafriri (1971).

Theorem 5.1.1. *Let X be a Banach space such that there is a bounded linear projection from X onto any of its closed subspace. Then X is isomorphic to a Hilbert space.*

A uniform boundedness argument (cf. Davis, Dean and Singer 1968) shows that the assumption implies the existence of a $\lambda < \infty$ so that there is a projection of norm $\leqslant \lambda$ from X onto any of its finite-dimensional subspaces. A compactness argument (cf. Lindenstrauss 1963) shows that if $d(Y,\ell_2^n) \leqslant \mu$ for every n-dimensional subspace of Y of X, $n = 2,3,\ldots$, for some μ then the distance of X itself from a Hilbert space is $\leqslant \mu$. Hence the main point in the proof of Theorem 5.1.1 is to show that if there is a projection of norm $\leqslant \lambda$ from X onto any of its finite-dimensional subspaces then $d(Y,\ell_2^n) \leqslant \varphi(\lambda)$ for a suitable function φ and every finite-dimensional subspace Y of X. This is done as follows. Let $Y \subset X$ with $\dim Y = n$ and put $\alpha = d(Y,\ell_2^n)$. There is a projection Q from X onto Y with $\|Q\| \leqslant \lambda$. By Theorem 1.1 (Dvoretzky's theorem) there is a subspace $Z \subset (I-Q)X$ with $d(Z,\ell_2^n) \leqslant 2$. Let T be an operator from Y onto Z with $\|T\| \leqslant 1$ and $\|T^{-1}\| \leqslant 2\alpha$. By the assumption there is a projection P of norm $\leqslant \lambda$ from X onto its subspace $\{y + c\lambda^2 Ty\colon y \in Y\}$, where c is a constant to be chosen later. By doing some elementary linear algebra one produces from P, Q and T an operator S from ℓ_2^n onto Y so that $\|S\| \cdot \|S^{-1}\| < \alpha$ unless $\alpha \leqslant c_1\lambda^2$ provided c and c_1 are chosen properly. This proves the desired result with $\varphi(\lambda) = c_1\lambda^2$. A "local" version of Theorem 5.1.1 is proved in Figiel, Lindenstrauss and Milman (1977).

It follows from Theorem 5.1.1 that in any space not close to a Hilbert space there are subspaces on which there are no "good" (i.e., with small norm) projections. There are many known concrete examples where the lack of good projections on specific subspaces can be verified. We mention just one such example which is quite surprising. It was proved by Bourgain (1981b) by using harmonic analysis that there is a λ_0 and subspaces Y_n of ℓ_1^n with $d(Y_n,\ell_1^{k_n}) \leqslant \lambda_0$ so that the norm of the best projection from ℓ_1^n onto Y_n tends to ∞ with n. This example shows how different the isomorphic situation can be from the isometric one (or even the almost isometric one). It is trivial to verify that if $T: \ell_1^k \to \ell_1^n$ is an isometry (into) then there is a projection of norm 1 from ℓ_1^n onto $T\ell_1^k$. It was proved by Dor (1975) that even if T is just an isomorphism with $d(T\ell_1^k,\ell_1^k) = \lambda < \sqrt{2}$ then there is a projection from ℓ_1^n onto $T\ell_1^k$ with norm $\leqslant \varphi(\lambda)$ for a suitable $\varphi(\lambda)$ (which tends to 1 as $\lambda \to 1$).

Nevertheless, there are also general results which ensure existence of good projections in non-Euclidean spaces on certain special subspaces (especially Euclidean subspaces). It was shown by Kadec and Pelczynski (1962) that any subspace of $L_p(0,1)$, $2 < p < \infty$ which is isomorphic to a Hilbert space is complemented. This result was generalized and also put in a quantitative form by Maurey (1974a). He proved that if $Y \supset X$ with Y of type 2 and X isomorphic to a Hilbert space then there is a projection P from Y onto X with $\|P\| \leqslant cT_2(Y)d_X$. Even if X is isometric to a Hilbert space the projection cannot however have a too small norm. If I_n denotes the identity operator in ℓ_2^n then $\gamma_p(I_n)$ (see (2.3.4) for the definition of γ_p) was evaluated by Gordon, Lewis and Retherford (1973). They showed that $\gamma_p(I_n) \sim \sqrt{p}$ as $n \to \infty$ (if $p > 2$; $\sqrt{p'}$ if $1 < p < 2$).

From the result of Figiel, Lindenstrauss and Milman mentioned after (3.1.17) and the Figiel and Tomczak-Jaegermann result (4.1.10) it follows that there is a $c > 0$ and a function $\varphi(\lambda)$ so that if $\dim X = n$ there is a subspace Y of X of dimension $k \geqslant c\log n$ with $d(Y,\ell_2^k) \leqslant 2$ and a projection P from X onto Y with $\|P\| \leqslant \varphi\big(K(X)\big)$. In general (i.e., without a K convexity assumption) it is unknown whether there is a $\lambda < \infty$ and a function $k(n)$ tending to ∞ with n so that any space X with $\dim X = n$ has a $k(n)$-dimensional subspace on which there is a projection of norm $\leqslant \lambda$.

If we assume that X has a 1-unconditional basis the preceding question becomes meaningless; by definition X is rich with good projections. However, the following result of Tzafriri (1974) is of interest (and may be valid without the assumption on the basis). For every sequence $\{X_n\}_{n=1}^{\infty}$ of normed spaces with $\dim X_n = n$ and a 1-unconditional basis there is a $\lambda < \infty$ and a p equal to either 1,2 or ∞ so that the following holds. For every integer k there is an $n = n(k)$ and an operator $T : \ell_p^k \to X_n$ with $d(\ell_p^k, T\ell_p^k) \leqslant \lambda$ and so that there is a projection of norm $\leqslant \lambda$ from X_n onto $T\ell_p^k$.

5.2. *Random normed spaces and existence of bases*

The set of n-dimensional spaces X over the reals with the (as usual multiplicative) Banach–Mazur distance is clearly a compact contractible metric space $\mathcal{K}_n$ called the Banach–Mazur compactum. $\mathcal{K}_n$ can, of course, also be considered as the space of all equivalence classes of symmetric convex bodies in $\mathbb{R}^n$ (two such sets are in the same class iff they are affinely equivalent). The study of the geometric structure of $\mathcal{K}_n$ leads to many natural questions, several of which are still open. Already estimating the diameter of $\mathcal{K}_n$ is by no means easy. It follows from John's result (2.1.4) that for all $X, Y \in \mathcal{K}_n$

$$d(X,Y) \leqslant d(X,\ell_2^n)d(\ell_2^n,Y) \leqslant n. \tag{5.2.1}$$

How much can (5.2.1) be improved? It happens that it is very difficult to find examples in $X, Y \in \mathcal{K}_n$ such that $d(X,Y)$ is much larger than $\sqrt{n}$, and therefore the problem was open for a long time. Gluskin (1981a) settled this question by proving the following.

Theorem 5.2.1. *There is an absolute positive constant c so that*

$$\operatorname{diameter} \mathcal{K}_n \geqslant cn. \tag{5.2.2}$$

Gluskin did not exhibit explicitly spaces $X_n, Y_n \in \mathcal{K}_n$ so that $d(X_n, Y_n) \geqslant cn$. He established their existence by probabilistic means. It is instructive to follow the reasoning which led Gluskin to his result. Assume that $X = Y = \ell_1^n$ and consider $T_\omega : X \to Y$ operators given by matrices whose n^2 entries are mutually independent standard Gaussian variables on some probability space. Such operators have long been considered in local theory and it is simple and well known how to compute the distribution of $\|T_\omega\|$ (see also the next subsection). It turns out that with high probability $\|T_\omega\|\|T_\omega^{-1}\| \geqslant cn$. This means that "computed randomly" the distance from ℓ_1^n to itself is cn ! This led Gluskin to the idea of creating a "random ℓ_1^n" so that the actual distance between such two random spaces is (with high probability)

$\geqslant cn$. The "random ℓ_1^n" Gluskin chose to consider were the convex hulls of $2n$ randomly chosen pairs of antipodal points on S^{n-1}. Observe that the convex hull of n such pairs gives the unit ball of ℓ_1^n itself. For some technical reasons it is more convenient to add to the $2n$ random pairs the extreme points of the standard ℓ_1^n ball. Thus one usually takes the unit ball of the Gluskin spaces to be

$$\operatorname{Conv}\{K_{\ell_1^n}, \pm x_j,\ 1 \leqslant j \leqslant 2n\}, \tag{5.2.3}$$

where the $\{x_j\}_{j=1}^{2n}$ are chosen on S^{n-1} independently of each other with the distribution given by the rotation invariant measure on S^{n-1}. The verification that with high probability the distance between two "Gluskin spaces" is $\geqslant cn$ is ingenious and involves volume estimates and precise estimates on the size of ε-nets in spaces of operators. The basic idea is to estimate for a fixed $T \in \mathrm{SL}_n$ the quantity

$$\eta(T) = \operatorname{Prob}\{(X_n, Y_n)\colon X_n, Y_n \text{ Gluskin spaces}, \|T\|_{X_n \to Y_n} \cdot \|T^{-1}\|_{Y_n \to X_n} \leqslant cn\}.$$

The estimate one gets is independent of T and (for a suitable $c > 0$) is so small that even if we multiply it by the cardinality of a very fine net in SL_n we get a small number. It is then easy to show that if X_n and Y_n are such that for all T in the net $\|T\|_{X_n \to Y_n} \cdot \|T^{-1}\|_{Y_n \to X_n} \geqslant cn$ then $d(X_n, Y_n) \geqslant cn/2$.

The fact that we took exactly $2n$ pairs $\{\pm x_j\}$ on S^{n-1} is not essential. For various purposes one varies the number of points chosen randomly. It is possible to take fewer (say $n/2$) pairs or more (say n^k with fixed k) pairs. The paper of Gluskin (1981a) led very quickly to the appearance of many papers in which its methods were streamlined and extended (see, e.g., Szarek 1983, 1986, Mankiewicz 1984, 1988) and applied to the solution of many other open problems.

The following two results were proved by Szarek (1986):

(i) There exists an $X \in \mathcal{K}_{2n}$ so that $d(X, Y) \geqslant c\sqrt{n}$ for every $Y \in \mathcal{K}_{2n}$ which admits also multiplication with complex scalars (i.e., Y is an n-dimensional complex space considered as a $2n$-dimensional real space).

(ii) There is a complex n-dimensional Y so that $d_{\mathbb{C}}(Y, \overline{Y}) \geqslant cn$ where $\overline{Y}$ is the same space as Y but with complex multiplication $\odot$ defined by $\lambda \odot y = \bar{\lambda} y$, and $d_{\mathbb{C}}$ denotes the Banach–Mazur distance with respect to transformation in $\mathrm{GL}_n(\mathbb{C})$. Note that as real spaces Y and $\overline{Y}$ are isometric.

A result similar to (ii) was proved independently by Bourgain (1986b). He used the Gluskin method and a glueing technique to produce an infinite-dimensional separable complex Y so that Y is not isomorphic to $\overline{Y}$.

Some further interesting consequences of Gluskin's method involve the notion of a basis. But first we need a definition. Let $\{x_i\}_{i=1}^n$ be an algebraic basis of some $X \in \mathcal{K}_n$. The basis constant of $\{x_i\}_{i=1}^n$ is defined to be

$$\max_{1 \leqslant k \leqslant n} \|P_k\|, \quad \text{where } P_k\left(\sum_{i=1}^n \lambda_i x_i\right) = \sum_{i=1}^k \lambda_i x_i. \tag{5.2.4}$$

Note that the order of $\{x_i\}_{i=1}^n$ is important here and that the basis constant is less or equal to the unconditional constant of the basis defined in (2.1.8). The unconditional constant of a basis is up to a factor 2 equal to $\max_\sigma \|P_\sigma\|$ where $P_\sigma\left(\sum_{i=1}^n \lambda_i x_i\right) = \sum_{i\in\sigma} \lambda_i x_i$ and the maximum is taken over all the 2^n subsets σ of $\{1,\dots,n\}$. Trivial examples show that the unconditional constant of a basis can be significantly larger than the basis constant.

In the setting of infinite-dimensional separable spaces X there is a notion of a Schauder basis. This is defined to be a sequence $\{x_i\}_{i=1}^\infty$ of non-zero vectors which spans X and for which the basis constant (defined in obvious analogy to (5.2.4) as $\sup_{1\leqslant k<\infty} \|P_k\|$) is finite. A famous example of Enflo (1973) shows that a separable space need not have a Schauder basis. (For a discussion of Schauder bases, Enflo's example and related notions we refer to Lindenstrauss and Tzafriri 1977, 1979.) Enflo's example does not however settle the question of existence of good bases in finite-dimensional spaces. This question was settled first using Gluskin's technique by Gluskin (1981b) himself and independently by Szarek (1983). They showed that

$$\exists X \in \mathcal{K}_n \text{ such that any basis in } X \text{ has a basis constant } \geqslant c\sqrt{n}, \tag{5.2.5}$$

where c is an absolute constant (Gluskin's estimate was slightly weaker: $c\sqrt{n/\log n}$). Note that by John's result (2.1.4) any space in $\mathcal{K}_n$ has a basis with basis constant $\leqslant n^{1/2}$. The proof of (5.2.5) consists of showing that Gluskin's method can be used to produce a space X of dimension n so that any projection on a subspace of dimension about $n/2$ has norm $\geqslant c\sqrt{n}$. (This fact is also relevant in connection with the problem discussed at the end of the previous subsection.)

It is worthwhile to mention that using Gluskin's method and variations on the glueing technique introduced by Bourgain, Szarek (1987) was able to construct a separable infinite-dimensional space which has the so-called bounded approximation property but still fails to have a Schauder basis.

Since the unconditional constant of a basis may be much larger than the basis constant it is natural that it is much easier to construct finite-dimensional spaces which fail to have a good unconditional basis than to prove (5.2.5). Spaces failing to have a good unconditional basis were constructed much earlier and in fact nice concrete spaces fail to have such a basis. This fact was proved first in Gordon and Lewis (1973). They noticed that it follows trivially from the definition of an absolutely summing norm that if $X \in \mathcal{K}_n$ has a λ-unconditional basis then for every $T: X \to Y$ there is a factorization $T = U \cdot V$, $V: X \to \ell_1^n$, $U: \ell_1^n \to Y$ with $\|U\|\,\|V\| \leqslant \lambda\pi_1(T)$. In other words (in the terminology of (2.3.4)) we get that $\gamma_1(T) \leqslant \lambda\pi_1(T)$. This led them to define a constant now called the Gordon and Lewis constant of X

$$\mathrm{gl}(X) = \sup\{\pi_1(T)/\gamma_1(T)\colon\ T: X \to \ell_2\}. \tag{5.2.6}$$

Thus $\mathrm{gl}(X)$ is at most the unconditional constant of any basis of X and is also clearly bounded by the distance of X to a subspace of $L_1(0,1)$. Gordon and Lewis verified that if C_p^n is the space of operators on ℓ_2^n which was mentioned in (5.1.3) then $\mathrm{gl}(C_p^n) \approx d_{C_p^n} \approx d_{\ell_p^n}$, $1 \leqslant p \leqslant \infty$. In particular, the unconditional constant of any basis of the n^2-dimensional space C_∞^n is at least of order $n^{1/2} = (\dim C_\infty^n)^{1/4}$. A direct proof of this fact was later given by Schütt (1978) who showed that in C_p^n (and many other spaces of operators) the basis with the smallest unconditional constant is the natural one consisting of matrices having only one non-zero entry. Spaces X for which $\mathrm{gl}(X)$ is of order $(\dim X)^{1/2}$ (obviously the largest possible) were constructed by Figiel, Kwapien and Pelczynski (1977) using a random method (see also Figiel and Johnson 1980).

Finally, we discuss yet another notion. This is a very natural notion from the algebraic (as well as geometric) point of view and is related to the concept of a symmetric basis. We call a subgroup G of GL_n rich if

$$T \in \mathrm{GL}_n; \qquad TS = ST \text{ for all } S \in G \Rightarrow T = \lambda I, \tag{5.2.7}$$

where I is the identity matrix. A normed space $X \in \mathcal{K}_n$ is said to have enough symmetries if it has a rich group of isometries. Since the group generated by the permutations of the coordinates and reflections with respect to the coordinate hyperplanes is evidently rich, it is clear that a space with a 1-symmetric basis (see (2.1.9)) has enough symmetries. It is also easy to verify that the spaces C_p^n, $1 \leqslant p \leqslant \infty$, have enough symmetries (though as stated above they fail even to have a good unconditional basis for $p \neq 2$). For a general $X \in \mathcal{K}_n$, Garling and Gordon (1971) defined the symmetry constant of X as follows:

$$\mathrm{sym}(X) = \inf\{d(X, Y) \colon Y \in \mathcal{K}_n \text{ with enough symmetries}\} \tag{5.2.8}$$

or, clearly equivalently,

$$\mathrm{sym}(X) = \inf\{\lambda\colon \exists G \subset \mathrm{SL}_n, G \text{ rich}, \|T\| \leqslant \lambda \text{ for all } T \in G\}. \tag{5.2.9}$$

The norm appearing in (5.2.9) is, of course, the norm of operators from X to itself. By the remarks above the symmetry constant of a space is less or equal to the symmetric basis constant of any basis in X. Spaces with a good unconditional basis need not, however, have a small symmetry constant. This was shown by Garling and Gordon. It is perhaps worthwhile to recall here their simple argument since it exhibits again the convenience of working with ideal norms and clarifies how the symmetry concept is often used.

By the duality relation $\pi_1^* = \gamma_\infty$ (see (2.3.14)) it follows immediately from the relevant definitions that for every $X \in \mathcal{K}_n$ we have

$$\lambda(X)^{-1} = \inf\{\pi_1(T)\colon\ T : X \to X,\ \mathrm{trace}\, T = 1\}. \tag{5.2.10}$$

By taking $T = n^{-1} I_X$ where I_X is the identity operator of X in (5.2.10) we get

$$n \leqslant \lambda(X)\pi_1(I_X). \tag{5.2.11}$$

We mention in passing that in view of the existence of an Auerbach basis (cf. (2.2.6)) $\pi_1(I_X) \leqslant n$ for every X with $\dim X = n$. It was proved by Deschaseaux

(1973) and Garling (1974) that $\pi_1(I_X) = n$ if and only if $X = \ell_\infty^n$. In view of (5.2.11) this is a stronger statement than the Goodner and Nachbin result, mentioned in section 5.1 above, that $X = \ell_\infty^n$ is characterized by $\lambda(X) = 1$.

Assume now that there is a rich group G of operators on X of norm $\leqslant \alpha$. We may clearly assume that G is compact (usually it is even finite) and let μ be its normalized Haar measure. Let $T\colon X \to X$ be any operator with trace 1. By (5.2.7) we get that $n^{-1}I_X = \int_G STS^{-1}\,\mathrm{d}\mu(S)$ and hence $\pi_1(I_X) \leqslant n\alpha^2\pi_1(T)$. Consequently

$$\lambda(X) \cdot \pi_1(I_X) \leqslant n(\mathrm{sym}(X)^2). \tag{5.2.12}$$

For spaces with enough symmetries (5.2.11) and (5.2.12) show that $\lambda(X)\pi_1(I) = n$ and this gives a convenient way to calculate $\lambda(\ell_p^n)$ (see (5.1.12)). Also, it is easy to deduce from (5.2.12) that the $2n$-dimensional space (with a 1-unconditional basis) $\ell_\infty^n \oplus \ell_2^n$ has a symmetry constant of order $n^{1/4}$ (and both sides of (5.2.12) are of order $n^{3/2}$ for this X).

Examples of spaces X with $\mathrm{sym}(X)$ of the order of $(\dim X)^{1/2}$ (clearly the largest possible value) were first constructed by Mankiewicz (1984), using the Gluskin method.

5.3. *Random operators and the distances between spaces of some special families*

Many results and computations in local theory are done by using probabilistic techniques. We have seen this, for example, in the discussion of the embedding theorems in section 3 and in several other places. Also many of the computations of distances are done via probabilistic arguments. Such computations in rather general situations were carried out before Gluskin introduced his method for constructing examples (or perhaps one should say "counterexamples") which was explained in the previous subsection. It seems that the first use of probabilistic arguments in computing distances was that done by Tomczak-Jaegerman (1978) while evaluating the distance between the spaces C_p^n (see (5.1.3)). Her paper was followed by the papers Davis, Milman and Tomczak-Jaegermann (1981), and Benyamini and Gordon (1981) in which a more systematic study was carried out and some general classes of spaces were considered. The general procedure is the following. Let $X, Y \in \mathcal{K}_n$. We first map in a suitable way X and Y onto $\mathbb{R}^n$, i.e., choose convenient coordinates in X and Y (or in the terminology of section 4 choose suitable positions for K_X and K_Y). Once this is done we compare TK_X with K_Y where T ranges over the orthogonal group SO_n. More precisely we compute the averages of $\|T\|_{X\to Y}$ and $\|T^{-1}\|_{Y\to X}$ with respect to normalized Haar measure on SO_n. The product of these expectations often gives a much better estimate for the distance than can be obtained by explicit choices of T. For the computation of the average over the orthogonal group it is often useful to use an inequality of Marcus and Pisier (1981) which allows one to pass from averages over SO_n to the average of norms of matrices whose entries are independent standard Gaussians. The average of the norm of Gaussian matrices is estimated by using an inequality of Chevet (1978) which is a consequence of a basic result concerning Gaussian variables, called Slepian's lemma.

We mention now some results obtained by the technique outlined above. In both papers mentioned above the following is proved

$$d(X,Y) \leqslant c\sqrt{n}\,(T_2(X)+T_2(Y^*)) \tag{5.3.1}$$

whenever $X, Y \in \mathcal{K}_n$ and c is a universal constant. By using a better Euclidean structure (namely the one given by (4.1.13)) Bourgain and Milman (1986) improved on (5.3.1) by showing that

$$d(X,Y) \leqslant c\,(d_X \cdot T_2(Y^*) + d_Y \cdot T_2(X)) \tag{5.3.2}$$

for every $X, Y \in \mathcal{K}_n$. For Y a subspace of $L_p(0,1)$, $1 < p \leqslant 2$, and X a subspace of $L_q(0,1)$, $2 \leqslant q < \infty$, (5.3.2) gives formula (5.1.5) which was mentioned in section 5.1. Another result of Bourgain and Milman in the same paper is

$$d(X,X^*) \leqslant \gamma(n) d_X^{5/3} \leqslant \gamma(n) n^{5/6}, \tag{5.3.3}$$

where $\gamma(n)$ denotes a factor of the form $c(\log n)^a$ (such factors will appear often below and we shall use the same notation throughout this subsection. The constants c and a may be different at different places). It may very well be that the constant $\frac{5}{6}$ in (5.3.3) can be reduced to $\frac{1}{2}$. This is also the case with other formulas below where an exponent strictly between $\frac{1}{2}$ and 1 appears. We know of no formula of a general nature (like (5.3.3)) where the best exponent is strictly between $\frac{1}{2}$ and 1.

Among other results obtained by the same method we mention the following (cf. Davis, Milman and Tomczak-Jaegermann 1981) which has an obvious geometric flavour. Assume that K_X and K_{Y^*} have "few" (say n^a where $1 < n = \dim X = \dim Y$) extreme points. Then

$$d(X,Y) \leqslant c(a)(n \log n)^{1/2}. \tag{5.3.4}$$

Note the interesting fact that if we assume that K_Y, rather than K_{Y^*}, has few extreme points then (5.3.4) no longer holds in view of Gluskin's spaces.

Benyamini and Gordon (1981) did not restrict themselves to computations of distances. They considered factorizations of an arbitrary operator $T: X \to Z$ through another space Y and computed the quantity

$$\gamma_Y(T) = \inf\{\|U\|\,\|V\|:\ U: X \to Y,\ V: Y \to Z,\ T = VU\},$$

by the same method of using random U and V. Note that if $X = Z$, $T =$ identity and $\dim X = \dim Y$ then $\gamma_Y(T) = d(X,Y)$, and also that for T, X and Z general and $Y = L_p(0,1)$ the norm $\gamma_Y(T)$ is the ideal norm $\gamma_p(T)$ (2.3.4). The flavor of their results can be gotten from the following example:

$$\gamma_{\ell_q^m}(Id_{\ell_p^n}) \leqslant c(n^{1/p-1/2} + n^{1/2} m^{-1/q}), \quad 1 \leqslant p \leqslant 2 \leqslant q,\ n \leqslant m. \tag{5.3.5}$$

Much work was done recently on estimating the maximal distance from a general $X \in \mathcal{K}^n$ to ℓ_1^n (or equivalently via duality to ℓ_∞^n). This problem which is of obvious geometric interest turns out to be surprisingly hard. The methods used for treating this problem are also (at least in part) probabilistic in nature but much more sophisticated than those used for the results mentioned earlier in this subsection. Bourgain and Szarek (1988) proved that

$$\max\{d(X,\ell_\infty^n)\colon X \in \mathcal{K}_n\} = \mathrm{o}(n). \tag{5.3.6}$$

For proving this they proved the strong version of the Dvoretzky and Rogers lemma (see (4.1.14)). Szarek and Talagrand, using a new combinatorial-geometric technique improved on (5.3.6) and showed that

$$d(X,\ell_1^n) \leqslant cn^{7/8}, \quad X \in \mathcal{K}_n \tag{5.3.7}$$

On the other hand by using ideas related to Wigner's semicircle law, as well as Gluskin's method, Szarek (1990) proved that

$$\max\{d(X,\ell_1^n)\colon X \in \mathcal{K}_n\} \geqslant c\sqrt{n}\log n\ . \tag{5.3.8}$$

Note that (5.3.8) gives the first example of a pair of spaces of which one of them is concrete so that their distance is essentially larger than $\sqrt{n}$. In connection with (5.3.8) it is interesting to record the following, somewhat surprising, result from Bourgain and Szarek (1988),

$$\max\{d(X,\ell_1^n \oplus \ell_2^n)\colon X \in \mathcal{K}_{2n}\} \leqslant c\sqrt{n}. \tag{5.3.9}$$

A more detailed survey concerning the $d(X,\ell_1^n)$ problem is given in Szarek (1991b).

We turn now to estimates on the distances $d(X,Y)$ where both X and Y belong to special classes of spaces. It was proved independently by Gluskin (1979) and Tomczak-Jaegermann (1979b) that if X and Y are n-dimensional spaces with 1-symmetric bases then $d(X,Y) \leqslant \gamma(n)\sqrt{n}$. The proof of this result is similar to the method outlined in the beginning of this subsection (in this case the choice of coordinates is obvious) with one difference. One does not use random matrices in SO_n but rather divides the coordinates into k sets of n/k elements each and use "block random matrices" consisting of k blocks on the main diagonal of $\frac{n}{k} \times \frac{n}{k}$ random orthogonal matrices (i.e., elements of $\mathrm{SO}_{n/k} \times \mathrm{SO}_{n/k} \times \cdots \times \mathrm{SO}_{n/k}$ with k factors; if n/k is not an integer we replace it by $[n/k]$ and add another random block of an appropriate size). Note that for $k = n$ the block random matrices reduce to the identity matrix. By using the Chevet estimate and an appropriate choice of $k = k(n)$ (depending on X and Y) the result of Gluskin and Tomczak-Jaegermann follows. In results of this type it is often very tricky to remove the "small" factor $\gamma(n)$ if that is at all possible. In the present case by using ingenious deterministic changes of the random matrices Tomczak-Jaegermann (1983) was able to prove the following.

Theorem 5.3.1. *Let X and Y be n-dimensional spaces with a* 1*-symmetric basis. Then*

$$d(X,Y) \leqslant c\sqrt{n} \tag{5.3.10}$$

($c = 2^{18}$ will do).

In spite of the substantial concrete information which has already been accumulated about calculating distances our understanding of this subject is far from satisfactory even among spaces with 1-symmetric bases.

The class of spaces with a 1-symmetric basis can be parametrized in a nice way and this allows the formulation of results and problems on these spaces in a concrete analytic form. Let $\{x_i\}_{i=1}^n$ be a 1-symmetric basis of X, normalized so that $\|x_i\| = 1$. Assume we know the function

$$\varphi_X(k) = \left\| \sum_{i=1}^k x_i \right\|, \quad 1 \leqslant k \leqslant n. \tag{5.3.11}$$

Then for any $x = \sum_{i=1}^k \lambda_i x_i$ in X we can calculate $\|x\|$ up to a factor of size $2\log n$. In order to parametrize in a convenient way the functions φ_X of (5.3.11) we introduce a class of functions on $[0,1]$

$$\mathscr{L} = \{f\colon [0,1] \to R,\ f(0) = 0,\ |f(s) - f(t)| \leqslant \tfrac{1}{2}(t-s),\ 0 \leqslant s < t \leqslant 1\}. \tag{5.3.12}$$

One easily checks that for every $f \in \mathscr{L}$ there is an X with a 1-symmetric basis so that

$$\varphi_X(k) = n^{f(\log k/\log n)} \cdot k^{1/2} \tag{5.3.13}$$

and conversely, every such X defines by this formula a function at the points $\log k/\log n$, $1 \leqslant k \leqslant n$ which can be extended to an $f \in \mathscr{L}$ (whose value at every t is determined up to an additive term of size $c/\log n$). In other words (5.3.13) defines an equivalence between the (dimension-free) set $\mathscr{L}$ and the 1-symmetric bases of length n which is essentially (i.e., up to $\sim \log n$) one to one for every n. In this correspondence the spaces ℓ_p^n correspond to the linear functions in $\mathscr{L}$ (with ℓ_2^n corresponding to $f \equiv 0$).

If f and g are both monotone (either increasing or decreasing) functions in $\mathscr{L}$ the distance between the corresponding n-dimensional 1-symmetric spaces was calculated by Gluskin (1983). It is shown that up to $\gamma(n)$ factors the distance is given by $n^{\beta(f,g)}$ where $\beta(f,g)$ is a suitable constant. The distance in this case is obtained by using the block random matrices mentioned above (cf. also Lindenstrauss and Szankowski 1986b). However, for general f and g in $\mathscr{L}$ the order of magnitude of the distance is unknown. What is known (see Lindenstrauss and Szankowski 1987) is that there are examples of f and g in $\mathscr{L}$ where block random matrices do not give the right answer. For example for some specific f and g in $\mathscr{L}$ the orthogonal matrix generated by the incidence matrix of projective planes (if $n = p^2 + p + 1$ with p prime) gives the right distance n^β for a suitable β (as usual up to a $\gamma(n)$ factor) between the symmetric spaces corresponding to f and g. In this example the estimate on the distance gotten by block-random matrices is at least $n^{\beta+\delta}$ for some fixed $\delta > 0$. This example illustrates clearly the limitation of the random method for computing distances.

In this connection it is worthwhile to introduce the notion of weak distance defined by Tomczak-Jaegermann (1984). Let $X, Y \in \mathcal{K}_n$, we want to measure how far we can distinguish between X and Y just by computing the "natural" parameters of both spaces. More precisely, we put

$$\mathrm{wd}(X,Y) = \sup\left\{\frac{\alpha(I_X)}{\alpha(I_Y)}, \frac{\alpha(I_Y)}{\alpha(I_X)}\right\}, \tag{5.3.14}$$

where the sup is taken over all possible operator ideal norms α. An alternative way to define $\mathrm{wd}(X,Y)$ is to use the following quantity

$$q(X,Y) = \inf\left\{\int_\Omega \|T_\omega\|\,\|S_\omega\|\,\mathrm{d}\omega\colon\ S_\omega : X \to Y,\ T_\omega : Y \to X,\ \int_\Omega T_\omega S_\omega\,\mathrm{d}\omega = I_X\right\} \tag{5.3.15}$$

where the infimum is over all measure spaces Ω and all choices of (measurable) T_ω and S_ω. Of course, in the case we consider here (i.e., finite-dimensional spaces), the infimum could be taken just over finite measure spaces Ω. In computing $q(X,Y)$ we measure how well I_X can be averaged by operators factoring through Y (instead of just factoring I_X itself through Y which yields $d(X,Y)$). The connection between $q(X,Y)$ and $\mathrm{wd}(X,Y)$ is easy to establish:

$$\mathrm{wd}(X,Y) = \max(q(X,Y),\ q(Y,X)). \tag{5.3.16}$$

Obviously $\mathrm{wd}(X,Y) \leqslant d(X,Y)$, and it is easy to see that $\mathrm{wd}(X,\ell_2^n) = d(X,\ell_2^n)(= d_X)$ for all $X \in \mathcal{K}_n$. Tomczak-Jaegermann (1984) showed that $\mathrm{wd}(X,\ell_p^n) \leqslant c\sqrt{n}$, $X \in \mathcal{K}_n$, $1 \leqslant p \leqslant \infty$; this result should be compared with (5.3.8) above. In the same paper it is shown that the weak distance between (at least "most") pairs of Gluskin spaces of dimension n is at most $c\sqrt{n}$.

Rudelson (1992) proved that there is a constant C so that

$$\mathrm{wd}(X,Y) \leqslant Cn^{17/18}(\log n)^{7/2} \tag{5.3.17}$$

for every $X, Y \in \mathcal{K}_n$. The proof uses, among other tools, Chevet's inequality on the norm of Gaussian operators and the selection techniques of Bourgain and Tzafriri (1987a). An important step in the argument is the proof of the fact that if the volume ratio of $X \in \mathcal{K}_n$ is close to being maximal (i.e., $\mathrm{vr}(X) \geqslant a\sqrt{n}$ for some constant a), then X has a k-dimensional subspace with $k = c\sqrt{n}/\log n$ whose distance from ℓ_∞^k is at most $C\log n$ for suitable constants c and C.

In Lindenstrauss and Szankowski (1986b) the weak distance was evaluated between an arbitrary pair of spaces with a 1-symmetric basis (up to a $\gamma(n)$ factor). The answer is of the form $n^{\eta(f,g)}$ where $\eta(f,g)$ is an explicit formula involving $f, g \in \mathcal{L}$. The operators T and S for which the infimum of (5.3.15) (or (5.3.18)) is achieved are "random Gaussian operators with lacunas". They consist of 0 outside a $k \times l$ submatrix (of the $n \times n$ matrix) for a suitable k and l. Inside this submatrix the elements $a_{i,j}$ are of the form $c\theta_{i,j}\gamma_{i,j}$ where $\theta_{i,j}$ are Bernoulli variables taking the value 1 with probability p (for a suitable p) and 0 with probability $1-p$ while the $\gamma_{i,j}$ are standard Gaussian variables. Thus in this case the random method can be said to give a complete answer.

We turn now to the question of uniqueness of symmetric bases which is closely related to a part of the discussion above. From the isometric point of view we have uniqueness except for a few isolated cases in low dimensions (see, e.g., the book of Rolewicz 1972, Th. IX 8.3): if X has 1-symmetric bases $\{u_i\}_{i=1}^n$ and $\{v_i\}_{i=1}^n$ of norm 1 then the map which sends $\sum_{i=1}^n \lambda_i u_i$ to $\sum_{i=1}^n \lambda_i v_i$ is an isometry (an example of an exceptional case is ℓ_∞^2 which is isometric to ℓ_1^2). However, the isomorphic version of the same question is more complex. The question can be phrased as follows. Assume X and Y are spaces with 1-symmetric bases $\{x_i\}_{i=1}^n$ and $\{y_i\}_{i=1}^n$. Let $T : X \to Y$ be defined by $T\sum_{i=1}^n \lambda_i x_i = \sum_{i=1}^n \lambda_i y_i$. Is it true that there exists a function f so that $\|T\|\,\|T^{-1}\| \leqslant f\big(d(X,Y)\big)$? This question was posed by Johnson et al. (1979) who gave a positive answer in a special case. Some further partial positive results were given in Schütt (1981), Lindenstrauss and Szankowski (1986b), and Bourgain, Kalton and Tzafriri (1989). In particular Schütt proved that if X is far from a Hilbert space the answer is affirmative. More precisely, if $d_X \geqslant n^\delta$ for some $\delta > 0$ then for a suitable function g, $\|T\|\,\|T^{-1}\| \leqslant g\big(d(X,Y),\delta\big)$. In general the answer to the problem stated above turned out to be negative. Gowers (1991) constructed spaces X and Y as above with $d(X,Y) \leqslant c$ and $\|T\|\,\|T^{-1}\| \geqslant C\log\log n$. It is still open whether $d(X,Y) \leqslant c$ implies $\|T\|\;\|T^{-1}\| \leqslant c_1(\log n)^a$ for suitable c_1 and a.

Finally, we mention a result concerning the distance between two spaces having a 1-unconditional basis. It was proved in Lindenstrauss and Szankowski (1986a) that there is a constant $\gamma \leqslant \frac{2}{3}$ so that for every $\varepsilon > 0$ there is a $c(\varepsilon)$ such that

$$d(X,Y) \leqslant c(\varepsilon)n^{\gamma+\varepsilon}, \quad X,Y \in \mathcal{K}_n \text{ both having a 1-unconditional basis.} \tag{5.3.18}$$

The constant γ obtained in the proof is given by a quite complicated expression. It is certainly in the interval $\left[\frac{5}{9},\frac{2}{3}\right]$ and numerical calculations seem to show that it is $\frac{5}{9}$. However, as we remarked above, the right constant should be $\frac{1}{2}$.

6. Applications to classical convexity theory in $\mathbb{R}^n$

6.1. *Isomorphic symmetrization and its applications*

In several places in the preceding sections we showed how the functional analytic approach yielded results concerning, e.g., geometric inequalities, which are in the mainstream of the subject matter of classical convexity theory in $\mathbb{R}^n$. This applies in particular to the subsections dealing with almost spherical sections, the Legendre ellipsoid and topics related to the inverse Brunn and Minkowski inequality. In the present section we shall present further results in the same spirit.

We start by presenting a detailed outline of the proof due to Milman (1988b) of Theorem 4.3.1. It is somewhat more convenient to prove here the following result which is easily seen to be equivalent to Theorem 4.3.1 (see also (4.3.9)): For every symmetric convex body K in $\mathbb{R}^n$ ($n = 2, 3, \ldots$) there is an ellipsoid $\mathscr{E}$ so that for every symmetric compact convex set A in $\mathbb{R}^n$

$$C^{-n}\operatorname{Vol}(\mathscr{E}+A) \leqslant \operatorname{Vol}(K+A) \leqslant C^n \operatorname{Vol}(\mathscr{E}+A), \tag{6.1.1}$$

$$C^{-n}\operatorname{Vol}(\mathscr{E}^\circ+A) \leqslant \operatorname{Vol}(K^\circ+A) \leqslant C^n \operatorname{Vol}(\mathscr{E}^\circ+A), \tag{6.1.2}$$

where C is an absolute constant.

We shall produce $\mathscr{E}$ from K by a procedure we call "isomorphic symmetrization". This terminology is used because it is a symmetrization procedure but unlike the usual procedures of this kind (which are discussed, e.g., in the next subsection) no natural parameter is precisely preserved under a single operation (and therefore under the entire procedure). What happens in our context is that the total change (for the entire procedure) of some parameters can be controlled. The name "isomorphic" is used also because at the end of the procedure we end up not with an ellipsoid but with a body with a fixed distance from an ellipsoid.

Before describing the operation itself let us point out what we get after applying it once and show why this will yield the desired result. We pass in a single step from the body K to another symmetric convex body K_1 in $\mathbb{R}^n$ so that

$$d_{X_{K_1}} \leqslant c_1 \log^3(1+d_{X_K}), \tag{6.1.3}$$

$$\mathrm{e}^{-c_2 n \log^{-2}(1+d_{X_K})} \leqslant \frac{\operatorname{Vol}(K_1+A)}{\operatorname{Vol}(K+A)} \leqslant \mathrm{e}^{c_2 n \log^{-2}(1+d_{X_K})}, \tag{6.1.4}$$

$$\mathrm{e}^{-c_2 n \log^{-2}(1+d_{X_K})} \leqslant \frac{\operatorname{Vol}(K_1^\circ+A)}{\operatorname{Vol}(K^\circ+A)} \leqslant \mathrm{e}^{c_2 n \log^{-2}(1+d_{X_K})}, \tag{6.1.5}$$

where c_1 and c_2 are absolute positive constants, A an arbitrary compact convex and symmetric set in $\mathbb{R}^n$ and X_K denotes $\mathbb{R}^n$ with the norm $\|\cdot\|_K$ whose unit ball is K.

Let now λ_0 be any number so that $\lambda_0 > c_1 \log^3(1+\lambda_0)$. We iterate the operation t times until we get for the first time a convex body K_t so that $d_{X_{K_t}} \leqslant \lambda_0$. It is trivial to check using (6.1.3) (noting that $d_{X_K} \leqslant n^{1/2}$) that

$$\sum_{j=0}^{t-1} \log^{-2}(1+d_{X_{K_j}}) \leqslant c_3, \tag{6.1.6}$$

where c_3 is again an absolute constant (and $K = K_0$).

It follows from (6.1.4), (6.1.5) and (6.1.6) that if $\mathscr{E}$ is an ellipsoid so that $\mathscr{E} \subset K_t \subset \lambda_0 \mathscr{E}$ then $\mathscr{E}$ satisfies (6.1.1) and (6.1.2) for suitable constant C.

It remains to produce K_1 from K. We introduce in $\mathbb{R}^n$ an Euclidean norm $|\cdot|$ which is the ℓ norm corresponding to $\|\cdot\|_K$ (see (4.1.9)). Recall that if

$$M_K = \int_{S^{n-1}} \|x\|_K \, \mathrm{d}\sigma_n(x),$$

where σ_n is the rotation invariant measure on the unit sphere S^{n-1} of $|\cdot|$, then by (4.1.10)

$$M_K M_{K^\circ} \leqslant c_1 \log(1 + d_{X_K}). \tag{6.1.7}$$

We normalize the ℓ norm so that $M_K = 1$. It is easy to check that

$$|x| \leqslant c_4 n \|x\|_K, \quad x \in \mathbb{R}^n. \tag{6.1.8}$$

We denote the unit ball of $|\cdot|$ by B. We recall next the notion of covering numbers (see (4.3.6)) and a well-known inequality connecting it to M_{K° due to Sudakov (1971)

$$N(K, \alpha B) \leqslant \mathrm{e}^{c_5 n (M_{K^\circ}/\alpha)^2} \tag{6.1.9}$$

for all $\alpha > 0$ and some absolute c_5. There is a dual inequality to (6.1.9) due to Pajor and Tomczak-Jaegermann (1985) which states

$$N(B, \alpha K) \leqslant c^{c_6 n (M_K/\alpha)^2}. \tag{6.1.10}$$

This was originally proved by duality from (6.1.9). Later Talagrand found an easy direct argument which proves (6.1.10) (see, e.g., Bourgain, Lindenstrauss and Milman 1989a).

With these preliminaries it is easy to describe K_1 and prove that it has the desired properties. We put

$$K_1 = \mathrm{Con}\left((K \cap \lambda B) \cup \mu^{-1} B\right), \tag{6.1.11}$$

where

$$\lambda = \log(1 + d_{X_K}) \cdot M_{K^\circ}, \qquad \mu = \log(1 + d_{X_K}) \cdot M_K. \tag{6.1.12}$$

Clearly $d_{X_{K_1}} \leqslant \lambda\mu$ and thus (6.1.3) is a consequence of (6.1.7) and (6.1.12). Because of obvious duality considerations it remains to verify that (6.1.4) holds. To establish this we note that, by the Brunn–Minkowski theorem, the even function $\varphi : \mathbb{R}^n \to \mathbb{R}$ defined by $\varphi(x) = \mathrm{Vol}\left((x + \lambda B) \cap K + A\right)$ satisfies $\varphi(x) = \frac{1}{2}\left(\varphi(x) + \varphi(-x)\right) \leqslant \varphi(0)$. (Here A is an arbitrary compact convex symmetric set.) Hence

$$\mathrm{Vol}(K + A) \leqslant N(K, \lambda B)\,\mathrm{Vol}\left((\lambda B \cap K) + A\right) \leqslant N(K, \lambda B)\,\mathrm{Vol}(K_1 + A). \tag{6.1.13}$$

A direct simple computation (using just the definition of covering numbers) shows that if L is a convex body and β a scalar so that $L \subset \beta K$ then

$$N\left(\mathrm{Con}(K \cup L), K\left(1+\tfrac{1}{n}\right)\right) \leqslant 2\beta n N(L, K). \tag{6.1.14}$$

Hence (by recalling (6.1.8) and using $L = \mu^{-1}B$ in (6.1.14))

$$\begin{aligned} \mathrm{Vol}(K_1 + A) &\leqslant N\left(\mathrm{Con}(K \cup \mu^{-1}B), K\left(1+\tfrac{1}{n}\right)\right) \mathrm{Vol}\left(K\left(1+\tfrac{1}{n}\right) + A\right) \\ &\leqslant 2c_4 n^2 \left(1+\tfrac{1}{n}\right)^n N(B, \mu K)\, \mathrm{Vol}(K + A). \end{aligned} \tag{6.1.15}$$

The inequality (6.1.4) is an immediate consequence of (6.1.9), (6.1.10), (6.1.12), (6.1.13) and (6.1.15).

Remarks.

(1) For a non-symmetric convex body K in $\mathbb{R}^n$ it is possible to find an ellipsoid $\mathscr{E}$ so that (6.1.1) holds for every compact convex set A. For a proof of this we have simply to replace K and A by the symmetric sets $\frac{1}{2}(K + (-K))$ and $\frac{1}{2}(A + (-A))$ and use the Rogers and Shephard (1957) inequality that

$$\mathrm{Vol}\, K \leqslant \mathrm{Vol}\, \tfrac{1}{2}(K + (-K)) \leqslant 2^{-n} \binom{2n}{n} \mathrm{Vol}\, K,$$

which reduces the general case to the symmetric one.

(2) The operation of passing from K to K_1 (see (6.1.11)) is called convex surgery.

Several geometric consequences of Theorem 4.3.1 were pointed out in section 4.3. We shall present some further consequences here. These results, taken from Milman (1991) require besides Theorem 4.3.1 arguments of the type used in the proofs of the results in section 3.1 on almost spherical sections.

Theorem 6.1.1. *Let K be a symmetric convex body in $\mathbb{R}^n$. Then there exist $U, V \in \mathrm{GL}_n$ such that if we put*

$$K_1 = K + UK, \qquad K_2 = K_1^\circ + VK_1^\circ, \tag{6.1.16}$$

then there is an ellipsoid $\mathscr{E}$ so that $\mathscr{E} \subset K_2 \subset C\mathscr{E}$ where C is an absolute constant.

In other words K_2 is an "isomorphic ellipsoid". The proof of Theorem 6.1.1 actually shows that if the Euclidean structure in $\mathbb{R}^n$ is chosen so that its unit ball is the M ellipsoid of K (i.e., the ellipsoid satisfying (6.1.1) and (6.1.2)) then one may take as U and V in (6.1.16) "most" (i.e., a set of large Haar measure) pairs in $\mathrm{SO}_n \times \mathrm{SO}_n$.

Note that the result of Kashin (1977) we mentioned in (3.1.20) states (if we pass to the dual) that if K is the unit cube $Q_n = [-1, 1]^n$ then $K_1 = K + UK$ is already an isomorphic ellipsoid (for most $U \in \mathrm{SO}_n$). This fact is not true in general. If K

is Q_n° (i.e., the crosspolytope) then at least $N = n/\log n$ summands are needed in order that $N^{-1}\sum_{j=1}^N U_jK$ becomes an isomorphic ellipsoid. This point and related topics will be discussed in the next subsection.

A result closely related to Theorem 6.1.1 is the following. Let K be a convex symmetric body in $\mathbb{R}^n$ and let B be the unit ball of the Euclidean norm in $\mathbb{R}^n$ (which need not be related at all to K). Put for $l = 1, 2, \ldots$

$$r_l(K) = \max\left\{r : rB \subset l^{-1}\sum_{j=1}^{l} U_jK,\ U_j \in \mathrm{SO}_n\right\}. \tag{6.1.17}$$

Then

$$r_2(K)r_3(K^\circ) \geqslant c, \tag{6.1.18}$$

where c is an absolute positive constant.

Note that the result of Kashin mentioned above is contained in (6.1.18). Indeed if $K = Q_n$ then since $\sum_{j=1}^3 U_jK^\circ$ has at most $6n$ extreme points (and is contained in $3B$) an easy computation of volumes shows that $r_3(K^\circ) \leqslant c_1/\sqrt{n}$. Hence, by (6.1.18), $r_2(Q_n) \geqslant c_3\sqrt{n}$. But clearly $\frac{1}{2}(U_1Q_n + U_2Q_n) \subset \sqrt{n}B$ and thus $Q_n + U_1^{-1}U_2Q_n$ is an "isomorphic ellipsoid". Also Theorem 6.1.1 can be deduced from (6.1.18).

6.2. Zonoids and Minkowski sums, approximation and symmetrization

A zonotope Z in $\mathbb{R}^n$ is by definition a Minkowski sum $\sum_{j=1}^N I_j$ of segments. Equivalently, Z is the affine image of an N-dimensional parallelopiped. Clearly Z is a convex polytope which is symmetric with respect to the sum of centers of the $\{I_j\}_{j=1}^N$. We shall assume below that the center of Z is the origin. The polar Z° of Z is an n-dimensional section of the unit ball of ℓ_1^N. A zonoid in $\mathbb{R}^n$ is a body which is the limit (in the sense of the Hausdorff metric) of zonotopes. It is a symmetric convex body. The polars of zonoids in $\mathbb{R}^n$ are exactly the unit balls of the n-dimensional subspaces of $L_1(0,1)$. In dimension 2 any symmetric convex body is a zonoid. For $n \geqslant 3$ the unit ball of ℓ_p^n is a zonoid if and only if $2 \leqslant p \leqslant \infty$ (cf. Dor 1976). A nice exposition of the basic properties of zonotopes and zonoids is Bolker (1969); for recent results on the structure of zonoids we refer to the article on this topic in the present volume.

The Euclidean ball B_n is a zonoid and a natural question is the following. Given n and $\varepsilon > 0$ how many segments $N = N(n, \varepsilon)$ are needed so as to obtain a zonotope $\sum_{j=1}^N I_j$ satisfying

$$(1-\varepsilon)B_n \subset \sum_{j=1}^{N} I_j \subset (1+\varepsilon)B_n. \tag{6.2.1}$$

By considering the support functionals we observe that (6.2.1) is equivalent to

$$(1-\varepsilon)|x| \leqslant \sum_{j=1}^{N} |\langle x_j, x\rangle| \leqslant (1+\varepsilon)|x| , \quad x \in \mathbb{R}^n, \tag{6.2.2}$$

where I_j is the segment $[-x_j, x_j]$ and $|x|$ denotes the Euclidean norm whose unit ball is B_n. Thus the problem of finding $N(n,\varepsilon)$ is equivalent to that of embedding ℓ_2^n into ℓ_1^N which was treated in section 3.1. In particular (3.1.19) means that $N(n,\varepsilon) \leqslant c(\varepsilon)n$ which is quite surprising geometrically. Moreover, it was noted in Bourgain, Lindenstrauss and Milman (1989a) that with that estimate on N the segments $\{I_j\}_{j=1}^N$ can be chosen to have all the same length. As a function of n alone (for fixed ε) the linear dependence of N on n is certainly optimal. If we are interested in N as a function of both ε and n the optimal result is not known. The upper estimate obtained from the results of section 3 is $N \leqslant c\varepsilon^{-2}n$ while the known lower estimates are $N \geqslant n/4\varepsilon^{1/2}$ (Betke and McMullen 1983) and $N \geqslant cn^2/(1+n\varepsilon)$ (Bourgain, Lindenstrauss and Milman 1989a).

For this specific problem it is of geometric interest to get also precise estimates of $N(n,\varepsilon)$ as a function of ε for a fixed n ($\geqslant 3$). One reason for this is the following. Recall that by a classical formula (going back to Cauchy) the surface area $s(K)$ of the convex body K in $\mathbb{R}^n$ is given by

$$s(K) = \alpha_n \int_{S^{n-1}} \mathrm{Vol}_{n-1}\left(P(K,u)\right) \mathrm{d}\sigma_n(u), \tag{6.2.3}$$

where $P(K,u)$ is the orthogonal projection of K on the hyperplane orthogonal to u, σ_n is the rotation invariant measure on S^{n-1} and α_n is a normalization factor. In the paper mentioned above Betke and McMullen noticed that the smallest $N(n,\varepsilon)$ for which (6.2.1) holds is also the smallest N for which there are $\{u_j\}_{j=1}^N$ in S^{n-1} and positive scalars $\{\lambda_j\}_{j=1}^N$ so that

$$\left| s(K) - \sum_{j=1}^{N} \lambda_j \,\mathrm{Vol}_{n-1} P(K,u_j)\right| \leqslant \varepsilon s(K) \tag{6.2.4}$$

for any convex body K in $\mathbb{R}^n$.

By using spherical harmonics it was proved in the Bourgain, Lindenstrauss and Milman paper mentioned above that if (6.2.1) (or equivalently, (6.2.4)) holds then

$$N(n,\varepsilon) \geqslant c(n)\varepsilon^{-2(n-1)/(n+2)}. \tag{6.2.5}$$

The proof of (6.2.5) is close in spirit to the proof of some results in the theory of irregularity of distribution (see Beck and Chen 1987). Linhart (1989) showed that a result in the book of Beck and Chen can be used to give also an estimate in the reverse direction which is better than ε^{-2}. By also using the approach of the theory of irregularity of distribution (of mixing a probabilistic method with a deterministic one) Bourgain and Lindenstrauss obtained later a stronger result than Linhart's which essentially shows that (6.2.5) is best possible, i.e., that (6.2.1) holds with

$$N(n,\varepsilon) \leqslant c(n)\big(\varepsilon^{-2}|\log\varepsilon|\big)^{(n-1)/(n+2)} \tag{6.2.6}$$

(see Bourgain and Lindenstrauss 1988).

In the context of this discussion the question whether the $\{I_j\}$ in (6.2.1) can be chosen to be of equal length (or equivalently, whether the $\{\lambda_j\}$ in (6.2.4) can be chosen independently of j) is a delicate question. It turns out that indeed (6.2.6) is valid also if one wants segments $\{I_j\}$ of equal length [this was proved for $n \leqslant 6$ by Wagner (1992) and for general n by Bourgain and Lindenstrauss (1992)].

Consider now an arbitrary zonoid K in $\mathbb{R}^n$. One may ask, just as in (6.2.1), how many summands are needed in order to approximate K up to ε by a zonotope $\sum_{j=1}^{N} I_j$. By passing to the dual (i.e., to the support functionals) one realizes at once that this question is equivalent to the question of embedding n-dimensional subspaces of $L_1(0,1)$ into ℓ_1^N which we discussed in section 3.2. Thus by the results of Bourgain, Lindenstrauss and Milman quoted there (see (3.2.7)) we get that there is a zonotope $\sum_{j=1}^{N} I_j$ containing $(1-\varepsilon)K$ and contained in $(1+\varepsilon)K$ with

$$N \leqslant c\varepsilon^{-2} n \log n. \tag{6.2.7}$$

Moreover, if the norm induced by K is strictly convex, then N can be estimated by γn (i.e., is linear in n) where γ depends just on ε and the modulus of convexity of the norm induced by K. It is not known if the $\log n$ term is really needed in (6.2.7) in the general case. There are also results concerning the dependence of N on ε for fixed n which generalize (6.2.6). It is shown in Bourgain and Lindenstrauss (1988) that (6.2.6) holds for a general zonoid K if $n \leqslant 4$ (with a slightly different logarithmic factor) while for $n > 4$ one gets that up to a logarithmic factor $N \leqslant c(n)\varepsilon^{-2(n-2)/n}$ for a general zonoid.

Another way in which results related to (6.2.1) were generalized is by considering Minkowski sums of convex sets which are not necessarily segments. We mention first a result on general Minkowski sums. Assume that $\{K_\alpha\}_{\alpha\in A}$ is a finite set of symmetric convex bodies in $\mathbb{R}^n$ and let $K = \sum_{\alpha\in A} K_\alpha$ be their Minkowski sum. Then for every $\varepsilon > 0$ there is a subset $\{\alpha_j\}_{j=1}^{N}$ of A and positive scalars $\{\lambda_j\}_{j=1}^{N}$ so that

$$(1-\varepsilon)K \subset \sum_{j=1}^{N} \lambda_j K_{\alpha_j} \subset (1+\varepsilon)K, \quad N \leqslant c\varepsilon^{-2}|\log\varepsilon| n^2. \tag{6.2.8}$$

This fact was observed in Bourgain, Lindenstrauss and Milman (1988) and is proved by a direct adaptation of the empirical distribution method of Schechtman (1987). It is not known if N in (6.2.8) can always be chosen to depend even linearly on n. This however is the case in the following special case which is of particular interest. Let K be a symmetric compact convex set in $\mathbb{R}^n$. The set

$$\int_{\mathrm{SO}_n} UK\,\mathrm{d}\mu(U), \tag{6.2.9}$$

where μ is the Haar measure on SO_n is well defined (either by approximating the integral by finite Minkowski sums or by considering the support functional which is a regular numerical integral). Obviously the set in (6.2.9) is rotation invariant, i.e., it is λB_n for some λ (actually $\lambda =$ mean width of K). It is proved in Bourgain, Lindenstrauss and Milman (1988) that for every K as above and $\varepsilon > 0$ there exist orthogonal transformations $\{U_j\}_{j=1}^N$ so that

$$(1-\varepsilon)\lambda B_n \subset N^{-1}\sum_{j=1}^{N} U_j K \subset (1+\varepsilon)\lambda B_n, \quad N \leqslant c\varepsilon^{-2}n. \tag{6.2.10}$$

[In the paper cited above the N had an additional factor $|\log\varepsilon|$. Schmuckenschläger (1991) showed how this term can be eliminated by the method of Schechtman (1989).]

In general (e.g., if K is a segment) the dependence of N on n in (6.2.10) cannot be better than linear. However if K is a convex body in $\mathbb{R}^n$ in a special position (e.g., so that B_n is the John ellipsoid of K) then (6.2.10) holds with $N \leqslant c\varepsilon^{-2}n/\log n$. The factor $n/\log n$ is optimal. If $\sum_{j=1}^N K_j$ is any Minkowski sum whose summands K_j are all crosspolytopes (i.e., affine images of the unit ball of ℓ_1^n) which is between say $\frac{1}{2}B_n$ and $2B_n$ then $N \geqslant cn/\log n$.

The results on Minkowski sums have analogues concerning symmetrizations of convex sets. We start with Minkowski symmetrization (called also Blaschke symmetrization). Let K be a convex body in $\mathbb{R}^n$ and let $u \in S^{n-1}$. The Minkowski symmetrization of K with respect to u is defined to be $\frac{1}{2}(K + \pi_u K)$ where π_u is the reflection with respect to the hyperplane orthogonal to u, i.e.,

$$\pi_u(x) = x - 2\langle x, u\rangle u, \quad x \in \mathbb{R}^n.$$

By a "random" Minkowski symmetrization we understand the operation above where u is chosen randomly on S^{n-1} with respect to the rotation invariant measure. We can now state the following result from Bourgain, Lindenstrauss and Milman (1988).

Theorem 6.2.1. *Let K be a convex body in $\mathbb{R}^n$ and let $\varepsilon > 0$. If $n > n_0(\varepsilon)$ and if we perform on K, $N = Cn\log n + c(\varepsilon)n$ mutually independent random Minkowski symmetrizations we obtain with probability $1-\exp(-\tilde{c}(\varepsilon)n)$ a body $\widetilde{K}$ which satisfies*

$$(1-\varepsilon)\lambda(K)B_n \subset \widetilde{K} \subset (1+\varepsilon)\lambda(K)B_n,$$

where $\lambda(K)$ is the mean width of K and $c(\varepsilon)$, $\tilde{c}(\varepsilon)$ are positive constants depending only on $\varepsilon > 0$.

The proof of Theorem 6.2.1 uses ideas related to the proof of Dvoretzky's theorem via the concentration of measure techniques. Here this concentration of measure phenomenon is used for SO_n as well as for S^{n-1}. Theorem 6.2.1 is in some sense related to the fact discovered by Diaconis and Shashahani (1986) that the product of $\frac{1}{2}n \log n + cn$ random reflections π_u yields a random orthogonal transformation on $\mathbb{R}^n$. However, the methods used in the proofs of these results are completely different.

The best known symmetrization procedure in convexity is the Steiner symmetrization. The Steiner symmetrization $\sigma_u K$ of a convex body $K \subset \mathbb{R}^n$ with respect to $u \in S^{n-1}$ consists by definition of all points of the form $x + \lambda u$ where $x \in P(K,u)$ (having the same meaning as in (6.2.3)) and $|\lambda| \leqslant \frac{1}{2}$ length $(x+\mathbb{R}u)\cap K$. It is well known, since the beginning of the century, that by suitably doing repeated Steiner symmetrizations of a convex body K one gets a sequence of convex bodies which converges in the Hausdorff metric to an Euclidean ball having the same volume as K. Mani (1986) proved that if the directions of symmetrizations are chosen randomly then almost surely we get such a sequence. There are many known strikingly elegant consequences of this property of the Steiner symmetrization, mostly to prove that in certain contexts the Euclidean ball is an optimal body. Originally this method was used for verifying the isoperimetric inequality. However, up to recently, little was known on the number of symmetrizations needed in order to obtain from an arbitrary K a body with a given Hausdorff distance from a ball. Probably the only estimate known until recently was the one given by Hadwiger (1952) which was superexponential in n. It was proved by Bourgain, Lindenstrauss and Milman (1989b) that there are absolute positive constants c and c_0 having the following property. Let K be a convex body in $\mathbb{R}^n$ whose volume is equal to that of the Euclidean ball B_n. Then there are $\{u_j\}_{j=1}^N$ is S^{n-1} with $N \leqslant c_0 n \log n$ so that

$$c^{-1}B_n \subset \prod_{j=1}^{N} \sigma_{u_j} K \subset cB_n \ . \tag{6.2.10}$$

The proof shows also that with $N \leqslant c_0 n \log n + c(\varepsilon)n$ symmetrizations one can ensure that

$$c^{-1}B_n \subset \prod_{j=1}^{N} \sigma_{u_j} K \subset (1+\varepsilon)B_n. \tag{6.2.11}$$

It seems that in (6.2.11) one should be able to replace c^{-1} by $(1-\varepsilon)$ but the proof does not show this.

Since the Steiner symmetrization is a nonlinear operation it is hard to treat directly the analytic behaviour of the iteration of such maps. The proof of (6.2.10) (or (6.2.11)) leans heavily on Theorem 6.2.1. One should point out also that the directions which one gets from the proof of (6.2.10) are not random.

6.3. Some additional results

We collect here a few geometric results on convex bodies in $\mathbb{R}^n$ which were proved by functional analytic tools. Most of these results are connected to some topics which were already considered above.

In section 4.2 we mentioned the problem whether there is an absolute constant $\delta > 0$ so that for every symmetric convex body K in $\mathbb{R}^n$ of volume 1 there is a hyperplane H through the origin so that $\mathrm{Vol}_{n-1}(K \cap H) \geqslant \delta$. We would like to mention now some known estimates on the volumes of central sections of special convex bodies.

Hensley (1980) proved that for $Q_n = \left[-\frac{1}{2}, \frac{1}{2}\right]^n$, the cube of volume 1 in $\mathbb{R}^n$, one has for every hyperplane H through the origin $1 \leqslant \mathrm{Vol}_{n-1}(Q_n \cap H) \leqslant 5$. (Actually the lower bound was first obtained by Hadwiger 1972.) He conjectured that the right upper bound is $\sqrt{2}$ and this was proved by Ball (1986b). Vaaler (1979) showed that for every subspace E of $\mathbb{R}^n$ of dimension k $\mathrm{Vol}_k(Q_n \cap E) \geqslant 1$. Meyer and Pajor (1988) proved that for every subspace $E \subset \mathbb{R}^n$ with $\dim E = k$ the function

$$\mathrm{Vol}_k(E \cap K_{\ell_p^n}) / \mathrm{Vol}_k(K_{\ell_p^k})$$

is a monotone increasing function of p on $[1, \infty]$. By comparing a general p to $p = 2$ one deduces

$$\mathrm{Vol}_k(E \cap K_{\ell_p^n}) \geqslant \mathrm{Vol}_k(K_{\ell_p^k}), \quad 2 \leqslant p \leqslant \infty \tag{6.3.1}$$

(for $p = \infty$ this is Vaaler's result), and

$$\mathrm{Vol}_k(E \cap K_{\ell_p^n}) \leqslant \mathrm{Vol}_k(K_{\ell_p^k}), \quad 1 \leqslant p \leqslant 2. \tag{6.3.2}$$

In the same paper they obtained good lower bounds for $\mathrm{Vol}_k(E \cap K_{\ell_1^n})$. In particular they showed that for every hyperplane H

$$\mathrm{Vol}_{n-1}(H \cap K_{\ell_1^n}) \geqslant \frac{\sqrt{n}}{4^{n-1}} \cdot \binom{2n \quad 2}{n-1} \cdot \frac{2^{n-1}}{(n-1)!} \tag{6.3.3}$$

with equality if and only if $H = \left\{x\colon \sum_{i=1}^n \theta_i x_i = 0,\ \theta_i = \pm 1 \text{ for all } i\right\}$.

Recently, Ball (1989) has extended his result from hyperplanes to general subspaces. For every $E \subset \mathbb{R}^n$ of dimension $n-k$

$$\mathrm{Vol}_{n-k}(Q_n \cap E) \leqslant (\sqrt{2})^k. \tag{6.3.4}$$

It is perhaps also worthwhile to mention the following result of McMullen (1984) (which was used in the Meyer and Pajor paper). Let E be a k-dimensional subspace of $\mathbb{R}^n$ and $E^\perp$ its orthogonal complement. Then

$$\mathrm{Vol}_k\,(P_E Q_n) = \mathrm{Vol}_{n-k}\,(P_{E^\perp} Q_n) \tag{6.3.5}$$

where P_E and $P_{E^\perp}$ denote the orthogonal projections on the corresponding subspaces.

An inequality in classical convexity theory which fits well in the local theory of normed spaces is Urysohn's inequality which states (in our usual notation)

$$(\operatorname{Vol} K/\operatorname{Vol} B_n)^{1/n} \leqslant M_{K^\circ} = \int_{S^{n-1}} \|x\|_{K^\circ}\, \mathrm{d}\sigma_n(x). \tag{6.3.6}$$

This inequality was generalized by Milman and Pajor (1989) as follows. For every pair of symmetric convex bodies K_1 and K_2 in $\mathbb{R}^n$

$$(\operatorname{Vol} K_1/\operatorname{Vol} K_2)^{1/n} \leqslant \left(1+\tfrac{1}{n}\right)(\operatorname{Vol} K_1)^{-1}\int_{K_1} \|x\|_{K_2}\, \mathrm{d}x. \tag{6.3.7}$$

To check that (6.3.7) reduces to (6.3.6) if $K_1 = B_n$ and $K_2 = K^\circ$ note that

$$M_{K^\circ} = \left(1+\tfrac{1}{n}\right)(\operatorname{Vol} B_n)^{-1}\int_{B_n} \|x\|_{K^\circ}\, \mathrm{d}x$$

and use the Blaschke–Santaló inequality (4.3.5). By using the inverse Blaschke–Santaló inequality (4.3.4), one gets from (6.3.6)

$$(\operatorname{Vol} K_2/\operatorname{Vol} K_1)^{1/n} \leqslant c(\operatorname{Vol} K_1^\circ)^{-1}\int_{K_1^\circ} \|x\|_{K_2^\circ}\, \mathrm{d}x. \tag{6.3.8}$$

The same paper of Milman and Pajor also contains a proof of the following sharper form of Urysohn's inequality.

$$(\operatorname{Vol} K/\operatorname{Vol} B_n)^{1/n} \leqslant \mathrm{e}^{1/n}\exp\left((\operatorname{Vol} B_n)^{-1}\int_{B_n} \log\|x\|_{K_0}\, \mathrm{d}x\right). \tag{6.3.9}$$

We end this survey by mentioning the elegant recent solution to the plank problem, due to Ball (1991b). A plank in a (finite- or infinite-dimensional) Banach space is the region between two parallel closed hyperplanes. In other words it is a set of the form $\{x\colon |\langle x^*, x\rangle - m| \leqslant w\}$ for some $x^* \in X^*$ of norm 1, and some scalars m and $w(>0)$. The number w is called for obvious reasons the half width of the plank. What Ball showed is that if the unit ball of X is covered by a (finite or countable) collection of planks then the sum of their width's is at least 2. For X a Hilbert space the same result was proved long ago by Bang (1951) (and even in that case the result is far from obvious). The analytic statement of Ball's result is the following.

Theorem 6.3.1. *Let $\{x_i\}_{i=1}^\infty$ be a sequence of unit vector in a normed space X. Let $\{m_i\}_{i=1}^\infty$ be a sequence of reals and let $\{w_i\}_{i=1}^\infty$ be a sequence of non-negative reals with $\sum_i w_i \leqslant 1$. Then there is an $x^* \in X^*$ with $\|x^*\| \leqslant 1$ so that*

$$|\langle x^*, x_i\rangle - m_i| \geqslant w_i \quad \textit{for all } i. \tag{6.3.10}$$

References

Alon, N., and V.D. Milman
[1983] Embedding of ℓ^k_∞ in finite-dimensional Banach spaces, *Israel J. Math.* **45**, 265–280.

Amir, D.
[1986] *Characterizations of Inner Product Spaces*, Operator Theory, Advances and Applications, Vol. 20 (Birkhäuser, Basel).

Amir, D., and V.D. Milman
[1980] Unconditional and symmetric sets in n-dimensional normed spaces, *Israel J. Math.* **37**, 3–20.
[1985] A quantitative finite-dimensional Krivine theorem, *Israel J. Math.* **50**, 1–12.

Ball, K.
[1986a] Isometric problems in l_p and sections of convex sets, Ph.D. Dissertation, Trinity College, Cambridge.
[1986b] Cube slicing in R^n, *Proc. Amer. Math. Soc.* **97**, 465–473.
[1988] Logarithmically concave functions and sections of convex sets in R^n, *Studia Math.* **88**, 69–84.
[1989] Volumes of sections of cubes and related problems, in: *Lecture Notes in Mathematics*, Vol. 1376 (Springer, Berlin) pp. 251–263.
[1991a] Normed spaces with a weak Gordon Lewis property, in: *Lecture Notes in Mathematics*, Vol. 1470 (Springer, Berlin) pp. 36–47.
[1991b] The plank problem for symmetric bodies, *Invent. Math.* **104**, 535–543.

Bang, T.
[1951] A solution to the plank problem, *Proc. Amer. Math. Soc.* **2**, 990–993.

Beck, A.
[1962] A convexity condition in Banach spaces and the strong law of large numbers, *Proc. Amer. Math. Soc.* **13**, 329–334.

Beck, J., and W. Chen
[1987] *Irregularities of Distribution*, Cambridge Tracts in Math., Vol. 89.

Bennett, G., L.E. Dor, V. Goodman, W.B. Johnson and C.M. Newman
[1977] On uncomplemented subspaces of L_p, $1<p<2$, *Israel J. Math.* **26**, 178.

Benyamini, Y., and Y. Gordon
[1981] Random factorization of operators between Banach spaces, *J. Anal. Math.* **39**, 45–74.

Betke, U., and P. McMullen
[1983] Estimating the size of convex bodies by projections, *J. London Math. Soc.* **27**, 525–538.

Blaschke, W.
[1918] Über affine Geometrie XIV, *Ber. Sachs. Akad. Wiss.* **70**, 72–75.

Bolker, E.D.
[1969] A class of convex bodies, *Trans. Amer. Math. Soc.* **145**, 323–346.

Borell, C.
[1975] The Brunn–Minkowski inequality in Gauss spaces, *Invent. Math.* **30**, 207–216.
[1979] On the integrability of Banach space valued Walsh polynomials, in: *Lecture Notes in Mathematics*, Vol. 721 (Springer, Berlin) pp. 1–3.

Bourgain, J.
[1981a] A remark on the finite dimensional P_λ spaces, *Studia Math.* **72**, 285–289.
[1981b] A counterexample to a complementation problem, *Compositio Math.* **43**, 133–144.
[1984a] On martingale transforms in finite dimensional lattices with an appendix on the K-convexity constant, *Math. Nachr.* **119**, 41–53.
[1984b] Subspaces of L^∞_N arithmetical diameter and Sidon sets, in: *Lecture Notes in Mathematics*, Vol. 1153 (Springer, Berlin) pp. 96–127.

[1986a] On high dimensional maximal functions associated to convex bodies, *Amer. J. Math.* **108**, 1467–1476.
[1986b] Real isomorphic complex Banach spaces need not be complex isomorphic, *Proc. Amer. Math. Soc.* **96**, 221–226.
[1988] On finite dimensional homogeneous Banach spaces, in: *Lecture Notes in Mathematics,* Vol. 1317 (Springer, Berlin) pp. 232–238.
[1991] On the distribution of polynomials on high dimensional convex sets, in: *Lecture Notes in Mathematics,* Vol. 1469 (Springer, Berlin) pp. 127–137.

Bourgain, J., and J. Lindenstrauss
[1988] Distribution of points on spheres and approximation by zonotopes, *Israel J. Math.* **64**, 25–31.
[1989] Almost Euclidean sections in spaces with a symmetric basis, in: *Lecture Notes in Mathematics,* Vol. 1376 (Springer, Berlin) pp. 278–288.
[1992] Approximating the sphere by a sum of segments of equal length, *J. Discrete Comput. Geom.*, to appear.

Bourgain, J., and V.D. Milman
[1985] Sections euclidiennes et volume des corps convexes symétriques, *C.R. Acad. Sci. Paris* **300**, 435–438.
[1986] Distances between normed spaces, their subspaces and quotient spaces, *Integral Equations and Operator Theory* **9**, 31–46.
[1987] New volume ratio properties for convex symmetric bodies in R^n, *Invent. Math.* **88**, 319–340.

Bourgain, J., and S.J. Szarek
[1988] The Banach Mazur distance to the cube and the Dvoretzky–Rogers factorization, *Israel J. Math.* **62**, 169–180.

Bourgain, J., and L. Tzafriri
[1987a] Invertibility of large submatrices with applications to the geometry of Banach spaces and harmonic analysis, *Israel J. Math.* **57**, 137–224.
[1987b] Complements of subspaces of l_p^n, $p \geqslant 1$, which are determined uniquely, in: *Lecture Notes in Mathematics,* Vol. 1267 (Springer, Berlin) pp. 39–52.
[1990] Embedding l_p^k in subspaces of L_p, for $p > 2$, *Israel J. Math.* **72**, 321–340.

Bourgain, J., J. Lindenstrauss and V.D. Milman
[1988] Minkowski sums and symmetrizations, in: *Lecture Notes in Mathematics,* Vol. 1317 (Springer, Berlin) pp. 44–66.

Bourgain, J., A. Pajor, S.J. Szarek and N. Tomczak-Jaegermann
[1989] On the duality problem for entropy numbers of operators, in: *Lecture Notes in Mathematics,* Vol. 1376 (Springer, Berlin) pp. 50–63.

Bourgain, J., N.J. Kalton and L. Tzafriri
[1989] Geometry of finite dimensional subspaces and quotients of L_p, in: *Lecture Notes in Mathematics,* Vol. 1376 (Springer, Berlin) pp. 138–175.

Bourgain, J., J. Lindenstrauss and V.D. Milman
[1989a] Approximation of zonoids by zonotopes, *Acta Math.* **162**, 73–141.
[1989b] Estimates related to Steiner symmetrizations, in: *Lecture Notes in Mathematics,* Vol. 1376 (Springer, Berlin) pp. 264–273.

Carl, B., and I. Stephani
[1990] *Entropy, Compactness and the Approximation of Operators,* Cambridge Tracts in Math., Vol. 98.

Chevet, S.
[1978] Series de variables aleatoires Gaussiens a valeurs dans E/F, *Sem. Maurey Schwartz,* Exp. 19 (Ecole Polytechnique, Paris).

Davis, W.J., and P. Enflo
[1977] The distance of symmetric spaces from l_p^n, in: *Lecture Notes in Mathematics,* Vol. 604 (Springer, Berlin) pp. 25–29.

Davis, W.J., and B. Maurey
[1977] The distance of a symmetric space from l_p^n, in: *Proc. Int. Conf. Operator Ideals* (Teubner, Leipzig) pp. 69–79.

Davis, W.J., D.W. Dean and I. Singer
[1968] Complemented subspaces and Λ systems in Banach spaces, *Israel J. Math.* **6**, 303–309.

Davis, W.J., V.D. Milman and N. Tomczak-Jaegermann
[1981] The distance between certain n-dimensional spaces, *Israel J. Math.* **39**, 1–15.

Deschaseaux, J.P.
[1973] Une charactérisation de certaines espaces vectoriel normés de dimension finie par leur constante de Macphail, *C.R. Acad. Sci. Paris* **276**, 1349–1351.

Diaconis, P., and M. Shahshahani
[1986] Products of random matrices as they arise in the study of random walks on groups, *Contemp. Math.* **50**, 183–195.

Dor, L.E.
[1975] On projections in L_1, *Ann. of Math.* **102**, 474–483.
[1976] Potentials and isometric embeddings in L_1, *Israel J. Math.* **24**, 260–268.

Dvoretzky, A.
[1959] A theorem on convex bodies and applications to Banach spaces, *Proc. Nat. Acad. Sci. U.S.A.* **45**, 223–226.
[1961] Some results on convex bodies and Banach spaces, in: *Proc. Sympos. Linear Spaces*, Jerusalem, pp. 123–160.

Dvoretzky, A., and C.A. Rogers
[1950] Absolute and unconditional convergence in normed linear spaces, *Proc. Nat. Acad. Sci. U.S.A.* **36**, 192–197.

Enflo, P.
[1972] Banach spaces which can be given an equivalent uniformly convex norm, *Israel J. Math.* **13**, 281–288.
[1973] A counter example to the approximation property, *Acta Math.* **130**, 309–317.

Figiel, T.
[1983] Local theory of Banach spaces and some operator ideals, in: *Proc. Int. Congress*, Warsaw, pp. 961–976.

Figiel, T., and W.B. Johnson
[1980] Large subspaces of l_∞^n and estimates of the Gordon Lewis constants, *Israel J. Math.* **37**, 92–112.

Figiel, T., and N. Tomczak-Jaegermann
[1979] Projections onto Hilbertian subspaces of Banach spaces, *Israel J. Math.* **33**, 155–171.

Figiel, T., S. Kwapien and A. Pelczynski
[1977] Sharp estimates for the constants of local unconditional structure in Minkowski spaces, *Bull. Acad. Polon. Sci. Ser. Sci. Math.* **25**, 1221–1226.

Figiel, T., J. Lindenstrauss and V.D. Milman
[1977] The dimension of almost spherical sections of convex bodies, *Acta Math.* **129**, 53–94.

Garling, D.J.H.
[1970] Absolutely p-summing operators in Hilbert spaces, *Studia Math.* **38**, 319–331.
[1974] Operators with large trace and a characterization of l_∞^n, *Proc. Cambridge Philos. Soc.* **76**, 413–414.

Garling, D.J.H., and Y. Gordon
[1971] Relation between some constants associated with finite dimensional Banach spaces, *Israel J. Math.* **9**, 346–361.

Geiss, S.
[1992] Antisymmetric tensor products of absolutely p-summing operators, *J. Approx. Theory* **68**, 223–246.

Giesy, D.P.
[1966] On a convexity condition in normed linear spaces, *Trans. Amer. Math. Soc.* **125**, 114–146.

Gilbert, J.E.
[1979] Nikišin–Stein theory and factorization with applications, *Proc. Symp. Pure Math.*, Vol. 35, part 2 (Amer. Math. Soc., Providence, RI) pp. 233–267.

Gluskin, E.D.
[1979] On the estimate of distance between finite dimensional symmetric spaces (in Russian), *Issled. Lin. Oper. Teor. Funk.* **92**, 268–273.
[1981a] The diameter of the Minkowskii compactum is approximately equal to n (in Russian), *Funct. Anal. Appl.* **15**, 72–73.
[1981b] Finite dimensional analogues of spaces without basis (in Russian), *Dokl. Akad. Nauk USSR* **216**, 1046–1050.
[1983] On distances between some symmetric spaces, *J. Soviet Math.* **22**, 1841–1846.
[1986] Probability in the geometry of Banach spaces, in: *Proc. Int. Congr., Berkeley*, Vol. 2, pp. 924–938.

Gluskin, E.D., N. Tomczak-Jaegermann and L. Tzafriri
[1992] Subspaces of l_p^N of small dimension, *Israel J. Math.*, to appear.

Goodner, D.A.
[1950] Projections in normed linear spaces, *Trans. Amer. Math. Soc.* **69**, 89–108.

Gordon, Y.
[1968] On the projection and Macphail constants of l_n^p spaces, *Israel J. Math.* **6**, 295–302.
[1988] Gaussian processes and almost spherical sections of convex bodies, *Ann. Probab.* **16**, 180–188.

Gordon, Y., and D.R. Lewis
[1974] Absolutely summing operators and local unconditional structure, *Acta Math.* **133**, 27–48.

Gordon, Y., D.R. Lewis and J.R. Retherford
[1973] Banach ideals of operators and applications, *J. Funct. Anal.* **14**, 85–129.

Gowers, W.T.
[1989] Symmetric block bases in finite-dimensional normed spaces, *Israel J. Math.* **68**, 193–219.
[1991] A finite dimensional normed space with two non-equivalent symmetric bases, Manuscript.

Gromov, M.
[1967] On a geometric conjecture of Banach, *Izv. Akad. Nauk USSR* **31**, 1105–1114.

Gromov, M., and V.D. Milman
[1984] Brunn theorem and a concentration of volume phenomenon for symmetric convex bodies, *GAFA Seminar Notes*, Tel Aviv University.

Grothendieck, A.
[1956a] Résumé de la théorie métrique des produits tensoriels topologiques, *Bol. Soc. Mat. São-Paulo* **8**, 1–79.
[1956b] Sur certaines classes de suites dans les espaces de Banach et le theorème de Dvoretzky–Rogers, *Bol. Soc. Mat. São-Paulo* **8**, 81–110.

Grünbaum, B.
[1960] Projection constants, *Trans. Amer. Math. Soc.* **95**, 451–456.

Gurari, V.E., M.I. Kadec and V.E. Mačaev
[1966] On the distance between isomorphic L_p spaces of finite dimension (in Russian), *Math. Sb.* **70**, 481–489.

Hadwiger, H.
[1952] Einfache Herleitung der isoperimetrischen Ungleichung für abgeschlossene Punktmengen, *Math. Ann.* **124**, 158–160.
[1972] Gitterperiodische Punktmengen und Isoperimetrie, *Monatsh. Math.* **76**, 410–418.

Hensley, D.
[1980] Slicing convex bodies and bounds of slice area in terms of the body's covariance, *Proc. Amer. Math. Soc.* **79**, 619–625.

Hoffmann-Jørgensen, J.
[1974] Sums of independent Banach space valued random variables, *Studia Math.* **52**, 159–186.

James, R.C.
[1964] Uniformly non-square Banach spaces, *Ann. of Math.* **80**, 542–550.
[1972] Some self-dual properties of normed linear spaces, *Ann. of Math. Studies* **69**, 159–175.
[1974] A non reflexive space that is uniformly nonoctahedral, *Israel J. Math.* **18**, 145–155.
[1978] Non reflexive spaces of type 2, *Israel J. Math.* **30**, 1–13.

James, R.C., and J. Lindenstrauss
[1974] The octahedral problem for Banach spaces, in: *Proc. Seminar on Random Series, Convex Sets and Banach Spaces* (Aarhus Univ., Denmark) pp. 100–120.

John, F.
[1937] Polar correspondence with respect to convex regions, *Duke Math. J.* **3**, 355–369.
[1948] Extremum problems with inequalities as subsidiary conditions, in: *Courant Anniversary Volume* (Interscience, New York) pp. 187–204.

Johnson, W.B., and G. Schechtman
[1982] Embedding l_p^m into l_1^n, *Acta Math.* **149**, 71–85.
[1991] On the distance of subspaces of l_p^n to l_p^k, *Trans. Amer. Math. Soc.* **324**, 319–319.
[1992] Computing p-summing norms with few vectors, *Israel J. Math.*, to appear.

Johnson, W.B., B. Maurey, G. Schechtman and L. Tzafriri
[1979] Symmetric structure in Banach spaces, *Mem. Amer. Math. Soc.* **217**.

Junge, M.
[1991] Computing (p, q) summing norms with n vectors, Manuscript.

Kadec, M.I., and A. Pelczynski
[1962] Bases lacunary sequences and complemented subspaces in the spaces L_p, *Studia Math.* **21**, 161–176.

Kadec, M.I., and M.G. Snobar
[1971] Certain functions of the Minkowski compactum, *Mat. Zametki* **10**, 453–458.

Kadec, V.M.
[1982] On two dimensional universal Banach spaces (in Russian), *C.R. Bulg. Acad. Sci.* **35**, 1331–1333.

Kahane, J.P.
[1968] *Some Random Series of Functions* (Heath Math. Monographs). Second edition: Cambridge Studia Adv. in Math., Vol. 5, 1985.

Kakutani, S.
[1939] Some characterization of Euclidean spaces, *Japan J. Math.* **16**, 93–97.

Kashin, B.S.
[1977] Sections of some finite-dimensional sets and classes of smooth functions (in Russian), *Izv. Akad. Nauk. SSSR Ser. Mat.* **41**, 334–351.

Knothe, H.
[1957] Contributions to the theory of convex bodies, *Michigan Math. J.* **4**, 39–52.

Kolmogoroff, A.N., and V.M. Tichomirov
[1959] ε-entropy and ε-capacity of sets in function spaces (in Russian), *Uspekhi Mat. Nauk* **14**, 3–86.

König, H.
[1980] Type constants and $(q, 2)$-summing norms defined by n vectors, *Israel J. Math.* **37**, 130–138.

König, H., and D.R. Lewis
[1987] A strict inequality for projection constants, *J. Funct. Anal.* **73**, 328–332.

König, H., and V.D. Milman
[1987] On the covering numbers of convex bodies, in: *Lecture Notes in Mathematics*, Vol. 1267 (Springer, Berlin) pp. 82–95.

König, H., and N. Tomczak-Jaegermann
[1990] Bounds for projection constants and 1-summing norms, *Trans. Amer. Math. Soc.* **320**, 799–823.

Krivine, J.L.
[1976] Sous-espaces de dimension finis des espaces de Banach réticulés, *Ann. of Math.* **104**, 1–29.

Kwapien, S.
[1972] Isomorphic characterizations of inner product spaces by orthogonal series with vector valued coefficients, *Studia Math.* **44**, 583–595.
[1976] A theorem on the Rademacher series with vector valued coefficients, in: *Lecture Notes in Mathematics,* Vol. 526 (Springer, Berlin) pp. 157–158.

Larman, D.G., and P. Mani
[1975] Almost ellipsoidal sections and projections of convex bodies, *Math. Proc. Cambridge Philos. Soc.* **77**, 529–546.

Ledoux, M., and M. Talagrand
[1991] *Probability in Banach Spaces*, Ergeb. Math. Grenzgeb., 3. Folge, Vol. 23 (Springer, Berlin).

Lemberg, H.
[1981] Nouvelle démonstration d'un théorème de J.L. Krivine sur la finie representation de l_p dans un espace de Banach, *Israel J. Math.* **39**, 341–348.

Lévy, P.
[1951] *Problèmes Concrets d'Analyse Fonctionnelle* (Gauthier-Villars, Paris).

Lewis, D.R.
[1979] Ellipsoids defined by Banach ideal norms, *Mathematika* **26**, 18–29.
[1988] An upper bound for the projection constant, *Proc. Amer. Math. Soc.* **103**, 1157–1160.

Lindenstrauss, J.
[1963] On the modulus of smoothness and divergent series in Banach spaces, *Michigan Math. J.* **10**, 241–252.
[1992] Almost spherical sections, their existence and their applications, *Jber. Deutsch. Math.-Vereinig.*, Jubiläumstagung 1990 (Teubner, Stuttgart) pp. 39–61.

Lindenstrauss, J., and A. Pelczynski
[1968] Absolutely summing operators in L_p spaces and their applications, *Studia Math.* **29**, 275–326.

Lindenstrauss, J., and A. Szankowski
[1986a] On the Banach–Mazur distance between spaces having an unconditional basis, in: *North-Holland Math. Studies,* Vol. 122 (North-Holland, Amsterdam) pp. 119–136.
[1986b] The weak distance between Banach spaces with a symmetric basis, *J. Reine Angew. Math.* **373**, 108–147.
[1987] The relation between the distance and the weak distance for spaces with a symmetric basis, in: *Lecture Notes in Mathematics,* Vol. 1267 (Springer, Berlin) pp. 21–38.

Lindenstrauss, J., and L. Tzafriri
[1971] On the complemented subspaces problem, *Israel J. Math.* **9**, 263–269.
[1977] *Classical Banach spaces, Vol. I,* Ergeb. Math. Grenzgeb., Vol. 92 (Springer, Berlin).
[1979] *Classical Banach Spaces, Vol. II*, Ergeb. Math. Grenzgeb., Vol. 97 (Springer, Berlin).

Linhart, J.
[1989] Approximation of a ball by zonotopes using uniform distribution on the sphere, *Arch. Math.* **53**, 82–86.

Lozanovskii, G.
[1969] Certain Banach lattices (in Russian), *Sibirsk. Mat. Ž.* **10**, 584–599.

Mani, P.
[1986] Random Steiner symmetrizations, *Studia Sci. Math. Hungar.* **21**, 373–378.

Mankiewicz, P.

[1984] Finite dimensional spaces with symmetry constant of order $\overline{n}$, *Studia Math.* **79**, 193–200.

[1988] Subspace mixing properties of operators in R^n with application to Gluskin spaces, *Studia Math.* **88**, 51–67.

Mankiewicz, P., and N. Tomczak-Jaegermann

[1991] A solution of the finite dimensional homogeneous Banach space problem, *Israel J. Math.* **75**, 129–160.

Marcus, M., and G. Pisier

[1981] *Random Fourier Series with Applications to Harmonic Analysis*, Ann. of Math. Studies, Vol. 101 (Princeton Univ. Press, Princeton, NJ).

Maurey, B.

[1974a] Un théorème de prolongement, *C.R. Acad. Sci. Paris* **279**, 329–332.

[1974b] Théorèmes de factorisation pour les opérateurs linéaires à valeurs dans les espaces L^p, *Astérisque* **11**, 1–163.

[1983] Sous espaces l^p des espaces de Banach, Bourbaki Seminar 82/83, *Astérisque* **105–106**, 199–215.

Maurey, B., and G. Pisier

[1976] Séries de variables aleatoires vectorielles independantes et propriétés geometriques des espaces de Banach, *Studia Math.* **58**, 45–90.

Mazur, S., and S. Ulam

[1932] Sur les transformationes isométriques d'espaces vectoriels, *C.R. Acad. Sci. Paris* **194**, 946–948.

McMullen, P.

[1984] Volume of projections of unit cubes, *Bull. London Math. Soc.* **16**, 278–280.

Meyer, M., and A. Pajor

[1988] Sections of the unit ball of l_p^n, *J. Funct. Anal.* **80**, 109–123.

[1989] On Santaló's inequality, in: *Lecture Notes in Mathematics*, Vol. 1376 (Springer, Berlin) pp. 261–263.

[1990] On the Blaschke–Santalo inequality, *Arch. Math.* **55**, 82–93.

Milman, V.D.

[1971] New proof of the theorem of Dvoretzky on sections of convex bodies, *Funct. Anal. Appl.* **5**, 28–37.

[1985a] Almost Euclidean quotient spaces of subspaces of finite dimensional normed spaces, *Proc. Amer. Math. Soc.* **94**, 445–449.

[1985b] Geometrical inequalities and mixed volumes in the local theory of Banach spaces, *Astérisque* **131**, 373–400.

[1986a] The concentration phenomenon and linear structure of finite-dimensional normed spaces, in: *Proc. Int. Congr., Berkeley*, Vol. 2, pp. 961–974.

[1986b] Inegalité de Brunn–Minkowski inverse et applications à la théorie locale des espaces normés, *C.R. Acad. Sci. Paris* **302**, 25–28.

[1988a] The heritage of P. Lévy in geometrical functional analysis, *Astérisque* **157–158**, 273–302.

[1988b] Isomorphic symmetrizations and geometric inequalities, in: *Lecture Notes in Mathematics*, Vol. 1317 (Springer, Berlin) pp. 107–131.

[1988c] A few observations on the connection between local theory and some other fields, in: *Lecture Notes in Mathematics*, Vol. 1317 (Springer, Berlin) pp. 283–289.

[1991] Some applications of duality relations, in: *Lecture Notes in Mathematics*, Vol. 1469 (Springer, Berlin) pp. 13–40.

Milman, V.D., and A. Pajor

[1989] Isotropic position and inertia ellipsoids and zonoids of the unit ball of a normed n-dimensional space, in: *Lecture Notes in Mathematics*, Vol. 1376 (Springer, Berlin) pp. 64–104.

Milman, V.D., and G. Pisier
[1986] Banach spaces with a weak cotype 2 property, *Israel J. Math.* **54**, 139–158.

Milman, V.D., and G. Schechtman
[1986] *Asymptotic Theory of Finite-Dimensional Normed Spaces,* Lecture Notes in Mathematics, Vol. 1200 (Springer, Berlin).

Milman, V.D., and M. Sharir
[1979] A new proof of the Maurey–Pisier theorem, *Israel J. Math.* **33**, 73–87.

Milman, V.D., and H. Wolfson
[1978] Minkowski spaces with extremal distance from Euclidean spaces, *Israel J. Math.* **29**, 113–130.

Nachbin, L.
[1950] A theorem of the Hahn Banach type for linear transformations, *Trans. Amer. Math. Soc.* **68**, 28–46.

Nikišin, E.M.
[1970] Resonance theorems and superlinear operators, *Russian Math. Surveys* **25**, 125–187.

Nordlander, G.
[1961] On sign-independent and almost sign-independent convergence in normed linear spaces, *Ark. Mat.* **4**, 287–296.

Pajor, A., and N. Tomczak-Jaegermann
[1985] Remarques sur les nombres d'entropie d'un opérateur et de son transposé, *C.R. Acad. Sci. Paris* **301**, 733–746.

Palmon, O.
[1992] The only convex body with extremal distance from the ball is the simplex, *Israel J. Math.* **79**, to appear.

Pelczynski, A.
[1967] A characterization of Hilbert–Schmidt operators, *Studia Math.* **28**, 355–360.
[1980] Geometry of finite dimensional Banach spaces and operator ideals, in: *Notes in Banach Spaces* (Univ. of Texas Press, Austin, TX) pp. 81–181.

Pelczynski, A., and S.J. Szarek
[1991] On parallelopipeds of minimal volume containing a convex symmetric body in R^n, *Proc. Cambridge Philos. Soc.* **109**, 125–148.

Persson, A., and A. Pietsch
[1969] p-nukleare und p-integrale Abbildungen in Banachräumen, *Studia Math.* **33**, 19–62.

Petty, C.M.
[1961] Centroid surfaces, *Pacific J. Math.* **11**, 1535–1547.

Phillips, R.S.
[1940] A characterization of Euclidean spaces, *Bull. Amer. Math. Soc.* **46**, 930–933.

Pietsch, A.
[1967] Absolut p-summierende Abbildungen in normierte Raümen, *Studia Math.* **28**, 333–353.
[1978] *Operator Ideals* (Deutscher Verlag der Wissenschaften, Berlin).

Pisier, G.
[1975] Martingales with values in uniformly convex spaces, *Israel J. Math.* **20**, 326–350.
[1978] Sur les espaces de Banach de dimension finie a distance extremal d'un espace euclidean, Exp. 16, *Seminaire d'Analyse Fonctionnelle* (Ecole Polytechnique, Paris).
[1980] Un théorème sur les opérateurs linéaires entre espaces de Banach qui se factorisent par un espace de Hilbert, *Ann. Sci. École Norm. Sup.* **13**, 23–43.
[1981] Remarques sur un résultat non publié de B. Maurey, Exp. 5, *Seminaire d'Analyse Fonctionnelle* (Ecole Polytechnique, Paris).
[1982] Holomorphic semi-groups and the geometry of Banach spaces, *Ann. of Math.* **115**, 375–392.
[1983] On the dimension of the l_p^n subspaces of Banach spaces, for $1 \leq p < 2$, *Trans. Amer. Math. Soc.* **276**, 201–211.

[1986] *Factorization of Linear Operators and the Geometry of Banach Spaces*, CBMS, Vol. 60 (American Math. Soc., Providence, RI).
[1989a] A new approach to several results of V. Milman, *J. Reine Angew. Math.* **393**, 115–131.
[1989b] *The Volume of Convex Bodies and Banach Space Geometry*, Cambridge Tracts in Math., Vol. 94.

Pisier, G., and Q. Xu
[1987] Random series in the real interpolation spaces between the spaces v_p, in: *Lecture Notes in Mathematics*, Vol. 1267 (Springer, Berlin) pp. 185–209.

Rogers, C.A., and G.C. Shephard
[1957] The difference body of a convex body, *Arch. Math.* **8**, 220–233.

Rolewicz, S.
[1972] *Metric Linear Spaces* (Monografie Mat., Warsaw).

Rudelson, M.
[1992] Manuscript, in preparation.

Rutovitz, D.
[1965] Some parameters associated with finite dimensional spaces, *J. London Math. Soc.* **40**, 241–255.

Schechtman, G.
[1987] More on embedding subspaces of L_p in l_r^n, *Comput. Math.* **61**, 159–170.
[1989] A remark concerning the dependence on ε in Dvoretzky's theorem, in: *Lecture Notes in Mathematics*, Vol. 1376 (Springer, Berlin) pp. 274–277.

Schmuckenschläger, M.
[1991] On the dependence on ε in a theorem of J. Bourgain, J. Lindenstrauss and V. Milman, in: *Lecture Notes in Mathematics*, Vol. 1469 (Springer, Berlin) pp. 166–173.

Schütt, C.
[1978] Unconditionality in tensor products, *Israel J. Math.* **31**, 209–216.
[1981] On the uniquencss of symmetric bases in finite dimensional Banach spaces, *Israel J. Math.* **40**, 97–117.

Stein, E.M.
[1961] On limits of sequences of operators, *Ann. of Math.* **74**, 140–170.

Sudakov, V.N.
[1971] Gaussian random processes and measures of solid angles in Hilbert spaces, *Soviet Math. Dokl.* **12**, 412–415.

Szarek, S.J.
[1978] On Kashin's almost Euclidean orthogonal decomposition of l_1^n, *Bull. Acad. Polon. Sci.* **26**, 691–694.
[1983] The finite dimensional basis problem, with an appendix on nets of Grassman manifold, *Acta Math.* **151**, 153–179.
[1986] On the existence and uniqueness of complex structure of spaces with 'few' operators, *Trans. Amer. Math. Soc.* **293**, 339–353.
[1987] A Banach space without a basis which has the bounded approximation property, *Acta Math.* **159**, 81–97.
[1990] Spaces with large distance to l_∞^n and random matrices, *Amer. J. Math.* **112**, 899–942.
[1991a] Computing summing norms and type constants on few vectors, *Studia Math.* **98**, 147–156.
[1991b] On the geometry of the Banach Mazur compactum, in: *Lecture Notes in Mathematics*, Vol. 1470 (Springer, Berlin) pp. 48–59.

Szarek, S.J., and M. Talagrand
[1989] An isomorphic version of the Sauer–Shelah lemma and the Banach–Mazur distance to the cube, in: *Lecture Notes in Mathematics*, Vol. 1376 (Springer, Berlin) pp. 105–112.

Szarek, S.J., and N. Tomczak-Jaegermann
[1980] On nearly Euclidean decompositions of some classes of Banach spaces, *Compositia Math.* **40**, 367–385.

Talagrand, M.
[1990] Embedding subspaces of L_1 into l_1^N, *Proc. Amer. Math. Soc.* **108**, 363–369.
Tomczak-Jaegermann, N.
[1978] The Banach Mazur distance between the trace classes C_p^n, *Proc. Amer. Math. Soc.* **72**, 305–308.
[1979a] Computing 2-summing norms with few vectors, *Ark. Mat.* **17**, 273–277.
[1979b] On the Banach Mazur distance between symmetric spaces, *Bull. Acad. Polon. Sci.* **27**, 273–276.
[1983] The Banach Mazur distance between symmetric spaces, *Israel J. Math.* **46**, 40–66.
[1984] The weak distance between Banach spaces, *Math. Nachr.* **119**, 291–307.
[1989] *Banach–Mazur Distance and Finite-Dimensional Operator Ideal,* Pitman Monographs, Vol. 38 (Pitman, London).
Tzafriri, L.
[1974] On Banach spaces with unconditional bases, *Israel J. Math.* **17**, 84–93.
Ullrich, D C.
[1988] An extension of the Kahane–Khintchine inequality, *Bull. Amer. Math. Soc.* **18**, 52–54.
Vaaler, J.D.
[1979] A geometric inequality with applications to linear forms, *Pacific J. Math.* **83**, 543–553.
Von Neumann, J.
[1942] Approximative properties of matrices of high order rank, *Portugal. Math.* **3**, 1–62.
Wagner, G.
[1992] On a new method for constructing good point sets on spheres, *J. Discrete Comput. Geom.*, to appear.
Zippin, M.
[1981a] The range of a projection of small norm in l_1^n, *Israel J. Math.* **39**, 349–358.
[1981b] The finite dimensional P_λ spaces with small λ, *Israel J. Math.* **39**, 359–365. Errata: **48** (1984) 255–256.

CHAPTER 4.6

Nonexpansive Maps and Fixed Points

Pier Luigi PAPINI

Dipartimento di Matematica dell' Università, Piazza di Porta San Donato 5, I-40127 Bologna, Italy

Contents

HANDBOOK OF CONVEX GEOMETRY
Edited by P.M. Gruber and J.M. Wills

1. Introduction

The literature concerning fixed point theory is so wide, that it is almost impossible to choose the most relevant results and condense them in a few pages, even if we limit the discussion to nonexpansive mappings in Banach spaces. Here we want to indicate one line of research in the area, along with a few side paths and generalizations, trying to point out the great role convexity plays in this context. For any path we indicate at least a reference: often a survey paper on the subject (not necessarily the most complete or the latest one); other times a paper containing enough references, or a meaningful result (not necessarily the deepest, or the first, or the latest one of some importance). We have no pretension to put the results (and their authors) into their right historical place or to quote the original source.

When not otherwise indicated, we shall deal with Banach spaces. Our main reference will be the recent book by Goebel and Kirk (1990).

1.1. Main definitions

We list a few definitions. The reader must be alerted that the terminology in use is not always standard: sometimes the same term is used to denote different classes of mappings.

Let $(X, \|\cdot\|)$ be a Banach space over the real field $\mathbb{R}$; we shall consider mappings, defined on nonempty subsets of X, into X. A map T, defined on $\operatorname{dom}(T) = A$, is said to be *k-Lipschitz* if we have, for some $k \in \mathbb{R}$,

$$\|Tx - Ty\| \leq k\|x - y\| \quad \text{for all } x, y \text{ in } A . \tag{1}$$

We say that T is a (strict) *contraction* if (1) holds for some $k \in [0, 1)$, i.e., when we have

$$\|Tx - Ty\| \leq k\|x - y\| \quad \text{for all } x, y \text{ in } A \text{ and some } k \in [0, 1) . \tag{2}$$

We say that T is a *nonexpansive* map if we have

$$\|Tx - Ty\| \leq \|x - y\| \quad \text{for all } x, y \text{ in } A . \tag{3}$$

Half way between these two classes of maps, we have the *generalized contractive* maps: a map T belongs to such class if

$$\|Tx - Ty\| < \|x - y\| \quad \text{for all } x, y \text{ in } A . \tag{$2\frac{1}{2}$}$$

Now set $F(T) = \{x \in X\colon Tx = x\}$.

We are interested in the following problem: under some assumption for A, can we say that for a map T of one of these types we have $F(T) \neq \emptyset$? In case this question has a positive answer, another important problem in this context is the following: how can we approximate elements in $F(T)$? When $F(T)=\emptyset$, we say that T is *fixed point free*.

We shall say that a set C (or a family $\mathbb{C}$ of sets) has the *fixed point property*, abbreviated FPP, for a class of mappings, when any self map of C (respectively of any $C \in \mathbb{C}$) from that class has at least one fixed point. We shall abbreviate by FPPNM the fact that a set or a class $\mathbb{C}$ of sets has FPP for nonexpansive maps. We shall often consider the class containing all weakly compact, convex sets, which we shall abbreviate by WCC; also we shall abbreviate bounded closed and convex by BCC.

2. Some examples

We list a few *examples of fixed point free maps*. They can be divided into two groups: the first one contains four simple examples, in $X=\mathbb{R}$, of maps with $\text{dom}(T) \subset \mathbb{R}$; they show that also in very nice spaces, a self map of a set C which is nonexpansive (or also, of a more special type) can be fixed point free, when one of the following assumptions is not satisfied for C: closure, convexity, boundedness. The second group contains examples which are somehow more elaborated, but which can be considered almost "classical".

Example 1. Set $T: x \to -x$, with $\text{dom}(T)=\{-1,1\}$; T satisfies (3) but not $(2\frac{1}{2})$.

Example 2. Let $Tx = x + 1/x$, with $\text{dom}(T)=[1,\infty)$. T satisfies $(2\frac{1}{2})$ but not (2).

Example 3. A translation in $\mathbb{R}$ is an affine isometry. A similar example can be given for the 1-dimensional torus.

Example 4. Let $Tx = \frac{1}{2}x + \frac{1}{2}$, with $\text{dom}(T)=\{x \in \mathbb{R}: \|x\|<1\}$.

Example 5 (Goebel and Kirk 1990, *p*. 30)**.** In c_0, the map sending $x=(x_1, x_2, \ldots, x_n, \ldots)$ in $Tx=(1, x_1, \ldots, x_{n-1}, \ldots)$ is a self map of the unit ball which is an (affine) isometry.

Example 6 (Goebel and Kirk 1990, *p*. 12)**.** Let $X=C[0,1]$ and $C=\{f \in X: 0 \leqslant f(x) \leqslant 1, f(0)=0, f(1)=1\}$. The map T sending $f(x)$ to $xf(x)$ is a self map of C satisfying (3).

Example 7 (Lin 1987)**.** Let $X=\ell^2$ and $C=\{x \in \ell^2: x=\Sigma_{i=1}^{\infty} \alpha_i e_i, \Sigma_{i=1}^{\infty} \alpha_i^2 \leqslant 1, \alpha_i \searrow 0\}$. For $x \in A$ set $g(x)=\max(a_1, 1-\|x\|)e_1 + \Sigma_{i=2}^{\infty} \alpha_{i-1} e_i$ and $Tx = g(x)/\|g(x)\|$. T is k-Lipschitz with $k \leqslant 20$; moreover, T satisfies the following property: for every $\delta>0$, there exists an $N_0 \in \mathbb{N}$ such that for every $x \in C$, for all $n>N_0$, $\|T^{n+1}(x)-T^n(x)\|<\delta$.

Example 8 (Rosenholtz 1976)**.** Let $X=c_0$; consider the following self map of its unit ball: for $x=(x_1, x_2, \ldots, x_n, \ldots)$, $Tx=(1, a_1x_1, a_2x_2, \ldots)$, where the sequence $\{a_n\}$ is such that:

(a) $0<a_i<1$ for every $i \in \mathbb{N}$;

(b) the sequence of partial products, $p_n = \Pi_{j=1}^n a_j$, is bounded away from 0.

For example, we can take $a_n=(2^n+1)/(2^n+2)$. Then T satisfies $(2\frac{1}{2})$; moreover, it is an affine homeomorphism onto its image and its inverse is Lipschitz.

Example 8 bis (Totik 1983). Let $X=c_0$; for $x=(x_1, x_2, \ldots, x_n, \ldots)$, let

$$Tx = \left(1, \frac{x_2}{2}+\frac{1}{2}, \frac{2x_3}{3}+\frac{1}{3}, \ldots, \frac{(n-1)x_n}{n}+\frac{1}{n}, \ldots\right).$$

T is affine, maps the unit ball of c_0 into its boundary and satisfies $(2\frac{1}{2})$.

Example 9 (Goebel and Kirk 1990, *p.* 85). Renorm c_0 by $|x| = \|x^+\|_\infty = \|x^-\|_\infty$ (x^+ and x^- denote the positive and the negative part of x). The dual of $(c_0, |\cdot|)$ is isometrically isomorphic to $(\ell^1, |\cdot|)$, with $|x| = \max(\|x^+\|, \|x^-\|)$. The set $C = \{x \in \ell^1\colon x_i \geq 0; \Sigma_i x_i \leq 1\}$ is w^*-compact and convex, while $T: C \to C$, $x \to (1-\Sigma_i x_i, x_1, x_2, \ldots)$ is an (affine) isometry.

Example 10 (Goebel and Kirk 1990, *p.* 202). In ℓ^2, if $x=(x_1, x_2, \ldots, x_n, \ldots)$ and $\varepsilon>0$, set $Tx=(\varepsilon(1-\|x\|)x_1, x_1, x_2, \ldots)$; this map is a $(1+\varepsilon)$-Lipschitz self map of the unit ball.

Example 11 (Goebel and Kirk 1990, *p.* 122). Let

$$C = \left\{f \in L^1\colon 0 \leq f \leq 2 \text{ almost everywhere}, \int_0^1 f(x)\,\mathrm{d}x = 1\right\}.$$

Set

$$Tx = \begin{cases} \inf(2f(2t), 2) & \text{if } 0 \leq t \leq \frac{1}{2}, \\ \sup(2f(2t-1)-2, 0) & \text{if } \frac{1}{2} < t \leq 1. \end{cases}$$

T is an isometric self map of C, which is convex and weakly compact.

The last example can be considered to be among the most important (and quoted) ones; for some comments and improvements see Sine (1982).

3. Some results (and some history)

Fixed point theory has – in some sense – a very long history: we may refer, e.g., to Roux (1978a,b) for some historical hints.

As Bolzano already showed in 1817, if $f:[a,b] \to \mathbb{R}$ is continuous and $f(a) \cdot f(b) < 0$, then there exists some $c \in [a,b]$ such that $f(c)=0$: in some sense this is one of the first fixed point results. The thesis fails if continuity of f or connectedness of its domain are not assumed.

In 1910, Brouwer proved the following result: if T is a continuous self map of a convex, compact subset of $\mathbb{R}^n$, then $F(T) \neq \emptyset$.

In 1930, Schauder proved that in any Banach space, for a continuous map T mapping a convex compact set C into itself we have $F(T) \neq \emptyset$. Also, the same is true if we "shift" compactness from the (closed) set C to the map T.

If we want to work with infinite dimensional Banach spaces, the assumption of compactness is in general a very strong one: for example, in this case the interior of a compact set must be empty. If we want to drop this assumption, we may strengthen the assumption of continuity for T. In fact, in 1922 Banach proved the following: if T is a self map of any C and satisfies (2), then $F(T) \neq \emptyset$. Unfortunately, assumption (2) is rather strong; in many cases arising also in applications, T satisfies (3) but not (2): but this is not enough to imply existence of fixed points for T [neither $(2\frac{1}{2})$ in general suffices]. So we may think at maps satisfying (3), mapping into itself some weakly compact, closed set: this is a much milder assumption with respect to compactness. For example, closed bounded and convex sets in any reflexive Banach space are weakly compact. Thus the role of convexity arises, as a condition to be added to closedness and boundedness (which are quite reasonable ones) to replace compactness in infinite dimensional spaces. Moreover, closure and weak closure for convex sets are equivalent.

In some sense, we can try to divide fixed point theorems into two main classes: those which have a more topological appeal, dealing with continuous functions and stressing the role of compactness; results in this area call to mind the names of Tychonoff, Lefschetz, KyFan (1935, 1942, 1961) and many others. The second class contains results using convexity (so weaker forms of compactness), and dealing with nonexpansive maps: related names of beginners are, among others, those of Brouwer, Göhde, Kirk, Karlovitz. In fact, much work concerning this second class of problems was done around 1965; a major later achievement was the discovery of Example 11 fifteen years later. Here we are going to discuss results of the second type, which are related to functional analysis (mainly, to geometry of Banach spaces).

3.1. Main results

Many results concerning FPPNM for WCC sets have been given for "good" classes of spaces: for Hilbert spaces (in 1965), then for uniformly convex spaces (1965), for reflexive spaces with normal structure (1965), for uniformly smooth spaces (1978–1979) (see Goebel and Kirk 1990). But also several spaces which are "bad" from the geometrical point of view have similar properties; a detailed account of some results of this kind concerning, e.g., the spaces c_0, ℓ^1, ℓ^∞ and L^∞ is given in Goebel and Kirk (1990) and in Aksoy and Khamsi (1990): the last book is – in some sense – a complement to Goebel and Kirk (1990).

Concerning the techniques used, to prove the result in Hilbert spaces concepts from the theory of monotone operators and best approximation were used. Then some special tools were developed. Instead of giving proofs concerning some of the numerous "classical" results of this type, we want to indicate briefly some of the main notions used to prove FPPNM in large classes of spaces.

3.2. *Approximate fixed points*

Though the FPPNM for WCC sets does not hold in any Banach space, the following weaker property is always satisfied. For every nonexpansive self map of a BCC subset A of X we have the "almost fixed point property", i.e.,

$$\inf\{\|x - Tx\|: x \in A\} = 0 . \tag{AFPP}$$

In Hilbert spaces, it is not difficult to see that for nonexpansive mappings and A as before, AFPP implies FPP (see also Goebel and Kirk 1990, p. 109). More generally, the following fact is true in a reflexive (but not in every) Banach space: if A is closed and convex, then A has AFPP for nonexpansive mappings if and only if it is linearly bounded (see section 4.5 for the definition).

A sequence $\{x_n\}$ of elements with $\lim_{n\to\infty}\|x_n - Tx_n\| = 0$ is called an approximate fixed point sequence. To obtain sequences of this type, we may consider, e.g., fixed points of $T_n = (1 + 1/n)T$. These sequences are useful to prove FPPNM, for example, in spaces satisfying the so-called "Opial condition", or in which the duality mapping possesses some kind of continuity (see again Goebel and Kirk 1990).

3.3. *Asymptotic centers and minimal invariant sets*

Given a bounded sequence $\{x_n\}$ in X, and a subset K, the number $\inf\{\limsup\|y - x_n\|: y \in K\}$ is called *asymptotic radius* of $\{x_n\}$ from K. A point $\bar{y} \in K$ is called an *asymptotic center* for $\{x_n\}$ from K if such infimum is attained at $\bar{y}$.

Note that existence (and also uniqueness) of asymptotic centers, from many "good" sets, is assured, e.g., in uniformly convex spaces. This fact, together with AFPP, implies results concerning FPPNM.

Another kind of methods is based on the following fact. Suppose K is a BCC set and T is a nonexpansive self map of K. A standard application of Zorn's lemma yields the following: K contains a BCC subset K_0 which is *minimal invariant* for T, i.e., no proper BCC subset of K_0 is invariant for T. A set of this type possesses nice, simple properties (see Goebel and Kirk 1990, p. 33). Moreover, if for some reason we can say that K_0 is a singleton, then we may conclude that T has a fixed point. By using these arguments, it is possible to obtain fixed point theorems for K when every subset A in K which is not a singleton has the following property, called normal structure (see Goebel and Kirk 1990, §4):

$$\text{there exists } x \in A \text{ such that } \sup\{\|x - y\|: y \in A\} < \text{diameter}(A) . \tag{NS}$$

Note that if a BCC set A is minimal invariant for T, then also the closed convex hull of $T(A)$ shares the same property. Also, it is possible to construct minimal invariant sets by using asymptotic centers. If A is compact, then points which are "half-way" between points whose distance is equal to diameter(A) play an important role.

Now assume X to be reflexive; in this case, the intersection of nonempty, closed, bounded subsets is nonempty, which gives nonemptiness properties for some invariant minimal sets; then we get FPPNM if all nonempty BCC subsets of X satisfy (NS) (this is always true, e.g., when X is uniformly convex).

For a result gluing together some of these notions, see Goebel and Kirk (1990, p. 91).

For results concerning (NS) and related conditions see, e.g., Nelson, Singh and Whitfield (1987). This notion is very important and really connected with FPPNM, as shown by Kassay (1986): in fact, a reflexive Banach space has (NS) for BCC subsets of X if and only if it has FPP for a class of mappings satisfying a condition more general than (3).

4. Some generalizations

The conditions indicated in the previous section can be generalized in many ways. We recall some of them here, which appear to be among the most popular. We shall give some indications concerning a single generalization among those possible; of course, we could also consider two or more generalizations: the class of mappings, the class of spaces, the class of sets considered for dom(T), and so on.

4.1. *Generalizing the class of mappings*

Concerning definitions (2) and ($2\frac{1}{2}$) and their analogues, the long list of generalizations considered by Rhoades in a few papers, starting from 1977 (see, e.g., Rhoades 1987, and Kincses and Totik 1990), is far from being exhaustive; note that these definitions are usually considered in a general metric space. The definition of nonexpansiveness (3) can be given similar generalizations; for some of them, see, e.g., Massa and Roux (1978).

In the following definitions, T is a map from $A \subset X$ into X ($A \neq \emptyset$). Of course, when a generalization of condition (3) is indicated, a similar generalization for condition (2) or ($2\frac{1}{2}$) is possible.

For any (possible) fixed point p of T, we have

$$\|Tx - p\| \leq \|x - p\| \quad \text{for all } x \text{ in } A\,, \tag{qne}$$

T is *quasi-nonexpansive*.

$$\|Tx - Ty\| \leq \max(\|x - y\|, \|Tx - y\|, \|Ty - x\|, \|Tx - x\|, \|Ty - y\|) \quad \text{for all } x, y \text{ in } A\,, \tag{gqn}$$

these maps are sometimes called *generalized nonexpansive*, or also *quasi-nonexpansive*.

$$\|Tx - Ty\| \leq a_1\|x - y\| + a_2\|Tx - y\| + a_3\|Ty - x\| + a_4\|Tx - x\| + a_5\|Ty - y\| \quad \text{for all } x, y \text{ in } A\,, \tag{gne}$$

where $\Sigma_{i=1}^{5} a_i = \Sigma_{i=1}^{5} |a_i| = 1$; T is *generalized nonexpansive*.

There exists a $k \in \mathbb{R}^+$ such that

$$\|T^n x - T^n y\| \leq k\|x - y\| \quad \text{for all } n \in \mathbb{N}\,, \tag{uk-L}$$

T is *uniformly k-Lipschitz*.

In general, for T (uk-L) with $k > 1$, we may have $F(T) = \emptyset$; but nice results can be given (for k in some interval $(1, 1 + \varepsilon]$) under simple assumptions on X and C: see, e.g., Nelson, Singh and Whitfield (1987).

For some results concerning maps satisfying (gne) (or more particular conditions of that type), or (gqn), see, e.g., Massa and Roux (1978); for maps of type (qne), see Roux (1978b).

Given a "measure of noncompactness" α in X, we may consider the following classes of maps: there exists a $k \in \mathbb{R}^+$ such that for any $D \subset \text{dom}(T)$ we have

$$\alpha(T(D)) \leq k\alpha(D)\,, \tag{4}$$

T is a *k-Lipschitz α-contraction*.

In general the cases $k \in [0, 1)$ and $k = 1$ are considered. When we use the usual Kuratowski or Hausdorff measure of noncompactness, we speak of k-set or, respectively, of k-ball contractions. For $k = 1$, we also speak of α-nonexpansive maps.

For results concerning maps which satisfy (4) with $k < 1$ [or the analogue of condition $(2\frac{1}{2})$, but defined in terms of α], see Akhmerov et al. (1982); for results concerning $k = 1$, see Petryshyn (1973).

A result concerning maps which "reduce" the measure of nonconvexity of a set, was given in Cano (1990).

Moreover, there are also (almost, weakly, locally, asymptotically, firmly, strongly, . . .) nonexpansive, condensing, hemicontractive, semicontractive, demicontractive, pseudo contractive, . . . maps!

4.2. *Generalizing the space*

Note that completeness of X (or closure of C) is a very important assumption in our context: without such an assumption fixed point theory becomes weak, see, e.g., Borwein (1983).

As stated at the beginning, we do not intend to consider here explicitly nonnormed spaces; but of course, many interesting facts could be considered in more general settings (locally convex spaces, metric spaces, . . .); we mention some reference papers. For results concerning locally convex spaces, we refer to

Naimpally, Singh and Whitfield (1982). For Hausdorff topological vector spaces, and also for multi-valued maps, we refer to Browder (1976) and Hadžić (1984).

Note that in most cases convexity plays a key role. For example, many authors considered these problems in convex metric spaces, i.e., in metric spaces such that for any pair x, y, it is possible to find some z such that $d(x, z) + d(y, z) = d(x, y)$. It is also possible to define strictly convex and uniformly convex metric spaces; for results in these spaces see Sastry et al. (1987).

More generally, it is possible to define spaces X with a convexity structure (CS); for results in these spaces see Naimpally, Singh and Whitfield (1984). In fact, a space of this type is essentially a convex subset of some normed space. Also, we may think at a more abstract notion of convexity, defined by means of a family of sets, containing $\emptyset$ and X, and closed with respect to intersection: for results in this setting see Kirk (1983) [a condition similar to (NS) can be used].

4.3. *Families of maps and multivalued maps*

Multivalued maps occur also in applications; in general, fixed point theory for these maps is done under the assumption that the map considered has closed and convex, or compact values. In this setting nonexpansiveness has the following meaning: if we denote by $H(A, D)$ the Hausdorff distance between two sets A and D, T must satisfy the condition

$$H(Tx, Ty) \leq \|x - y\| \quad \text{for all } x, y \text{ in } \mathrm{dom}(T). \tag{mne}$$

For these maps, FPP is the following: there exists some $x \in \mathrm{dom}(T)$ such that $x \in T(x)$. For different contraction definitions as well as for some other results see Borisovich et al. (1984).

Common fixed points for a family of nonexpansive (or more general) maps were studied. Some nice results have been given, e.g., under the assumption that the family commutes and forms a semigroup. For results of this type, see, e.g., Goebel and Kirk (1990, §15). The literature contains also many simple results concerning common fixed points for pairs of maps.

4.4. *Fixed points and nonconvex sets*

Concerning the possibility of extending some results about nonexpansive mappings which hold for BCC sets to starshaped (or also to more general) sets, we refer, e.g., to Chandler and Faulkner (1980). A rather general fixed point result in Hilbert spaces was given in Goebel and Schöneberg (1977). Examples of "strange" sets with the fixed point property were given in Goebel and Kuczumow (1979).

4.5. *Fixed points and unbounded sets*

Let us drop the boundedness assumption for A; for example, assume that $T: A \to A$ is a nonexpansive mapping and A is closed convex and unbounded.

Assume now that X is a Hilbert space. If A is also *linearly unbounded*, i.e., A has an unbounded intersection with some line in X, then it is not difficult to construct a fixed point free, self map $T : A \rightarrow A$; also, the set $A = \{x \in \ell^1 \colon \|x\| \leqslant 1$ for all $i\}$ is linearly bounded and the map $T : (x_1, x_2, x_3, \ldots) \rightarrow (1, x_1, x_2, \ldots)$ is an isometry, mapping A into itself, which is fixed point free. Rather unexpectedly, the following was proved to be true (see Goebel and Kirk 1990, pp. 130–132):
– let A be a convex and linearly bounded subset of a real Hilbert space. Then K has the FPPNM if and only if A is bounded.

For a similar result, in general Banach spaces, see Shafrir (1990).

Concerning linearly bounded sets, we recall the following fact (Allen 1986):
– a Banach space is finite dimensional if and only if every unbounded convex set contains a ray.

Concerning fixed points for maps on (convex) cones, we refer to Lafferriere and Petryshyn (1989).

For a general reference about fixed points for some kinds of maps on unbounded sets, see Marino and Pietramala (1992).

5. A few miscellaneous results

Let A be again a BCC set. Though the existence of fixed points for nonexpansive self-mappings of A is not always assured, "most" maps (also in a rather strong sense) have FPP: see De Blasi and Myjak (1989).

Fixed point property is a highly instable property: see, e.g., Goebel and Kirk (1990, p. 96).

We also indicate the following fact (see Lin 1986): for T nonexpansive, the set $\bigcap_{n=1}^{\infty} T^n(A)$, A BCC, is not always convex, also in Hilbert spaces.

5.1. *Iteration and approximation*

The simplest way of convergence concerning fixed points we may consider is the convergence of $T^{(n)}(x)$, for x in dom(T), hopefully in a "monotone" way, to some element of $F(T)$; we may study strong or weak convergence (or convergence in some other sense). Unfortunately, this convergence does not occur frequently, so we are led to consider other processes by using "averages" with the iterates of T, $x_n = T^n x$, the iterates of $\alpha I + (1 - \alpha)T$ and so on.

In general, in infinite dimensional spaces the convergence of these methods is not guaranteed (see Goebel and Kirk 1990, p. 104). For a discussion concerning this topic, see also Roux (1978b) or Kirk (1983); for some more results on convergence see Bruck (1983).

5.2. *The structure of the fixed point set*

The fixed set of a nonexpansive mapping in general is neither convex, nor connected, nor weakly closed: see Gruber (1975) and Sine (1982). But when X is strictly convex, $F(T)$ is (closed and) convex for T in most classes considered: see

Goebel and Kirk (1990, pp. 34–35 and p. 117). More precisely (see Khamsi 1989), a space X is strictly convex if and only if for any nonexpansive mapping T on any convex set, $F(T)$ is convex. For another characterization of strict convexity in terms of nonexpansive mappings, see Müller and Reinermann (1979), where it was proved that convexity of a set is implied by the validity of a property slightly stronger than FPPNM.

In Lin and Sternfeld (1985), a characterization of compactness for convex sets in terms of FPP for Lipschitz maps was given. A similar result concerning starshapedness and finite dimensional spaces was indicated in Müller and Reinermann (1979).

5.3. Some more particular spaces, sets, and maps

Results in spaces with a richer structure were considered; for example, for Banach lattices we refer to Nelson, Singh and Whitfield (1987), and Aksoy and Khamsi (1990).

We recall that an isometry is a map T (from A to X) such that $\|Tx - Ty\| = \|x - y\|$ for all x, y in A. For results concerning fixed points for isometries see Lau (1980). In particular, if X is strictly convex, then X has FPP for isometries and WCC sets (compare with Example 11 in section 2; see also section 4.5 and the remark after Problem 1 bis in section 6.1).

Also results for other classes of sets were considered: balls [see Nadler (1981) for Euclidean spaces], etc. We recall that some topological results concerning fixed points and polyhedra (in finite dimensional spaces) were indicated by Thomeier (1982).

5.4. Changing the space and preserving FPP

We may consider the following question (which is related to FPP for k-Lipschitz mappings): assume that X has FPPNM for a class of subsets; let Y be a Banach space, isomorphic to X, with a Banach–Mazur distance d from X not too large: can we say that Y has again FPPNM, for the same class of sets, for d in some interval $[1, 1 + \varepsilon]$, $\varepsilon \in \mathbb{R}^+$? For some results of this type see Goebel and Kirk (1990, §14).

Now let X and Y have FPPNM; we may ask when this property is preserved if we construct a product space. For results in this direction see Kuczumow (1990).

6. Some other general facts

We conclude this article with the following remarks:
– The existing literature contains, together with some remarkable results, many simple remarks or exercises (making the reputation of the area rather low!).
– Fixed point theory stimulates "pure" mathematicians (working, e.g., in metric geometry and in nonlinear analysis) to introduce new properties and to study new classes of spaces.

– Fixed point theory plays a great role: it is a good tool to prove some theorems usually taught in a course of "calculus", it has relations with some important classes of mappings and it is quite important for applications in many fields.

6.1. A few open problems

Some of the main problems in the area stand still unsolved, notwithstanding many efforts and recent progress. We quote a few among them.

Problem 1. Does every reflexive (or every reflexive, strictly convex) Banach space have FPPNM for WCC sets?

Problem 1 bis. Does any superreflexive Banach space have FPPNM for WCC sets? We recall that superreflexive spaces have FPP for isometries and WCC sets: see Aksoy and Khamsi (1990, p. 99).

Problem 2. Does every reflexive (or every superreflexive) Banach lattice have FPPNM for WCC sets?

Problem 3. Assume that a WCC set C has FPPNM. Does C have FPP for nonexpansive multivalued maps which are compact, convex valued?

For a discussion of the above problems and other ones, see Nelson, Singh and Whitfield (1987); see also Reich (1983).

6.2. Some general references

For a few very simple applications, we may refer to Wagner (1982); for a discussion of the role fixed points play in nonlinear functional analysis, see, e.g., Browder (1976). For applications to operator equations, one may refer to Akhmerov et al. (1982). For applications to approximation theory, see Singh (1985).

As we said at the beginning, we considered here some "functional" aspects of fixed point theory. There are many general works related to these and to the more topological aspects of the theory, while many books on more general subjects contain chapters dedicated to this topic. As general references, we may indicate the books and the bibliography by Istrăţescu (1981, 1985, 1989): unfortunately, not all these works are easily accessible. We quote also the books by Eisenack and Fenske (1978), and Goebel and Reich (1984).

Acknowledgement

We are indebted to E. Casini, E. Maluta and C. Zanco for some useful discussions.

References

Akhmerov, R.R., M.I. Kamenskii, S.I. Potapov and B.N. Sadovskii
[1982] Condensing operators, *J. Soviet Math.* **18**, 551–592.

Aksoy, A.G., and M.A. Khamsi
[1990] *Nonstandard Methods in Fixed Point Theory* (Springer, Berlin).

Allen, C.S.
[1986] A characterization of dimension, *Amer. Math. Monthly* **93**, 635–636.

Borisovich, Yu.G., B.D. Gel'man, A.D. Mikshkis and V.V. Obukhovskii
[1984] Multivalued mappings, *J. Soviet Math.* **24**, 719–791.

Borwein, J.M.
[1983] Completeness and the contraction principle, *Proc. Amer. Math. Soc.* **87**, 246–250.

Browder, F.E.
[1976] Nonlinear operators and nonlinear equations of evolution in Banach spaces, in: *Nonlinear Functional Analysis,* Proc. Symposia in Pure Math., Vol. 18, Part 2, ed. F.E. Browder (Amer. Math. Soc., Providence, RI).

Bruck, R.E.
[1983] Asymptotic behaviour of nonexpansive mappings, in: *Fixed Points and Nonexpansive Maps,* Contemp. Math., Vol. 18, ed. R.C. Sine (Amer. Math. Soc., Providence, RI) pp. 1–47.

Cano, J.
[1990] A measure of non-convexity and another extension of Schauder's theorem, *Bull. Math. Soc. Sci. Roumanie (N.S.)* **34**(82), 3–6.

Chandler, E., and G. Faulkner
[1980] A fixed point theorem for non-expansive condensing mappings, *J. Austral. Math. Soc. A* **29**, 393–398.

De Blasi, F.S., and J. Myjak
[1989] Sur la porosité de l'ensemble des contractions non linéaires sans points fixes, *C.R. Acad. Sci. Paris Sér. I Math.* **308**, 51–54.

Eisenack, G., and C. Fenske
[1978] *Fixpunkttheorie* (Wissenschaftsverlag Bibliographisches Institut, Mannheim).

Goebel, K., and W.A. Kirk
[1990] *Topics in Metric Fixed Point Theory* (Cambridge Univ. Press, London).

Goebel, K., and T. Kuczumow
[1979] Irregular convex sets with fixed-point property for nonexpansive mappings, *Colloq. Math.* **40**, 259–264.

Goebel, K., and S. Reich
[1984] *Uniform Convexity, Hyperbolic Geometry, and Nonexpansive Mappings* (Marcel Dekker, New York).

Goebel, K., and R. Schöneberg
[1977] Moons, bridges, birds... and nonexpansive mappings in Hilbert space, *Bull. Austral. Math. Soc.* **17**, 463–466.

Gruber, P.M.
[1975] Fixpunktmengen von Kontraktionen in endlichdimensionalen normierten Räumen, *Geom. Dedicata* **4**, 179–198.

Hadžić
[1984] *Fixed Point Theory in Topological Vector Spaces* (Inst. of Mathematics, Novi Sad).

Istrăţescu, V.I.
[1981] *Fixed Point Theory – An Introduction* (Reidel, Dordrecht).
[1985] *Bibliography. Fixed Point Theory 1836–1985* (3 volumes) (Kiel).
[1989] *Fixed Point Theory* (Kiel).

Kassay, G.
[1986] A characterization of reflexive Banach spaces with normal structure, *Boll. Un. Mat. Ital. A (6)* **5**, 273–276.

Khamsi, M.A.
[1989] On normal structure, fixed point property and contractions of type (γ), *Proc. Amer. Math. Soc.* **106**, 995–1001.

Kincses, J., and V. Totik
[1990] Theorems and counterexamples on contractive mappings, *Math. Balkanica* **4**, 69–90.

Kirk, W.A.
[1983] Fixed point theory for nonexpansive mappings II, in: *Fixed Points and Nonexpansive Maps*, Contemp. Math., Vol. 18, ed. R.C. Sine (Amer. Math. Soc., Providence, RI) pp. 121–140.

Kuczumow, T.
[1990] Fixed point theorems in product spaces, *Proc. Amer. Math. Soc.* **108**, 727–729.

Lafferriere, B., and W.V. Petryshyn
[1989] New positive fixed point and eigenvalue results for P_γ-compact maps and some applications, *Nonlinear Anal.* **13**, 1427–1439.

Lau, A.T.-M.
[1980] Sets with fixed point property for isometric mappings, *Proc. Amer. Math. Soc.* **79**, 388–392.

Lin, P.-K.
[1986] On the core of nonexpansive mappings, *Houston J. Math.* **12**, 537–540.
[1987] A uniformly asymptotically regular mapping without fixed points, *Canad. Math. Bull.* **30**, 481–483.

Lin, P.-K., and Y. Sternfeld
[1985] Convex sets with the Lipschitz fixed point property are compact, *Proc. Amer. Math. Soc.* **93**, 633–639.

Marino, G., and P. Pietramala
[1992] Fixed points and almost fixed points for mappings defined on unbounded sets in Banach spaces, *Atti Sem. Mat. Fis. Univ. Modena* **40**, 1–9.

Massa, S., and D. Roux
[1978] A fixed point theorem for generalized nonexpansive mappings, *Boll. Un. Mat. Ital. A (5)* **15**, 624–634.

Müller, G., and J. Reinermann
[1979] Eine Characterisierung strikt konvexer Banach-Räume über Fixpunktsatz für nichtexpansive Abbildungen, *Math. Nachr.* **93**, 239–247.

Nadler Jr, S.B.
[1981] Examples of fixed point free maps from cells onto larger cells and spheres, *Rocky Mountain J. Math.* **11**, 319–325.

Naimpally, S.A., K.L. Singh and J.H.M. Whitfield
[1982] Fixed points and close-to-normal structure in locally convex spaces, in: *Nonlinear Analysis and Applications,* eds S.P. Singh and J.H. Burry (Marcel Dekker, New York) pp. 203–221.
[1984] Fixed points in convex metric spaces, *Math. Japon.* **29**, 585–597.

Nelson, J.L., K.L. Singh and J.H.M. Whitfield
[1987] Normal structures and nonexpansive mappings in Banach spaces, in: *Nonlinear Analysis,* ed. Th.M. Rassias (World Scientific, Singapore) pp. 433–492.

Petryshyn, W.V.
[1973] Fixed point theorems for various classes of 1-set contractive and 1-ball contractive mappings in Banach spaces, *Trans. Amer. Math. Soc.* **182**, 323–352.

Reich, S.
[1983] Some problems and results in fixed point theory, in: *Topological Methods in Nonlinear Functional Analysis,* Contemp. Math., Vol. 21, eds S.P. Singh, S. Thomeier and B. Watson (Amer. Math. Soc., Providence, RI) pp. 179–187.

Rhoades, B.E.
[1987] Contractive definitions, in: *Nonlinear Analysis,* ed. Th.M. Rassias (World Scientific, Singapore) pp. 513–526.

Rosenholtz, I.
[1976] On a fixed point problem of D.R. Smart, *Proc. Amer. Math. Soc.* **55**, 252.

Roux, D.
[1978a] Generalizzazioni del teorema di Brower a spazi a infinite dimensioni, in: *Applicazioni del Teorema di Punto Fisso all'Analisi Economica* (Accad. Naz. Lincei, Roma) pp. 73–86.
[1978b] Teoremi di punto fisso per applicazioni contrattive, in: *Applicazioni del Teorema di Punto Fisso all'Analisi Economica* (Accad. Naz. Lincei, Roma) pp. 89–110.

Sastry, K.P.R., S.V.R. Naidu, I.H.N. Rao and K.P.R. Rao
[1987] Geometry of metric linear spaces with applications to fixed point theory, *Tamkang J. Math.* **18**, 331–340.

Shafrir, I.
[1990] The approximate fixed point property in Banach and hyperbolic spaces, *Israel J. Math.* **71**, 211–223.

Sine, R.
[1982] Remarks on the example of Alspach, in: *Nonlinear Analysis and Applications,* eds S.P. Singh and J.H. Burry (Marcel Dekker, New York) pp. 237–241.

Singh, K.L.
[1985] Applications of fixed points to approximation theory, in: *Approximation Theory and Applications,* ed. S.P. Singh (Pitman, London) pp. 198–213.

Thomeier, S.
[1982] Some remarks and examples concerning the fixed point property of polyhedra, in: *Nonlinear Analysis and Applications,* eds S.P. Singh and J.H. Burry (Marcel Dekker, New York) pp. 271–278.

Totik, V.
[1983] On two open problems of contractive mappings, *Publ. Inst. Math. (Beograd) (N.S.)* **34**(48), 239–242.

Wagner, C.H.
[1982] A generic approach to iterative methods, *Math. Mag.* **55**, 259–273.

CHAPTER 4.7

Critical Exponents

Vlastimil PTÁK

Institute of Mathematics, Czechoslovak Academy of Sciences, Žitná 25, 11567 Praha 1, Czechoslovakia

Contents

HANDBOOK OF CONVEX GEOMETRY
Edited by P.M. Gruber and J.M. Wills

1. Motivation

The notion of the critical exponent of a Banach space has its origin in considerations concerning convergence of the iterative process

$$x_{r+1} = Ax_r + y$$

where A is a bounded linear operator on a Banach space E and y a given vector in E.

It is a known fact that this iterative procedure converges for every choice of the initial vector x_0 and every right-hand side y if and only if $|A|_\sigma$, the spectral radius of A, is less that one. This will be the case if (and only if) some power of the operator A has norm less than one; for practical purposes it is, of course, important to know how far we have to go in the sequence

$$|A|, |A^2|, |A^3|, \ldots$$

to find a value less than one provided such a term exists.

The spectral radius of a scalar multiple λA being $|\lambda A|_\sigma = |\lambda||A|_\sigma$ it is obvious that the question is only meaningful if we impose some restriction on the norm of A. Thus we will restrict ourselves to operators of norm not exceeding 1. In conformity with general usage, operators of norm $\leqslant 1$ will be called contractions.

Obviously it would be desirable to find a bound for the number of steps needed to reach a power $|A^n| < 1$ if such a power exists: in other words we are looking for a number q with the following property: if $|A|_\sigma < 1$ then $|A^m|$ will be < 1 already for some exponent $m \leqslant q$. Expressed in its negative form this would mean the following: if the first q powers $|A|, |A^2|, \ldots, |A^q|$ are all $\geqslant 1$ there is no hope of finding a value less than one in the sequence $|A^m|$. Thus the convergence of the iterative process $x_{r+1} = Ax_r + y$ could be tested on the basis of the behaviour of the first q steps only; in other words (formulated in a somewhat loose manner) the process either starts converging before the qth step or it does not converge at all.

These remarks should be sufficient to motivate the following.

Definition 1.1. Let E be a Banach space. The critical exponent q of E may be defined as the smallest integer possessing one of the following equivalent properties.

(1) If A is a bounded linear operator on E such that

$$1 = |A| = |A^q| \quad \text{then} \quad |A|_\sigma = 1.$$

(2) If A is a contraction on E then $|A|_\sigma < 1$ if and only if $|A^q| < 1$.

(3) If A is a contraction on E and $|A^m| < 1$ for some m then already $|A^q| < 1$.

We say that the critical exponent of E is infinite if there is no integer that satisfies one of the above properties.

There is another characterization of the critical exponent which puts into evidence the geometric character of the notion and makes it evident that the existence of the critical exponent is a fairly delicate matter. Before stating the definition in this form it will be convenient to recall some facts about the spectral radius.

We shall consider the more general situation of a Banach algebra A. Let us recall the formula relating the spectral radius of an element $a \in A$ to the norms of the iterates of a:

$$|a|_\sigma = \lim |a^r|^{1/r} = \inf |a^r|^{1/r}.$$

In particular this formula yields a criterion for the spectral radius to be less than one: the following three assertions about an element $a \in A$ are equivalent:

(i) the spectral radius of a is smaller than one,

(ii) the norms $|a^r|$ are less than one for large r,

(iii) there exists an m for which $|a^m| < 1$.

Consequently, there exists, for each $a \in A$ with $|a|_\sigma < 1$, an exponent m for which $|a^m| < 1$; let us denote by $m(a)$ the smallest exponent with this property.

In terms of the function $m(\cdot)$ the definition may be restated as follows

Definition 1.1 (Continued)**.** (4) The critical exponent of the Banach space E is the maximum of $m(A)$ as A ranges over all contractions $A \in B(E)$ such that $|A|_\sigma < 1$.

It is, in particular, this form of the definition which reveals the geometric substance of the notion: the set on which the maximum of $m(\cdot)$ is to be taken,

$$\{A \in B(E)\colon |A| \leqslant 1, |A|_\sigma < 1\},$$

is not compact – not even in the case of a finite-dimensional E. This helps to explain the somewhat unexpected fact that there exist finite-dimensional Banach spaces E whose critical exponent is infinite.

If E is a Banach space of finite dimension, it is possible to express the definition of the critical exponent in terms of the behaviour of vectors in E. Given a contraction A on E and a vector $x \in E$ of norm one consider the sequence

$$1 = |x| \geqslant |Ax| \geqslant |A^2 x| \geqslant \cdots.$$

For a sequence of this type we either have

$$|A^n x| = 1 \quad \text{for all } n \quad \text{or} \quad \lim |A^n x| < 1.$$

Clearly $|A|_\sigma = 1$ if and only if there exists an x of norm one such that all terms of the above sequence are equal to one. In the case $|A|_\sigma < 1$ there will be, for each x of norm one, a string of ones in the sequence considered this string being followed by values less than a fixed number $\alpha < 1$. This may be used in order to reformulate the definition in yet another form.

Definition 1.1 (Continued). (5) In the case of a finite-dimensional Banach space E the critical exponent may be defined as the maximum length of a string of ones in the sequence

$$|x| \geqslant |Ax| \geqslant |A^2x| \geqslant \cdots$$

as A ranges over all contractions in E with $|A|_\sigma < 1, x$ being an arbitrary vector in $E, |x| = 1$.

A simple compactness argument using the finite-dimensionality of E makes it possible to reduce this form of the definition to that one stated in (4). On the other hand, the finite dimensionality of E does not guarantee the existence of a finite critical exponent: in view of (5), to produce an example of a finite-dimensional E with infinite critical exponent it suffices to construct a finite-dimensional E with the following property: given any q, there exists a contraction A on E with spectral radius < 1 and a vector $x \in E$ such that

$$|x| = |Ax| = \cdots = |A^q x| = 1.$$

This is, indeed, possible – a moment's reflection shows, however, that this is a fairly delicate problem in geometry of Banach spaces.

The definition of the critical exponent in its qualitative form as stated above is of theoretical interest only. The formulation in (2) admits, however, a quantitative refinement. It is to be expected that the following question would have a closer relation to reality. If A is a contraction and if we find that the norm of A^q is very close to one then it is reasonable to expect that this is caused by the presence of an eigenvalue of A the modulus of which is close to one. This question may be given a quantitative form. Find

$$\inf\{|T|_\sigma\colon |T| \leqslant 1,\ |T^q| \geqslant r\} = g(r);$$

this would mean the following: if A is a contraction then $|A|_\sigma \geqslant g(|A^q|)$. In this manner the norm of the qth power of a contraction A is of decisive importance for the convergence of the iterative process

$$1 + A + A^2 + \cdots.$$

Indeed, we either find that the value of $|A^q|$ is appreciably less than one and we have convergence at least as fast as that of a geometric series with quotient $|A^q|$ or else, the value of $|A^q|$ is close to one; in this second case the above estimate $|A|_\sigma \geqslant g(|A^q|)$ shows that poor convergence may be expected, the spectral radius being also close to one.

The problem of computing the function $g(r)$ seems to become more tractable if it is replaced by the inverse maximum problem: compute

$$f(r) = \sup\{|T^2|\colon |T| \leqslant 1, |T|_\sigma \leqslant r\}.$$

In a manner of speaking f is a function inverse to g; if we restrict ourselves to finite-dimensional Banach spaces E, the statement that q is the critical exponent of E finds its reflection in the fact that $f(r) < 1$ for $r < 1$. The computation of f is a considerably more complicated task than establishing the assertion that q is the critical exponent of E. Thus far the only case where the function f was computed is the case of n-dimensional Hilbert space and $q = n$ (Pták 1968); in fact f was computed in the sense that we can identify the operator T for which the maximum is attained.

The definition of the critical exponent was motivated by considerations concerning convergence of iterative processes and was formulated, accordingly, for Banach spaces. It is obvious that the definition is a particular case of a more general situation which might be described as follows.

We are given a set K and assume that we have, for the set K, a meaningful notion of interior (taken algebraically or topologically). Furthermore, let $\mathscr{C}$ be a class of mappings. It is natural to call a $T \in \mathscr{C}$ a contraction if $TK \subset K$. It is, of course, necessary to restrict our considerations to mappings related to the notion of interior in a natural manner; the interior should be preserved by mappings of the class $\mathscr{C}$, in particular, the inclusion $TK \subset K$ should imply $T(\operatorname{int} K) \subset \operatorname{int} K$ or $T(K) \subset \operatorname{bd} K$. A mapping $T \in \mathscr{C}$ will be called stable if there exists an exponent m such that T^m maps K into its interior. For each stable $T \in \mathscr{C}$ we denote by $m(T)$ the smallest integer m for which

$$T^m K \subset \operatorname{int} K.$$

The critical exponent of K with respect to the class $\mathscr{C}$, denoted by $q(K, \mathscr{C})$ is then defined as

$$q(K, \mathscr{C}) = \sup m(T)$$

as T ranges over all stable contractions $T \in \mathscr{C}$. It is easy to see that the critical exponent of a Banach space E is a particular case of this general concept in the following situation. We take, for K, the closed unit cell of E and consider the class $\mathscr{C}$ of all bounded linear mappings of E into itself; the requirement that $TK \subset K$ is equivalent to the inequality $|T| \leqslant 1$. Stable mappings will then be those whose spectral radius is less than one.

If we adopt the convention $\inf \emptyset = \infty$ we may state the definition of the critical exponent $q(K, \mathscr{C})$ of K with respect to the class $\mathscr{C}$ as follows

$$q(K, \mathscr{C}) = \sup \inf \{m\colon T^m K \subset \operatorname{int} K\}$$

the supremum being taken over all contractions $T \in \mathscr{C}$.

It is possible to consider also a local version of this concept

$$\bar{q}(K, \mathscr{C}) = \sup \sup \inf \{m\colon T^m x \in \operatorname{int} K\}$$

the supremum being taken over all contractions T and all $x \in K$ for which $\inf\{m\colon T^m x \in \operatorname{int} K\} < \infty$.

In the rest of this paper we shall consider a fixed convex set K in a Banach space E. By a projective transformation of E we mean a mapping of the form

$$x \to (f(x)+d)^{-1}(Ax+b),$$

where A is a bounded linear operator in E (not necessarily invertible), $b \in E$, d a scalar and f a bounded linear functional on E. The transformation is called affine if $f = 0$ and $d \neq 0$.

The projective, affine and linear critical exponent of K will be defined as $q(K, \mathscr{C})$ if $\mathscr{C}$ is taken to be the class of projective, affine and (bounded) linear transformations of E respectively. The corresponding local notions are defined in an analogous manner.

This chapter is organized as follows. The preceding section explains the motivation. Section 2 is intended as an epitome of the history of the subject. The case of n-dimensional Hilbert space forms the subject matter of the independent section 3 – not only is the theory more complete than in other cases but section 3 also contains a discussion of the quantitative version of the problem. The quantitative version, in its turn, has interesting connections with the theory of complex functions. Another case that deserves a separate section is that of the n-dimensional l^∞ space. The methods used there are entirely different and of purely combinatorial character. Some of the combinatorial results used there are of independent interest.

The concluding section is devoted to the discussion of open problems.

2. History

The definition of the critical exponent appears first in the author's paper (Pták 1967). There is an earlier paper of Mařík and Pták (1960) the main result of which – although formulated in other terms – may be interpreted as the statement that the critical exponent of the n-dimensional l^∞ space is $n^2 - n + 1$.

The quantitative version of the problem was first formulated in the author's papers (Pták 1967a, 1968) where the maximum problem

$$f(r) = \max\{|T^n|\colon |T| \leqslant 1, |T|_\sigma \leqslant r\}$$

was solved for operators T on n-dimensional Hilbert space.

As soon as the definition was formulated in its full generality it became obvious that there is a host of interesting questions that can be asked in this context.

In a lecture in Seattle (1961) the author suggested investigating the critical exponent as an integer-valued characteristic of convex bodies. The subject aroused interest among the specialists gathered for the convexity symposium and the subsequent discussions resulted in the solution of two basic questions. Grünbaum disproved a conjecture that then seemed natural: The critical exponent of a finite dimensional Banach space cannot be smaller than its dimension.

Indeed, Grünbaum showed that the l^1 sum of a four-dimensional and an eight-dimensional Hilbert space is a counterexample. Grünbaum and Danzer computed the critical exponent for a number of interesting polytopes with increasing number of faces and finally Danzer constructed a finite-dimensional Banach space with infinite critical exponent. The thesis of Perles (1964) written under the supervision of Grünbaum brought a number of interesting ideas and results.

The methods used by Mařík and the author to investigate the critical exponent of finite-dimensional l^∞ spaces are purely combinatorial. The theory may be reduced to the study of powers of directed graphs (Mařík and Pták 1960; see also Pták 1958, and Pták and Sedláček 1958). Two of the main results of the paper (Mařík and Pták 1960), theorems (1,14) and (1,15) were generalized to lattices by Perles. The graph-theoretical background of the l^∞ results was reformulated by Ljubič and Tabačnikov in the form of statements about subharmonic functions on oriented graphs; the technical parts of the paper of Mařík and the author appear thus in a broader context and in a more natural light, in particular the proof that the critical exponent of n-dimensional l^∞ space is $n^2 - n + 1$ assumes thus a somewhat less technical form (Ljubič and Tabačnikov 1969b). In a subsequent paper Kiržner and Tabačnikov (1971) prove a fairly general existence (finiteness) theorem for the critical exponent: if the unit sphere of the (finite-dimensional) Banach space may be imbedded in an algebraic manifold of a certain kind, then the critical exponent of the space is finite.

3. Hilbert space

The paper (Pták 1962) of the author contains the definition of the critical exponent of a Banach space and the proof that the critical exponent of the n-dimensional (complex) Hilbert space H_n equals n. The proof is based on the following proposition.

Proposition 3.1. *If A is a contraction on H_n and if $|A^n| = 1$ then there exists a nontrivial subspace $\mathcal{M} \subset H_n$ invariant with respect to A such that $A|\mathcal{M}$ is an isometry.*

The critical exponent of H_n cannot be smaller than n since the finite shift operator S_n given by the n by n matrix

$$S_n = \begin{pmatrix} 0 & \dots & 0 & 0 \\ 1 & \dots & 0 & 0 \\ \vdots & \ddots & \vdots & \vdots \\ 0 & \dots & 1 & 0 \end{pmatrix}$$

satisfies

$$|S_n| = |S_n^2| = \dots = |S_n^{n-1}| = 1$$

and $S_n^n = 0$. A number of authors presented simpler proofs (Flanders 1974, Goldberg and Zwas 1974, Wimmer 1974). Young gave a proof of the following more general proposition.

Proposition 3.2. *Suppose $A_1, \ldots, A_n$ are n commuting contractions on n-dimensional Hilbert space. If the spectral radii of each A_j are less than one then $|A_1 \cdots A_n| < 1$.*

Perles investigated l^1 sums of two Hilbert spaces. Denote by $X^{m,n}(R)$ the direct sum of two real Hilbert spaces of dimensions m and n equipped with the norm

$$|x \oplus y| = |x| + |y|.$$

Analogously, $X^{m,n}(C)$ will denote the same construction with complex Hilbert spaces. Then $q(X^{m,n}(F)) = n + 1$ if either

$$F = R, \qquad m > 1 \quad \text{and} \quad n > \binom{m+1}{2}$$

or if

$$F = C, \qquad m > 1 \quad \text{and} \quad n > m^2.$$

This result furnishes examples of Banach spaces whose critical exponents are less than their dimension. The smallest known examples of this type are

$$q(X^{2,4}(R)) - 5,$$

$$q(X^{2,5}(C)) = 6.$$

The space $X^{1,2}(R)$ whose unit ball is a double cone is an example of a three-dimensional space with the same critical exponent as three-dimensional real Hilbert space.

In 1968 the author presented in (Pták 1967a, 1968) the solution of the following problem.

Problem 3.3. Compute the maximum of $|T^n|$ as T ranges over the set of all linear operators T on n-dimensional Hilbert space such that $|T| \leqslant 1$ and $|T|_\sigma \leqslant r$.

The solution consisted in the identification of the operator where the maximum is attained. We can speak of *the* maximal operator since Dostál (1978a) was able to show, by a careful analysis of the proof in (Pták 1968), that the maximal operator is uniquely determined up to unitary equivalence.

The method adopted by the author for the solution of the extremal problem consisted in dividing the problem into two stages.

The first maximum problem. We replace the constraint $|T|_\sigma \leqslant r$ by a more stringent one and solve the corresponding extremal problem. Take a polynomial

p of degree n with all zeros in the disk $|z| \leqslant r$ and compute the maximum of $|T^n|$ on the set of all linear operators on n-dimensional Hilbert space such that

$$|T| \leqslant 1 \quad \text{and} \quad p(T) = 0.$$

Clearly $p(T) = 0$ implies $|T|_\sigma \leqslant r$. For reasons that will become obvious later we write the condition $p(T) = 0$ in the equivalent form $\varphi(T) = 0$ where φ is the Blaschke product obtained from p as follows if $p(z) = \prod(z - \alpha_j)$ then

$$\varphi(z) = \prod \frac{z - \alpha_j}{1 - \alpha_j^* z}.$$

It turns out (Pták 1968) that there exists an operator $T(\varphi)$ defined on an n-dimensional Hilbert space such that $|T| = 1$ and $\varphi(T(\varphi)) = 0$ which maximizes $|T^n|$ under these constraints. Sz.-Nagy (1969) observed that $T(\varphi)$ also solves the following more general maximum problem.

Problem 3.4. Let h be a function holomorphic in a neighbourhood of the spectrum of φ. The maximum of $|h(T)|$ on the set of all linear operators T on n-dimensional Hilbert space with $|T| \leqslant 1$ and $\varphi(T) = 0$ is attained at $T(\varphi)$.

It follows from the Cayley–Hamilton theorem that $|T|_\sigma \leqslant r$ if and only if $\varphi(T) = 0$ for some Blaschke product φ of length n with all zeros in $|z| \leqslant r$. Thus it suffices to solve *the second maximum problem.*

Problem 3.5. Compute the maximum of

$$|h(T(\varphi))|$$

as φ ranges over the set of all Blaschke products of length n with all zeros in $|z| \leqslant r$.

In the case $h(z) = z^n$ the author was able to show (Pták 1968) that the maximum is attained for the function

$$\varphi(z) = \left(\frac{z - r}{1 - rz}\right)^n.$$

In the general case of an arbitrary h the problem remains open.

To return to the first maximum problem, let us mention that its first solution (Pták 1968) was based on a careful use of the properties of the cone of nonnegative definite matrices, in particular of the fact that its extreme rays are of the form aa^*, a being a column vector of length n. Although the first paper (Pták 1968) only deals with the case $h(z) = z^n$, the proof actually works for an arbitrary h (Pták 1984).

It turned out later that the result may also be obtained using methods of the theory of complex functions. The von Neumann inequality and a theorem of Sarason yield the following.

If $h \in H^\infty$ and φ is a Blaschke product of length n with zeros $\alpha_1, \dots, \alpha_n \in D$ then

$$|h(T(\varphi))| = |h + \varphi H^\infty|$$

the norm being taken in the quotient space $H^\infty/\varphi H^\infty$. The connections are carefully explained in Pták (1984).

The solution of the first maximum problem is thus the norm of the class of h in the quotient space H^∞/pH^∞. The second maximum problem admits then the following reformulation.

Problem 3.5'. Among all polynomials p of degree n whose zeros are all contained in the disk $\{z: |z| \leqslant r\}$ find one which maximizes the norm $|h + pH^\infty|_{H^\infty/pH^\infty}$.

The original result of the author says that, for $h(z) = z^n$, the extremal polynomial can be taken to be $(z - r)^n$.

It is not unnatural to conjecture that this polynomial will also do for h of the form $h(z) = z^m$ for $m > n$ or even for a wider class of functions in H^∞. The method of proof used by the author for the function $h(z) = z^n$ does not seem to extend even to the case $h(z) = z^{n+1}$.

A recent result of Hayashi (1987) shows that, for an arbitrary h, the zeros of the worst polynomial must lie on the circle $\{z: |z| = r\}$. In fact, the constraint for the polynomial p may be even more general.

Using the Schur algorithm, Hayashi proves the following maximum principle.

Theorem. *Let F be a compact subset of the open unit disk and let h be a fixed function in H^∞. Consider the function*

$$p \to |h - pH^\infty|_{H^\infty/pH^\infty}$$

as p varies over all polynomials of degree n with all zeros in F. If this function is not constant then it attains its maximum at p only if all the zeros of p lie on the boundary of F.

In the context of the theory of C^* algebras the idea of the critical exponent was investigated by Kato (1988).

The quantitative version of the theory of the critical exponent is discussed in detail in the survey article (Pták 1967b, 1982), the connections with the theory of complex function is analysed in Pták (1984). In the book of Belickii and Ljubič (1984) these questions are treated in the broader context of a discussion of norms on linear spaces.

4. Polytopes

The theory of the critical exponent of a polytope is essentially combinatorial; to explain this, we limit ourselves to the case of the n-dimensional cube. The case of a general polytope is not substantially different although more complicated technically. Comments on the generalizations are given at the end of this discussion.

For our purposes it is convenient to view a directed graph as an additive set-valued function since we will have to study its iterates. More precisely, an oriented graph φ on a set N is a mapping that assigns to each subset $A \subset N$ a subset $\varphi(A) \subset N$ such that $\varphi(\emptyset) = \emptyset$ and $\varphi(A_1 \cup A_2) = \varphi(A_1) \cup \varphi(A_2)$. The set $\varphi(A)$ is the set of endpoints of all arrows starting in A. Then, for $k \geqslant 1, \varphi^k(A)$ is the set of those points in N that can be reached by a path consisting of k arrows.

Now let N be the set $N = \{1, 2, \ldots, n\}$. Given an n by n matrix M, i.e., a complex function defined on $N \times N$ we define the graph corresponding to M as follows

$$j \in \varphi(A) \quad \text{iff} \quad a_{ij} \neq 0 \quad \text{for some } i \in A.$$

Clearly, for a nonnegative matrix M, the graph corresponding to M^k is the kth iterate of the graph corresponding to M. In the theory of the critical exponent the relation between the iterates of a complex matrix M and the iterates of the corresponding φ will be of decisive importance.

We formulate first a general theorem on the behaviour of the iterates of a finite graph. A graph φ is said to be indecomposable if it has no nontrivial invariant subset, in other words, if there exists no nonvoid proper subset $A \subset N$ with $\varphi(A) \subset A$. A cyclic partition (of N with respect to φ) of length k is a system of k disjoint subsets $R, \varphi(R), \ldots, \varphi^{k-1}(R)$ such that $\varphi^k(R) = R$ and

$$N = R \cup \varphi(R) \cup \cdots \cup \varphi^{k-1}(R).$$

The maximal possible length of a cyclic partition is called the index of imprimitivity of φ. If no proper cyclic partition exists and if φ is indecomposable we say that φ is primitive.

The following theorem (Mařík and Pták 1960, Pták 1958, Pták and Sedláček 1958) shows that the class of all indecomposable graphs may be divided into two subclasses according to the behaviour of the iterated mappings.

Theorem 4.1. *Let φ be an indecomposable graph. Then the following two cases are possible.*

(1) *The mappings $\varphi, \varphi^2, \ldots, \varphi^n$ are all indecomposable; then φ^v is indecomposable for every v and $\varphi^p(A) = N$ for every nonvoid A as soon as $p \geqslant n^2 - 2n + 2$.*

(2) *There exists a $k \leqslant n$ such that φ^k is decomposable; then φ^v is decomposable for infinitely many v. If φ^v is decomposable there exists a divisor $d > 1$ of v and a cyclic partition of length d.*

The index of imprimitivity of φ equals the greatest common divisor of all lengths of cycles of φ.

In particular, if $Q \subset N$ and $\varphi^v(Q) \subset Q$ then Q is the union of some of the members of the cyclic partition, so that $\varphi^d(Q) = Q$.

If φ is primitive, we have $\varphi^k(i) = N$ for every $i \in N$ as soon as $k \geqslant n^2 - 2n + 2$. (For simplicity, we write i instead of $\{i\}$.) It is important to know that the bound $n^2 - 2n + 2$ is sharp; indeed, there is a primitive graph whose $(n^2 - 2n + 1)$th iterate is not the full graph. In fact, there is only one graph of this property: an n-cycle with exactly one bypass. More precisely, the following theorem can be proved (Pták 1958).

Theorem 4.2. *Let $n > 2$ and let φ be primitive. Set $m = n^2 - 2n + 1$. Suppose that $j \in N$ and $\varphi^m(j) \neq N$. Then it is possible to arrange the elements of N into a sequence $j_1, \ldots, j_n$ in such a manner that $j_1 = j, \varphi(j_r) = \{j_{r+1}\}$ for $r = 1, 2, \ldots, n-1$ and $\varphi(j_n) = \{j_1, j_2\}$. This graph satisfies the identity $\varphi^m(x) = N$ for every $x \in N$ different from j and $\varphi^m(j) = N \backslash \{j\}$.*

When applied to nonnegative matrices these results represent the combinatorial substance of the classical theorems proved first by Frobenius (1912) by direct methods.

Theorem 4.3. *Let A be an indecomposable nonnegative matrix. Then the following conditions are equivalent:*
(1) *A is primitive,*
(2) *all iterates of A are indecomposable,*
(3) *the matrices $A, A^2, \ldots, A^n$ are indecomposable,*
(4) *all powers A^v are positive for $v \geqslant n^2 - 2n + 2$,*
(5) *some power of A is positive,*
(6) *A^v is primitive for every v.*

Frobenius only proved the existence of an exponent v for which A^v is positive. The lower bound $n^2 - 2n + 2$ for v was given without proof by Wielandt (1950). He also produces an example to show that this bound cannot be sharpened. In fact, Theorem 4.2 shows that there is combinatorially only one type of matrix where the bound is attained.

In the case of imprimitivity, the matrix may be transformed by a suitable permutation of rows and columns to a block cyclic form

$$A = \begin{pmatrix} 0 & A_{12} & 0 & \ldots & 0 \\ 0 & 0 & A_{23} & \ldots & 0 \\ \vdots & \vdots & \vdots & \ddots & \vdots \\ 0 & 0 & 0 & \ldots & A_{h-1,h} \\ A_{h,1} & 0 & 0 & \ldots & 0 \end{pmatrix}$$

If $\varepsilon^h = 1$, consider the matrix

$$D = \begin{pmatrix} \varepsilon I_1 & 0 & \ldots & 0 \\ 0 & \varepsilon I_2 & \ldots & 0 \\ \vdots & \vdots & \ddots & \vdots \\ 0 & 0 & \ldots & \varepsilon^h I_h \end{pmatrix},$$

the I_j being unit matrices of appropriate sizes; since $\varepsilon A = D^{-1}AD$ the spectrum of A is invariant with respect to rotation by the angle $2\pi/h$.

The corresponding statement about nonnegative matrices may be stated as follows.

Theorem 4.4. *Let A be indecomposable nonnegative. Suppose A has exactly h eigenvalues of modulus $|A|_\sigma$. Then the whole spectrum of A is invariant with respect to rotation by the angle $2\pi/h$. All eigenvalues of modulus $|A|_\sigma$ are simple. The number h is the smallest integer for which A^h decomposes into primitive matrices. For any v the iterate A^v decomposes into exactly (v,h) indecomposable matrices.*

Similar results, though not in this simple form, were obtained independently about the same time by Holladay and Varga (1958). It seems that our approach to oriented graphs as join-endomorphisms in $\exp N$ is more suitable for investigations of iterated mappings.

The application of the combinatorial set-up described above to nonnegative matrices is more or less straightforward; only the combinatorial structure of the iterates is relevant there. Considerably more sophisticated methods have to be used in the theory of the critical exponent where the existence of an eigenvalue of modulus one has to be proved.

Now take a complex n by n matrix A and consider it as a linear operator on the space E of (column) vectors of length n equipped with the norm $|(x_1,\ldots,x_n)^{\mathrm{T}}| = \max|x_j|$. The operator norm of A will be $|A| = \max_i \sum_k |a_{ik}|$. Given a vector $x = (x_1,\ldots,x_n)^{\mathrm{T}}$ such that $|x| \leqslant 1$, let $P(x)$ be the set of those indices $j \in N$ for which $|x_j| = 1$. The basic relation between the geometry of E and the combinatorial structure of A is the following inclusion:

$$\text{if} \quad |A| \leqslant 1 \text{ and } |x| \leqslant 1 \quad \text{then} \quad P(x) \supset \varphi(P(Ax)).$$

This fact may be used as follows. If A is a contraction and if $|A^m| = 1$ for some m, there exists a vector x of norm one such that $|A^m x| = 1$. Thus $P(A^j x)$ is nonvoid for $j = 0,1,\ldots,m$ and we obtain a string of inclusions of the form $P(A^j x) \supset \varphi(P(A^{j+1}x))$; it is not difficult to realize that this series of inclusions will yield significant information about A provided it is long enough. Indeed, the assertion that $n^2 - n + 1$ is the critical exponent of E is nothing more than the statement that $n^2 - n + 1$ inclusions of this type permit the conclusion that A has an eigenvalue of modulus 1.

The case of a general convex set requires a somewhat more complicated combinatorial structure; in particular, the family of all subsets of N has to be replaced by the lattice of all faces of the set.

We consider a lattice L with 0 and 1 and a group G of automorphisms of L. For $a,b \in L$ we write $a \sim b$ if there exists a $\sigma \in G$ such that $b = \sigma a$. Let φ be a join-endomorphism of L which commutes with all elements of G. An element $a \in L$ is said to be periodic if $\varphi^r(a) \sim a$ for some $r > 0$; a is almost periodic if it may be written as the join of a finite number of periodic elements. Let us remark that, in the case that G is finite, both periodicity and almost periodicity reduce to the condition $\varphi^r(a) = a$ for some $r > 0$. This equivalence is easily proved by iterating the condition $a = \sigma\varphi^r(a)$. Denote by $\zeta(\varphi)$ the smallest nonnegative integer s much that $\varphi^s(a)$ is almost periodic for all $a \in L$. If L is finite, then $\zeta(\varphi)$ is the smallest nonnegative integer s such that $\varphi^s = \varphi^{s+t}$ for some $t > 0$.

The theorem about primitive graphs has a natural counterpart in this more general situation (Perles 1964, 1967). The following conditions are imposed on L.

(1) The set of all non-zero joint-irreducible elements of L splits into k congruence classes modulo G,

(2) if $a \in L$ and $\sigma, \tau \in G$ then either $\sigma a = \tau a$ or $\sigma a \vee \tau a = 1$.

Let d be the largest integer r such that there exists r elements $a_1, \ldots, a_r \in L$ with

$$0 < a_1 < a_2 < \cdots < a_r < 1.$$

Under these assumptions Perles proves the following generalization of the primitivity theorem.

Theorem 4.5. *The characteristic $\zeta(\varphi)$ is bounded by $dk - d + 1$. If $d \geqslant 2$ then the following conditions are equivalent*

(1) $\zeta(\varphi) = dk - d + 1$,

(2) *all the joint-irreducible elements of L are atoms and there exist k atoms $p_0, p_1, \ldots, p_{k-1}$ such that $p_i \sim p_j$ and two elements $\sigma, \tau \in G$ with the following properties:*

$$\varphi(p_{i-1}) = p_i, \quad \textit{for } 1 \leqslant i \leqslant k-1,$$
$$\varphi(p_{k-1}) = \sigma p_0 \vee \tau p_1,$$
$$\varphi^{dk-d}(p_0) \neq 1, \qquad \varphi^{dk-d+1}(p_0) = 1.$$

In applications to convex sets, the lattice L will be the lattice of all faces of the convex set K (the empty set and K itself will also be considered as faces of K). If P is a projective mapping such that $PK \subset K$ we define a mapping $\bar{P}$ from L into L by declaring, for each $A \in L, \bar{P}(A)$ to be the smallest face containing $P(A)$. It is easy to show that $\bar{P}$ is a join-endomorphism of L and that $(\bar{P})^r = \overline{P^r}$ for each $r \geqslant 0$.

In what follows we collect theorems obtained for general convex sets by Perles. Both the methods as well as the results are generalizations of the theorem that $n^2 - n + 1$ is the critical exponent of n-dimensional l^∞ space. The more general estimate $dk - d + 1$ puts into evidence the role played by the dimension and the geometrical shape of the convex set. In the theorem of Mařík and Pták both d and k are equal to n.

To simplify the statements we introduce the following notation: given a convex set K, we denote by $\mathcal{A}(K), \mathcal{P}(K)$ and $\mathcal{L}(K)$ respectively the set of all affine, projective and linear transformations mapping K into itself.

Theorem 4.6. *Let K be a convex polyhedral cone,* $\dim K = m+d+1$ *with* $\dim(K \cap -K) = m$ *and* $d \geqslant 0$. *If K has k $(m+1)$-faces or k $(m+d)$-faces then*

$$\bar{q}(K,A) \leqslant \zeta(\bar{A}) \leqslant dk - d + \tfrac{1}{2}(1+(-1)^{dk}) \quad \textit{for all } A \in \mathcal{L}(A).$$

Theorem 4.7. *Let K be a d-dimensional bounded convex polytope with k vertices or k $(d-1)$-faces. Then*

$$\bar{q}(K,P) \leqslant \zeta(\bar{P}) \leqslant dk - d + \tfrac{1}{2}(1+(-1)^{dk}) \quad \textit{for all } P \in \mathcal{P}(\mathcal{K}).$$

Theorem 4.8. *Let K be a centrally symmetric bounded convex polytope of dimension d. Let $A \in \mathcal{L}(K)$. If K has $2k$ vertices or $2k$ $(d-1)$-faces, then*

$$\bar{q}(K,A) \leqslant \zeta(\bar{A}) \leqslant dk - k + 1.$$

If d is odd, k is even and if $k \leqslant 2d$, then $\zeta(\bar{A}) \leqslant dk - d$.

The estimate in Theorem 4.7 is sharp. Indeed, the following holds.

Theorem 4.9. *Let $k-1 = d = 1$ or $k > d \geqslant 2$. Then there exists a d-dimensional bounded convex polytope K with k vertices, containing* 0 *in its interior and a regular $A \in \mathcal{L}(K)$ such that*

$$q(K,A) = dk - d + \tfrac{1}{2}(1+(-1)^{dk}).$$

For the other two estimates stated above there are similar sharpness results (Perles 1967) except that, this time, more stringent conditions have to be imposed on d and k.

In the case of a polytope it is also possible to give lower bounds for the critical exponent in terms of its dimension. The following result of Perles (1967) shows that there is a bound independent of the shape of the polytope.

Theorem 4.10. *Let K be a bounded convex polytope of dimension d. Then there exists a regular $A \in \mathcal{A}(K)$ such that $q(K,A) = d+1$.*

Of course, in terms of linear mappings this result may be reformulated as follows:

Theorem 4.11. *There exists a point $t \in \operatorname{int} K$ and a regular $A \in \mathcal{L}(K-t)$ such that $q(K-t,A) = d+1$.*

In his thesis Pham Dinh-Tao (1981) throws some more light on the case of n-dimensional l^∞ space by looking at the dual space, n-dimensional l^1 space. In the early days of the theory (in the sixties), Grünbaum computed the critical exponent for a number of polytopes. These results were never published; now they are covered by the work of Perles.

5. Open problems

A moment's reflection shows – in particular if we consider the definition of the critical exponent in its form under (4) – that its existence is a rather surprising fact. It seems that, among all norms on a given vector space, the norms with a finite critical exponent form an exception.

Let $\mathcal{K}^d$ be the family of all centrally symmetric convex bodies in R^d. Taken in the Hausdorff metric, $\mathcal{K}^d$ becomes a locally compact metric space. Denote by Q_r^d the subset of $\mathcal{K}^d$ consisting of all K for which $q(K) \geqslant r$. Clearly $Q_r^d \supset Q_{r+1}^d \supset \cdots$ and $Q_d^d \neq \mathcal{K}^d$. The set Q_r^d is an F_σ set in $\mathcal{K}^d$. Define

$$q(d) = \inf\{r\colon Q_r^d \text{ is meagre in } \mathcal{K}^d\}.$$

Using these notions, we may formulate the following:

Problem 1. Is $q(d)$ finite? If so, is it bounded as a function of d?

Problem 2. Describe the topological properties of the set of all norms in $\mathcal{K}^d$ whose critical exponent equals m. Is it thin in some sense?

This is an almost unexplored area. Some partial results are contained in the thesis of Perles (1964).

A treacherously simple problem which is still unsolved is the following:

Problem 3. Does there exist an infinite-dimensional Banach space with a finite critical exponent?

More important problems arise of course in finite dimensions. We collect first problems for general Banach spaces.

It would be desirable to find sufficiently general criteria for the finiteness of the critical exponents together with reasonable bounds. In a somewhat simplified form we formulate this as Problem 4.

Problem 4. Characterize Banach spaces whose critical exponent is finite.

A general qualitative result that covers both the case of the cube as well of the sphere was obtained by Kiržner and Tabačnikov (1971). In this generality the bounds obtained cannot be expected to be sharp.

Problem 5. Determine the asymptotic behaviour of the function $M(n)$ where $M(n)$ is the minimum of the critical exponents of all n-dimensional Banach spaces.

Since for a Hilbert space the critical exponent equals its dimension, we have $M(n) \leqslant n$ for every n. Grünbaum and Perles have shown that $M(n)$ can be considerably less than n. Hence even the following weaker version of the preceding problem is interesting.

Problem 6. Is $\liminf M(n)$ infinite?

In other words, does there exist a sequence of Banach spaces E_n with $\dim E_n$ tending to infinity and such that the critical exponents of all E_n lie below a certain finite bound?

A deeper investigation of the function f of section 3 immediately suggests the following problems.

Problem 7. Let E be a Banach space, q a natural number, p a positive number, $p < 1$. Suppose a_0 is the operator which realizes the maximum of $|a^q|$ subject to constraints $|a| \leqslant 1$ and $|a|_\sigma \leqslant p$. Does it follow that $|a_0| = 1$ and $|a_0|_\sigma = p$? Clearly, at least one of the two equalities must hold.

To clarify the meaning of the constraints $|a| \leqslant 1, |a|_\sigma \leqslant p$ it would be useful to solve:

Problem 8. Let E be a Banach space. For each $0 < p < 1$ give a description of the set

$$C(p) = \{a \in B(E): |a| = 1, |a|_\sigma = p\}.$$

Is it always nonvoid?

A closely related problem is:

Problem 9. Let E be a Banach space, q a natural number; for each positive $p < 1$ set

$$f(p) = \sup |a^q|,$$

the supremum being taken over al linear operators a on E subject to the constraints $|a| \leqslant 1, |a|_\sigma \leqslant p$. Is f a strictly increasing function of p? Is f continuous?

The critical exponent of the n-dimensional l^1 or l^∞ space is $n^2 - n + 1$ (Dostál 1979), for l^2 it equals n (Pták 1962). It remains to solve:

Problem 10. Determine the critical exponent of the n-dimensional l^p space.

Apart from the interesting and ingenious investigations of Perles (1964, 1967) who proved the existence and obtained estimates for some p, little progress was made.

The case of the Hilbert space is both interesting and important. Let us conclude with the following intriguing problem.

Problem 11. Let H be an n-dimensional Hilbert space. Let p be a positive number, $0 < p < 1$, and denote by $D(p)$ the disk $\{z: |z| \leqslant p\}$. Let ψ be a given polynomial [or a function holomorphic on a neighbourhood of $D(p)$]. Let $F(p)$ be the set of all polynomials whose roots lie in $D(p)$. For each $\varphi \in F(p)$ we know that

$$\max\{|\psi(a)|: a \in B(H), |a| \leqslant 1, \varphi(a) = 0\} = |\psi(T(\varphi))|.$$

Find the polynomial $\varphi \in F(p)$ which realizes the maximum of the function

$$\varphi \to |\psi(T(\varphi))| \quad \text{on } F(p).$$

The classical result of the present author says that, for $\psi(z) = z^n$ (n being the dimension) this maximum is attained for the polynomial $\varphi(z) = (z-p)^n$.

The quantitative question solved for Hilbert spaces is also meaningful in the general case. To be quite modest, it would be interesting to start by investigating the following problem.

Problem 12. Consider the n-dimensional l^∞ space E, set $q = n^2 - n + 1$. Compute, for $r < 1$, the maximum

$$\sup\{|T^q|: |T| \leqslant 1, |T|_\sigma \leqslant r\}.$$

It would be sufficient to identify the operator (operators) for which the maximum is attained.

References

Asplund, E., and V. Pták
[1971] A minimax inequality for operators and a related numerical range, *Acta Math. (Uppsala)* **126**, 53–62.

Belickii, G.R., and U.I. Ljubič
[1984] *Norms of Matrices and their Applications* (in Russian) (Naukova Dumka, Kiev).

Dostál, Z.
[1978a] Uniqueness of the operator attaining $C(H_{n,r,n})$, *Časopis Pêst. Mat.* **103**, 236–243.
[1978b] Polynomials of the eigenvalues and powers of matrices, *Comment. Math. Univ. Carolin.* **19**(3), 459–469.
[1978c] Critical exponent of operators with constrained spectral radius, *Comment. Math. Univ. Carolin.* **19**, 315–318.
[1979] Negative powers and the spectrum of matrices, *Comment. Math. Univ. Carolin.* **20**(1), 19–27.
[1980a] Norms of iterates and the spectral radius of matrices, *Comment. Math. Univ. Carolin.* **105**, 256–260.

[1980b] A note on estimates of the spectral radius of symmetric matrices, *Comment. Math. Univ. Carolin.* **21**(2), 333–340.

Flanders, H.
[1974] On the norm and spectral radius, *Linear and Multilear Algebra* **2**, 239–240.

Frobenius, G.
[1912] Über Matrizen aus nicht-negativen Elementen, *Sitzungsber. Preuss. Akad. Wiss.* **23**, 456–477.

Goldberg, M., and G. Zwas
[1974] On matrices having equal spectral radius and spectral norm, *Linear Algebra Appl.* **8**, 427–434.

Hayashi, E.
[1987] A maximum principle for quotient norms in H^{∞}, *Proc. Amer. Math. Soc.* **99**, 323–327.

Holladay, J.C., and R.S. Varga
[1958] On powers of non-negative matrices, *Proc. Amer. Math. Soc.* **9**, 631–634.

Kato, Y.
[1988] Pták type theorem for C^*-algebras, *Arch. Math.* **50**, 550–552.

Kiržner, V., and M.I. Tabačnikov
[1971] On critical exponents of norms in n-dimensional spaces, *Siberian Math. J.* **12**, 672–675.

Ljubič, U.I., and M.I. Tabačnikov
[1969a] On a theorem of Mařík and Pták, *Siberian Math. J.* **10**, 470–473.
[1969b] Subgarmoničeskie funkcii na orientovannom grafe, *Siberian Mat. J.* **10**, 600–613.

Mařík, J., and V. Pták
[1960] Norms, spectra and combinatorial properties of matrices, *Czech. Math. J.* **85**, 181–196.

Perles, M.
[1964] Critical exponents of convex bodies, Ph.D. Thesis (Hebrew with English Summary), Hebrew University, Jerusalem.
[1967] Critical exponents of convex sets, in: *Proc. Colloq. Convexity, Copenhagen 1965* (Københavns Univ. Mat. Inst.) pp. 221–228.

Pham, Dinh-Tao
[1981] Contribution à la théorie de normes et ses applications à l'analyse numérique, Thèse, Université de Grenoble.

Potapov, V.P.
[1955] La structure multiplicative des fonctions matricielles J-contractives, *Trudy Moskov. Mat. Obšč.* **4**, 125–236.

Pták, V.
[1958] On a combinatorial theorem and its applications to nonnegative matrices, *Czech. Math. J.* **83**, 487–495.
[1962] Norms and the spectral radius of matrices, *Czech. Math. J.* **87**, 553–557.
[1967a] Rayon spectral norme des itérés d'un opérateur et exposant critique, *C.R. Acad. Sci. Paris* **265**, 257–259.
[1967b] Critical exponents, in: *Proc. Colloq. Convexity, Copenhagen 1965* (Københavns Univ. Mat. Inst.) pp. 244–248.
[1968] Spectral radius, norms of iterates and the critical exponent, *Linear Algebra Appl.* **1**, 245–260.
[1976a] Isometric parts of operators and the critical exponent, *Časopis Pěst. Mat.* **101**, 383–388.
[1976b] The spectral radii of an operator and its modulus, *Comment. Math. Univ. Carolin.* **17**, 273–279.
[1978] An infinite companion matrix, *Comment. Math. Univ. Carolin.* **19**, 447–458.
[1979a] A maximum problem for matrices, *Lin. Alg. Appl.* **28**, 193–204.
[1979b] Critical exponents, in: *Proc. Fourth Conf. on Operator Theory, Timisoara*, pp. 320–329.
[1980] A lower bound for the spectral radius, *Proc. Amer. Math. Soc.* **80**, 435–440.

[1982] Universal estimates of the spectral radius, in: *Proc. Semester on Spectral Theory,* Banach Center Publ., Vol. 8 (Publishing House of the Polish Academy of Sciences, Warszawa) pp. 373–387.
[1983a] Uniqueness in the first maximum problem, *Manuscripta Math.* **42**, 101–104.
[1983b] Biorthogonal systems and the infinite companion matrix, *Lin. Alg. Appl.* **49**, 57–78.
[1984] A maximum problem for operators, *Časopis Pêst. Mat.* **109**, 168–193.
[1985a] Extremal operators and oblique projections, *Časopis Pêst. Mat.* **110**, 343–350.
[1985b] Isometries in H^2, generating functions and extremal problems, *Časopis Pêst. Mat.* **110**, 33–57.
[1986] An extremal problem for operators, *Lin. Alg. Appl.* **84**, 213–226.

Pták, V., and J. Sedláček
[1958] On the index of imprimitivity of nonnegative matrices, *Czech. Math. J.* **83**, 495–501.

Pták, V., and N.J. Young
[1980] Functions of operators and the spectral radius, *Lin. Alg. Appl.* **29**, 357–392.

Sarason, D.
[1967] Generalized interpolation in H^∞, *Trans. Amer. Math. Soc.* **127**, 179–203.

Sz.-Nagy, B.
[1969] Sur la norme des fonctions de certains opérateurs, *Acta Math. Acad Sci. Hungar.* **20**, 331–334.

Wielandt, H.
[1950] Unzerlegbare, nicht negative Matrizen, *Math. Z.* **52**, 642–648.

Wimmer, H.
[1974] Spektralradius und Spektralnorm, *Czech. Math. J.* **99**, 501–502.

Young, N.J.
[1978] Analytic programmes in matrix algebras, *Proc. London Math. Soc.* **36**, 226–242.
[1978b] Norms of matrix powers, *Comment. Math. Univ. Carolin.* **19**, 415–430.
[1979a] Norms of powers of matrices with constrained spectrum, *Lin. Alg. Appl.* **23**, 227–244.
[1979b] Matrices which maximise any analytic function, *Acta Math. Acad. Sci. Hungar.* **34**, 239–243.
[1980] Norm and spectral radius for algebraic elements of a Banach algebra, *Math. Proc. Cambridge Philos. Soc.* **88**, 129–133.

CHAPTER 4.8

Fourier Series and Spherical Harmonics in Convexity

H. GROEMER

Department of Mathematics, The University of Arizona, Tucson, AZ 85721, USA

Contents

HANDBOOK OF CONVEX GEOMETRY
Edited by P.M. Gruber and J.M. Wills

In 1901, Hurwitz published a short paper which showed that the isoperimetric inequality for plane domains can be deduced from simple properties of Fourier series. This paper marks the beginning of the use of Fourier series for purely geometric purposes. In a subsequent article Hurwitz pursued this subject further and for the first time used spherical harmonics for proving geometric results. A few years later there appeared a short note of Minkowski where it is shown that spherical harmonics can be used to prove an interesting characterization of three-dimensional convex bodies of constant width in terms of the perimeter of their projections. After Hurwitz and Minkowski had convincingly demonstrated the usefulness of Fourier series and spherical harmonics expansions for geometric investigations there appeared a large number of mathematical papers that deal with this subject. The principal applications of these analytic methods in convexity are still focused on the two types of problems that have already been considered by Hurwitz and Minkowski, namely geometric inequalities and uniqueness theorems (and a kind of combination of these two topics that has more recently emerged under the title of stability theorems). Fourier series and spherical harmonics have turned out to be appealing and often surprisingly powerful tools for proving geometric theorems. In fact, there are many results that can, at present, not be proved by any other means. The disadvantage, from a geometric point of view, of such proofs is that they offer hardly any possibilities for intuitive geometric interpretations.

The present article is a survey of major geometric results that have been obtained by the use of Fourier series or spherical harmonics. Occasionally some proofs are outlined and in all cases pertinent references are given. Not all areas are thoroughly covered. For example, since this volume is devoted to the theory of convex sets we hardly ever mention applications of Fourier series and spherical harmonics to the geometry of non-convex sets although the extension from the convex case to certain non-convex sets is sometimes rather straightforward. Some theorems can be proved by Fourier series or spherical harmonics and also by other means. In these cases we usually will discuss only the first possibility and not the methods and the literature associated with possible alternatives. Another subject area that will be largely neglected concerns "second-generation theorems", i.e., theorems that are consequences of theorems that can be proved by the use of Fourier series or spherical harmonics.

1. Notations and basic concepts

1.1. Convex bodies

We let $\mathbf{E}^d$ denote the Euclidean d-dimensional space and always assume that $d \geqslant 2$. The usual inner product of points x, y of $\mathbf{E}^d$ will be denoted by (x, y) and the norm of x by $|x|$. The class of all *convex bodies* (non-empty compact convex sets) in $\mathbf{E}^d$ will be denoted by $\mathbb{C}^d$. A convex body will be called a *central body* if it is centrally symmetric with respect to the origin o of $\mathbf{E}^d$. For any $K \in \mathbb{C}^d$ we let

$V(K)$, $S(K)$, and $D(K)$ denote, respectively, the volume, surface area, and diameter of K. We use the traditional notation $W_i(K)$ to denote the ith mean projection measure (Quermaßintegral) of K, and $V(K_1, \ldots, K_d)$ to denote the mixed volume of the convex bodies $K_1, \ldots, K_d$. If $d = 2$ the mixed volume $V(K_1, K_2)$ is also called the *mixed area* and denoted by $A(K_1, K_2)$. The unit sphere in $\mathbf{E}^d$ with center at o will be denoted by S^{d-1} and the corresponding unit ball by B^d. We let σ_d denote the surface area and κ_d the volume of B^d. In order to simplify the formulation of some of our statements we often refer to plane convex bodies as *convex domains*. Moreover, depending on the context, the words *circle* and *polygon* may either mean the respective convex curves or the corresponding convex domains bounded by these curves. If K is a convex domain we usually write $A(K)$ [instead of $V(K)$] for the area, and $P(K)$ [instead of $S(K)$] for the perimeter of K.

The support function of a convex body $K \in \mathbb{C}^d$ will be denoted by $h_K(u)$, or simply by $h(u)$. It will always be clear from the context whether $h(u)$ is to be considered as a function on S^{d-1}, or a function on $\mathbf{E}^d$ that is positively homogeneous of degree 1. If in $\mathbf{E}^2$ the usual (x, y)-coordinate system is given and $K \in \mathbb{C}^2$, any $u \in S^1$ is uniquely determined by the angle, say ω, between the positive x-axis and the vector u. Consequently, if $d = 2$ it is convenient to view h_K as a function of ω and to denote it by $h_K(\omega)$ or $h(\omega)$. If $K \in \mathbb{C}^d$ the mean width $\bar{w}(K)$ of K is defined by

$$\bar{w}(K) = \frac{1}{\sigma_d} \int_{S^{d-1}} (h_K(u) + h_K(-u))\, \mathrm{d}\sigma(u)\,.$$

It is well known that $\bar{w}(K) = (2/\kappa_d)W_{d-1}(K)$. In the case $d = 2$ this can also be written as $\bar{w}(K) = P(K)/\pi$. Among the various special points that can be associated with a convex body K we need only the centroid and the Steiner point of K. For the history, the general properties, and a rather complete list of references (up to 1971) regarding the Steiner point see Schneider (1972b). We denote the Steiner point of K by $z(K)$ and remark that it can be defined by

$$z(K) = \frac{1}{\kappa_d} \int_{S^{d-1}} u h(u)\, \mathrm{d}\sigma(u)\,.$$

The ball of diameter $\bar{w}(K)$ with center at $z(K)$ will be called the *Steiner ball* (if $d = 2$ *Steiner disc*) of K.

If $K, L \in \mathbb{C}^d$ the Hausdorff distance between K and L will be denoted by $\delta(K, L)$, and the L_2-distance by $\delta_2(K, L)$. Hence,

$$\delta(K, L) = \max\{|h_L(u) - h_K(u)|\colon u \in S^{d-1}\}\,,$$

and

$$\delta_2(K, L) = \left(\int_{S^{d-1}} (h_L(u) - h_K(u))^2\, \mathrm{d}\sigma(u)\right)^{1/2},$$

where $d\sigma(u)$ denotes the surface area differential on S^{d-1}. See Vitale (1985), Groemer and Schneider (1991), and chapter 1.4, section 1.3 for inequalities relating δ and δ_2.

We freely use the standard definitions and theorems of the theory of convex sets without giving specific references. The pertinent material can be found in books like Bonnesen and Fenchel (1934), Eggleston (1958), or Leichtweiß (1980).

1.2. The Laplace–Beltrami operator

Many applications of spherical harmonics to problems concerning geometric inequalities depend on the Laplace–Beltrami operator. We describe here a definition and some of the important properties of this operator in the special case when the underlying manifold is a sphere. If f is a function on $\mathbf{E}^d \backslash \{o\}$ we let $f_{S^{d-1}}$ denote the restriction of f to S^{d-1}. The Laplace operator Δ is defined by

$$\Delta = \sum_{i=1}^{d} \frac{\partial^2}{\partial x_i^2} ,$$

and can be applied to any twice differentiable function on an open subset of $\mathbf{E}^d$. The analogue of the Laplace operator for functions on S^{d-1} is the *Laplace–Beltrami operator* which will be denoted by Δ_o. If g is a twice differentiable function on S^{d-1}, the function $g(x/|x|)$ is the radial extension of g to $\mathbf{E}^d \backslash \{o\}$ which is constant on each half-line starting at o. Using this extension we can define Δ_o by

$$\Delta_o g(x) = \Delta g(x/|x|)_{S^{d-1}} .$$

If f is a twice differentiable function on $\mathbf{E}^d \backslash \{o\}$ that is positively homogeneous of degree 1 (for example, if it is the support function of a sufficiently smooth convex body) it is easy to calculate that

$$\Delta_o f_{S^{d-1}} = (\Delta f - (d-1)f)_{S^{d-1}} .$$

Another important feature of Δ_o is the formula

$$\int_{S^{d-1}} g(u)\, \Delta_o h(u)\, d\sigma(u) = \int_{S^{d-1}} h(u)\, \Delta_o g(u)\, d\sigma(u) , \tag{1}$$

which holds for all pairs of functions g, h on S^{d-1} that are twice continuously differentiable. For proofs of these formulas and for more details regarding Δ and Δ_o see Seeley (1966). For geometric applications of particular importance is the possibility to express the mean projection measure $W_{d-2}(K)$ of any sufficiently smooth $K \in \mathcal{C}^d$ in terms of Δh_K or $\Delta_o h_K$. The pertinent formula is (see Bonnesen and Fenchel 1934, pp. 62–63)

$$W_{d-2}(K) = \frac{1}{d(d-1)} \int_{S^{d-1}} h_K(u)\, \Delta h_K(u)\, \mathrm{d}\sigma(u)$$

$$= \frac{1}{d} \int_{S^{d-1}} h_K(u)\Big(h_K(u) + \frac{1}{d-1}\, \Delta_o h_K(u)\Big)\, \mathrm{d}\sigma(u)\,. \tag{2}$$

Further informative discussions of these operators with regard to the theory of convex sets can be found in the articles of Dinghas (1940) and Berg (1969).

1.3. Fourier series

We list here some of the definitions, notations, and a few facts about Fourier series that will be used. Proofs of these theorems can be found in the standard textbooks on this subject such as Sz.-Nagy (1965) or Zygmund (1977). We consistently use Fourier series in their real form although it would of course be possible to employ (as some authors have done) their complex version.

Let f be a real valued integrable function on $[-\pi, \pi]$. The *Fourier coefficients* of f are defined by

$$a_0 = \frac{1}{2\pi} \int_{-\pi}^{\pi} f(x)\, \mathrm{d}x\,, \qquad a_k = \frac{1}{\pi} \int_{-\pi}^{\pi} f(x) \cos kx\, \mathrm{d}x\,,$$

$$b_0 = 0\,, \qquad b_k = \frac{1}{\pi} \int_{-\pi}^{\pi} f(x) \sin kx\, \mathrm{d}x\,,$$

and the series

$$\sum_{k=0}^{\infty} (a_k \cos kx + b_k \sin kx)$$

is called the *Fourier series* of f. To indicate that this series is the Fourier series of f we write

$$f(x) \sim \sum_{k=0}^{\infty} (a_k \cos kx + b_k \sin kx)\,.$$

Of particular importance for our purpose is *Parseval's equation*

$$\int_{-\pi}^{\pi} f(x)^2\, \mathrm{d}x = \pi\Big(2a_0^2 + \sum_{k=1}^{\infty} (a_k^2 + b_k^2)\Big)\,,$$

which holds for any square integrable function on $[-\pi, \pi]$. More generally, if g is another such function and

$$g(x) \sim \sum_{k=0}^{\infty} (c_k \cos kx + d_k \sin kx)\,,$$

then an application of Parseval's equation to $f + g$ yields immediately

$$\int_{-\pi}^{\pi} f(x)g(x)\,\mathrm{d}x = \pi\Big(2a_0c_0 + \sum_{k=1}^{\infty} (a_kc_k + b_kd_k)\Big)\,.$$

This relation will be referred to as the *generalized Parseval's equation*.

1.4. Spherical harmonics

We now introduce some definitions, notations, and basic facts about spherical harmonics on S^{d-1}. Proofs and further results can be found in the standard literature dealing with this subject, specifically in the monographs of Erdélyi et al. (1953), Müller (1966), Berg (1969), and in the books of Hochstadt (1986, chapter 6) and Stein and Weiss (1971, chapter IV, §2). Schneider (1992a) lists in an appendix those results on spherical harmonics that are used for the geometric applications in this book. (But no further references to this book will be given here, since it was not yet available.)

We define the *inner product* $\langle f, g\rangle$ of two square integrable real valued functions f, g on S^{d-1} by:

$$\langle f, g\rangle = \int_{S^{d-1}} f(u)g(u)\,\mathrm{d}\sigma(u)\,,$$

and let $\|f\| = \langle f, f\rangle^{1/2}$ denote the norm of f. The functions f and g are said to be *orthogonal* if $\langle f, g\rangle = 0$.

A homogeneous polynomial $p(x)$ in d variables with real coefficients will be called a *harmonic polynomial* if $\Delta p(x) = 0$. The restriction of a harmonic polynomial of degree n to S^{d-1} is called a *spherical harmonic of order n and dimension d*. It is easy to see that in the case $d = 2$, with the points on S^1 identified by the corresponding angle ω, the spherical harmonics of dimension 2 and order n are exactly the functions $a \cos n\omega + b \sin n\omega$. If $N(d, n)$ denotes the maximum number of linearly independent spherical harmonics of dimension d and order n, then

$$N(d, n) = \frac{2n + d - 2}{n + d - 2}\binom{n + d - 2}{d - 2}\,.$$

It can be shown that any two spherical harmonics of different orders are orthogonal. Furthermore, for any given d it clearly is possible to select from the set of all d-dimensional spherical harmonics of order n a subset of $N(d, n)$ mutually orthogonal functions (none of which is identically 0). If this is done successively for $n = 0, 1, \ldots$ it leads to a sequence $P_0, P_1, \ldots$ of spherical harmonics with the property that any two terms are orthogonal and that for any n it contains $N(d, n)$ linearly independent spherical harmonics of order n. Such a sequence will be called a *standard sequence*. If f is an integrable real-valued function on S^{d-1} and if a standard sequence $P_0, P_1, \ldots$ is given we associate with

f the series

$$\sum_{k=0}^{\infty} c_k P_k , \tag{3}$$

where

$$c_k = \frac{\langle f, P_k \rangle}{\|P_k\|^2} . \tag{4}$$

We call the series (3) the *harmonic expansion* of f with respect to the given standard sequence $P_0, P_1, \ldots$ and write

$$f \sim \sum_{k=0}^{\infty} c_k P_k . \tag{5}$$

It can be shown that any standard sequence $P_0, P_1, \ldots$ is complete in the sense that for any square integrable function g on S^{d-1} the conditions $\langle P_i, g \rangle = 0$ (for all i) imply that $g = 0$ almost everywhere. Equivalently, this fact can be expressed in terms of *Parseval's equation*

$$\|f\|^2 = \sum_{k=0}^{\infty} c_k^2 \|P_k\|^2 ,$$

which holds for all square integrable functions f on S^{d-1} satisfying (5). More generally, and in analogy to the case of Fourier series, if in addition to (5) one has $g \sim \Sigma_{k=0}^{\infty} d_k P_k$, then

$$\langle f, g \rangle = \sum_{k=0}^{\infty} c_k d_k \|P_k\|^2 . \tag{6}$$

It sometimes is convenient to combine in (3) all terms of the same order. If Q_n denotes the sum of all terms $c_k P_k$ of order n this leads to a series of the form $\Sigma_{k=0}^{\infty} Q_n$, where each Q_n is a spherical harmonic of order n. We call this series the *condensed harmonic expansion* of f. When dealing with such condensed harmonic expansions it is not necessary to refer to the standard sequence $P_0, P_1, \ldots$ that was used to construct this expansion since it is easy to see that any two standard sequences yield the same functions Q_n. As before we write

$$f \sim \sum_{n=0}^{\infty} Q_n \tag{7}$$

and remark that the corresponding Parseval's equation becomes

$$\|f\|^2 = \sum_{n=0}^{\infty} \|Q_n\|^2 .$$

Similarly as in the case of (6) if $f \sim \Sigma_{n=0}^{\infty} U_n$ and $g \sim \Sigma_{n=0}^{\infty} V_n$ are condensed expansions, then

$$\langle f, g \rangle = \sum_{n=0}^{\infty} \langle U_n, V_n \rangle . \tag{8}$$

Both (6) and (8) will be referred to as *generalized Parseval's equations.*

In most cases it is rather difficult to decide whether the expansion of a given function f on S^{d-1} converges pointwise to f. It is known (see Kubota 1925a or Müller 1966) that for a continuous f the series is (uniformly on S^{d-1}) Abelian summable to f. From this fact it follows that any continuous f can be uniformly approximated by finite sums of spherical harmonics. One can also prove that the harmonic expansion of f converges uniformly to f if this function is sufficiently often differentiable. This follows, for example, from estimates given by Schneider (1967).

Of interest is the behavior of spherical harmonics with respect to the Laplace–Beltrami operator. It can be shown that if Q_n is a spherical harmonic of order n, then

$$\Delta_o Q_n = -n(n + d - 2)Q_n . \tag{9}$$

This relation reveals in particular that the spherical harmonics are eigenfunctions of the Laplace–Beltrami operator. Relation (9) can be used to find the harmonic expansion of $\Delta_o f$ if an expansion of f is given. Indeed, if $\Delta_o f$ exists and is continuous and if U_n is a spherical harmonic of dimension d and order n it follows from (1) and (9) that

$$\langle \Delta_o f, U_n \rangle = \langle f, \Delta_o U_n \rangle = -n(d + n - 2)\langle f, U_n \rangle .$$

Consequently, (7) implies that

$$\Delta_o f \sim -\sum_{n=0}^{\infty} n(d + n - 2)Q_n . \tag{10}$$

In the case $d = 2$ (with f considered as a function of the angle ω) this relation implies that $\Delta_o f = \mathrm{d}^2 f/\mathrm{d}\omega^2$.

For geometric applications a very useful result for spherical harmonics is the *Funk–Hecke theorem*. If Q_n is any d-dimensional spherical harmonic of order n and g a bounded integrable function on $[-1, 1]$, this theorem states that

$$\int_{S^{d-1}} g((u, v))Q_n(u)\,\mathrm{d}\sigma(u) = c_{d,n}(g)Q_n(v) , \tag{11}$$

where $c_{d,n}(g)$ depends only on d, n, and g. For proofs and explicit representations of $c_{d,n}(g)$ see the monographs on spherical harmonics mentioned above. (The formula is usually only stated for continuous functions on S^{d-1} but the extension to the case of bounded integrable functions follows without difficulties from standard approximation procedures of integration theory.)

2. Geometric applications of Fourier series

This section is a survey of results in the theory of plane convex sets that have been proved by the use of Fourier series. It focuses mainly on theorems that have no natural extension to the d-dimensional situation and on results that provide good illustrations of the general methods under the simplifying assumption that $d=2$.

2.1. The work of Hurwitz on the isoperimetric inequality

The proof of Hurwitz (1901) of the isoperimetric inequality is the earliest and probably best known example of the use of Fourier series for the purpose of proving a geometric problem, namely the isoperimetric inequality $P(C)^2 - 4\pi A(C) \geqslant 0$, where C is a (sufficiently smooth) simple closed curve of length $P(C)$ that encloses a region of area $A(C)$.

Because of its historical importance we outline here the proof of Hurwitz. Assuming, as one may, that the length of C is 2π one wishes to prove that

$$\pi - A(C) \geqslant 0 .$$

If s denotes the length of an arc in C (measured from some given initial point) and if $x(s)$, $y(s)$ are the Euclidean coordinates of the point on C corresponding to this value of s one may set

$$x(s) \sim \sum_{k=0}^{\infty} (a_k \cos ks + b_k \sin ks) , \qquad y(s) \sim \sum_{k=0}^{\infty} (c_k \cos ks + d_k \sin ks) , \tag{12}$$

Using the generalized Parseval's equation one immediately finds

$$A(C) = \int_0^{2\pi} x(s)y'(s)\,\mathrm{d}s = \pi \sum_{k=0}^{\infty} k(a_k d_k - b_k c_k) ,$$

where the derivative is taken with respect to s. Moreover, since $(x')^2 + (y')^2 = 1$, we have, again using Parseval's equation,

$$\pi = \tfrac{1}{2} \int_0^{2\pi} ((x')^2 + (y')^2)\,\mathrm{d}s = \tfrac{1}{2}\pi \sum_{k=0}^{\infty} k^2(a_k^2 + b_k^2 + c_k^2 + d_k^2) .$$

Hence,

$$\pi - A(C) = \tfrac{1}{2}\pi \sum_{k=0}^{\infty} \Big((k^2 - k)(a_k^2 + b_k^2 + c_k^2 + d_k^2) + k((a_k - d_k)^2 + (b_k + c_k)^2) \Big)$$

and it follows that $\pi - A(C) \geqslant 0$. It is easily seen that equality holds if and only if C is a circle.

This proof of the isoperimetric inequality, sometimes in a modified version that uses the complex form of the Fourier series, has been reproduced frequently in educationally oriented articles and textbooks as an example for the applicability of Fourier series in geometry. Lebesgue (1906) has pointed out that essentially the same proof yields the isoperimetric inequality under the sole assumption that C is a rectifiable Jordan curve; see also Sz.-Nagy (1965). Another variation of the above proof of Hurwitz is due to Fisher, Ruoff and Shilleto (1985). It proceeds from the case of polygons to the general case.

In a subsequent paper, Hurwitz (1902) presented a different proof of the isoperimetric inequality based on the Fourier expansion of the radius of curvature. This approach is similar to that in the following section but requires regularity assumptions that can be avoided if one uses the support function. Various other results of Hurwitz (l.c.) which he derived from the expansion of the radius of curvature will be (selectively) mentioned in section 2.4.

2.2. The Fourier expansion of the support function and mixed area inequalities

Although the method discussed in the previous section is of interest the most important series for geometric purposes is the Fourier expansion of the support function. The reason for this is the fact that important geometric data, like area, perimeter, mixed area, and Steiner point can be succinctly expressed in terms of the coefficients of this Fourier series. Moreover, some of the most interesting features regarding this expansion can be generalized to higher dimensions.

If K is a convex domain with support function $h(\omega)$ and

$$h(\omega) \sim \sum_{k=0}^{\infty} (a_k \cos k\omega + b_k \sin k\omega)\,,$$

then

$$A(K) = \pi a_0^2 - \tfrac{1}{2}\pi \sum_{k=2}^{\infty} (k^2 - 1)(a_k^2 + b_k^2)\,, \tag{13}$$

$$P(K) = 2\pi a_0\,, \tag{14}$$

and the Steiner point is given by

$$z(K) = (a_1, b_1)\,. \tag{15}$$

Furthermore, if L is another convex domain whose support function, say $g(\omega)$, has the Fourier expansion $\Sigma(c_k \cos k\omega + d_k \sin k\omega)$, then the mixed area of K and L is given by the formula

$$A(K, L) = \pi a_0 c_0 - \tfrac{1}{2}\pi \sum_{k=2}^{\infty} (k^2 - 1)(a_k c_k + b_k d_k)\,, \tag{16}$$

and the L_2-distance between K and L can be expressed in terms of the Fourier coefficients by

$$\delta_2(K, L)^2 = 2\pi(a_0 - c_0)^2 + \pi \sum_{k=1}^{\infty} \left((a_k - c_k)^2 + (b_k - d_k)^2\right). \tag{17}$$

The proof of these equalities follows immediately from the definition of $z(K)$ and δ_2, the generalized Parseval's equation and the well-known formulas

$$P(K) = \int_{-\pi}^{\pi} h(\omega)\,\mathrm{d}\omega\,, \tag{18}$$

$$A(K) = \tfrac{1}{2}\int_{-\pi}^{\pi} \left(h(\omega)^2 - h'(\omega)^2\right) \mathrm{d}\omega\,, \tag{19}$$

and

$$A(K, L) = \tfrac{1}{2}\int_{-\pi}^{\pi} \left(h(\omega)g(\omega) - h'(\omega)g'(\omega)\right) \mathrm{d}\omega\,,$$

where the derivatives are taken with respect to ω.

As a straightforward consequence of (13), (14), (15), and (17) one obtains the following strengthened version of the isoperimetric inequality

$$P(K)^2 - 4\pi A(K) \geq 6\pi\delta_2(K, C(K))^2\,, \tag{20}$$

where $C(K)$ denotes the Steiner disc of K. Furthermore, if one uses instead of (13) the equality (16) one finds (after a slightly more involved calculation) the following generalization of the isoperimetric inequality:

$$A(K, L)^2 - A(K)A(L) \geq \tfrac{3}{2}\bar{w}(L)^2 A(K)\delta_2(K^*, L^*)^2\,, \tag{21}$$

where K^* and L^* denote homothetic copies of K and L (respectively) of mean width 1 and with coincident Steiner points. Similar proofs of the inequalities $P(K^2) - 4\pi A(K) \geq 0$ and $A(K, L)^2 - A(K)A(L) \geq 0$ have been given repeatedly; for example, by Blaschke (1914), Görtler (1937a), and Dinghas (1940). Bol (1939) has shown how a more elaborate application of Fourier series leads to strengthened forms of the isoperimetric inequality and the mixed area inequality of a type that has first been considered by Bonnesen (see chapter 1.4, sections 2.1 and 2.2). Using spherical harmonics instead of Fourier series one can generalize both (20) and (21) to the higher dimensional case (see section 3.7). There also exist generalizations of this kind of inequalities to non-convex domains with the property that the concept of the support function can be suitably generalized. Such results are proved in the articles of Geppert (1937), Bol (1939), and in the work of Gericke (1940a) who employs systematically complex Fourier series. Further inequalities of this type that are more or less obvious consequences of the

Fourier expansion of the support function can be found in the survey article of Groemer (1990); see also Letac (1983) and Fisher (1984).

Gericke (1941) introduced (for the two-dimensional case) a support function with respect to a given Minkowskian geometry in the plane and has studied the development of this function with respect to a system of orthogonal functions that appear as eigenfunctions of a certain differential equation. Using such series he proved generalizations of the isoperimetric inequality, the mixed area inequality and some related estimates.

We conclude this section by mentioning area estimates of the following kind:

$$A(K) \leqslant \tfrac{1}{2} \int_0^{2\pi} h(\omega)h(\omega + \alpha)\,\mathrm{d}\omega\ ,$$

where h is the support function of a given domain K. This inequality is due to Heil (1972) who used the Fourier expansion of h. For the corresponding integral with h replaced by the width of K in the direction ω and $\alpha = \frac{1}{2}\pi$, an earlier proof of an inequality of this kind (also based on Fourier series) has been given by Chernoff (1969). Chakerian (1979) used Fourier series to obtain substantial generalizations of this inequality involving the integral over the product of the support functions of two different domains. If $\alpha = \frac{1}{2}\pi$ the above inequality can also be interpreted as an estimate of the mean value of the area of circumscribed rectangles of K. From this point of view it can be generalized to higher dimensions. For example, Schneider (1972a) has used spherical harmonics to show that the mean value (with respect to normalized Haar measure on the rotation group of $\mathbf{E}^3$) of the surface area of the circumscribed boxes of a three-dimensional convex body K is at least $6S(K)/\pi$. He also remarks that corresponding results can be proved for $\mathbf{E}^d$ and for other types of circumscribed polytopes. See also Groemer (1992b) for work on circumscribed cylinders and Lutwak (1977) for an approach to this subject area that does not depend on the use of Fourier series or spherical harmonics.

2.3. *Circumscribed polygons and rotors*

Let K be a convex domain and P a polygon (always assumed to be convex). P will be called a *tangential polygon* of K, and K an *osculating domain* in P, if $K \subset P$ and every side of P has a nonempty intersection with K. Thus, for the purpose of our discussion, tangential polygons are circumscribed, osculating domains are inscribed. We discuss here how Fourier expansions have been used to derive interesting results on osculating domains in various types of tangential polygons. Most of these results have first been proved by Meissner (1909) and some of them have been occasionally rediscovered or presented in the context of related investigations. (See, e.g., Cieślak and Góźdź 1987 and Focke 1969.) Meissner's work is primarily based on the Fourier series expansion of the radius of curvature and consequently requires regularity assumptions. Using the Fourier series of the support function one can prove the same results without any regularity assumptions.

A polygon will be said to be *equiangular* if all its interior angles at the vertices are equal. We begin by citing a theorem that characterizes, in terms of the Fourier coefficients associated with the support function, those convex domains that have the property that all tangential equiangular polygons have the same perimeter.

Let K be a convex domain with support function $h(\omega)$ and assume that

$$h(\omega) \sim \sum_{k=0}^{\infty} (a_k \cos k\omega + b_k \sin k\omega) \tag{22}$$

and that n is an integer not less than 3. Then every tangential equiangular n-gon of K has the same perimeter, say P_n°, if and only if

$$a_n = a_{2n} = a_{3n} = \cdots = 0\,, \qquad b_n = b_{2n} = b_{3n} = \cdots = 0\,.$$

If this condition is satisfied, then P_n° and the perimeter of K are related by the equation

$$P_n^{\circ} = \frac{n}{\pi}\left(\tan \frac{\pi}{n}\right) P(K)\,.$$

To convey the idea of proof of this theorem set $\varepsilon_n = 2\pi/n$ and let $g(\omega)$ denote the side of a tangential equiangular n-gon of K, which is contained in the support line of K corresponding to the angle ω. Let $\lambda(g(\omega))$ denote the length of $g(\omega)$. Then the other sides of this polygon are $g(\omega + \varepsilon_n)$, $g(\omega + 2\varepsilon_n), \ldots, g(\omega + (n-1)\varepsilon_n)$ and elementary geometry yields that

$$\lambda(g(\omega + j\varepsilon_n)) = \frac{1}{\sin \varepsilon_n}\,[h(\omega + (j-1)\varepsilon_n) - 2h(\omega + j\varepsilon_n)\cos \varepsilon_n + h(\omega + (j+1)\varepsilon_n)]\,.$$

Summation of these expressions shows that the perimeter, say $P_n(\omega)$, of this tangential n-gon is given by

$$P_n(\omega) = 2 \tan \tfrac{1}{2}\varepsilon_n \sum_{j=0}^{n-1} h(\omega + j\varepsilon_n)\,.$$

Expressing h as a Fourier series the relation $P_n(\omega) = P_n^{\circ}$ enables one to deduce an identity in ω that leads immediately to the above conditions on the Fourier coefficients. (It is not difficult to prove that $h(\omega)$ is sufficiently regular to justify these operations, cf. Blaschke 1914.)

We next consider convex domains with the property that all their tangential equiangular n-gons are regular (but not necessarily of equal size). Using the conditions $\lambda(g(\omega + j\varepsilon_n)) = \lambda(g(\omega))$ and the above representation of $\lambda(g(\omega + j\varepsilon_n))$ one can derive the following characterization of these domains in terms of the Fourier series of the support function: Every equiangular tangential n-gon of a convex domain K is regular if and only if the Fourier coefficients a_k, b_k of the Fourier expansion of the support function of K have the property that $a_k = 0$ and $b_k = 0$ for all k that are not congruent to 0, 1, or -1 modulo n.

A convex domain K will be called a *rotor* in a polygon Q if for every rotation ρ there is a translation vector p_ρ such that $\rho K + p_\rho$ is an osculating domain in K. For example, rotors in rhombi (with equally distanced parallel sides) are exactly convex domains of constant width. Clearly, K is a rotor in a regular n-gon Q exactly if all tangential equiangular n-gons are regular and have equal perimeters. Hence we can state the following theorem: A convex domain K is a rotor in a regular n-gon Q if and only if the Fourier coefficients a_k, b_k of the Fourier series of the support function of K have the property that $a_k = b_k = 0$ for all $k > 0$ that are not congruent to ± 1 modulo n. Furthermore, all rotors in Q have the same perimeter, namely $(\pi/n)\cot(n/\pi)P(Q)$.

As a consequence of these results it can easily be deduced that for every $n \geqslant 3$ there are non-circular rotors in regular n-gons, and that there exist convex domains which are not rotors in regular n-gons but have property that all their tangential n-gons are regular.

We now consider rotors in not necessarily regular convex polygons. Most of the results that will be described here have first been found by Fujiwara (1915). A more recent presentation is due to Schaal (1962); see also Hayashi (1918), Kamenezki (1947), Focke (1969) and the discussion of the pertinent literature by Schneider (1971a), and Chakerian and Groemer (1983). In the study of rotors it is advantageous to admit also certain "unbounded polygons". We call a convex domain a *polygonal domain* if it is the intersection of finitely many closed half-planes, has nonempty interior, and a boundary that does not consist of only one line or of two half-lines with a common initial point. In other words, we exclude those domains (half-planes and "wedges") for which every convex body is a rotor. It is clear how to generalize the concepts of a side, a tangential polygon, an osculating domain, and a rotor to the case of polygonal domains. Of particular importance are *triangular domains*, i.e., polygonal domains whose boundary is contained in three lines without any two of them being parallel. If α_2, α_3 are interior angles of a triangular domain T we define a third angle α_1 by $\alpha_1 = \pi - (\alpha_2 + \alpha_3)$ and call $\alpha_1, \alpha_2, \alpha_3$ the associated angles of T. (In the case when T is an ordinary triangle these are of course the three interior angles of T.) The principal result relating rotors in triangular domains to the Fourier series of their support function can be formulated as follows: Let T be a triangular domain with associated angles $\alpha_1, \alpha_2, \alpha_3$ and let r denote the radius of the osculating circle of T. If at least one of the angles α_i is an irrational multiple of π, then T has only one rotor, namely its osculating circle. If all angles α_i are rational multiples of π one may set $\alpha_i = \pi g_i/N$ $(i = 1, 2, 3)$, where g_1, g_2, g_3, and N are integers such that $\gcd(g_1, g_2, g_3) = 1$ and $N > 2$. In this case T has infinitely many rotors and a convex domain K with support function $h(\omega)$ is a rotor in T if and only if the Fourier expansion (22) has the property that $a_0 = r$ and $a_k = b_k = 0$ for all $k > 1$ that are not congruent to ± 1 modulo N', where $N' = N$ if all g_i are odd, and $N' = 2N$ if at least one g_i is even. Combined with the earlier mentioned result about rotors in regular polygons it follows that every rotor in a triangular domain is also a rotor in a regular polygon.

These facts can be used to deduce results about rotors in arbitrary polygonal domains. To formulate these results it is convenient to introduce the following definitions. If P and Q are two polygonal domains we say that Q is *derived* from P if every side of Q contains a side of P. Any polygonal domain derived from a rhombus will be called a *rhombic domain*. Domains of this kind are either rhombi, strips, or "half-strips" (bounded by a line segment and two parallel half-lines). Obviously, a convex domain is a rotor in a rhombic domain exactly if it is of constant width. It is easy to see that a convex domain K is a rotor in a non-rhombic polygonal domain P if and only if it is a rotor in every triangular domain derived from P. Moreover, if P has a rotor it has an osculating circle (since the radius of the osculating circle of each derived triangular domain is the same, namely a_0). As a consequence of this and the previously stated characterization of rotors in triangular domains we immediately obtain the fact that a non-rhombic polygonal domain Q has a non-circular rotor if and only if it has an osculating circle and all its interior angles are rational multiples of π. Moreover we can state that every polygonal domain that has a non-circular rotor is derived either from a rhombus or a regular polygon, and that every rotor in any polygonal domain is also a rotor in some regular polygon.

For a detailed investigation of rotors bounded by circular arcs in triangles see Fujiwara (1919). Fourier expansions associated with domains of constant width have already been studied by Hurwitz (1902) and have been frequently used to prove various properties of such domains, see Nakajima (1920). More recent studies regarding this subject have been published by Tennison (1976), Fisher (1987), Cieślak and Góźdź (1989), and Góźdź (1990). It is an interesting but difficult problem to find rotors of minimal area in regular n-gons. Regarding the role of Fourier series for this and a related minimum problem see Fujiwara and Kakeya (1917), Fujiwara (1919), Focke (1969), and Klötzler (1975).

Nakajima (1920) considered the problem of characterizing those strictly convex domains with the property that there is a wedge such that under rotation in that wedge (so that it always is an osculating domain of the wedge) the length of the inner boundary curve between the two contact points is constant. Using the Fourier expansion of the support function he showed, under regularity assumptions, that such a domain must either be a circle (if the angle of the wedge is irrational) or it must have n-fold rotational symmetry (for some integer n). A very special case of this problem concerning tangential equilateral triangles has been studied by Kubota (1920). Nakajima (l.c.) also considered a number of variations of this problem and even certain generalizations to convex bodies in $\mathbf{E}^3$.

Using Fourier series Su (1927) investigated the domain bounded by the path traced out by any given point of a convex domain under a full rotation in a wedge (so that it is always tangential). In particular he showed that its area (suitably signed according to orientation) is minimal if and only if this point is the Steiner point of the domain.

Rotors in $\mathbf{E}^d$, where the situation is quite different, will be discussed in section 3.6.

2.4. Other geometric applications of Fourier series

We start by listing some of the results that are already contained in the classical paper of Hurwitz (1902).

If $A_o(K)$ denotes the absolute value of the signed area (according to orientation) of the evolute of the boundary curve of a convex domain K, then

$$P(K)^2 - 4\pi A(K) \leqslant \pi A_o(K) .$$

Hurwitz also found sharp inequalities for the mean value of the squared width and the squared radius of curvature of convex domains (however, these estimates can also be obtained by straightforward applications of Hölder's inequality). For further applications of Fourier series along these lines, involving also the Steiner point and the pedal curve of a given domain, see Meissner (1909), Hayashi (1924), and Gericke (1940a). Another accomplishment of Hurwitz is a proof, based on Fourier series, of Crofton's formula

$$\tfrac{1}{2}P(K)^2 - \pi A(K) = \int\int (\alpha - \sin \alpha)\, dx\, dy ,$$

where α is the angle between two support lines of the convex domain K meeting at a point (x, y) and the integration is extended over all points (x, y) outside K. Finally, we mention that he established an upper bound for the isoperimetric deficit in terms of the integral of the squared radius of curvature of the domain and that he gave estimates for the perimeter of a convex curve in terms of the maximum and minimum of the radius of curvature.

Several results of Hurwitz have been reestablished by Dinghas (1940) with slightly different proofs and with a view towards generalizations to the d-dimensional case. Gericke (1940a,b) used Fourier series to study properties of moments of inertia (also for $d = 3$).

Fourier series have occasionally been employed to determine under which circumstances the equality sign holds in a given geometric inequality. For example, if K is a convex domain and K_n the inscribed convex n-gon of maximum area, then it has been shown by Sas (see Fejes Tóth 1972) that $A(K_n) \geqslant (nA(K)/2\pi)\sin(2\pi/n)$. Fejes Tóth (l.c.) has employed a Fourier series to determine that in this relation the equality sign holds exactly if K is an ellipse. For related problems of this kind see also Schneider (1971c), together with the supplementary work of Florian and Prachar (1986), and the pertinent remarks of Florian in chapter 1.6 of this Handbook.

Görtler (1937a) derived a condition on the respective Fourier coefficients of the support functions of two domains with the property that the area of their Minkowski sum is not only translation but also rotation invariant. Moreover, he discusses the problem under which circumstances the area of a linear combination of the form $\lambda K + (1-\lambda)L$ (where $0 \leqslant \lambda \leqslant 1$) is a concave function of λ (not the square root of this combination, as guaranteed by the Brunn–Minkowski Theorem). Another study of this problem has been published by Geppert (1937), who also investigates the corresponding three-dimensional problem for the surface area using the expansion of the support function in terms of spherical harmonics.

Meissner (1909) used Fourier series to prove various results about the centroid and Steiner point of a given convex domain. For example, he established necessary and sufficient conditions on the Fourier coefficients in order that the centroid of K is the same as the Steiner point of K and proved the theorem of Steiner that the area of the pedal curve of a convex domain is minimal if it is taken with respect to the Steiner point. Regarding the former theorem see also Kubota (1918).

As already proved by Archimedes the surface area of a ball equals the lateral surface area of a circumscribed cylinder and the ratio of the volume of a ball to the volume of a circumscribed cylinder is 2 : 3. Knothe (1957) used Fourier series to show that each of these properties characterizes balls. (The statement regarding the surface area has been proved by Firey (1959) without the use of Fourier series.) An interesting inequality in Knothe's work regarding the mean value of the lateral surface area of circumscribed cylinders has been generalized to the d-dimensional case by Groemer (1992b).

Fourier series have been used to prove theorems of the type of the four vertex theorem and of theorems concerning the existence of support lines with particular properties. Investigations of this kind can be found in the articles of Meissner (1909), Hayashi (1926), and Fisher (1987).

There exists a variety of results concerning the properties of the Fourier coefficients that are associated with convex curves through the relations (12). For example, if a convex curve has perimeter 2π and if $n \geqslant 2$, then $a_n^2 + b_n^2 + c_n^2 + d_n^2 \leqslant 8/\pi n^4$. Results of this type are proved in the articles of de Vries (1969), Wegmann (1975), Hall (1983, 1985) and the papers listed by these authors as references.

There are several articles that concern themselves with the composition of domains in terms of certain "basic domains". Analytically this means the representation of support functions (suitably generalized for certain non-convex domains) in terms of functions having particular properties. One such representation is the Fourier series expansion of the support function, where the terms are interpreted geometrically. For work along these lines see Görtler (1937b), Gericke (1940a, 1941), and Inzinger (1949). The work of Görtler (1938) and Gericke (1940b) contains also generalizations of such concepts to the case $d = 3$.

We finally mention that various problems concerning polygons have been solved by the use of "finite Fourier series", i.e., trigonometric sums of the same type as partial sums of Fourier series. See Blaschke (1916b), Schoenberg (1950) and Fisher, Ruoff and Shilleto (1985).

3. Geometric applications of spherical harmonics

3.1. The harmonic expansion of the support function

If $u = (u_1, u_2, \ldots, u_d)$ is a point of $\mathbf{E}^d$, then the functions $\Phi_i(u) = u_i$, restricted to S^{d-1}, form a collection of $N(d, 1) = d$ linearly independent spherical harmonics of order 1. Hence, if h is the support function of some $K \in \mathbb{C}^d$ it has a condensed harmonic expansion of the form

$$h(u) \sim \alpha_0 + (\beta_1 u_1 + \cdots + \beta_d u_d) + \sum_{n=2}^{\infty} Q_n(u)\,.$$

The definition of the mean width and (4) show that $\alpha_0 = \frac{1}{2}\bar{w}(K)$. Furthermore,

$$\beta_i = \frac{1}{\|u_i\|^2} \int_{S^{d-1}} h(u)u_i \,\mathrm{d}\sigma(u) = \frac{1}{\kappa_d} \int_{S^{d-1}} h(u)u_i \,\mathrm{d}\sigma(u)\,,$$

and it follows that the point $(\beta_1, \beta_2, \ldots, \beta_d)$ is actually the Steiner point $z(K)$ of K. Hence the condensed harmonic expansion of h can always be written in the form

$$h(u) \sim \tfrac{1}{2}\bar{w}(K) + (u, z(K)) + \sum_{n=2}^{\infty} Q_n(u)\,. \tag{23}$$

As a consequence of (2), (10), and the generalized version of Parseval's equation, we find that for any (sufficiently smooth) $K \in \mathbb{C}^d$ whose condensed expansion of the support function is written in the form (23) we have

$$W_{d-2}(K) = \frac{\kappa_d}{4}\,\bar{w}(K)^2 - \frac{1}{d(d-1)} \sum_{n=2}^{\infty} (n-1)(n+d-1)\|Q_n\|^2\,. \tag{24}$$

Using less elementary techniques the same relation can be proved without any smoothness assumptions (cf. Goodey and Groemer 1990).

In one form or another the formulas of this section appear (with or without proofs) in many articles devoted to the application of spherical harmonics in geometry. Besides the seminal paper of Hurwitz (1902) we mention in particular Blaschke (1916b), Kubota (1925a), and Dinghas (1940) where one can also find relations between the radii of curvature and harmonic series.

Another immediate but interesting consequence of (23) and Parseval's equation is the following fact: If $K \in \mathbb{C}^d$ and X ranges over all homothetic copies of some given convex body in $\mathbf{E}^d$, then $\delta_2(K, X)$, considered as a function of X, is minimal if and only if X has the same Steiner point and mean width as K. Arnold (1989) has studied in detail the problem of the best L_2 approximation of one convex body by another and, in particular, the relationship to the corresponding Fourier expansions of the pertinent support functions in the case $d = 2$.

3.2. *Minkowski's Theorem on convex bodies of constant width*

If $K \in \mathbb{C}^d$ and $u \in S^{d-1}$ we let K_u denote the orthogonal projection of K onto the $(d-1)$-dimensional linear subspace of $\mathbf{E}^d$ that is orthogonal to u. Clearly, if K is of constant width, then $\bar{w}(K_u)$ does not depend on u. Minkowski (1904–1906) showed in the case $d = 3$ and under smoothness assumptions that the converse of this statement is also true. Calling $P(K_u)$ the perimeter of K in the direction u he formulated his theorem by saying that convex bodies of constant perimeter are convex bodies of constant width.

To describe the essential features of Minkowski's proof of this theorem let us assume that some convex body $K \in \mathbb{C}^3$ with support function h is given and let us set $P(K_u)/\pi = c$. Then, if h is sufficiently regular, we may write

$$h(v) + h(-v) - c = \sum_{n=0}^{\infty} Q_n(v) \tag{25}$$

and assume that this condensed harmonic expansion converges uniformly. Minkowski shows by a direct calculation involving the Legendre polynomials that appear in the explicit definition of spherical harmonics in the case $d = 3$ that

$$\int_{S(u)} Q_n(v)\, \mathrm{d}\sigma_u(v) = \beta_n Q_n(u)\,, \tag{26}$$

where $S(u)$ denotes the great circle $\{w\colon\ w \in S^2,\ (w, u) = 0\}$ and $\mathrm{d}\sigma_u$ indicates integration with respect to arc length in $S(u)$. The factor β_k depends only on k and is not zero if k is even. It follows that for all $u \in S^2$

$$2(P(K_u) - \pi c) = \int_{S_u} (h(v) + h(-v) - c)\, \mathrm{d}\sigma_u(v) = \sum_{k=0}^{\infty} \beta_k Q_k(u) = 0\,.$$

Hence, $\beta_k Q_k(u) = 0$ (for all u). Since $h(v) + h(-v) - c$ is an even function it is obvious that $Q_k(u) = 0$ if k is odd, and since $\beta_k \neq 0$ if k is even one can deduce that in this case again $Q_k(u) = 0$. Thus (25) shows that $h(v) + h(-v) - c = 0$ and K must be of constant width. For references concerning different proofs of this theorem and related results see Chakerian and Groemer (1983).

3.3. *More about functions on the sphere with vanishing integrals over great circles*

In this section we always assume that $d \geq 3$. It is clear how to generalize the theorem and proof of Minkowski to the d-dimensional situation. $P(K_u)/\pi$ needs to be replaced by the mean width of K_u and the analogue of the great circle $S(u)$ on S^2 is the $(d-2)$-dimensional sphere $S(u) = \{w\colon\ w \in S^{d-1},\ (w, u) = 0\}$ with surface area differential $\mathrm{d}\sigma_u$. Analytically the critical part of the above proof of Minkowski's theorem is obviously the following theorem: If f is an even continuous function on S^{d-1}, then

$$\int_{S(u)} f(v)\,\mathrm{d}\sigma_u(v) = 0 \quad (\text{for all } u \in S^{d-1}) \;\Rightarrow\; f = 0\,. \tag{27}$$

(If one does not require continuity the conclusion is $f = 0$ a.e.)

Results of this type are of course among the basic facts concerning spherical Radon transforms but we restrict ourselves here to the role of spherical harmonics in this area. It has been noted repeatedly (e.g., by Schneider 1970b and Falconer 1983) that this result can be obtained in full generality by elementary analysis as a limiting case of the Funk–Hecke formula (11) by setting $g(x) = 1$ for $|x| \leqslant \varepsilon$ and 0 elsewhere and letting ε tend to 0. Aside from Minkowski's work early proofs of this theorem based on spherical harmonics, usually with some restrictions regarding the dimension and regularity, have been given by Funk (1913) and Blaschke (1916b); see also Bonnesen and Fenchel (1934). For more recent proofs and generalizations see Schneider (1970a,b), and Falconer (1983). Regarding the situation when the integration is not extended over "great circles" but over $(d-2)$-dimensional subspheres of a given radius, see Schneider (1969). A geophysical application of this theorem has been discussed by Backus (1964). Concerning the associated stability problem to estimate the size of an even f (on S^3) if the corresponding integrals over all great circles are small and f has a certain degree of regularity, see Campi (1981). The latter article contains also an investigation of a discrete analogue of this stability problem.

We now describe some geometric consequences of (27). Minkowski's theorem which marked the beginning of research in this area has been generalized and amplified in various directions. Nakajima (1930) noted that a straightforward generalization of the proof of Minkowski shows that two convex bodies in $\mathbf{E}^3$ whose corresponding orthogonal projections onto all planes have the same perimeter must be "equiwide". Generalizing work of Campi (1986) for the case $d = 3$, Goodey and Groemer (1990) find explicit stability estimates for the deviation of a convex body K in $\mathbf{E}^d$ from a body of constant width if for some constant c and all u the condition $|\bar{w}(K_u) - c| \leqslant \varepsilon$ is satisfied. They also give stability estimates for the (d-dimensional version of the) result of Nakajima just mentioned and note that the problem can be interpreted as a result for first order projection bodies.

The following consequence of (27) has already been noted by Funk (1913) (for $d = 3$); see also Kubota (1920) and Schneider (1970b). If $K \in \mathbb{C}^d$ is a central body with the property that all sections with hyperplanes through o have the same $(d-1)$-dimensional volume, then K must be a ball. More generally, Petty (1961) and Falconer (1983) proved the following result: Let K, L be central convex bodies in $\mathbf{E}^d$ and let $K(u)$, $L(u)$ denote the respective intersections of K and L with a hyperplane through o that is orthogonal to u. If for every $u \in S^{d-1}$ the bodies $K(u)$ and $L(u)$ have the same $(d-1)$-dimensional volume, then $K = L$. The proof is a rather straightforward application of (27) and the fact that according to the volume formula for polar coordinates one has for every $u \in S^{d-1}$

$$v_{d-1}(K(u)) = \frac{1}{d-1} \int_{S(u)} r(w)^{d-1}\,\mathrm{d}\sigma_u(w)\,,$$

where v_{d-1} denotes the $(d-1)$-dimensional volume, r is the radial function of K, and $S(u)$, $\mathrm{d}\sigma_u$ are defined as in (26). Another generalization (for $d=3$) of the above theorem about bodies with constant area of the cross-sections has been proved by Kubota (1920).

Schneider (1980) has used (27) to show that a convex body with the property that all hyperplane sections through one of its points are congruent must be a ball. We finally mention that Berwald (1937) has employed (27) to give a characterization of three-dimensional convex bodies of constant brightness that is reminiscent of Minkowski's characterization of convex bodies of constant width. Schneider (1970a) has proved corresponding results for all $d \geqslant 3$.

3.4. Projections of convex bodies and related matters

As in the preceding section it is again assumed that $d \geqslant 3$ and that K_u is defined as in section 3.2. We follow the often used convention to denote the ith mean projection measures, if $\mathbf{E}^{d-1}$ is the underlying space, by W'_i. One of the basic problems in convexity is to obtain information on K if $W'_i(K_u)$ is given (for some $i \in \{0, 1, \ldots, d-2\}$ and all $u \in S^{d-1}$). As evidenced by examples of non-spherical convex bodies of constant width [i.e., constant $W'_{d-2}(K_u)$], or constant brightness [i.e., constant $W'_0(K_u)$] K is in general not completely determined by $W'_i(K_u)$. A useful concept for the formulation of problems in this area is obtained by noting that $W'_i(K_u)$, considered as a function of u, is the support function (restricted to S^{d-1}) of a central convex body. This body is called the *projection body* of order $d-i-1$ of K. Projection bodies of order $d-1$ are frequently referred to simply as projection bodies (without specifying an order) or as *zonoids*. See chapter 4.10 of this Handbook or Schneider and Weil (1983) for more information regarding these concepts.

A fundamental result in the theory of convex sets states that for $i<d-1$ and within the class of proper central convex bodies (i.e., central bodies with o as interior point) the relationship between convex bodies and their projection bodies is injective. In other words, if K and L are proper central convex bodies in $\mathbf{E}^d$ and if for some $i \in \{0, 1, \ldots, d-2\}$ and all $u \in S^{d-1}$ it is true that $W'_i(K_u) = W'_i(L_u)$, then $K = L$. Already Minkowski's work on convex bodies of constant width (section 3.2) can easily be modified to qualify as a result of this kind (for $d=3$, $i=1$), and Blaschke (1916b) has pointed out that spherical harmonics can be used to settle under some regularity assumptions the case $d=3$, $i=0$. In full generality this result has first been proved, again with the aid of spherical harmonics, by Aleksandrov (1937).

To indicate the role of spherical harmonics in this connection let us consider the case $i=0$. It is fairly easy to see that for every $u \in S^{d-1}$

$$W'_0(K_u) = \int_{S^{d-1}} |(u, v)| \,\mathrm{d}\lambda(u)\,,$$

where λ is a Borel measure on S^{d-1} that is derived in the natural way from the spherical image correspondence between subsets of ∂K and S^{d-1}. Thus, taking differences and using the fact that within the class of proper central bodies K is uniquely determined by λ and that λ is even, one sees that the problem can be solved by showing that for every signed even measure μ on S^{d-1} the relation

$$\int_{S^{d-1}} |(u, v)| \, \mathrm{d}\mu(u) = 0 \tag{28}$$

(for all $v \in S^{d-1}$) implies that $\mu = 0$. If (28) is multiplied by a spherical harmonic Q_n of order n and integrated one obtains

$$\int_{S^{d-1}} \left(\int_{S^{d-1}} |(u, v)| Q_n(u) \, \mathrm{d}\sigma(u) \right) \mathrm{d}\mu(u) = 0 \, .$$

But the Funk–Hecke formula (11) shows that in the case when $g(x) = |x|$

$$\int_{S^{d-1}} |(u, v)| Q_n(v) \, \mathrm{d}\sigma(v) = c_{d,n} Q_n(u) \, , \tag{29}$$

with $c_{d,n}$ depending only on d and n and $c_{d,n} \neq 0$ if n is even. (See Petty 1961 and Schneider 1967 for explicit evaluations of $c_{d,k}$.) Hence, for all spherical harmonics of order n we have

$$\int_{S^{d-1}} Q_n(v) \, \mathrm{d}\mu(v) = 0 \, .$$

(If n is odd this is trivially true.) Using uniform approximations of continuous functions on S^{d-1} by finite sums of spherical harmonics one finds immediately that

$$\int_{S^{d-1}} g(v) \, \mathrm{d}\mu(v) = 0$$

for every continuous g. With the aid of well-known results of integration theory one deduces that this is only possible if $\mu = 0$. In a more general context this theorem has been proved by Schneider (1970a,b). For further literature on this subject, see the pertinent references of Goodey and Howard (1990).

More difficult is the problem of finding stability estimates, i.e., to estimate the distance (in terms of suitable metrics) between centrally symmetric bodies K and L if $W_i'(K_u)$ and $W_i'(L_u)$ do not differ very much. In the case $i = 0$ such stability estimates have been found for $d = 3$ by Campi (1986, 1988) and (independently) in the general case by Bourgain (1988), and by Bourgain and Lindenstrauss (1988a,b). In these proofs spherical harmonics are used to find estimates for the deviation of the surface area measures of the two bodies. From such estimates the desired result follows then from known stability estimates for surface area measures (see chapter 1.8, section 6). For $i = d - 2$ (first order projection bodies) it is easier to prove stability estimates. The problem which reduces to the study of relations as discussed in section 3.3 has been investigated, again with the aid of spherical harmonics, by Campi (1986) (for $d=3$) and by Goodey and Groemer (1990) without any dimension restriction. Apparently no stability estimates are available for projection bodies whose order is different from $d-1$ and 1.

Schneider and Weil (1970) used spherical harmonics to investigate to which extent a central convex body is uniquely determined by the $(d-1)$-dimensional volumes $V(K_u)$ if u belongs to a suitable subset of S^{d-1}.

We add several comments on zonoids. Schneider (1967) has considered the problem (originally posed by Shephard) whether the assumption $K, L \in \mathbb{C}^d$ and $W_i'(L_u) < W_i'(K_u)$ for some $i \in \{0, 1, \ldots, d-2\}$ and all $u \in S^{d-1}$ imply that $V(L) < V(K)$. If K is a zonoid, then this is true but in general it is not (even if K and L are centrally symmetric). For the proof of the latter statement Schneider uses spherical harmonics to show that any even function of class C^k with $k \geqslant d+2$ has a representation of the form $\int |(u, v)| g(u)\, d\sigma(u)$ with a continuous g and the integration extending over S^{d-1}. Goodey (unpublished) has suggested further refinements of this theorem; see also Schneider (1970a). The same representation theorem has been used by Schneider (1975) to show that there are zonoids which are not ellipsoids but whose duals are also zonoids. For further results that involve spherical harmonics and zonoids see Schneider (1967, 1970a) and Bourgain, Lindenstrauss and Milman (1989).

Although the following concepts and results do not directly concern projections the methods of proof resemble those for projections. Petty (1961) has used spherical harmonics to prove an interesting uniqueness theorem concerning sections of a centrally symmetric body with half-spaces. Let K be a central body in $\mathbf{E}^d$, and for any $u \in S^{d-1}$ let $H^+(u)$ denote the closed half-space associated with u in the sense that its boundary plane is orthogonal to u, contains o and is such that $u \in H^+(u)$. It can be shown that the centroids of $H^+(u) \cap K$ form the boundary of a convex body, say $C(K)$. This convex body, which is also centrally symmetric with respect to o, is called the *centroid body* of K. Petty has shown that if K and L are central bodies in $\mathbf{E}^d$ with the property that $C(K) = C(L)$, then $K = L$. If r denotes the radial function of K the support function of $C(K)$ can be written in the form

$$h_{C(K)}(u) = \frac{1}{(d+1)v(C)} \int_{S^{d-1}} |(u, x)| r^{d+1}(x)\, d\sigma(x)\,.$$

As Petty notes, from this representation and (29) one can easily deduce the desired uniqueness result. As a consequence of Petty's result one obtains that if a three-dimensional centrally symmetric convex body of uniform density $\frac{1}{2}$ will float in stable equilibrium in any orientation in a liquid of density 1, then it must be a ball. Independently of Petty's work (but by the same means) this was also proved by Falconer (1983) and again (under an additional assumption) by Gilbert (1991).

Ôishi (1920) has used spherical harmonics to prove that a convex body in $\mathbf{E}^3$ whose orthogonal projections onto all planes are centrally symmetric must itself be centrally symmetric and Kubota (1922) has employed spherical harmonics to derive Cauchy's surface area formula (in $\mathbf{E}^3$) and the fact that the total mean curvature is a constant multiple of the mean width. However, these results may be more naturally established without spherical harmonics. Another investigation regarding Cauchy's formula that depends also on spherical harmonics is due to Groemer (1991). It concerns a stability question that arises if non-isotropic averages of the projections of a convex body are taken into consideration.

Schneider (1977) employed spherical harmonic to prove two characterizations of balls in terms of projections. One of these characterizations shows that a convex body is a ball if and only if it is homothetic to its first order projection body.

3.5. Functions on the sphere with vanishing integrals over hemispheres

If one applies the Funk–Hecke theorem in the case when $g(x)=1$ for $x \geqslant 0$ and $g(x)=0$ elsewhere one obtains easily the following theorem: If f is a continuous odd function on S^{d-1} and if

$$\int_H f(u)\,\mathrm{d}\sigma = 0$$

for every hemisphere H in S^{d-1}, then $f=0$. Schneider (1970a,b) and Falconer (1983) have proved this result within the framework of more general investigations and have applied it for geometric purposes. Nakajima (1920), Ungar (1954), and Schneider (1970a) have also dealt with the more general situation when the integration is extended over spherical caps of a given radius.

We now discuss several geometric results that are consequences of this theorem. The following theorem (for $d=3$) has already been mentioned by Funk (1915): If a convex body $K \in \mathbb{C}^d$ has the property that there is a point p such that every hyperplane through p divides K into two parts of equal volume, then it must be centrally symmetric. As noted by Schneider (1970a) and Falconer (1983) this theorem follows immediately from the above fact about odd functions on S^{d-1} by observing that the equality of the volumes means the same as

$$\int_H (r(u)^d - r(-u)^d)\,\mathrm{d}\sigma(u) = 0\,,$$

where r denotes the radial function of K. Blaschke (1917) has shown that from this theorem one can easily deduce that a convex body $K \in \mathbb{C}^3$ must be centrally symmetric if there is a $p \in K$ such that the intersection of K with any plane through p has p as centroid. Funk (1915) has also considered the more general question to which extent a body in $\mathbf{E}^3$ is determined by its "half-volumes". A more elaborate study of this problem including certain non-convex sets and stability considerations has been published by Campi (1984).

In analogy to this theorem for the volume Schneider (1970a) has shown that if for a given $K \in \mathbb{C}^d$ there is a point p such that every hyperplane through p divides K into two parts of equal surface area, then it must be centrally symmetric with respect to p. He also proved a similar result for integrals of suitable functions of the principal curvatures of a sufficiently smooth convex surface. If $K \in \mathbb{C}^d$ and $u \in S^{d-1}$ we call the set of points with outer normal vector x such that $\langle x, u \rangle < 0$ the *illuminated portion* of K. Kubota (1920) (for $d = 3$ and under smoothness assumptions) and Schneider (1970a) (for arbitrary d and without additional conditions) have proved the following characterization of centrally symmetric bodies. If $K \in \mathbb{C}^d$ has the property that for every $u \in S^{d-1}$ the surface area of the illuminated portion of K in the direction u is the same as the surface area of the illuminated portion in the direction $-u$, then K must be centrally symmetric. Schneider (l.c.) remarks also that a similar theorem can be proved for the affine surface area (assuming that K is sufficiently regular). Anikonov and Stepanov (1981) have used spherical harmonics to show that a convex body K in $\mathbf{E}^3$ with sufficiently regular boundary is up to translations uniquely determined if (for all directions) the values of a suitable linear combination of the area of their orthogonal projections and the surface area of the corresponding illuminated portion of K are given. They also prove a stability statement regarding this uniqueness result.

3.6. Rotors in polytopes

Since in section 2.3 the major results regarding two-dimensional rotors have already been described, we discuss here only the case when $d > 2$. Similarly as in the two-dimensional case we call an intersection of finitely many closed half-spaces of $\mathbf{E}^d$ a d-dimensional *polytopal set* if it has nonempty interior and if the normal vectors of its facets, i.e., of its (possibly unbounded) $(d-1)$-dimensional boundary faces, are linearly dependent. If P is a d-dimensional polytopal set and $K \in \mathbb{C}^d$, then K is called an *osculating body* in P if $K \subset P$ and every facet of P has a nonempty intersection with K. The body K is called a *rotor* in P if for every (proper) rotation ρ of $\mathbf{E}^d$ there is a translation vector p_ρ such that $\rho K + p_\rho$ is an osculating body in P. Finally, we say that a polytopal set Q is *derived* from P if every facet of Q contains a facet of P. The condition regarding the linear dependence of the unit vectors of the facets of a polytopal set has been imposed so that it cannot happen that every convex body is a rotor. Obvious examples of rotors are balls that are contained in a polytopal set whose facets meet the ball, or convex bodies of constant width 1 in a slab of width 1 or a cube of side length 1.

Meissner (1918) has used spherical harmonics to investigate rotors in three-dimensional regular polytopes. A complete description of all polytopal sets that have non-spherical rotors and characterizations of the corresponding rotors have been presented in an exhaustive study of Schneider (1971a). He first shows that every polytopal set with a rotor must have an osculating ball and the diameter of this ball is the mean width of the rotor. This ball can be obtained from any given rotor by rotational symmetrization (Drehmittelung). To describe the main result one has to introduce a particular polytopal set C_o in $\mathbf{E}^3$. This set (a cone with square cross-section) is defined by $C_o = \{x \in \mathbf{E}^3 \colon (x, q_i) \leq 0,\ i = 0, 1, 2, 3\}$, where $q_0 = (\sqrt{6}, 0, 1)$, $q_1 = (-\sqrt{6}, 0, 1)$, $q_2 = (0, \sqrt{6}, 1)$, $q_3 = (0, -\sqrt{6}, 1)$. Schneider's principal result can be formulated as follows:

Let $P \in \mathbb{C}^d$ be a polytopal set that has a non-spherical rotor K with support function h and let ΣQ_n denote the condensed harmonic expansion of h. Then, one of the following conditions must be satisfied:

(i) P is derived from a parallelotope whose distances between opposite facets are equal and K is of constant width;

(ii) $d = 3$, P is a (regular) tetrahedron and $h = Q_0 + Q_1 + Q_2 + Q_5$;

(iii) $d = 3$, P is derived from a (regular) octahedron, but is not a tetrahedron or derived from a parallelotope, and $h = Q_0 + Q_1 + Q_5$;

(iv) $d = 3$, P is congruent to C_o, and $h = Q_0 + Q_1 + Q_4$;

(v) $d > 3$, P is a regular simplex, and $h = Q_0 + Q_1 + Q_2$.

In addition, Schneider (l.c.) pointed out that all the listed polyhedral sets do in fact have non-spherical rotors and that convex bodies whose respective support functions have the indicated harmonic expansions and whose mean width is the diameter of an osculating ball are rotors in the corresponding polyhedral sets.

Let P be a polyhedral set with an osculating sphere of radius r and with facets that are determined by unit normal vectors $p_1, p_2, \ldots, p_n$. The proof of the theorem just mentioned is based on the fact that a convex body with support function h is a rotor in P exactly if every relation of the form $\alpha_1 p_1 + \alpha_2 p_2 + \cdots + \alpha_n p_n = 0$ implies that for every rotation ρ of $\mathbf{E}^d$

$$\alpha_1(h(\rho p_1) - r) + \alpha_2(h(\rho p_2) - r) + \cdots + \alpha_n(h(\rho p_n) - r) = 0 .$$

Using the harmonic expansion of h Schneider's proof proceeds by showing that in the case of a non-spherical rotor only the possibilities listed above satisfy these conditions. Case (iv) is the most bothersome in this analysis. See also the remarks in the article of Schneider (1970b) regarding the above theorem.

3.7. *Inequalities for mean projection measures and mixed volumes*

The original idea to use spherical harmonics for proving inequalities for certain mean projection measures is due to Hurwitz (1902) (in the case $d = 3$ and with regularity assumptions). We first consider the inequality

$$W_{d-1}^2(K) - \kappa_d W_{d-2}(K) \geq 0 , \tag{30}$$

which is valid for all $K \in \mathbb{C}^d$ and where equality holds if and only if K is a ball. Actually, one can show that

$$W_{d-1}^2(K) - \kappa_d W_{d-2}(K) \geq \frac{d+1}{d(d-1)} \delta_2(K, B(K))^2 , \tag{31}$$

where $B(K)$ is the Steiner ball of K. Equality holds in (31) if and only if the support function h of K has a harmonic expansion of the form

$$h(u) \sim Q_0 + Q_1 + Q_2 ,$$

with Q_i having order i. As pointed out in the previous section, in the case $d \geqslant 4$ the class of these bodies coincides with the class of rotors in simplexes.

A proof of (31), together with the condition for equality, can be obtained in complete analogy to the proof of (20). In other words, it is an immediate consequence of (23), (24), Parseval's equation, and the fact that

$$h_{B(K)}(u) \sim \tfrac{1}{2}\bar{w}(K) + (z(K), u)$$

and therefore

$$\delta(K, B(K))^2 = \sum_{n=2}^{\infty} \|Q_n\|^2 .$$

(Since all functions appearing in (31) depend continuously on K it suffices to give a proof under the assumption that the support function of K is twice continuously differentiable.) This proof is a refinement of the original arguments of Hurwitz (1902) who considers only the case $d = 3$. The proof of Hurwitz of (30) has been often reproduced (occasionally with minor changes). See, e.g., Geppert (1937) and Kubota (1925a) (who does not restrict the dimension to 3). Although most authors have observed that the left-hand side of (30) can be expressed as a series of positive terms, the interpretation as an L_2-distance has not been realized. This necessitates in most cases the imposition of regularity assumptions to characterize the case of equality in (30). Essentially the same proof of (31) as indicated above has been published by Goodey and Groemer (1990). Another approach to prove (31), based on ideas of the following section, has been used by Schneider (1989). Concerning different kinds of sharpened versions of (30) (for $d = 3$, and also proved with the aid of spherical harmonics), see Bol (1939) and Wallen (1991). Dinghas (1940) used spherical harmonics to find an upper bound of $W_{d-1}(K)^2 - \kappa_d W_{d-2}(K)$ that is analogous to an estimate of Hurwitz for $d = 2$.

Assuming that $K \in \mathbb{C}^d$, $L \in \mathbb{C}^d$, and writing $V_{1,1}(K, L)$ to denote the mixed volume $V(K, L, B^d, \ldots, B^d)$ one can use a slightly more sophisticated argument to derive from the expansion of the support functions of K and L as series of spherical harmonics the inequality

$$V_{1,1}(K, L)^2 - V_{1,1}(K, K)V_{1,1}(L, L) \geqslant \frac{d+1}{d(d-1)} \bar{w}(L)^2 V_{1,1}(K, K)\delta(K^*, L^*) , \tag{32}$$

where K^* and L^* denote homothetic copies of K and L, respectively, that have coincident Steiner points and mean width 1. If $d = 2$, (32) is the same as (21), and in the case $K = B^d$, (32) yields again (31). The weaker inequality

$$V_{1,1}(K, L)^2 - V_{1,1}(K, K)V_{1,1}(L, L) \geqslant 0 \tag{33}$$

is one of the classical inequalities of Minkowski (in the case $d = 3$). A proof of

(33) based on spherical harmonics has first been given (under some regularity assumptions) by Kubota (1925b); see also Geppert (1937) and Dinghas (1940). The stronger version (32) has been proved by Schneider (1989) and Goodey and Groemer (1990). Schneider (1990) has used this inequality to prove stability statements for the Aleksandrov–Fenchel inequality (cf. chapter 1.4, section 4.3) and the mean curvature of convex surfaces.

As shown by Groemer and Schneider (1991), inequality (30) can be used to prove stability versions of a whole array of inequalities for the mean projection measures including the classical isoperimetric inequality. Fuglede (1986, 1989) has given proofs of stability statements for the isoperimetric inequality that involve the harmonic expansion of the radial function of the given body. The inequalities in Fuglede (1989) are in a certain sense best possible. For a more detailed description of these results, see chapter 1.4, section 5.2. Groemer (1992a) has used the spherical harmonics expansion of the support functions of two convex bodies to prove a strong stability version of the Brunn–Minkowski Theorem for W_{d-2}. For the special case of this theorem (without a stability statement) concerning the convex body $\frac{1}{2}(K+(-K))$ spherical harmonics have already been used by Kubota (1925b). See also the remark at the end of the following section.

3.8. Wirtinger's inequality and its applications

We discuss here Wirtinger's inequality and some of its generalizations for functions on S^{d-1}. The significance of this inequality for the topic of our survey lies in the fact that its proof depends on the use of Fourier series or spherical harmonics and that some geometric inequalities are straightforward consequences of Wirtinger's inequality. Although all of these geometric inequalities can also be proved directly the approach through Wirtinger's inequality is of some interest.

For the sake of completeness we consider also the case $d=2$. Let f be a real-valued function on $(-\infty, \infty)$ of period 2π. Furthermore, let us assume that f' exists and is square integrable, and that

$$\int_0^{2\pi} f(x)\,\mathrm{d}x = 0\,.$$

Wirtinger's inequality states that under these assumptions

$$\int_0^{2\pi} f(x)^2\,\mathrm{d}x \leq \int_0^{2\pi} f'(x)^2\,\mathrm{d}x\,, \tag{34}$$

with equality if and only if $f(x) = a\cos x + b\sin x$ (a, b constant).

From the Fourier expansion of f one sees immediately that Wirtinger's inequality is a trivial consequence of Parseval's inequality. If one applies Wirtinger's inequality in the case $f = h - P(K)/2\pi$ where h is the support function of a convex domain K and recalls (18) and (19) one obtains immediately the isoperimetric inequality. As shown by Blaschke (1916b) it is also easy to derive the mixed area inequality (21) from Wirtinger's inequality. Wallen (1987) has shown how Wirtinger's inequality can be used to derive a Bonnesen-type inequality for the mixed areas of two convex domains (see chapter 1.4, section 2.2, and section 2.2 above).

We now describe some generalizations of Wirtinger's inequality for functions on S^{d-1}. Let f be such a function and assume that $\Delta_o f$ exists, is square integrable, and that

$$\int_{S^{d-1}} f(u)\,\mathrm{d}\sigma(u) = 0\,.$$

Then,

$$\int_{S^{d-1}} f(u)^2\,\mathrm{d}\sigma(u) + \frac{1}{d-1}\int_{S^{d-1}} f(u)\,\Delta_o f(u)\,\mathrm{d}\sigma(u) \leqslant 0\,. \tag{35}$$

Equality holds exactly if f is a spherical harmonic of order 1.

Using the condensed harmonic expansion of f one obtains this inequality as a straightforward consequence of (10) and the generalized form of Parseval's equality. In a slightly different form (and for $d=3$) (35) has been proved by Blaschke (1916b) who has also pointed out that in the case $f = h - \bar{w}(K)/2$, where h is the support function of a convex body K, Minkowski's inequality (30) follows immediately from (35) and the representation (2) of W_{d-2}. Note that in the case $d=2$, (35) reduces to (34) (after an obvious integration by parts). For arbitrary dimensions several forms of Wirtinger's inequality have been proved by Dinghas (1940). He uses these results to establish a strengthened version of (30) and, generalizing a result of Hurwitz, to derive an upper bound for the left-hand sides of (30) in terms of an integral over the squared mean curvature. He also derives (33) and a sharpened version of this inequality.

Schneider (1989) has proved the following more flexible version of Wirtinger's inequality: For any two constants a, b with $a \leqslant 2db$, $b \geqslant 0$, the conditions

$$\int_{S^{d-1}} f(u)\,\mathrm{d}\sigma(u) = 0\,, \qquad \int_{S^{d-1}} f(u)u\,\mathrm{d}\sigma(u) = 0$$

imply that

$$a\int_{S^{d-1}} f(u)^2\,\mathrm{d}\sigma(u) + b\int_{S^{d-1}} f(u)\,\Delta_o f(u)\,\mathrm{d}\sigma(u) \leqslant 0\,.$$

From this generalization of Wirtinger's inequality Schneider derives not only the strengthened form (32) of Minkowski's inequality for mixed volumes but also a stability statement for area measures (see chapter 1.8, section 6). Moreover, in a recent article Schneider (1992b) used this inequality to prove a stability version of the Brunn–Minkowski Theorem for the mean projection measures and more general types of mixed volumes.

3.9. Other geometric applications of spherical harmonics

We start by describing an important result regarding the Steiner point. From the definition of the Steiner point it follows immediately that $z(K)$, considered as a mapping from $\mathbb{C}^d$ onto $\mathbf{E}^d$, is continuous (with respect to the Hausdorff metric on $\mathbb{C}^d$) and that $z(K+L)=z(K)+z(L)$ (for all $K, L \in \mathbb{C}^d$). Moreover, if μ is a proper motion of $\mathbf{E}^d$, then $z(\mu K)=\mu z(K)$. Using approximations of support functions by spherical harmonics it has been shown by Schneider (1971b) that any mapping from $\mathbb{C}^d$ onto $\mathbf{E}^d$ that has these three properties must be the Steiner point. An earlier result of this type for the case $d=2$ has been obtained by Shephard (1968) using Fourier series. The following property of the Steiner point has also been proved by Schneider (1971b): Assume that $p \in K \in \mathbb{C}^d$. If for every $u \in S^{d-1}$ the Steiner point of the projection K_u coincides with p_u, then K is centrally symmetric with respect to p.

Another subject area where spherical harmonics play an essential role concerns mappings of $\mathbb{C}^d$ into itself. Such a mapping is called an *endomorphism* of $\mathbb{C}^d$ if it is with respect to Minkowski addition additive, with respect to the Hausdorff metric continuous, and if it commutes with all proper rigid motions of $\mathbf{E}^d$. Among various other results Schneider (1974) has proved the following theorem: The only surjective endomorphisms of $\mathbb{C}^d$ are given by $K \rightarrow \lambda\mu(K-z(K))+z(K)$, where $z(K)$ is the Steiner point, $\lambda>0$, and μ is either the identity motion, or (if $d \geqslant 3$) a reflection in the origin, or (if $d=2$) an element of SO(2).

Since it is possible (under certain conditions and with respect to suitable norms) to approximate functions on the sphere by sums of spherical harmonics it is natural to try to use this fact for the approximation of arbitrary convex bodies by others that have desirable smoothness properties. Based on this idea Schneider (1984) has shown that any convex body can be approximated arbitrarily closely (in the Hausdorff metric) by convex bodies with algebraic support functions and everywhere positive Gaussian curvature. He also proved (as a special case of more general results) that if the given convex body is of constant width the approximating body can be selected so that it has the same property. Fillmore (1969) has used spherical harmonics to construct convex sets of constant width that have certain prescribed symmetry properties.

Blaschke (1916a) has considered the following problem related to a result of Monge. Let K be a convex body in $\mathbf{E}^3$ and let us call a cone that is the intersection of three half-space with mutually orthogonal boundary planes an *orthogonal cone*. If K is an ellipsoid, then the set of apices of all orthogonal cones with K as osculating body is a sphere. Using spherical harmonics Blaschke has shown that the converse of this fact is also true. Regarding the generalization of this result to $\mathbf{E}^d$ see Schneider (1970b) and Burger (1990), where a stability version of this theorem is proved. Blaschke (l.c.) discusses also an analogous characterization of elliptic paraboloids.

Recently, Goodey and Weil (1992) have studied the problem of recovering a convex body K in $\mathbf{E}^d$ ($d \geqslant 3$) from the average of sections with random k-flats, where k is a fixed number between 1 and $d-1$. The average is defined in terms of an integral mean of the support functions of the convex bodies that arise as the intersections of K with the k-flats. Based on an evaluation of the coefficients in a particular case of the Funk–Hecke formula they show that for $k=2$ such a reconstruction of the body from its mean value of the intersections is indeed possible.

We finally mention the application of spherical harmonics in the construction of dense sphere packings and the use of multidimensional Fourier series in the geometry of numbers. We list the survey of Sloane (1982) as a good reference for the former subject area and Gruber and Lekkerkerker (1987) for the latter.

Acknowledgements

The more recent developments regarding geometric applications of spherical harmonics, and consequently this survey, have been substantially influenced by the work of Professor Rolf Schneider. On a more personal level I wish to thank him for pointing out to me several passages that needed corrective changes and for proposing some additions to the cited literature. I also would like to thank Professors William Firey and Erwin Lutwak for suggesting several improvements of this article. Finally, I gratefully acknowledge financial support from the National Science Foundation (Research Grant DMS 8922399).

References

Aleksandrov, A.D.
[1937] On the theory of mixed volumes. New inequalities between mixed volumes and their applications (in Russian), *Mat. Sb.* **44**, 1205–1238.

Anikonov, Yu.E., and V.N. Stepanov
[1981] Uniqueness and stability of the solution of a problem of geometry in the large (in Russian), *Mat. Sb. (N.S.)* **116**(158), 539–546, 607 [*Math. USSR-Sb.* **44**, 483–490].

Arnold, R.
[1989] Zur L^2-Bestapproximation eines konvexen Körpers durch einen bewegten konvexen Körper, *Monatsh. Math.* **108**, 277–293.

Backus, G.
[1964] Geographical interpretation of measurements of average phase velocities of surface waves over great circular and semi circular paths, *Bull. Seism. Soc. Amer.* **54**, 571–610.

Berg, C.
[1969] Corps convexes et potentiels sphériques, *Danske Vid. Selsk. Mat-Fys. Medd.* **37**, 1–64.

Berwald, L.
[1937] Integralgeometrie 25. Über Körper konstanter Helligkeit, *Math. Z.* **42**, 737–738.

Blaschke, W.
[1914] Beweise zu Sätzen von Brunn–Minkowski über die Minimaleigenschaft des Kreises, *Jber. Deutsch. Math.-Vereinig.* **23**, 210–234.
[1916a] Eine kennzeichnende Eigenschaft des Ellipsoids und eine Funktionalgleichung auf der Kugel, *Ber. Verh. Sächs. Akad. Leipzig* **68**, 129–136.

[1916b] *Kreis und Kugel* (Veit & Co., Leipzig). 2nd Ed.: Walter de Gruyter, Berlin, 1956.
[1917] Über affine Geometrie IX: Verschiedene Bemerkungen und Aufgaben, *Verh. Sächs. Akad. Leipzig* **69**, 412–420.

Bol, G.
[1939] Zur Theorie der konvexen Körper, *Jber. Deutsch. Math.-Vereinig.* **49**, 113–123.

Bonnesen, T., and W. Fenchel
[1934] *Theorie der konvexen Körper,* Ergeb. Math., Bd. 3, (Springer, Berlin) [*Theory of Convex Bodies,* BCS Assoc. Moscow, ID, 1987].

Bourgain, J.
[1988] Remarques sur les zonoïdes (projection bodies, etc.), *Séminaire d'Analyse Fonctionnelle,* 1985/1986/1987, Publ. Math. Univ. Paris VII, Vol. 28 (Paris) pp. 171–186.

Bourgain, J., and J. Lindenstrauss
[1988a] Nouveaux résultats sur les zonoïdes et les corps de projection, *C.R. Acad. Sci. Paris* **306**, 377–380.
[1988b] Projection Bodies, in: *Lecture Notes in Mathematics*, Vol. 1317 (Springer, Berlin) pp. 250–270.

Bourgain, J., J. Lindenstrauss and V.D. Milman
[1989] Approximation of zonoids by zonotopes, *Acta Math.* **162**, 73–141.

Burger, T.
[1990] Stabilitätsfragen bei konvexen Körpern, Diplomarbeit Univ. Freiburg i. Brg.

Campi, S.
[1981] On the reconstruction of a function on a sphere by its integrals over great circles, *Boll. Un. Mat. Ital. C(5)* **18**, 195–215.
[1984] On the reconstruction of a star-shaped body from its 'half-volumes', *J. Austral. Math. Soc. A* **37**, 243–257.
[1986] Reconstructing a convex surface from certain measurements of its projections, *Boll. Un. Math. Ital. (6)* **5B**, 945–959.
[1988] Recovering a centred convex body from the areas of its shadows: a stability estimate, *Ann. Mat. Pura Appl.* **151**, 289–302.

Chakerian, G.D.
[1979] Geometric inequalities for plane convex bodies, *Canad. Math. Bull.* **22**, 9–16.

Chakerian, G.D., and H. Groemer
[1983] Convex bodies of constant width, in: *Convexity and its Applications,* eds P.M. Gruber and J. M. Wills (Birkhäuser, Basel) pp. 49–96.

Chernoff, P.R.
[1969] An area-width inequality for convex curves, *Amer. Math. Monthly* **76**, 34–35.

Cieślak, W., and S. Góźdź
[1987] On curves which bound special convex sets, *Serdica* **13**, 281–286.
[1989] Properties of finite systems of convex curves, *Istanbul Üniv. Fen Fak. Mat. Der.* **48**, 99–108.

De Vries, H.L.
[1969] Über Koeffizientenprobleme bei Eilinien und über die Heinzsche Konstante, *Math. Z.* **112**, 101–106.

Dinghas, A.
[1940] Geometrische Anwendungen der Kugelfunktionen, *Nachr. Ges. Wiss. Göttingen Math.-Phys. Kl.* **1**, 213–235.

Eggleston, H.G.
[1958] *Convexity* (Cambridge Univ. Press, Cambridge).

Erdélyi, A., W. Magnus, F. Oberhettinger and F. Tricomi
[1953] *Higher Transcendental Functions,* Vol. 2 (McGraw-Hill, New York).

Falconer, K.J.
[1983] Applications of a result on spherical integration to the theory of convex sets, *Amer. Math. Monthly* **90**, 690–693.

Fejes Tóth, L.
[1972] *Lagerungen in der Ebene auf der Kugel und im Raum*, Grundl. Math. Wiss., Bd. 65 (Springer, Berlin).

Fillmore, J.R.
[1969] Symmetries of surfaces of constant width, *J. Differential Geom.* **3**, 103–110.

Firey, Wm.J.
[1959] A note on a theorem of Knothe, *Michigan Math. J.* **6**, 53–54.

Fisher, J.C.
[1984] Fourier series and geometric inequalities, unpublished manuscript.
[1987] Curves of constant width from a linear viewpoint, *Math. Mag.* **60**, 131–140.

Fisher, J.C., D. Ruoff and J. Shilleto
[1985] Perpendicular polygons, *Amer. Math. Monthly* **92**, 23–37.

Florian, A., and K. Prachar
[1986] On the Diophantine equation $\tan(k\pi/m) = k \tan(\pi/m)$, *Monatsh. Math.* **102**, 263–266.

Focke, J.
[1969] Symmetrische n-Orbiformen kleinsten Inhalts, *Acta Math. Acad. Sci. Hungar.* **20**, 39–68.

Fuglede, B.
[1986] Stability in the isoperimetric problem, *Bull. London Math. Soc.* **18**, 599–605.
[1989] Stability in the isoperimetric problem for convex or nearly spherical domains in R^n, *Trans. Amer. Math. Soc.* **314**, 619–638.

Fujiwara, M.
[1915] Über die einem Vielecke eingeschriebenen und umdrehbaren geschlossenen Kurven, *Sci. Rep. Tôhoku Univ.* **4**, 43–55.
[1919] Über die innen-umdrehbare Kurve eines Vielecks, *Sci. Rep. Tôhoku Univ.* **8**, 221–246.

Fujiwara, M., and S. Kakeya
[1917] On some problems of maxima and minima for the curve of constant breadth and the in-revolvable curve of the equilateral triangle, *Tôhoku Math. J.* **11**, 92–110.

Funk, P.
[1913] Über Flächen mit lauter geschlossenen geodätischen Linien, *Math. Ann.* **74**, 278–300.
[1915] Über eine geometrische Anwendung der Abelschen Integralgleichung, *Math. Ann.* **77**, 129–135.

Geppert, H.
[1937] Über den Brunn–Minkowskischen Satz, *Math. Z.* **42**, 238–254.

Gericke, H.
[1940a] Stützbare Bereiche in komplexer Fourierdarstellung, *Deutsch. Math.* **5**, 279–299.
[1940b] Über stützbare Flächen und ihre Entwicklung nach Kugelfunktionen, *Math. Z.* **46**, 55–61.
[1941] Zur Relativgeometrie ebener Kurven, *Math. Z.* **47**, 215–228.

Gilbert, E.N.
[1991] How things float, *Amer. Math. Monthly* **98**, 201–216.

Goodey, P.R., and H. Groemer
[1990] Stability results for first order projection bodies, *Proc. Amer. Math. Soc.* **109**, 1103–1114.

Goodey, P.R., and R. Howard
[1990] Processes of flats induced by higher dimensional processes, *Adv. in Math.* **80**, 92–109.

Goodey, P.R., and W. Weil
[1992] The determination of convex bodies from the mean of random sections, preprint.

Görtler, H.
[1937a] Zur Addition beweglicher ebener Eibereiche, *Math. Z.* **42**, 313–321.

[1937b] Erzeugung Stützbarer Bereiche I, *Deutsch. Math.* **2**, 454–466.
[1938] Erzeugung Stützbarer Bereiche II, *Deutsch. Math.* **3**, 189–200.

Góźdź, S.
[1990] Barbier type theorems for plane curves, *Mathematiche (Catania)* **55**, 369–377.

Groemer, H.
[1990] Stability properties of geometric inequalities, *Amer. Math. Monthly* **97**, 382–394.
[1991] Stability properties of Cauchy's surface area formula, *Monatsh. Math.* **112**, 43–60.
[1992a] On the stability of a Brunn–Minkowski type inequality, to appear, in *Exposition. Math.*.
[1992b] On circumscribed cylinders of convex sets, preprint.

Groemer, H., and R. Schneider
[1991] Stability estimates for some geometric inequalities, *Bull. London Math. Soc.* **23**, 67–74.

Gruber, P.M., and C.G. Lekkerkerker
[1987] *Geometry of Numbers* (North-Holland, Amsterdam).

Hall, R.R.
[1983] On an inequality of E. Heinz, *J. Anal. Math.* **42**, 185–198.
[1985] A class of isoperimetric inequalities, *J. Anal. Math.* **45**, 169–180.

Hayashi, T.
[1918] On a certain functional equation, *Sci. Rep. Tôhoku Univ.* **7**, 1–32.
[1924] On Steiner's curvature centroid, *Sci. Rep. Tôhoku Univ.* **13**, 109–132.
[1926] Some geometrical applications of Fourier series, *Rend. Circ. Mat. Palermo* **50**, 96–102.

Heil, E.
[1972] Eine Verschärfung der Bieberbachschen Ungleichung und einige andere Abschätzungen für ebene konvexe Bereiche, *Elem. Math.* **27**, 4–8.

Hochstadt, H.
[1986] *The Functions of Mathematical Physics* (Dover Publ., New York).

Hurwitz, A.
[1901] Sur le problème des isopérimètres, *C.R. Acad. Sci. Paris* **132**, 401–403 [*Math. Werke,* 1. Bd. (Birkhäuser, Basel, 1932) pp. 490–491].
[1902] Sur quelques applications géométriques des séries Fourier, *Ann. Sci. École Norm. Sup. (3)* **19**, 357–408 [*Math. Werke,* 1. Bd. (Birkhäuser, Basel, 1932) pp. 509–554].

Inzinger, R.
[1949] Stützbare Bereiche, trigonometrische Polynome und Defizite höherer Ordnung, *Monatsh. Math.* **53**, 302–323.

Kamenezki, M.
[1947] Solution of a geometric problem of L. Lusternik (in Russian), *Uspekhi Mat. Nauk II* **2**, 199–202.

Klötzler, R.
[1975] Beweis einer Vermutung über n-Orbiformen kleinsten Inhalts, *Z. Angew. Math. Mech.* **55**, 557–570.

Knothe, H.
[1957] Inversion of two theorems of Archimedes, *Michigan Math. J.* **4**, 53–56.

Kubota, T.
[1918] Über die Schwerpunkte der konvexen geschlossenen Kurven und Flächen, *Tôhoku Math. J.* **14**, 20–27.
[1920] Einige Probleme über konvex-geschlossene Kurven und Flächen. *Tôhoku Math. J.* **17**, 351–362.
[1922] Beweise einiger Sätze über Eiflächen, *Tôhoku Math. J.* **21**, 261–264.
[1925a] Über die konvex-geschlossenen Mannigfaltigkeiten im n-dimensionalen Raume, *Sci. Rep. Tôhoku Univ.* **14**, 85–99.
[1925b] Über die Eibereiche im n-dimensionalen Raume, *Sci. Rep. Tôhoku Univ.* **14**, 399–402.

Lebesgue, H.
[1906] *Leçons sur Les Series Trigonometriques* (Gauthier-Villars, Paris).

Leichtweiß, K.
[1980] *Konvexe Mengen* (Springer, Berlin).

Letac, G.
[1983] Mesures sur le circle et convexes du plan, *Ann. Sci. Univ. Clermont-Ferrand II* **76**, 35–65.

Lutwak, E.
[1977] Mixed width-integrals of convex bodies, *Israel J. Math.* **28**, 249–253.

Meissner, E.
[1909] Über die Anwendung von Fourier-Reihen auf einige Aufgaben der Geometrie und Kinematik, *Vierteljahresschr. Naturforsch. Ges. Zürich* **54**, 309–329.
[1918] Über die durch reguläre Polyeder nicht stützbaren Körper, *Vierteljahresschr. Naturforsch. Ges. Zürich* **63**, 544–551.

Minkowski, H.
[1904–1906] Über die Körper konstanter Breite (in Russian), *Mat. Sb.* **25**, 505–508. German translation: *Gesammelte Abhandlungen,* 2. Bd. (Teubner, Leipzig, 1911) pp. 277–279.

Müller, C.
[1966] *Spherical Harmonics,* Lecture Notes in Mathematics, Vol. 17 (Springer, Berlin).

Nakajima, S.
[1920] On some characteristic properties of curves and surfaces, *Tôhoku Math. J.* **18**, 277–287.
[1930] Eiflächenpaare gleicher Breiten und gleicher Umfänge, *Japan. J. Math.* **7**, 225–226.

Ôishi, K.
[1920] A note on the closed convex surfaces, *Tôhoku Math. J.* **18**, 288–290.

Petty, C.M.
[1961] Centroid surfaces, *Pacific J. Math.* **11**, 1535–1547.

Schaal, H.
[1962] Prüfung einer Kreisform mit Hohlwinkel und Taster, *Elem. Math.* **17**, 33–37.

Schneider, R.
[1967] Zu einem Problem von Shephard über die Projektionen konvexer Körper, *Math. Z.* **101**, 71–82.
[1969] Functions on a sphere with vanishing integrals over certain subspheres, *J. Math. Anal. Appl.* **26**, 381–384.
[1970a] Über eine Integralgleichung in der Theorie der konvexen Körper, *Math. Nachr.* **44**, 55–75.
[1970b] Functional equations connected with rotations and their geometric applications, *Enseign. Math.* **16**, 297–305.
[1971a] Gleitkörper in konvexen Polytopen, *J. Reine Angew. Math.* **248**, 193–220.
[1971b] On Steiner points of convex bodies, *Israel J. Math.* **9**, 241–249.
[1971c] Zwei Extremalaufgaben für konvexe Bereiche, *Acta Math. Hungar.* **22**, 379–383.
[1972a] The mean surface area of the boxes circumscribed about a convex body, *Ann. Polon. Math.* **25**, 325–328.
[1972b] Krümmungsschwerpunkte konvexer Körper I, *Abh. Math. Sem. Hamburg* **37**, 112–132.
[1974] Equivariant endomorphisms of the space of convex bodies, *Trans. Amer. Math. Soc.* **194**, 53–78.
[1975] Zonoids whose polars are zonoids, *Proc. Amer. Math. Soc.* **50**, 365–368.
[1977] Rekonstruktion eines konvexen Körpers aus seinen Projektionen, *Math. Nachr.* **79**, 325–329.
[1980] Convex bodies with congruent sections, *Bull. London Math. Soc.* **12**, 52–54.
[1984] Smooth approximations of convex bodies, *Rend. Circ. Mat. Palermo 11* **33**, 436–440.
[1989] Stability in the Aleksandrov–Fenchel–Jessen theorem, *Mathematika* **36**, 50–59.
[1990] A stability estimate for the Aleksandrov–Fenchel inequality with an application to mean curvature, *Manuscripta Math.* **69**, 291–300.

[1992a] *Convex Bodies: The Brunn–Minkowski Theory* (Cambridge Univ. Press, Cambridge).
[1992b] On the general Brunn–Minkowski theorem, preprint.

Schneider, R., and W. Weil
[1970] Über die Bestimmung eines konvexen Körpers durch die Inhalte seiner Projektionen, *Math. Z.* **116**, 338–348.
[1983] Zonoids and related topics, in: *Convexity and its Applications*, eds P.M. Gruber and J. M. Wills (Birkhäuser, Basel) pp. 296–317.

Schoenberg, I.J.
[1950] The finite Fourier series and elementary geometry, *Amer. Math. Monthly* **58**, 390–404.

Seeley, R.T.
[1966] Spherical harmonics, *Amer. Math. Monthly* **73**, 115–121.

Shephard, G.C.
[1968] A uniqueness theorem for the Steiner point of a convex region, *J. London Math. Soc.* **43**, 439–444.

Sloane, N.J.A.
[1982] Recent bounds for codes, sphere packings and related problems obtained by linear programming and other methods, *Contemp. Math.* **9**, 153–185.

Stein, E.M., and G. Weiss
[1971] *Introduction to Fourier Analysis on Euclidean Spaces* (Princeton Univ. Press, Princeton, NJ).

Su, B.
[1927] On Steiner's curvature centroid, *Japan. J. Math.* **4**, 195–201.

Sz.-Nagy, B.
[1965] *Introduction to Real Functions and Orthogonal Expansions* (Oxford Univ. Press, New York).

Tennison, R.L.
[1976] Smooth curves of constant width, *Math. Gaz.* **60**, 270–272.

Ungar, P.
[1954] Freak theorem about functions on a sphere, *J. London Math. Soc.* **29**, 100–103.

Vitale, R.A.
[1985] L_p metrics for compact convex sets, *J. Approx. Theory* **45**, 280–287.

Wallen, L.J.
[1987] All the way with Wirtinger: a short proof of Bonnesen's inequality, *Amer. Math. Monthly* **94**, 440–442.
[1991] An abstract theorem of Bonnesen type with applications to mixed areas, preprint.

Wegmann, R.
[1975] Extremalfiguren für eine Klasse von isoperimetrischen Problemen, *Math. Nachr.* **69**, 173–190.

Zygmund, A.
[1977] *Trigonometric Series* (Cambridge Univ. Press, London).

CHAPTER 4.9

Zoonoids and Generalisations

Paul GOODEY

Department of Mathematics, The University of Oklahoma, 601 Elm Avenue, Norman, OK 73019, USA

Wolfgang WEIL

Mathematisches Institut II, TH Karlsruhe, Englerstrasse 2, D-76131 Karlsruhe, Germany

Contents

HANDBOOK OF CONVEX GEOMETRY
Edited by P.M. Gruber and J.M. Wills

1. Introduction

Because of the linear structure of the space $\mathcal{K}^d$ of convex bodies (see chapter 1.9) it is a natural problem to study convex bodies that are composed (in the sense of Minkowski addition) of simpler ones. The simplest non-trivial convex bodies are the line segments, and finite sums of line segments in $\mathbb{R}^d$ (the *zonotopes*) comprise a class of convex polytopes with interesting symmetry properties. These bodies found attention quite early, in fact it was the Russian crystallographer Fedorov, who invented them in the last century. This chapter, however, concentrates on analytic aspects of centrally symmetric convex bodies. A first step in this direction involves the study of *zonoids*, which are limits of zonotopes in the Hausdorff metric. Zonoids are of particular interest in convex geometry because they occur as images of a natural operation, namely as *projection bodies*. In addition, zonoids allow one to extend some analytic notions, properties and formulae in an easy way to more general convex sets (*generalised zonoids*) or even to all centrally symmetric convex bodies. The analytic point of view was already emphasised by Blaschke (1916, 1923) whose results have been cited in a slightly incorrect form in Bonnesen and Fenchel (1934, sections 19 and 61). Blaschke (1916, pp. 154–155) remarked on the connection between zonoids and the spherical Radon transform. Zonoids were also implicitly present in the work of Aleksandrov (1937) on projections of convex bodies. In particular, Aleksandrov was the first to give a general proof of the uniqueness property in the basic integral equation. This was rediscovered later a number of times by various authors. Interest in zonoids arose from some surprising connections between convex geometry, analysis (positive definite functions, Radon transforms), functional analysis (vector measures, subspaces of L^1), and stochastic geometry (point processes). A theory of zonoids gradually emerged in the last 25 years with increasing interest in recent times. Many of the results which will be described in the following are fairly new.

Because of the analytic context of this chapter, there will not be a separate section on zonotopes. Some of their properties, in particular those which have an analytic form, will be mentioned at appropriate places. For more details, the reader is referred to the survey articles of Bolker (1969), and Schneider and Weill (1983), as well as to chapters 2.3 and 3.7 of this Handbook. Further information, applications and references can be found in the above surveys.

In contrast, this chapter will focus its attention on some of the more recent results in zonoid theory and explain their origin.

2. Basic definitions and properties

Most convex bodies in the following are centrally symmetric and it may be assumed that this centre is the origin of $\mathbb{R}^d$. These are called the *centred* bodies and the corresponding class is denoted by $\mathcal{K}_0^d$.

A *zonotope* is a finite sum of line segments. As shown by Blaschke (1923, p. 250) and Coxeter (1963, section 2.8) for $d = 3$, and in full generality by Bolker

(1969) and Schneider (1970a), zonotopes are characterised (among all polytopes) by the fact that all their 2-dimensional faces have a centre of symmetry. The support function of a centred zonotope K is the sum of the support functions of centred line segments and therefore takes the form

$$h(K; u) = \sum_{i=1}^{n} |\langle u, v_i \rangle| \rho(v_i), \quad u \in S^{d-1}, \tag{2.1}$$

with different unit vectors v_i (unique up to reflections) and positive weights $\rho(v_i)$ (here, $\langle \cdot, \cdot \rangle$ denotes the Euclidean scalar product in $\mathbb{R}^d$). *Zonoids* are limits, in the Hausdorff metric, of zonotopes. For a zonoid K, the sum (2.1) transforms into an integral involving an even measure, i.e., one which assigns equal measure to antipodal sets.

Theorem 2.1. *A convex body K is a (centred) zonoid if and only if*

$$h(K; u) = \int_{S^{d-1}} |\langle u, v \rangle| \rho_K(\mathrm{d}v) \tag{2.2}$$

for all $u \in S^{d-1}$, where ρ_K is a positive even measure on S^{d-1}.

This result is proved in Bolker (1969), Schneider (1970a), Lindquist (1975a), and Matheron (1975, p. 99). For one direction, let the support function of K obey (2.2). Then, ρ_K can be written as the weak limit of a sequence of even measures ρ_k with finite support on S^{d-1}. Each ρ_k generates a zonotope K_k through (2.1), and the support functions $h(K_k; \cdot)$ converge pointwise to $h(K; \cdot)$. Thus, K is a zonoid. For the other direction, it suffices to show that the set $\mathscr{Z}^d$ of convex bodies whose support function has a representation (2.2) is closed. This follows from a standard compactness argument for measures on the sphere.

The characterisation of surface area measures of convex bodies shows that centred zonoids (with inner points) are precisely the projection bodies of convex bodies (with inner points). Therefore, a result of Aleksandrov (1937, §8), which is described in more detail in section 5, shows that the measure ρ_K in (2.2) is unique. This result can be formulated in terms of signed measures.

Theorem 2.2. *If μ is an even signed measure on S^{d-1} with*

$$\int_{S^{d-1}} |\langle u, v \rangle| \mu(\mathrm{d}v) = 0$$

for all $u \in S^{d-1}$, then $\mu \equiv 0$.

Theorem 2.2 has been rediscovered a number of times, e.g., by Petty (1961), Rickert (1967a,b), and Matheron (1974a, 1975). Matheron's proof is of interest since he relates the integral equation (2.2) to an integral representation of Lévy (1937) for infinitely divisible probability distributions.

Earlier than Aleksandrov, Blaschke gave two analytic proofs of Theorem 2.2 for $d=3$ and smooth measures μ (Blaschke 1916, pp. 152 and 154–155). The one proof (as well as the proofs of Petty and Rickert) uses spherical harmonics, this method was generalised to uniqueness problems of a similar type by Schneider (1970a), and applied to a number of problems in convex geometry [some of these results were rediscovered by Falconer (1983)].

The other proof of Blaschke is based on the obvious equivalence between the uniqueness of ρ_K, and the fact that the vector space spanned by the functions $u \mapsto |\langle u, v\rangle|$, $v \in S^{d-1}$, is dense in the Banach space $C_e(S^{d-1})$ of even continuous real functions on S^{d-1}. A short and elementary proof of this latter fact was given by Choquet (1969a,b). Blaschke showed (for $d=3$) somewhat more, namely that any smooth function $f \in C_e(S^{d-1})$ has a representation

$$f(u) = \int_{S^{d-1}} |\langle u, v\rangle| g(v) \lambda_{d-1}(\mathrm{d}v) \tag{2.3}$$

with some function $g \in C_e(S^{d-1})$. Here, λ_{d-1} is the spherical Lebesgue measure on S^{d-1} with total measure $\omega_{d-1} = \lambda_{d-1}(S^{d-1})$. Exact smoothness conditions (in arbitrary dimensions) were given by Schneider (1967) (again using spherical harmonics), a related result is presented in section 4. For C^∞-functions, these results give the following lemma.

Lemma 2.3. *For each $f \in C_e^\infty(S^{d-1})$ there is a (unique) $g \in C_e^\infty(S^{d-1})$ with*

$$f(u) = \int_{S^{d-1}} |\langle u, v\rangle| g(v) \lambda_{d-1}(\mathrm{d}v), \quad u \in S^{d-1}.$$

Further information on the spherical harmonics aspects of these results, in particular their connection with the Funk–Hecke theorem, can be found in chapter 4.8.

The results described so far hold true in dimensions $d \geqslant 2$. However, the planar case $d=2$ plays a special role, since every centrally symmetric polygon $K \subset \mathbb{R}^2$ is a zonotope and hence the set $\mathscr{Z}^2$ of (centred) planar zonoids coincides with $\mathscr{K}_0^2$. For this reason, some of the previous and the following results are trivial in the case $d=2$, while other considerations only make sense for $d \geqslant 3$. Therefore, it will mostly be assumed that $d \geqslant 3$ in the following. Some particular instances where the two-dimensional case is of interest will be mentioned separately.

Although subtraction of line segments does not seem to be a very natural process, there are bodies $K \in \mathscr{K}_0^d$ whose support functions have the integral representation (2.2) with a signed even measure ρ_K. Such bodies are called *generalised zonoids* (Weil 1976a). Specific examples of generalised zonoids which are not zonoids are given in Schneider (1970a).

It is reasonable to expect that zonoids, as limits of zonotopes, will also exhibit a high degree of symmetry and therefore will not be dense among the centrally symmetric bodies. Since the set $\mathscr{Z}^d$ of centred zonoids is closed, this follows, of course, from Schneider's above-mentioned examples of bodies $K \not\in \mathscr{Z}^d$. It is however also a consequence of a general decomposition theorem of Shephard (1964a). Shephard's result implies that any polytope which can be approximated by zonotopes (and hence is a zonoid), in fact has to be a zonotope itself. Therefore, a centrally symmetric polytope with non-symmetric faces, for example an octahedron, is not a zonoid. In a similar vein, Schneider (1970a) showed that the only polytopal generalised zonoids are the zonotopes. See section 7 for an approach to this result which uses Radon transforms. In contrast to the above observations, the generalised zonoids are dense in the centrally symmetric bodies. In fact, Lemma 2.3 shows that any sufficiently smooth centrally symmetric body has to be a generalised zonoid.

Theorem 2.4. *The generalised zonoids are dense in the class of centrally symmetric bodies.*

This denseness result led Weil (1976b) to show that all centrally symmetric bodies have a generating distribution. Its definition involves the *cosine transform* $T: C_e^\infty(S^{d-1}) \to C_e^\infty(S^{d-1})$ defined by:

$$(Tf)(u) = \int_{S^{d-1}} |\langle u, v\rangle| f(v)\lambda_{d-1}(\mathrm{d}v) . \tag{2.4}$$

It follows from Theorem 2.2 that T is invertible and from Lemma 2.3 that T is a continuous bijection of $C_e^\infty(S^{d-1})$ to itself. Now $C_e^\infty(S^{d-1})$ is complete and metrisable and so the Open Mapping Theorem shows that T^{-1} is a continuous mapping of $C_e^\infty(S^{d-1})$ to itself. Fubini's theorem together with (2.4) implies that

$$\int_{S^{d-1}} (Tf)(u)g(u)\lambda_{d-1}(\mathrm{d}u) = \int_{S^{d-1}} f(u)(Tg)(u)\lambda_{d-1}(\mathrm{d}u) . \tag{2.5}$$

This duality result, together with the above continuity properties of T, shows that T can be extended to a continuous bijection $T: D_e(S^{d-1}) \to D_e(S^{d-1})$ on the dual space $D_e(S^{d-1}) = [C_e^\infty(S^{d-1})]'$ of even distributions on S^{d-1}, endowed with the strong topology. So, for $\rho \in D_e(S^{d-1})$, $T\rho$ is defined by:

$$(T\rho)(f) = \rho(Tf) , \quad f \in C_e^\infty(S^{d-1}) .$$

Therefore, the inverse $T^{-1}\rho$ exists for every $\rho \in D_e(S^{d-1})$, and in particular the *generating distribution* $\rho_K = T^{-1}h(K; \cdot)$ of a body $K \in \mathcal{K}_0^d$ exists and satisfies

$$\rho_K(f) = \int_{S^{d-1}} h(K; u)(T^{-1}f)(u)\lambda_{d-1}(\mathrm{d}u) \tag{2.6}$$

for $f \in C_e^\infty(S^{d-1})$. Because of Theorem 2.2, ρ_K can be extended to a linear functional on the space of all functions $f \in C_e(S^{d-1})$ of the form

$$f(u) = \int_{S^{d-1}} |\langle u, v\rangle| \rho_f(\mathrm{d}v) ,$$

with ρ_f an even measure. This extension of (2.6) is defined by:

$$\rho_K(f) = \int_{S^{d-1}} h(K; u)\rho_f(\mathrm{d}u) .$$

In particular, this gives

$$\rho_K(|\langle u, \cdot \rangle|) = h(K; u) \tag{2.7}$$

for all centrally symmetric bodies K.

Consequently, there is a hierarchy of centrally symmetric convex bodies corresponding to the nature of the generating distribution. Zonotopes are the bodies whose generating distribution is an atomic measure, zonoids are those for which it is a positive measure, and the generalised zonoids correspond to the case of signed measures. The smoothness results related to (2.3) give upper bounds on the order of the generating distributions.

The continuity properties of T and T^{-1} imply that the generating distribution ρ_K depends continuously on the body K (with respect to the Hausdorff metric). However, T^{-1} is not continuous on the vector space of even signed measures (supplied with the weak* topology) and so, for a sequence of generalised zonoids converging to a generalised zonoid, the generating measures need not converge. On the other hand, for zonoids K_i, K, we have $K_i \to K$ if and only if $\rho_{K_i} \to \rho_K$ weakly.

For generalised zonoids $K, K_1, \ldots, K_d$, the basic geometric functionals, the intrinsic volumes $V_j(K)$, $j = 0, \ldots, d$, the mixed volume $V(K_1, \ldots, K_d)$, the surface area measures $S_j(K; \cdot)$, $j = 0, \ldots, d-1$, and the mixed surface area measure $S(K_1, \ldots, K_{d-1}; \cdot)$ can be expressed in terms of the generating measures (see chapters 1.2 and 1.8 for the definitions and properties of these functionals). To explain the results, let $D_k(u_1, \ldots, u_k)$ (for $u_1, \ldots, u_k \in S^{d-1}$) be the absolute value of the determinant of $u_1, \ldots, u_k$ (computed in an appropriate k-dimensional space). The partial mapping $S : (S^{d-1})^{d-1} \to S^{d-1}$ is defined, for all linearly independent $(u_1, \ldots, u_{d-1}) \in (S^{d-1})^{d-1}$, by $S(u_1, \ldots, u_{d-1}) = u_d$, where u_d is orthogonal to $u_1, \ldots, u_{d-1}$ and $u_1, \ldots, u_{d-1}, u_d$ are positively oriented. D_k and S are measurable mappings. Let $S\mu$ denote the image measure of a (signed) measure μ on $(S^{d-1})^{d-1}$ under S.

Theorem 2.5. *For generalised zonoids $K, K_1, \ldots, K_d$, we have*

$$S(K_1, \ldots, K_{d-1}; \cdot) = \frac{2^d}{(d-1)!} S\left[\int_{(\cdot)} D_{d-1}\, \mathrm{d}(\rho_{K_1} \times \cdots \times \rho_{K_{d-1}})\right], \tag{2.8}$$

$$S_j(K; \cdot) = \frac{2^{j+1}(d-1)^{d-j-1}}{(d-1)!\,\omega_{d-2}^{d-j-1}} S\left[\int_{(\cdot)} D_{d-1}\, \mathrm{d}(\rho_K^j \times \lambda_{d-1}^{d-j-1})\right],$$

$$j = 0, \ldots, d-1 , \tag{2.9}$$

$$V(K_1, \ldots, K_d) = \frac{2^d}{d!} \int_{S^{d-1}} \cdots \int_{S^{d-1}} D_d(u_1, \ldots, u_d) \rho_{K_1}(du_1) \cdots \rho_{K_d}(du_d) , \tag{2.10}$$

and

$$V_j(K) = \frac{2^j}{j!} \int_{S^{d-1}} \cdots \int_{S^{d-1}} D_j(u_1, \ldots, u_j) \rho_K(du_1) \cdots \rho_K(du_j) , \quad j = 0, \ldots, d . \tag{2.11}$$

These formulae have been proved in Weil (1976a); (2.11) was also obtained for zonoids by Matheron (1975). Note that (2.9)–(2.11) all follow from (2.8). Furthermore, (2.9) can be used to give an answer to the natural question, which signed measures are generating measures of generalized zonoids. Although generating measures are not necessarily positive, they do have to satisfy certain positivity conditions which turn out to be characteristic.

Theorem 2.6. *The even signed measures ρ on S^{d-1} which are generating measures of generalised zonoids are precisely those for which*

$$S\left[\int_{(\cdot)} D_{d-1} \, d(\rho^j \times \lambda_{d-1}^{d-j-1})\right] \geq 0$$

for $j = 1, \ldots, d-1$.

This result is due to Weil (1976a) and follows from (2.9) and a theorem in Weil (1974). The latter characterises support functions in the vector space $\mathscr{L}$ of differences of support functions by the positivity of their surface area measures.

A different characterisation of generating measures of generalised zonoids was given in Weil (1982), using projections. The projection ΠK of a zonoid K onto a subspace of arbitrary dimension is again a zonoid. This implies that ΠK, for a generalised zonoid K, is again a generalised zonoid. Weil (1982) showed how to obtain the generating measure $\rho_{\Pi K}$ of ΠK from that of the original body K. Defining the projection $\Pi \rho$ of a measure on S^{d-1} in a natural way, it turns out that $\rho_{\Pi K} = \Pi \rho_K$. This was used in Weil (1982) to obtain a characterisation of generating measures of generalised zonoids, which is different from Theorem 2.6. The following special case seems to be the most interesting; it was obtained earlier by Lindquist (1975b).

Theorem 2.7. *A function $f \in C_e(S^{d-1})$ is the density of the generating measure of a generalised zonoid if and only if*

$$\int_{S^{d-2}} |\langle u, v \rangle|^2 f(v) \lambda_{d-2}(dv) \geq 0$$

for all $(d-2)$-dimensional subspheres $S^{d-2} \subset S^{d-1}$ and all $u \in S^{d-2}$.

Of course, it is an interesting problem to extend these results to general centrally symmetric bodies. For the formulae of Theorem 2.5 this can be done in an indirect way by approximation (see Weil 1976b). Also the above-mentioned characterisation by projections can be extended to distributions. However, it seems to be difficult to formulate Theorem 2.6 for distributions and, up to now, no generating distributions which are not signed measures are known explicitly. For example, it is an open problem to characterise generating distributions of polytopes (other than zonotopes) or even to give the generating distribution of an octahedron.

3. Analytic characterisations of zonoids

The following two analytic descriptions of zonoids are of a more functional analytic nature and will not be discussed further.

Theorem 3.1. *The zonoids in $\mathbb{R}^d$ are (up to translations) precisely the ranges of nonatomic $\mathbb{R}^d$-valued measures.*

Theorem 3.2. *A body $K \in \mathcal{K}_0^d$ with inner points is a zonoid if and only if the normed space $(\mathbb{R}^d, h(K; \cdot))$ is isometric to a subspace of $L_1 = L_1([0, 1])$.*

Details and references can be found in Schneider and Weil (1983). These, as well as some of the following characterisations, admit certain infinite-dimensional generalisations [some references are also given in Schneider and Weil (1983)].

Other analytic aspects of these normed spaces $(\mathbb{R}^d, h(K; \cdot))$ involve characterisations of zonoids K by sets of inequalities which are to be satisfied by the norm $\| \cdot \| = h(K; \cdot)$. A function $f : \mathbb{R}^d \to \mathbb{R}$ is called *positive definite* if

$$\sum_{i,j=1}^{n} f(x_i - x_j) w_i w_j \geq 0 \tag{3.1}$$

for $n \in \mathbb{N}$, all $x_1, \ldots, x_n \in \mathbb{R}^d$ and all real numbers $w_1, \ldots, w_n$. If (3.1) is merely assumed to hold under the additional assumption that $\Sigma\, w_i = 0$, then f is *conditionally positive definite*. The function is said to be of *negative type*, if for every $t > 0$ the function e^{-tf} is positive definite. A real normed vector space $(V, \| \cdot \|)$ is called *hypermetric* if

$$\sum_{i,j=1}^{n} w_i w_j \|x_i - x_j\| \leq 0 \tag{3.2}$$

for $n \in \mathbb{N}$, all $x_1, \ldots, x_n \in V$ and all integers $w_1, \ldots, w_n$ satisfying $\Sigma\, w_i = 1$.

The following theorem summarises results of Lévy (1937, pp. 219–223), Schoenberg (1938), Bolker (1969), Choquet (1969a, pp. 55–59, 1969b, p. 173), and Witsenhausen (1973).

Theorem 3.3. *For a centrally symmetric body K (with inner points), the following are equivalent*:

(a) *K is a zonoid*,
(b) *$h(K;\cdot)$ is of negative type*,
(c) *$e^{-h(K;\cdot)}$ is positive definite*,
(d) *$(\mathbb{R}^d, \|\cdot\|)$ with $\|\cdot\| = h(K;\cdot)$ is hypermetric*,
(e) *$-h(K;\cdot)$ is conditionally positive definite.*

The proof of the equivalence of (b), (c), (d), and (e) is fairly easy, as it is to show that the support function of a zonoid K has one (and hence all) of these properties. For the implication from (b), (c), (d), or (e) to (a), all authors refer to Lévy's (1937, pp. 219–223) general results on infinitely divisible probability distributions. Since (2.2) is an integral representation of Choquet type, a more direct but still complicated proof using Choquet's theorem is possible.

Some equivalences of this and the aforementioned type for zonoids generalise to convex bodies K, which, for fixed $1 \leqslant p \leqslant 2$, satisfy

$$h(K; u) = \left[\int_{S^{d-1}} |\langle u, v \rangle|^p \rho(\mathrm{d}v) \right]^{1/p} \quad \text{for } u \in \mathbb{R}^d ,$$

for some even positive measure ρ on S^{d-1} (Lévy 1937, Herz 1963, Choquet 1969a, Bretagnolle, Dacunha Castelle and Krivine 1966). It seems however that these bodies, for $p > 1$, are of much less geometric interest.

Theorem 3.3 characterises zonoids among centrally symmetric bodies by certain systems of inequalities which are satisfied by their support functions. For polytopes there are much simpler characteristic inequalities (Witsenhausen 1978, Assouad 1980).

Theorem 3.4. *A polytope K is a zonotope if and only if the norm $\|\cdot\| = h(K;\cdot)$ satisfies Hlawka's inequality*

$$\|x\| + \|y\| + \|z\| + \|x + y + z\| \geqslant \|x + y\| + \|y + z\| + \|z + x\|$$

for all $x, y, z \in \mathbb{R}^d$.

The one direction follows from the more general results in Theorem 3.3. For the other direction, one first observes that Hlawka's inequality for $h(K;\cdot)$ implies the same inequality for all directional derivatives $h_v(K;\cdot)$, $v \in S^{d-1}$. Since this means that $h_v(K;\cdot)$ is (up to a linear summand) an even function, all faces of K are symmetric, and hence K is a zonotope.

Assouad (1980) also gives another norm inequality (the 7-*polygonal inequality*) which characterises (the support functions of) zonotopes. From a result of Weil (1982), which will be discussed later in more detail, it is clear that none of these inequalities characterises zonoids in the same manner, at least not in all dimensions. More precisely, if there is a characterisation of zonoids by a finite number of inequalities for the support function, the number k of points involved has to increase with d. It is however open, whether the Hlawka inequality characterises zonoids in $\mathbb{R}^3$.

The question of finding a minimal set of inequalities for the support function which characterise zonoids, or, in other words, the problem to characterise the essential inequalities in (3.1) or (3.2) can be formulated in terms of positive linear functionals. The set $\mathscr{Z}^d$ of centred zonoids is a closed convex cone. Theorems 2.1 and 2.2 show that this cone is simplicial in the sense of Choquet [see, for example, Alfsen (1971)] and that its extremal rays comprise the centred line segments. The map $K \mapsto h(K; \cdot)$ is an isometric isomorphism from the cone $\mathscr{Z}^d$ onto the cone $\mathscr{S} \subset C_e(S^{d-1})$ of functions of the form (2.2) (*zonoidal functions*). One can formulate questions about $\mathscr{Z}^d$ or $\mathscr{S}$ in terms of the dual cone $\mathscr{S}^*$. The dual of $C_e(S^{d-1})$ is the space $\mathscr{M}_e(S^{d-1})$ of even signed measures on S^{d-1}. So

$$\mathscr{S}^* = \{ \mu \in \mathscr{M}_e(S^{d-1}) \colon\ \mu(h) \geq 0 \text{ for all } h \in \mathscr{S} \} .$$

It follows that

$$\mathscr{S}^* = \left\{ \mu \in \mathscr{M}_e(S^{d-1}) \colon \int_{S^{d-1}} |\langle x, v \rangle| \, \mu(\mathrm{d}v) \geq 0 \text{ for all } x \in S^{d-1} \right\} .$$

Characterisations of $\mathscr{S}$, $\mathscr{S}^*$, and $\mathscr{Z}^d$ are closely connected. The Hahn–Banach theorem implies that K is a zonoid if and only if $\mu(h(K; \cdot)) \geq 0$ for all $\mu \in \mathscr{S}^*$. Of course, it is enough to consider a subset $\mathscr{N}$ of $\mathscr{S}^*$ which contains the extreme rays in its closure and therefore *generates* $\mathscr{S}^*$ (in the sense that $\mathscr{S}^*$ is the smallest closed convex cone containing $\mathscr{N}$).

Theorem 3.5. *Let $\mathscr{N}$ be a generating subset of $\mathscr{S}^*$. Then a centred convex body K is a zonoid if and only if $\mu(h(K; \cdot)) \geq 0$ for all $\mu \in \mathscr{N}$.*

Similar observations will be used in section 7 to find some further characterisations of zonoids and other related classes of convex bodies. It should however be pointed out that these characterisations use rather "big" subsets $\mathscr{N}$ of $\mathscr{S}^*$, namely dense ones. It would of course be preferable to use "small" sets $\mathscr{N}$, e.g., the extremal rays of $\mathscr{S}^*$ themselves. However, the characterisation of these extremal rays appears to be a completely open problem. In view of Theorem 3.3, the inequalities in (3.2) correspond to certain discrete measures in $\mathscr{S}^*$, and thus these questions can also be formulated in terms of a minimal set of inequalities.

4. Centrally symmetric bodies and the spherical Radon transform

The cosine transform T is closely related to the (spherical) *Radon transform* $R : C_e^\infty(S^{d-1}) \to C_e^\infty(S^{d-1})$ defined by:

$$(Rf)(u) = \int_{u^\perp \cap S^{d-1}} f(v)\nu^{u^\perp}(dv)\,, \quad u \in S^{d-1}\,, \tag{4.1}$$

where $\nu^{u^\perp}$ is the invariant probability measure on $u^\perp \cap S^{d-1}$. The important properties of R can be found in Helgason (1980, 1984). In particular, the Radon transform satisfies a duality equation analogous to (2.5) and so, like T, can be extended to distributions, resulting in a continuous bijection, mapping $D_e(S^{d-1})$ to itself. Historically, the injectivity of R (for $d=3$) is due to Minkowski (1911); other proofs can be found in Funk (1913), and Bonnesen and Fenchel (1934, p. 137). In case $d \geqslant 3$, Helgason (1959) gives inversion formulae for R; see Petty (1961) and Schneider (1969) for other proofs of the injectivity of R.

The relationship between T and R seems to have first been noted (in case $d=3$) by Blaschke (1916), and is studied in Petty (1961), Schneider (1969, 1970c), and Goodey and Weil (1992a). It can be expressed quite concisely using the Laplace–Beltrami operator Δ on S^{d-1}. Berg (1969), in his solution of the Christoffel problem, showed that, for any convex body K,

$$((d-1)^{-1}\Delta + 1)h(K;\cdot) = S_1(K;\cdot) \tag{4.2}$$

as distributions. In the case of a centred line segment K with endpoints $\pm u \in S^{d-1}$, (4.1) and (4.2) show that

$$(((d-1)^{-1}\Delta + 1)|\langle u,\cdot\rangle|)(f) = 2\omega_{d-2}(d-1)^{-1}(Rf)(u)\,, \tag{4.3}$$

where $f \in C_e^\infty(S^{d-1})$. So, if $\Box = (\Delta + d - 1)/2\omega_{d-2}$, then (4.3) is equivalent to

$$\Box T = R\,. \tag{4.4}$$

This formula, relating T and R, facilitates the application of results from the theory of Radon transforms to the geometry of centrally symmetric convex bodies and vice versa.

R is known (Helgason 1980, 1984) to be an invertible intertwining operator, so

$$T^{-1} = R^{-1}\Box = \Box R^{-1}\,. \tag{4.5}$$

There are specific inversion formulae for R (Helgason 1980, 1984, Semyanistyi 1961, Strichartz 1981). For even dimensions d there are polynomials p_d such that

$$R^{-1} = p_d(\Delta)R \tag{4.6}$$

(Helgason 1959), in odd dimensions the inversion formulae are more complicated, some formulations involving $\Delta^{1/2}$.

The above inversion formulae show that the hierarchy of centrally symmetric sets described in section 2 has another interpretation in terms of first order surface area measures. This follows from the definition of generating distributions ρ_K and

(4.5) which give

$$R\rho_K = RT^{-1}h(K;\cdot) = \Box h(K;\cdot) = ((d-1)/2\omega_{d-2})S_1(K;\cdot)\,. \tag{4.7}$$

So the zonoids (respectively generalised zonoids) are the bodies whose first surface area measures are Radon transforms of positive (signed) measures. Results of the form (4.7) appeared in Weil (1976a).

It was mentioned in section 2, that all sufficiently smooth bodies are generalised zonoids. Schneider (1967) showed that smoothness of order d yields a generalised zonoid whose generating measure has a continuous density. For many purposes, the natural setting for the Radon transform is L_2 space, and here Schneider's calculations show that smoothness of order $\frac{1}{2}d$ implies ρ_K has an L_2 density (see Goodey and Weil 1992a). In fact, similar calculations can be found in Strichartz (1981) where he analyses the range of R on different Sobolev spaces. These results give the following theorem, where, as usual, $C_e^k(S^{d-1})$ denotes the subspace of k times continuously differentiable functions in $C_e(S^{d-1})$.

Theorem 4.1. *If K is a centrally symmetric body with $h(K;\cdot) \in C_e^k(S^{d-1})$, then K is a generalised zonoid if $k = \frac{1}{2}(d+5)$.*

This shows that the generating distributions of centrally symmetric bodies are of order at most k. Further small improvements in k can be found when the residue of $d \bmod 4$ is known. But it is not known what are the best values of k, or if k might even be independent of d.

Radon transforms provide a technique for studying the Christoffel problem for centrally symmetric bodies. This problem, for arbitrary convex bodies, was solved independently by Berg (1969) and Firey (1967, 1968) (see chapter 1.8). The following is an outline of a short approach to Berg's solution in the case of centrally symmetric bodies [see Goodey and Weil (1992a) for more details]. For such bodies, the Christoffel problem asks for conditions on a positive measure $\mu \in \mathcal{M}_e(S^{d-1})$ which guarantee that there is a $K \in \mathcal{K}_0^d$ with $\mu = S_1(K;\cdot)$. If it were possible to find a function $f_d \in L_1(S^{d-1})$ with

$$Rf_d = |\langle u, \cdot\rangle| \tag{4.8}$$

(for fixed $u \in S^{d-1}$), then f_d must be rotationally symmetric in the sense that there is a $g_d \in L_1([0,1])$ with $f_d = g_d(|\langle u,\cdot\rangle|)$. Then (2.7), (4.7) and (4.8) yield

$$Rh(K;\cdot) = R\rho_K(|\langle\cdot,\cdot\rangle|) = \frac{(d-1)}{2\omega_{d-2}} R\int_{S^{d-1}} g_d(|\langle\cdot, v\rangle|)S_1(K; \mathrm{d}v)\,,$$

and therefore, because of the injectivity of R,

$$h(K;\cdot) = \frac{(d-1)}{2\omega_{d-2}} \int_{S^{d-1}} g_d(|\langle\cdot, v\rangle|)S_1(K; \mathrm{d}v)\,.$$

It follows that the positive measures $\mu \in \mathcal{M}_e(S^{d-1})$ for which

$$\int_{S^{d-1}} g_d(|\langle \cdot, v\rangle|)\mu(dv)$$

is convex are precisely those that are the first surface area measures of convex bodies. Explicit expressions for the functions f_d, or equivalently g_d, can be obtained from the observation that (4.8) is equivalent to the Fredholm integral equation

$$(1-r^2)^{1/2} = 2(\omega_{d-3}/\omega_{d-2}) \int_0^1 g_d(sr)(1-s^2)^{(d-4)/2}\, ds$$

if $d \geqslant 3$ (and a simpler one in the case $d = 2$). Solutions g_d are given recursively by

$$g_2(r) = (1-r^2)^{1/2},$$

$$g_3(r) = 1 + \frac{r}{2} \log_e \frac{1-r}{1+r},$$

and

$$g_{d+2}(r) = \frac{r}{d-1} g_d'(r) + g_d(r), \quad d \geqslant 2.$$

These are essentially the even parts of Berg's (1969) functions.

Since zonotopes are characterised by a simple symmetry property of their faces, one would expect a correspondingly simple characterisation of zonoids K. In fact, Blaschke (1923, p. 250) [see also Blaschke and Reidemeister (1922, pp. 81–82)] and Bolker (1971) asked for a "local" condition. In analytic terms, such a condition would imply that a centrally symmetric body K with the property that for each $v \in S^{d-1}$ there is a neighborhood $U_v \subset S^{d-1}$ of v and a zonoid K_v with

$$h(K;\cdot) = h(K_v;\cdot) \quad \text{on } U_v \tag{4.9}$$

must itself be a zonoid. Weil (1977) used Lemma 2.3 to construct counterexamples for all dimensions $d \geqslant 3$. He then reformulated the problem, asking whether a *zonal characterisation* is possible, that is, whether (4.9) characterises zonoids, if instead of a neighborhood U_v of v, a neighborhood E_v of the equator $v^\perp$ (a *zone*) on S^{d-1} is considered. A proof of this conjecture in even dimension was given by Panina (1988, 1989) based on some involved techniques from combinatorial integral geometry. The following outlines a direct approach using the Radon transform R (see Goodey and Weil 1992a).

Theorem 4.2. *Let K be a centrally symmetric convex body such that for any $v \in S^{d-1}$ there is a zone E_v, containing the equator $v^\perp$, and a zonoid $K_v = K(E_v)$ such that*

$$h(K;u) = h(K_v;u) \quad \text{for all } u \in E_v .$$

If d is even, then K is a zonoid.

In order to show that ρ_K is a positive distribution, it suffices to prove that to each $v \in S^{d-1}$ there is a small neighbourhood U_v (a *cap*) such that $\rho_K(g) \geq 0$ for all positive test functions g supported on U_v. U_v is chosen so that the orthogonal zone $U_v^\perp$ is contained in E_v. Then by eqs. (4.4)–(4.6)

$$T^{-1}g = \Box R^{-1}g = \Box^2 p_d(\Delta)Tg ,$$

which is supported on $U_v^\perp \subset E_v$. Then

$$\begin{aligned}\rho_K(g) &= \int_{S^{d-1}} h(K;u)(T^{-1}g)(u)\lambda_{d-1}(\mathrm{d}u) = \int_{E_v} h(K;u)(T^{-1}g)(u)\lambda_{d-1}(\mathrm{d}u) \\ &= \int_{E_v} h(K_v;u)(T^{-1}g)(u)\lambda_{d-1}(\mathrm{d}u) \\ &= \rho_{K_v}(g) \geq 0 .\end{aligned}$$

It is clear that, in the above argument, the particular form of the inversion formula (4.6) was not important. The essential step was the following support property of the cosine transform, which follows from (4.6) [and (4.4) and (4.5)].

Lemma 4.3. *Let d be even. Assume that $g \in C_e(S^{d-1})$ is such that Tg is supported by a (symmetric) cap C. Then the support of g is contained in the orthogonal zone $C^\perp$. Equivalently, assume $g \in C_e(S^{d-1})$ is such that $Tg \equiv 0$ on $C^\perp$ for some cap C. Then $g \equiv 0$ on C.*

There is an interesting connection between these results, which hold for all even d, and a result of Schneider and Weil (1970), which holds for all odd d and which will be described in the next section. The latter has the following analytic formulation.

Lemma 4.4. *Let d be odd and let $g \in C_e(S^{d-1})$ be such that $g \equiv 0$ on a cap C and $Tg \equiv 0$ on $C^\perp$. Then $g \equiv 0$ on S^{d-1}.*

Schneider and Weil (1970) also show, by means of a counterexample, that this result is false in even dimensions. This can now be seen as a consequence of Lemma 4.3 as follows. Let d be even and $g \in C_e(S^{d-1})$ such that $Tg \equiv 0$ on a zone $C^\perp$. Then Lemma 4.3 implies that $g \equiv 0$ on the cap C. So if Lemma 4.4 would hold, $g \equiv 0$ on S^{d-1}. But, of course, there are non-zero functions $g \in C_e(S^{d-1})$ with $T_g \equiv 0$ on a zone $C^\perp$ (this follows, e.g., from Lemma 2.3). Therefore, Lemma 4.4 cannot hold for even d.

This argument is of interest since it can also be used in the reverse direction to conclude from Lemma 4.4 that there is no support theorem analogous to Lemma 4.3 for the cosine transform T (and hence also not for the Radon transform R) in odd dimensions (Goodey and Weil 1992a). Nevertheless, the question of a zonal characterisation of zonoids in odd dimensions remains open.

The related questions about the support properties of functions whose Radon transform is supported by a zone were considered by Quinto (1983) and more recently by Helgason (1990).

5. Projections onto hyperplanes

For an arbitrary convex body K and $u \in S^{d-1}$, let $v_j^{(d-1)}(K; u)$ be the jth intrinsic volume of the projection of K onto the hyperplane $u^\perp$ and let $v_{d-1}(K; u) = v_{d-1}^{(d-1)}(K; u)$. $v_j^{(d-1)}(K; u)$ is given in terms of the jth surface area measure $S_j(K; \cdot)$ by

$$v_j^{(d-1)}(K; u) = \frac{\binom{d-1}{j}}{2\kappa_{d-j-1}} \int_{S^{d-1}} |\langle u, v\rangle| S_j(K; \mathrm{d}v) ,$$

where κ_i denotes the i-dimensional volume of the unit i-ball; note that $\omega_{d-1} = d\kappa_d$. Since $S_j(K; \cdot)$determines a convex body K of dimension $\geqslant j+1$ uniquely (up to a translation), the following is an equivalent formulation of Theorem 2.2 (the version that Aleksandrov proved).

Theorem 5.1. *If, for $j \in \{1, \ldots, d-1\}$, K, L are centred convex bodies (of dimension $\geqslant j+1$), such that $v_j^{(d-1)}(K; \cdot) = v_j^{(d-1)}(L; \cdot)$, then $K = L$.*

The following result of Schneider and Weil (1970), which was mentioned in the previous section, shows cases in which the assumption $v_j^{(d-1)}(K; \cdot) = v_j^{(d-1)}(L; \cdot)$ in Theorem 5.1 can be weakened.

Theorem 5.2. *Let d be odd and, for $j \in \{1, \ldots, d-1\}$, let K, L be centred convex bodies (of dimension $\geqslant j+1$) that both have a vertex with a common interior outward normal $u \in S^{d-1}$. Assume $v_j^{(d-1)}(K; \cdot) = v_j^{(d-1)}(L; \cdot)$ on a zone containing $u^\perp$. Then $K = L$.*

As was already mentioned, this result is false in even dimensions, and also the vertex condition on both bodies cannot be suppressed. Moreover, for any nonempty symmetric open set $A \subset S^{d-1}$, there are two different bodies $K, L \in \mathcal{K}_0^d$ with $v_j^{(d-1)}(K; \cdot) = v_j^{(d-1)}(L; \cdot)$ outside A. However, for polytopes K, L there is a stronger result (Schneider 1970b).

For zonoids K, the projections frequently behave in an intuitive manner, so much so that their properties are sometimes characteristic of zonoids. The following paragraphs describe such characteristic properties and the instances of less intuitive behaviour, in the case of non-zonoidal bodies.

A natural starting point for this discussion is contained in the works of Petty (1967) and Schneider (1967), in which they investigated a problem posed by Shephard (1964b). Shephard asked, if K_1 and K_2 are centrally symmetric bodies with

$$v_{d-1}(K_1; u) > v_{d-1}(K_2; u) \tag{5.1}$$

for all $u \in S^{d-1}$ is it necessarily the case that $V(K_1) > V(K_2)$? First, note that the restriction to centrally symmetric sets is necessary. This follows most easily from a result of Petty (1967).

Theorem 5.3. *If K is any convex body with interior points, then the family of bodies L with*

$$v_{d-1}(L; \cdot) = v_{d-1}(K; \cdot)$$

has (up to a translation) only one centrally symmetric member, and this body has greater volume than any non-centrally symmetric member.

The following result was obtained independently by Petty (1967) and Schneider (1967).

Theorem 5.4. *If K_1 and K_2 are centrally symmetric convex bodies satisfying* (5.1) *and if K_1 is a zonoid, then $V(K_1) > V(K_2)$.*

Both authors showed that, in order to obtain the desired volume inequality, some restrictions must be placed on the bodies K_1 or K_2. The following general result is due to Schneider (1967).

Theorem 5.5. *For every sufficiently smooth centrally symmetric convex body K_2, which is not a zonoid, there is a centrally symmetric convex body K_1 satisfying* (5.1) *and for which $V(K_1) < V(K_2)$.*

The analogous questions for central sections of symmetric bodies also yield some interesting and unexpected results (Larman and Rogers 1975, Ball 1986, 1988, Lutwak 1988). For example, in dimensions $d \geq 10$ one can construct a cube C and a ball K such that the central sections of K have greater volume than the parallel central sections of C and yet $V(C) > V(K)$.

Weil (1976b) gave a characterisation of zonoids using properties similar to those investigated by Schneider. It uses the mixed volumes $V(L; K, d-1) = V(L, K, \ldots, K)$ of convex bodies K, L.

Theorem 5.6. *Let L be a centrally symmetric convex body. Then L is a zonoid if and only if*

$$V(L; K_1, d-1) \leq V(L; K_2, d-1)$$

whenever $K_1, K_2 \in \mathcal{K}_0^d$ *satisfy*

$$v_{d-1}(K_1; \cdot) \leq v_{d-1}(K_2; \cdot) .$$

This result is a simple consequence of Theorem 3.5 since any (full dimensional) measure $\mu \in \mathcal{S}^*$ is the difference of surface area measures,

$$\mu = S_{d-1}(K_2; \cdot) - S_{d-1}(K_1; \cdot) .$$

Modifications and generalisations are obtained if one of the measures $S_{d-1}(K_1; \cdot)$, $S_{d-1}(K_2; \cdot)$ is assumed to be the spherical Lebesgue measure (and hence the body is a ball), if the $(d-1)$st surface area measures are replaced by jth surface area measures, or if discrete measures are used. There are also analogous characterisations of generalised zonoids and arbitrary centrally symmetric bodies. These modifications and analogues can be found in Goodey (1977), Weil (1979), and Schneider and Weil (1983).

For a convex body K, $v_j^{(d-1)}(K; u)$ is the support function of a zonoid $\Pi_j K$ which is called the *jth projection body* of K. The corresponding map $K \mapsto \Pi_j K$ defines the projection operator Π_j, $j = 1, \ldots, d-1$. Theorem 5.1 shows that Π_j is injective on $\mathcal{K}_0^d$. Π_j and (on $\mathcal{K}_0^d$) Π_j^{-1} are also continuous with respect to the Hausdorff metric. Schneider's result above shows that Π_{d-1}^{-1} is not monotonic with respect to set inclusion. So these inverse functions appear to demonstrate a quite complicated behaviour. In fact, of all the transformations Π_j and Π_j^{-1} for $j \in \{1, \ldots, d-1\}$, only Π_1 is uniformly continuous. Most of these observations arise quite easily although the non-uniform continuity of Π_1^{-1} is less obvious, see Goodey (1986). If one places bounds on both the inradius and the circumradius of the bodies concerned, it is then natural to investigate the stability of Π_j^{-1}. The stability of Π_{d-1}^{-1} was recently demonstrated by Bourgain and Lindenstrauss (1988a). The stability of Π_1^{-1} was established by Campi (1986), and Goodey and Groemer (1990) (see chapter 1.4). We note that in view of (4.7), the latter is just a stability result for the spherical Radon transform R.

In a rather different direction it is interesting to study the range of the operators Π_j. These operators obviously have smoothing properties and, as has been seen, tend to increase the symmetry of the sets to which they are applied. Iterations of Π_j therefore lead to a decreasing sequence of sets $\Pi_j^k(\mathcal{K}_0^d)$, $k = 1, 2, \ldots$. No description of the limit sets $\Pi_j^\infty(\mathcal{K}_0^d)$ seems to be known. In this connection, the determination of all bodies K which are homothetic to their projection body $\Pi_j K$ (respectively to $\Pi_j^2 K$) is also of interest. Only some partial results are known. Weil (1971) showed that the polytopal members of $\Pi_{d-1}^\infty(\mathcal{K}_0^d)$ are precisely the polytopes K which are homothetic to $\Pi_{d-1}^2 K$. He then described these polytopes as direct sums of (possibly degenerate) symmetric polygons. A corresponding characterisation of polytopes K with $\Pi_{d-1} K = cK$ is also given. Schneider (1977) showed that $\Pi_1 K = cK$ implies K is a ball.

6. Projection functions on higher rank Grassmannians

This section is concerned with projections onto subspaces E of dimension k with $1<k<d-1$. Here again the focus of attention will be the intrinsic volumes $v_j^{(k)}(K; E)$ of these projections for centrally symmetric bodies K, $j \in \{0, \ldots, k\}$. $v_j^{(k)}(K; \cdot)$ is a continuous function on the compact Grassmannian manifold L_k^d of k-dimensional subspaces of $\mathbb{R}^d$. The *kth projection function* $v_k(K; \cdot) = v_k^{(k)}(K; \cdot)$ will again play a special role. To simplify the connections with earlier results, even functions on S^{d-1} are identified with functions on L_1^d. It is also convenient to make use of the orthogonality operator which transforms a function f on L_k^d into a function $f^{\perp}$ on L_{d-k}^d by $f^{\perp}(E) = f(E^{\perp})$ and, similarly, a measure ρ on L_k^d into a measure $\rho^{\perp}$ on L_{d-k}^d.

In the case of generalised zonoids K, a measure $\rho_k(K; \cdot)$ on L_k^d, the *kth projection generating measure* of K, can be introduced as the image of

$$\int_{(\cdot)} D_k \, \mathrm{d}\rho_K^k$$

(notation as in Theorem 2.5) under the mapping S_k which assigns to each (linearly independent) k vectors $u_1, \ldots, u_k \in S^{d-1}$ the subspace $E \in L_k^d$ spanned by $u_1, \ldots, u_k$. This terminology for $\rho_k(K; \cdot)$ is justified by the fact that

$$v_k(K; E) = \frac{2^k}{k!} \int_{L_k^d} |E, F| \rho_k(K; \mathrm{d}F) \tag{6.1}$$

for all $E \in L_k^d$. Here $|E, F|$ denotes the absolute value of the determinant of the orthogonal projection of E onto F. In the case $k = d-1$ when $E = u^{\perp}$ and $F = v^{\perp}$ it is clear that $|E, F| = |\langle u, v \rangle|$ [and $\rho_{d-1}(K; \cdot)$ reduces to $((d-1)!/2^d) \times S_{d-1}(K: \cdot)^{\perp}$]. Also, for $k=1$, one has $E = u$, $F = v$ and $\rho_1(K; \cdot) = \rho_K$. So (6.1) is a generalisation of the integral equation (2.2). The uniqueness result, analogous to Theorem 2.2, for $2 \leq k \leq d-2$ [a conjecture of Matheron (1974a, 1975)], was disproved by Goodey and Howard (1990). This lack of uniqueness is the cause of many of the differences that occur in the study of these higher rank Grassmannians.

It is important to note that there is no smoothness result, analogous to that for support functions, which guarantees that a (sufficiently smooth) function on L_k^d is a difference of projection functions. In fact, such differences are not dense in $C(L_k^d)$, since, if they were, the same would be true of differences of projection functions of generalised zonoids, but the measures constructed in Goodey and Howard (1990) are orthogonal to all these functions.

Another way in which differences from the case $k = d-1$ (and $k=1$) occur can be seen by considering the Radon transforms $R_{i,j}$ on the Grassmannians. For $1 \leq i, j \leq d-1$, $R_{i,j} : C(L_i^d) \to C(L_j^d)$ is defined by:

$$R_{i,j} f(E) = \int_{L_i^d(E)} f(F) \nu_i^{(E)}(\mathrm{d}F) \, .$$

Here $L_i^d(E)$ is the submanifold of L_i^d which consists of all $F \in L_i^d$ which contain (respectively are contained in) E, and $\nu_i^{(E)}$ is the invariant probability measure on $L_i^d(E)$. There is an obvious orthogonality relation, for functions on L_i^d,

$$R_{i,j} f(E) = R_{d-1,d-j} f^{\perp}(E^{\perp}), \quad E \in L_j^d . \tag{6.2}$$

Exactly as for the spherical transform R (which can be identified with $R_{1,d-1}$ and $R_{d-1,1}$), the Radon transforms $R_{i,j}$, $1 \leq i, j \leq d-1$ [and the orthogonality relation (6.2)], can be extended to distributions. In particular, for a finite (signed) measure ρ on L_i^d, $R_{i,j}\rho$ is a (signed) measure on L_j^d defined as

$$R_{i,j}\rho = \int_{L_i^d} \nu_j^{(F)} \rho(\mathrm{d}F) .$$

From (6.2), the definition of jth projection bodies, and a relation between projection generating measures and surface area measures [see (6.4)], one easily gets

$$R_{j,d-1} v_j(K; \cdot) = \frac{\kappa_j \kappa_{d-1-j}}{\binom{d-1}{j}\kappa_{d-1}}\; h(\Pi_j K; \cdot)^{\perp} \tag{6.3}$$

for all convex bodies K. Alternatively, (6.3) also follows from Hadwiger's characterisation of intrinsic volumes. Now for $i<j$, $R_{i,j}$ is injective if and only if $i+j \leq d$ (Grinberg 1986, Gelfand, Graev and Rosu 1984). So $R_{k,d-1}$ is injective only in the case $k=1$. It follows that $v_1(K; \cdot)$ on L_1^d can be retrieved from $h(\Pi_1 K; \cdot)$ (see Firey 1970, Chakerian 1967). For $2 \leq k \leq d-2$, $R_{k,d-1}$ is injective when restricted to projection functions of centrally symmetric bodies; but it is not known whether this is true for arbitrary projection functions. In addition, it is known that there are bodies K such that if L is centrally symmetric, then

$$v_k(K; \cdot) \neq v_k(L; \cdot) .$$

Notice that this is in contrast to the cases $k=1$, $d-1$; for the latter, see Theorem 5.3. In a positive direction, it can be shown that if P is a polytope and K is any convex body such that

$$R_{k,d-1} v_k(K; \cdot) = R_{k,d-1} v_k(P; \cdot) ,$$

then K is a polytope and $v_k(K; \cdot) = v_k(P; \cdot)$. Of course this is an area where there are many open problems. At the heart of the matter is the unfortunate fact that there is no useful characterisation of projection functions. Some properties of projection functions are studied in Busemann, Ewald and Shephard (1963) and Shephard (1964c,d).

It would be desirable to construct the analogue of (6.1) for arbitrary centrally symmetric convex bodies, thus obtaining a distribution $\rho_k(K; \cdot)$. But the difficulty arises in showing that such a distribution is defined on functions of the form $|E, \cdot|$.

It is an open problem whether (or under which conditions) for a convex body K and $j \in \{2, \ldots, d-2\}$ there is a body L with $v_j(K;\cdot) = v_{d-j}(L;\cdot)^{\perp}$. McMullen (1984, 1987) studied the centred polytopes K that fulfill $v_j(K;\cdot) = v_{d-j}(K;\cdot)^{\perp}$ for all $j \in \{1, \ldots, d-1\}$. By (6.3), the latter equation is equivalent to $\Pi_j K = \Pi_{d-j} K$. For $j=1$, this means $2K = \Pi_{d-1} K$ and hence the result of Weil (1971), described in the last section, leads to a description of these polytopes too.

In the case of generalised zonoids K,

$$\binom{d}{j} S_j(K;\cdot) = d\kappa_{d-j} R_{d-j,1} \rho_j^{\perp}(K;\cdot)\,. \tag{6.4}$$

This is analogous to and a direct consequence of the earlier mentioned formulae for zonoids (Weil 1976a). These projection generating measures also arise in a number of formulae in (translative) integral geometry (Goodey and Weil 1987, 1992c, Weil 1990; see chapter 5.1). In particular, kinematic formulae for projection functions are obtained with the help of these measures (Goodey and Weil 1992b).

A Crofton formula for projection functions is an example of such a result. The homogeneous space of k-flats (affine k-dimensional subspaces of $\mathbb{R}^d$) is denoted by E_k^d and μ_k denotes the (suitably normalised) invariant measure on E_k^d.

Theorem 6.1. *For a centred convex body K and $1 \leq j < k \leq d$,*

$$\int_{E_k^d} v_j(K \cap E;\cdot)\,\mu_k(\mathrm{d}E) = c_{djk} R_{d+j-k,j} v_{d+j-k}(K;\cdot)\,,$$

with some explicitly given constant c_{djk}.

7. Classes of centrally symmetric bodies

In earlier sections it was observed that there is a hierarchy of centrally symmetric bodies based on the nature of their generating distributions. In addition, (4.7) shows that, for $K \in \mathcal{K}_0^d$, the first surface area measure $S_1(K;\cdot)$ can be expressed in terms of $R\rho_K$. Consequently, this hierarchy can be expressed in terms of the nature of the inverse Radon transform of $S_1(K;\cdot)$. It is natural to compare these various classes of centred bodies (zonotopes, zonoids, generalised zonoids) with other more geometrically defined classes.

Such a comparison can be obtained by considering projections. All projections of zonoids are zonoids. Conversely, if all three-dimensional projections of a polytope P are zonotopes, then P is a zonotope. This follows directly from the symmetry property of the 2-faces and was used by Witsenhausen (1978) in his proof of Theorem 3.4. It is therefore natural to ask whether the analogous result for zonoids is true. Weil (1982) showed that this is not the case. Indeed, if K_j^d denotes the class of convex bodies in $\mathcal{K}_0^d$ (of dimension d) for which all projections on j-spaces are zonoids, then all inclusions between these classes

K_j^d, $j = 2, \ldots, d$, are strict. Here

$$\mathsf{K}_d^d = \{K \in \mathscr{Z}^d \colon \dim K = d\}$$

and

$$\mathsf{K}_2^d = \{K \in \mathscr{K}_0^d \colon \dim K = d\} .$$

It would be interesting to have more information about these classes.

The zonoids satisfy (6.1) with a positive projection generating measure $\rho_k(K;\cdot)$ on L_k^d. So another natural generalisation is to consider, for $k = 1, \ldots, d-1$, the class $\mathsf{K}(k)$ of convex bodies $K \in \mathscr{K}_0^d$, for which there is a positive measure $\rho_k(K;\cdot)$ on L_k^d such that

$$v_k(K; E) = \int_{\mathsf{L}_k^d} |E, F| \rho_k(K; \mathrm{d}F) \tag{7.1}$$

for all $E \in \mathsf{L}_k^d$. Then $\mathsf{K}(1) = \mathscr{Z}^d$ and $\mathsf{K}(d-1) = \mathscr{K}_0^d$, which implies

$$\mathsf{K}(1) \subset \mathsf{K}(k) \subset \mathsf{K}(d-1) .$$

Based on the assumption that the measure in (7.1) is unique [the Matheron conjecture, which was disproved in Goodey and Howard (1990)], it was conjectured in Weil (1982) (see also Schneider and Weil 1983) that, for $j = 2, \ldots, d$,

$$\{K \in \mathsf{K}(d-j+1) \colon \dim K = d\} = \{K \in \mathsf{K}_j^d \colon \dim K = d\} .$$

Although this is true for $j = 2, d$, it was shown to be false for all other j in Goodey and Weil (1991). They considered the polytopal members of these classes. On the right side there are only zonotopes. But if $K \in \mathscr{K}(d-j+1)$, then [compare (6.4)]

$$S_{d-j+1}(K;\cdot) = \frac{d\kappa_{j-1}}{\binom{d}{j-1}} R_{j-1,1} \rho_{d-j+1}^{\perp}(K;\cdot) , \tag{7.2}$$

which is the Radon transform of a positive measure. For polytopes K the support of $S_{d-j+1}(K;\cdot)$ comprises the spherical images of the $(d-j+1)$-faces. (7.2) implies that $S_{d-j+1}(K;\cdot)$ is uniformly distributed on great $(j-2)$-spheres. This, in turn, means that the polytopal members of $\mathsf{K}(d-j+1)$ are the polytopes with centrally symmetric $(d-j+2)$-faces. However, McMullen (1970) constructed non-zonotopal polytopes with centrally symmetric facets. Any such polytope is in $\mathsf{K}(d-2)$ but not in K_3^d.

It is however possible to obtain some positive results about the classes $\mathsf{K}(j)$ by using functional analytic techniques similar to those mentioned in section 3. If $\mathscr{M}(\mathsf{L}_j^d)$ denotes the signed measures on L_j^d and

$$S_j^* = \left\{ \rho \in \mathcal{M}(\mathsf{L}_j^d) \colon \int_{\mathsf{L}_k^d} |E, F| \rho(\mathrm{d}F) \geq 0 \text{ for all } E \in \mathsf{L}_j^d \right\},$$

then the following extension of Theorem 3.5 can be found in Goodey and Weil (1991).

Theorem 7.1. *For $1 \leq j \leq d-1$, let N be a set of measures such that S_j^* is the closed convex hull of N. Then $K \in \mathsf{K}(j)$ if and only if*

$$\int_{\mathsf{L}_j^d} v_j(K; E) \rho(\mathrm{d}E) \geq 0$$

for all $\rho \in N$.

By appropriate choice of N one finds that $K \in \mathsf{K}(j)$ if and only if

$$\sum_{i=1}^{n} v_j(K; E_i) \leq V_j(K) \max_{F \in \mathsf{L}_j^d} \sum_{i=1}^{n} |E_i, F|$$

for all $E_1, \ldots, E_n \in \mathsf{L}_j^d$ and all $n = 1, 2, \ldots$.

The nonlinear analogues of Theorem 5.6 lead one to investigate classes such as $\mathsf{Z}(j, k)$ for $1 \leq j \leq d-1$, $1 \leq k \leq d-j$, which comprise those $K \in \mathcal{K}_0^d$ with $\dim K \geq j+1$ and for which

$$V(K, j; L, k; B, d-j-k) \leq V(K, j; M, k; B, d-j-k)$$

whenever $L, M \in \mathcal{K}_0^d$ satisfy

$$v_k^{(d-j)}(L; \cdot) \leq v_k^{(d-j)}(M; \cdot) \quad \text{on } \mathsf{L}_{d-j}^d .$$

For $j = 1$, Weil (1979) showed that $\mathsf{Z}(1, k) = \mathcal{Z}^d$ for $1 \leq k \leq d-1$. It would be interesting to know if, for fixed $j > 1$, the classes $\mathsf{Z}(j, k)$ are independent of k. In any case

$$\{K \in \mathsf{K}(j) \colon \dim K \geq j+1\} \subset \mathsf{Z}(j, k), \quad 1 \leq k \leq d-j .$$

It is also shown in Goodey and Weil (1991) that $K \in \mathsf{Z}(j, 1)$ if and only if $S_j(K; \cdot)$ is the Radon transform $R_{d-j,1}$ of a positive measure on L_{d-j}^d.

8. Zonoids in integral and stochastic geometry

Any positive even measure on S^{d-1} generates a zonoid by way of (2.2). Using the natural identification between the spaces S^{d-1}, L_1^d and L_{d-1}^d, one obtains zonoids corresponding to any line or hyperplane measure. This relation extends to positive measures on E_1^d and E_{d-1}^d under some stationarity (or translational invariance) properties. It is even possible, to associate a zonoid to quite general 1- or $(d-1)$-dimensional sets via the tangent spaces. It is clear that in such a way problems in integral and stochastic geometry can sometimes be formulated in terms of associated zonoids and therefore results from convex geometry can be applied to such general (random) sets too. This section surveys some of these cross connections and gives appropriate references for further reading. Finally, attention is focussed on a geometric problem for zonoids which in turn is motivated by applications in integral and stochastic geometry. Results of an integral geometric or stochastic nature are to be found in chapters 5.1 and 5.2.

Translation invariant hyperplane or line measures are at the heart of Combinatorial Integral Geometry. In this field, the connection with zonoids has been particularly emphasised by Ambartzumian (1987, 1990) and Panina (1988, 1989). The relationship to Hilbert's 4th problem on projective metrics is surveyed in Alexander (1988).

There are various situations where zonoids arise from probability measures. Random points on the sphere and their (random) determinants are considered by Vitale (1988, 1991). Schneider (1982a,b) uses inequalities for zonoids to solve some extremal problems for random lines and random hyperplanes. Similar extremal properties and uniqueness results for point processes of lines or hyperplanes are studied by Matheron (1974b, 1975), Janson and Kallenberg (1981), Mecke (1981), Thomas (1984), and Weil (1987); for point processes of fibres, hypersurfaces and other more general sets, see Mecke (1981), Mecke and Nagel (1980), Mecke and Stoyan (1980), Pohlmann, Mecke and Stoyan (1981), Schneider (1987), and Wieacker (1986, 1989); for random mosaics, see Mecke (1987). In some of these references, explicit use is made of geometric properties of the zonoids generated by the probability measures, in other cases zonoids are not mentioned, although the results allow a geometric interpretation. See also Stoyan, Kendall and Mecke (1987) and Mecke et al. (1990), for surveys.

A slightly different occurrence of zonoids stems from measures which are associated with rather general (random or non-random) sets, for example, surface area measures. Here, zonoids appear as projection bodies or other associated bodies. The most general and systematic treatment of this aspect is due to Wieacker (1986, 1989).

A typical example of the occurrence of zonoids in applied stochastic problems arises when the integral equation (2.2) takes the form

$$\theta(u) = c \int_{S^2} |\langle u, v \rangle| P(\mathrm{d}v) . \tag{8.1}$$

Here $\theta(u)$ is the *intersection density* of a (random) field of 1-dimensional sets in $\mathbb{R}^3$ with a plane with normal u (or of a (random) field of 2-dimensional sets in $\mathbb{R}^3$ with a line in direction u). P is the distribution of the tangential or normal directions of the field (the *directional distribution*). Usually, the 3-dimensional structure is observed in a series of finitely many sections (with directions

$u_1, \ldots, u_n$), from which the corresponding values $\theta(u_1), \ldots, \theta(u_n)$ are determined (more precisely, estimated). The question is to give the approximate shape of P.

Because of the special structure of zonoids, it is a non-trivial problem to approximate a zonoid K (respectively its generating measure ρ_K) on the basis of finitely many support values $\pm h(K; u_1), \ldots, \pm h(K; u_n)$. By intersection of the corresponding supporting halfspaces one gets a centred polytope P_n. In general, however, this will not be a zonotope. Interpolation and smoothing procedures can be used to produce a smooth function agreeing with $h(K; \cdot)$ at $\pm u_1, \ldots, \pm u_n$. This smooth function will be the support function $h(K_n; \cdot)$ of a generalised zonoid. But again, K_n need not be a zonoid, and then the inversion $T^{-1}h(K_n; \cdot)$ of the integral equation (which is possible with spherical harmonics expansion) leads to the signed measure ρ_{K_n} which might be far from ρ_K. This instability of T^{-1} on the set $\mathscr{Z}^d_{\text{sign}}$ of centred generalised zonoids with respect to the weak topology on $\mathscr{M}_e(S^{d-1})$ has apparently been overlooked in a number of practically orientated papers (Hilliard 1962, Philofsky and Hilliard 1969, Kanatani 1984). Functional analytic approaches for a solution are described in Coleman (1989). The 2-dimensional version of this problem is, of course, very different, since here T^{-1} is stable. In fact, any centred symmetric body $K' \subset \mathbb{R}^2$ with $h(K'; u_i) = h(K; u_i)$, $i = 1, \ldots, n$, can be used for the approximation of ρ_K. A practical procedure of this type has been described in Rataj and Saxl (1989).

The estimation problem described above is obviously connected with the problem of approximating a zonoid by zonotope (with a given maximum number of segments, say). Theoretical results in this direction have been obtained by Betke and McMullen (1983), Bourgain and Lindenstrauss (1988b), Bourgain, Lindenstrauss and Milman (1989), and Linhart (1989). In this connection, it is interesting to mention the following simple result: let $K \subset \mathbb{R}^d$ be a zonoid and let $u_1, \ldots, u_n \in S^{d-1}$ be given directions. Then there exists a zonotope P_n with $h(P_n; u_i) = h(K; u_i)$, $i = 1, \ldots, n$. In fact, a linear programming argument (or, as J. Bourgain pointed out to us, Carathéodory's theorem) shows that P_n can be chosen to be a sum of n segments.

References

Aleksandrov, A.D.

[1937] Zur Theorie der gemischten Volumina von konvexen Körpern, II. Neue Ungleichungen zwischen den gemischten Volumina und ihre Anwendungen (in Russian), *Mat. Sb. N.S.* **2**, 1205–1238.

Alexander, R.

[1988] Zonoid theory and Hilbert's fourth problem, *Geom. Dedicata* **28**, 199–211.

Alfsen, E.M.

[1971] *Compact Convex Sets and Boundary Integrals* (Springer, Berlin).

Ambartzumian, R.V.

[1987] Combinatorial integral geometry, metrics, and zonoids, *Acta Appl. Math.* **9**, 3–27.

[1990] *Factorization Calculus and Geometrical Probability* (Cambridge Univ. Press, Cambridge).

Assouad, P.
[1980] Charactérisations de sous-espaces normés de L^1 de dimension finie, *Seminaire d'analyse fonctionelle (École Polytechnique Palaiseau), 1979-1980*, exposé No 19.

Ball, K.M.
[1986] Cube slicing in R^n, *Proc. Amer. Math. Soc.* **97**, 456–473.
[1988] Some remarks on the geometry of convex sets, in: *Geometric Aspects of Functional Analysis,* eds J. Lindenstrauss and V.D. Milman, Lecture Notes in Mathematics, Vol. 1317 (Springer, New York) pp. 224–231.

Berg, C.
[1969] Corps convexes et potentiels sphériques, *Danske Vid. Selsk., Mat.-Fys. Medd.* **37**(b), 1–64.

Betke, U., and P. McMullen
[1983] Estimating the sizes of convex bodies from projections, *J. London Math. Soc.* **27**, 525–538.

Blaschke, W.
[1916] *Kreis und Kugel* (Veit, Leipzig). 2nd Ed.: De Gruyter, Berlin, 1956.
[1923] *Vorlesungen über Differentialgeometrie, II. Affine Differentialgeometrie* (Springer, Berlin).

Blaschke, W., and K. Reidemeister
[1922] Über die Entwicklung der Affingeometrie, *Jber. Deutsch. Math.-Vereinig.* **31**, 63–82.

Bolker, E.D.
[1969] A class of convex bodies, *Trans. Amer. Math. Soc.* **145**, 323–346.
[1971] The zonoid problem, *Amer. Math. Monthly* **78**, 529–531.

Bonnesen, T., and W. Fenchel
[1934] *Theorie der konvexen Körper* (Springer, Berlin).

Bourgain, J., and J. Lindenstrauss
[1988a] Projection bodies, in: *Geometric Aspects of Functional Analysis*, eds J. Lindenstrauss and V.D. Milman, Lecture Notes in Mathematics, Vol. 1317 (Springer, New York) pp. 250–270.
[1988b] Distribution of points on spheres and approximation by zonotopes, *Israel J. Math.* **64**, 25–31.

Bourgain, J., J. Lindenstrauss and V. Milman
[1989] Approximation of zonoids by zonotopes, *Acta Math.* **162**, 73–141.

Bretagnolle, J., D. Dacunha Castelle and J.L. Krivine
[1966] Lois stables et espaces L^p, *Ann. Inst. H. Poincaré N.S.* **2**, 231–259.

Busemann, H., G. Ewald and G.C. Shephard
[1963] Convex bodies and convexity on Grassmann cones I–IV, *Math. Ann.* **151**, 1–41.

Campi, S.
[1986] Reconstructing a convex surface from certain measurements of its projections, *Boll. Un. Mat. Ital. B* **5**, 945–959.

Chakerian, G.D.
[1967] Sets of constant relative width and constant relative brightness, *Trans. Amer. Math. Soc.* **129**, 26–37.

Choquet, G.
[1969a] *Lectures on Analysis,* Vol. III (W.A. Benjamin, Reading, MA).
[1969b] Mesures coniques, affines et cylindriques, *Symposia Math., Vol. II (INDAM, Rome 1968)* (Academic Press, London).

Coleman, R.
[1989] Inverse problems, *J. Microscopy* **153**, 233–248.

Coxeter, H.S.M.
[1963] *Regular Polytopes* (Macmillan, New York, 2nd edition).

Falconer, K.J.
[1983] Applications of a result on spherical integration to the theory of convex sets, *Amer. Math. Monthly* **90**, 690–693.

Firey, W.J.
[1967] The determination of convex bodies from their mean radius of curvature functions, *Mathematika* **14**, 1–13.
[1968] Christoffel's problem for general convex bodies, *Mathematika* **15**, 7–21.
[1970] Convex bodies with constant outer p-measure, *Mathematika* **17**, 21–27.

Funk, P.
[1913] Über Flächen mit lauter geschlossenen geodätischen Linien, *Math. Ann.* **74**, 278–300.

Gelfand, I.M., M.I. Graev and R. Rosu
[1984] The problem of integral geometry and intertwining operators for a pair of real Grassmannian manifolds, *J. Operator Theory* **12**, 359–383.

Goodey, P.R.
[1977] Centrally symmetric convex sets and mixed volumes, *Mathematika* **24**, 193–198.
[1986] Instability of projection bodies, *Geom. Dedicata* **20**, 295–305.

Goodey, P.R., and H. Groemer
[1990] Stability results for first order projection bodies, *Proc. Amer. Math. Soc.* **109**, 1103–1114.

Goodey, P.R., and R. Howard
[1990] Processes of flats induced by higher dimensional processes, *Adv. in Math.* **80**, 92–109.

Goodey, P.R., and W. Weil
[1987] Translative integral formulae for convex bodies, *Aequationes Math.* **34**, 64–77.
[1991] Centrally symmetric convex bodies and Radon transforms on higher order Grassmannians, *Mathematika* **38**, 117–133.
[1992a] Centrally symmetric convex bodies and the spherical Radon transform, *J. Differential Geom.* **35**, 675–688.
[1992b] Integral geometric formulae for projection functions, *Geom. Dedicata* **41**, 117–126.
[1992c] The determination of convex bodies from the mean of random sections, *Math. Proc. Camb. Phil. Soc.*, to appear.

Grinberg, E.
[1986] Radon transforms on higher rank Grassmannians, *J. Differential Geom.* **24**, 53–68.

Helgason, S.
[1959] Differential operators on homogeneous spaces, *Acta Math.* **102**, 239–299.
[1980] *The Radon Transform* (Birkhäuser, Boston).
[1984] *Groups and Geometric Analysis* (Academic Press, Orlando, FL).
[1990] The totally-geodesic Radon transform on constant curvature spaces, in: *Integral Geometry and Tomography*, eds E. Grinberg and E.T. Quinto, Contemp. Math., Vol. 113, pp. 141–149.

Herz, C.S.
[1963] A class of negative definite functions, *Proc. Amer. Math. Soc.* **14**, 670–676.

Hilliard, J.E.
[1962] Specification and measurement of microstructural anisotropy, *Trans. Amer. Inst. Mining Metallurg. Engrg.* **224**, 1201–1211.

Janson, S., and O. Kallenberg
[1981] Maximizing the intersection density of fibre processes, *J. Appl. Probab.* **18**, 820–828.

Kanatani, K.
[1984] Stereological determination of structural anisotropy, *Internat. J. Engrg. Sci.* **22**, 531–546.

Larman, D.G., and C.A. Rogers
[1975] The existence of a centrally symmetric convex body with central sections that are unexpectedly small, *Mathematika* **22**, 164–175.

Lévy, P.
[1937] *Théorie de l'Addition des Variables Aléatoires* (Gauthier-Villars, Paris).

Lindquist, N.F.
[1975a] Approximation of convex bodies by sums of line segments, *Portugal. Math.* **34**, 233–240.
[1975b] Support functions of central convex bodies, *Portugal. Math.* **34**, 241–252.

Linhart, J.
[1989] Approximation of a ball by zonotopes using uniform distribution on the sphere, *Arch. Math.* **53**, 82–86.

Lutwak, E.
[1988] Intersection bodies and dual mixed volumes, *Adv. in Math.* **71**, 232–261.

Matheron, G.
[1974a] Un theoreme d'unicite pour les hyperplans poissoniens, *J. Appl. Probab.* **11**, 184–189.
[1974b] Hyperplans poissoniens et compacts de Steiner, *Adv. in Appl. Probab.* **6**, 563–579.
[1975] *Random Sets and Integral Geometry* (Wiley, New York).

McMullen, P.
[1970] Polytopes with centrally symmetric faces, *Israel J. Math.* **8**, 194–196.
[1984] Volumes of projections of unit cubes, *Bull. London Math. Soc.* **16**, 278–280.
[1987] Volumes of complementary projections of convex polytopes, *Monatsh. Math.* **104**, 265–272.

Mecke, J.
[1981] Formulas for stationary planar fibre processes III – Intersections with fibre systems, *Math. Operationsforsch. Statist. Ser. Statist.* **12**, 201–210.
[1987] Extremal properties of some geometric processes, *Acta Appl. Math.* **9**, 61–69.

Mecke, J., and W. Nagel
[1980] Stationäre räumliche Faserprozesse und ihre Schnittzahlrosen, *Elektron. Informationsverarb. Kybernet.* **16**, 475–483.

Mecke, J., and D. Stoyan
[1980] Formulas for stationary planar fibre processes I – General theory, *Math. Operationsforsch. Statist. Ser. Statist.* **11**, 267–279.

Mecke, J., R. Schneider, D. Stoyan and W. Weil
[1990] *Stochastische Geometrie* (Birkhäuser, Basel).

Minkowski, H.
[1911] Theorie der konvexen Körper, insbesondere Begründung ihres Oberflächenbegriffs, *Ges. Abh.*, Vol. II (Teubner, Leipzig) pp. 131–229.

Panina, Y.
[1988] The representation of an n-dimensional body in the form of a sum of $(n-1)$-dimensional bodies (in Russian), *Izv. Akad. Nauk Armjan. SSR Ser. Mat.* **23**, 385–395 [*Soviet J. Contemp. Math. Anal.* **23**, 91–103].
[1989] Convex bodies integral representations, in: *Geobild '89*, eds A. Hübler, W. Nagel, B.D. Ripley and G. Werner, Mathematical Res., Vol. 51 (Akademie–Verlag, Berlin) pp. 201–204.

Petty, C.M.
[1961] Centroid surfaces, *Pacific J. Math.* **11**, 1535–1547.
[1967] Projection bodies, in: *Proc. Coll. Convexity, Copenhagen 1965* (Københavns Univ. Mat. Inst.) pp. 234–241.

Philofsky, E.M., and J.E. Hilliard
[1969] On the measurements of the orientation distribution of lineal and areal arrays, *Quart. Appl. Math.* **27**, 79–86.

Pohlmann, S., J. Mecke and D. Stoyan
[1981] Formulas for stationary surface processes, *Math. Operationsforsch. Statist. Ser. Statist.* **12**, 429–440.

Quinto, E.T.
[1983] The invertibility of rotation invariant Radon transforms, *J. Math. Anal. Appl.* **91**, 510–522.

Rataj, J., and I. Saxl
[1989] Analysis of planar anisotropy by means of the Steiner compact, *J. Appl. Probab.* **26**, 490–502.

Rickert, N.W.
[1967a] Measures whose range is a ball, *Pacific J. Math.* **23**, 361–371.

[1967b] The range of a measure, *Bull. Amer. Math. Soc.* **73**, 560–563.

Schneider, R.

[1967] Zu einem Problem von Shephard über die Projektionen konvexer Körper, *Math. Z.* **101**, 71–82.

[1969] Functions on a sphere with vanishing integrals over certain subspheres, *J. Math. Anal. Appl.* **26**, 381–384.

[1970a] Über eine Integralgleichung in der Theorie der konvexen Körper, *Math. Nachr.* **44**, 55–75.

[1970b] On the projections of a convex polytope, *Pacific J. Math.* **32**, 799–803.

[1970c] Functional equations connected with rotations and their geometric applications, *Enseign. Math.* **16**, 297–305.

[1977] Rekonstruktion eines konvexen Körpers aus seinen Projektionen, *Math. Nachr.* **79**, 325–329.

[1982a] Random hyperplanes meeting a convex body, *Z. Wahrscheinlichkeitsth. Verw. Geb.* **61**, 379–387.

[1982b] Random polytopes generated by anisotropic hyperplanes, *Bull. London Math. Soc.* **14**, 549–553.

[1987] Geometric inequalities for Poisson processes of convex bodies and cylinders, *Results Math.* **11**, 165–185.

Schneider, R., and W. Weil

[1970] Über die Bestimmung eines konvexen Körpers durch die Inhalte seiner Projektionen, *Math. Z.* **116**, 338–348.

[1983] Zonoids and related topics, in: *Convexity and its Applications,* eds P. Gruber and J.M. Wills (Birkhäuser, Basel) pp. 296–317.

Schoenberg, I.J.

[1938] Metric spaces and positive definite functions, *Trans. Amer. Math. Soc.* **44**, 522–536.

Semyanistyi, V.I.

[1961] Homogeneous functions and some problems of integral geometry in spaces of constant curvature (in Russian), *Dokl. Akad. Nauk SSSR* **136**, 228–291 [*Soviet Math. Dokl.* **2**, 59–62].

Shephard, G.C.

[1964a] Approximation problems for convex polyhedra, *Mathematika* **11**, 9–18.

[1964b] Shadow systems of convex bodies, *Israel J. Math.* **2**, 229–236.

[1964c] Convex bodies and convexity on Grassmann cones VI: The projection functions of a simplex, *J. London Math. Soc.* **39**, 307–319.

[1964d] Convex bodies and convexity on Grassmann cones VII: Projection functions of vector sums of convex sets, *J. London Math. Soc.* **39**, 417–423.

Stoyan, D., W.S. Kendall and J. Mecke

[1987] *Stochastic Geometry and its Applications* (Akademie-Verlag/Wiley, Berlin/New York).

Strichartz, R.

[1981] L^p estimates for Radon transforms in Euclidean and non-Euclidean spaces, *Duke Math. J.* **48**, 699–727.

Thomas, C.

[1984] Extremum properties of the intersection densities of stationary Poisson hyperplane processes, *Math. Operationsforsch. Statist. Ser. Statist.* **15**, 443–449.

Vitale, R.A.

[1988] An alternate formulation of mean value for random geometric figures, *J. Microscopy* **151**, 197–204.

[1991] Expected absolute random determinants and zonoids, *Ann. Appl. Probab.* **1**, 293–300.

Weil, W.

[1971] Über die Projektionenkörper konvexer Polytope, *Arch. Math.* **22**, 664–672.

[1974] Über den Vektorraum der Differenzen von Stützfunktionen konvexer Körper, *Math. Nachr.* **59**, 353–369.

[1976a] Kontinuierliche Linearkombination von Strecken, *Math. Z.* **148**, 71–84.

[1976b] Centrally symmetric convex bodies and distributions, *Israel J. Math.* **24**, 352–367.
[1977] Blaschkes Problem der lokalen Charakterisierung von Zonoiden, *Arch. Math.* **29**, 655–659.
[1979] Centrally symmetric convex bodies and distributions, II, *Israel J. Math.* **32**, 173–182.
[1982] Zonoide und verwandte Klassen konvexer Körper, *Monatsh. Math.* **94**, 73–84.
[1987] Point processes of cylinders, particles, and flats, *Acta Appl. Math.* **9**, 103–136.
[1990] Iterations of translative integral formulae and nonisotropic Poisson processes of particles, *Math. Z.* **205**, 531–549.

Wieacker, J.A.
[1986] Intersections of random hypersurfaces and visibility, *Probab. Theory Related Fields* **71**, 405–433.
[1989] Geometry inequalities for random surfaces, *Math. Nachr.* **142**, 73–106.

Witsenhausen, H.S.
[1973] Metric inequalities and the zonoid problem, *Proc. Amer. Math. Soc.* **40**, 517–520.
[1978] A support characterization of zonotopes, *Mathematika* **25**, 13–16.

CHAPTER 4.10

Baire Categories in Convexity

Peter M. GRUBER

Abteilung für Analysis, Technische Universität Wien, Wiedner Hauptstraße 8–10, A-1040 Wien, Austria

Contents

HANDBOOK OF CONVEX GEOMETRY
Edited by P.M. Gruber and J.M. Wills

1. Introduction and basic definitions

Let $\mathcal{K} = \mathcal{K}(\mathbb{E}^d)$ be the space of *(proper) convex bodies* in $\mathbb{E}^d$; these are compact convex subsets of $\mathbb{E}^d$ with non-empty interior. It is a natural question to ask for a tool to distinguish between "large" and "small" subsets of $\mathcal{K}$ or of more general spaces. For a discussion of this problem see chapter 1.9. Here we consider the topological tool of Baire categories which in recent years gave rise to a large number of partly rather surprising convexity results.

A topological space X is called *Baire* if each of its meager sets has a dense complement; a subset of X is *meager* or of *first (Baire) category* if it is a countable union of nowhere dense sets. All other sets are called *non-meager* or of *second (Baire) category*; in particular, a set with meager complement is called *residual.* When speaking of *most, typical* or *generic* elements of a Baire space we mean all elements with a meager set of exceptions. One version of the category theorem of Baire (1899) (see also Osgood 1900) shows that any complete metric and any locally compact space is Baire, compare Holmes (1975) and Oxtoby (1971). The notion of Baire categories was introduced for the investigation of real functions and has ever since proved a valuable and frequently used tool in real and functional analysis.

Let the space $\mathcal{C} = \mathcal{C}(\mathbb{E}^d)$ of all (non-empty) compact subsets of $\mathbb{E}^d$ be endowed with its natural topology (from our point of view). It is induced by, for example, the *Hausdorff metric* δ^{H}. (Let $C, D \in \mathcal{C}$. Then $\delta^{\mathrm{H}}(C, D)$ is the maximum Euclidean distance which a point of one of the sets C, D can have from the other set.) A more general version of the selection theorem of Blaschke (1916) (see, e.g., Falconer 1985) shows that $\mathcal{C}$ is a boundedly compact complete metric space. Its subspace $\mathcal{K}$ is locally compact. Thus $\mathcal{C}$ and $\mathcal{K}$ both are Baire.

Let $\dim, \mathrm{bd}, \mathrm{int}, \mathrm{conv}$ stand for dimension, boundary, interior and convex hull. S^{d-1} denotes the $(d-1)$-dimensional unit sphere in $\mathbb{E}^d$ and $\|\cdot\|$ the Euclidean norm.

2. A typical proof of a Baire category type result in convexity

In the following a Baire category type result in convexity together with its proof will be presented. Although almost trivial, the proof displays the main characteristics of many proofs of such results in convexity. Sometimes it is only due to the formulation of the proofs that these characteristics are not easily recognizable.

The *affine surface* area $A(C)$ of a convex body $C \in \mathcal{K}$ (with boundary) of (differentiability) class $\mathcal{C}^2$ was introduced by Blaschke (1923) in the context of affine differential geometry. His definition was extended to all convex bodies by Leichtweiss (1988). Using properties of this more general notion of affine surface area due to Leichtweiss we will prove the following result.

Theorem 1. *For most convex bodies $C \in \mathcal{K}$ the extended affine surface area $A(C)$ is* 0.

Proof. First, for $n = 1, 2, \ldots$, let

$$\mathcal{K}_n = \{C \in \mathcal{K}: A(C) \geqslant 1/n\}. \tag{1}$$

Then clearly

$$\{C \in \mathcal{K}: A(C) > 0\} \subset \bigcup \mathcal{K}_n. \tag{2}$$

Second, we show that

$$\mathcal{K}_n \text{ is nowhere dense in } \mathcal{K}. \tag{3}$$

To see this it is sufficient to show that $\mathcal{K}_n$ is closed and has empty interior. The closedness of $\mathcal{K}_n$ follows from the definition of $\mathcal{K}_n$ since the extended affine surface area is upper semicontinuous on $\mathcal{K}$ by a result of Leichtweiss (1988). The interior of $\mathcal{K}_n$ must be empty. Otherwise $\mathcal{K}_n$ contains polytopes (which are dense in $\mathcal{K}$). Since, again by Leichtweiss (1988), each polytope has extended affine surface area 0, this contradicts (1).

Third, (3) implies that

$$\bigcup \mathcal{K}_n \text{ is meager.}$$

Together with (2) this yields Theorem 1. See section 4 for a different proof. □

Some comments are in order. If one wants to prove that most convex bodies have a certain property $\mathcal{P}$, then in general one proceeds as in the proof of Theorem 1. First, one represents the set of convex bodies not having property $\mathcal{P}$ as a countable union of sets of convex bodies $\mathcal{K}_n$. Second, it is shown that each $\mathcal{K}_n$ is nowhere dense in $\mathcal{K}$. For this purpose one may prove first that $\mathcal{K}_n$ is closed, using some semicontinuity property of a function related to $\mathcal{P}$, and then show that $\mathcal{K}_n$ has empty interior. In general the latter is the most difficult step. It can be achieved by constructing a set of convex bodies having a, possibly, stronger property – depending on $\mathcal{K}_n$ – than property $\mathcal{P}$ which is dense in $\mathcal{K}$. Third, the first two steps together show that all convex bodies have property $\mathcal{P}$ with the possible exception of those contained in the meager set $\bigcup \mathcal{K}_n$.

Since many Baire category type results exhibit rather unexpected features of typical convex bodies, a first – and often difficult – problem is to discover them.

3. Boundary properties of arbitrary convex bodies

A boundary point x of a convex body C is *smooth* if there is a unique supporting hyperplane of C containing x. Call C *smooth* if all its boundary points are smooth. This is equivalent to the condition that C is of class $\mathscr{C}^1$. A finite-dimensional version of a result of Mazur (1933) (see Holmes 1975) is the following.

Theorem 2. *For any $C \in \mathcal{K}$ most $x \in \operatorname{bd} C$ are smooth.*

This is in accordance with a measure-theoretic result of Reidemeister (1921) and its more precise form due to Anderson and Klee (1952) which says that the set of non-smooth points in $\operatorname{bd} C$ are of σ-finite $(d-2)$-dimensional (Hausdorff) measure. Aleksandrov (1939) proved that the points where $\operatorname{bd} C$ is not twice differentiable is of $(d-1)$-dimensional measure 0. As may be seen from Theorem 7 below, there is no corresponding category result to Aleksandrov's theorem. A result of Asplund (1968) contains Theorem 2 as a special case. Asplund's theorem was refined by several authors. In order to state a result of Preiss and Zajiček (1984a,b) we need to introduce a concept related to Baire categories due to Dolženko (1967) which in essence goes back to Denjoy.

Let X be a metric space and $S \subset X$. The *porosity* of S at $x \in S$ is the limit superior as $\varepsilon \to +0$ of $\varphi(\varepsilon)/\varepsilon$ where $\varphi(\varepsilon)$ is the supremum of the radii of the open balls disjoint from S whose centers have distance at most ε from x. S is *porous* if it has positive porosity at any of its points and *σ-porous* if it is a countable union of porous sets. By *nearly all* elements of X all elements of X are meant, except those in a σ-porous set.

A porous set is nowhere dense in X and if X is a metric Baire space, any σ-porous set is meager. It has been proved by Gandini and Zucco (1991) that in $\mathcal{K}$ there are meager subsets which are not σ-porous. If μ is a measure on X and X and μ satisfy certain conditions, then any measurable σ-porous set has measure 0. Such conditions may be obtained from Zaanen (1967, chapter 8). In particular this holds for $X = \mathbb{E}^d$ and $\mu =$ Lebesgue measure, according to the density theorem of Lebesgue. Preiss and Zajiček (1984a,b) proved the following result; for a refinement see their paper (1984b).

Theorem 3. *For any $C \in \mathcal{K}$ nearly all $x \in \operatorname{bd} C$ are smooth.*

We state the following theorem without proof. It is related to a result of Ewald, Larman and Rogers (1970) in the same way as Theorem 2 is related to the result of Reidemeister and Anderson and Klee, respectively.

Theorem 4. *For any $C \in \mathcal{K}$ the set of $u \in S^{d-1}$ parallel to line segments on $\operatorname{bd} C$ is meager.*

For results on the face-function of a convex body see Klee and Martin (1971).

4. Smoothness and strict convexity

In the following several results for convex bodies are quoted. In some cases analogous results for norms or convex functions can be found in the literature; see, for example, Gruber (1977) and Fabian, Zajiček and Zizler (1982). In another case the result originally was given for convex function, compare Klima and Netuka (1981).

Theorem 5. *Most $C \in \mathcal{K}$ are smooth, i.e., $C \in \mathcal{C}^1$, and strictly convex.*

This result constitutes the first example of a Baire category result on the space of convex bodies. Theorem 5 was first discovered by Klee (1959), but completely forgotten. It was independently rediscovered by Gruber (1977) and (concerning the smoothness part) by Schneider, Choquet (unpublished) and Howe (1982). This led to the great interest in Baire type results since and, in particular, to the following investigations of smoothness properties.

First, the author (1977) showed that most convex bodies are not of class $\mathcal{C}^2$ and are not "very strictly convex". Then Klima and Netuka (1981) proved that most convex bodies are not of class $\mathcal{C}^{1+\varepsilon}$ for any $\varepsilon > 0$ at "many" points. (They state their result in terms of convex functions f and show that for a typical f the partial derivatives do not satisfy a Hoelder condition of the form $|f_{,i}(x) - f_{,i}(y)| \leqslant \text{const}\|x - y\|^{\tau}$ for any $\tau > 0$.)

Second, Schneider (1979) and Zamfirescu (1980c,d) proved that a typical convex body has quite unexpected curvature properties. Without giving precise definitions we state the following results of Zamfirescu:

Theorem 6. *For most $C \in \mathcal{K}$, each $x \in \operatorname{bd} C$ and each tangent direction t of C at x the lower sectional curvature of $\operatorname{bd} C$ at x in direction t is 0 or the upper sectional curvature is ∞ or both.*

Thus for most $C \in \mathcal{K}$ at any $x \in \operatorname{bd} C$ and any tangent direction of $\operatorname{bd} C$ at x for which the sectional curvature of $\operatorname{bd} C$ exists, it is 0. By Aleksandrov's (1939) theorem cited in section 3 this holds for almost all $x \in \operatorname{bd} C$ and any tangent direction. By the definition of a generalized affine surface area due to Schütt and Werner, this implies that for most $C \in \mathcal{K}$ the generalized affine surface area is 0. Since it was recently shown that the notions of generalized surface area due to Schütt–Werner and Leichtweiß coincide, we obtain again Theorem 1.

Theorem 7. *For most $C \in \mathcal{K}$ at most $x \in \operatorname{bd} C$ and any tangent direction t of C at x the lower sectional curvature of C at x in direction t is 0 and the upper sectional curvature is ∞.*

For a thorough discussion of the relation of the sectional curvatures in opposite tangent directions in the spirit of Theorems 6, 7, see Zamfirescu (1988b).

Third, Zamfirescu (1987) gave the following refinement of Theorem 5 (but see the remark in section 3).

Theorem 8. *Nearly all $C \in \mathcal{K}$ are smooth and strictly convex.*

Fourth, a consequence of Theorem 6 and of Aleksandrov's theorem mentioned before implies that for most $C \in \mathcal{K}$ the set $F(C)$ of *farthest points* of C has $(d-1)$-dimensional measure 0 where $x \in \operatorname{bd} C$ is called a farthest point of C if there is a $p \in \mathbb{E}^d$ such that $\|p-x\| = \sup\{\|p-y\|: y \in C\}$. Results of Schneider and Wieacker (1981) and Zamfirescu (1988b) yield the next theorem:

Theorem 9. *For most $C \in \mathcal{K}$ the set $F(C)$ has Hausdorff dimension* 0 *and is meager.*

Zamfirescu (1988b) also considers nearest points of $\operatorname{bd} C$ for $x \in \operatorname{int} C$.

5. Geodesics

A *geodesic segment* on the boundary of a convex body C is a continuous curve on $\operatorname{bd} C$ connecting two points of $\operatorname{bd} C$ and having minimum length among all such curves. For any two points of $\operatorname{bd} C$ there is at least one geodesic segment connecting them. A *geodesic* is a continuous curve on $\operatorname{bd} C$ which locally consists of geodesic segments. For C of class $\mathscr{C}^2$ the concepts of geodesics in convexity and differential geometry coincide, as follows from a result of Siegel (1957). The standard treatise on geodesics in convexity is Aleksandrov (1948).

The next three theorems are due to Zamfirescu (1982b, 1992):

Theorem 10. *For most $C \in \mathcal{K}$ through most $x \in \operatorname{bd} C$ there passes no geodesic. If $d = 3$ then for most $C \in \mathcal{K}$ and any $x \in \operatorname{bd} C$ in most tangent directions of* $\operatorname{bd} C$ *at x there starts no geodesic.*

Theorem 11. *For most $C \in \mathcal{K}(\mathbb{E}^3)$ and any $n = 1, 2, \ldots,$ there is a dense set of pairs (x, t) where $x \in \operatorname{bd} C$ and t is tangent vector of length* 1 *of C at x such that there is a geodesic of length $2n$ with midpoint x and direction t at x.*

Theorem 12. *For most $C \in \mathcal{K}(\mathbb{E}^3)$ there are geodesics of arbitrary lengths without self-intersections.*

Theorem 13. *On most $C \in \mathcal{K}(\mathbb{E}^3)$ there is no closed geodesic, even admitting self-intersections.*

The latter result was obtained by Gruber (1991). It contrasts well-known classical theorems on the existence of closed geodesics on sufficiently often differentiable convex surfaces, see, e.g., Lyusternik and Schnirel'man (1929), Klingenberg (1976) and Hingston (1984). For a weaker earlier version of Theorem 13 see Gruber (1988b).

In a paper not yet published Gruber (1992b) derived the following result.

Theorem 14. *For $C \in \mathcal{K}$ each point $x \in \operatorname{bd} C$ is connected with nearly all points $y \in \operatorname{bd} C$ by a unique geodesic segment.*

The proof of this result was achieved through several consecutive steps. First the author (1988b) showed that for most $C \in \mathcal{K}$ most pairs $(x, y) \in \operatorname{bd} C \times \operatorname{bd} C$ are connected by a unique geodesic segment. Then Zamfirescu noted that this holds in case $d = 3$ for any $C \in \mathcal{K}$. The next step was Zamfirescu's (1991c) proof of Theorem 14 for $d = 3$.

6. Billiards

A *billiard table* C in $\mathbb{E}^d$ is a smooth convex body. A *billiard ball* is a point in C which moves with unit velocity along a straight line in C until it hits $\operatorname{bd} C$ where it is reflected in the usual way. The curve described by a billiard ball is a (*billiard*) *trajectory*. It clearly can be described by the sequence of *vertices* $\dots, p_{-1}, p_0, p_1, \dots$, i.e., the points where it hits $\operatorname{bd} C$. The *phase space* $\operatorname{ph} C$ of C consists of all pairs (p, v) where $p \in \operatorname{bd} C$ and $v \in S^{d-1}$ points from p into $\operatorname{int} C$. Obviously, each trajectory can also be described by a sequence $\dots, (p_{-1}, v_{-1}), (p_0, v_0), (p_1, v_1), \dots$ in $\operatorname{ph} C$. The following Theorems 15–18 are due to the author (1990).

Call a compact convex set $K \subset \operatorname{int} C$ a *caustic* of the billiard table C if any trajectory which touches K once touches K again after each reflection. Caustics were investigated by Minasian (1973), Lazutkin (1979) and Turner (1982). They are related to the eigenfunction problem of the Laplace operator.

Theorem 15. *Most billiard tables contain no caustic.*

In $\mathbb{E}^2$ there is a dense set of billiard tables containing caustics, whereas for $d \geqslant 3$ we conjecture that only the ellipsoids contain caustics and their caustics are precisely the confocal ellipsoids. A partial positive result in this direction has been obtained by Berger (1990).

Halpern (1977) showed that there are billiard tables C in $\mathbb{E}^2$ containing a trajectory with vertices $\dots, p_{-1}, p_0, p_1, \dots$, say, the length of which is finite, i.e.,

$$\|p_0 - p_1\| + \|p_1 - p_2\| + \cdots < +\infty.$$

We then say that the trajectory *terminates* on $\operatorname{bd} C$.

Theorem 16. *On most billiard tables no trajectory terminates on the boundary.*

Theorem 17. *Let C be a billiard table. Then the set of elements $(p, v) \in \operatorname{ph} C$ such that the trajectory starting at p in direction v terminates on $\operatorname{bd} C$ is meager.*

Density results for billiards were given, among others, by Zemlyakov and Katok (1975).

Theorem 18. *For most billiard tables C in $\mathbb{E}^2$ for most $(p, v) \in \operatorname{ph} C$ the trajectory starting at p in direction v is dense in C.*

For further results see Gruber (1990).

7. Normals, mirrors and diameters

The problem of intersecting normals of a convex body of class $\mathscr{C}^1$ has attracted some interest, see, e.g., Heil (1985). A surprising result in this context is due to Zamfirescu (1982a, 1984c):

Theorem 19. *For most $C \in \mathcal{K}$ (which are smooth by Theorem 5) most $x \in \mathbb{E}^d$ lie on infinitely many normals of* bd C.

$x \in \mathbb{E}^d$ *sees the mirror image of* $y \in \mathbb{E}^d$ with respect to bd C where $C \in \mathcal{K}$ is smooth if there is a point $z \in \text{bd}\, C$ such that the interior normal of bd C at z bisects the angle of the triangle xyz at z. A result of Zamfirescu (1982a) shows that "most mirrors are magic":

Theorem 20. *For most $C \in \mathcal{K}(\mathbb{E}^2)$ (which are smooth by Theorem 5) most $x \in \mathbb{E}^2$ can see infinitely many mirror images with respect to* bd C *of any given point y in* $\mathbb{E}^2$.

A *diameter* of a convex body C is a line segment with endpoints $x, y \in \text{bd}\, C$, such that there is a parallel strip containing C with x, y on its boundary hyperplanes. For results on diameters see Hammer and Sobczyk (1953) and Kosinski (1958). The following results are taken from Bárány and Zamfirescu (1990). Theorem 22 was proved earlier by Zamfirescu (1984b) for $d = 2$.

Theorem 21. *For most $C \in \mathcal{K}$ on most diameters each point belongs to infinitely many other diameters.*

Theorem 22. *For most $C \in \mathcal{K}$ most points belong to infinitely many diameters.*

8. Approximation of convex bodies by polytopes

Let δ be a metric or another measure of deviation on the space $\mathcal{K}$ of convex bodies or on subspaces of $\mathcal{K}$ such as $\mathcal{K}_o = \{C \in \mathcal{K}\colon C = -C\}$. Examples are the Hausdorff metric δ^{H}, the *symmetric difference metric* δ^{S} and the (multiplicative) *Banach–Mazur distance* δ^{BM} (not a metric) on $\mathcal{K}_o$. δ^{S} and δ^{BM} are defined by

$$\delta^{\mathrm{S}}(C, D) = V(C \triangle D) \quad \text{for } C, D \in \mathcal{K},$$

$$\delta^{\mathrm{BM}}(C, D) = \inf\{1 + \varepsilon \geqslant 1\colon C \subset l(D) \subset (1 + \varepsilon)C, l : \mathbb{E}^d \to \mathbb{E}^d \text{ linear}\}$$
$$\text{for } C, D, \in \mathcal{K}_0.$$

It is of interest to investigate for a convex body

$$\delta(C, \mathfrak{Q}) = \inf\{\delta(C, P)\colon P \in \mathfrak{Q}\}$$

where $\mathfrak{Q}$ is a subset of the set of all polytopes in $\mathcal{K}$ such as $\mathcal{P}_n, \mathcal{P}_{(n)}, \mathcal{P}^{\mathrm{i}}_n, \mathcal{P}^{\mathrm{c}}_n, \mathcal{P}^{\mathrm{i}}_{(n)}, \mathcal{P}^{\mathrm{c}}_{(n)}$. Here n, resp. (n) means at most n vertices, resp. facets and i, resp. c stands for inscribed, resp. circumscribed. For surveys see Fejes Tóth (1953) and Gruber (1983b) and, for random approximation, Schneider (1988).

The following result of Gruber (1983a) applies to many approximation problems.

Theorem 23. *Let X be a Baire space. Then the following hold.*

(i) *Let* $\alpha_1, \alpha_2, \ldots > 0$ *and let* $\varphi_1, \varphi_2, \ldots : X \to [0, +\infty[$ *be upper semicontinuous functions such that* $\{x \in X : \varphi_n(x) = \mathrm{o}(\alpha_n) \text{ as } n \to \infty\}$ *is dense in X. Then for most* $x \in X$ *the inequality* $\varphi_n(x) < \alpha_n$ *holds for infinitely many indices n.*

(ii) *Let* $\beta_1, \beta_2, \ldots \geqslant 0$ *and let* $\psi_1, \psi_2, \ldots : X \to \mathbb{R}^+$ *be lower semicontinuous functions such that* $\{x \in X : \beta_n = \mathrm{o}(\psi_n(x)) \text{ as } n \to \infty\}$ *is dense in X. Then for most* $x \in X$ *the inequality* $\beta_n < \psi_n(x)$ *holds for infinitely many indices n.*

Next we consider several types of problems and cite in each case a characteristic result.

The following result is a consequence of Theorem 23 and of Theorem 4 in the author's paper (1992a, I).

Theorem 24. *Let* $C \in \mathcal{K} \cap \mathcal{C}^2$ *have positive Gaussian curvature and let* $\varphi, \psi : \mathbb{N} \to \mathbb{R}^+$ *such that* $1/n^{2/(d-1)} = \mathrm{o}(\varphi(n)), \psi(n) = \mathrm{o}(1)$ *as* $n \to \infty$. *Then for most sequences* (x_n) *in* $\mathrm{bd}\, C$ *(i.e., for most elements of the compact product space* $(\mathrm{bd}\, C)^\infty$ *endowed with the product topology)*

$$\delta^{\mathrm{H}}(C, \mathrm{conv}\{x_1, \ldots, x_n\}) \leqslant \varphi(n) \quad \text{for infinitely many indices } n,$$

$$\delta^{\mathrm{H}}(C, \mathrm{conv}\{x_1, \ldots, x_n\}) \geqslant \psi(n) \quad \text{for infinitely many indices } n.$$

Second, in analogy to similar results for $\delta^{\mathrm{H}}, \delta^{\mathrm{S}}$ due to Schneider and Wieacker (1981) and Gruber and Kenderov (1982) the next one follows from Theorem 23 and Theorem 1 in Gruber (1992a, I) where $\mathcal{P}_{on} = \mathcal{K}_o \cap \mathcal{P}_{2n}$.

Theorem 25. *Let* $\varphi, \psi : \mathbb{N} \to \mathbb{R}^+$ *be such that* $0 < \varphi(n) < \psi(n) = \mathrm{o}(1/n^{2/(d-1)})$ *as* $n \to \infty$. *Then for most* $C \in \mathcal{K}_0$

$$\delta^{\mathrm{BM}}(C, \mathcal{P}_{0n}) \leqslant 1 + \varphi(n) \quad \text{for infinitely many indices } n,$$

$$\delta^{\mathrm{BM}}(C, \mathcal{P}_{0n}) \geqslant 1 + \psi(n) \quad \text{for infinitely many indices } n.$$

Third, for $C \in \mathcal{K}$ let $E_n(C)$ be the mathematical expectation of the volume $V(\mathrm{conv}\{x_1, \ldots, x_n\})$ where $x_1, \ldots, x_n$ are n independently and uniformly distributed points in C. Sorger (1987) (for $d = 2$) and Bárány and Larman (1988) (for general d) proved the following result.

Theorem 26. *Let* $\varphi, \psi : \mathbb{N} \to \mathbb{R}^+$ *be such that* $1/n^{2/(d+1)} = \mathrm{o}(\varphi(n))$, $\psi(n) = \mathrm{o}((\log n)^{d+1}/n^{2/(d+1)})$ *as* $n \to \infty$. *Then for most* $C \in \mathcal{K}$

$$V(C) - E_n(C) \leqslant \varphi(n) \quad \text{for infinitely many indices } n,$$
$$V(C) - E_n(C) \geqslant \psi(n) \quad \text{for infinitely many indices } n.$$

Fourth, Kenderov (1980, 1983) and the author and Kenderov (1982) found a uniqueness result:

Theorem 27. *For most* $C \in \mathcal{K}(\mathbb{E}^2)$ *and* $n = 3, 4, \ldots$, *there are unique polygons* $P_n, Q_n \in \mathcal{P}_n$ *such that*

$$\delta^{\mathrm{H}}(C, P_n) = \delta^{\mathrm{H}}(C, \mathcal{P}_n), \qquad \delta^{\mathrm{S}}(C, Q_n) = \delta^{\mathrm{S}}(C, \mathcal{P}_n).$$

Using a more general definition for δ^{H} an analogous result was derived by Zhivkov (1982). Most probably Theorem 27 can be extended to many other cases and, in particular, to higher dimensions, but the proofs might be difficult.

9. Points of contact

The *circumsphere* $B^{\mathrm{c}}(C)$ of $C \in \mathcal{K}$ is the (unique) Euclidean ball of minimum radius containing C, an *insphere* $B^{\mathrm{i}}(C)$ is a (not necessarily unique) Euclidean ball of maximum radius contained in C. The *minimal shell* $S(C)$ is the (unique) shell between two concentric Euclidean spheres of minimum difference of radii containing $\operatorname{bd} C$. The *minimum circumscribed* or *Loewner ellipsoid* $E^{\mathrm{c}}(C)$, resp. the *maximum inscribed ellipsoid* $E^{\mathrm{i}}(C)$ is the unique ellipsoid containing, resp. contained in C of minimum, resp. maximum volume. The following result is due to Zamfirescu (1980b) (i), Zucco (1989, 1990) (ii), and Gruber (1988) (iii), where # stands for cardinal number.

Theorem 28. *For most* $C \in \mathcal{K}$ *the following hold:*
(i) $\# \operatorname{bd} C \cap \operatorname{bd} B^{\mathrm{c}}(C) = \# \operatorname{bd} C \cap \operatorname{bd} B^{\mathrm{i}}(C) = d + 1$,
(ii) $\# \operatorname{bd} C \cap \operatorname{bd} S(C) = d + 2$,
(iii) $\# \operatorname{bd} C \cap \operatorname{bd} E^{\mathrm{c}}(C) = \# \operatorname{bd} C \cap \operatorname{bd} E^{\mathrm{i}}(C) = d(d+3)/2$.

For further results related to (iii) and an application to experimental designs see Gruber (1988). See also Zucco (1992), Peri and Zucco (1992) and Peri (1992).

For a series of different results on "touching" convex bodies, resp. on the "order of contact" with supporting hyperplanes see Zamfirescu (1981a, 1985b, 1988b).

10. Shadow boundaries

The *shadow boundary* $S(C, u)$ of $C \in \mathcal{K}$ in the direction $u \in S^{d-1}$ is the set of all $x \in \operatorname{bd} C$ such that the line $\{x + \lambda u \colon \lambda \in \mathbb{R}\}$ supports C. It is *sharp* if for any $x \in S(C, u)$ the line $\{x + \lambda u \colon \lambda \in \mathbb{R}\}$ intersects C only at x.

The next result, due to Gruber and Sorger (1989), contrasts a measure-theoretic theorem of Steenaertz (1985).

Theorem 29. *For most pairs* $(C,u) \in \mathcal{K} \times S^{d-1}$ *the shadow boundary* $S(C,u)$ *is sharp, has Hausdorff dimension* $d-2$ *and infinite* $(d-2)$*-dimensional Hausdorff measure.*

Corresponding results for illumination from point sources are also mentioned there. While the case $d > 3$ requires difficult tools from geometric measure theory, the proof for $d = 3$ is rather simple. This case also follows from Zamfirescu's (1988c) result on illumination parallel to $(d-2)$-dimensional subspaces. A corresponding measure theorem was announced by Larman and Mani (1992).

Let $C \in \mathcal{K}$ and $x \in \mathbb{E}^d \backslash C$. Then the *shadow boundary* $S(C,x)$ of C with respect to illumination from x consists of all $y \in C$ such that the line $\{(1-\lambda)x+\lambda y\colon \lambda \in \mathbb{R}\}$ supports C. Zamfirescu (1991b) proved the following.

Theorem 30. *For most* $C \in \mathcal{K}$ $(d \geqslant 3)$ *no shadow boundary* $S(C,x)$ *is contained in a hyperplane.*

Related to this is the following unpublished result of the author.

Theorem 31. *For most norms* $|\cdot|$ *in* $\mathbb{E}^d$ $(d \geqslant 3)$ *there is no linear subspace* L *of* $\mathbb{E}^d$ *with* $2 \leqslant \dim L \leqslant d-1$ *for which there exists a projection* $p : \mathbb{E}^d \to L$ *of norm* $|p|(:= \sup\{|p(x)|\colon x \in \mathbb{E}^d, |x| \leqslant 1\}) = 1$.

11. Metric projections

Given $C \in \mathcal{K}$ the (single-valued) *metric projection* or *nearest point mapping* $p_C\colon \mathbb{E}^d \to C$ is defined for $x \in \mathbb{E}^d$ by $p_C(x) = y$ where $y \in C$ is the unique point with $\|x-y\| = \min\{\|x-z\|\colon z \in C\}$. p_C is (*Stolz*- or *Frechet*-) *differentiable* at $x \in \mathbb{E}^d \backslash C$ if there is a linear mapping $l : \mathbb{E}^d \to \mathbb{E}^d$ such that for any $\varepsilon > 0$

$$\|p_C(y) - p_C(x) - l(y-x)\| \leqslant \varepsilon \|x-y\|$$

for all $y \in \mathbb{E}^d$ sufficiently close to x.

Asplund (1973) proved that for any $C \in \mathcal{K}$ the metric projection p_C is differentiable almost everywhere on $\mathbb{E}^d \backslash C$. Zajiček (1983b) found a convex body $C \in \mathcal{K}(\mathbb{E}^2)$ for which p_C is not differentiable at most points of $\mathbb{E}^2 \backslash C$. A result of Zamfirescu (1989b,c) shows that this is typical:

Theorem 32. *For most* $C \in \mathcal{K}$ *the metric projection* p_C *is not differentiable at most points of* $\mathbb{E}^d \backslash C$.

In the case $d = 2$ more precise results are known. For results on "farthest" points see Wieacker (1988) and De Blasi and Myjak (1991a).

12. Miscellaneous results for typical convex bodies

A typical convex body C is quite "asymmetric":

Theorem 33. *For most $C \in \mathcal{K}$ the group of affinities mapping C onto itself consists of the identity mapping alone.*

See Gruber (1988a). A similar result holds for norms.

The *Steiner symmetral* of $C \in \mathcal{K}$ with respect to a hyperplane H is the union of all line segments of the following form: let L be a line orthogonal to H and meeting C and consider the line segment $C \cap L$ translated along L such that its midpoint is in H. The basic problem of tomography for convex bodies C is to distinguish different C's and to reconstruct C from its Steiner symmetrals. For results of this type see Giering (1962) and Gardner (1983). Volčič and Zamfirescu (1989) showed the following result (formulated as a problem by Gruber 1987).

Theorem 34. *Let H, K be two hyperplanes in $\mathbb{E}^d$. Then most $C \in \mathcal{K}$ are uniquely determined by their Steiner symmetrals in H and K.*

The *critical exponent* of a norm $|\cdot|$ on $\mathbb{E}^d$ is the (unique) number $q = q(|\cdot|)$ (if it exists) such that for any linear mapping $l : \mathbb{E}^d \to \mathbb{E}^d$ the condition $|l| = |l^q| = 1$ implies that the spectral radius of l is 1. The spectral radius is the maximum of the absolute values of the eigenvalues of l; see Pták (1967). Perles (1967) proved the following result.

Theorem 35. *For most norms $|\cdot|$ in $\mathbb{E}^2$ the inequality $q(|\cdot|) \leqslant 3$ holds.*

A *lattice* L in $\mathbb{E}^d$ consists of all integer linear combinations of d linearly independent vectors. L provides a *lattice packing* of a convex body C if the bodies $C + l : l \in L$ have pairwise disjoint interiors. The *density* of the lattice packing is, roughly speaking, the ratio of the set covered by the bodies of the packing and the whole space. Two bodies $C, C + l, l \neq o$, are *neighbours* if $C \cap (C + l) \neq \emptyset$. In any lattice packing of a convex body C the number of neighbours is at most $3^d - 1$ by a result of Minkowski and if the packing has maximum density, it is at least $d(d+1)$ according to Swinnerton-Dyer, see, e.g., Erdős, Gruber and Hammer (1989). The following result is due to Gruber (1986).

Theorem 36. *For most $C \in \mathcal{K}$ the number of neighbours of C in any of its lattice packings of maximum density is at most $2d^2$.*

A recent result in the same context is due to Fejes Tóth and Zamfirescu (1992).

Typical convex curves on smooth convex surfaces in $\mathbb{E}^3$ are smooth; see Zamfirescu (1987b).

Silin (1991) considered functions with values in $\mathcal{K}$ in the context of problems of optimal control.

The space of all compact convex sets in $\mathbb{E}^d$ is "convex". Hence its "convex" subsets may be investigated. A first step in this direction was made by Schwarz and Zamfirescu (1987).

13. Starbodies, starsets and compact sets

A *star set* S is a compact set in $\mathbb{E}^d$ for which there exists a point $k \in S$ such that for each $x \in S$ the line segment $kx = \{(1-\lambda)k + \lambda x: \lambda \in [0,1]\}$ is contained in S. The set of all such points k form the *kernel* $\ker S$ of S. The kernel is compact and convex. In our context a *star body* is a star set S with $\ker S \in \mathcal{K}$.

The star bodies whose kernels contain a fixed convex body K, say, and the star sets form closed subspaces of the space of all compact sets in $\mathbb{E}^d$ and thus are Baire.

Zamfirescu (1989a) discovered the following properties of a generic star body:

Theorem 37. *Let $K \in \mathcal{K}$ be given. Then for most star bodies S with $K \subset \ker S$ the following hold:*

(i) $K = \ker S$,

(ii) $\operatorname{bd} S$ *is not differentiable at most of its points,*

(iii) $\operatorname{bd} S$ *is differentiable at almost all of its points and the tangent hyperplanes support K.*

The following results on star bodies and compact sets show that a typical such set is quite small. For the concept of porosity see section 3.

From Zamfirescu (1988a) and Gruber and Zamfirescu (1990) the following results are taken.

Theorem 38. *Most star sets S in $\mathbb{E}^d$ have the following properties:*

(i) $\ker S$ *consists of a single point k, say,*

(ii) $\{(x-k)/\|x-k\|: x \in S\backslash\{k\}\}$ *is a dense meager subset of S^{d-1} of cardinality c,*

(iii) *S has Hausdorff dimension* 1, *but is of non-σ-finite* 1*-dimensional Hausdorff measure.*

Ostaszewski (1974) and Gruber (1983a, 1989) proved the next theorem.

Theorem 39. *Most compact sets C in $\mathbb{E}^d$ have the following properties:*

(i) *C is totally disconnected and perfect,*

(ii) *C has Hausdorff and lower entropy dimension* 0 *and upper entropy dimension d,*

(iii) *for any $x \in C$ and $0 < \varepsilon < 1$ there are arbitrary small $\sigma > 0$ such that the shell $\{y \in \mathbb{E}^d: \varepsilon\sigma \leqslant \|y - x\| \leqslant \sigma\}$ is disjoint from C; thus C has porosity* 1 *at any of its points.*

Wieacker (1988) thoroughly studied the boundary structure of the convex hull of generic compact sets. Our last theorem presents one of his many results.

Theorem 40. *For most compact sets C in $\mathbb{E}^d$,* conv C *is a convex body of class $\mathscr{C}^1$ but not of class $\mathscr{C}^2$.*

For other results belonging to this subsection see De Blasi and Myjak (1991b).

Acknowledgement

For their helpful comments I am obliged to F.J. Schnitzer, T. Zamfirescu and A. Zucco.

References

Aleksandrov, A.D.
[1939] Almost everywhere existence of the second differential of a convex function and some properties of convex surfaces connected with it, *Učen. Zap. Leningr. Gos. Univ. Ser. Mat. Nauk* **6**, 3–35.
[1948] *Die innere Geometrie der konvexen Flächen* (Goz. Izdat. Tehn. Teor. Lit., Moscow–Leningrad) [Akademie-Verlag, Berlin, 1955].

Anderson, R.D., and V. Klee
[1952] Convex functions and upper semi-continuous collections, *Duke Math. J.* **19**, 349–357.

Asplund, E.
[1968] Frechet differentiability of convex functions, *Acta Math.* **121**, 31–47.
[1973] Differentiability of the metric projection in finite-dimensional Euclidean space, *Proc. Amer. Math. Soc.* **38**, 218–219.

Baire, R.
[1899] Sur les fonctions de variables réelles, *Ann. Mat. Pura Appl. (3)* **3**, 1–122.

Bárány, I., and D.G. Larman
[1988] Convex bodies, economic cap coverings, random polytopes, *Mathematika* **35**, 274–291.

Bárány, I., and T. Zamfirescu
[1990] Diameters in typical convex bodies, *Canad. J. Math.* **62**, 50–61.

Beer, G.A.
[1980] On closed starshaped sets and Baire categories, *Proc. Amer. Math. Soc.* **78**, 555–558.

Berger, M.
[1990] Sur les caustiques de surfaces en dimension 3, *C.R. Acad. Sci. Paris* **311**, 333–336.

Blaschke, W.
[1916] *Kreis und Kugel* (Göschen, Leipzig). Later editions: Chelsea, New York, 1949; De Gruyter, Berlin, 1956.
[1923] *Vorlesungen über Differentialgeometrie II* (Springer, Berlin).

De Blasi, F.S., and J. Myjak
[1991a] Ambiguous loci of the farthest distance mapping from compact convex sets, Manuscript.
[1991b] Ambiguous loci of the nearest point mapping in Banach spaces, Manuscript.

Dolženko, E.
[1967] Boundary properties of arbitrary functions, *Izv. Akad. Nauk SSSR Ser. Mat.* **31**, 3–14.

Erdős, P., P.M. Gruber and J. Hammer
[1989] *Lattice Points* (Longman Scientific, Harlow).

Ewald, D.G., D. Larman and C.A. Rogers
[1970] The directions of the line segments and of the r-dimensional balls on the boundary of a convex body in Euclidean space, *Mathematika* **17**, 1–20.

Fabian, M., L. Zajiček and V. Zizler
[1982] On residuality of the set of rotund norms on a Banach space, *Math. Ann.* **258**, 349–351.

Falconer, K.J.
[1985] *The Geometry of Fractal Sets* (Cambridge Univ. Press, Cambridge).

Fejes Tóth, G., and T. Zamfirescu
[1992] For most convex disks thinnest covering is not lattice like, Manuscript.

Fejes Tóth, L.
[1953] *Lagerungen in der Ebene, auf der Kugel und im Raum* (Springer, Berlin). 2nd Ed.: 1972.

Gandini, P.M., and A. Zucco
[1991] A nowhere dense but not porous set in the space of convex bodies, *Rend. Accad. Naz. Sci. XL Mem. Mat. (5)* **15**, 213–218.

Gardner, R.J.
[1983] Symmetrals and X-rays of planar convex bodies, *Arch. Math.* **41**, 183–189.

Giering, O.
[1962] Bestimmung von Eibereichen und Eikörpern durch Steiner-Symmetrisierungen, *Bayer. Akad. Wiss. Math.-Natur. Kl. Sitzungsber.*, 225–253.

Gruber, P.M.
[1977] Die meisten konvexen Körper sind glatt, aber nicht zu glatt, *Math. Ann.* **229**, 259–266.
[1981] Approximation of convex bodies by polytopes, *C.R. Acad. Bulgare Sci.* **34**, 621–622.
[1983a] In most cases approximation is irregular, *Rend. Sem. Mat. Univ. Politec. Torino* **41**, 19–33.
[1983b] Approximation of convex bodies, in: *Convexity and its Applications,* eds P.M. Gruber and J.M. Wills (Birkhäuser, Basel) pp. 131–162.
[1985] Results of Baire category type in convexity, in: *Discrete Geometry and Convexity,* Ann. New York Acad. Sci., Vol. 440 (New York Acad. Sci., New York) pp. 162–169.
[1986] Typical convex bodies have surprisingly few neighbours in densest lattice packings, *Studia Sci. Math. Hungar.* **21**, 163–173.
[1987] Radons Beiträge zur Konvexität/Radon's contribution to convexity, in: *J. Radon: Collected Works,* Vol. 1 (Österr. Akad. Wiss./Birkhäuser, Vienna/Basel) pp. 330–342.
[1988a] Minimal ellipsoids and their duals, *Rend. Circ. Mat. Palermo (2)* **37**, 35–64.
[1988b] Geodesics on typical convex surfaces, *Atti Acc. Naz. Lincei, Cl. Sci. Fis. Mat. Natur.* **82**, 651–659.
[1989] Dimension and structure of typical compact sets, continua and curves, *Monatsh. Math.* **108**, 149–164.
[1990] Convex Billiards, *Geom. Dedicata* **33**, 205–226.
[1991] A typical convex surface contains no closed geodesic!, *J. Reine Angew. Math.* **416**, 195–205.
[1992a] Asymptotic estimates for best and stepwise approximation of convex bodies I, II, *Forum Math.*, to appear.
[1992b] in preparation.

Gruber, P.M., and P. Kenderov
[1982] Approximation of convex bodies by polytopes, *Rend. Circ. Mat. Palermo (2)* **31**, 195–225.

Gruber, P.M., and H. Sorger
[1989] Shadow boundaries of typical convex bodies, measure properties, *Mathematika* **36**, 142–152.

Gruber, P.M., and T. Zamfirescu
[1990] Generic properties of compact starshaped sets, *Proc. Amer. Math. Soc.* **108**, 207–214.

Halpern, B.
[1977] Strange billiard tables, *Trans. Amer. Math. Soc.* **232**, 297–305.

Hammer, P.C., and A. Sobczyk
[1953] Planar line families II, *Proc. Amer. Math. Soc.* **4**, 341–349.

Heil, E.
[1985] Concurrent normals and critical points under weak smoothness assumptions, in: *Discrete Geometry and Convexity,* Ann. New York Acad. Sci., Vol. 440 (New York Acad. Sci., New York) pp. 170–178.

Hingston, N.
[1984] Equivalent Morse theory and closed geodesics, *J. Differential Geom.* **19**, 85–116.

Holmes, R.B.
[1975] *Geometric Functional Analysis and its Applications* (Springer, Berlin).

Howe, R.
[1982] Most convex functions are smooth, *J. Math. Econ.* **9**, 37–39.

Kenderov, P.
[1980] Approximation of plane convex compacta by polygons, *C.R. Acad. Bulgare Sci.* **33**, 889–891.
[1983] Polygonal approximation of plane convex compacta, *J. Approx. Theory* **38**, 221–239.

Klee, V.
[1959] Some new results on smoothness and rotundity in normed linear spaces, *Math. Ann.* **139**, 51–63.

Klee, V., and M. Martin
[1971] Semicontinuity of the face-function of a convex set, *Comment. Math. Helv.* **46**, 1–12.

Klima, V., and I. Netuka
[1981] Smoothness of a typical convex function, *Czechoslovak Math. J.* **31**(106), 569–572.

Klingenberg, W.
[1976] Existence of infinitely many closed geodesics, *J. Differential Geom.* **11**, 299–308.

Kosinski, A.
[1958] On a problem of Steinhaus, *Fund. Math.* **46**, 47–59.

Larman, D., and P. Mani
[1992] Almost all shadow boundaries are almost smooth, Manuscript.

Lazutkin, V.F.
[1979] The existence of caustics for billiards in a convex domain, *Izv. Akad. Nauk SSSR* **37**, 186–216 [*Math. USSR Izv.* **7**, 185–214].

Leichtweiss, K.
[1988] Über einige Eigenschaften der Affinoberfläche beliebiger konvexer Körper, *Results Math.* **13**, 255–282.

Lyusternik, L.A., and L.G. Schnirel'man
[1929] Sur le problème de trois géodésiques fermées sur les surfaces de genre 0, *C.R. Acad. Sci. Paris* **189**, 269–271.

Mazur, S.
[1933] Über konvexe Mengen in linearen normierten Räumen, *Studia Math.* **4**, 70–84.

McMullen, P.
[1979] Problem 54, contribution to 'Problems in geometric convexity' (P.M. Gruber, and R. Schneider), in: *Contributions to Geometry,* eds J. Tölke and J.M. Wills (Birkhäuser, Basel) pp. 255–278.

Minasian
[1973] in: *Introduction to Ergodic Theory,* by Ya.G. Sinai (Princeton Univ. Press, Princeton, 1977).

Minkowski, H.
[1896] *Geometrie der Zahlen* (Teubner, Leipzig). Reprinted: Johnson, New York, 1968.

Osgood, W.F.
[1900] Zweite Note über analytische Funktionen mehrer Veränderlichen, *Math. Ann.* **53**, 461–463.

Ostaszewski, A.J.
[1974] Families of compact sets and their universals, *Mathematika* **21**, 116–127.

Oxtoby, J.C.
[1971] *Measure and Category* (Springer, Berlin).

Peri, C.
[1992] On the minimal convex shell of a convex body, *Canad. Math. Bull.*, to appear.

Peri, C., and A. Zucco
[1992] On the minimal convex annulus of a planar convex body.

Perles, M.A.
[1967] Critical exponents of convex sets, in: *Proc. Coll. Convexity, Copenhagen, 1965* (Mat. Inst. Univ., København) pp. 221–228.

Preiss, D., and L. Zajiček
[1984a] Stronger estimates of smallness of sets of Frechet nondifferentiability of convex functions, *Suppl. Rend. Circ. Mat. Palermo (2)* **3**, 219–223.
[1984b] Frechet differentiation of convex functions in a Banach space with separable dual, *Proc. Amer. Math. Soc.* **91**, 202–204.

Pták, V.
[1967] Critical exponents, in: *Proc. Coll. Convexity, Copenhagen, 1965* (Mat. Inst. Univ., København) pp. 244–248.

Reidemeister, K.
[1921] Über die singularen Randpunkte eines konvexen Körpers, *Math. Ann.* **83**, 116–118.

Schneider, R.
[1979] On the curvature of convex bodies, *Math. Ann.* **240**, 177–181.
[1988] Random approximation of convex sets, *J. Microscopy* **151**, 211–227.

Schneider, R., and J.A. Wieacker
[1981] Approximation of convex bodies by polytopes, *Bull. London Math. Soc.* **13**, 149–156.

Schwarz, T., and T. Zamfirescu
[1987] Typical convex sets of convex sets, *J. Austral. Math. Soc. (A)* **43**, 287–290.

Siegel, C.L.
[1957] Integralfreie Variationsrechnung, *Nachr. Akad. Wiss. Göttingen, Math.-Phys. Kl.*, 81–86; *Ges. Abh.*, Vol. III (Springer, Berlin, 1966) pp. 264–269.

Silin, D.B.
[1991] On a typical property of convex sets, *Mat. Zametki* **49**.

Sorger, H.
[1987] Eigenschaften konvexer Körper und Verwandtes, Ph.D. Thesis, Techn. Univ. Vienna.

Steenaertz, P.
[1985] Mittlere Schattengrenzenlänge konvexer Körper, *Results Math.* **8**, 54–77.

Swinnerton-Dyer, H.P.F.
[1953] Extremal lattices of convex bodies, *Proc. Cambridge Philos. Soc.* **49**, 161–162.

Turner, Ph.H.
[1982] Convex caustics for billiards in R^2 and R^3, in: *Convexity and Related Combinatorial Geometry*, eds D.C. Kay and M. Breen (Marcel Dekker, New York) pp. 85–105.

Volčič, A., and T. Zamfirescu
[1989] Ghosts are scarce, *J. London Math. Soc. (2)* **40**, 171–178.

Wieacker, J.A.
[1988] The convex hull of a typical compact set, *Math. Ann.* **282**, 637–644.

Yost, D.
[1991] Irreducible convex sets, *Mathematika* **38**, 134–153.

Zaanen, A.C.
[1967] *Integration* (North-Holland, Amsterdam).

Zajiček, L.
[1976] Sets of σ-porosity and sets of σ-porosity (q), *Časopis Pest. Mat.* **101**, 350–359.
[1979] On the differentiation of convex functions in finite and infinite dimensional spaces, *Czechoslovak Math. J.* **29**(104), 340–348.
[1983a] Differentiability of the distance function and points of multi-valuedness of the metric projection in Banach space, *Czechoslovak Math. J.* **33**(108), 292–308.
[1983b] On differentiation of metric projections in finite dimensional Banach spaces, *Czechoslovak Math. J.* **33**(108), 325–336.
[1984a] On the Fréchet differentiability of distance functions, *Rend. Circ. Mat. Palermo (2)* **5**, 161–165.
[1984b] A generalization of an Ekeland–Lebourg theorem and the differentiability of distance functions, *Rend. Circ. Mat. Palermo (2)* **3**, 403–410.

Zamfirescu, T.
[1980a] Spreads, *Abh. Math. Sem. Univ. Hamburg* **50**, 238–253.
[1980b] Inscribed and circumscribed circles to convex curves, *Proc. Amer. Math. Soc.* **80**, 455–457.
[1980c] Nonexistence of curvature in most points of most convex surfaces, *Math Ann.* **252**, 217–219.
[1980d] The curvature of most convex surfaces vanishes almost everywhere, *Math. Z.* **174**, 135–139.
[1981a] Intersections of tangent convex curves, *J. Austral. Math. Soc. A* **31**, 456–458.
[1981b] Most monotone functions are singular, *Amer. Math. Monthly* **28**, 46–49.
[1981c] On continuous families of curves VI, *Geom. Dedicata* **10**, 205–217.
[1982a] Most convex mirrors are magic, *Topology* **21**, 65–69.
[1982b] Many endpoints and few interior points of geodesics, *Invent. Math.* **69**, 253–257.
[1984a] Typical monotone continuous functions, *Arch. Math.* **42**, 151–156.
[1984b] Intersecting diameters in convex bodies, *Ann. Discrete Math.* **20**, 311–316.
[1984c] Points on infinitely many normals to convex surfaces, *J. Reine Angew. Math.* **350**, 183–187.
[1985a] Using Baire categories in geometry, *Rend. Sem. Mat. Univ. Politec. Torino* **43**, 67–88.
[1985b] Convex curves in gear, *Acta Math. Hungar.* **46**, 297–300.
[1987a] How many sets are porous?, *Proc. Amer. Math. Soc.* **100**, 382–387.
[1987b] Typical convex curves on convex surfaces, *Monatsh. Math.* **103**, 241–247.
[1987c] Nearly all convex bodies are smooth and strictly convex, *Monatsh. Math.* **103**, 57–62.
[1987d] Typical convex sets of convex sets, *J. Austral. Math. Soc. A* **43**, 287–290.
[1988a] Typical starshaped sets, *Aequationes Math.* **36**, 188–200.
[1988b] Curvature properties of typical convex surfaces, *Pacific J. Math.* **131**, 191–207.
[1988c] Too long shadow boundaries, *Proc. Amer. Math. Soc.* **103**, 586–590.
[1989a] Description of most starshaped surfaces, *Math. Proc. Cambridge Philos. Soc.* **106**, 245–251.
[1989b] Nondifferentiability properties of the nearest point mapping, *J. Analyse Math.* **54**, 90–98.
[1989c] The nearest point mapping is single valued nearly everywhere, *Arch. Math.*, to appear.
[1991a] Baire categories in convexity, *Atti Sem. Mat. Fis. Univ. Modena* **39**, 139–164.
[1991b] On two conjectures of Franz Hering about convex surfaces, *J. Discrete Comput. Math.* **6**, 171–180.
[1991c] Conjugate points on convex surfaces, *Mathematika* **38**, 312–317.
[1992] Long geodesics on convex surfaces.

Zemlyakov, A.N., and A.B. Katok
[1975] Topological transitivity of billiards in polygons, *Mat. Zametki* **18**, 291–300 [*Math. Notes* **18**, 760–764].

Zhivkov, N.V.
[1982] Plane polygonal approximation of bounded convex sets, *C.R. Acad. Bulgare Sci.* **35**, 1631–1634.

Zucco, A.
[1989] Minimal annulus of a convex body, *Arch. Math.* **52**, 92–94.
[1990] The minimal shell of a typical convex body, *Proc. Amer. Math. Soc.* **109**, 797–802.
[1992] The generic contact of convex bodies with circumscribed homothets of a convex surface, *Discrete Comput. Geom.* **7**, 319–323.

Part 5
Stochastic Aspects of Convexity

CHAPTER 5.1

Integral Geometry

Rolf SCHNEIDER and John A. WIEACKER

Mathematisches Institut der Universität Freiburg, Albertstrasse 23b, D-79104 Freiburg, Germany

Contents

HANDBOOK OF CONVEX GEOMETRY
Edited by P.M. Gruber and J.M. Wills

Integral geometry is concerned with the study, computation, and application of invariant measures on sets of geometric objects. It has its roots in some questions on geometric probabilities. The early development, where the names of Crofton, Sylvester, Poincaré, Lebesgue and others play a role, is subsumed in the book of Deltheil (1926); see also Stoka (1968). Integral geometry, as considered here, was essentially promoted by Wilhelm Blaschke and his school in the mid-thirties; lectures of Herglotz (1933) mark the beginning of this period. Standard sources are the books by Blaschke (1955) (first published in 1935 and 1937; see also vol. 2 of his collected works, Blaschke 1985), Santaló (1953), Hadwiger (1957 chapter 6), and in particular the comprehensive work of Santaló (1976).

From its very beginning, and more so in the work of Hadwiger, integral geometry was closely connected to the geometry of convex bodies. In the following, we restrict ourselves essentially to those parts of integral geometry which are related to convexity. In contrast to the existing monographs, we prefer a measure-theoretic approach; in particular, Federer's (1959) curvature measures and the area measures related to the theory of mixed volumes play an essential role. The article will, therefore, be different in spirit from the books listed above.

For different views on integral geometry and some more recent developments, we refer to the books of Matheron (1975) and Ambartzumian (1982, 1990). Some connections to convexity appear there, too, but these cannot be taken into consideration in the present article. To a certain amount, this article is continued by chapter 5.2 on "Stochastic Geometry".

1. Preliminaries: Spaces, groups, and measures

In this section, we give a brief account of some notation, concepts and results concerning the main spaces occurring in the integral geometry of Euclidean spaces.

We work in d-dimensional real Euclidean vector space $\mathbb{E}^d$ with the usual scalar product $\langle \cdot, \cdot \rangle$, the Euclidean norm $\| \cdot \|$, the induced topology and the corresponding σ-algebra $\mathcal{B}(\mathbb{E}^d)$. (Generally, $\mathcal{B}(X)$ denotes the σ-algebra of Borel subsets of a topological space X.) For $m \in \mathbb{N} \cup \{0\}$, the m-dimensional Hausdorff (outer) measure λ_m on $\mathbb{E}^d$ is defined by

$$\lambda_m(A) := 2^{-m}\kappa_m \lim_{\delta \to 0} \inf\{\sum_{j=1}^{\infty}(\operatorname{diam} M_j)^m\colon\ A \subset \bigcup_{j=1}^{\infty} M_j, \operatorname{diam} M_j \leqslant \delta\}$$

whenever $A \subset \mathbb{E}^d$, where κ_m is the volume of the unit ball B^m in $\mathbb{E}^m$ and diam denotes the diameter. The restriction of λ_m to $\mathcal{B}(\mathbb{E}^d)$ is a measure which coincides for $m = d$ with d-dimensional Lebesgue measure and for $m = 0$ with the counting measure. Moreover, the restriction of λ_m to the σ-algebra of Borel sets of an m-dimensional C^1-submanifold of $\mathbb{E}^d$ coincides with the classical measures of arc length, surface area, etc., used in differential geometry. In particular, the restriction of λ_{d-1} to $\mathcal{B}(S^{d-1})$ is the spherical Lebesgue measure on S^{d-1}, the unit sphere of

$\mathbb{E}^d$. The measure λ_m is obviously invariant under Euclidean isometries. A detailed investigation of Hausdorff measures can be found in Federer (1969).

The group G_d of rigid motions, that is, of all orientation preserving isometries of $\mathbb{E}^d$, will be endowed with the compact-open topology which is generated by the family

$$\{\{g \in G_d\colon\ gK \subset U\}\colon\ K \subset \mathbb{E}^d \text{ compact}, U \subset \mathbb{E}^d \text{ open}\}.$$

With this topology, G_d is a topological group, locally compact, σ-compact and Hausdorff, and the action of G_d on $\mathbb{E}^d$ is continuous (see, e.g., Kobayashi-Nomizu 1963); moreover, there is no other topology sharing all these properties. In the following, all subgroups of G_d will be endowed with the trace of this topology. The most important of them are the group of translations, which may be identified with $\mathbb{E}^d$, and the compact group $\mathrm{SO}(d)$ of rotations, which may be identified with the group of orthogonal $(d \times d)$-matrices with determinant 1 considered as a topological subspace of $\mathbb{R}^{d^2}$. In particular, G_d may be identified with the semidirect product $\mathbb{E}^d \times \mathrm{SO}(d)$ with the product topology, the identifying mapping being

$$\gamma : \mathbb{E}^d \times \mathrm{SO}(d) \to G_d, \qquad \gamma(t, \rho) := \beta_{t,\rho}$$

where $\beta_{t,\rho}(x) := \rho x + t$ for $x \in \mathbb{E}^d$. Recall that the law of composition of the semidirect product $\mathbb{E}^d \times \mathrm{SO}(d)$ is given by

$$(t_1, \rho_1)(t_2, \rho_2) := (t_1 + \rho_1 t_2, \rho_1 \rho_2).$$

All these groups have a left Haar measure (nonzero regular Borel measure invariant under left translations), unique up to normalization, which is also invariant under right translations. The Haar measure on $\mathbb{E}^d$ normalized so that the unit cube has measure 1 coincides with Lebesgue measure. We shall denote by ν the Haar measure on $\mathrm{SO}(d)$ satisfying $\nu(\mathrm{SO}(d)) = 1$. Finally, μ will be the Haar measure on G_d with $\mu(\{g \in G_d\colon\ g0 \in C\}) = 1$ for a unit cube C. Thus μ is the image measure of $\lambda_d \otimes \nu$ under the map γ.

The set $\mathcal{F}^d$ of all closed subsets of $\mathbb{E}^d$ is endowed with the topology generated by the family

$$\{\mathcal{F}_U\colon\ U \subset \mathbb{E}^d \text{ open}\} \cup \{\mathcal{F}^K\colon\ K \subset \mathbb{E}^d \text{ compact}\}$$

where

$$\mathcal{F}_A := \{F \in \mathcal{F}^d\colon\ F \cap A \neq \emptyset\}, \qquad \mathcal{F}^A := \{F \in \mathcal{F}^d\colon\ F \cap A = \emptyset\}$$

for $A \subset \mathbb{E}^d$. With this topology, sometimes called the hit-or-miss topology (or topology of closed convergence), $\mathcal{F}^d$ is compact, Hausdorff and separable, and the natural action of G_d on $\mathcal{F}^d$ given by

$$(g, F) \mapsto gF := \{gx\colon\ x \in F\}, \quad F \in \mathcal{F}^d,$$

is continuous. A sequence $(F_n)_{n \in \mathbb{N}}$ in $\mathcal{F}^d \backslash \{\emptyset\}$ converges to $F \in \mathcal{F}^d \backslash \{\emptyset\}$ if and only if each $x \in F$ is the limit of some sequence $(x_n)_{n \in \mathbb{N}}$ with $x_n \in F_n$ for all

$n \in \mathbb{N}$ and, for each sequence $(n_k)_{k\in\mathbb{N}}$ in $\mathbb{N}$, each convergent sequence $(y_k)_{k\in\mathbb{N}}$ with $y_k \in F_{n_k}$ for all $k \in \mathbb{N}$ satisfies $\lim y_k \in F$. Using this criterion, it is easy to see that the map $(F,F') \mapsto F \cup F'$ from $\mathcal{F}^d \times \mathcal{F}^d$ into $\mathcal{F}^d$ is continuous. Clearly, the σ-algebra $\mathcal{B}(\mathcal{F}^d)$ is generated by the family $\{\mathcal{F}_U\colon U \subset \mathbb{E}^d \text{ open}\}$, as well as by the family $\{\mathcal{F}^K\colon K \subset \mathbb{E}^d \text{ compact}\}$. Since for each compact $K \subset \mathbb{E}^d$ the set $\{(F_1,F_2) \in \mathcal{F}^d \times \mathcal{F}^d\colon F_1 \cap F_2 \in \mathcal{F}^K\}$ is open in $\mathcal{F}^d \times \mathcal{F}^d$, the map $(F_1,F_2) \mapsto F_1 \cap F_2$ is measurable. Similarly, since for each open $U \subset \mathbb{E}^d$ the set $\{F \in \mathcal{F}^d\colon \operatorname{bd} F \in \mathcal{F}_U\}$ is open in $\mathcal{F}^d$, the map $F \mapsto \operatorname{bd} F$ is measurable. More details about $\mathcal{F}^d$ can be found in Matheron (1975) and Ripley (1976). On $\mathcal{K}^d$, the set of non-empty compact convex subsets of $\mathbb{E}^d$, the trace of the hit-or-miss topology coincides with the topology induced by the Hausdorff metric δ^{H}.

The space $\mathcal{E}_q^d$ of all q-flats (q-dimensional affine subspaces) of $\mathbb{E}^d$ is considered as a topological subspace of $\mathcal{F}^d$ and is locally compact, σ-compact and Hausdorff. Moreover, with the natural action of G_d, it is a homogeneous space and has a Haar measure which is unique up to normalization. We shall denote by μ_q the Haar measure on $\mathcal{E}_q^d$ satisfying $\mu_q(\{E \in \mathcal{E}_q^d\colon E \cap B^d \neq \emptyset\}) = \kappa_{d-q}$. For some computations, the following representation is useful. Let $E_q \in \mathcal{E}_q^d$ with $0 \in E_q$ be fixed and let $E_q^\perp$ be the orthogonal complement of E_q. Further, define a map

$$\gamma_q : E_q^\perp \times \mathrm{SO}(d) \to \mathcal{E}_q^d, \qquad \gamma_q(t,\rho) := \rho(E_q + t).$$

Then γ_q is surjective and continuous, and μ_q is the image measure of the product measure $(\lambda_{d-q}\,|_{\mathcal{B}(E_q^\perp)}) \otimes \nu$ under γ_q. Hence, for any μ_q-integrable function f on $\mathcal{E}_q^d$ we have

$$\int_{\mathcal{E}_q^d} f \,\mathrm{d}\mu_q = \int_{\mathrm{SO}(d)} \int_{E_q^\perp} f \circ \gamma_q(t,\rho)\,\mathrm{d}\lambda_{d-q}(t)\,\mathrm{d}\nu(\rho).$$

2. Intersection formulae

The most familiar type of integral-geometric formulae refers to the intersection of a fixed and a "moving" geometric object. For example, the principal kinematic formula for convex bodies provides an explicit expression for the measure of all positions of a moving convex body K' in which it meets a fixed convex body K. Crofton's intersection formula does the same for the invariant measure of the set of all k-dimensional flats meeting a convex body. The functionals of convex bodies appearing in the results, the quermassintegrals or intrinsic volumes, can in turn replace the characteristic functions (or Euler characteristic) in the integrations with respect to invariant measures. The resulting formulae can be further generalized, since they are valid in local versions, namely for curvature measures. In this form we shall now present these intersection formulae.

For a convex body $K \in \mathcal{K}^d$, we denote by $\Phi_0(K,\cdot),\ldots,\Phi_{d-1}(K,\cdot)$ the Federer curvature measures of K, as introduced in chapter 1.8, and we write $\Phi_d(K,\cdot) := \lambda_d(K \cap \cdot)$. Thus, $\Phi_i(K,\cdot)$ is a finite measure on $\mathcal{B}(\mathbb{E}^d)$. We refer to the cited article for its geometric meaning.

Theorem 2.1. *For convex bodies $K, K' \in \mathcal{K}^d$, Borel sets $\beta, \beta' \in \mathcal{B}(\mathbb{E}^d)$, and for $j \in \{0, \dots, d\}$,*

$$\int_{G_d} \Phi_j(K \cap gK', \beta \cap g\beta') \, \mathrm{d}\mu(g) = \sum_{k=j}^{d} \alpha_{djk} \Phi_{d+j-k}(K, \beta) \Phi_k(K', \beta') \tag{2.1}$$

with

$$\alpha_{djk} = \frac{\binom{k}{j} \kappa_k \kappa_{d+j-k}}{\binom{d}{k-j} \kappa_j \kappa_d} = \frac{\Gamma\left(\frac{k+1}{2}\right) \Gamma\left(\frac{d+j-k+1}{2}\right)}{\Gamma\left(\frac{j+1}{2}\right) \Gamma\left(\frac{d+1}{2}\right)}.$$

Theorem 2.1 contains, in particular, the *complete system of kinematic formulae*, namely

$$\int_{G_d} V_j(K \cap gK') \, \mathrm{d}\mu(g) = \sum_{k=j}^{d} \alpha_{djk} V_{d+j-k}(K) V_k(K'). \tag{2.2}$$

Here

$$V_j(K) = \Phi_j(K, \mathbb{E}^d) = \frac{\binom{d}{j}}{\kappa_{d-j}} W_{d-j}(K) \tag{2.3}$$

is the jth intrinsic volume of K, and $W_m(K)$ is the mth quermassintegral of K. Of course, formula (2.2) could equivalently be written in terms of $W_0, \dots, W_d$. The intrinsic volume V_0 is equal to the Euler characteristic χ, hence a special case of (2.2) is the *principal kinematic formula*

$$\int_{G_d} \chi(K \cap gK') \, \mathrm{d}\mu(g) = \frac{1}{\kappa_d} \sum_{k=0}^{d} \frac{\kappa_k \kappa_{d-k}}{\binom{d}{k}} V_{d-k}(K) V_k(K'). \tag{2.4}$$

Observe that the left side is equal to $\mu(\{g \in G_d \colon K \cap gK' \neq \emptyset\})$.

It is an essential feature of the kinematic formulae (2.1) that on the right side the bodies K and K' appear separated. This simplifying effect is a consequence of the integration over the group of rigid motions. If the integration extends only over the group of translations, the result can only be expressed in terms of measures that depend simultaneously on K and K':

Theorem 2.2. *For convex bodies $K, K' \in \mathcal{K}^d$, Borel sets $\beta, \beta' \in \mathcal{B}(\mathbb{E}^d)$, and for $j \in \{0, \dots, d\}$,*

$$\begin{aligned} &\int_{\mathbb{E}^d} \Phi_j(K \cap (K' + t), \beta \cap (\beta' + t)) \, \mathrm{d}\lambda_d(t) \\ &\quad = \Phi_j(K, \beta) \Phi_d(K', \beta') + \sum_{k=j+1}^{d-1} \Phi_k^{(j)}(K, K', \beta \times \beta') \\ &\qquad + \Phi_d(K, \beta) \Phi_j(K', \beta') \end{aligned} \tag{2.5}$$

with unique maps $\Phi_k^{(j)}: \mathcal{K}^d \times \mathcal{K}^d \times \mathcal{B}(\mathbb{E}^d \times \mathbb{E}^d) \to \mathbb{R}$ *having the following properties:*

(a) $\Phi_k^{(j)}(K, K', \cdot)$ *is a finite measure,*

(b) *the map* $(K, K') \mapsto \Phi_k^{(j)}(K, K', \cdot)$ *from* $\mathcal{K}^d \times \mathcal{K}^d$ *into the space of finite measures on* $\mathbb{E}^d \times \mathbb{E}^d$ *with the weak topology is continuous,*

(c) $\Phi_k^{(j)}(\cdot, K, \alpha)$ *and* $\Phi_k^{(j)}(K, \cdot, \alpha)$ *are additive* ($K \in \mathcal{K}^d, \alpha \in \mathcal{B}(\mathbb{E}^d \times \mathbb{E}^d)$),

(d) $\Phi_k^{(j)}(\cdot, K, \cdot \times \beta)$ *is positively homogeneous of degree* k, $\Phi_k^{(j)}(K, \cdot, \beta \times \cdot)$ *is positively homogeneous of degree* $d + j - k$ ($K \in \mathcal{K}^d, \beta \in \mathcal{B}(\mathbb{E}^d)$).

In contrast to Theorem 2.1, the result of Theorem 2.2 on the translative case is mainly of a qualitative character, since no explicit representation of the measure $\Phi_k^{(j)}(K, K', \cdot)$ is available, except in special cases (see below).

A counterpart to formula (2.1), with the "moving convex body" gK' replaced by a "moving flat", is given by the following result.

Theorem 2.3. *For a convex body* $K \in \mathcal{K}^d$, *a number* $k \in \{0, \ldots, d\}$, *a Borel set* $\beta \in \mathcal{B}(\mathbb{E}^d)$, *and for* $j \in \{0, \ldots, k\}$,

$$\int_{\mathscr{E}_k^d} \Phi_j(K \cap E, \beta \cap E) \, \mathrm{d}\mu_k(E) = \alpha_{djk} \Phi_{d+j-k}(K, \beta) \tag{2.6}$$

(*with* α_{djk} *as in Theorem* 2.1).

Again, some special cases are worth noting. With $\beta = \mathbb{E}^d$, we obtain *Crofton's intersection formula*

$$\int_{\mathscr{E}_k^d} V_j(K \cap E) \, \mathrm{d}\mu_k(E) = \alpha_{djk} V_{d+j-k}(K). \tag{2.7}$$

The case $j = 0$ of (2.6),

$$\alpha_{d0k} \Phi_{d-k}(K, \beta) = \int_{\mathscr{E}_k^d} \Phi_0(K \cap E, \beta \cap E) \, \mathrm{d}\mu_k(E), \tag{2.8}$$

interprets the curvature measure Φ_{d-k}, up to a constant factor, as an integral-geometric mean value of the "Gaussian" curvature measure Φ_0. The specialization $\beta = \mathbb{E}^d$ yields

$$\begin{aligned}\alpha_{d0k} V_{d-k}(K) &= \int_{\mathscr{E}_k^d} \chi(K \cap E) \, \mathrm{d}\mu_k(E) \\ &= \mu_k(\{E \in \mathscr{E}_k^d \colon K \cap E \neq \emptyset\}).\end{aligned} \tag{2.9}$$

We give some hints to the literature and to proofs. The principal kinematic formula goes back, in different degrees of generality, to Blaschke, Santaló, Chern and Yien. One finds references in Blaschke (1955), Santaló (1953) and, in particular, Santaló (1976). In the latter book, formulae of type (2.2) are proved for domains with smooth boundaries, where the V_j are expressed as curvature integrals. For convex bodies (and, more generally, for sets of the convex ring, see section 5 below), Hadwiger (1950, 1951, 1956) has proved (2.2) and other integral-geometric formulae in an elegant way, making use of his axiomatic characterization of the linear combinations of the quermassintegrals. This method is also employed in chapter 6 of his book, Hadwiger (1957). From there (but with different notations) we quote a general version of the principal kinematic formula for convex bodies.

Theorem 2.4. *If $\varphi : \mathcal{K}^d \to \mathbb{R}$ is an additive and continuous function, then*

$$\int_{G_d} \varphi(K \cap gK')\, \mathrm{d}\mu(g) = \frac{1}{\kappa_d} \sum_{k=0}^{d} \frac{\kappa_k \kappa_{d-k}}{\binom{d}{k}} \varphi_{d-k}(K) V_k(K')$$

for $K, K' \in \mathcal{K}^d$, where

$$\varphi_{d-k}(K) := \frac{1}{\alpha_{d0k}} \int_{\mathcal{E}_k^d} \varphi(K \cap E)\, \mathrm{d}\mu_k(E).$$

From (2.4) and (2.9) it is clear that for $\varphi = \chi$ this reduces to the principal kinematic formula.

A short proof of formula (2.4) was also given by Mani-Levitska (1988).

Theorems 2.1 and 2.3 in their general forms for curvature measures are due to Federer (1959), who proved them for sets of positive reach. A shorter proof for this general version of (2.1) was given by Rother and Zähle (1990). For convex bodies, considerably simpler approaches are possible. Schneider (1978a) gave a proof by a method similar to that of Hadwiger, proving and applying an axiomatic characterization of the curvature measures. A slightly simpler proof was given in Schneider (1980b), using uniqueness results for Lebesgue measures and external angles in the case of convex polytopes, and approximation to obtain the general case. A method of Federer (1959) to deduce (2.6) from (2.1) was extended in Schneider (1980b) to obtain a common generalization of Theorems 2.1 and 2.3, namely a kinematic formula for a fixed convex body and a moving convex cylinder.

A still different approach to Theorem 2.1, this time via the translative case and thus leading also to Theorem 2.2, was followed in Schneider and Weil (1986). We briefly sketch the main ideas. After showing the measurability of the map $g \mapsto \Phi_j(K \cap gK', \beta \cap g\beta')$, one may write

$$\int_{G_d} \Phi_j(K \cap gK', \beta \cap g\beta')\, \mathrm{d}\mu(g) = \int_{\mathrm{SO}(d)} I(\rho)\, \mathrm{d}\nu(\rho)$$

with

$$I(\rho) := \int_{\mathbb{E}^d} \Phi_j(K \cap (\rho K' + t), \beta \cap (\rho\beta' + t))\, \mathrm{d}\lambda_d(t)$$

for $\rho \in \mathrm{SO}(d)$. Assuming first that K and K' are polytopes one obtains, by direct computation and for ν-almost all ρ, the formula

$$I(\rho) = \Phi_j(K,\beta)\Phi_d(K',\beta') + \Phi_d(K,\beta)\Phi_j(K',\beta')$$

$$+ \sum_{k=j+1}^{d-1} \sum_{F\in\mathcal{F}_{d+j-k}(K)} \sum_{F'\in\mathcal{F}_k(K')} \gamma(F,\rho F',K,\rho K')[F,\rho F']$$

$$\times \lambda_{d+j-k}(\beta \cap F)\lambda_k(\beta' \cap F').$$

Here $\mathcal{F}_m(K)$ denotes the set of m-faces of the polytope K. The number $\gamma(F,\rho F',K,\rho K')$ is the external angle of the polytope $K \cap (\rho K' + t)$ at its face $F \cap (\rho F' + t)$, where $t \in \mathbb{E}^d$ is chosen so that relint $F\cap$ relint$(\rho F' + t) \neq \emptyset$. Finally, $[F,\rho F']$ is a number depending only on the relative positions of the affine hulls of F and $\rho F'$. (The assumption of general relative position made in Schneider and Weil (1986) is superfluous.) Choosing for ρ the identity, we see that the equality for $I(\rho)$ proves the translative formula (2.5) for polytopes and at the same time gives an explicit representation, in this case, for the measures $\Phi_k^{(j)}(K,K',\cdot)$ appearing in Theorem 2.2. The general assertions of Theorem 2.2 are then obtained by approximation. The proof of Theorem 2.1 next requires the computation of the integral

$$\int_{SO(d)} \gamma(F,\rho F',K,\rho K')[F,\rho F'] \,\mathrm{d}\nu(\rho).$$

This can conveniently be achieved in an indirect way, using the uniqueness of spherical Lebesgue measure to show that the integral must be proportional to the product of the external angles of K at F and of K' at F'. This yields formula (2.1) for polytopes, except that the numerical values of the coefficients have to be determined by an additional argument. The proof is then completed by approximation. Details are found in Schneider and Weil (1986). There one also finds a result similar to Theorem 2.2 which holds for convex bodies and translates of convex cylinders, and as a special case a translative Crofton formula for curvature measures.

We add some remarks on translative formulae in special cases. For $j = d$ and $j = d-1$, formula (2.5) follows from general formulae of measure theory; see Groemer (1977, 1980a), Schneider (1981b). The global case of (2.5), that is, the case $\beta = \beta' = \mathbb{E}^d$, can be written in the form

$$\int_{\mathbb{E}^d} V_j(K \cap (K'+t)) \,\mathrm{d}\lambda_d(t)$$

$$= V_j(K)V_d(K') + \sum_{k=j+1}^{d-1} V_{k,d+j-k}^{(j)}(K,K') + V_d(K)V_j(K') \tag{2.10}$$

with $V_{k,d+j-k}^{(j)}(K,K') := \Phi_k^{(j)}(K,K',\mathbb{E}^d\times\mathbb{E}^d)$. In special cases, some more information on the functionals $V_{k,d+j-k}^{(j)}(K,K')$ is available. Investigations referring to the cases

$d = 2$ and $d = 3$ are found in Blaschke (1937), Berwald and Varga (1937), Miles (1974) (cf. also Firey 1977). For $j = 0$ one obtains (see Groemer 1977 for some extensions)

$$\int_{\mathbb{E}^d} V_0(K \cap (K' + t)) \, \mathrm{d}\lambda_d(t)$$

$$= \lambda_d(\{t \in \mathbb{E}^d \colon K \cap (K' + t) \neq \emptyset\})$$

$$= V_d(K + \check{K}') = \sum_{k=0}^{d} \binom{d}{k} V(\underbrace{K, \ldots, K}_{k}, \underbrace{\check{K}', \ldots, \check{K}'}_{d-k}), \tag{2.11}$$

where $\check{K}' := \{-x \colon x \in K'\}$ and V denotes the mixed volume. Mixed volumes also appear in the following translative Crofton formula. Let $k \in \{1, \ldots, d-1\}$, $E_k \subset \mathbb{E}^d$ a k-dimensional linear subspace and $E_k^\perp$ its orthogonal complement; let B_k denote a k-dimensional unit ball in E_k. Then

$$\int_{E_k^\perp} V_j(K \cap (E_k + t)) \, \mathrm{d}\lambda_{d-k}(t)$$

$$= \frac{1}{\kappa_{k-j}} \binom{d}{k-j} V(\underbrace{K, \ldots, K}_{d+j-k}, \underbrace{B_k, \ldots, B_k}_{k-j}) \tag{2.12}$$

for $j = 0, \ldots, k$ (see Schneider 1981a).

The functionals $V^{(j)}_{k,d+j-k}(K, K')$ appearing in (2.10) can be expressed as integrals of mixed volumes, in the form

$$V^{(j)}_{k,d+j-k}(K, K')$$

$$= c_{djk} \int_{\mathscr{E}^d_{d-j}} V(\underbrace{K \cap E, \ldots, K \cap E}_{k-j}, \underbrace{\check{K}', \ldots, \check{K}'}_{d+j-k}) \, \mathrm{d}\mu_{d-j}(E) \tag{2.13}$$

with

$$c_{djk} = \binom{d}{j} \binom{d}{k-j} \frac{\kappa_d}{\kappa_j \kappa_{d-j}}.$$

The special case $d = 3$, $k = 2$, $j = 1$ appears in Berwald and Varga (1937). The general case, and its extension to corresponding formulae for cylinders, is due to Goodey and Weil (1987). They also have different representations for $V^{(j)}_{k,d+j-k}$ in the case of centrally symmetric bodies, involving measures on Grassmannians or projection bodies.

The kinematic formula (2.2) can be iterated. Since on the right side of (2.2) there appear only intrinsic volumes, which can serve as integrands on the left side, one can use induction to obtain the formula

$$\int_{G_d} \cdots \int_{G_d} V_j(K_0 \cap g_1 K_1 \cap \cdots \cap g_m K_m)\, \mathrm{d}\mu(g_1) \cdots\, \mathrm{d}\mu(g_m)$$

$$= \sum_{\substack{k_0, \ldots, k_m = j \\ k_0 + \cdots + k_m = dm + j}}^{d} c^{(j)}_{k_0, k_1, \ldots, k_m} V_{k_0}(K_0) V_{k_1}(K_1) \cdots V_{k_m}(K_m) \tag{2.14}$$

for $K_0, K_1, \ldots, K_m \in \mathcal{K}^d$, $m \in \mathbb{N}$, $j \in \{0, \ldots, d\}$, with explicitly known constants $c^{(j)}_{k_0, k_1, \ldots, k_m}$ (see, e.g., Streit 1970). Iterated versions of the translative formula (2.10) are easy to obtain for $d = 2$ (Blaschke 1937, Miles 1974) and for $j = d - 1$, d (Streit 1973, 1975), but not so in the general case. The latter has been investigated by Weil (1990). He showed that one has an expression

$$\int_{\mathbb{E}^d} \cdots \int_{\mathbb{E}^d} V_j(K_0 \cap (K_1 + t_1) \cap \cdots \cap (K_m + t_m))\, \mathrm{d}\lambda_d(t_1) \cdots\, \mathrm{d}\lambda_d(t_m)$$

$$= \sum_{\substack{k_0, \ldots, k_m = j \\ k_0 + \cdots + k_m = dm + j}}^{d} V^{(j)}_{k_0, k_1, \ldots, k_m}(K_0, K_1, \ldots, K_m), \tag{2.15}$$

by which a variety of mixed functionals $V^{(j)}_{k_0, k_1, \ldots, k_m}$ is introduced. Weil investigated the properties of these functionals and showed, in particular, that, for fixed j, they can be computed from the finitely many functionals $V^{(j)}_{k_1, \ldots, k_d}$, where $k_1, \ldots, k_d \in \{j, \ldots, d\}$ and $k_1 + \cdots + k_d = d(d-1) + j$. However, explicit geometric descriptions are only known in special cases. The mixed functionals $V^{(j)}_{k_0, k_1, \ldots, k_m}$ satisfy in turn integral-geometric formulae; see Weil (1990) and also the short survey in Weil (1989b).

As a by-product of studies in translative integral geometry, Goodey and Weil (1987) and Weil (1990) obtained some Crofton-type formulae for mixed volumes, among them, for $j \in \{1, \ldots, d \quad 1\}$,

$$\int_{\mathcal{E}^d_j} V(\underbrace{K \cap E, \ldots, K \cap E}_{j}, \underbrace{L, \ldots, L}_{d-j})\, \mathrm{d}\mu_j(E)$$

$$= \binom{d}{j}^{-2} \frac{\kappa_j \kappa_{d-j}}{\kappa_d} V_d(K) V_{d-j}(L) \tag{2.16}$$

for $K, L \in \mathcal{K}^d$ and

$$\int_{\mathcal{E}^d_{d-j+1}} \int_{\mathcal{E}^d_{j+1}} V(\underbrace{K \cap E, \ldots, K \cap E}_{j}, \underbrace{L \cap F, \ldots, L \cap F}_{d-j})$$

$$\times\, \mathrm{d}\mu_{j+1}(E)\, \mathrm{d}\mu_{d-j+1}(F)$$

$$= \frac{d(d-1)\alpha_{d0(j+1)}}{4\binom{d}{j}\kappa_{d-2}} V(\Pi K, \Pi L, B^d, \ldots, B^d) \tag{2.17}$$

for centrally symmetric convex bodies $K, L \in \mathcal{K}^d$, where ΠK denotes the projection body of K.

3. Minkowski addition and projections

The formulae of the preceding section refer to the intersection of a fixed convex body and a moving convex set. Similar formulae exist for other geometric operations, namely Minkowski addition and projection. Some global formulae of this type are immediate consequences of the principal kinematic formula, and we mention these first.

For convex bodies $K, K' \in \mathcal{K}^d$ and a rotation $\rho \in \mathrm{SO}(d)$, we integrate the trivial relation

$$V_d(K + \rho K') = \int_{\mathbb{E}^d} \chi(K \cap (\rho \check{K}' + t)) \, \mathrm{d}\lambda_d(t)$$

over $\mathrm{SO}(d)$ and then use (2.4) to get

$$\int_{\mathrm{SO}(d)} V_d(K + \rho K') \, \mathrm{d}\nu(\rho) = \int_{G_d} \chi(K \cap g\check{K}') \, \mathrm{d}\mu(g)$$

$$= \sum_{k=0}^{d} \alpha_{d0k} V_k(K) V_{d-k}(K'). \tag{3.1}$$

Replacing K by $K + \varepsilon B^d$, expanding and comparing coefficients of equal powers of ε, we obtain

$$\int_{\mathrm{SO}(d)} V_j(K + \rho K') \, \mathrm{d}\nu(\rho) = \sum_{k=0}^{j} \beta_{djk} V_k(K) V_{j-k}(K') \tag{3.2}$$

with

$$\beta_{djk} = \frac{\binom{d-k}{j-k} \kappa_{d-k} \kappa_{d+k-j}}{\binom{d}{j-k} \kappa_{d-j} \kappa_d}. \tag{3.3}$$

More generally, in (3.1) we may write V_k, V_{d-k} as mixed volumes (with numerical factors), replace K and K' by Minkowski combinations of convex bodies, expand both sides and compare terms of equal degrees of homogeneity. Thus we obtain

$$\int_{\mathrm{SO}(d)} V(K_1, \ldots, K_m, \rho K_{m+1}, \ldots, \rho K_d) \, \mathrm{d}\nu(\rho)$$

$$= \frac{1}{\kappa_d} V(K_1, \ldots, K_m, \underbrace{B^d, \ldots, B^d}_{d-m}) V(\underbrace{B^d, \ldots, B^d}_{m}, K_{m+1}, \ldots, K_d). \tag{3.4}$$

To treat local versions of these and further formulae, we use the generalized curvature measures Θ_j (see chapter 1.8). Recall that they can be defined by

$$\lambda_d(M_\varepsilon(K,\eta)) = \frac{1}{d}\sum_{j=0}^{d-1} \varepsilon^{d-j}\binom{d}{j}\Theta_j(K,\eta) \tag{3.5}$$

for $K \in \mathcal{K}^d$, $\eta \in \mathcal{B}(\Sigma)$, and $\varepsilon > 0$. Here $\Sigma = \mathbb{E}^d \times S^{d-1}$, and the set $M_\varepsilon(K,\eta)$ is defined as follows. For $x \in \mathbb{E}^d$, we denote by $p(K,x)$ the unique point in $K \in \mathcal{K}^d$ nearest to x and by $r(K,x) := \|x - p(K,x)\|$ its distance from x. If $x \notin K$, the unit vector pointing from $p(K,x)$ to x is defined by $u(K,x) := (x - p(K,x))/r(K,x)$. Thus, for $x \in \mathbb{E}^d \setminus K$, the pair $(p(K,x), u(K,x))$ is a *support element* of K, by which we mean that $p(K,x)$ is a boundary point of K and $u(K,x)$ is an exterior unit normal vector to K at this point. The set of all support elements of K is denoted by $\operatorname{Nor} K$. Now the set $M_\varepsilon(K,\eta)$ appearing in (3.5) is defined by

$$M_\varepsilon(K,\eta) := \{x \in \mathbb{E}^d\colon\ 0 < r(K,x) \leqslant \varepsilon,\ (p(K,x), u(K,x)) \in \eta\}. \tag{3.6}$$

Special cases of the generalized curvature measures $\Theta_0(K,\cdot), \ldots, \Theta_{d-1}(K,\cdot)$ are Federer's curvature measures and the area measures of lower order, which in the literature appear in two different normalizations:

$$\Theta_j(K, \beta \times S^{d-1}) = C_j(K,\beta) = d\binom{d}{j}^{-1}\kappa_{d-j}\Phi_j(K,\beta), \tag{3.7}$$

$$\Theta_j(K, \mathbb{E}^d \times \omega) = S_j(K,\omega) = d\binom{d}{j}^{-1}\kappa_{d-j}\Psi_j(K,\omega) \tag{3.8}$$

for $\beta \in \mathcal{B}(\mathbb{E}^d)$ and $\omega \in \mathcal{B}(S^{d-1})$. In the following, it seems preferable to formulate the local formulae in terms of Θ_j, C_j, S_j instead of the renormalized versions.

For sets $\eta, \eta' \subset \Sigma$ we define

$$\eta * \eta' := \{(x + x', u) \in \Sigma\colon\ (x,u) \in \eta,\ (x',u) \in \eta'\}.$$

This operation includes the behaviours of sets of boundary points and of normal vectors of convex bodies under addition: if $\eta \subset \operatorname{Nor} K$ and $\eta' \subset \operatorname{Nor} K'$, then $\eta * \eta' \subset \operatorname{Nor}(K + K')$. For $\beta, \beta' \subset \mathbb{E}^d$ and $\omega, \omega' \subset S^{d-1}$ we have

$$(\beta \times \omega) * (\beta' \times \omega') = (\beta + \beta') \times (\omega \cap \omega').$$

The following theorem contains a rather general local version of (3.2).

Theorem 3.1. *If $K, K' \in \mathcal{K}^d$ are convex bodies, $\eta \subset \operatorname{Nor} K$ and $\eta' \subset \operatorname{Nor} K'$ are Borel sets of support elements, and $j \in \{0, \ldots, d-1\}$, then*

$$\int_{\mathrm{SO}(d)} \Theta_j(K + \rho K', \eta * \rho\eta')\, \mathrm{d}\nu(\rho) = \frac{1}{d\kappa_d}\sum_{k=0}^{j}\binom{j}{k}\Theta_k(K,\eta)\Theta_{j-k}(K',\eta'). \tag{3.9}$$

Special cases are

$$\int_{\mathrm{SO}(d)} C_j(K+\rho K', \beta+\rho\beta')\,\mathrm{d}\nu(\rho) = \frac{1}{d\kappa_d}\sum_{k=0}^{j}\binom{j}{k} C_k(K,\beta)C_{j-k}(K',\beta') \tag{3.10}$$

for Borel sets $\beta \subset \operatorname{bd} K$, $\beta' \subset \operatorname{bd} K'$, and

$$\int_{\mathrm{SO}(d)} S_j(K+\rho K', \omega\cap\rho\omega')\,\mathrm{d}\nu(\rho) = \frac{1}{d\kappa_d}\sum_{k=0}^{j}\binom{j}{k} S_k(K,\omega)S_{j-k}(K',\omega') \tag{3.11}$$

for $\omega, \omega' \in \mathcal{B}(S^{d-1})$.

In the same way as formula (3.4) was deduced from (3.1) we may derive, from the case $j = d-1$ of formula (3.11), a rotational mean value formula for the mixed area measure S, namely

$$\begin{aligned}
&\int_{\mathrm{SO}(d)} S(K_1,\dots,K_m,\rho K_{m+1},\dots,\rho K_{d-1},\omega\cap\rho\omega')\,\mathrm{d}\nu(\rho)\\
&\quad = \frac{1}{d\kappa_d} S(K_1,\dots,K_m,\underbrace{B^d,\dots,B^d}_{d-1-m},\omega)\\
&\quad\quad \times S(\underbrace{B^d,\dots,B^d}_{m},K_{m+1},\dots,K_{d-1},\omega')
\end{aligned} \tag{3.12}$$

for $K_1,\dots,K_{d-1} \in \mathcal{K}^d$ and $\omega, \omega' \in \mathcal{B}(S^{d-1})$.

The case $j = d-1$ of (3.9) can also be used to obtain more general formulae involving arbitrary functions. For example, let $v(K,x)$ be the exterior unit normal vector of the convex body K at x if x is a regular boundary point of K (otherwise, $v(K,x)$ remains undefined); let $f : S^{d-1} \to \mathbb{R}$ be a nonnegative measurable function. Then

$$\begin{aligned}
&\int_{\mathrm{SO}(d)} \int \mathbf{1}_{\beta+\rho\beta'}(x) f(v(K+\rho K',x))\,\mathrm{d}C_{d-1}(K+\rho K',x)\,\mathrm{d}\nu(\rho)\\
&\quad = \sum_{k=0}^{d-1}\binom{d-1}{k}\int \mathbf{1}_{\beta}(x) f(u)\,\mathrm{d}\Theta_k(K,(x,u))C_{d-1-k}(K',\beta')
\end{aligned} \tag{3.13}$$

for $K, K' \in \mathcal{K}^d$ and Borel sets $\beta \subset \operatorname{bd} K$, $\beta' \subset \operatorname{bd} K'$; here $\mathbf{1}$ denotes the indicator function.

Some of the rotational mean value formulae can be specialized to yield projection formulae. In the following, E denotes a fixed k-dimensional linear subspace of

$\mathbb{E}^d$, where $k \in \{1, \ldots, d-1\}$. The image of a set $A \subset \mathbb{E}^d$ under orthogonal projection onto E is denoted by $A \mid E$. The mixed volume in a k-dimensional linear subspace will be denoted by $v^{(k)}$. If $K_1, \ldots, K_k \in \mathcal{K}^d$ are convex bodies and if $U \subset E^\perp$ is a convex body with $\lambda_{d-k}(U) = 1$, then one shows in the theory of mixed volumes that

$$v^{(k)}(K_1 \mid E, \ldots, K_k \mid E) = \binom{d}{k} V(K_1, \ldots, K_k, \underbrace{U, \ldots, U}_{d-k}).$$

Hence, from (3.4) we can infer that

$$\int_{\mathrm{SO}(d)} v^{(k)}(K_1 \mid \rho E, \ldots, K_k \mid \rho E)\, \mathrm{d}\nu(\rho) = \frac{\kappa_k}{\kappa_d} V(K_1, \ldots, K_k, \underbrace{B^d, \ldots, B^d}_{d-k}). \tag{3.14}$$

A special case can be written in the form

$$\int_{\mathrm{SO}(d)} V_j(K \mid \rho E)\, \mathrm{d}\nu(\rho) = \beta_{d(d+j-k)j} V_j(K), \tag{3.15}$$

valid for $j \in \{0, \ldots, k\}$, and further specialization gives

$$\int_{\mathrm{SO}(d)} \lambda_k(K \mid \rho E)\, \mathrm{d}\nu(\rho) = \frac{\kappa_{d-k}\kappa_k}{\binom{d}{k}\kappa_d} V_k(K) = \frac{\kappa_k}{\kappa_d} W_{d-k}(K). \tag{3.16}$$

(This explains the name "quermassintegral", since the k-dimensional measure of the projection, $\lambda_k(K \mid \rho E)$, can in German be called a "Quermaß".)

The general formula (3.15) is often called *Kubota's integral recursion.* The case $k = d-1$ of (3.16) is *Cauchy's surface area formula.* The case $k = 1$ of (3.16) shows that

$$V_1(K) = \frac{d\kappa_d}{2\kappa_{d-1}} \bar{b}(K), \tag{3.17}$$

where $\bar{b}$ is the mean width.

From the local formula (3.9) one may also derive a local formula for projections. For $\eta \subset \Sigma$ we write

$$\eta \mid E := \{(x \mid E, u)\colon\ (x, u) \in \eta \ \text{and}\ u \in E\}.$$

Theorem 3.2. *If $K \in \mathcal{K}^d$ is a convex body, $\eta \subset \mathrm{Nor}\, K$ is a Borel set of support elements, and $E \subset \mathbb{E}^d$ is a k-dimensional linear subspace ($k \in \{1, \ldots, d-1\}$), then*

$$\int_{\mathrm{SO}(d)} \Theta_j^{(k)}(K \mid \rho E, \eta \mid \rho E)\, \mathrm{d}\nu(\rho) = \frac{k\kappa_k}{d\kappa_d} \Theta_j(K, \eta) \tag{3.18}$$

for $j \in \{0, \ldots, k-1\}$, where $\Theta_j^{(k)}$ is the generalized curvature measure taken with respect to the subspace ρE.

By specialization, we obtain the formulae

$$\int_{\mathrm{SO}(d)} C_j^{(k)}(K \mid \rho E, \beta \mid \rho E)\, \mathrm{d}\nu(\rho) = \frac{k\kappa_k}{d\kappa_d} C_j(K,\beta) \tag{3.19}$$

for Borel sets $\beta \subset \operatorname{bd} K$, and

$$\int_{\mathrm{SO}(d)} S_j^{(k)}(K \mid \rho E, \omega \cap \rho E)\, \mathrm{d}\nu(\rho) = \frac{k\kappa_k}{d\kappa_d} S_j(K,\omega) \tag{3.20}$$

for $\omega \in \mathcal{B}(S^{d-1})$; here $C_j^{(k)}$ and $S_j^{(k)}$ are computed in ρE.

In a similar way as (3.14) was deduced, one may obtain a corresponding mean value formula for mixed area measures. With E as above, we have

$$\int_{\mathrm{SO}(d)} s^{(k)}(K_1 \mid \rho E, \ldots, K_{k-1} \mid \rho E, \omega \cap \rho E)\, \mathrm{d}\nu(\rho)$$

$$= \frac{k\kappa_k}{d\kappa_d} S(K_1, \ldots, K_{k-1}, \underbrace{B^d, \ldots, B^d}_{d-k}, \omega) \tag{3.21}$$

for $\omega \in \mathcal{B}(S^{d-1})$, where $s^{(k)}$ denotes the mixed area measure in ρE.

We give some hints to the literature. Rotation integrals for Minkowski addition of type (3.2) first appear in Hadwiger (1950), obtained in a different way; see also Hadwiger (1957, section 6.2.4). Theorems 3.1 and 3.2 are due to Schneider (1986); the special cases (3.11), (3.20) were proved before by Schneider (1975a) and the special cases (3.10), (3.19) by Weil (1979b), but in different and more indirect ways. These formulae are essential tools for the formulae of the next section. The equation (3.13) was applied by Papaderou-Vogiatzaki and Schneider (1988) to a question on geometric collision probabilities. The projection formulae (3.15) are classical, see, e.g., Hadwiger (1957).

Finally, we mention a formula that combines intersection and projection. Such a formula exists for a moving convex cylinder meeting a fixed convex body; one projects the intersection orthogonally into a generating subspace of the cylinder. Let a convex body $K \in \mathcal{K}^d$, a q-dimensional linear subspace E of $\mathbb{E}^d$, where $q \in \{0, \ldots, d-1\}$, and a convex body $C \subset E^\perp$ be given. Then $Z = C + E$ is a convex cylinder with generating subspace E. For $j \in \{0, \ldots, q\}$ we have

$$\int_{\mathrm{SO}(d)} \int_{E^\perp} V_j((K \cap \rho(Z+t)) \mid \rho E)\, \mathrm{d}\lambda_{d-q}(t)\, \mathrm{d}\nu(\rho)$$

$$= \sum_{k=j}^{d+j-q} \gamma_{djkq} V_k(K) V_{d+j-q-k}(C) \tag{3.22}$$

with

$$\gamma_{djkq} = \frac{\binom{q}{j} \kappa_q \kappa_k \kappa_{d-k}}{\binom{d}{k} \kappa_{q-j} \kappa_d \kappa_j}.$$

A proof (with different notations) can be found in Schneider (1981a); the case where C is a ball in $E^\perp$ was treated earlier by Matheron (1976).

4. Distance integrals and contact measures

The principal kinematic formula (2.4), now written in terms of quermassintegrals, thus

$$\int_{G_d} \chi(K \cap gK')\, \mathrm{d}\mu(g) = \frac{1}{\kappa_d} \sum_{k=0}^{d} \binom{d}{k} W_k(K) W_{d-k}(K'), \tag{4.1}$$

refers to the set of rigid motions g for which $K \cap gK' \neq \emptyset$. One can also integrate over the complementary set of motions if one introduces suitable functions of the distance between K and gK'.

The distance, $r(K, L)$, of a compact set $K \subset \mathbb{E}^d$ and a closed set $L \subset \mathbb{E}^d$ is defined by

$$r(K, L) := \min \{\| x - y \| : x \in K, y \in L\}.$$

Now let $f : [0, \infty) \to [0, \infty)$ be a measurable function for which $f(0) = 0$ and

$$M_k(f) := k \int_0^\infty f(r) r^{k-1}\, \mathrm{d}r < \infty \qquad \text{for } k = 1, \ldots, d.$$

Then, for convex bodies $K, L \in \mathcal{K}^d$,

$$\int_{G_d} f(r(K, gL))\, \mathrm{d}\mu(g)$$

$$= \frac{1}{\kappa_d} \sum_{k=1}^{d} \sum_{j=d+1-k}^{d} \binom{d}{k} \binom{k}{d-j} M_{k+j-d}(f) W_k(K) W_j(L). \tag{4.2}$$

This was first proved by Hadwiger (1975a). Similar results for moving flats are due to Bokowski, Hadwiger and Wills (1976). In his proof of (4.2), Hadwiger assumed monotonicity for f and then deduced the result from his axiomatic characterization of the quermassintegrals. In a more direct way, (4.2) can be obtained as follows. First assume that f is the indicator function of the interval $(a, b]$, where $0 \leqslant a < b$. Then

$$\int_{G_d} f(r(K, gL))\, \mathrm{d}\mu(g)$$

$$= \mu(\{g \in G_d\colon\ (K + bB^d) \cap gL \neq \emptyset\})$$

$$- \mu(\{g \in G_d\colon\ (K + aB^d) \cap gL \neq \emptyset\})$$

$$= \frac{1}{\kappa_d} \sum_{k=0}^{d} \binom{d}{k} [W_k(K + bB^d) - W_k(K + aB^d)] W_{d-k}(L)$$

by (4.1). The application of the Steiner formula for the quermassintegral W_k of a parallel body now leads to (4.2) for functions f of the special type considered.

The extension to more general functions is then achieved by standard arguments of integration theory. By essentially this method, different generalizations of (4.2) were proved by Schneider (1977) and Groemer (1980b).

Local versions of (4.2) make sense in different ways. For example, one may integrate only over those rigid motions g for which a pair of points realizing the distance of K and gL, or the direction of the difference of these points, belongs to a specified set. First we state a simpler result of the latter type. For a convex body K and a closed convex set L, let $x \in K$ and $y \in L$ be points at distance $r(K,L)$. Then the unit vector pointing from K to L is defined by $u(K,L) := (y-x)/r(K,L)$. This vector is unique, although the pair (x,y) is not necessarily unique. If $\alpha, \beta \subset \mathcal{B}(S^{d-1})$ are Borel sets on the unit sphere, if

$$M(K,L;\alpha,\beta) := \{g \in G_d\colon\ K \cap gL = \emptyset \text{ and } u(K,gL) \in \alpha \cap g_0\check{\beta}\},$$

where $g_0 \in \mathrm{SO}(d)$ denotes the rotation part of $g \in G_d$, and if f is as in (4.2), then

$$\int_{M(K,L;\alpha,\beta)} f(r(K,gL))\,\mathrm{d}\mu(g) = \frac{1}{d^2\kappa_d}\sum_{k,j=0}^{d-1}\binom{d}{k}\binom{d-k}{j} M_{d-k-j}(f) S_k(K,\alpha) S_j(L,\beta). \tag{4.3}$$

Formulae of this type were first obtained by Hadwiger (1975b). A short proof of (4.3), using (3.11), was given by Schneider (1977). Since (4.3) can be interpreted as involving the indicator functions of α and β, it is not surprising that further generalizations, involving more general functions, are possible.

We describe some rather general formulae, due mainly to Weil (1979a,b, 1981). These concern integrals of the types

$$\int_{K\cap gL=\emptyset} f(g)\,\mathrm{d}\mu(g) \quad\text{and}\quad \int_{K\cap E=\emptyset} h(E)\,\mathrm{d}\mu_q(E),$$

for functions f and h depending in different ways on the geometric situation. For given convex bodies $K, L \in \mathcal{K}^d$, Weil (1979b) established a decomposition of the form

$$\mu\,|_{\{g\colon\ K\cap gL=\emptyset\}} = \int_0^\infty \mu^{(r)}(K,L;\cdot)\,\mathrm{d}r, \tag{4.4}$$

where $\mu^{(r)}(K,L;\cdot)$ is a finite Borel measure concentrated on the set of rigid motions g for which $r(K,gL) = r$. He deduced that, for a μ-integrable function f on G_d,

$$\int_{K\cap gL=\emptyset} f(g)\,\mathrm{d}\mu(g) = \int_0^\infty \int_{\mathrm{SO}(d)} \int_{\mathbb{E}^d} f(\gamma(t,\rho))\,\mathrm{d}C_{d-1}(K + rB^d + \rho\check{L}, t)\,\mathrm{d}\nu(\rho)\,\mathrm{d}r. \tag{4.5}$$

From this, the following result can be derived.

Theorem 4.1. *Let* $f : (0,\infty) \times S^{d-1} \times \mathrm{SO}(d) \to \mathbb{R}$ *be a measurable function for which the integrals in* (4.6) *are finite; then*

$$\int_{K\cap gL=\emptyset} f(r(K,gL),u(K,gL),g_0)\,\mathrm{d}\mu(g)$$

$$= \sum_{j=0}^{d-1} \binom{d-1}{j} \int_0^\infty \int_{\mathrm{SO}(d)} \int_{S^{d-1}} f(r,u,\rho)\,\mathrm{d}S_j(K+\rho\check{L},u)\,\mathrm{d}\nu(\rho) r^{d-j-1}\,\mathrm{d}r. \tag{4.6}$$

For more restricted functions, the integration over the rotation group on the right side disappears:

Theorem 4.2. *Let* $f : (0,\infty) \times S^{d-1} \times S^{d-1} \to \mathbb{R}$ *be a measurable function for which the integrals in* (4.7) *are finite, then*

$$\int_{K\cap gL=\emptyset} f(r(K,gL),u(K,gL),g_0^{-1}u(gL,K))\,\mathrm{d}\mu(g)$$

$$= d\kappa_d \sum_{j=0}^{d-1}\sum_{k=0}^{j} \binom{d-1}{j}\binom{j}{k}$$

$$\times \int_0^\infty \int_{S^{d-1}} \int_{S^{d-1}} f(r,u,v)\,\mathrm{d}S_k(K,u)\,\mathrm{d}S_{j-k}(L,v) r^{d-j-1}\,\mathrm{d}r. \tag{4.7}$$

The following theorem contains a counterpart for variable q-flats instead of a moving convex body.

Theorem 4.3. *If* $q \in \{0,\ldots,d-1\}$ *and if* $h : (0,\infty) \times S^{d-1} \to \mathbb{R}$ *is a measurable function for which the integrals in* (4.8) *are finite, then*

$$\int_{K\cap E=\emptyset} h(r(K,E),u(K,E))\,\mathrm{d}\mu_q(E)$$

$$= (d-q)\kappa_{d-q} \sum_{j=0}^{d-q-1} \binom{d-q-1}{j} \int_0^\infty \int_{S^{d-1}} h(r,u)\,\mathrm{d}S_j(K,u) r^{d-q-j-1}\,\mathrm{d}r. \tag{4.8}$$

One also has a translative version of Theorem 4.2:

Theorem 4.4. *If* $h : (0,\infty) \times S^{d-1} \to \mathbb{R}$ *is a measurable function for which the integrals in* (4.9) *are finite, then*

$$\int_{K\cap(L+t)=\emptyset} h(r(K,L+t),u(K,L+t))\, \mathrm{d}\lambda_d(t)$$

$$= \sum_{i,k=0}^{d-1} \binom{d-1}{i+k}\binom{i+k}{i}$$

$$\times \int_0^\infty \int_{S^{d-1}} h(r,u)\, \mathrm{d}S(\underbrace{K,\dots,K}_{i},\underbrace{L,\dots,L}_{k},\underbrace{B^d,\dots,B^d}_{d-1-i-k},u) r^{d-1-i-k}\, \mathrm{d}r. \tag{4.9}$$

So far, we have integrated functions that involve the positive distance of a convex body K from a moving convex set gL and the unit vector pointing from K to gL, but not the boundary points realizing the distance. If one wants to take the latter into account, the difficulty arises that a pair (x,y) of points realizing the distance of K and gL, where $K\cap gL=\emptyset$, is in general not unique. However, for μ-almost all $g\in G_d$, the distance of the disjoint convex bodies K and gL is realized by a unique pair. This was proved in Schneider (1978b), and the corresponding result for flats in Schneider (1978a). Once this is known, one can prove analogues of some of the results above. For a convex body K and a closed convex set L with $K\cap L=\emptyset$ define $x(K,L):=x$ if $x\in K$ is such that $\| x-y \| = r(K,L)$ for some $y\in L$ and the pair (x,y) is unique.

Theorem 4.5. *Let $f:(0,\infty)\times \operatorname{bd} K\times \operatorname{bd} L\to \mathbb{R}$ be a measurable function for which the integrals in* (4.10) *are finite, then*

$$\int_{K\cap gL=\emptyset} f(r(K,gL),x(K,gL),g^{-1}x(gL,K))\, \mathrm{d}\mu(g)$$

$$= d\kappa_d \sum_{j=0}^{d-1}\sum_{k=0}^{j} \binom{d-1}{j}\binom{j}{k}$$

$$\times \int_0^\infty \int_{\operatorname{bd} K}\int_{\operatorname{bd} L} f(r,x,y)\, \mathrm{d}C_k(K,x)\, \mathrm{d}C_{j-k}(L,y) r^{d-j-1}\, \mathrm{d}r. \tag{4.10}$$

Theorem 4.6. *If $q\in\{0,\dots,d-1\}$ and if $h:(0,\infty)\times \operatorname{bd} K\to\mathbb{R}$ is a measurable function for which the integrals in* (4.11) *are finite, then*

$$\int_{K\cap E=\emptyset} h(r(K,E),x(K,E)) d\mu_q(E)$$

$$= (d-q)\kappa_{d-q} \sum_{j=0}^{d-q-1} \binom{d-q-1}{j} \int_0^\infty \int_{\operatorname{bd} K} h(r,x)\, \mathrm{d}C_j(K,x) r^{d-q-j-1}\, \mathrm{d}r. \tag{4.11}$$

It should be clear by now that common generalizations of these results are possible by extending them to support elements, generalized curvature measures, and convex cylinders.

The integrals involving the distance of a convex body and a moving convex set are closely related to contact measures. The contact measure of two convex bodies K and L is a measure which is concentrated on the set of rigid motions g for which K and gL are in contact, that is, touch each other, and which is derived from the Haar measure on G_d in the following natural way. For convex bodies $K, L \in \mathcal{K}^d$ we define

$$G_0(K,L) := \{g \in G_d\colon\ gL \text{ touches } K\}.$$

(gL touches K if $K \cap L \neq \emptyset$, but gL and K can be separated weakly by a hyperplane.) If $K \cap L = \emptyset$, there is a unique translation $\tau = \tau(K,L)$ by a vector of length $r(K,L)$ (namely, $-r(K,L)u(K,L)$) such that $K \cap \tau L \neq \emptyset$. If now $\alpha \in \mathcal{B}(G_d)$ is a Borel set in the motion group and if $\varepsilon > 0$, we define

$$A_\varepsilon(K,L,\alpha) := \{g \in G_d\colon\ 0 < r(K,gL) \leqslant \varepsilon,\ \tau(K,gL) \circ g \in \alpha\}.$$

This set is a Borel set, and for its Haar measure one obtains

$$\mu(A_\varepsilon(K,L,\alpha)) = \frac{1}{d}\sum_{j=1}^{d-1} \varepsilon^{d-j}\binom{d}{j}\int_{\mathrm{SO}(d)} C_j(K + \rho\check{L}, T(\alpha,\rho))\, \mathrm{d}\nu(\rho)$$

with

$$T(\alpha,\rho) := \{t \in \mathbb{E}^d\colon\ \gamma(t,\rho) \in \alpha \cap G_0(K,L)\}.$$

Hence, the limit

$$\varphi(K,L,\alpha) := \lim_{\varepsilon \to 0} \frac{1}{\varepsilon}\mu(A_\varepsilon(K,L,\alpha))$$

exists and is given by

$$\varphi(K,L,\alpha) = \int_{\mathrm{SO}(d)} C_{d-1}(K + \rho\check{L}, T(\alpha,\rho))\, \mathrm{d}\nu(\rho). \tag{4.12}$$

Thus, $\varphi(K,L,\cdot)$ is a finite Borel measure on G_d, which is concentrated on the set $G_0(K,L)$ of rigid motions bringing L into contact with K. It is called the *contact measure* of K and L and was introduced in this way by Weil (1979a), who extended and unified formerly treated special cases. Using a different approach, Weil (1979b) also showed that this contact measure is the weak limit

$$\varphi(K,L,\cdot) = w - \lim_{r \to 0} \mu^{(r)}(K,L,\cdot),$$

where $\mu^{(r)}(K,L,\cdot)$ is defined by the disintegration (4.4).

For suitable sets of rigid motions defined by special touching conditions of geometric significance, the contact measure can be expressed in terms of curvature measures. For $K, K' \in \mathcal{K}^d$, $\omega, \omega' \in \mathcal{B}(S^{d-1})$, $\beta, \beta' \in \mathcal{B}(\mathbb{E}^d)$ and $\varepsilon > 0$ we define the following sets. $M_0(K,K';\omega,\omega')$ is the set of rigid motions $g \in G_0(K,K')$ for which the unit normal vector u, pointing from K to gK', of a separating hyperplane of K and gK' satisfies $u \in \omega \cap g_0\check{\omega}'$. $L_0(K,K';\beta,\beta')$ is the set of motions $g \in G_0(K,K')$ for which $\beta \cap \operatorname{bd} K \cap g(\beta' \cap \operatorname{bd} K') \neq \emptyset$. Further,

$$\begin{aligned} &M_\varepsilon(K,K';\omega,\omega') \\ &\quad := \{g \in G_d\colon\ 0 < r(K,gK') \leqslant \varepsilon,\ u(K,gK') \in \omega,\ u(gK',K) \in g_0\omega'\}, \\ &L_\varepsilon(K,K';\beta,\beta') \\ &\quad := \{g \in G_d\colon\ 0 < r(K,gK') \leqslant \varepsilon,\ x(K,gK') \in \beta,\ x(gK',K) \in g\beta'\}. \end{aligned}$$

(Observe that the latter set is only defined up to a set of μ-measure zero.) The sets $M_0(K,K';\omega,\omega'), L_0(K,K';\beta,\beta')$ are $\varphi(K,K',\cdot)$-measurable, and

$$\begin{aligned} &\varphi(K,K',M_0(K,K';\omega,\omega')) \\ &\quad = \lim_{\varepsilon\to 0}\frac{1}{\varepsilon}\mu(M_\varepsilon(K,K';\omega,\omega')) \\ &\quad = \frac{1}{d\kappa_d}\sum_{j=0}^{d-1}\binom{d-1}{j}S_j(K,\omega)S_{d-1-j}(K',\omega'). \end{aligned} \tag{4.13}$$

Here the first equality follows from the definition of the contact measure and the second from a special case of (4.3) and thus of Theorem 4.2. Further,

$$\begin{aligned} &\varphi(K,K',L_0(K,K';\beta,\beta')) \\ &\quad = \lim_{\varepsilon\to 0}\frac{1}{\varepsilon}\mu(L_\varepsilon(K,K';\beta,\beta')) \\ &\quad = \frac{1}{d\kappa_d}\sum_{j=0}^{d-1}\binom{d-1}{j}C_j(K,\beta)C_{d-1-j}(K',\beta'), \end{aligned} \tag{4.14}$$

from a special case of Theorem 4.5. A common generalization of (4.13) and (4.14), involving generalized curvature measures, can be obtained if the touching conditions are formulated in terms of Borel sets of support elements.

In a similar way, contact measures for a convex body and a moving q-flat can be treated. Let $K \in \mathcal{K}^d$ and $q \in \{0,\dots,d-1\}$ be given. Proceeding in obvious analogy to the above, one constructs a natural measure $\varphi_q(K,\cdot)$ on $\mathcal{B}(\mathcal{E}_q^d)$, concentrated on the set of q-flats touching K, and finds that it is given by

$$\varphi_q(K,\alpha) = \frac{1}{d\kappa_d}\int_{\mathrm{SO}(d)} C_{d-q-1}^{(d-q)}(K \mid \rho L_q^\perp, T_q(\alpha,\rho))\,\mathrm{d}\nu(\rho), \tag{4.15}$$

where $\alpha \in \mathcal{B}(\mathcal{E}_q^d), L_q \subset \mathbb{E}^d$ is a fixed q-dimensional linear subspace, and

$$T_q(\alpha,\rho) := \{t \in \rho L_q^\perp\colon\ \rho L_q + t \in \alpha,\ \rho L_q + t \text{ touches } K\}.$$

We mention only one result analogous to (4.13), (4.14), this time formulated for generalized curvature measures. For $\eta \in \mathcal{B}(\Sigma)$ and $\varepsilon > 0$, let $N^q_\varepsilon(K,\eta)$ be the set of all q-flats $E \in \mathcal{E}^d_q$ for which there exist points $x \in K$ and $y \in E$ for which $0 < \|x - y\| = r(K,E) \leqslant \varepsilon$ and $(x, u(K,E)) \in \eta$. Further, let $N^q_0(K,\eta)$ be the set of all q-flats E touching K at a point x and lying in a supporting hyperplane of K with outer unit normal vector u such that $(x,u) \in \eta$. Then $N^q_0(K,\eta)$ is $\varphi_q(K,\cdot)$-measurable (though not necessarily a Borel set, see Burton 1980), and

$$\varphi_q(K, N^q_0(K,\eta)) = \lim_{\varepsilon \to 0} \frac{1}{\varepsilon} \mu_q(N^q_\varepsilon(K,\eta))$$

$$= \frac{(d-k)\kappa_{d-k}}{d\kappa_d} \Theta_{d-q-1}(K,\eta). \tag{4.16}$$

The latter limit relation gives a direct geometric interpretation of the generalized curvature measures. For η of the special kind $\mathbb{E}^d \times \omega$ one obtains a representation of $S_{d-q-1}(K,\omega)$, and for $\eta = \beta \times S^{d-1}$ one of $C_{d-q-1}(K,\beta)$. These special cases are due to Firey (1972) and Schneider (1978a), respectively; the general case was mentioned in Schneider (1980a, Theorem 4.12).

In a common generalization of the cases of touching convex bodies and touching flats, one may define a contact measure for a convex body and a, possibly unbounded, closed convex set; see Weil (1989a), where similar results are obtained.

In the literature, results on contact measures have been studied in connection with so-called *collision* or *touching probabilities*. For example, let d-dimensional convex bodies $K, K' \in \mathcal{K}^d$ and subsets β, β' of their respective boundaries be given. Let K' undergo random motion in such a way that it touches K. What is the probability that the bodies touch at a point belonging to the prescribed boundary sets? (Of course, the same motions are applied to K' and β'.) A reasonable way to make this question precise is to choose the completion of the probability space

$$(G_0(K,K'), \mathcal{B}(G_0(K,K')), \varphi(K,K',\cdot)/\varphi(K,K',G_0(K,K')))$$

as an underlying probabilistic model. If this is done and if β, β' are Borel sets, then (4.14) yields the value

$$\frac{\sum_{j=0}^{d-1} \binom{d-1}{j} C_j(K,\beta) C_{d-1-j}(K',\beta')}{\sum_{j=0}^{d-1} \binom{d-1}{j} C_j(K,K) C_{d-1-j}(K',K')}$$

for the probability of a collision at the preassigned boundary sets.

Contributions to this field of touching probabilities are due to Firey (1974, 1979), McMullen (1974), Molter (1986), Schneider (1975a,b, 1976, 1978b, 1980b), Schneider and Wieacker (1984), and Weil (1979a,b, 1981, 1982, 1989a).

5. Extension to the convex ring

Several of the integral-geometric formulae considered earlier in this article are not restricted to convex bodies. According to the class of sets envisaged, extensions require, say, methods of differential geometry or geometric measure theory. A class of sets which, at our present stage of considerations, is technically easy to treat, but which on the other hand is sufficiently general for applications (see, e.g., chapter 5.2), is provided by the convex ring $\mathcal{R}^d$. This is the set of all finite unions of convex bodies in $\mathbb{E}^d$ (for formal reasons, we assume that also $\emptyset \in \mathcal{R}^d$). A natural and useful extension of the curvature measures, and thus of the intrinsic volumes, to the convex ring is achieved if one exploits their additivity property.

Recall that a function $\varphi : \mathcal{R}^d \to A$ from $\mathcal{R}^d$ into some Abelian group A is called *additive* if $\varphi(\emptyset) = 0$ and

$$\varphi(K \cup L) + \varphi(K \cap L) = \varphi(K) + \varphi(L) \tag{5.1}$$

for all $K, L \in \mathcal{R}^d$ (compare chapter 3.6). For such a function, the *inclusion–exclusion principle* says that

$$\varphi(K_1 \cup \cdots \cup K_m) = \sum_{r=1}^{m} (-1)^{r-1} \sum_{i_1 < \cdots < i_r} \varphi(K_{i_1} \cap \cdots \cap K_{i_r}) \tag{5.2}$$

for $K_1, \ldots, K_m \in \mathcal{R}^d$. In particular, the values of the function φ on $\mathcal{R}^d$ are uniquely determined by its values on $\mathcal{K}^d$. For a more concise notation, let $S(m)$ be the set of nonempty subsets of $\{1, \ldots, m\}$ and write $|v| := \operatorname{card} v$ for $v \in S(m)$ as well as

$$K_v := K_{i_1} \cap \cdots \cap K_{i_r} \quad \text{for } v = \{i_1, \ldots, i_r\} \in S(m)$$

if $K_1, \cdots, K_m$ are given. Then (5.2) can be written in the form

$$\varphi(K_1 \cup \cdots \cup K_m) = \sum_{v \in S(m)} (-1)^{|v|-1} \varphi(K_v). \tag{5.3}$$

Let $j \in \{0, 1, \ldots, d-1\}$. The generalized curvature measure Θ_j is additive on $\mathcal{K}^d$, that is, it satisfies

$$\Theta_j(K \cup L, \cdot) + \Theta_j(K \cap L, \cdot) = \Theta_j(K, \cdot) + \Theta_j(L, \cdot)$$

whenever $K, L \in \mathcal{K}^d$ are such that $K \cup L$ is convex (see chapter 1.8). One can show that the map Θ_j can be extended from $\mathcal{K}^d \times \mathcal{B}(\Sigma)$ to $\mathcal{R}^d \times \mathcal{B}(\Sigma)$ such that $\Theta_j(\cdot, \eta)$ is additive for each $\eta \in \mathcal{B}(\Sigma)$. If then K is a set of the convex ring, represented as $K = K_1 \cup \cdots \cup K_m$ with convex bodies $K_1, \ldots, K_m$, we have

$$\Theta_j(K, \cdot) = \sum_{v \in S(m)} (-1)^{|v|-1} \Theta_j(K_v, \cdot), \tag{5.4}$$

which shows that $\Theta_j(K, \cdot)$ is a finite signed measure on $\mathcal{B}(\Sigma)$. The possibility of additive extension could be deduced from a general theorem of Groemer (1978), using the weak continuity of the generalized curvature measures. The following approach proceeds in a more explicit way and yields additional information.

For $K \in \mathcal{R}^d$ and $q, x \in \mathbb{E}^d$ the *index* of K at q with respect to x is defined by

$$j(K,q,x) := \begin{cases} 1 - \lim_{\delta \to +0} \lim_{\varepsilon \to +0} \chi(K \cap B(x, \|x-q\| - \varepsilon) \cap B(q,\delta)) & \text{if } q \in K, \\ 0 \quad \text{if } q \notin K, \end{cases}$$

where χ denotes the Euler characteristic and $B(z,\rho)$ is the closed ball with centre z and radius ρ. Then $j(\cdot, q, x)$ is additive, and for convex K one simply has

$$j(K,q,x) = \begin{cases} 1 & \text{if } q = p(K,x), \\ 0 & \text{else} \end{cases}$$

(recall that $p(K,x)$ is the point of K nearest to x). Next, for $K \in \mathcal{R}^d$, a Borel set $\eta \in \mathcal{B}(\Sigma)$, a number $\varepsilon > 0$ and for $x \in \mathbb{E}^d$ one defines

$$c_\varepsilon(K,\eta,x) := \sum_* j(K \cap B(x,\varepsilon), q, x),$$

where the sum $\sum_*$ extends over the points $q \in \mathbb{E}^d$ for which $q \neq x$ and $(q,u) \in \eta$ for $u := (x-q)/\|x-q\|$ (only finitely many summands are not zero). If K is convex, then $c_\varepsilon(K,\eta,\cdot)$ is the indicator function of the set $M_\varepsilon(K,\eta)$ defined by (3.6). The function $c_\varepsilon(\cdot,\eta,x)$ is additive on $\mathcal{R}^d$, hence the function $\mu_\varepsilon(\cdot,\eta)$ defined by

$$\mu_\varepsilon(K,\eta) := \int_{\mathbb{E}^d} c_\varepsilon(K,\eta,x)\, \mathrm{d}\lambda_d(x)$$

for $K \in \mathcal{R}^d$ is additive, too. From (3.5) it follows that

$$\mu_\varepsilon(K,\eta) = \frac{1}{d} \sum_{j=0}^{d-1} \varepsilon^{d-j} \binom{d}{j} \sum_{v \in S(m)} (-1)^{|v|-1} \Theta_j(K_v, \eta),$$

if $K = \bigcup_{i=1}^m K_i$ with $K_i \in \mathcal{K}^d$. Since the left side does not depend on the special representation of K, the same is true for

$$\Theta_j(K,\eta) := \sum_{v \in S(m)} (-1)^{|v|-1} \Theta_j(K_v,\eta).$$

This defines the generalized curvature measures $\Theta_0, \ldots, \Theta_{d-1}$ on the convex ring $\mathcal{R}^d$. The defining Steiner type formula

$$\int_{\mathbb{E}^d} c_\varepsilon(K,\eta,x)\, \mathrm{d}\lambda_d(x) = \frac{1}{d} \sum_{j=0}^{d-1} \varepsilon^{d-j} \binom{d}{j} \Theta_j(K,\eta)$$

is the immediate generalization of (3.5), with the measure of the local parallel set $M_\varepsilon(K,\eta)$ replaced by the integral of the additive extension $c_\varepsilon(K,\eta,\cdot)$ of its indicator function.

The measures $C_j(K,\cdot)$, $\Phi_j(K,\cdot)$ on $\mathcal{B}(\mathbb{E}^d)$ and $S_j(K,\cdot)$, $\Psi_j(K,\cdot)$ on $\mathcal{B}(S^{d-1})$ are now obtained by specialization, verbally in the same way as in (3.7), (3.8). One also defines $\Phi_d(K,\beta) := \lambda_d(K \cap \beta)$ for $\beta \in \mathcal{B}(\mathbb{E}^d)$.

In the way described here, the additive extensions of the generalized curvature measures to the convex ring were constructed in Schneider (1980a).

Once this additive extension has been achieved, the generalization of some of the integral-geometric formulae is immediate. We demonstrate this for Theorem 2.1. Let $K \in \mathcal{R}^d$ be a set of the convex ring. It has a representation $K = K_1 \cup \cdots \cup K_m$ with $K_1,\ldots,K_m \in \mathcal{K}^d$. First let $K' \in \mathcal{K}^d$ be a convex body. For $j \in \{0,\ldots,d\}$ and $g \in G_d$ we have

$$\begin{aligned}\Phi_j(K \cap gK',\cdot) &= \Phi_j((K_1 \cap gK') \cup \cdots \cup (K_m \cap gK'),\cdot)\\ &= \sum_{v \in S(m)} (-1)^{|v|-1} \Phi_j(K_v \cap gK',\cdot)\end{aligned}$$

by (5.4). Now (2.1) yields, for $\beta,\beta' \in \mathcal{B}(\mathbb{E}^d)$,

$$\begin{aligned}&\int_{G_d} \Phi_j(K \cap gK', \beta \cap g\beta')\, \mathrm{d}\mu(g)\\ &\quad= \sum_{v \in S(m)} (-1)^{|v|-1} \int_{G_d} \Phi_j(K_v \cap gK', \beta \cap g\beta')\, \mathrm{d}\mu(g)\\ &\quad= \sum_{v \in S(m)} (-1)^{|v|-1} \sum_{k=j}^{d} \alpha_{djk} \Phi_{d+j-k}(K_v,\beta)\Phi_k(K',\beta')\\ &\quad= \sum_{k=j}^{d} \alpha_{djk} \Phi_{d+j-k}(K,\beta)\Phi_k(K',\beta').\end{aligned}$$

In a second step, K' can be replaced by a set of the convex ring, in precisely the same way. Thus formula (2.1) is valid if K and K' are elements of the convex ring.

In a strictly analogous way, Theorem 2.3 extends to the convex ring. Extensions are also possible for Theorem 2.2, and for Theorem 2.4 if φ is additive on $\mathcal{R}^d$ and continuous on $\mathcal{K}^d$. The role of additivity for the extension of integral-geometric formulae to the convex ring was mainly emphasized by Hadwiger (1957).

The interpretation of the generalized curvature measures for convex bodies that is given by (4.16) can be extended to the measures Θ_{d-q-1} on the convex ring, if the measure of the set $N_\varepsilon^q(K,\eta)$ is replaced by the μ_q-integral of a suitable multiplicity function. This can be achieved using a more general version of the index introduced above; see Schneider (1988b). There one also finds an extension of the projection formula (3.18) to the convex ring, where a notion of tangential projections with multiplicities plays a role.

6. Translative integral geometry and auxiliary zonoids

Some translative integral-geometric formulae for convex bodies have already been mentioned in section 2. In an essentially different way, convex bodies play an unexpected and useful role in translative integral geometry for quite general rectifiable sets. This is due to the fact that each finite measure on the space of m-dimensional linear subspaces of $\mathbb{E}^d$ induces, in a natural way, a pseudo-norm on $\mathbb{E}^d$ which is the support function of a zonoid. Thus one can exploit classical results on convex bodies to treat several extremal problems of translative integral geometry and of stochastic geometry. For applications of the latter type we refer to chapter 5.2. Auxiliary zonoids in the sense to be discussed were first introduced by Matheron (1975) (although a special result of this type occurs already in Blaschke 1937). We present the method in a rather general version.

By $\mathcal{L}_m^d$ we denote the space of m-dimensional linear subspaces of $\mathbb{E}^d$, topologized as usual (as a subspace of $\mathcal{F}^d$, the space of closed subsets of $\mathbb{E}^d$, see section 1). For $m \in \{1, \ldots, d-1\}$, a subspace $L \in \mathcal{L}_m^d$ and $x \in \mathbb{E}^d$ we write $r_L(x) := r(L, \{x\})$ for the distance of x from L. Then r_L is the support function of the convex body $B^d \cap L^\perp$. Hence, if τ is some finite (Borel) measure on $\mathcal{L}_m^d$, then there is a unique convex body $\Pi^m(\tau)$ such that

$$h(\Pi^m(\tau), \cdot) = \tfrac{1}{2} \int_{\mathcal{L}_m^d} r_L(\cdot)\, \mathrm{d}\tau(L), \tag{6.1}$$

where h denotes the support function. Since each $B^d \cap L^\perp$ is a zonoid, $\Pi^m(\tau)$ is a zonoid (see chapter 4.9). For the mean width of $\Pi^m(\tau)$ one finds (using Fubini's theorem and (56) on p. 217 of Hadwiger 1957)

$$\bar{b}(\Pi^m(\tau)) = \frac{(d-m)\kappa_{d-m}}{d\kappa_d} \frac{\kappa_{d-1}}{\kappa_{d-m-1}} \tau(\mathcal{L}_m^d). \tag{6.2}$$

A second zonoid is obtained by putting $\Pi_m(\tau) := \Pi^{d-m}(\tau^\perp)$, where $\tau^\perp$ is the image measure of τ under the map $L \mapsto L^\perp$ from $\mathcal{L}_m^d$ on to $\mathcal{L}_{d-m}^d$. In the case $m = d-1$, $h(\Pi_m(\tau), \cdot)$ is essentially the spherical Radon transform of $h(\Pi^m(\tau), \cdot)$ (see, e.g., Wieacker 1984, Lemma 1).

Since stationary (i.e., translation invariant) measures on the space $\mathcal{E}_m^d$ of m-flats, or more generally on the space of m-dimensional surfaces, under weak assumptions induce finite measures on $\mathcal{L}_m^d$, the foregoing simple construction has far-reaching consequences.

A subset $M \subset \mathbb{E}^d$ is called *m-rectifiable* $(1 \leqslant m \leqslant d-1)$ if it is the image of some bounded subset of $\mathbb{E}^m$ under a Lipschitz map, and *countably m-rectifiable* if it is the union of a countable family of m-rectifiable sets (see Federer 1969 for details). If M is countably m-rectifiable and λ_m-measurable, then there are countably many m-dimensional C^1-submanifolds $N_1, N_2, \ldots$ of $\mathbb{E}^d$ such that $\lambda_m(M \setminus \bigcup_{i \in \mathbb{N}} N_i) = 0$. Suppose that, moreover, $\lambda_m(M) < \infty$ and let $T_x N_i$ denote the tangent space of N_i at $x \in N_i$. Then, defining

$$\tau_M(S) := \lambda_m\Big(\bigcup_{i \in \mathbb{N}} \{x \in M \cap N_i \colon\ T_x N_i \in S\}\Big) \tag{6.3}$$

for $S \in \mathcal{B}(\mathcal{L}_m^d)$, we get a finite measure τ_M on $\mathcal{L}_m^d$, which can be shown to depend only on M. Hence, we can define

$$\Pi^m(M) := \Pi^m(\tau_M), \qquad \Pi_m(M) := \Pi_m(\tau_M).$$

Since $\tau_M(\mathcal{L}_m^d) = \lambda_m(M)$, (6.2) implies that the mean widths of the two zonoids $\Pi^m(M)$ and $\Pi_m(M)$ are, up to numerical factors, equal to $\lambda_m(M)$. As an example, if $K \in \mathcal{K}^d$ is a convex body with interior points, then $\operatorname{bd} K$ is countably $(d-1)$-rectifiable and $\Pi^{d-1}(\operatorname{bd} K)$ is the usual projection body of K. If K is a line segment, then $\Pi_1(K)$ is a translate of K. Clearly, if M and M' are countably m-rectifiable and λ_m-measurable subsets of $\mathbb{E}^d$ with finite λ_m-measure, then

$$\Pi^m(M \cup M') + \Pi^m(M \cap M') = \Pi^m(M) + \Pi^m(M'), \tag{6.4}$$

and the same relation holds for Π_m.

The preceding construction can be considerably generalized. Let

$$\mathcal{LC}_m := \{F \in \mathcal{F}^d\colon\ F \cap K \text{ is countably } m\text{-rectifiable } \forall K \in \mathcal{K}^d\}$$

be the space of locally countably m-rectifiable closed sets. If θ is a stationary σ-finite measure on $\mathcal{LC}_m$ and $\lambda_m(\cdot \cap [0,1]^d)$ is θ-integrable, then there is a unique convex body $\Pi^m(\theta)$ satisfying

$$\lambda_d(A) h(\Pi^m(\theta), x) = \int_{\mathcal{LC}_m} h(\Pi^m(F \cap A), x)\, \mathrm{d}\theta(F) \tag{6.5}$$

for all $x \in \mathbb{E}^d$, whenever $A \in \mathcal{B}(\mathbb{E}^d)$ and $\lambda_d(A) < \infty$. The same holds for Π_m instead of Π^m.

The following lemma (Wieacker 1989) is often useful in exploiting the translation invariance in the proof of integral-geometric formulae for stationary σ-finite measures.

Lemma 6.1. *Let $\mathcal{T} \subset \mathcal{F}^d$ be a translation invariant measurable subset, θ a stationary σ-finite measure on $\mathcal{T}$, and ξ a measure on $\mathbb{E}^d$. If the map $K \mapsto \xi(K)$ is measurable on the space of nonempty compact subsets of $\mathbb{E}^d$, then*

$$\int_{\mathcal{T}} \xi(F)\, \mathrm{d}\theta(F) = \int_{\mathcal{T}} \int_{\mathbb{E}^d} \xi((F \cap]0,1]^d) + x)\, \mathrm{d}\lambda_d(x)\, \mathrm{d}\theta(F).$$

The general construction of auxiliary zonoids described here includes some special cases appearing in the literature.

Example 6.1. Let M be a countably m-rectifiable closed set with $\lambda_m(M) < \infty$, and let θ be the image measure of λ_d under the map $x \mapsto M + x$ from $\mathbb{E}^d$ into $\mathcal{LC}_m$. Then one easily shows that $\Pi^m(M) = \Pi^m(\theta)$.

Example 6.2. Let θ be a stationary measure on $\mathcal{LC}_m$ which is concentrated on $\mathcal{E}_m^d$ and locally finite in the sense that $\theta(\mathcal{F}_K \cap \mathcal{E}_m^d) < \infty$ for each compact subset K of $\mathbb{E}^d$. Then there is a unique finite measure $\theta^\dagger$ on $\mathcal{L}_m^d$ such that

$$\theta(A) = \int_{\mathcal{L}_m^d} \int_{L^\perp} \mathbf{1}_A(L + x)\, \mathrm{d}\lambda_{d-m}(x)\, \mathrm{d}\theta^\dagger(L) \tag{6.6}$$

for each Borel subset A of $\mathcal{L}^d_m$, where $\mathbf{1}_A$ denotes the indicator function of A. This is a consequence of Proposition 3.2.2 and its corollary in Matheron (1975) (an extension to convex cylinders is treated in Schneider 1987, Lemma 3.3). Hence, by (6.1), we may associate with each locally finite, stationary measure θ on $\mathcal{E}^d_m$ the zonoids $\Pi^m(\theta^\dagger)$ and $\Pi_m(\theta^\dagger)$, and it turns out that $\Pi^m(\theta) = \Pi^m(\theta^\dagger)$ and $\Pi_m(\theta) = \Pi_m(\theta^\dagger)$. From (6.1) and (6.3) it is now easy to see that, for a locally finite stationary measure θ on $\mathcal{E}^d_{d-1}$ and $x \in \mathbb{E}^d$, we have $2h(\Pi^{d-1}(\theta), x) = \theta(\mathcal{F}_{[0,x]} \cap \mathcal{E}^d_{d-1})$. We shall see that this is a special case of a more general result.

Example 6.3. Let θ be a stationary measure on the space of non-empty compact sets with $\theta(\{K \in \mathcal{K}^d\colon K \cap [0,1]^d \neq \emptyset\}) < \infty$, and let Q be the set of all $K \in \mathcal{K}^d$ the circumsphere of which has centre 0. Then there is a unique finite measure $\theta^\sharp$ on Q such that θ is the image measure of $\theta^\sharp \otimes \lambda_d$ under the map $(K,x) \mapsto K + x$ from $Q \times \mathbb{E}^d$ into $\mathcal{K}^d$. Now, if θ is concentrated on $\mathcal{L}\mathcal{C}_m \cap \mathcal{K}^d$ and if $\lambda_m(\cdot \cap [0,1]^d)$ is θ-integrable, then Example 1 shows that

$$h(\Pi^m(\theta), x) = \int_Q h(\Pi^m(K), x)\, \mathrm{d}\theta^\sharp(K) \tag{6.7}$$

for all $x \in \mathbb{E}^d$. Thus, $\Pi^m(\theta)$ is in a certain sense the $\theta^\sharp$-integral of the function which associates with each $K \in \mathcal{L}\mathcal{C}_m \cap Q$ the convex body $\Pi^m(K)$. A formula analogous to (6.7) and a similar remark hold for $\Pi_m(\theta)$.

Now we turn to the relations between intersection problems of translative integral geometry and associated convex bodies, in particular to *Poincaré-type formulae* and extremal problems. We need some more notation. For $i = 1, \ldots, n$ let L_i be an m_i-dimensional linear subspace of $\mathbb{E}^d$. By $D(L_1, \ldots, L_n)$ we denote the $(m_1 + \cdots + m_n)$-dimensional volume of $K_1 + \cdots + K_n$, where $K_i \subset L_i$ is a compact set of λ_{m_i}-measure one; clearly $D(L_1, \ldots, L_n)$ depends only on $L_1, \ldots, L_n$. Further, if M_i is a countably m_i-rectifiable Borel subset of $\mathbb{E}^d$ with $\lambda_{m_i}(M_i) < \infty$, for $i = 0, \ldots, n$, and if $m := m_0 + \cdots + m_n \geqslant nd$, then we use the notation

$$I(M_0, m_0; \ldots; M_n, m_n)$$

$$:= \int_{\mathbb{E}^d} \cdots \int_{\mathbb{E}^d} \lambda_{m-nd}(M_0 \cap (M_1 + y_1) \cap \cdots \cap (M_n + y_n))\, \mathrm{d}\lambda_d(y_1) \cdots \mathrm{d}\lambda_d(y_n).$$

The following theorem shows that the value of this translative intersection integral depends only on the measures $\tau_{M_0}, \ldots, \tau_{M_n}$ defined by (6.3).

Theorem 6.2. *For $i = 0, \ldots, n < d$, let M_i be a countably m_i-rectifiable Borel subset of $\mathbb{E}^d$ with $\lambda_{m_i}(M_i) < \infty$, and suppose that $m := m_0 + \cdots + m_n \geqslant nd$. Then*

$$I(M_0, m_0; \ldots, M_n, m_n)$$

$$= \int_{\mathcal{L}^d_{m_0}} \cdots \int_{\mathcal{L}^d_{m_n}} D(L_0^\perp, \ldots, L_n^\perp)\, \mathrm{d}\tau_{M_0}(L_0) \cdots \mathrm{d}\tau_{M_n}(L_n).$$

For the proof, one first assumes that $m = nd$ and applies Federer's area formula to the Lipschitz map $f : M_0 \times \cdots \times M_n \to (\mathbb{E}^d)^n$ defined by $f(x_0, \ldots, x_n) := (x_0 - x_1, \ldots, x_0 - x_n)$, observing that

$$\lambda_0(M_0 \cap (M_1 + y_1) \cap \cdots \cap (M_n + y_n)) = \lambda_0(f^{-1}(\{(y_1, \ldots, y_n)\}))$$

in this case. To prove the general case, let M_{n+1} be some m_{n+1}-dimensional cube in $\mathbb{E}^d$, where $m_{n+1} := (n+1)d - m$, and compute the integral

$$\int_{\mathrm{SO}(d)} I(M_0, m_0; \ldots; M_n, m_n; \rho(M_{n+1}), m_{n+1}) \, \mathrm{d}\nu(\rho)$$

with the formula obtained in the first case.

Very special cases of this statement go back to Berwald and Varga (1937) (a detailed proof of the general assertion can be found in Wieacker 1984).

For a countably $(d-1)$-rectifiable Borel subset M of $\mathbb{E}^d$ with $\lambda_{d-1}(M) < \infty$, Theorem 6.2 implies

$$2h(\Pi^{d-1}(M), x) = I([0, x], 1; M, d-1).$$

Hence, if θ is a translation invariant σ-finite measure on $\mathcal{LC}_{d-1}$ and $\lambda_{d-1}(\cdot \cap [0,1]^d)$ is θ-integrable, we infer from Lemma 6.1 that

$$2h(\Pi^{d-1}(\theta), x) = \int_{\mathcal{LC}_{d-1}} \lambda_0([0, x] \cap F) \, \mathrm{d}\theta(F)$$

for all $x \in \mathbb{E}^d$.

The connection between translative integral geometry of general surfaces and the theory of convex bodies is now established by the observation that in some cases the integral in Theorem 6.2 can be expressed as a mixed volume of auxiliary convex bodies. We give two typical examples.

Theorem 6.3. *Let $\theta_1, \ldots, \theta_m$ be translation invariant σ-finite measures on $\mathcal{LC}_{d-1}$ satisfying $\theta_i(\mathcal{F}_K) < \infty$ for all $K \in \mathcal{K}^d$ and $i = 1, \ldots, m$. If $\lambda_{d-1}(\cdot \cap [0,1]^d)$ is θ_i-integrable for $i = 1, \ldots, m$, then*

$$\int_{\mathcal{LC}_{d-1}} \cdots \int_{\mathcal{LC}_{d-1}} \lambda_{d-m}(F_1 \cap \cdots \cap F_m \cap A) \, \mathrm{d}\theta_1(F_1) \cdots \mathrm{d}\theta_m(F_m)$$

$$= \frac{d!}{(d-m)!\kappa_{d-m}} V(\Pi^{d-1}(\theta_1), \ldots, \Pi^{d-1}(\theta_m), B^d, \ldots, B^d) \lambda_d(A)$$

for each bounded Borel subset A of $\mathbb{E}^d$.

For the proof, we take $m_0 = \cdots = m_n = d-1$ in Theorem 6.2; then the multiple integral there can be considered as an integral over $(S^{d-1})^{n+1}$, with τ_{M_i} corresponding to the generating measure of the zonoid $\Pi^{d-1}(M_i)$. Hence, the statement of Theorem 6.3 follows from Lemma 6.1, known results about zonoids, and the linearity of the mixed volume in each argument (more details in Wieacker 1986). A very special case of this theorem was proved by Goodey and Woodcock (1979).

As a consequence of Theorem 6.3, each inequality for mixed volumes leads to an inequality for the integral on the left-hand side in Theorem 6.3. For example ($\theta = \theta_1 = \cdots = \theta_m$),

$$\int_{\mathcal{LE}_{d-1}} \cdots \int_{\mathcal{LE}_{d-1}} \lambda_{d-m}(F_1 \cap \cdots \cap F_m \cap A)\, \mathrm{d}\theta(F_1) \cdots \mathrm{d}\theta(F_m)$$

$$\leqslant \frac{d!\kappa_d}{(d-m)!\kappa_{d-m}} \left(\frac{\kappa_{d-1}}{d\kappa_d} \int_{\mathcal{LE}_{d-1}} \lambda_{d-1}(F \cap [0,1]^d)\, \mathrm{d}\theta(F) \right)^m \lambda_d(A),$$

with equality if and only if $\Pi^{d-1}(\theta)$ is a ball (which, for instance, is the case if θ is rigid motion invariant).

If $\Pi^{d-1}(\theta)$ has interior points, then by Minkowski's theorem there is a uniquely determined centrally symmetric convex body $\Psi^{d-1}(\theta)$, centred at the origin, the area measure of which is the generating measure of $\Pi^{d-1}(\theta)$. This associated convex body appears in the following intersection formula.

Theorem 6.4. *Suppose that η and θ are translation invariant σ-finite measures on $\mathcal{LE}_m$ and $\mathcal{LE}_{d-1}$, respectively, and that $\lambda_m(\cdot \cap [0,1]^d)$ is η-integrable and $\lambda_{d-1}(\cdot \cap [0,1]^d)$ is θ-integrable. Then*

$$\int_{\mathcal{LE}_{d-1}} \int_{\mathcal{LE}_m} \lambda_{m-1}(F_1 \cap F_2 \cap A)\, \mathrm{d}\eta(F_1)\, \mathrm{d}\theta(F_2)$$

$$= 2dV(\Psi^{d-1}(\theta), \ldots, \Psi^{d-1}(\theta), \Pi_m(\eta))\lambda_d(A)$$

for each bounded Borel set $A \subset \mathbb{E}^d$.

The proof can be found in Wieacker (1989). Since the mixed volume in Theorem 6.4 is essentially the Minkowski area of $\Psi^{d-1}(\theta)$ with respect to $\Pi_m(\eta)$, the isoperimetric inequality of Minkowski geometry shows that, on the set of all convex bodies K of given positive volume, the integral

$$\int_{\mathcal{LE}_m} \lambda_{m-1}(F \cap \operatorname{bd} K)\, \mathrm{d}\eta(F)$$

attains a minimum at K if and only if K is homothetic to $\Pi_m(\eta)$.

Further extremal problems in a similar spirit are treated in Schneider (1982, 1987), Wieacker (1984, 1986, 1989). Applications to stochastic processes of geometric objects are described in chapter 5.2.

7. Lines and flats through convex bodies

The present section is devoted to an entirely different facet of the close relations between integral geometry and convex bodies, this time of a more classical type. We consider various questions related to flats meeting a convex body. In this context, it is often convenient to formulate the results in terms of random flats, probabilities of geometric events, and expectations of geometric random variables. In essence, however, the results to be discussed here are either interpretations of integral-geometric identities, or inequalities obtained by methods from convex geometry. For a survey on related investigations of a more probabilistic flavour, see chapter 5.2.

Let $K \in \mathcal{K}^d$ be a convex body. A *random r-dimensional flat through K* (or *meeting K*) is a measurable map X from some probability space into the space $\mathcal{E}_r^d$ of r-flats such that $X \cap K \neq \emptyset$ with probability 1. The random flat X is called *uniform* if its distribution, which is a probability measure on $\mathcal{B}(\mathcal{E}_r^d)$, can be obtained from a translation invariant measure on $\mathcal{B}(\mathcal{E}_r^d)$ by restricting it to the flats meeting K and by normalizing the restriction to a probability measure. If the distribution of X can be derived, in this way, from a rigid motion invariant measure on $\mathcal{B}(\mathcal{E}_r^d)$, then X is called an *isotropic uniform random flat*, often abbreviated by IUR flat. In the case $r = 0$, both notions coincide, and we talk of a *uniform random point in K*.

In the following, the central setup will be that of a given finite number of independent random flats through K. These flats determine other geometric objects as well as geometric functions, and one may ask for various probabilities, expectations, or distributions connected with these. Integral-geometric identities may be useful to transform the problem, and often methods from convex geometry then lead to sharp inequalities.

We exclude from the following survey convex hulls of $d+1$ or more random points; for these, chapter 5.2 gives a unified treatment.

Let $K \in \mathcal{K}^d$ be a convex body with interior points. First we consider an isotropic uniform random line X through K. Let σ_K denote the length of the random secant $X \cap K$. The random variable σ_K has found considerable interest in the literature. The moments of its distribution are essentially (up to a factor involving the surface area of K) the *chord power integrals* of K, defined by

$$I_k(K) := \frac{d\kappa_d}{2} \int_{\mathcal{E}_1^d} \lambda_1(L \cap K)^k \, \mathrm{d}\mu_1(L) \tag{7.1}$$

for $k \in \mathbb{N}_0$ (where $0^0 := 0$). (The factor before the integral occurs because in the older literature a different normalization of μ_1 is chosen.) In particular, (2.7) shows that

$$I_0(K) = \frac{\kappa_{d-1}}{2} S(K), \tag{7.2}$$

where S denotes the surface area, and

$$I_1(K) = \frac{d\kappa_d}{2} V_d(K). \tag{7.3}$$

From relation (7.5) below it follows that

$$I_{d+1}(K) = \frac{d(d+1)}{2} V_d(K)^2. \tag{7.4}$$

For the ball B^d_ρ of radius ρ one has

$$I_k(B^d_\rho) = \frac{2^{k-1}\pi^{d-1/2}k\Gamma(\frac{1}{2}k)}{\Gamma(\frac{1}{2}d)\Gamma(\frac{1}{2}(k+d+1))}\rho^{k+d-1};$$

see Santaló (1986). There one also finds the representation

$$I_k(K) = k(k-1)\int_0^\infty \ell^{k-2}M_K(\ell)\,\mathrm{d}\ell$$

where $M_K(\ell)$ denotes the (suitably normalized) kinematic measure of the set of all line segments of length ℓ contained in K.

A number of inequalities satisfied by the chord power integrals I_k are known; for these and for references, see Santaló (1976, pp. 48 and 238, 1986); see also Hadwiger (1957, section 6.4.6), and Voss (1984).

Blaschke (1955, p. 52), posed the following questions. If positive numbers $c_0, c_1, c_2, \ldots$ are given, what are the necessary and sufficient conditions in order that there exists a convex domain $K \in \mathcal{K}^2$ for which $I_k(K) = c_k$ for $k \in \mathbb{N}_0$? If K exists, to what extent is it determined by the numbers c_k? Mallows and Clark (1970) constructed two noncongruent convex polygons with the same chord length distribution. Gates (1982) showed how triangles and quadrangles may be reconstructed from their chord length distributions. A thorough study of the plane case was made by Waksman (1985). He was able to show that a convex polygon which is generic (in a precise sense, roughly saying that the polygon is sufficiently asymmetric) can, in fact, be reconstructed from its chord length distribution. Some more information on the distribution of the chord length for general convex bodies is contained in papers by Sulanke (1961), Gečiauskas (1987).

A classical integral-geometric transformation (see (7.9) below) relates the chord power integrals to distance power integrals of point pairs:

$$\int_K\int_K \|x_1 - x_2\|^k\,\mathrm{d}x_1\,\mathrm{d}x_2 = \frac{2}{(d+k)(d+k+1)}I_{d+k+1}(K) \tag{7.5}$$

for $k = -d+1, -d+2$; here $\mathrm{d}x_i$ stands for $\mathrm{d}\lambda_d(x_i)$. Formula (7.5) goes back to Crofton for $d = 2$ and $k = 0$; the general case was proved independently by Chakerian (1967) and Kingman (1969). In a generalization of (7.5) obtained by Piefke (1978b), the integrations on the left side may be extended over two different convex bodies, and the integrand can be of a more general form. For example,

$$\int_K\int_K f(\|x_1 - x_2\|)\,\mathrm{d}x_1\,\mathrm{d}x_2 = d\kappa_d\int_{\mathcal{E}^d_1} g(\lambda_1(L\cap K))\,\mathrm{d}\mu_1(L),$$

if f is, say, continuous and g is determined from

$$g''(t) = f(t)t^{d-1} \quad \text{for } t \geqslant 0, \qquad g(0) = g'(0) = 0.$$

Piefke (1978a) also showed how the probability densities of the secant length σ_K and of the distance between two independent uniform random points in K can be computed from each other. This together with results of Coleman (1969) and Hammersley (1950) yields the explicit distributions of both random variables in the case of a 3-dimensional cube and a d-dimensional ball. Many special results and references on the distance between two random points in plane regions can be found in the paper by Sheng (1985); see also Gečiauskas (1976).

From the viewpoint of applications, an isotropic uniform random line through K need not be the most natural type of random line meeting the convex body K. We mention two other and equally natural ways of generating random lines through K. Let X_1, X_2 be independent uniform random points in K. With probability 1, they span a unique line. The random line through K thus generated is called λ-random. In this context, an isotropic uniform random line through K has been called μ-random. Another simple generation of random lines through K proceeds as follows. Choose a uniform random point in K and through that point a line with direction given by a unit vector which is chosen at random, independent from the point and with uniform distribution on the sphere S^{d-1}. The random line thus obtained is called ν-random. If the probability distribution of a μ-random, ν-random, λ-random line through K is denoted, respectively, by P_μ, P_ν, P_λ, then P_λ and P_ν are absolutely continuous with respect to P_μ, and for the corresponding Radon-Nikodym derivatives one has

$$\frac{\mathrm{d}P_\nu}{\mathrm{d}P_\mu}(L) = \frac{\kappa_{d-1}}{d\kappa_d}\frac{S(K)}{V_d(K)}\lambda_1(L \cap K), \tag{7.6}$$

$$\frac{\mathrm{d}P_\lambda}{\mathrm{d}P_\mu}(L) = \frac{\kappa_{d-1}}{d(d+1)}\frac{S(K)}{V_d(K)^2}\lambda_1(L \cap K)^{d+1}. \tag{7.7}$$

Essentially, these results can be found in Kingman (1965, 1969).

These and some other different types of random lines through convex bodies, as well as the induced random secants, were investigated in papers by Kingman (1965, 1969), Coleman (1969), Enns and Ehlers (1978, 1980, 1988), and Ehlers and Enns (1981); see also Warren and Naumovich (1977).

Enns and Ehlers (1978) had conjectured that, for all convex bodies K with given volume, the expectation of the length of a ν-random secant of K is maximal precisely when K is a ball. This was proved, independently, by Davy (1984), Schneider (1985), and Santaló (1986). All three authors establish the following more general result. For $k \in \mathbb{N}$, let $M_k(K)$ denote the expectation of $\lambda_1(L \cap K)^k$, where L is a ν-random line through K. If B denotes a ball with $V_d(K) = V_d(B)$, then

$$M_k(K) \begin{cases} \leqslant M_k(B) & \text{for } 1 \leqslant k < d, \\ = M_k(B) & \text{for } k = d, \\ \geqslant M_k(B) & \text{for } k > d. \end{cases} \tag{7.8}$$

Equality holds for $k \neq d$ only if K is a ball. The proof makes use of (7.5) and of the following result, which can be obtained by Steiner symmetrization.

Theorem 7.1. *Let f be a decreasing measurable function on $(0, \infty)$ such that $x^{d-1} \times |f(x)|$ is integrable over all finite intervals. Then among all convex bodies K with* *fixed volume, the double integral*

$$\int_K \int_K f(\|x - y\|) \, dx \, dy$$

achieves its maximum value for the ball (and only for the ball, if f is strictly decreasing).

This was proved by Carleman (1919) (see also Blaschke 1918) for $d = 2$; extensions are due to Groemer (1982), Davy (1984) (where the above formulation appears), Pfiefer (1982, 1990).

Some of the results for random lines through a convex body K extend to random r-flats through K, for $r \in \{1, \ldots, d-1\}$. Here the following integral-geometric transformation, going back to Blaschke and Petkantschin is useful (see Santaló 1976, p. 201, but also Kingman 1969 and Miles 1971, 1979). For any integrable function $f : (\mathbb{E}^d)^{r+1} \to \mathbb{R}$,

$$\begin{aligned}\int_{\mathbb{E}^d} \cdots \int_{\mathbb{E}^d} f(x_1, \ldots, x_{r+1}) \, dx_1 \cdots dx_{r+1} \\ = c_{dr}(r!)^{d-r} \int_{\mathscr{E}_r^d} \int_E \cdots \int_E f(x_1, \ldots, x_{r+1}) \lambda_r(\operatorname{conv}\{x_1, \ldots, x_{r+1}\})^{d-r} \\ \times \, dx_1^E \cdots dx_{r+1}^E \, d\mu_r(E),\end{aligned} \tag{7.9}$$

where

$$c_{dr} = \frac{\omega_d \cdots \omega_{d-r+1}}{\omega_1 \cdots \omega_r}, \qquad \omega_m := m\kappa_m,$$

and dx_i^E indicates integration with respect to r-dimensional Lebesgue measure in the r-flat E.

If we choose a uniform random point in K and through that point independently a random r-flat with uniformly distributed direction (specified by an element of the Grassmannian $\mathscr{L}_r^d$), then we obtain a ν-random r-flat meeting K. With respect to the distribution of an isotropic uniform random r-flat through K, the distribution of a ν-random r-flat has a density proportional to $\lambda_r(E \cap K)$ for $E \in \mathscr{E}_r^d$. Such r-weighted r-flats through K, as they have been called, play an important role in stereology (Davy and Miles 1977, Miles and Davy 1976). Some of the inequalities of (7.8) can be extended to ν-random r-flats, see Schneider (1985).

Now we consider a finite number of independent uniform random flats through K. The case of points we mention only briefly. An extensive study of the case of up to $d+1$ points was made by Miles (1971). For more points, in particular the asymptotic behaviour of the convex hull of n points for $n \to \infty$, we refer to chapter 5.2 and to the survey by Schneider (1988a). In some of the treatments, integral-geometric transforms of type (7.9) play an essential role. We mention two further problems on random points in convex bodies where integral-geometric methods have been applied. Hall (1982) derived a formula for the probability that three uniform random points in the ball B^d form an acute triangle. If B^d is replaced by an arbitrary convex body K, Hall conjectured that the corresponding probability is maximized when K is a ball. The following problem was treated by Affentranger (1990). If $m+1$ $(2 \leqslant m \leqslant d-1)$ independent uniform random points in a convex body K are given, they determine, with probability one, a unique $(m-1)$-dimensional sphere C_{m-1} containing the points. Affentranger showed that the probability that $C_{m-1} \subset K$ is maximized if and only if K is a ball; for this case, the explicit value was computed.

For a thorough investigation of uniform random flats, we refer to Miles (1969). Here we consider only a few, mostly later, results. Let $E_1, \ldots, E_s$ be independent isotropic uniform random flats through K, where $2 \leqslant s \leqslant d$,

$$1 \leqslant \dim L_i = r_i \leqslant d-1 \quad \text{for } 1, \ldots, s,$$

$$m := r_1 + \cdots + r_s - (s-1)d \geqslant 0.$$

With probability 1, the intersection $X := E_1 \cap \cdots \cap E_s$ is a flat of dimension m. It is a matter of integral geometry to compute the distribution of this random m-flat. For example, the probability $p(K)$ that X meets K can be expressed in terms of quermassintegrals of K, and from the Aleksandrov–Fenchel inequalities it then follows that $p(K)$ is maximal precisely when K is a ball; see Schneider (1985), also Santaló (1976, III.14.2), and Miles (1969, p. 231). As another example, consider the case $d=2$, $s=2$, $r_1=r_2=1$. The distribution of the intersection point X of two independent IUR lines through $K \in \mathcal{K}^2$ is given by

$$\operatorname{Prob}\{X \in \beta\} = \frac{2}{L(K)^2}\left[\pi \lambda_2(\beta \cap K) + \int_{\beta \setminus K} (\omega(x) - \sin \omega(x))\, \mathrm{d}\lambda_2(x)\right]$$

for $\beta \in \mathcal{B}(\mathbb{E}^2)$ (cf. Santaló 1976, I.4.3). Here $L(K)$ is the perimeter of K and $\omega(x)$ denotes the angle between the two supporting rays of K emanating from $x \in \mathbb{E}^2 \setminus K$. In particular, the distribution of X is uniform inside K. The probability that $X \in K$ is given by $2\pi V_2(K)/L(K)^2 \leqslant \frac{1}{2}$, by the isoperimetric inequality.

More generally, we may consider n independent IUR lines through K and ask for the distribution of the number of intersection points inside K. Only some partial results are known, see Sulanke (1965), Gates (1984).

For random flats through K which generally do not intersect, for dimensional reasons, one can study probabilities connected to the closeness of the flats. For example, n lines meeting K may be called *ρ-close*, for some given number $\rho > 0$, if there is a point in K at distance at most ρ from each of the lines. Hadwiger and Streit (1970) determined the probability that n independent IUR lines through $K \in \mathcal{K}^3$ are ρ-close, and they treated similar questions for points and planes. The proof makes use of iterated kinematic formulae for cylinders. Next, let L_1, L_2 be two independent IUR flats through $K \in \mathcal{K}^d$, with $\dim L_1 = r$, $\dim L_2 = s$, and $r+s \leqslant d-1$. With probability 1, there is a unique pair of points $x_1 \in L_1$, $x_2 \in L_2$ with smallest distance. Let $p(K)$ denote the probability that $x_1, x_2 \in K$. Then $p(K)$ is maximal if and only if K is a ball. This was proved for $d=3$, $r=s=1$ by Knothe (1937), for $r+s=d-1$ by Schneider (1985), and in general by Affentranger (1988).

New problems arise if we consider uniform random flats through K which are not necessarily isotropic. For example, let $H_1, \ldots, H_n$ be independent, identically distributed uniform random hyperplanes through K. There is an even probability measure φ on the unit sphere determining the orientation of H_i (φ is the distribution of the unit normal vector of H_i if K is a ball). We assume that φ is not concentrated on a great subsphere. For $2 \leqslant n \leqslant d$, let $p_n(K, \varphi)$ denote the probability that $H_1 \cap \cdots \cap H_n$ meets K. Let ω be the rotation invariant probability measure on S^{d-1}. Then

$$p_n(K, \omega) \leqslant p_n(B^d, \omega)$$

(Miles 1969), a special case of a result mentioned above. The inequality

$$p_n(B^d, \varphi) \leqslant p_n(B^d, \omega) \tag{7.10}$$

was conjectured by Miles (1969, p. 224), and proved by Schneider (1982). There it is also proved that, for given φ,

$$p_d(K, \varphi) \leqslant p_d(M_\varphi, \varphi)$$

for a convex body M_φ which is unique up to homothety. For $d = 2$, one obtains $p_2(K, \varphi) \leqslant \frac{1}{2} = p_2(B^2, \omega)$, with equality for a unique homothety class of convex bodies determined by φ. However, for $d \geqslant 3$ it is shown that $p_d(K, \varphi) > p_d(B^d, \omega)$ is possible, and the maximum value of $p_d(K, \varphi)$ remains unknown. The probabilities $p_n(K, \varphi)$ also appear in the following formula. If $H_1, \ldots, H_n$ are as above (but allowing arbitrary $n \geqslant 1$), these hyperplanes determine, in the obvious way, a decomposition of the interior of K into relatively open convex cells. The random variable ν_k is defined as the number of k-dimensional cells of this decomposition ($k = 0, \ldots, d$). Then its expectation is given by

$$E(\nu_k) = \sum_{j=d-k}^{d} \binom{j}{d-k} \binom{n}{j} p_j(K, \varphi),$$

as shown by Schneider (1982), extending special results of Santaló. In particular, if the hyperplanes are isotropic, then

$$E(\nu_k) = \sum_{j=d-k}^{d} \binom{j}{d-k} \binom{n}{j} \frac{j!}{2^j} \frac{V_j(K)}{V_1(K)^j},$$

which becomes maximal if and only if K is a ball.

Interesting new intersection phenomena arise for lower-dimensional flats. For example, the extremal property of the isotropic distribution expressed by (7.10) does not necessarily extend. Mecke (1988a,b) considers the case of two independent, identically distributed r-dimensional flats meeting the ball B^d in $\mathbb{E}^d$ for $d = 2r$ ($r \geqslant 2$). With probability 1, the two flats have a unique intersection point X. Mecke was able to find all uniform distributions for which the probability $\mathrm{Prob}\{X \in B^d\}$ becomes maximal; they are not rotation-invariant.

References

Affentranger, F.
[1988] Pairs of non-intersecting random flats, *Probab. Theory Related Fields* **79**, 47–50.
[1990] Random spheres in a convex body, *Arch. Math.* **55**, 74–81.

Ambartzumian, R.V.
[1982] *Combinatorial Integral Geometry* (Wiley, Chichester).
[1990] *Factorization Calculus and Geometric Probability* (Cambridge Univ. Press, Cambridge).

Berwald, L., and O. Varga
[1937] Integralgeometrie 24, Über die Schiebungen im Raum, *Math. Z.* **42**, 710–736.

Blaschke, W.
[1918] Eine isoperimetrische Eigenschaft des Kreises, *Math. Z.* **1**, 52–57.
[1937] Integralgeometrie 21, Über Schiebungen, *Math. Z.* **42**, 399–410.
[1955] *Vorlesungen über Integralgeometrie* (VEB Deutsch. Verl. d. Wiss., Berlin, 3rd ed.). First edition: Part I, 1935; Part II, 1937.
[1985] Gesammelte Werke, Vol. 2: *Kinematik und Integralgeometrie*, eds W. Burau, S.S. Chern, K. Leichtweiß, H.R. Müller, L.A. Santaló, U. Simon and K. Strubecker (Thales, Essen).

Bokowski, J., H. Hadwiger and J.M. Wills
[1976] Eine Erweiterung der Croftonschen Formeln für konvexe Körper, *Mathematika* **23**, 212–219.

Burton, G.R.
[1980] Subspaces which touch a Borel subset of a convex surface, *J. London Math. Soc.* **21**, 167–170.

Carleman, T.
[1919] Über eine isoperimetrische Aufgabe und ihre physikalischen Anwendungen, *Math. Z.* **3**, 1–7.

Chakerian, G.D.
[1967] Inequalities for the difference body of a convex body, *Proc. Amer. Math. Soc.* **18**, 879–884.

Coleman, R.
[1969] Random paths through convex bodies, *J. Appl. Probab.* **6**, 430–441.

Davy, P., and R.E. Miles
[1977] Sampling theory for opaque spatial specimens, *J. Roy. Statist. Soc. Ser. B* **39**, 56–65.

Davy, P.J.
[1984] Inequalities for moments of secant length, *Z. Wahrscheinlichkeitsth. Verw. Geb.* **68**, 243–246.

Deltheil, R.
[1926] *Probabilités Géométriques* (Gauthier-Villars, Paris).

Ehlers, P.F., and E.G. Enns
[1981] Random secants of a convex body generated by surface randomness, *J. Appl. Probab.* **18**, 157–166.

Enns, E.G., and P.F. Ehlers
[1978] Random paths through a convex region, *J. Appl. Probab.* **15**, 144–152.
[1980] Random paths originating within a convex region and terminating on its surface, *Austral. J. Statist.* **22**, 60–68.
[1988] Chords through a convex body generated from within an embedded body, *J. Appl. Probab.* **25**, 700–707.

Federer, H.
[1959] Curvature measures, *Trans. Amer. Math. Soc.* **93**, 418–491.
[1969] *Geometric Measure Theory* (Springer, Berlin).

Firey, W.J.
[1972] An integral-geometric meaning for lower order area functions of convex bodies, *Mathematika* **19**, 205–212.
[1974] Kinematic measures for sets of support figures, *Mathematika* **21**, 270–281.
[1977] Addendum to R.E. Miles' paper on the fundamental formula of Blaschke in integral geometry, *Austral. J. Statist.* **19**, 155–156.
[1979] Inner contact measures, *Mathematika* **26**, 106–112.

Gates, J.
[1982] Recognition of triangles and quadrilaterals by chord length distribution, *J. Appl. Probab.* **19**, 873–879.
[1984] Bounds for the probability of complete intersection of random chords in a circle, *J. Appl. Probab.* **21**, 419–424.

Gečiauskas, E.
[1976] Distribution of distance within a convex region. I, $x \leq d$ (in Russian), *Litovsk. Mat. Sb.* **16**, 105–111 [1977, *Lith. Math. J.* **16**, 546–551].
[1987] Geometrical parameters of the chord length distribution of a convex domain (in Russian), *Litovsk. Mat. Sb.* **27**, 255–257.

Goodey, P.R., and W. Weil
[1987] Translative integral formulae for convex bodies, *Aequationes Math.* **34**, 64–77.

Goodey, P.R., and M.M. Woodcock
[1979] Intersections of convex bodies with their translates, in: *The Geometric Vein,* eds C. Davis, B. Grünbaum and F.A. Sherk (Springer, New York) pp. 289–296.

Groemer, H.
[1977] On translative integral geometry, *Arch. Math.* **29**, 324–330.
[1978] On the extension of additive functionals on classes of convex sets, *Pacific J. Math.* **75**, 397–410.
[1980a] The average measure of the intersection of two sets, *Z. Wahrscheinlichkeitsth. Verw. Geb.* **54**, 15–20.
[1980b] The average distance between two convex sets, *J. Appl. Probab.* **17**, 415–422.
[1982] On the average size of polytopes in a convex set, *Geom. Dedicata* **13**, 47–62.

Hadwiger, H.
[1950] Einige Anwendungen eines Funktionalsatzes für konvexe Körper in der räumlichen Integralgeometrie, *Monatsh. Math.* **54**, 345–353.
[1951] Beweis eines Funktionalsatzes für konvexe Körper, *Abh. Math. Sem. Univ. Hamburg* **17**, 69–76.
[1956] Integralsätze im Konvexring, *Abh. Math. Sem. Univ. Hamburg* **20**, 136–154.
[1957] *Vorlesungen über Inhalt, Oberfläche und Isoperimetrie* (Springer, Berlin).
[1975a] Eine Erweiterung der kinematischen Hauptformel der Integralgeometrie, *Abh. Math. Sem. Univ. Hamburg* **44**, 84–90.
[1975b] *Eikörperrichtungsfunktionale und kinematische Integralformeln,* Studienvorlesung (Manuskript) (Universität Bern).

Hadwiger, H., and F. Streit
[1970] Über Wahrscheinlichkeiten räumlicher Bündelungserscheinungen, *Monatsh. Math.* **74**, 30–40.

Hall, G.R.
[1982] Acute triangles in the n-ball, *J. Appl. Probab.* **19**, 712–715.

Hammersley, J.M.
[1950] The distribution of distance in a hypersphere, *Ann. of Math. Statist.* **21**, 447–452.

Herglotz, G.
[1933] *Geometrische Wahrscheinlichkeiten,* Vorlesungsausarbeitung (Mimeographed notes) (Göttingen) 156 pp.

Kingman, J.F.C.
[1965] Mean free paths in a convex reflecting region, *J. Appl. Probab.* **2**, 162–168.
[1969] Random secants of a convex body, *J. Appl. Probab.* **6**, 660–672.

Knothe, H.
[1937] Über Ungleichungen bei Sehnenpotenzintegralen, *Deutsch. Math.* **2**, 544–551.

Kobayashi, S., and K. Nomizu
[1963] *Foundations of Differential Geometry,* Vol. I (Interscience, New York).

Mallows, C.L., and J.M.C. Clark
[1970] Linear-intercept distributions do not characterize plane sets, *J. Appl. Probab.* **7**, 240–244.

Mani-Levitska, P.
[1988] A simple proof of the kinematic formula, *Monatsh. Math.* **105**, 279–285.

Matheron, G.
[1975] *Random Sets and Integral Geometry* (Wiley, New York).
[1976] La formule de Crofton pour les sections épaisses, *J. Appl. Probab.* **13**, 707–713.

McMullen, P.
[1974] A dice probability problem, *Mathematika* **21**, 193–198.

Mecke, J.
[1988a] Random r-flats meeting a ball, *Arch. Math.* **51**, 378–384.
[1988b] An extremal property of random flats, *J. Microscopy* **151**, 205–209.

Miles, R.E.
[1969] Poisson flats in Euclidean spaces, Part I: A finite number of random uniform flats, *Adv. in Appl. Probab.* **1**, 211–237.
[1971] Isotropic random simplices, *Adv. in Appl. Probab.* **3**, 353–382.
[1974] The fundamental formula of Blaschke in integral geometry and geometric probability, and its iteration, for domains with fixed orientations, *Austral. J. Statist.* **16**, 111–118.
[1979] Some new integral geometric formulae, with stochastic applications, *J. Appl. Probab.* **16**, 592–606.

Miles, R.E., and P. Davy
[1976] Precise and general conditions for the validity of a comprehensive set of stereological formulae, *J. Microscopy* **107**, 211–226.

Molter, U.M.
[1986] Tangential measure on the set of convex infinite cylinders, *J. Appl. Probab.* **23**, 961–972.

Papaderou-Vogiatzaki, I., and R. Schneider
[1988] A collision probability problem, *J. Appl. Probab.* **25**, 617–623.

Pfiefer, R.E.
[1982] The extrema of geometric mean values, Dissertation, Univ. of California, Davis.
[1990] Maximum and minimum sets for some geometric mean values, *J. Theoret. Probab.* **3**, 169–179.

Piefke, F.
[1978a] Beziehungen zwischen der Sehnenlängenverteilung und der Verteilung des Abstandes zweier zufälliger Punkte im Eikörper, *Z. Wahrscheinlichkeitsth. Verw. Geb.* **43**, 129–134.
[1978b] Zwei integralgeometrische Formeln für Paare konvexer Körper, *Z. Angew. Math. Phys.* **29**, 664–669.

Ripley, B.D.
[1976] The foundations of stochastic geometry, *Ann. Probab.* **4**, 995–998.

Rother, W., and M. Zähle
[1990] A short proof of a principal kinematic formula and extensions, *Trans. Amer. Math. Soc.* **321**, 547–558.

Santaló, L.A.
[1953] *Introduction to Integral Geometry* (Hermann, Paris).
[1976] *Integral Geometry and Geometric Probability* (Addison-Wesley, Reading, MA).

[1986] On the measure of line segments entirely contained in a convex body, in: *Aspects of Mathematics and its Applications,* ed. J.A. Barroso (North-Holland, Amsterdam) pp. 677–687.

Schneider, R.

[1975a] Kinematische Berührmaße für konvexe Körper, *Abh. Math. Sem. Univ. Hamburg* **44**, 12–23.

[1975b] Kinematische Berührmaße für konvexe Körper und Integralrelationen für Oberflächenmaße, *Math. Ann.* **218**, 253–267.

[1976] Bestimmung eines konvexen Körpers durch gewisse Berührmaße, *Arch. Math.* **27**, 99–105.

[1977] Eine kinematische Integralformel für konvexe Körper, *Arch. Math.* **28**, 217–220.

[1978a] Curvature measures of convex bodies, *Ann. Mat. Pura Appl.* **116**, 101–134.

[1978b] Kinematic measures for sets of colliding convex bodies, *Mathematika* **25**, 1–12.

[1980a] Parallelmengen mit Vielfachheit und Steinerformeln, *Geom. Dedicata* **9**, 111–127.

[1980b] Curvature measures and integral geometry of convex bodies, *Rend. Sem. Mat. Univ. Politec. Torino* **38**, 79–98.

[1981a] Crofton's formula generalized to projected thick sections, *Rend. Circ. Mat. Palermo* **30**, 157–160.

[1981b] A local formula of translative integral geometry, *Arch. Math.* **36**, 149–156.

[1982] Random hyperplanes meeting a convex body, *Z. Wahrscheinlichkeitsth. Verw. Geb.* **61**, 379–387.

[1985] Inequalities for random flats meeting a convex body, *J. Appl. Probab.* **22**, 710–716.

[1986] Curvature measures and integral geometry of convex bodies II, *Rend. Sem. Mat. Univ. Politec. Torino* **44**, 263–275.

[1987] Geometric inequalities for Poisson processes of convex bodies and cylinders, *Results Math.* **11**, 165–185.

[1988a] Random approximation of convex sets, *J. Microscopy* **151**, 211–227.

[1988b] Curvature measures and integral geometry of convex bodies III, *Rend. Sem. Mat. Univ. Politec. Torino* **46**, 111–123.

Schneider, R., and W. Weil

[1986] Translative and kinematic integral formulae for curvature measures, *Math. Nachr.* **129**, 67–80.

Schneider, R., and J.A. Wieacker

[1984] Random touching of convex bodies, in: *Proc. Conf. Stochastic Geom., Geom. Statist., Stereology, Oberwolfach, 1983,* eds R.V. Ambartzumian and W. Weil (Teubner, Leipzig) pp. 154–169.

Sheng, T.K.

[1985] The distance between two random points in plane regions, *Adv. in Appl. Probab.* **17**, 748–773.

Stoka, M.I.

[1968] *Géométrie Intégrale* (Gauthier-Villars, Paris).

Streit, F.

[1970] On multiple integral geometric integrals and their applications to probability theory, *Canad. J. Math.* **22**, 151–163.

[1973] Mean-value formulae for a class of random sets, *J. Roy. Statist. Soc. Ser. B* **35**, 437–444.

[1975] Results on the intersection of randomly located sets, *J. Appl. Probab.* **12**, 817–823.

Sulanke, R.

[1961] Die Verteilung der Sehnenlängen an ebenen und räumlichen Figuren, *Math. Nachr.* **23**, 51–74.

[1965] Schnittpunkte zufälliger Geraden, *Arch. Math.* **16**, 320–324.

Voss, K.

[1984] Integrals of chord length powers for planar convex figures, *Elektron. Informationsverarb. Kybern.* **20**, 488–494.

Waksman, P.
[1985] Plane polygons and a conjecture of Blaschke's, *Adv. in Appl. Probab.* **17**, 774–793.
Warren, R., and N. Naumovich
[1977] Relative frequencies of random intercepts through convex bodies, *J. Microscopy* **110**, 113–120.
Weil, W.
[1979a] Berührwahrscheinlichkeiten für konvexe Körper, *Z. Wahrscheinlichkeitsth. Verw. Geb.* **48**, 327–338.
[1979b] Kinematic integral formulas for convex bodies, in: *Contributions to Geometry, Proc. Geometry Symp., Siegen, 1978,* eds J. Tölke and J.M. Wills (Birkhäuser, Basel) pp. 60–76.
[1981] Zufällige Berührung konvexer Körper durch q-dimensionale Ebenen, *Resultate Math.* **4**, 84–101.
[1982] Inner contact probabilities for convex bodies, *Adv. in Appl. Probab.* **14**, 582–599.
[1989a] Collision probabilities for convex sets, *J. Appl. Probab.* **26**, 649–654.
[1989b] Translative integral geometry, in: *Geobild '89,* eds A. Hübler, W. Nagel, B.D. Ripley and G. Werner, Math. Research, Vol. 51 (Akademie-Verlag, Berlin) pp. 75–86.
[1990] Iterations of translative integral formulae and nonisotropic Poisson processes of particles, *Math. Z.* **205**, 531–549.
Wieacker, J.A.
[1984] Translative Poincaré formulae for Hausdorff rectifiable sets, *Geom. Dedicata* **16**, 231–248.
[1986] Intersections of random hypersurfaces and visibility, *Probab. Theory Related Fields* **71**, 405–433.
[1989] Geometric inequalities for random surfaces, *Math. Nachr.* **142**, 73–106.

CHAPTER 5.2

Stochastic Geometry

Wolfgang WEIL

Mathematisches Institut II, TH Karlsruhe, Englerstrasse 2, D-76131 Karlsruhe, Germany

John A. WIEACKER

Mathematisches Institut der Universität Freiburg, Albertstrasse 23b, D-79104 Freiburg, Germany

Contents

HANDBOOK OF CONVEX GEOMETRY
Edited by P.M. Gruber and J.M. Wills

Preliminaries

The roots of Stochastic Geometry can be traced back to the famous *needle problem* of Buffon in 1733. He asked for the probability that a needle of length L, randomly thrown onto a grid of parallel lines in the plane (with distance $D > L$), hits one of the lines. By using a suitable parametrisation of the needle and a subsequent elementary integration, Buffon showed this probability to be $2L/\pi D$ [he published this result only in 1777; see Miles and Serra (1978), for further historical remarks]. The potential danger in using parametrisations of geometrical objects, when dealing with problems of probabilistic type, was pointed out by an example of Bertrand in 1888 (it is usually referred to as Bertrand's paradox, although the apparent contradiction has a simple reason). Bertrand considered the probability for a random chord of the unit circle to be longer than $\sqrt{3}$. According to three different parametrisations of secants, he obtained three different answers using the corresponding uniform measures in the parameter space. Of course, the different solutions belong to different experiments to obtain a random secant of the unit circle. One of the solutions is a natural one from the mathematical point of view, since it is related to the (up to a normalising constant) unique motion invariant measure on the space $\mathscr{E}_1^2$ of lines in the plane $\mathbb{R}^2$. It is remarkable that similar problems occurred in $\mathbb{R}^3$ in stereological applications. Here, expectation formulae for random two-dimensional sections through solid particles in $\mathbb{R}^3$ are used in practice, but again they depend on the performance of the experiment. The methods frequently used to slice a particle randomly are different from the one obtained from the motion invariant measure on $\mathscr{E}_2^3$ (see section 8).

In view of Bertrand's paradox it seemed natural to connect probabilistic questions of geometric type to measures invariant under the group of rigid motions. This obviously spanned a bridge to Integral Geometry, and for a long time Geometrical Probability was just viewed as an application of integral geometric formulae (see chapter 5.1). A fairly complete overview of this period with numerous further references is given by Santaló (1976). In this survey, we do not try to copy most of the material which is already described in Santaló's book, but concentrate on the numerous new aspects and results in Geometrical Probability and Stochastic Geometry.

The direct application of integral geometric measures forces some limitations on the probabilistic problems to be considered. First, only a finite number of random objects are allowed. More seriously, the shapes of the objects have to be fixed, only their position and orientation may be random. Finally, since the invariant measure μ on the motion group G^d is infinite, reference sets have to be introduced in order to obtain compact subsets of G^d on which then μ can be normalised to give a probability measure. Consequently, the resulting distributions of the random geometric objects have limited invariance properties with respect to rigid motions. For example, the question, "Is the triangle spanned by three uniformly distributed random points in the plane more likely to be acute or obtuse?" only makes sense, if the points are chosen from a given bounded set $K \subset \mathbb{R}^2$. But of course, then the result will depend on the shape of K.

These problems were overcome, when *random sets* were introduced and combined with the already existing notion of *point processes* (this event also marked the transition from Geometrical Probability to Stochastic Geometry). Two models of random sets were presented independently by Kendall (1974) and Matheron (1972). Kendall's approach is slightly more general, but Matheron's model of *random closed sets* (RACS) is easier to follow, and found more applications in practical fields, like Image Analysis and Stereology (see Matheron 1975). Of course, for applications in stochastic processes, a more general notion of random set is necessary.

Since our goal is in applications of convexity, we limit ourselves to random sets in the class $\mathcal{K}^d$ of convex bodies, the convex ring $\mathcal{R}^d$, and the extended convex ring $\mathcal{S}^d$,

$$\mathcal{S}^d = \{K \subset \mathbb{R}^d \colon K \cap K' \in \mathcal{R}^d \text{ for all } K' \in \mathcal{K}^d\} ,$$

as well as to point processes on these set classes. More general RACS and point processes are treated in section 6, since there convex bodies appear as secondary notions. Moreover, we will concentrate on results connected with convexity. Therefore, point processes of flats are not treated as a separate topic but included in sections 6 and 7, and many results on processes of flats which are of a nonconvex nature do not appear here. We refer to Stoyan, Kendall and Mecke (1987) for further information and references, also concerning statistical questions and applications. References are also found in Mecke et al. (1990), a book which is close to some parts of the following presentation.

We close this introductory section with some notation that will be used in the following. Besides the set classes $\mathcal{K}^d$, $\mathcal{R}^d$, and $\mathcal{S}^d$, which we have already introduced, we need the systems $\mathcal{F}^d$ and $\mathcal{C}^d$ of all closed and compact subsets of $\mathbb{R}^d$, respectively. Moreover, $\mathcal{P}^d$ denotes the polytopes in $\mathcal{K}^d$. We further use the groups G^d and SO^d with their Haar measures μ and ν, and the homogeneous spaces $\mathcal{E}_k^d$ of affine k-subspaces and $\mathcal{L}_k^d$ of linear k-subspaces with their invariant measures μ_k and ν_k. For details, and in particular for the topologies used on these spaces, we refer to chapter 5.1.

If X is a topological space, we denote by $\mathcal{B}(X)$ its Borel σ-algebra. All the above-mentioned spaces are supplied with their Borel σ-algebra, and measurability always refers to this Borel structure.

As in chapter 5.1, V_j will denote the jth intrinsic volume, and Φ_j the jth curvature measure, $j = 0, \ldots, d$. Here $\Phi_d(K, \cdot)$ is the Lebesgue measure λ_d restricted to K. We will also write V for the volume V_d, S for the surface area $2V_{d-1}$, W for the mean width (which is proportional to V_1), and χ for the Euler characteristic V_0. In the planar case, we use A for the area and L for the perimeter. For a polyhedral set Q, f_i denotes the number of i-faces. Also, in $\mathbb{R}^3$ we use the integral mean curvature M instead of V_1. κ_d is the volume of the unit ball $B^d \subset \mathbb{R}^d$.

For m-dimensional sets $A \subset \mathbb{R}^d$ with appropriate regularity properties, we use λ_m to denote the m-dimensional Hausdoff measure (on A). If A belongs to $\mathcal{R}^d$ or

$\mathcal{S}^d$, this coincides with $\Phi_m(A, \cdot)$. Also, for $m = d-1$, and if A is the boundary of a set K in $\mathcal{R}^d$ or $\mathcal{S}^d$, we have $\lambda_{d-1} = 2\Phi_{d-1}(K, \cdot)$ on A.

Throughout the article, the letters $\mathbb{P}$ and $\mathbb{E}$ will denote probability measures and expectations.

1. Random points in a convex body

Random points, i.e., $\mathbb{R}^d$-valued random variables, are the simplest random objects in geometry. They can be used in different ways, to generate a great variety of random geometrical objects (random planes, random segments, random polytopes, random tessellations, etc.), and they have applications in several domains, for instance, in statistics, computer geometry and pattern analysis (see, e.g., Eddy and Gale 1981, Dwyer 1988, Ronse 1989, Grenander 1973, 1977). There is a very extensive literature on random points in a convex body. From the viewpoint of convex geometry it seems that the most important investigations in this field are those concerning the convex hull of random points in a convex body. Here we shall only consider the case where the random points are independently and uniformly distributed in the body, results concerning other generating procedures can be found in the excellent surveys of Schneider (1988) and Buchta (1985).

Let $K \subset \mathbb{R}^d$ be a convex body with interior points, and let $X_1, \ldots, X_n$ be n independently and uniformly distributed random points in K. Independently means that the joint distribution $\mathbb{P}$ of $X_1, \ldots, X_n$ is given by the product measure $\mathbb{P}_{X_1} \otimes \cdots \otimes \mathbb{P}_{X_n}$ of the distributions $\mathbb{P}_{X_i}$ of X_i. Uniformly means that the random points X_i all have the same distribution $\mathbb{P}_{X_i} = \Phi_d(K, \cdot)/V(K)$. Then the convex hull $Q_n := \operatorname{conv}\{X_1, \ldots, X_n\}$ is a random polytope, i.e., a random element of $\mathcal{P}^d$ (see section 4 for more details). For any measurable nonnegative function $g : \mathcal{P}^d \to [0, \infty]$, $g \circ Q_n$ is a random variable and we write $\mathbb{E}_n(g)$ for the expected value of $g \circ Q_n$. Some functions g are of particular interest, for instance, for $d \geq 3$, the volume V, the surface area S, the mean width W or the number f_i of i-dimensional faces. For $d = 2$, interesting functions are the area A, the perimeter L, and the vertex or edge number $f_0 = f_1$.

In most cases, the explicit computation of $\mathbb{E}_n(g)$ for one of the above-mentioned functionals g is complicated, even for simple convex bodies K. For example, for a tetrahedron K in $\mathbb{R}^3$, $\mathbb{E}_n(V)$ is still unknown, and the formulae derived by Buchta (1984b) for the explicit computation of $\mathbb{E}_n(A)$ in the case of a planar polygon K may give an idea of the difficulties which occur in this type of computation. Nevertheless, a number of explicit results are known in the case where K is the unit ball B^d of $\mathbb{R}^d$ (Hostinsky 1925, Kingman 1969, Buchta and Müller 1984, Affentranger 1988), and in the plane some results concerning the distribution and the moments of $A(Q_3)$ have been obtained in the cases where K is a triangle, a parallelogram or an ellipse (Reed 1974, Alagar 1977, Henze 1983). For an arbitrary convex body K in $\mathbb{R}^d$, some relations between the rth normalised moment $M_r(n, K) := \mathbb{E}_n(V^r)/V(K)^r$ of the volume of Q_n (an affine invariant parameter of K) and other expected values are easy to obtain; for example, the identity

$$\mathbb{P}(Q_n \text{ is a } d\text{-simplex}) = \binom{n}{d+1} M_{n-d-1}(d+1, K)$$

[in the planar case this relates an old problem of Sylvester to the computation of $M_1(3, K)$], or Efron's identity $\mathbb{E}_n(f_0) = n(1 - M_1(n-1, K))$. Both of them are direct consequences of Fubini's theorem. A further relation of this type is the identity $2\mathbb{E}_{d+2}(V) = (d+2)\mathbb{E}_{d+1}(V)$ proved by Buchta (1986). From the geometrical point of view, the most significant result concerning $M_r(n, K)$ up to now is probably the following theorem of Groemer (1974) (see also Schöpf 1977), which provides a characterisation of ellipsoids.

Theorem 1.1. *Let K be a convex body in $\mathbb{R}^d$ with interior points, and let $n > d$, $r \in \mathbb{N}$. Then $M_r(n, K)$ is minimal if and only if K is an ellipsoid.*

We give a brief outline of Groemer's proof. Using the affine invariance of $M_r(n, K)$, the existence of a "minimum body" can be proved with standard compactness arguments. For $Y = (y_1, \ldots, y_n) \in (\mathbb{R}^{d-1})^n$ and $Z = (z_1, \ldots, z_n) \in \mathbb{R}^n$, define:

$$V(Y, Z) = V(\mathrm{conv}\{(y_1, z_1), \ldots, (y_n, z_n)\}),$$

where (y_i, z_i) is considered as a point of $\mathbb{R}^d$. From the convexity of the map $Z \mapsto V(Y, Z)$, we infer that

$$\int_{|z_1 - p_1| \leqslant a_1} \cdots \int_{|z_n - p_n| \leqslant a_n} V(Y, Z)^r \, \mathrm{d}z_1 \cdots \mathrm{d}z_n \geqslant \int_{|z_1| \leqslant a_1} \cdots \int_{|z_n| \leqslant a_n} V(Y, Z)^r \, \mathrm{d}z_1 \cdots \mathrm{d}z_n$$

whenever $Y \in ((\mathbb{R}^{d-1})^n$, $p_1, \ldots, p_n \in \mathbb{R}$ and $a_1, \ldots, a_n > 0$, with equality if and only if $(p_1, \ldots, p_n) = (0, \ldots, 0)$. Now, if K is not an ellipsoid, then there is a line L such that the set $Q(K, L)$ of the midpoints of the segments $K \cap (L + x)$, $x \in K - L$, is not contained in a hyperplane. Hence, for the convex body K' obtained from K by Steiner symmetrisation with respect to a hyperplane orthogonal to L, the above inequality implies $M_r(n, K') < M_r(n, K)$.

The convexity of the map $Z \mapsto V(Y, Z)$ has been used recently by Dalla and Larman (1991) to prove that, in the plane, $M_1(n, K)$ is maximal if K is a triangle, thus extending an old result of Blaschke (1917). However, the conjecture that in higher dimensions $M_1(n, K)$ is maximal if and only if K is a simplex, is still open.

More general results have been obtained for the asymptotic behaviour of $\mathbb{E}_n(g)$ as n tends to infinity. Problems of this type (in the plane) were first investigated in two classical papers of Rényi and Sulanke (1963, 1964). Up to now, most of the explicit computations of the asymptotic value of $\mathbb{E}_n(g)$ are based on (extensions or modifications of) their method, except in the case $g = W$ where an identity of Efron (1965) leads to a simpler proof. These results are collected in the following theorem, as long as they concern a large class of convex bodies and hold in arbitrary dimension. We use $\approx$ to denote asymptotic equality, as n tends to infinity.

Theorem 1.2. *Let K be a convex body in $\mathbb{R}^d$ with interior points. If* bd K *is of class C^3 and has positive Gauss–Kronecker curvature k everywhere, then*

$$\mathbb{E}_n(W) = W(K) - c_1(d) \int_{\partial K} k^{(d+2)/(d+1)}(x)\, \mathrm{d}\lambda_{d-1}(x) \left(\frac{n}{V(K)}\right)^{-2/(d+1)} + \mathrm{O}(n^{-3/(d+1)}) ,$$

$$\mathbb{E}_n(V) = V(K) - c_2(d) \int_{\partial K} k^{1/(d+1)}(x)\, \mathrm{d}\lambda_{d-1}(x) \left(\frac{n}{V(K)}\right)^{-2/(d+1)} + \mathrm{O}(n^{-3/(d+1)} \log^2 n) ,$$

$$\mathbb{E}_n(f_0) \approx c_2(d) \int_{\partial K} k^{1/(d+1)}(x)\, \mathrm{d}\lambda_{d-1}(x) \left(\frac{n}{V(K)}\right)^{(d-1)/(d+1)} ,$$

$$\mathbb{E}_n(f_{d-1}) \approx c_3(d) \int_{\partial K} k^{1/(d+1)}(x)\, \mathrm{d}\lambda_{d-1}(x) \left(\frac{n}{V(K)}\right)^{(d-1)/(d+1)} ,$$

with explicitly given constants $c_1(d)$, $c_2(d)$ and $c_3(d)$ depending only on d. If K is a polytope, then

$$W(K) - \mathbb{F}_n(W) \approx c_4(K) n^{-1/d} ,$$

with a constant $c_4(K)$ depending only on the shape of K in arbitrarily small neighbourhoods of the vertices of K, and if K is a simple polytope, then

$$\mathbb{E}_n(f_0) = \frac{d}{(d+1)^{d-1}} f_0(K) \log^{d-1} n + \mathrm{O}(\log^{d-2} n) ,$$

$$\mathbb{E}_n(f_{d-1}) = \frac{d^d}{d!} f_{d-1}(K) M_1(\Delta_{d-1}) \log^{d-1} n + \mathrm{O}(\log^{d-2} n) ,$$

$$\mathbb{E}_n(V) = V(K) - V(K) f_0(K) \frac{d}{(d+1)^{d-1}} \frac{\log^{d-1} n}{n} + \mathrm{O}\left(\frac{\log^{d-2} n}{n}\right) ,$$

where $M_1(\Delta_{d-1})$ is the normalised expected volume of a random simplex in a $(d-1)$-dimensional simplex Δ_{d-1}. [The value of $M_1(\Delta_{d-1})$ is affine invariant and hence does not depend on the special choice of the simplex Δ_{d-1}.]

The main idea in the proof of these results may be formulated as follows. Let g be a bounded real function on the set of all oriented $(d-1)$-dimensional polytopes contained in K. For any polytope $Q \subset K$, define

$$g(Q) = \sum_{F \text{ facet of } Q} g(F), \tag{1}$$

where the orientation of a facet F is given by the outer normal of Q at F (e.g., f_{d-1}, S and V are functions of this type). Suppose that $g \circ Q_n$ is integrable. Since Q_n is almost surely a simplicial polytope and the points are independently and uniformly distributed, $\mathbb{E}_n(g)$ can be expressed as in integral over K^d. The transformation of this integral into an integral over all hyperplanes meeting K via the Blaschke–Petkantschin identity [see Santaló (1976, p. 201) or chapter 5.1, section 7], leads to the relation

$$\mathbb{E}_n(g) = \frac{d\kappa_d}{2}\binom{n}{d} V(K)^{-d} \int_{\mathscr{E}^d_{d-1}} g_K(E)\left(1 - \frac{V(K_E)}{V(K)}\right)^{n-d} \mathrm{d}\mu_{d-1}(E) + \mathrm{o}(\varepsilon^n), \tag{2}$$

with some $\varepsilon < 1$. Here K_E is the part of K cut out by E with the smaller volume $V(K_E) < \frac{1}{2}V(K)$ (both parts cut out by E have almost surely a different volume). Also, g_K is given by

$$g_K(E) := (d-1)! \int_{K\cap E} \cdots \int_{K \cap E} g(\mathrm{conv}\{x_1, \ldots, x_d\}) \times \lambda_{d-1}(\mathrm{conv}\{x_1, \ldots, x_d\})\, \mathrm{d}\lambda_{d-1}(x_1) \cdots \mathrm{d}\lambda_{d-1}(x_d),$$

and the orientation of $\mathrm{conv}\{x_1, \ldots, x_d\}$ is given by the outer normal of the face $K \cap E$ of $K \backslash K_E$. From here on, one has to use the special properties of g and $\mathrm{bd}\, K$ to get a result expressed in terms of geometric parameters of K. For instance, if K is a smooth convex body in $\mathbb{R}^d$ (i.e., if $\mathrm{bd}\, K$ is of class C^3 and has positive curvature k everywhere), then a local Taylor approximation of $\mathrm{bd}\, K$ can be used to evaluate (2) in the case $g = f_{d-1}$ (Raynaud 1970, Wieacker 1978). Unfortunately, this local Taylor approximation is not good enough to obtain precise asymptotic results for $\mathbb{E}_n(V)$ or $\mathbb{E}_n(S)$, except in the case where K is a ball (Wieacker 1978; see also Affentranger 1992, and Meilijson 1990). In $\mathbb{R}^3$, the asymptotic behaviour of $\mathbb{E}_n(V)$ can easily be deduced from the asymptotic behaviour of $\mathbb{E}_n(f_2)$, since f_0 is a.s. related to f_2 via Euler's polyhedron theorem, and $\mathbb{E}_n(f_0)$ is related to $\mathbb{E}_{n-1}(V)$ via Efron's relation (Wieacker 1978). The asymptotic results concerning $\mathbb{E}_n(V)$ and $\mathbb{E}_n(f_0)$ in the case where K is smooth are due to Bárány (1992), who reduced the problem to the case of a ball (the reduction is not trivial). For a simple polytope K, (2) has been evaluated in the case where $g(F) = \eta_F^q \lambda_{d-1}(F)^q$ for some given $q \in \mathbb{N}$, where η_F is the distance between F and the supporting hyperplane of K parallel to F and oriented in the same direction. Some results about the position of the vertices of Q_n for large n

and an Efron-type argument show that one of these functionals has the same asymptotic behaviour as $\mathbb{E}_n(V)$, thus leading to the asymptotic value of $\mathbb{E}_n(V)$ and $\mathbb{E}_n(f_0)$ (Affentranger and Wieacker 1991; weaker results have been given by Dwyer 1988, and van Wel 1989). An extension of the above result on $\mathbb{E}_n(f_0)$ to arbitrary d-polytopes has been announced recently by Bárány and Buchta (1990). The computation of $\mathbb{E}_n(W)$ is due to Schneider and Wieacker (1980) for a smooth convex body (an extension to more general distributions was obtained by Ziezold 1984), and Schneider (1987a) for polytopes K.

Rényi and Sulanke (1963) obtained slightly more precise results for the planar case, since then $\mathbb{E}_n(f_{d-1})$ is also the expected number of vertices of Q_n. If K is a smooth convex body in the plane (i.e., bd K is of class C^3 and has positive curvature k everywhere), the asymptotic value of $\mathbb{E}_n(A)$ can be obtained from Theorem 1.2 via Efron's relation (Efron 1965). In this case, Rényi and Sulanke (1964) obtained the following relation by a direct computation:

$$\mathbb{E}_n(A) = A(K) - (\tfrac{2}{3})^{1/3}\Gamma(\tfrac{5}{3}) \int_{\partial K} k^{1/3}(x)\, \mathrm{d}\lambda_1(x) \left(\frac{n}{A(K)}\right)^{-2/3} + \mathrm{O}(n^{-1}) . \tag{3}$$

For a polygon K with interior points in the plane, Rényi and Sulanke (1963) obtained the more precise result

$$\mathbb{E}_n(f_0) = \tfrac{2}{3} f_0(K)(\log n + C) + c_5(K) + \mathrm{o}(1) , \tag{4}$$

where C is Euler's constant, and $c_5(K)$ is a constant depending only on K and given explicitly (see also Ziezold 1970). The asymptotic value of $\mathbb{E}_n(A)$ was computed by Rényi and Sulanke in the case where K is a square, and by Buchta (1984a) for arbitrary polygons in the plane. The method of Rényi and Sulanke has also been used to compute the asymptotic value of $\mathbb{E}_n(L)$ (which is essentially the mean width) in the planar case (Rényi and Sulanke 1964, Buchta 1984a). Since all these results concern only smooth convex bodies or polytopes, the following estimates, which hold for arbitrary convex bodies and are in a certain sense best possible (see Theorem 1.4), are useful.

Theorem 1.3. *For each convex body K in $\mathbb{R}^d$ with interior points there are positive constants $c_6(K)$, $c_7(K)$, $c_8(d)$, $c_9(d)$ such that for large n:*

(a) $c_6(K)n^{-2/(d+1)} \leqslant W(K) - \mathbb{E}_n(W) \leqslant c_7(K)n^{-1/d}$,

(b) $c_6(K)\,\dfrac{\log^{d-1} n}{n} \leqslant V(K) - \mathbb{E}_n(V) \leqslant c_7(K)n^{-2/(d+1)}$,

(c) $c_8(d)\log^{d-1} n \leqslant \mathbb{E}_n(f_i) \leqslant c_9(d)n^{(d-1)/(d+1)}$.

In this theorem, (a) is due to Schneider (1987a), (b) to Bárány and Larman (1988), and (c) to Bárány (1989). The proof of Bárány and Larman is based on the following interesting idea. For a convex body K in $\mathbb{R}^d$ and $\varepsilon > 0$ they define

$$K_\varepsilon := \{x \in K : V(K \cap H) \leq \varepsilon \text{ for some half-space } H \text{ with } x \in \text{bd } H\}, \tag{5}$$

and they show that

$$c_{10} V(K_{1/n}) < V(K) - \mathbb{E}_n(V) < c_{11}(d) V(K_{1/n}) \tag{6}$$

if $V(K) = 1$ and n is large enough. One of the main steps in the proof of this relation is the construction of an economic cap covering for K_ε. As a consequence, they obtain the first inequality in (b) from a lower bound for $V(K_{1/n})$, while the second one follows from Theorem 1.1 and the above-mentioned results for a ball. Similarly, Bárány deduced (c) from the relation

$$c_{12}(d) n V(K_{1/n}) \leq \mathbb{E}_n(f_i) \leq c_{13}(d) n V(K_{1/n}) \tag{7}$$

for $i = 0, \ldots, d-1$, $V(K) = 1$, and n sufficiently large.

Bárány (1989) obtained also an analogue of (7) for the intrinsic volume V_j, $j = 1, \ldots, d-1$. The following theorem is proved in Bárány and Larman (1988) and Bárány (1989). Here, $h(n) = \Theta(f(n))$ means that $h(n) = O(f(n))$ and $f(n) = O(h(n))$, as $n \to \infty$.

Theorem 1.4. *Let K be a convex body in $\mathbb{R}^d$ with interior points. If* bd K *is of class C^3 and has positive Gauss–Kronecker curvature k everywhere, then*

$$\mathbb{E}_n(f_i) = \Theta(n^{(d-1)/(d+1)}) \quad \text{for } i = 0, \ldots, d-1,$$

$$V_j(K) - \mathbb{E}_n(V_j) = \Theta(n^{-2/(d+1)}) \quad \text{for } j = 1, \ldots, d.$$

If K is a polytope, then

$$\mathbb{E}_n(f_i) = \Theta(\log^{d-1} n) \quad \text{for } i = 0, \ldots, d-1,$$

$$V(K) - \mathbb{E}_n(V) = \Theta\left(\frac{\log^{d-1} n}{n}\right),$$

$$V_j(K) - \mathbb{E}_n(V_j) = \Theta(n^{-1/(d-j+1)}) \quad \text{for } j = 1, \ldots, d-1.$$

For a smooth convex body a more precise result concerning $\mathbb{E}_n(V_j)$ (with a sketch of the proof) and a conjecture concerning the asymptotic value of $\mathbb{E}_n(f_i)$ can be found in Bárány (1992). Another interesting theorem proved by Bárány (1989) concerns the expected Hausdorff distance $\delta^{\mathrm{H}}(K, Q_n)$ of K and Q_n: if bd K is of class C^2 with positive Gauss–Kronecker curvature everywhere, then

$$\mathbb{E}(\delta^{\mathrm{H}}(K, Q_n)) = \Theta\left(\left(\frac{\log n}{n}\right)^{2/(d+1)}\right).$$

The preceding theorems show that the shape of K strongly influences the rate of convergence and that with respect to the asymptotic behaviour of f_i, W and V, the polytopes and the smooth convex bodies are in a certain sense extreme cases. However, the relation between the boundary structure of K and the rate of convergence of the expected values considered here seems to be fairly complicated, and for convex bodies which are neither smooth nor polytopes rather less is known. As a consequence of a general theorem of Gruber (1983) and the previous results, it turns out that for most convex bodies (in the sense of Baire categories, see chapter 4.10) the asymptotic behaviour of $W(K) - \mathbb{E}_n(W)$, $V(K) - \mathbb{E}_n(V)$ and $\mathbb{E}_n(f_i)$ is highly irregular. For instance, most convex bodies K have the property, that for any $\varepsilon > 0$, there are strictly increasing sequences $(p_i)_{i\in\mathbb{N}}$ and $(q_i)_{i\in\mathbb{N}}$ in $\mathbb{N}$ such that

$$W(K) - \mathbb{E}_{p_i}(W) < p_i^{-(2/(d+1))+\varepsilon} \quad \text{and} \quad W(K) - \mathbb{E}_{q_i}(W) > q_i^{-(1/d)-\varepsilon}$$

for all $i \in \mathbb{N}$ (see, e.g., Schneider 1987a). Similar results for $V(K) - \mathbb{E}_n(V)$ and $\mathbb{E}_n(f_i)$ can be found in Bárány and Larman (1988), and in Bárány (1989). Since most convex bodies are strictly convex and have a boundary of class C^1, this shows that, in the range between C^1 and C^3, a small change of the smoothness properties may strongly influence the rate of convergence (see also Bárány and Larman 1988, Theorem 4). An interesting result concerning a special type of convex bodies that are neither smooth nor polytopes is due to Dwyer (1990). He proved that in the case where K is a product of lower dimensional balls, $K = B^{d_1} \times \cdots \times B^{d_k}$, with $d_1 = \cdots = d_m > d_{m+1} \geq \cdots \geq d_k$ and $d_1 + \cdots + d_k = d$, we have

$$\mathbb{E}_n(V) = \Theta(n^{(d_1-1)/(d_1+1)} \log^{m-1} n) . \tag{8}$$

Here the main idea, previously used by Bentley et al. (1978) and by Devroye (1980), is the following. Suppose that the random points $x_1, \ldots, x_n$ are independently and uniformly distributed in K. If d orthogonal hyperplanes are chosen through x_1, they divide K into 2^d convex sets. Let w be the probability content of the smallest of these 2^d sets, then the probability that x_1 is a vertex of $\operatorname{conv}\{x_1, \ldots, x_n\}$ is bounded above by $2^d(1-w)^{n-1}$. On the other hand, if $\tilde{w}$ is the probability content of a closed halfspace bounded by a hyperplane through x_1, then the probability that x_1 is a vertex of $\operatorname{conv}\{x_1, \ldots, x_n\}$ is bounded below by $(1-\tilde{w})^{n-1}$.

While the previous methods use essentially analytical and geometrical tools, a more stochastical approach due to Groeneboom (1988) provided very strong results in the case where K is a polygon or unit disk in the plane, among others a central limit theorem for f_0 and the asymptotic behaviour of the variance of f_0.

All results mentioned so far concern the case where the points are randomly distributed in the interior of the convex body K. The case where some of the points (or all of them) are randomly chosen on the boundary of K and the remaining ones in the interior, was investigated by several authors (Miles 1971a,

Mathai 1982, Buchta, Müller and Tichy 1985, Affentranger 1988, Müller 1989, 1990). Some further results concern the asymptotic behaviour as the dimension tends to infinity (Miles 1971a, Ruben 1977, Mathai 1982, Buchta 1986, Bárány and Füredi 1988). Almost sure approximation of convex bodies by random polytopes (or more generally of smooth curves) in the plane was treated by Drobot (1982), Stute (1984), and Schneider (1988).

Further results and an extensive literature concerning other problems about random points may be found in the books of Santaló (1976) and Hall (1988). More recent contributions are due to Affentranger (1989, 1990, 1992), Dette and Henze (1989, 1990), Dwyer (1990), Meilijson (1990), Bárány and Vitale (1992), and Carnal and Hüsler (1991).

2. Random flats intersecting a convex body

A natural generalisation of the notion of a random point in a convex body is the notion of a random flat meeting a convex body. Such random flats have already been considered in chapter 5.1, section 7. Here we shall only mention a few asymptotic results which are closely related to the theory of Poisson processes of hyperplanes (see section 4, for the notion of Poisson processes). Some of them are in a certain sense dual to the results of the preceding section. Instead of considering the convex hull of random points, one may also consider the intersection of random closed halfspaces generated by random hyperplanes. Since a random polyhedral set generated in this way may be viewed as the solution set of a finite system of random linear inequalities, this type of random polyhedral sets is of interest in the average case analysis of linear programming algorithms (concerning this aspect of the problem, see, e.g., Prékopa 1972, Schmidt and Mattheiss 1977, Kelly and Tolle 1981, Borgwardt 1987, Buchta 1987a,b).

Let K and C be convex bodies in $\mathbb{R}^d$, C being contained in the interior of K, and let $\mathcal{H} := \{H \in \mathscr{E}^d_{d-1}\colon H \cap K \neq \emptyset, H \cap C = \emptyset\}$ be the set of all hyperplanes meeting K but not C. A *random hyperplane* X in $\mathcal{H}$ is a measurable map from some probability space into the space $\mathscr{E}^d_{d-1}$ such that $X \in \mathcal{H}$ almost surely. X is thus a special random closed set (see section 4). The random hyperplane X is called uniform (respectively uniform and isotropic) if its probability distribution is obtained from a translation invariant (motion invariant) measure on $\mathscr{B}(\mathscr{E}^d_{d-1})$ after restriction to $\mathscr{B}(\mathcal{H})$ and normalisation. For a random hyperplane X in $\mathcal{H}$ we shall denote by X^+ the random closed halfspace bounded by X and containing C. In the following, $X_1, \ldots, X_n$ are independent, identically distributed random hyperplanes in $\mathcal{H}$, and Q_n is the random polyhedral set $X_1^+ \cap \cdots \cap X_n^+$. As in the preceding section, $\mathbb{E}_n(f_0)$ denotes the expected number of vertices of Q_n.

The random polyhedral set Q_n was first investigated by Rényi and Sulanke (1968) in the case where $d=2$ and the random hyperplanes are uniform and isotropic. They proved that $\mathbb{P}(Q_n \not\subseteq K) = \mathrm{O}(\gamma^n)$, $0<\gamma<1$, as $n \to \infty$. Further, in the case where the boundary of C is smooth enough with positive and bounded curvature k, they proved that

$$\mathbb{E}_n(f_0) = (\tfrac{2}{3})^{1/3}\Gamma(\tfrac{5}{3})\left(\frac{n}{L_2 - L_1}\right)^{1/3} \int_0^{L_1} k^{2/3}(x)\,\mathrm{d}\lambda_1(x) + \mathrm{O}(1)$$

as $n \to \infty$. Here, L_1 and L_2 are the perimeters of C and K. In the case where C is a convex polygon with $f_0(C)$ vertices, they obtained

$$\mathbb{E}_n(f_0) = \tfrac{2}{3} f_0(C) \log n + \mathrm{O}(1)$$

for $n \to \infty$. While Rényi and Sulanke gave a direct proof of these results, Ziezold (1970) showed that they can be deduced from the corresponding results (Theorem 1.2) in the preceding section by means of a duality relation. However, the reduction of such results for random intersections to analogous results for random convex hulls is difficult, in general. In particular, the image distribution is a rather complicated one, and in some cases a direct proof may be easier. Similar results for higher dimensions and more general situations have been obtained recently by Kaltenbach (1990). In the case where the boundary of C is of class C^3 with positive Gauss–Kronecker curvature k, one of his results implies

$$\mathbb{E}_n(f_0) = c(d)\left(\frac{n}{W(K) - W(C)}\right)^{(d-1)/(d+1)} \int_{\partial K} k^{d/(d+1)}(x)\,\mathrm{d}\lambda_{d-1}(x) + \mathrm{O}(n^{(d-2)/(d+1)})$$

for $n \to \infty$, with an explicitly given constant $c(d)$ depending only on the dimension d. Kaltenbach also investigated the behaviour of other functionals like the volume of the part of Q_n contained in a ball centred at the origin and containing K.

The case where C reduces to the origin and K is the unit ball is of particular interest. For uniform and isotropic random hyperplanes in $\mathbb{R}^d$, Schmidt (1968) proved that $\mathbb{F}_n(f_0)$ converges as $n \to \infty$. The limit was computed by Rényi and Sulanke (1968) in the case $d = 2$, and by Sulanke and Wintgen (1972) for arbitrary dimension. Schneider (1982) considered the case where the distribution of the random hyperplanes comes from a translation invariant, but not necessarily isotropic measure τ on $\mathscr{E}^d_{d-1}$ with $\tau(A) = \mathbb{P}(X \in A)$, for $A \in \mathscr{B}(\mathscr{H})$. Under the assumption that the random hyperplanes are not almost surely parallel to a line, he proved that

$$\lim_{n\to\infty} \mathbb{E}_n(f_0) = 2^{-d} d!\, V_d(\Pi^{d-1}(\tau))V_d(\Pi^{d-1}(\tau)^*)\,,$$

where $\Pi^{d-1}(\tau)$ is a zonoid associated with τ (see chapter 5.1, section 6, for details) and $\Pi^{d-1}(\tau)^*$ is the polar body of $\Pi^{d-1}(\tau)$. Since the right-hand side is essentially the volume product of $\Pi^{d-1}(\tau)$, this implies that

$$2^d \le \lim_{n\to\infty} \mathbb{E}_n(f_0) \le 2^{-d} d!\, \kappa_d^2\,,$$

with equality on the left if and only if $\Pi^{d-1}(\tau)$ is a parallelotope and equality on the right if and only if $\Pi^{d-1}(\tau)$ is an ellipsoid (in particular, this is the case when the random hyperplanes are isotropic). Here, the lower bound and the corresponding equality case follow from results of Reisner (1985, 1986), while the upper bound is a consequence of the Blaschke–Santaló inequality (the equality case was treated by Saint Raymond 1981). In particular it follows that the limit is maximal in the isotropic case. These results coincide with the corresponding results for stationary Poisson hyperplane processes (see Theorem 7.2), and in fact the asymptotic behaviour of the above-described model is closely related to the behaviour of a stationary Poisson hyperplane process. A comparison of both models may be found in Kaltenbach (1990).

Related problems concerning almost sure approximation of planar convex bodies with smooth boundaries by circumscribed polytopes generated by independently and identically distributed random tangent lines are studied in Carlsson and Grenander (1967), and Schneider (1988).

3. Random convex bodies

A *random compact set* or a *random convex body* X is a random element of the measurable spaces $\mathscr{C}^d$ or $\mathscr{K}^d$, i.e., X is given by a probability measure on $\mathscr{C}^d$ or $\mathscr{K}^d$, which we call the *distribution* $\mathbb{P}_X$ of X.

Since the spaces $\mathscr{C}^d$ and $\mathscr{K}^d$ carry a linear structure, i.e., they are convex cones with respect to Minkowski addition and multiplication by nonnegative scalars, results for random elements X of $\mathscr{C}^d$ or $\mathscr{K}^d$, analogous to the classical results for real random variables, can be expected. Here, the main difference is that $\mathscr{K}^d$ can be embedded into a Banach space, e.g., into the space $C(S^{d-1})$ of continuous functions on S^{d-1} (see chapter 1.9), using the support function. Therefore, results for $C(S^{d-1})$-valued random elements can be transferred to random convex bodies. This is not possible directly for random compact sets, since the semigroup $\mathscr{C}^d$ is not embeddable into a group. Therefore, and in view of the goals of this Handbook, the considerations in this section will concentrate on random convex bodies, results for nonconvex sets will be mentioned at the end.

Some of the usual probabilistic notions like joint distribution and (stochastic) independence transfer to random compact sets immediately. For others, like the expectation, we need additional explanations. If the random compact set X is viewed as a measurable mapping from a basic probability space $(\Omega, \mathscr{A}, \mathbb{P})$ into $(\mathscr{C}^d, \mathscr{B}(\mathscr{C}^d))$, the *expectation* $\mathbb{E}X$ can be defined by:

$$\mathbb{E}X = \{\mathbb{E}\xi\colon\ \xi : \Omega \to \mathbb{R}^d \text{ integrable, } \xi(\omega) \in X(\omega) \text{ for almost all } \omega \in \Omega\}\,.$$

A measurable mapping $\xi : \Omega \to \mathbb{R}^d$ with $\xi(\omega) \in X(\omega)$, $\omega \in \Omega$, is called a selection of X, so $\mathbb{E}X$ is the set built by the mean vectors of all integrable selections of X. $\mathbb{E}X$ is compact if and only if $\mathbb{E}d(X, \{0\}) < \infty$. Moreover, if the probability space $(\Omega, \mathscr{A}, \mathbb{P})$ is nonatomic, $\mathbb{E}X$ is convex (see Aumann 1965). This indicates a disadvantage in the definition of $\mathbb{E}X$, due to the use of selections the expectation will depend on the structure of the underlying probability space and not on the distribution alone. To overcome this difficulty, we therefore assume in the following that $(\Omega, \mathcal{A}, \mathbb{P})$ is nonatomic (and fulfills $\mathbb{E}d(X, \{0\}) < 0$) (for more information about expectations, see Vitale 1988, 1990; different aspects of means for random sets are discussed in Stoyan 1989).

For a random compact set X, the convex hull conv X is a random convex body (see section 4) and then $\mathbb{E}(\text{conv}\, X)$ is a convex body, too. Moreover (in view of our assumptions), $\mathbb{E}X = \mathbb{E}(\text{conv}\, X)$. For a random convex body X, the support function h_X is a random element of $C(S^{d-1})$. Since the supremum norm $\|h_X\|_\infty$ obeys $\|h_X\|_\infty = d(X, \{0\})$, we have $\mathbb{E}\|h_X\|_\infty < \infty$. So the expectation $\mathbb{E}h_X$ exists in the (usual) weak sense, this means

$$\varphi(\mathbb{E}h_X) = \int_\Omega \varphi(h_{X(\omega)})\, \mathrm{d}\mathbb{P}(\omega)$$

for all linear functionals $\varphi \in C'(S^{d-1})$ (see Araujo and Giné 1980). As one would expect, $\mathbb{E}h_X$ is the support function of $\mathbb{E}X$,

$$\mathbb{E}h_X = h_{\mathbb{E}X} \,.$$

Thus if $X_1, X_2, \ldots$ is a sequence of random convex bodies, then $h_{X_1}, h_{X_2}, \ldots$ is a sequence of random elements of the Banach space $C(S^{d-1})$ and distributional properties of the random bodies $X_1, X_2, \ldots$ like independence or identical distribution transfer immediately to the sequence of (random) support functions $h_{X_1}, h_{X_2}, \ldots$. Therefore, results for Banach-space-valued random variables can be used. This is the key to most of the results mentioned in the following. For example, the strong law of large numbers in $C(S^{d-1})$ (Araujo and Giné 1980) immediately gives the following *Strong Law of Large Numbers* for random convex bodies.

Theorem 3.1. *Let $X_1, X_2, \ldots$ be a sequence of independent, identically distributed (i.i.d.) random convex bodies* (*such that* $\mathbb{E}d(X_1, \{0\}) < \infty$), *then almost surely*

$$\frac{1}{n}(X_1 + \cdots + X_n) \to EX_1$$

as $n \to \infty$.

This theorem was first obtained by Artstein and Vitale (1975) who thus initiated a variety of subsequent results of a similar nature. Variants of Theorem 3.1 and generalisations are due to Cressie (1978), Hess (1979), Giné, Hahn and Zinn (1983), Puri and Ralescu (1983) and Hiai (1984, 1985).

As a second basic result from probability theory which can be transferred to random convex bodies by the above method we present the *Central Limit Theorem*. To formulate it, we need the covariance Γ_X of a random element X in $C(S^{d-1})$. Γ_X is the mapping $\Gamma_X : C'(S^{d-1}) \times C'(S^{d-1}) \to \mathbb{R}$ defined by

$$\Gamma_X(\varphi, \psi) = \mathbb{E}[\varphi(X - \mathbb{E}X)\psi(X - \mathbb{E}X)], \quad \varphi, \psi \in C'(S^{d-1}).$$

For a random convex body X we set

$$\Gamma_X = \Gamma_{h_X}.$$

We also denote by $\underset{\mathcal{D}}{\rightarrow}$ the convergence in distribution.

Theorem 3.2. *Let $X_1, X_2, \ldots$ be a sequence of i.i.d. random convex bodies (such that $\mathbb{E}d(X_1, \{0\}) < \infty$), then*

$$n^{1/2} d\left(\frac{1}{n}(X_1 + \cdots + X_n), \mathbb{E}X_1\right) \underset{\mathcal{D}}{\rightarrow} \|Z\|_\infty$$

as $n \to \infty$, where Z is a centred Gaussian $C(S^{d-1})$-variable with $\Gamma_Z = \Gamma_{X_1}$.

This result was proved independently in Weil (1982) and Giné, Hahn and Zinn (1983), versions for more special distributions are due to Cressie (1979) and Lyashenko (1982) (also, unpublished manuscripts on the Central Limit Theorem of Eddy and Vitale are listed in Giné, Hahn and Zinn 1983). Both results, Theorems 3.1 and 3.2, hold as well for random compact sets X_i. The reason is that the summation of sets is a convexifying operation. This has been made precise in a theorem of Shapley–Folkmann–Starr (see, e.g., Arrow and Hahn 1971), which is used in Artstein and Vitale (1975), and Weil (1982).

Further limit theorems for random compact sets or random convex bodies, which will not be mentioned in detail, are a law of the iterated logarithm (Giné, Hahn and Zinn 1983), ergodic theorems (Hess 1979, Schürger 1983), and the characterisations of infinitely divisible and stable random bodies (Mase 1979, Giné and Hahn 1985a,b,c). For the latter, it is important to mention that the Gaussian case does not play the same important role for random convex bodies as in classical probability theory. The fact that the space $\mathcal{K}^d$ of convex bodies is a convex cone, implies that any Gaussian measure on $\mathcal{K}^d$ is degenerated (Lyashenko 1983, Vitale 1983a).

Further results concern extensions to closed sets, sets in Banach spaces and convexification of compact sets (Artstein and Hart 1981, Artstein and Hansen 1985, Puri and Ralescu 1985, Puri, Ralescu and Ralescu 1986). General surveys are given by Giné, Hahn and Zinn (1983), Vitale (1983b) and Cressie (1984).

For random convex bodies X, the usual geometric functionals (intrinsic volumes, mixed volumes) become real random variables and some of their relations transfer into expectation formulae. We mention only two of them, the *Brunn–Minkowski Theorem* for random convex bodies,

$$V^{1/d}(\mathbb{E}X) \geqslant \mathbb{E}V^{1/d}(X),$$

proved by Vitale (1990), and the *generalised Steiner formula* (Matheron 1975, Stoyan, Kendall and Mecke 1987)

$$\mathbb{E}V(X+K)=\sum_{k=0}^{d}\alpha_{d0k}\mathbb{E}V_k(X)V_{d-k}(K)\,. \tag{9}$$

Here, X is assumed to have a rotation invariant distribution and $K\in\mathcal{K}^d$ is arbitrary. Equation (9) is a simple consequence of

$$\mathbb{E}V(X+K)=\int_{\mathrm{SO}_d}\mathbb{E}V(\vartheta X+K)\,\mathrm{d}\nu(\vartheta)=\mathbb{E}\int_{G_d}\chi(gX\cap K)\,\mathrm{d}\mu(g)$$

and the *Principal Kinematic Formula* (see chapter 5.1). The coefficients α_{d0k} are also determined by the latter.

4. Random sets

A *random closed set* (RACS) X is a random element of $(\mathcal{F}^d,\mathcal{B}(\mathcal{F}^d))$, i.e., a measurable mapping $X:(\Omega,\mathcal{A},\mathbb{P})\to(\mathcal{F}^d,\mathcal{B}(\mathcal{F}^d))$, where $(\Omega,\mathcal{A},\mathbb{P})$ is an abstract probability space. The image measure of $\mathbb{P}$ under X is the distribution $\mathbb{P}_X$ of X, it is a probability measure on $(\mathcal{F}^d,\mathcal{B}(\mathcal{F}^d))$. Two random closed sets X, X' with the same distribution $\mathbb{P}_X=\mathbb{P}_{X'}$ are called *equivalent*. For a RACS X, there is an analogue of the classical distribution function for real random variables. This is the *capacity functional* T_X of X, defined on $\mathcal{C}^d$ by

$$C\mapsto T_X(C):=\mathbb{P}_X(\mathcal{F}_C)=\mathbb{P}(X\cap C\neq\emptyset)\,.$$

Here, we have used the abbreviation $\mathcal{F}_C=\{F\in\mathcal{F}^d\colon F\cap C\neq\emptyset\}$. The following uniqueness result is part of a more general theorem of Choquet (see also Kendall 1974 and Matheron 1975).

Theorem 4.1. *Two random closed sets X and X' have the same distribution if and only if $T_X=T_{X'}$.*

Other familiar probabilistic notions (joint distribution, independence, etc.) can be transferred to random closed sets X in the obvious way. Moreover, the following geometric transformations map random closed sets X (respectively X, Y) into random closed sets since they are either continuous or have a certain semi-continuity property, and hence are measurable (for details, see Matheron 1975):

$$(X,Y)\mapsto X\cup Y\,,\qquad (X,Y)\mapsto X\cap Y\,,$$

$$(\alpha,X)\mapsto\alpha X\,,\quad \alpha\in\mathbb{R}\,,\qquad (g,X)\mapsto gX\,,\quad g\in G^d\,,$$

$$(X,Y)\mapsto X+Y\quad\text{for } Y \text{ compact}\,,$$

$$X\mapsto\operatorname{conv}X\quad\text{for } X \text{ compact}\,,$$

$$X\mapsto\operatorname{bd}X\,.$$

If the distribution $\mathbb{P}_X$ of a random closed set X is concentrated on one of the measurable subsets $\mathcal{S}^d$, $\mathcal{C}^d$, $\mathcal{R}^d$, $\mathcal{K}^d$, $\mathcal{E}_k^d$ and $\mathcal{L}_k^d$, we will speak of a *random* $\mathcal{S}^d$*-set, random compact set, random* $\mathcal{R}^d$*-set, random convex body, random k-flat* and *random (k-dimensional) subspace*, respectively. Some of these notions have already been mentioned and used in previous sections; they all appear now as special cases of a RACS.

From a theoretical as well as practical point of view, the main interest is in random closed sets X that have certain invariance properties against geometric transformations $\varphi : \mathcal{F}^d \rightarrow \mathcal{F}^d$. We call X φ*-invariant*, if $\varphi(X)$ and X have the same distribution, i.e., if $\mathbb{P}_X$ is invariant under φ. In particular, X is called *stationary* if X is φ-invariant for all translations φ, and X is called *isotropic* if X is φ-invariant for all rotations φ.

Throughout this section, we assume that X is stationary. If we also require $X \neq \emptyset$ (almost surely), then X is almost surely unbounded. In view of the geometric aspect of this Handbook, it is therefore natural to concentrate on (stationary) random $\mathcal{S}^d$-sets X. The main goal in the following is to define mean values $D_j(X)$ of the intrinsic volumes V_j for X (we call these mean values *quermass densities*) and to transfer classical integral geometric formulae (see chapter 5.1) to such random $\mathcal{S}^d$-sets.

For a stationary random $\mathcal{S}^d$-set X and $j \in \{0, \ldots, d\}$, the curvature measure $\Phi_j(X, \cdot)$ is a random signed Radon measure. Also we may consider $V_j(X \cap K)$ for all $K \in \mathcal{K}^d$ and get a real random variable. In order that the expectations $\mathbb{E}\Phi_j(X, \cdot)$ and $\mathbb{E}V_j(X \cap K)$ exist, integrability conditions have to be fulfilled by X. Here we use a simple, but surely not optimal condition. For $K \in \mathcal{R}^d$ let $N(K)$ be the smallest number n such that $K = K_1 \cup \cdots \cup K_n$, with $K_i \in \mathcal{K}^d$. The mapping $N : \mathcal{R}_d \rightarrow \mathbb{N}_0$ is measurable. We now make the general assumption that

$$\mathbb{E}2^{N(X \cap K)} < \infty$$

for all $K \in \mathcal{K}^d$. This ensures that all expectations which appear in the following will exist.

For example, $\mathbb{E}\Phi_j(X, \cdot)$ is now again a signed Radon measure, and, because of the stationarity, translation invariant. Hence

$$\mathbb{E}\Phi_j(X, \cdot) = c\lambda_d ,$$

with a constant $c \in \mathbb{R}$, which can serve as quermass density.

Another approach could be to consider

$$\lim_{i \rightarrow \infty} \frac{\mathbb{E}V_j(X \cap K_i)}{V_d(K_i)} ,$$

where $K_i \nearrow \mathbb{R}^d$, if this limit exists and is independent of the sequence K_i.

As a third approach, one can use a set C_0 from a lattice tesselation of $\mathbb{R}^d$, e.g., $C_0 = [0, 1]^n$ (or any other box of unit volume), and subtract half the value of the boundary of C_0. More precisely, let

$$\partial^+ C_0 = \{x \in C_0 \colon \max_{i=1,\ldots,d} x_i = 1\}$$

be the "upper right" boundary of C_0. Since $\partial^+ C_0 \in \mathcal{R}^d$,

$$\mathbb{E}[V_j(X \cap C_0) - V_j(X \cap \partial^+ C_0)]$$

is another candidate for the quermass density of X.

The following result shows that all three approaches are equivalent.

Theorem 4.2. *Let X be a stationary $\mathcal{S}^d$-set. Then for $j = 0, \ldots, d$ there exists a number $D_j(X)$ such that*

$$\mathbb{E}\Phi_j(X, \cdot) = D_j(X) \cdot \lambda_d\,, \tag{10}$$

$$\lim_{r\to\infty} \frac{\mathbb{E}V_j(X \cap rK)}{V_d(rK)} = D_j(X) \quad \textit{for all } K \in \mathcal{K}^d \textit{ with } V_d(K) > 0\,, \tag{11}$$

and

$$\mathbb{E}[V_j(X \cap C_0) - V_j(X \cap \partial^+ C_0)] = D_j(X)\,. \tag{12}$$

We call $D_j(X)$ the *jth quermass density* of X. The following formulae for quermass densities are the counterpart of the two basic formulae from integral geometry, the *Principal Kinematic Formula* and the *Crofton Formula* (see chapter 5.1, in particular for the explicit value of the coefficients).

Theorem 4.3. *Let X be a stationary and isotropic $\mathcal{S}^d$-set and let $K \in \mathcal{K}^d$. Then, for $j = 0, \ldots, d$,*

$$\mathbb{E}V_j(X \cap K) = \sum_{k=j}^{d} \alpha_{djk} V_k(K) D_{d+j-k}(X)\,. \tag{13}$$

Theorem 4.4. *Let X be a stationary and isotropic $\mathcal{S}^d$-set and let $L \subset \mathbb{R}^d$ be a q-dimensional subspace. Then $X \cap L$ is also a stationary and isotropic $\mathcal{S}^d$-set (in L), and for $j = 0, \ldots, q$*

$$D_j(X \cap L) = \alpha_{djq} D_{d+j-q}(X)\,. \tag{14}$$

The proofs are similar. We give a short outline in the case of Theorem 4.3. From the Principal Kinematic Formula and Fubini's theorem we have

$$(V_d(rB^d))^{-1} \int_{G_d} \mathbb{E}V_j(X \cap rB^d \cap gK)\, \mathrm{d}\mu(g)$$

$$= (V_d(rB^d))^{-1} \sum_{k=j}^{d} \alpha_{djk} V_k(K) \mathbb{E}V_{d+j-k}(X \cap rB^d)\,.$$

The right side converges for $r \to \infty$, because of (11), towards the right side of (13). For the integration on the left side we may asymptotically (for $r \to \infty$) concentrate on those $g \in G^d$, for which $gK \subset rB^d$. For these g,

$$\mathbb{E}V_j(X \cap rB^d \cap gK) = \mathbb{E}V_j(X \cap gK) = \mathbb{E}V_j(X \cap K)\,,$$

hence the integrand is constant. The result therefore follows from

$$\lim_{r \to \infty} \frac{\mu(\{g \in G_d\colon\ gK \subset rB^d\})}{V_d(rB^d)} = 1\,.$$

These results were obtained in Weil and Wieacker (1984), and Weil (1984); for more general classes of sets, see Zähle (1986).

A theory of random closed sets can be developed in any suitable topological space, and in fact Theorem 4.1 is given in Matheron (1975) in this more general setting. We will use it in the particular case of the locally compact space $\overline{\mathcal{F}}^d = \mathcal{F}^d \backslash \{\emptyset\}$.

5. Point processes

For point processes on $\mathcal{F}^d$ two closely connected approaches exist. They can either be described geometrically as random collections of sets in $\mathcal{F}^d$ or analytically as locally finite random measures. We shortly survey both developments.

For the first approach, we call a set $\eta \subset \overline{\mathcal{F}}^d$ *locally finite*, if

$$\operatorname{card}\{F \in \eta\colon\ F \cap C \neq \emptyset\} < \infty$$

for all $C \in \mathcal{C}^d$. (Here card denotes the number of elements.) Let N be the class of all locally finite subsets $\eta \subset \overline{\mathcal{F}}^d$ and let $\mathcal{N}$ be the σ-algebra on N, generated by the "counting mappings" $\gamma_{\mathcal{F}} : \eta \mapsto \operatorname{card}(\eta \cap \mathcal{F})$, where $\mathcal{F}$ runs through the σ-algebra $\mathcal{B}(\overline{\mathcal{F}}^d)$. A (*simple*) *point process* X on $\mathcal{F}^d$ is then a measurable mapping $X : (\Omega, \mathcal{A}, \mathbb{P}) \to (N, \mathcal{N})$, where $(\Omega, \mathcal{A}, \mathbb{P})$ denotes an abstract probability space. The distribution $\mathbb{P}_X$ is the image of $\mathbb{P}$ under X. Two point processes X, X' on $\mathcal{F}^d$ with the same distribution are again called *equivalent*.

It should be emphasised that a realisation $X(\omega)$ of a point process X on $\mathcal{F}^d$ is a collection of sets, hence it has a spatial component but no temporal interpretation. Therefore, X can also be called a *random field of closed sets*. Due to the definition, sets $F \in X(\omega)$ can occur only once (i.e., with multiplicity one), therefore, these point processes X are called *simple*. Simple point processes X on $\mathcal{F}^d$ are just locally finite random subsets of $\overline{\mathcal{F}}^d$, since $\mathcal{N}$ coincides with the Borel-σ-algebra $\mathcal{B}(\overline{\mathcal{F}}^d)$ restricted to N (see Ripley 1976). Hence all tools and results which can be formulated generally for random sets immediately transpose to point processes. In particular, this shows how the geometric transformations from the list in the previous section act on point processes and also, stationarity and isotropy for point processes is defined.

More specifically, for two point processes X, X' on $\mathscr{F}^d$ the *union* (or *superposition*) $X \cup X'$ is again a point process on $\mathscr{F}^d$, and for a point process X on $\mathscr{F}^d$ and $\mathscr{F} \in \mathscr{B}(\overline{\mathscr{F}}^d)$ the *intersection* $X \cap \mathscr{F}$ is again a point process on $\mathscr{F}^d$ (the *restriction* of X to $\mathscr{F}$). The *section process* $X \cap F$, for $F \in \overline{\mathscr{F}}^d$ is of a different nature, it consists of the sets $F' \cap F$, $F' \in X$.

If the point process X is concentrated on one of the sets $\mathscr{C}^d$, $\mathscr{R}^d$ or $\mathscr{K}^d$, we call it a *particle process*, point processes on $\mathscr{E}_k^d$ are called *processes of flats* (*k-flats*), and processes on $\mathbb{R}^d$ are called *ordinary point processes*. Also, the meaning of *line process*, *hyperplane process*, *process of curves* (*fibre process*) is now evident.

The following uniqueness result for point processes follows from the general version of Theorem 4.1, which goes back to Choquet, Kendall, and Matheron (see Matheron 1975).

Theorem 5.1. *Two point processes X, X' on $\mathscr{F}^d$ have the same distribution if and only if*

$$\mathbb{P}(X \cap \mathscr{F} = \emptyset) = \mathbb{P}(X' \cap \mathscr{F} = \emptyset)$$

for all compact $\mathscr{F} \subset \overline{\mathscr{F}}^d$.

The set-theoretic approach described so far, has the disadvantage that some natural operations lead to point processes which cannot be described as (simple) random collections any more. For example, for a particle process X, the centres (Steiner point, centre of gravity, etc.) of the particles $K \in X$ build an ordinary point process $\tilde{X}$ on $\mathbb{R}^d$ where multiple points are possible. Here, the random measure approach is more appropriate; it also embeds the theory of point processes in the theory of random measures.

We call a Borel measure φ on $\overline{\mathscr{F}}^d$ a *locally finite counting measure* if

$$\varphi(\{F \in \overline{\mathscr{F}}^d : F \cap C \neq \emptyset\}) \in \mathbb{N}_0$$

for all $C \in \mathscr{C}^d$. Let M be the collection of all locally finite counting measures on $\overline{\mathscr{F}}^d$, and let $\mathscr{M}$ be the σ-algebra on M, generated by the "evaluation mappings" $\gamma_{\mathscr{F}} : \varphi \mapsto \varphi(\mathscr{F})$, where $\mathscr{F}$ runs through the σ-algebra $\mathscr{B}(\overline{\mathscr{F}}^d)$. $\varphi \in M$ is called *simple* if $\varphi(\{F\}) \leq 1$ for all $F \in \overline{\mathscr{F}}^d$. There is an isomorphism between the simple measures $\varphi \in M$ and the sets $\eta \in N$ given by the representation

$$\varphi = \sum_{F \in \eta} \delta_F ,$$

where δ_F denotes the Dirac measure in $F \in \overline{\mathscr{F}}^d$. This isomorphism also preserves the measurability structure, i.e., the σ-algebra $\mathscr{M}$, restricted to the simple measures, is isomorphic to the σ-algebra $\mathscr{N}$. As an extension of the definition given before, we therefore define a (*general*) *point process* X on $\mathscr{F}^d$ as a measurable mapping $X : (\Omega, \mathscr{A}, \mathbb{P}) \rightarrow (M, \mathscr{M})$, where $(\Omega, \mathscr{A}, \mathbb{P})$ denotes an abstract probability space.

It is clear that Theorem 5.1 is no longer true for general point processes (but general uniqueness results for random measures apply, see Kallenberg 1986). *Stationarity* and *isotropy* are defined for general point processes as in the simple case, based on the corresponding action of G^d on M.

In the following, we will concentrate on simple point processes without further mention; nonsimple processes can only occur as secondary processes (like the process of centres). We will however use both approaches to point processes simultaneously, i.e., we will not distinguish strictly between simple random measures in M and their corresponding random set in N. This allows us, e.g., to write $X(\mathcal{F})$ for the number of elements of X which lie in $\mathcal{F} \in \mathcal{B}(\overline{\mathcal{F}}^d)$, on one hand, and $F \in X$ for the elements F of X, on the other.

A basic notion for (simple or general) point processes X is the *intensity measure* Λ, a counterpart to the classical expectation for random variables. Λ is a measure on $\overline{\mathcal{F}}^d$ defined by

$$\Lambda(\mathcal{F}) = \mathbb{E}X(\mathcal{F}), \quad \mathcal{F} \in \mathcal{B}(\overline{\mathcal{F}}^d).$$

$\Lambda(\mathcal{F})$ is thus the mean number of sets of X lying in $\mathcal{F}$. We assume throughout that Λ is *locally finite*, i.e., obeys

$$\Lambda(\{F \in \overline{\mathcal{F}}^d\colon F \cap C \neq \emptyset\}) < \infty$$

for all $C \in \mathcal{C}^d$. For stationary (isotropic) X, Λ is translation (rotation) invariant.

A simple but rather useful result is the following. For indicator functions $f = 1_{\mathcal{F}}$ it is a direct consequence of the definition of the intensity measure, for general functions $f \geq 0$ it follows from the monotone convergence theorem of measure theory.

Theorem 5.2 (Campbell). *For a point process X on $\mathcal{F}^d$ and a measurable function $f : \mathcal{F}^d \to \mathbb{R}_+$,*

$$\omega \mapsto \sum_{F \in X(\omega)} f(F)$$

is measurable and

$$\mathbb{E} \sum_{F \in X} f(F) = \int_{\mathcal{F}^d} f \, \mathrm{d}\Lambda. \tag{15}$$

If X is a point process on $\mathcal{F}^d$, we may consider the *union set* Y_X defined by

$$Y_X = \bigcup_{F \in X} F.$$

Y_X is a RACS, and if X is stationary (isotropic), then Y_X is stationary (isotropic). The capacity functional of Y_X follows directly from the distribution of X, since

$$T_{Y_X}(C) = 1 - \mathbb{P}(Y_X \cap C = \emptyset) = 1 - \mathbb{P}(X \cap \mathcal{F}_C = \emptyset), \quad C \in \mathcal{C}^d.$$

The most important class of point processes is given by the Poisson processes. A point process X on $\mathcal{F}^d$ with intensity measure Λ is called a *Poisson process*, if $\operatorname{card}(X \cap \mathscr{F})$ is, for each $\mathscr{F} \in \mathscr{B}(\mathcal{F}^d)$ with $\Lambda(\mathscr{F}) < \infty$, a Poisson random variable with mean $\Lambda(\mathscr{F})$, i.e.,

$$\mathbb{P}(\operatorname{card}(X \cap \mathscr{F}) = k) = \mathrm{e}^{-\Lambda(\mathscr{F})} \frac{(\Lambda(\mathscr{F}))^k}{k!}, \quad k = 0, 1, 2, \ldots .$$

For the general theory of Poisson processes and their most important properties, see Daley and Vere-Jones (1988), or Karr (1986). Here, we only mention some existence and uniqueness results.

Theorem 5.3. *Let Λ be a locally finite measure on $\mathcal{F}^d$. Then there is (up to equivalence) a unique Poisson process X on $\mathcal{F}^d$ with intensity measure Λ. X is stationary (isotropic) if and only if Λ is translation invariant (rotation invariant).*

It follows from Theorem 5.1 that the condition

$$\mathbb{P}(\operatorname{card}(X \cap \mathscr{F}) = 0) = \mathrm{e}^{-\Lambda(\mathscr{F})}, \quad \mathscr{F} \in \mathscr{B}(\mathcal{F}^d),$$

already characterises a Poisson process. This can even be generalised slightly.

Theorem 5.4. *A point process X on $\mathcal{F}^d$ is a Poisson process (with intensity measure Λ) if and only if*

$$\mathbb{P}(X \cap \mathscr{F}_C = \emptyset) = \mathrm{e}^{-\Lambda(\mathscr{F}_C)}$$

for all $C \in \mathscr{C}^d$.

In particular, this means that the Poisson property of a point process X on $\mathcal{F}^d$ is already determined by the union set Y_X! For further results on Poisson processes of sets, see Matheron (1975) and Stoyan, Kendall and Mecke (1987).

In view of the theme of this Handbook, we are mainly interested in particle processes on $\mathscr{K}^d$ or $\mathscr{R}^d$. For such processes, the union set Y_X is a random $\mathscr{S}^d$-set.

We first give a decomposition of the intensity measure of a stationary point process X on $\mathscr{C}^d$. Let $z : \mathscr{C}^d \to \mathbb{R}^d$ be a mapping which supplies each set $C \in \mathscr{C}^d$ with a center $z(C)$ in a motion covariant manner. We will use as $z(C)$ the midpoint of the circumsphere of C (on $\mathscr{R}^d$, the Steiner point is another reasonable choice). z is easily seen to be continuous on $\mathscr{C}^d$. Let

$$\mathscr{C}_0^d = \{C \in \mathscr{C}^d : z(C) = 0\}$$

be the set of centred particles (the sets $\mathscr{R}_0^d$ and $\mathscr{K}_0^d$ are defined analogously). Then

$$\varphi : \mathscr{C}^d \to \mathscr{C}_0^d \times \mathbb{R}^d, \quad C \mapsto (C - z(C), z(C)),$$

is a homeomorphism. Let $\Lambda' = \Lambda \circ \varphi^{-1}$ be the image of Λ under φ, Λ' is thus a measure on $\mathscr{C}_0^d \times \mathbb{R}^d$. For stationary X, Λ' is translation invariant in the second coordinate, hence it is of the form

$$\Lambda' = \rho \otimes \lambda_d .$$

Since Λ was assumed to be locally finite, ρ is a finite measure on $\mathscr{C}_d^0$. In case $\Lambda \not\equiv 0$, ρ can be normalised to a probability measure.

Theorem 5.5. *Let X be a stationary point process on $\mathscr{C}^d$ with intensity measure $\Lambda \not\equiv 0$. Then there is a $\lambda \in (0, \infty)$ and a probability measure $\mathbb{P}_0$ on $\mathscr{C}_0^d$ with*

$$\Lambda = \lambda \cdot (\mathbb{P}_0 \otimes \lambda_d) \circ \varphi .$$

λ and $\mathbb{P}_0$ are uniquely determined by Λ. If X is isotropic, then $\mathbb{P}_0$ is rotation invariant.

We call λ the *intensity* of X and $\mathbb{P}_0$ the *shape distribution*. The interpretation of the latter is obvious, $\mathbb{P}_0$ is the distribution of a "typical" particle of X. For λ, the interpretation is given in the next theorem.

Theorem 5.6. *Let X be a stationary point process on $\mathscr{C}^d$. Then*

$$\lambda = \frac{1}{\kappa_d} \mathbb{E}\,\mathrm{card}\{C \in \mathscr{C}^d \colon C \in X, z(C) \in B^d\}$$

and

$$\lambda = \lim_{r \to \infty} \frac{1}{V_d(rK)} \mathbb{E}\,\mathrm{card}(X \cap \mathscr{F}_{rK})$$

for all $K \in \mathscr{K}^d$ with $V_d(K) > 0$.

The proof of the first equation follows from

$$\mathbb{E}\,\mathrm{card}(X \cap \{C \in \mathscr{C}^d \colon z(C) \in B^d\}) = \lambda \cdot \lambda_d(B^d)\mathbb{P}_0(\mathscr{C}_0^d) = \lambda \cdot \kappa_d .$$

To prove the second, we assume $K \in \mathscr{K}_0^d$. Then

$$\mathbb{E}\,\mathrm{card}(X \cap \mathscr{F}_{rK}) = \lambda \int_{\mathscr{C}_0^d} \lambda_d(A_r(C))\, \mathrm{d}\mathbb{P}_0(C),$$

where

$$A_r(C) = \{x \in \mathbb{R}^d \colon (C + x) \cap rK \neq \emptyset\} .$$

Since

$$\lim_{r\to\infty} \frac{\lambda_d(A_r(C))}{V_d(rK)} = 1$$

for all $C \in \mathscr{C}_0^d$, the result follows.

For Poisson processes further results are true.

Theorem 5.7. *Let $\lambda \in [0, \infty)$ and let $\mathbb{P}_0$ be a probability measure on $\mathscr{C}_0^d$. Then there is (up to equivalence) a unique stationary Poisson process X on $\mathscr{C}^d$ with intensity λ and shape distribution $\mathbb{P}_0$. X is isotropic, if and only if $\mathbb{P}_0$ is rotation invariant.*

An important property of (general) Poisson point processes is that the different points are independent. For a stationary Poisson process X on $\mathscr{C}^d$ this implies a simple but very useful procedure to simulate X, e.g., on a computer. First, an ordinary Poisson process $\tilde{X}$ of intensity λ in $\mathbb{R}^d$ is simulated (this involves the determination of a random number according to the appropriate Poisson distribution and afterwards a simulation of uniformly distributed points in a region). Then, to each point x of $\tilde{X}$ a random set X_x with distribution $\mathbb{P}_0$ is added independently. The resulting configuration is a realisation of X.

We now concentrate on processes X on $\mathscr{R}^d$ and aim to introduce quermass densities $D_j(X)$ of X. Again, we need an integrability condition. Let

$$\mathbb{E}2^{N(X_0)} < \infty ,$$

where X_0 is a random set with distribution $\mathbb{P}_{X_0} = \mathbb{P}_0$. For brevity, we sometimes denote the expectation $\mathbb{E}f(X_0)$ by $\bar{f}$.

In contrast to section 4, we now have the possibility of a direct definition of $D_j(X)$. We call $D_j(X) = \lambda \cdot \mathbb{E}V_j(X_0)$ the *jth quermass density* of X, $j = 0, \ldots, d$. The following theorem shows that the other approaches from section 4 lead to the same quantity.

Theorem 5.8. *Let X be a stationary point process on $\mathscr{R}^d$. Then, for $j = 0, \ldots, d$,*

$$\mathbb{E} \sum_{C\in X} \Phi_j(C, \cdot) = D_j(X) \cdot \lambda_d , \tag{16}$$

$$\lim_{r\to\infty} \frac{\mathbb{E}\sum_{C\in X} V_j(C \cap rK)}{V_d(rK)} = D_j(X) \quad \textit{for all } K \in \mathscr{K}^d \textit{ with } V_d(K) > 0 , \tag{17}$$

$$\mathbb{E} \sum_{C\in X} [V_j(C \cap C_0) - V_j(C \cap \partial^+ C_0)] = D_j(X) . \tag{18}$$

The proofs are easier here, since eq. (15) can be used. If X is ergodic, the second equation holds almost surely, i.e., without the expectation sign (Nguyen and Zessin 1979).

The transfer of the integral geometric formulae now proceeds without problems. We get the following versions of the Principal Kinematic Formula and the Crofton Formula.

Theorem 5.9. *Let X be a stationary and isotropic point process on $\mathscr{R}^d$, and let $K \in \mathscr{K}^d$. Then, for $j = 0, \ldots, d$,*

$$\mathbb{E} \sum_{C \in X} V_j(C \cap K) = \sum_{k=j}^{d} \alpha_{djk} V_k(K) D_{d+j-k}(X) . \tag{19}$$

Theorem 5.10. *Let X be a stationary and isotropic point process on $\mathscr{R}^d$, and let $L \subset \mathbb{R}^d$ be a q-dimensional subspace. Then $X \cap L$ is also a stationary and isotropic point process on $\mathscr{R}^q$ (in L), and for $j = 0, \ldots, q$*

$$D_j(X \cap L) = \alpha_{djq} D_{d+j-q}(X) . \tag{20}$$

Both formulae follow from

$$\mathbb{E} \sum_{C \in X} \Phi_j(C \cap K, A) = \sum_{k=j}^{d} \alpha_{djk} \Phi_k(K, A) D_{d+j-k}(X) , \quad A \in \mathscr{B}_d ,$$

which is a consequence of eq. (15) for $f(C) = \Phi_k(C, \cdot)$. Equation (19) is obtained with $A = \mathbb{R}^d$. For eq. (20), let $K = B^q$ be the unit ball in L and A the relative interior of B^q. Then

$$\mathbb{E} \sum_{C \in X} \Phi_j(C \cap K, A) = \mathbb{E} \sum_{C \in X \cap L} \Phi_j(C \cap K, A) = D_j(X \cap L) \cdot \kappa_q$$

and

$$\Phi_k(K, A) = 0 , \quad k = 0, \ldots, q-1 ; \qquad \Phi_q(K, A) = \kappa_q .$$

The last results have been derived in a more general setting, for point processes of cylinders, in Weil (1987). There, formulae are also given for stationary, nonisotropic processes. Point processes on more general classes of sets are treated in Zähle (1982, 1986).

For a stationary Poisson process X on $\mathscr{C}^d$, the union set Y_X is called a *Boolean model*. Here, the capacity functional T_{Y_X} can be calculated,

$$T_{Y_X}(C) = \mathbb{P}(Y_X \in \mathscr{F}_C) = 1 - \mathbb{P}(X \cap \mathscr{F}_C = \emptyset) = 1 - \mathrm{e}^{-\Lambda(\mathscr{F}_C)}$$

with

$$\begin{aligned}\Lambda(\mathscr{F}_C) &= \lambda \int_{\mathscr{C}_0^d} \int_{\mathbb{R}^d} 1_{\mathscr{F}_C}(K + x) \, \mathrm{d}\lambda_d(x) \, \mathrm{d}\mathbb{P}_0(K) \\ &= \lambda \int_{\mathscr{C}_0^d} \lambda_d(K + \check{C}) \, \mathrm{d}\mathbb{P}_0(K)\end{aligned}$$

(where $\check{C}$ is the set C reflected in the origin). If X is moreover an isotropic process on $\mathcal{K}^d$ and if $K \in \mathcal{K}^d$, we can simplify the latter formula because of (9).

Theorem 5.11. *Let X be a stationary and isotropic Poisson process on $\mathcal{K}^d$ and $K \in \mathcal{K}^d$. Then*

$$-\ln(1 - T_{Y_X}(K)) = \sum_{k=0}^{d} \alpha_{d0k} V_k(K) D_{d-k}(X) . \tag{21}$$

For isotropic Boolean models (in $\mathcal{S}^d$), a connection between the quermass densities of Y_X and those of X can also be given. The derivation of this result uses the additivity of V_j, the independence properties of Poisson processes X, and the corresponding product form of the moment measures of X.

Theorem 5.12. *Let X be a stationary Poisson process on $\mathcal{R}^d$, $j \in \{0, \ldots, d\}$ and $K \in \mathcal{K}^d$. Then*

$$\mathbb{E}V_j(Y_X \cap K) = \sum_{k=1}^{\infty} \frac{(-1)^{k-1}}{k!} \lambda^k \int_{\mathcal{R}_0^d} \cdots \int_{\mathcal{R}_0^d} F_j(K, K_1, \ldots, K_k) \, \mathrm{d}\mathbb{P}_0(K_1) \cdots \mathrm{d}\mathbb{P}_0(K_k) , \tag{22}$$

with

$$F_j(K, K_1, \ldots, K_k) = \int_{\mathbb{R}^d} \cdots \int_{\mathbb{R}^d} V_j(K \cap (K_1 + x_1) \cap \cdots \cap (K_k + x_k)) \, \mathrm{d}\lambda_d(x_1) \cdots \mathrm{d}\lambda_d(x_k) . \tag{23}$$

Moreover, for isotropic X, we have

$$F_j(K, K_1, \ldots, K_k) = \int_{G^d} \cdots \int_{G^d} V_j(K \cap g_1 K_1 \cap \cdots \cap g_k K_k) \, \mathrm{d}\mu(g_1) \cdots \mathrm{d}\mu(g_k) . \tag{24}$$

By iteration of the Principal Kinematic Formula it is therefore possible, in the isotropic case, to express $\mathbb{E}V_j(Y_X \cap K)$ by (22) and (24) in terms of the quermassintegrals of K and the quermass densities of X. Theorem 4.2 then implies corresponding formulae for the quermass densities $D_j(Y_X)$. Here we give only the latter formulae.

Corollary 5.13. *The quermass densities of the Boolean model Y_X fulfill*

$$D_d(Y_X) = 1 - \mathrm{e}^{-D_d(X)} ,$$

$$D_{d-1}(Y_X) = D_{d-1}(X) \, \mathrm{e}^{-D_d(X)} ,$$

and

$$D_j(Y_X) = e^{-D_d(X)}\left[D_j(X) + \sum_{k=2}^{d-j} \frac{(-1)^{k-1}}{k!} \sum_{\substack{m_1,\ldots,m_k=j+1 \\ m_1+\cdots+m_k=(k-1)d+j}}^{d-1} c_{m_1,\ldots,m_k}^{(j)} D_{m_1}(X) \cdots D_{m_k}(X) \right],$$

with

$$c_{m_1,\ldots,m_k}^{(j)} = \frac{d!\kappa_d}{j!\kappa_j} \prod_{i=1}^{k} \frac{m_i!\kappa_{m_i}}{d!\kappa_d}$$

for $j = 0, \ldots, d-2$.

The most interesting cases for applications are of course $d=2$ and $d=3$ and there the results look less complicated. For $d=2$ we use the notation A_A, L_A and χ_A to denote the area density, density of the boundary length and density of the characteristic of Y_X, as well as $\bar{A}$, $\bar{L}$ and $\bar{\chi}$ for the integrals of these functionals with respect to $\mathbb{P}_0$. In three dimensions a similar notation is used. The resulting formulae are then the following:

$$\begin{aligned} A_A &= 1 - e^{-\lambda \bar{A}}, \\ L_A &= \lambda \bar{L}\, e^{-\lambda \bar{A}}, \\ \chi_A &= e^{-\lambda \bar{A}}\left(\lambda \bar{\chi} - \frac{1}{4\pi} \lambda^2 \bar{L}^2 \right), \end{aligned} \tag{25}$$

for $d=2$ and

$$\begin{aligned} V_V &= 1 - e^{-\lambda \bar{V}}, \\ S_V &= \lambda \bar{S}\, e^{-\lambda \bar{V}}, \\ M_V &= e^{-\lambda \bar{V}}\left(\lambda \bar{M} - \frac{\pi^2}{32} \lambda^2 \bar{S}^2 \right), \\ \chi_V &= e^{-\lambda \bar{V}}\left(\lambda \bar{\chi} - \frac{1}{4\pi} \lambda^2 \bar{M}\bar{S} + \frac{\pi}{384} \lambda^3 \bar{S}^3 \right), \end{aligned} \tag{26}$$

for $d=3$.

Our approach in this latter part of the section followed that of Weil and Wieacker (1984), but similar considerations in different generality are due to Matheron (1975), Miles (1976), Davy (1976, 1978), A. Kellerer (1983, 1985), H. Kellerer (1984), and Zähle (1986). In the nonisotropic case, (22) and (23) can also be used, but then iterations of the translative version of the Principal Kinematic Formula are necessary. The resulting formulae for the quermass densities of Y_X look similar to those in Corollary 5.13 but involve mixed densities (Weil 1990). In two and three dimensions and for Poisson processes X on $\mathcal{K}^d$ with some symmetry conditions these mixed functionals can be expressed as mixed volumes of convex mean bodies associated with X. This allows the application of classical inequalities to formulate and solve some extremal properties of Boolean models (Weil 1988). For example, for a stationary Poisson process X on $\mathcal{K}^2$, the smallest value of the density χ_V of the Euler characteristic of the union set Y_X is obtained if X is almost surely a process of homothetic equilateral triangles (Betke and Weil 1991). It is open, whether these are the only extremal processes. If the particles of the Poisson process X are all convex, a similar formula holds, without isotropy conditions, for the density χ_V^+ (respectively χ_A^+) of the "lower points of convexity" (specific convexity number) of Y_X (Stoyan, Kendall and Mecke 1987, p. 78).

6. Random surfaces

In the preceding sections, convex bodies were used to construct random sets and point processes. Random surfaces are implicit in this theory, e.g., they occur as boundaries bd X of random $\mathcal{S}^d$-sets X. Here, in this section, we will discuss a different connection between convex geometry and random surfaces, which is based on the integral geometric results discussed in chapter 5.1, section 6, and in which convex bodies occur as a secondary notion associated with a random surface X. Therefore, more general random closed sets X will be considered that have realisations in the space $\mathcal{LC}_m$ of locally countable m-rectifiable closed subsets of $\mathbb{R}^d$.

A random closed set $X : (\Omega, \mathcal{A}, \mathbb{P}) \to \mathcal{F}^d$ is called a *random m-surface* (*random hypersurface* if $m = d - 1$ and *random curve* if $m = 1$) if $X \in \mathcal{LC}_m$ almost surely and $\mathbb{E}\lambda_m(X \cap K) < \infty$ for all $K \in \mathcal{K}^d$. If X is stationary, then its distribution $\mathbb{P}_X$ is a stationary (i.e., translation invariant) probability measure on $\mathcal{LC}_m$ and all results about stationary σ-finite measures on $\mathcal{LC}_m$ mentioned in chapter 5.1, section 6, apply also to X. In particular, we may associate with X the two auxiliary zonoids $\Pi^m(X) := \Pi^m(\mathbb{P}_X)$ and $\Pi_m(X) := \Pi_m(\mathbb{P}_X)$ defined there. The most important real parameter of a stationary random m-surface X is the density $D_m(X)$ defined by $\mathbb{E}\lambda_m(X \cap A) = D_m(X)\lambda_d(A)$, for $A \in \mathcal{B}(\mathbb{R}^d)$. This generalises the quermass density $D_m(X)$ used in previous sections for random $\mathcal{S}^d$-sets. It is natural to call $D_m(X)$ the (m-dimensional) surface area density of X; it is related to the above-mentioned zonoids by the identities

$$\tfrac{1}{2} D_m(X) = \frac{\kappa_{d-m-1}}{(d-m)\kappa_{d-m}} V_1(\Pi^m(X)) = \frac{\kappa_{m-1}}{m\kappa_m} V_1(\Pi_m(X)) . \tag{27}$$

The support functions $h(\Pi^m(X), \cdot)$ and $h(\Pi_m(X), \cdot)$ of the zonoids $\Pi^m(X)$ and $\Pi_m(X)$ are closely related to the "rose of length of orthogonal intersections" and the "rose of number of intersections" considered by Pohlmann, Mecke and Stoyan (1981). They describe in a certain sense mean first order properties of the random surface. For $u \in S^{d-1}$ we have

$$\mathbb{E}\lambda_{m-1}(X \cap A \cap L(u)) = 2h(\Pi_m(X), u)\lambda_{d-1}(A \cap L(u)) \tag{28}$$

whenever $A \in \mathcal{B}(\mathbb{R}^d)$ and $L(u)$ is a hyperplane orthogonal to u. Moreover, a slightly modified version of Theorem 6.4 in chapter 5.1 and the isoperimetric inequality for the Minkowski area relative to a convex body show that, among all convex bodies K with fixed positive volume, $\mathbb{E}\lambda_{m-1}(X \cap \operatorname{bd} K)$ attains a minimum if and only if K is homothetic to $\Pi_m(X)$. For $m < d-1$, there is a similar but more complicated interpretation of $h(\Pi^m(X), u)$ involving projected thick sections (Wieacker 1989). In the case $m = d-1$, where the role of $\Pi^m(X)$ is particularly important, there is a natural analogue of (28). We shall first consider this case.

Let X be a stationary random hypersurface. Then, denoting by $[a, b]$ the segment joining the points $a, b \in \mathbb{R}^d$, we have

$$h(\Pi^{d-1}(X), x) = \tfrac{1}{2}\mathbb{E}\operatorname{card}(X \cap [0, x]) \tag{29}$$

for each $x \in \mathbb{R}^d$ (see Wieacker 1986, for details). In particular, if $\Pi^{d-1}(X)$ is not degenerated, then (29) implies that the map $x \mapsto \frac{1}{2}\mathbb{E}\operatorname{card}(X \cap [0, x])$ is a norm in $\mathbb{R}^d$, the unit ball of which is the polar body $\Pi^{d-1}(X)^*$ of $\Pi^{d-1}(X)$. The behaviour of X in the Minkowski geometry corresponding to this norm is in some sense similar to the behaviour of a stationary and isotropic random hypersurface in Euclidean geometry. For a unit vector u, $h(\Pi^{d-1}(X), u)$ may also be viewed as the intersection density of X in the direction u. If the intersection density in a given direction is large and the random hypersurface is considered as opaque, then one may expect that the visible distance in the same direction is not too large. This observation leads to a second interpretation of $h(\Pi^{d-1}(X), u)$. More precisely, for $u \in S^{d-1}$ and $r \geq 0$ let $\varphi_u(r) := \mathbb{P}([0, ru] \cap X = \emptyset)$ be the probability that the visible distance from 0 in the direction u is at least r. A straightforward argument shows that the function φ_u is convex, and that the right derivative $\varphi_u'(0)$ of φ_u at the origin 0 exists. It turns out that

$$\varphi_u'(0) = -2h(\Pi^{d-1}(X), u)$$

for all $u \in S^{d-1}$. This shows that, for each $\varepsilon > 0$, $\Pi^{d-1}(X)$ is uniquely determined by the values of the capacity functional T_X of X on the set $\{[0, x]: \|x\| < \varepsilon\}$. For any opaque subset $A \subset \mathbb{R}^d$ let $S_A := \{y \in \mathbb{R}^d: [0, y] \cap A = \emptyset\}$ be the open star-shaped set of all points which are visible from the origin. Then, $\lambda_d(S_X)$ is a random variable and by Fubini's theorem we have

$$\begin{aligned}\mathbb{E}(\lambda_d(S_X)) &= \int_{\mathbb{R}^d} \mathbb{P}([0, x] \cap X = \emptyset)\, \mathrm{d}\lambda_d(x) \\ &= \int_{S^{d-1}} \int_0^\infty r^{d-1}\varphi_u(r)\, \mathrm{d}r\, \mathrm{d}\lambda_{d-1}(u)\,.\end{aligned}$$

A rough estimation of $\varphi_u(r)$ yields

$$\mathbb{E}(\lambda_d(S_X)) \geq \frac{1}{2^d(d+1)} V_d(\Pi^{d-1}(X)^*) .$$

Here the best numerical constant in the inequality is still unknown. Similar results for a stationary RACS (the boundary of which is a random hypersurface) and proofs may be found in Wieacker (1986). More precise results for special types of random hypersurfaces are given in Theorem 6.2 and in the next section. Visibility properties have been studied also in Serra (1982), and Yadin and Zacks (1985).

Random hypersurfaces may be used to generate lower-dimensional random sets. For instance, the intersection of a stationary random hypersurface with an m-dimensional C^1 submanifold (or more generally with some countably m-rectifiable Borel subset) M of $\mathbb{R}^d$ is an $(m-1)$-dimensional random closed subset of M. The expected λ_{m-1}-measure of this random set depends only on M and $\Pi^{d-1}(X)$.

Theorem 6.1. *Let X be a stationary random hypersurface. Then, for any m-dimensional submanifold M of class C^1, we have*

$$\mathbb{E}\lambda_{m-1}(X \cap A) = \frac{1}{\kappa_{m-1}} \int_A \int_{S(T_yM)} h(\Pi^{d-1}(X), x)\, \mathrm{d}\lambda_{m-1}(x)\, \mathrm{d}\lambda_m(y)$$

for each Borel subset A of M, where T_yM is the tangent space of M at y and $S(T_yM) := T_yM \cap S^{d-1}$.

Since the integral over $S(T_yM)$ is essentially the mean width of the orthogonal projection of $\Pi^{d-1}(X)$ on T_yM, Theorem 6.1 may be viewed as an analogue of (27). It should be also noticed that the integral over $S(T_yM)$ as a function of y is constant in M if $\Pi^{d-1}(X)$ is a ball (for instance if X is isotropic) or if M is an affine subspace of $\mathbb{R}^d$. In both cases, the measures $\mathbb{E}\lambda_{m-1}(X \cap \cdot)$ and λ_m, when restricted to $\mathscr{B}(M)$, are proportional. We may also consider the intersection of several random hypersurfaces. If a random m-surface Y is almost surely the intersection of $d-m$ independent stationary random hypersurfaces $X_1, \ldots, X_{d-m}$, then a slightly modified version of Theorem 6.3 in chapter 5.1 shows that $\Pi_m(Y)$ is essentially the mixed projection body of $\Pi^{d-1}(X_1), \ldots, \Pi^{d-1}(X_{d-m})$ and B^d $(m-1$ times).

The case where the random hypersurface X is generated by a Poisson process on $\mathscr{L}\mathscr{C}_{d-1}$ (see section 5) is of particular interest. Suppose that X is the union set X_Y where Y is a stationary Poisson process on $\mathscr{L}\mathscr{C}_{d-1}$ with intensity measure Λ. Then we have $\Pi^{d-1}(X_Y) = \Pi^{d-1}(\Lambda)$, where $\Pi^{d-1}(\Lambda)$ is defined as in chapter 5.1, section 6. For $\omega \in \Omega$ and $k \in \mathbb{N}$ let

$$X_Y^k(\omega) := \{x \in \mathbb{R}^d \colon Y(\omega)(\mathscr{F}_{\{x\}}) \geq d-k\}$$

be the set of points belonging to at least $d-k$ elements of Y. Then X_Y^k is a

k-dimensional random closed set and we have

$$D_k(X_Y^k) = V_{d-k}(\Pi^{d-1}(X_Y)) , \tag{30}$$

where the density $D_k(X_Y^k)$ is again defined by $\mathbb{E}\lambda_k(X_Y^k \cap A) = D_k(X_Y^k)\lambda_d(A)$, $A \in \mathcal{B}(\mathbb{R}^d)$. Hence, eq. (27) and the Minkowski–Fenchel–Aleksandrov inequalities for the intrinsic volumes imply

$$D_k(X_Y^k) \leq \binom{d}{k} \frac{\kappa_d}{\kappa_k} \left(\frac{\kappa_{d-1}}{d\kappa_d} D_{d-1}(X_Y) \right)^{d-k} , \tag{31}$$

with equality if and only if $\Pi^{d-1}(X_Y)$ is a ball (this is the case, for instance, if the Poisson process Y is isotropic). In the case where X is generated by a Poisson process of hyperplanes, (30) is due to Matheron (1975) and (31) is due to Thomas (1984) (the general case is treated in Wieacker 1986). These densities, and in particular $D_0(X_Y^0)$, measure in a certain sense the denseness of the process. If the elements of the Poisson process are the boundaries of nondegenerate convex bodies, more precise results can be obtained. For a particle process X, we denote by bd X the corresponding process of boundary sets. If X is a Poisson process then also bd X, and bd X has the same invariance properties as X.

Theorem 6.2. *Let Z be a stationary Poisson process of nondegenerate convex bodies with locally finite intensity measure Λ and such that $0 < \mathbb{E}\lambda_{d-1}(\mathrm{bd}\, Z \cap [0,1]^d) < \infty$. Then the conditional expectation $\mathbb{E}(\lambda_d(S_{X_Z}) \mid 0 \notin X_Z)$ fulfills*

$$\mathbb{E}(\lambda_d(S_{X_Z}) \mid 0 \notin X_Z) = d!V_d(\Pi^{d-1}(X_{\mathrm{bd}\, Z})^*)$$
$$\geq d!\kappa_d \left(\frac{\kappa_{d-1}}{d\kappa_d} D_{d-1}(X_{\mathrm{bd}\, Z}) \right)^{-d} ,$$

with equality if and only if $\Pi^{d-1}(X_{\mathrm{bd}\, Z})$ is a ball (in particular if Z is isotropic). Moreover, we have

$$4^d \leq D_0(X_{\mathrm{bd}\, Z}^0)\mathbb{E}(\lambda_d(S_{X_Z}) \mid 0 \notin X_Z) \leq d!\kappa_d^2 ,$$

with equality on the left if and only if $\Pi^{d-1}(X_{\mathrm{bd}\, Z})$ is a parallelotope, and equality on the right if and only if $\Pi^{d-1}(X_{\mathrm{bd}\, Z})$ is an ellipsoid.

The lower bound for the mean visible volume is a consequence of (27) and Jensen's inequality. In the second assertion, both inequalities and the corresponding equality cases follow as in section 2. Similar results have been obtained for the random mosaics generated by Poisson hyperplane processes. A common generalisation of these results for Poisson processes of cylinders and further inequalities involving the intensity measure are due to Schneider (1987b). Random surfaces dividing the space into convex polytopes (random mosaics), and in particular stationary Poisson hyperplane networks, are treated separately in section 7.

For stationary random m-surfaces with $1 < m < d-1$, it is generally difficult to get results of this type, even in the case where the random m-surface is generated by a Poisson process of m-flats. Here analytical methods seem to be more successful (see Matheron 1975, Mecke and Thomas 1986, Goodey and Howard 1990a,b). The case $m = 1$ is more accessible. In particular, if X is the union set of a stationary Poisson process of straight lines, then X is uniquely determined by $\Pi_1(X)$ (up to equivalence). Moreover, if we denote by Y the random hypersurface of all points which have distance r from at least one line of the process generating X, then Y is the union set of a stationary Poisson process of boundaries of cylinders, and we have

$$\Pi^{d-1}(Y) = 2\kappa_{d-2} r^{d-2} \Pi^1(X) .$$

Hence, Schneider's results about Poisson processes of cylinders (Schneider 1987b) lead to several stochastic interpretations of the parameters of $\Pi^1(X)$. For random curves a few inequalities have been obtained (Wieacker 1989). In the planar case, additional results are due to the fact that $\Pi^1(X)$ is obtained by rotating $\Pi_1(X)$ by the angle $\frac{1}{2}\pi$. For instance, if C is a fixed curve given as the image $C = f([0, 1])$ of some injective Lipschitzian function $f : [0, 1] \to \mathbb{R}^2$ and if Y is a stationary Poisson process of translates of C with finite intensity λ, then (30) may be combined with an inequality for $V_2(\Pi_1(X))$ to get

$$D_0(X_Y^0) \geqslant 2V_2(\operatorname{conv} C) .$$

Here equality holds, for instance, if C is a half circle. Further results about random curves and random surfaces of intermediate dimension which are closely related to convex geometry may be found in Wieacker (1989).

Concerning the classical part of the theory which is not particularly related to convex geometry, we refer to Stoyan, Kendall and Mecke (1987, chapter 9) and the literature quoted there. Random processes of Hausdorff rectifiable closed sets were first investigated in Zähle (1982). While the theory described here is based on first order tangential properties, a second order theory based on Federer's sets with positive reach is also due to Zähle (1986). A very different aspect of the theory of random surfaces is treated in Wschebor (1985).

7. Random mosaics

A hypersurface $F \in \mathscr{L}\mathscr{C}_{d-1}$ is called a *mosaic* if $\mathbb{R}^d \setminus F$ is a locally finite union of bounded, disjoint open convex sets the closures of which are called the cells of the mosaic. Here, "locally finite" means that each compact subset of $\mathbb{R}^d$ meets only a finite number of cells. From the definition it follows that the cells are convex polytopes, and the k-faces of these polytopes are called k-faces of the mosaic. A random hypersurface $X : (\Omega, \mathscr{A}, \mathbb{P}) \to \mathscr{L}\mathscr{C}_{d-1}$ is called a *random mosaic* if X is almost surely a mosaic (the terms "mosaic" and "random mosaic" are sometimes used in a more general sense). Since X is also uniquely determined by the random measure ξ_X satisfying

$$\xi_X(\omega, U) := \operatorname{card}\{K \in U:\ K \text{ is a cell of } X(\omega)\}$$

for each $\omega \in \Omega$ and each $U \in \mathcal{B}(\mathcal{K}^d)$, a random mosaic may also be viewed as a particle process (concentrated on the space $\mathcal{P}^d$). We call ξ_X the point process associated with the random mosaic X. Other particle processes connected with X are the processes of the k-faces, $k \in \{0, \ldots, d-1\}$. However, for $k \leqslant d-2$, they do not determine X uniquely.

Classical examples of random mosaics are obtained from a stationary Poisson point process Y with finite intensity in $\mathbb{R}^d$. We may, for instance, associate with each point $y \in Y(\omega)$ its Voronoi cell, i.e., the set of all points $z \in \mathbb{R}^d$ which fulfill $\|z-y\| \leqslant \|z-\tilde{y}\|$ for all $\tilde{y} \in Y(\omega)$. The random mosaic obtained in this way is called the *Voronoi mosaic* associated with the Poisson process Y, it is stationary and isotropic. This type of random mosaic and related models in $\mathbb{R}^2$ and $\mathbb{R}^3$ have been investigated, for instance, by Meijering (1953), Gilbert (1962), and Miles (1970). Examples of stationary random mosaics which are not necessarily isotropic are the nondegenerate Poisson hyperplane networks treated at the end of this section. The role of convex geometry in the theory of stationary random mosaics has two different aspects. On the one hand, each realisation of a random mosaic may be viewed as an aggregate of convex polytopes and hence as a natural object of study in convex geometry. On the other hand, a stationary random mosaic is a stationary random hypersurface, and hence the auxiliary zonoids introduced in the preceding section may be used to describe its behaviour in many situations. Here the emphasis will be on the relations between both aspects.

Let X be a stationary random mosaic and assume that the intensity measure of the associated point process ξ_X on $\mathcal{P}^d$ is locally finite. Thus, the expected number of cells meeting a compact set is finite. Then two random polytopes may be associated with X. On the one hand, the shape distribution $\mathbb{P}_0$ of ξ_X defined in section 5 is the distribution of a random set Q_X called the typical cell of the random mosaic. Typical k-faces, $k = 0, \ldots, d-1$, may be defined in a similar way. These notions have their origin in the theory of Palm measures, which proved very useful in the investigation of random mosaics (see, e.g., Mecke 1980, and Møller 1989). For $d = 2$ and $d = 3$ many relationships between mean values concerning the typical faces and other mean values concerning the stationary random mosaics are due to Mecke (1984a) (extensions to random mosaics with not necessarily convex cells may be found in Stoyan 1986 and Weiss and Zähle 1988). On the other hand, since X is stationary, the origin 0 belongs almost surely to exactly one cell of X denoted by R_X. R_X is a random polytope called the 0-cell of X. Between the distribution of the typical cell and the distribution of the 0-cell we have the relation

$$\int_\Omega f(R_X(\omega))\,\mathrm{d}\mathbb{P}(\omega) = \left(\int_{\mathcal{P}^d} V_d(K)\,\mathrm{d}\mathbb{P}_0(K)\right)^{-1} \int_{\mathcal{P}^d} f(K) V_d(K)\,\mathrm{d}\mathbb{P}_0(K)$$

for each measurable and translation invariant function $f \geq 0$ on $\mathcal{P}^d$ (for the special case of a stationary Poisson hyperplane network see the remarks concerning the number law and the volume law in Matheron 1975, section 6.2). On the other hand, if Λ denotes the intensity measure of ξ_X and f is a nonnegative measurable function on $\mathcal{P}^d$, then we also have

$$\int_{\mathcal{P}^d} f(K)\,\mathrm{d}\Lambda(K) = \int_\Omega V_d(R_X(\omega))^{-1} \int_{\mathbb{R}^d} f(R_X(\omega) + y)\,\mathrm{d}\lambda_d(y)\,\mathrm{d}\mathbb{P}(\omega)\,. \tag{32}$$

In fact, R_X is just the closure of the random open set S_X defined in the last section. Since the random closed set $Y := \mathbb{R}^d \setminus S_X$ is (up to equivalence) uniquely determined by its capacity functional T_Y (Theorem 4.1) and

$$T_Y(C) = 1 - \mathbb{P}(C \subset S_X) = T_X(\mathrm{conv}(\{0\} \cup C))$$

for each $C \in \mathscr{C}^d$, it follows that R_X is uniquely determined by the values of T_X on the set of convex bodies containing the origin and vice versa. Consequently, the zonoid $\Pi^{d-1}(X)$ associated with the random hypersurface X (see section 6) is uniquely determined by R_X, and from (32) we infer that

$$2\Pi^{d-1}(X) - \mathbb{E}(V_d(R_X)^{-1}\Pi^{d-1}(\mathrm{bd}\,R_X))\,, \tag{33}$$

where $\Pi^{d-1}(\mathrm{bd}\,R_X)$ is the zonoid associated with bd R_X (see chapter 5.1) and on the right-hand side the expectation of the random set $V_d(R_X)^{-1}\Pi^{d-1}(\mathrm{bd}\,R_X)$ is defined as in section 3. Since the mean width of $\Pi^{d-1}(X)$ is essentially the surface area density of X, this implies

$$D_{d-1}(X) = \mathbb{E}\left(\frac{V_{d-1}(R_X)}{V_d(R_X)}\right).$$

Further, since $\Pi^{d-1}(\mathrm{bd}\,R_X)$ is the projection body of R_X in the usual sense, (33) gives some information concerning the relation between R_X and the inverse projection body $\Psi^{d-1}(X)$ of $\Pi^{d-1}(X)$. $\Psi^{d-1}(X)$ is the unique centrally symmetric convex body centred at the origin, the surface area measure of which is the generating measure of $\Pi^{d-1}(X)$ (see chapter 4.9). The relation between both convex bodies is rather complicated because of the factor $V_d(R_X)^{-1}$ in (33).

Theorem 7.1. *If X is a stationary random mosaic in $\mathbb{R}^d$, then*

$$2^d V_d(\Psi^{d-1}(X))^{d-1} \geq \mathbb{E}(V_d(R_X))^{-1}\,,$$

with equality if and only if for some convex polytope K, X is almost surely the boundary of a tiling of $\mathbb{R}^d$ by translates of K, and in this case K is homothetic to $\Psi^{d-1}(X)$.

This inequality, which has no analogue in the general case of a stationary random surface, is of interest because lower bounds involving $V_d(\Psi^{d-1}(X))$ appear in several extremal problems related to Theorem 6.4 in chapter 5.1. Both theorems together and classical results about mixed volumes lead to further inequalities for random mosaics. For a simple rectifiable curve K, for instance, we get a sharp inequality involving the mean number of intersection points of X and K, the volume of the convex hull conv K of K and the expected volume of the 0-cell, namely

$$\mathbb{E}(\operatorname{card}(X \cap K))^d \geqslant \frac{d!V_d(\operatorname{conv} K)}{d^d \mathbb{E}(V_d(R_X))} .$$

Further results of this type and a proof of the theorem may be found in Wieacker (1989).

In the important special case of a nondegenerate stationary Poisson hyperplane network more precise results may be obtained. A random hypersurface is called a *stationary Poisson hyperplane network* if it is the union set of a stationary Poisson process of hyperplanes in $\mathbb{R}^d$ with locally finite intensity measure. Nondegenerate means that the intensity measure of the Poisson process is not concentrated on the set of hyperplanes parallel to a line. This implies that, for almost all realisations of the process, all cells are bounded. If X is a nondegenerate and stationary Poisson hyperplane network in $\mathbb{R}^d$, then we may use Theorem 5.4 to compute the capacity functional T_X of X. The intensity measure Λ of the underlying hyperplane process can be decomposed into a *directional distribution* $\mathbb{P}_0$ and an intensity λ similar to Theorem 5.5 (compare also Example 2 in chapter 5.1, section 6). If we represent $\mathbb{P}_0$ as an even measure on the unit sphere, $\Lambda_0 = \lambda\mathbb{P}_0$ is precisely the generating measure of $\Pi^{d-1}(X)$ and hence the area measure of $\Psi^{d-1}(X)$. Therefore, we get

$$T_X(K) = 1 - \exp(-dV(\Psi^{d-1}(X), \ldots, \Psi^{d-1}(X), K - K)) ,$$

for all $K \in \mathcal{K}^d$. Here we have used the notation $K - K := \{x - y\colon x, y \in K\}$. This result and the isoperimetric inequality for the Minkowski area of $\Psi^{d-1}(X)$ (relative to a convex body K), where equality holds if and only if $\Psi^{d-1}(X)$ and K are homothetic, lead to a characterisation of $\Psi^{d-1}(X)$ (up to a homothety) as the solution of an extremal problem.

Theorem 7.2. *Let X be a nondegenerate and stationary Poisson hyperplane network in $\mathbb{R}^d$. Then, among all convex bodies K with given volume and containing the origin in their interior, $\mathbb{P}(K \subset R_X)$ attains a maximum if and only if K is homothetic to $\Psi^{d-1}(X)$.*

From the viewpoint of convex geometry, nondegenerate stationary Poisson hyperplane networks are of particular interest. On the one hand, a nondegenerate stationary Poisson hyperplane network X is uniquely determined (up to equivalence) by the intensity measure of the generating Poisson process of hyperplanes, and this intensity measure is uniquely determined by the generating measure of the zonoid $\Pi^{d-1}(X)$. On the other hand, if Z is a nondegenerate zonoid centred at the origin, then we may use the generating measure of Z to construct a nondegenerate stationary Poisson hyperplane network X satisfying $\Pi^{d-1}(X)=Z$. Hence, if we do not distinguish between random closed sets having the same distribution, then the map $X \mapsto \Pi^{d-1}(X)$ provides a bijection between the space of nondegenerate stationary Poisson hyperplane networks and the space of nondegenerate zonoids centred at the origin. It readily follows that a nondegenerate stationary Poisson hyperplane network X is isotropic if and only if $\Pi^{d-1}(X)$ is a ball. Moreover, many geometrical parameters of a nondegenerate zonoid may be expressed in terms of parameters of the corresponding Poisson hyperplane network and conversely. We give a few examples. For a nondegenerate and stationary Poisson hyperplane network X and for $k=0,\ldots,d-1$ we shall denote by X^k the union set of all k-faces of the cells of X and by $\mathrm{skel}_k\, R_X$ the set of all k-extreme points of R_X. Then X^k is a random k-surface and from (30) we have $D_k(X^k)=V_{d-k}(\Pi^{d-1}(X))$ [moreover, if in inequality (31) Y is a stationary Poisson process of hyperplanes, then equality holds if and only if Y is isotropic]. The following theorem is an analogue of Theorem 6.2.

Theorem 7.3. *Let X be a nondegenerate and stationary Poisson hyperplane network in $\mathbb{R}^d$. Then*

$$\mathbb{E}(V_d(R_X)) = d!2^{-d}V_d(\Pi^{d-1}(X)^*) \geq d!\kappa_d\left(\frac{2\kappa_{d-1}}{d\kappa_d}\, D_{d-1}(X)\right)^{-d},$$

with equality if and only if X is isotropic. Moreover, for $k=0,\ldots,d-1$, we have

$$\begin{aligned}\mathbb{E}(\lambda_k(\mathrm{skel}_k\, R_X)) &= D_k(X^k)\mathbb{E}(V_d(R_X))\\ &= d!2^{-d}V_{d-k}(\Pi^{d-1}(X))V_d(\Pi^{d-1}(X)^*)\,.\end{aligned}$$

In particular, $(2^d/d)\mathbb{E}(\lambda_0(\mathrm{skel}_0\, R_X))$ is the volume product of $\Pi^{d-1}(X)$, and hence

$$2^d \leq \mathbb{E}(\lambda_0(\mathrm{skel}_0\, R_X)) \leq d!2^{-d}\kappa_d^2\,,$$

with equality on the left side if and only if $\Pi^{d-1}(X)$ is a parallelotope, and equality on the right side if and only if $\Pi^{d-1}(X)$ is an ellipsoid.

The last statement is closely related to Schneider's results about random polytopes generated by anisotropic hyperplanes (Schneider 1982; see also section 2). Many interesting relations of this type, for instance concerning the expected intrinsic volumes of Q_X, may be found in Matheron (1975, chapter 6). For a convex body K containing the origin, Kaltenbach (1990) studied the asymptotic behaviour of the conditional expectation $\mathbb{E}(V_d(R_X) \mid K \subset R_X)$ as the intensity measure of the Poisson process generating X tends to infinity in a suitable way.

This question is closely related to the problem considered in section 2, and also here the asymptotic behaviour strongly depends on the boundary structure of K.

Since we had to omit many contributions of continuing importance, some remarks concerning the literature about random mosaics are in order here. The survey of Miles (1972) and the collection of papers edited by Harding and Kendall (1974) give an idea of the progress in the sixties and in the early seventies. A detailed investigation of stationary and isotropic Poisson hyperplane networks can be found in the important papers of Miles (1961, 1971b). The treatment of the anisotropic case in the path-breaking work of Matheron (1975) was the starting point for many of the developments in this section and section 6. Random mosaics generated by hyperplanes were investigated by Mecke (1984b) (some of the relations proved there have an interesting deterministic analogue, as was shown by Schneider 1987c). A good account of the applications and the statistical analysis of random mosaics with many references is given in Stoyan, Kendall and Mecke (1987, chapter 10), see also Mecke et al. (1990). A unified exposition of the theory of random mosaics including results on Voronoi mosaics and Delauney mosaics (the dual of Voronoi mosaics) can be found in the paper of Møller (1989). For a different approach based on ergodic theory, see, for instance, Cowan (1980). Far-reaching generalisations for random cell complexes and generalised sets are due to Zähle (1988).

8. Stereology

Stereological problems are currently the main field of applications of stochastic geometry and stereological questions had a strong influence on the development of the theory described in some of the last sections. In the most general formulation, stereology deals with the determination (or estimation) of characteristic geometric properties of (usually three-dimensional) objects by investigations of sections, projections, intersections with test sets, or other transformed images. Such problems are inherent to most of the experimental sciences, whenever a direct investigation of a three-dimensional feature is not possible. Examples are biology, medicine, geology, metallurgy, forestry, to name only a few. Frequently, two-dimensional images are treated in a similar way with two- or one-dimensional test sets. Here similar problems occur in Image Analysis and Spatial Statistics, two fields which have also strong connections to stereology and stochastic geometry. Therefore, and for mathematical simplicity, we will formulate the following presentation in the general d-dimensional setting. We will however only sketch some of the basic stereological problems, for further details we refer to the literature (Weil 1983, Jensen et al. 1985, Stoyan, Kendall and Mecke 1987, Mecke et al. 1990, Stoyan 1990, and Baddeley 1991, are some of the more mathematically oriented references).

To mention a typical stereological example: the direct determination of the specific alveolar surface area S_V of the human (or animal) lung is practically impossible due to the complicated structure of the lung tissue. To estimate the quantity S_V, in pathology usually small (cubical) pieces of lung tissue are cut randomly in (parallel) thin slices which are then examined under the microscope. The resulting 2-dimensional image is then either treated directly or by imposing randomly placed grids of lines or segments. This allows the determination of the mean boundary length per unit area, L_A, of the planar microscopical image. Integral geometric formulae and their random versions now give the connection between the observed values of L_A and the quantity S_V which is to be estimated. The models of stochastic geometry also make precise the conditions under which certain estimators will work.

The stereological literature usually distinguishes between two dual mathematical idealisations of the practical problem, a so-called *designed-based* and a *model-based* approach. In the first, the object of investigation is assumed to be a fixed set (in $\mathbb{R}^3$) which is intersected by randomly chosen planes. In the second approach, the set itself is assumed to be random (with certain invariance properties). Then, sectioning planes with fixed orientations can be used. In the first case, integral geometric formulae can be applied directly after some probabilistic modifications. In the second case, the corresponding results for random sets or point processes have to be used. We shortly describe both situations.

Let $K_0 \subset \mathbb{R}^d$ be a convex body with inner points and let $K \subset K_0$, $K \in \mathcal{R}_d$. We assume that K_0 is a reference set of known shape and size (in applications K_0 usually is a cube or ball), whereas certain geometric quantities of K are to be estimated (the assumption $K \in \mathcal{R}_d$ is not a serious restriction for applications). For this purpose, a random q-dimensional section of K_0 is taken (i.e., a random q-flat X_q intersecting K_0) and the section $X_q \cap K$ is observed. If, for example, X_q is a *uniform isotropic random* flat, then the distribution of X_q is the invariant measure μ_q on the space $\mathscr{E}_q^d$ of q-flats, restricted to $\{E \in \mathscr{E}_q^d\colon E \cap K_0 \neq \emptyset\}$ and normalised. In that case, the *Crofton Formulae* (see chapter 5.1) lead directly to the following expectation formula for the intrinsic volumes,

$$\mathbb{E}V_j(K \cap X_q) = \frac{\alpha_{djq}}{\alpha_{d0q}} \frac{V_{d+j-q}(K)}{V_{d-q}(K_0)}, \tag{34}$$

$0 \leq j \leq q \leq d-1$. Thus,

$$\frac{\alpha_{d0q}}{\alpha_{djq}} V_{d-q}(K_0) V_j(K \cap X_q)$$

is an unbiased estimator of $V_{d+j-q}(K)$. A disadvantage of formula (34) is that it requires the determination of $V_{d-q}(K_0)$, on the other hand one quite often wants to estimate a quermassintegral of K per unit d-volume of K_0, i.e., an estimator for $V_{d+j-q}(K)/V_d(K_0)$. Here the above formula implies

$$\frac{\mathbb{E}V_j(K \cap X_q)}{\mathbb{E}V_q(K_0 \cap X_q)} = \alpha_{djq} \frac{V_{d+j-q}(K)}{V_d(K_0)}, \tag{35}$$

but of course $f(X_q) = V_j(K \cap X_q)/V_q(K_0 \cap X_q)$ is not an unbiased estimator of the right-hand side. If however, X_q is chosen to be a *volume-weighted random* flat (which is determined by a uniformly distributed random point in K_0 and an independently chosen uniform direction; see chapter 5.1 for more details), then

$$\mathbb{E}\,\frac{V_j(K \cap X_q)}{V_q(K_0 \cap X_q)} = \alpha_{djq}\,\frac{V_{d+j-q}(K)}{V_d(K_0)}\,, \tag{36}$$

i.e., the estimator $f(X_q)$ is unbiased.

A number of other formulae from integral geometry have similar stereological interpretations and, after a suitable normalisation of the corresponding invariant measure, give unbiased estimators for certain stereological quantities. As a further example, we mention only the formula for projected thick sections (see chapter 5.1), which is the appropriate model for microscopical images, and which allows the estimation of particle number, a quantity which is not directly accessible with ordinary planar sections.

The model-based approach is simpler since the results from sections 4 and 5 can be used directly, the difficulties with formulae (35) and (36) do not occur. The basic assumption is that the underlying structure is a bounded part of a realisation of a stationary and isotropic random set or particle point process X. Here, the quotients of intrinsic volumes are replaced by the quermass densities. Thus, for a fixed q-dimensional subspace $L \subset \mathbb{R}^d$, the Crofton formula

$$D_j(X \cap L) = \alpha_{djq} D_{d+j-q}(X)\,, \tag{37}$$

[formula (14) in section 4 and (20) in section 5] replaces (35) and (36). In this case, both sides of (37) are expectations. Formulae (10), (11) and (12) [respectively (16), (17) and (18)] show different possibilities for unbiased or asymptotically unbiased estimation of the quermass densities. For example, in the case of a random set X and a convex body K of volume 1, $\Phi_j(X, \operatorname{int} K)$ and $V_j(X \cap C_0) - V_j(X \cap \partial^+ C_0)$ are unbiased estimators of $D_j(X)$, whereas the estimator $r^{-d} V_j(X \cap rK)$ is asymptotically unbiased. If $V_j(X \cap K)$ is used instead of $\Phi_j(X, \operatorname{int} K)$, the effects of bd K (the so-called *edge effects*) lead to an error, the mean of which is given by (13). Moreover, if all the intrinsic volumes $V_j(X \cap K)$, $j = 0, \ldots, d$, are evaluated, the linear system

$$\mathbb{E} V_j(X \cap K) = \sum_{k=j}^{d} \alpha_{djk} V_k(K) D_{d+j-k}(X)\,, \quad j = 0, \ldots, d\,,$$

can be solved for the unknowns $D_i(X)$, $i = 0, \ldots, d$, and another set of unbiased estimators for the quermass densities results. Of course, for the stereological situation described earlier, these estimators are applied correspondingly to the section $X \cap L$. Also, there is no principal difference between random sets and particle processes, so overlapping particle systems can be treated in the same way with the results from section 5. We mention that the section formula (37) contains the so-called *Fundamental formulae of stereology*:

$$V_V = A_A = L_L = \chi_X \,, \qquad S_V = \frac{4}{\pi} L_A = 2\chi_L \,, \qquad M_V = 2\pi\chi_A \,, \tag{38}$$

where we have used the notation from the end of section 5.

An important problem for applications is the determination of mean particle quantities for overlapping particle systems X when only the union set $Y = \bigcup_{K \in X} K$ is observable. Here, the formulae (25) and (26) are applicable, provided the underlying assumption of a Boolean model Y (i.e., a stationary and isotropic Poisson process X) is realistic. This is the case, whenever the particles are independently and uniformly distributed in a certain region. If all the particles are simply connected (hence $\bar{\chi} = 1$) and if the quermass densities of Y on the left-hand side of (25) and (26) are estimated by the above-mentioned methods, (25) allows successively the estimation of $\lambda \bar{A}$, $\lambda \bar{L}$, and λ, and similarly (26) can be used.

Another important aspect of this method is that Boolean models Y are basic models of random sets Y which can be modified to match real set-valued data. The unknown parameters are then the intensity γ and the shape distribution $\mathbb{P}_0$. The method described above allows the estimation of γ and of some mean values of $\mathbb{P}_0$. It is an interesting problem to get more information on $\mathbb{P}_0$ (in particular, in the nonisotropic case). Here, integral geometric formulae, other than the basic ones, have to be developed.

We finally emphasise that the results described so far give mean values hence first-order information on random sets and point processes. This is due to the nature of the underlying integral geometric results. There are also some less geometric methods to obtain higher-order informations or distributions, but generally the determination of variances, e.g., is a major open problem.

References

Affentranger, F.

[1988] The expected volume of a random polytope in a ball, *J. Microscopy* **151**, 277–287.

[1989] Random circles in the d-dimensional unit ball, *J. Appl. Probab.* **26**, 408–412.

[1990] Random spheres in a convex body, *Arch. Math.* **55**, 74–81.

[1992] The convex hull of random points with spherically symmetric distributions, *Rend. Sem. Mat. Univ. Politec. Torino*, in print.

Affentranger, F., and J.A. Wieacker

[1991] On the convex hull of uniform random points in a simple d-polytope, *Discrete Comput. Geom.* **6**, 291–305.

Alagar, V.S.

[1977] On the distribution of a random triangle, *J. Appl. Probab.* **14**, 284–297.

Araujo, A., and E. Giné

[1980] *The Central Limit Theorem for Real and Banach Valued Random Variables* (Wiley, New York).

Arrow, K.J., and F.H. Hahn

[1971] *General Competitive Analysis* (Holden-Day, San Francisco).

Artstein, Z., and J.C. Hansen
[1985] Convexification in limit laws of random sets in Banach spaces, *Ann. Probab.* **13**, 307–309.
Artstein, Z., and S. Hart
[1981] Law of large numbers for random sets and allocation processes, *Math. Oper. Res.* **6**, 485–492.
Artstein, Z., and R.A. Vitale
[1975] A strong law of large numbers for random compact sets, *Ann. Probab.* **5**, 879–882.
Aumann, R.J.
[1965] Integrals of set-valued functions, *J. Math. Anal. Appl.* **12**, 1–12.
Baddeley, A.
[1991] Stereology, in: *Spatial Statistics and Digital Image Analysis* (National Academy Press, Washington, DC) pp. 181–216.
Bárány, I.
[1989] Intrinsic volumes and f-vectors of random polytopes, *Math. Ann.* **285**, 671–699.
[1992] Random polytopes in smooth convex bodies, *Mathematika* **39**, 81–92.
Bárány, I., and C. Buchta
[1990] On the convex hull of uniform random points in an arbitrary d-polytope, *Anz. Österreich. Akad. Wiss. Math.-Natur. Kl.* **127**, 25–27.
Bárány, I., and Z. Füredi
[1988] On the shape of the convex hull of random points, *Probab. Theory Related Fields* **77**, 231–240.
Bárány, I., and D.G. Larman
[1988] Convex bodies, economic cap coverings, random polytopes, *Mathematika* **35**, 274–291.
Bárány, I., and R.A. Vitale
[1992] Random convex hulls: floating bodies and expectations, *J. Approx. Theory*, to appear.
Bentley, J.L., H.T. Kung, M. Schkolnick and C.D. Thompson
[1978] On the average number of maxima in a set of vectors, *J. ACM* **25**, 536–543.
Betke, U., and W. Weil
[1991] Isoperimetric inequalities for the mixed area of plane convex sets, *Arch. Math.* **57**, 501–507.
Blaschke, W.
[1917] Über affine Geometrie XI: Lösung des 'Vierpunktproblems' von Sylvester aus der Theorie der geometrischen Wahrscheinlichkeiten, *Leipz. Ber.* **69**, 436–453.
Borgwardt, K.H.
[1987] *The Simplex Method; A Probabilistic Analysis* (Springer, Berlin).
Buchta, C.
[1984a] Stochastische Approximation konvexer Polygone, *Z. Wahrscheinlichkeitsth. Verw. Geb.* **67**, 283–304.
[1984b] Zufallspolygone in konvexen Vielecken, *J. Reine Angew. Math.* **347**, 212–220.
[1985] Zufällige Polyeder – Eine Übersicht, in: *Zahlentheoretische Analysis*, ed. E. Hlawka, Lecture Notes in Mathematics, Vol. 1114 (Springer, Berlin) pp. 1–13.
[1986] On a conjecture of R.E. Miles about the convex hull of random points, *Monatsh. Math.* **102**, 91–102.
[1987a] On the number of vertices of random polyhedra with a given number of facets, *SIAM J. Algebraic Discrete Methods* **8**, 85–92.
[1987b] On nonnegative solutions of random systems of linear inequalities, *Discrete Comput. Geom.* **2**, 85–95.
Buchta, C., and J. Müller
[1984] Random polytopes in a ball, *J. Appl. Probab.* **21**, 753–762.
Buchta, C., J. Müller and R.F. Tichy
[1985] Stochastical approximation of convex bodies, *Math. Ann.* **271**, 225–235.

Carlsson, S., and U. Grenander
[1967] Statistical approximation of plane convex sets, *Skand. Aktuarietidsskr.* **3/4**, 113–127.

Carnal, H., and J. Hüsler
[1991] On the convex hull of n random points on a circle, *J. Appl. Probab.* **28**, 231–237.

Cowan, R.
[1980] Properties of ergodic random mosaic processes, *Math. Nachr.* **97**, 89–102.

Cressie, N.
[1978] A strong limit theorem for random sets, *Adv. in Appl. Probab. Suppl.* **10**, 36–46.
[1979] A central limit theorem for random sets, *Z. Wahrscheinlichkeitsth. Verw. Geb.* **49**, 37–47.
[1984] Modelling sets, in: *Lecture Notes in Math.*, Vol. 1091 (Springer, New York) pp. 138–149.

Daley, D.J., and D. Vere-Jones
[1988] *An Introduction to the Theory of Point Processes* (Springer, New York).

Dalla, L., and D.G. Larman
[1991] Volumes of a random polytope in a convex set, in: *Applied Geometry and Discrete Mathematics. The Victor Klee Festschrift*, eds P. Gritzmann and B. Sturmfels, DIMACS Series in Discrete Mathem. and Comp. Science, Vol. 4 (AMS, ACM, New York) pp. 175–180.

Davy, P.
[1976] Projected thick sections through multidimensional particle aggregates, *J. Appl. Probab.* **13**, 714–722. Correction: **15** (1978) 456.
[1978] Stereology – A Statistical Viewpoint, Thesis, Australian National Univ., Canberra.

Dette, H., and N. Henze
[1989] The limit distribution of the largest nearest-neighbor link in the unit d-cube, *J. Appl. Probab.* **26**, 67–80.
[1990] Some peculiar boundary phenomena for extremes of the rth nearest neighbor links, *Statist. Probab. Lett.* **10**, 381–390.

Devroye, L.P.
[1980] A note on finding convex hulls via maximal vectors, *Inform. Process. Lett.* **11**, 53–56.

Drobot, V.
[1982] Probabilistic version of a curvature formula, *Ann. Probab.* **10**, 860–862.

Dwyer, R.A.
[1988] On the convex hull of random points in a polytope, *J. Appl. Probab.* **25**, 688–699.
[1990] Random convex hulls in a product of balls, *Probab. Theory Related Fields* **86**, 457–468.

Eddy, W.F., and J.D. Gale
[1981] The convex hull of a spherically symmetric sample, *Adv. in Appl. Probab.* **13**, 751–763.

Efron, B.
[1965] The convex hull of a random set of points, *Biometrika* **52**, 331–343.

Gilbert, E.N.
[1962] Random subdivision of space into crystals, *Ann. of Math. Statist.* **33**, 958–972.

Giné, E., and M.C. Hahn
[1985a] The Lévy–Khinchin representation for random compact convex subsets which are infinitely divisible under Minkowski addition, *Z. Wahrscheinlichkeitsth. Verw. Geb.* **70**, 271–287.
[1985b] Characterization and domains of attraction of p-stable random compact sets, *Ann. Probab.* **13**, 447–468.
[1985c] M-infinitely divisible random compact convex sets, in: *Lecture Notes in Mathematics*, Vol. 1153 (Springer, New York) pp. 226–248.

Giné, E., M.C. Hahn and J. Zinn
[1983] Limit theorems for random sets: an application of probability in Banach space results, in: *Lecture Notes in Mathematics*, Vol. 990 (Springer, New York) pp. 112–135.

Goodey, P., and R. Howard
[1990a] Processes of flats induced by higher dimensional processes, *Adv. in Math.* **80**, 92–109.
[1990b] Processes of flats induced by higher dimensional processes II, *Contemp. Math.* **113**, 111–119.

Grenander, U.
[1973] Statistical geometry: a tool for pattern analysis, *Bull. AMS* **79**, 829–856.
[1977] *Pattern Analysis. Lectures in Pattern Theory,* Vol. II (Springer, New York).

Groemer, H.
[1974] On the mean value of the volume of a random polytope in a convex set, *Arch. Math.* **25**, 86–90.

Groeneboom, P.
[1988] Limit theorems for convex hulls, *Probab. Theory Related Fields* **79**, 327–368.

Gruber, P.M.
[1983b] In most cases approximation is irregular, *Rend. Sem. Mat. Univ. Politec. Torino* **41**, 19–33.

Hall, P.
[1988] *Introduction to the Theory of Coverage Processes* (Wiley, New York).

Harding, E.F., and D.G. Kendall, eds
[1974] *Stochastic Geometry* (Wiley, New York).

Henze, N.
[1983] Random triangles in convex regions, *J. Appl. Probab.* **20**, 111–125.

Hess, C.
[1979] Théorème ergodique et loi forte des grandes nombres pour des ensembles aléatoires, *C.R. Acad. Sci., Ser. A* **288**, 519–522.

Hiai, F.
[1984] Strong laws of large numbers for multivalued random variables, in: *Lecture Notes in Mathematics,* Vol. 1091 (Springer, New York) pp. 160–172.
[1985] Convergence of conditional expectations and strong laws of large numbers for multivalued random variables, *Trans. Amer. Math. Soc.* **291**, 613–627.

Hostinsky, B.
[1925] Sur les probabilités géométriques, *Publ. Fac. Sci. Univ. Masaryk (Brno)* **50**, 1–26.

Jensen, E.B., A.J. Baddeley, H.J.G. Gundersen and R. Sundberg
[1985] Recent trends in stereology, *Internat. Statist. Rev.* **53**, 99–108.

Kallenberg, O.
[1986] *Random Measures* (Academic Press, London).

Kaltenbach, F.J.
[1990] Asymptotsches Verhalten zufälliger konvexer Polyeder, Thesis, Freiburg.

Karr, A.F.
[1986] *Point Processes and their Statistical Inference* (Marcel Dekker, New York).

Kellerer, A.M.
[1983] On the number of clumps resulting from the overlap of randomly placed figures in the plane, *J. Appl. Probab.* **20**, 126–135.
[1985] Counting figures in planar random configurations, *J. Appl. Probab.* **22**, 68–81.

Kellerer, H.G.
[1984] Minkowski functionals of Poisson processes, *Z. Wahrscheinlichkeitsth. Verw. Geb.* **67**, 63–84.

Kelly, D.G., and J.W. Tolle
[1981] Expected number of vertices of a random convex polyhedron, *SIAM J. Algebraic Discrete Methods* **2**, 441–451.

Kendall, D.G.
[1974] Foundations of a theory of random sets, in: *Stochastic Geometry*, eds E.F. Harding and D.G. Kendall (Wiley, New York) pp. 322–376.

Kingman, J.F.C.
[1969] Random secants of a convex body, *J. Appl. Probab.* **6**, 660–672.

Lyashenko, N.N.
[1982] Limit theorems for sums of independent, compact, random subsets of Euclidean space, *J. Soviet Math.* **20**, 2187–2196.
[1983] Statistics of random compact sets in Euclidean space, *J. Soviet Math.* **21**, 76–92.

Mase, S.
[1979] Random compact sets which are infinitely divisible with respect to Minkowski addition, *Adv. in Appl. Probab.* **11**, 222–224.

Mathai, A.M.
[1982] On a conjecture in geometric probability regarding asymptotic normality of a random simplex, *Ann. Probab.* **10**, 247–251.

Matheron, G.
[1972] Ensembles fermés aléatoires, ensembles semimarkoviens et polyèdres poissoniens, *Adv. in Appl. Probab.* **4**, 508–541.
[1975] *Random Sets and Integral Geometry* (Wiley, New York).

Mecke, J.
[1980] Palm methods for stationary random mosaics, in: *Combinatorial Principles in Stochastic Geometry*, ed. R.V. Ambartzumian (Armenian Academy of Sciences, Erevan) pp. 124–132.
[1984a] Parametric representation of mean values for stationary random mosaics, *Math. Operationsforsch. Stat. Ser. Statist.* **15**, 437–442.
[1984b] Random tessellations generated by hyperplanes, in: *Stochastic Geometry, Geometric Statistics, Stereology*, Proc. Conf. Oberwolfach, 1983, eds R.V. Ambartzumian and W. Weil (Teubner, Leipzig) pp. 104–109.

Mecke, J., and C. Thomas
[1986] On an extreme value problem for flat processes, *Comm. Statist. Stochastic Models* **2**, 273–280.

Mecke, J., R. Schneider, D. Stoyan and W. Weil
[1990] *Stochastische Geometrie* (Birkhäuser, Basel).

Meijering, J.L.
[1953] Interface area, edge length, and number of vertices in crystal aggregates with random nucleation, *Philips Res. Rep* **8**, 270–290.

Meilijson, I.
[1990] The expected value of some functions of the convex hull of a random set of points sampled in R^d, *Israel J. Math.* **72**, 341–352.

Miles, R.E.
[1961] Random Polytopes: The generalisation to n dimensions of the intervals of a Poisson process, Ph.D. Thesis, Cambridge Univ.
[1970] On the homogeneous planar Poisson point process, *Math. Biosci.* **6**, 85–127.
[1971a] Isotropic random simplices, *Adv. in Appl. Probab.* **3**, 353–382.
[1971b] Poisson flats in euclidean spaces; Part II: Homogeneous Poisson flats and the complementary theorem, *Adv. in Appl. Probab.* **3**, 1–43.
[1972] The random division of space, *Adv. in Appl. Probab. Suppl.* **4**, 243–266.
[1976] Estimating aggregate and overall characteristics from thick sections by transmission microscopy, *J. Microscopy* **107**, 227–233.

Miles, R.E., and J. Serra, eds
[1978] *Geometrical Probability and Biological Structures: Buffon's 200th anniversary*, Lecture Notes in Biomath., Vol. 23 (Springer, Berlin).

Møller, J.
[1989] Random tessellations in R^d, *Adv. in Appl. Probab.* **21**, 37–73.

Müller, J.S.
[1989] Über die mittlere Breite von Zufallspolyedern, *Probab. Theory Related Fields* **82**, 33–37.
[1990] Approximation of a ball by random polytopes, *J. Approx. Theory* **63**, 198–209.

Nguyen, X.X., and H. Zessin
[1979] Ergodic theorems for spatial processes, *Z. Wahrscheinlichkeitsth. Verw. Geb.* **48**, 133–158.

Pohlmann, S., J. Mecke and D. Stoyan
[1981] Stereological formulas for stationary surface processes, *Math. Operationsforsch. Stat. Ser. Statist.* **12**, 429–440.

Prékopa, A.
[1972] On the number of vertices of random convex polyhedra, *Period. Math. Hungar.* **2**, 259–282.

Puri, M.L., and D.A. Ralescu
[1983] Strong law of large number for Banach space-valued random sets, *Ann. Probab.* **11**, 222–224.
[1985] Limit theorems for random compact sets in Banach space, *Math. Proc. Cambridge Philos. Soc.* **97**, 151–158.

Puri, M.L., D.A. Ralescu and S.S. Ralescu
[1986] Gaussian random sets in Banach space, *Teor. Veroyatnost i Primenen.* **31**, 598–601.

Raynaud, H.
[1970] Sur l'enveloppe convexe des nuages de points aléatoires dans R^n, *J. Appl. Probab.* **7**, 35–48.

Reed, W.J.
[1974] Random points in a simplex, *Pacific J. Math.* **54**, 183–198.

Reisner, S.
[1985] Random polytopes and the volume product of symmetric convex bodies, *Math. Scand.* **57**, 386–392.
[1986] Zonoids with minimal volume-product, *Math. Z.* **192**, 339–346.

Rényi, A., and R. Sulanke
[1963] Über die konvexe Hülle von n zufällig gewählten Punkten, *Z. Wahrscheinlichkeitsth. Verw. Geb.* **2**, 75–84.
[1964] Über die konvexe Hülle von n zufällig gewählten Punkten. II, *Z. Wahrscheinlichkeitsth. Verw. Geb.* **3**, 138–147.
[1968] Zufällige konvexe Polygone in einem Ringgebiet, *Z. Wahrscheinlichkeitsth. Verw. Geb.* **9**, 146–157.

Ripley, B.D.
[1976] Locally finite random sets: Foundations for point process theory, *Ann. Probab.* **4**, 983–994.

Ronse, C.
[1989] A bibliography on digital and computational convexity (1961–1988), *IEEE Trans. Pattern Anal. Machine Intell.* **2**, 181–190.

Ruben, H.
[1977] The volume of a random simplex in an n-ball is asymptotically normal, *J. Appl. Probab.* **14**, 647–653.

Saint Raymond, J.
[1981] Sur le volume des corps convexes symétriques, *Publ. Math. Univ. Pierre et Marie Curie* **46**, exposé No. 11, 25 pp.

Santaló, L.A.
[1976] *Integral Geometry and Geometric Probability* (Addison-Wesley, Reading, MA).

Schmidt, B.K., and T.H. Mattheiss
[1977] The probability that a random polytope is bounded, *Math. Oper. Res.* **2**, 292–296.

Schmidt, W.
[1968] Some results in probabilistic geometry, *Z. Wahrscheinlichkeitsth. Verw. Geb.* **9**, 158–162.

Schneider, R.
[1982] Random polytopes generated by anisotropic hyperplanes, *Bull. London Math. Soc.* **14**, 549–553.
[1987a] Approximation of convex bodies by random polytopes, *Aequationes Math.* **32**, 304–310.
[1987b] Geometric inequalities for Poisson processes of convex bodies and cylinders, *Results Math.* **11**, 165–185.
[1987c] Tessellations generated by hyperplanes, *Discrete Comput. Geom.* **2**, 223–232.
[1988] Random approximation of convex sets, *J. Microscopy* **151**, 211–227.

Schneider, R., and J.A. Wieacker
[1980] Random polytopes in a convex body, *Z. Wahrscheinlichkeitsth. Verw. Geb.* **52**, 69–73.

Schöpf, P.
[1977] Gewichtete Volumsmittelwerte von Simplices, welche zufällig in einem konvexen Körper des R^n gewählt werden, *Monatsh. Math.* **83**, 331–337.

Schürger, K.
[1983] Ergodic theorems for subadditive superstationary families of convex compact random sets, *Z. Wahrscheinlichkeitsth. Verw. Geb.* **62**, 125–135.

Serra, J.P.
[1982] *Image Analysis and Mathematical Morphology* (Academic Press, London).

Stoyan, D.
[1986] On generalized planar random tessellations, *Math. Nachr.* **128**, 215–219.
[1989] On means, medians and variances of random compact sets, in: *Geobild '89*, eds A. Hübler, W. Nagel, B.D. Ripley and G. Werner, Math. Res., Vol. 51 (Akademie-Verlag, Berlin) pp. 99–104.
[1990] Stereology and stochastic geometry, *Internat. Statist. Rev.* **58**, 227–242.

Stoyan, D., W.S. Kendall and J. Mecke
[1987] *Stochastic Geometry and its Applications* (Akademie-Verlag, Berlin).

Stute, W.
[1984] Random approximation of smooth curves, *Mitt. Math. Sem. Giessen* **165**, 205–210.

Sulanke, R., and P. Wintgen
[1972] Zufällige konvexe Polyeder im N-dimensionalen euklidischen Raum, *Period. Math. Hungar.* **2**, 215–221.

Thomas, C.
[1984] Extremum properties of the intersection densities of stationary Poisson hyperplane processes, *Math. Operationsforsch. Stat. Ser. Statist.* **15**, 443–449.

van Wel, B.F.
[1989] The convex hull of a uniform sample from the interior of a simple d-polytope, *J. Appl. Probab.* **27**, 259–273.

Vitale, R.A.
[1983a] On Gaussian random sets, in: *Stochastic Geometry, Geometric Statistics, Stereology*, Proc. Conf. Oberwolfach, 1983, eds R.V. Ambartzumian and W. Weil (Teubner, Leipzig) pp. 222–224.
[1983b] Some developments in the theory of random sets, *Bull. Int. Stat. Inst.* **50**, 863–871.
[1988] An alternate formulation of the mean value for random geometric figures, *J. Microscopy* **151**, 197–204.
[1990] The Brunn–Minkowski inequality for random sets, *J. Multivariate Anal.* **33**, 286–293.

Weil, W.
[1982] An application of the central limit theorem for Banach-space-valued random variables to the theory of random sets, *Z. Wahrscheinlichkeitsth. Verw. Geb.* **60**, 203–208.

[1983] Stereology: A survey for geometers, in: *Convexity and its Applications,* eds P.M. Gruber and J.M. Wills (Birkhäuser, Basel) pp. 360–412.

[1984] Densities of quermassintegrals for stationary random sets, in: *Stochastic Geometry, Geometric Statistics, Stereology,* Proc. Conf. Oberwolfach, 1983, eds R.V. Ambartzumian and W. Weil (Teubner, Leipzig) pp. 233–247.

[1987] Point processes of cylinders, particles and flats. *Acta Appl. Math.* **9**, 103–136.

[1988] Expectation formulas and isoperimetric properties for non-isotropic Boolean models, *J. Microscopy* **151**, 235–245.

[1990] Iterations of translative integral formulae and nonisotropic Poisson processes of particles, *Math. Z.* **205**, 531–549.

Weil, W., and J.A. Wieacker

[1984] Densities for stationary random sets and point processes, *Adv. in Appl. Probab.* **16**, 324–346.

Weiss, V., and M. Zähle

[1988] Geometric measures for random curved mosaics of R^d, *Math. Nachr.* **138**, 313–326.

Wieacker, J.A.

[1978] Einige Probleme der polyedrischen Approximation, Diplomarbeit, Univ. Freiburg.

[1986] Intersections of random hypersurfaces and visibility, *Probab. Theory Related Fields* **71**, 405–433.

[1989] Geometric inequalities for random surfaces, *Math. Nachr.* **142**, 73–106.

Wschebor, M.

[1985] *Surfaces aléatoires,* Lecture Notes in Mathematics, Vol. 1147 (Springer, Berlin).

Yadin, M., and S. Zacks

[1985] The visibility of stationary and moving targets in the plane subject to a Poisson field of shadowing elements, *J. Appl. Probab.* **22**, 776–786.

Zähle, M.

[1982] Random processes of Hausdorff rectifiable closed sets, *Math. Nachr.* **108**, 49–72.

[1986] Curvature measures and random sets, II, *Probab. Theory Related Fields* **71**, 37–58.

[1988] Random cell complexes and generalised sets, *Ann. Probab.* **16**, 1742–1766.

Ziezold, H.

[1970] Über die Eckenzahl zufälliger konvexer Polygone, *Izv. Akad. Nauk Armjan. SSR V* **3**, 296–312.

[1984] The mean breadth of a random polytope in a convex body, *Z. Wahrscheinlichkeitsth. Verw. Geb.* **68**, 121–125.

Author Index

Subject Index

The index is based on the proposals of the contributors to the Handbook. Because of their heterogeneous standpoints it was necessary to homogenize the proposals. Names occurring in entries are always placed first, e.g., we cite “Aleksandrov–Fenchel inequality” instead of “inequality of Aleksandrov–Fenchel”. Well-known geometric notions are cited in general in the form in which they are used, e.g., we take “affine surface area” and not “surface area, affine”.